Unit Root Tests in Time Series Volume 2

Palgrave Texts in Econometrics
General Editor: **Kerry Patterson**

Titles include:

Simon P. Burke and John Hunter
MODELLING NON-STATIONARY TIME SERIES

Michael P. Clements
EVALUATING ECONOMETRIC FORECASTS OF ECONOMIC AND FINANCIAL VARIABLES

Lesley Godfrey
BOOTSTRAP TESTS FOR REGRESSION MODELS

Terence C. Mills
MODELLING TRENDS AND CYCLES IN ECONOMIC TIME SERIES

Kerry Patterson
A PRIMER FOR UNIT ROOT TESTING

Kerry Patterson
UNIT ROOT TESTS IN TIME SERIES VOLUME 1
Key Concepts and Problems

Kerry Patterson
UNIT ROOT TESTS IN TIME SERIES VOLUME 2
Extensions and Developments

Palgrave Texts in Econometrics

Series Standing Order ISBN 978- 1–4039–0172-9 hardback
978-1–4039–0173–6 paperback (*outside North America only*)

You can receive future titles in this series as they are published by placing a standing order. Please contact your bookseller or, in case of difficulty, write to us at the address below with your name and address, the title of the series and the ISBN quoted above.

Customer Services Department, Macmillan Distribution Ltd, Houndmills, Basingstoke, Hampshire RG21 6XS, England

---

Unit Root Tests in Time Series Volume 2

Extensions and Developments

Kerry Patterson

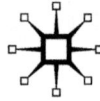

First published 2012 by
PALGRAVE MACMILLAN

Palgrave Macmillan in the UK is an imprint of Macmillan Publishers Limited, registered in England, company number 785998, of Houndmills, Basingstoke, Hampshire RG21 6XS.

Palgrave Macmillan in the US is a division of St Martin's Press LLC, 175 Fifth Avenue, New York, NY 10010.

Palgrave Macmillan is the global academic imprint of the above companies and has companies and representatives throughout the world.

Palgrave® and Macmillan® are registered trademarks in the United States, the United Kingdom, Europe and other countries

ISBN: 978–0–230–25026–0 hardback
ISBN: 978–0–230–25027–7 paperback

This book is printed on paper suitable for recycling and made from fully managed and sustained forest sources. Logging, pulping and manufacturing processes are expected to conform to the environmental regulations of the country of origin.

A catalogue record for this book is available from the British Library.

A catalog record for this book is available from the Library of Congress.

Contents

Detailed Contents

List of Tables

List of Figures

Symbols and Abbreviations (for Volumes 1 and 2)

$\hat{\tau}$	standard DF t-type test statistic, with no deterministic terms
$\hat{\tau}_\mu$	standard DF t-type test statistic using demeaned data
$\hat{\tau}_\beta$	standard DF t-type test statistic using detrended data
$\hat{\delta}$	$\equiv T(\hat{\rho} - 1)$, where $\hat{\rho}$ is the least squares (LS) estimator
$\hat{\delta}_\mu$	as in the specification of $\hat{\tau}_\mu$
$\hat{\delta}_\beta$	as in the specification of $\hat{\tau}_\beta$
$\Rightarrow_D$	convergence in distribution (weak convergence)
$\rightarrow_p$	convergence in probability
$\rightarrow$	tends to, for example ε tends to zero, $\varepsilon \rightarrow 0$
$\mapsto$	mapping
$\Rightarrow$	implies
$\sim$	is distributed as (or, from the context, left and right hand sides approach equality)
$\equiv$	definitional equality
$\neq$	not equals
$\approx$	approximately
$\Phi(z)$	the cumulative distribution function of the standard normal distribution
$\Re$	the set of real numbers; the real line ($-\infty$ to ∞)
$\Re^+$	the positive half of the real line
N^+	the set of nonnegative integers
ε_t	white noise unless explicitly excepted
$\prod_{j=1}^n x_j$	the product of x_j, $j = 1, ..., n$
$\sum_{j=1}^n x_j$	the sum of x_j, $j = 1, ..., n$
0^+	approach zero from above
$_-0$	approach zero from below
L	the lag operator, $L^j y_t \equiv y_{t-j}$
Δ	the first difference operator, $\Delta \equiv (1 - L)$
Δ_s	the s-th difference operator, $\Delta_s \equiv (1 - L^j)$
Δ^s	the s-th multiple of the first difference operator, $\Delta^s \equiv (1 - L)^s$
$\subset$	a proper subset of
$\subseteq$	a subset of
$\cap$	intersection of sets
$\cup$	union of sets
$\in$	an element of

$\lvert a \rvert$	the absolute value (modulus) of a
iid	independent and identically distributed
m.d.s	martingale difference sequence
$N(0, 1)$	the standard normal distribution, with zero mean and unit variance
niid	independent and identically normally distributed
$B(t)$	standard Brownian motion, that is with unit variance
$W(t)$	nonstandard Brownian motion

Preface

This book is the second volume of *Unit Root Tests in Time Series* and is subtitled 'Extensions and Developments'. The first volume was published in 2011 (Patterson, 2011), and subtitled 'Key Concepts and Problems' (referred to herein as *UR, Vol. 1*). Additionally, a third contribution, although the first in the sequence, entitled *A Primer for Unit Root Testing* (referred to herein as *A Primer*), was published in 2010 (Patterson, 2010), and completes the set. The books can be read independently depending on the reader's background and interests.

I conceived this project around ten years ago, recognising a need to present and critically assess the key developments in unit root testing, a topic that barely twenty years before was only occasionally referenced in the econometrics literature, although its importance had been recognised long before, for example in the modelling strategy associated with Box and Jenkins (1970). (See Mills (2011), for an excellent overview in the context of the development of modern time series analysis.)

An econometric or statistical procedure can be judged to be influential when it reaches into undergraduate courses as well as becoming standard practice in research papers and articles, and no serious empirical analysis of time series is now presented without reporting seemingly obligatory unit root tests. However, in the course of becoming standard practice many of the nuances and concerns about the unit testing framework, carefully detailed in the original research, may be overlooked, the job being done if the unit root 'tick box' has been checked. Undergraduate dissertations and projects, as well as PhD theses, involving time series, usually report Dickey-Fuller (DF) statistics or perhaps, depending on the available software, tests in the form due to Elliott, Rothenberg and Stock (1996), the 'ERS' tests. However, it can be the case that the analysis of the time series properties of the data is somewhat superficial. Faced with the task of analysing and modelling time series, I am struck by how careful one has to be and how many possibilities there are that complicate the question of interpreting the outcome of a unit root test. This is the starting point of this book that distinguishes it from Volume 1.

When considering what to include here I took the view that the topics must address the development of the standard unit root testing framework to situations that were likely to occur in practice. Moreover, whilst an overview of as many problems as possible would have been one organising principle, this approach would risk taking a 'menu'-type solution to potential problems and, instead, it would be preferable to address in detail some developments that have been particularly important in shaping the direction of research and application. This, then, was the guiding principle for organising the chapters in this volume although, having concluded this contribution as far as time will allow, I am as aware now as at the start of the project that there still remains so much of interest untold here. Whilst the testing procedure for a unit root seems straightforward, indeed simple enough, as noted, to be taught in an introductory econometrics or statistics course, some important problems have come to light, and the explicit dichotomy (unit root or no unit root) in standard tests seems a questionable simplicity.

Although much of the unit root framework has been developed in the econometrics literature, the presence of key articles in general statistical journals (for example, *The Annals of Statistics, Biometrika, The Journal of the Royal Statistical Society, The Journal of the American Statistical Association* and so on) indicates the wider interest and wider importance of the topic. Moreover, the techniques have been finding interesting applications outside economics, showing just how pervasive is the basic underlying problem. To give some illustrations, Stern and Kaufman (1999) analysed a number of time series related to global climate change, including temperatures for the northern and southern hemispheres, carbon dioxide, nitrous oxide and sulphate aerosol emissions. Wang et al. (2005) analysed the streamflow processes of 12 rivers in western Europe for nonstationarity for the twentieth century, streamflow being an issue of concern in the design of a flood protection system, a hydrological connection that includes Hurst (1951) and Hosking (1984). Aspects of social behaviour have been the subject of study, such as the nature of the gender gap in drink-driving (see Schwartz and Rookey, 2008).

I noted in Volume 1 that the research and literature on this topic has grown almost exponentially since Nelson and Plosser's (N&P) seminal article published in 1982. N&P applied the framework due to Dickey (1976) and Fuller (1976) to testing for a unit root in a number of macroeconomic time series. Whilst the problem of modelling nonstationary series was known well before 1982, and was routinely taken into account in the Box-Jenkins methodology, the focus of the N&P study was on the stochastic nature of the trend and its implications for economic policy and the interpretation of 'shocks'; this led to interest in the properties of economic time series as worthy of study in itself.

I also noted in Volume 1 that articles on unit root tests are amongst the most cited in economics and econometrics and have influenced the direction of economic research at a much wider level. A citation summary for articles based on univariate processes was included therein and showed that there has been a sustained interest in the topic over the last thirty years. Out of interest I have updated this summary and calculated the implied annual growth rate, with the 'top 5' on a citations basis presented in Table P1. The rate of growth is quite astonishing and demonstrates the continuing interest in the topic of unit roots;

Table P1   Number of citations of key articles on unit roots

Author(s)	February 2010	July 2011	p.a. growth
Dickey and Fuller (1979)	7,601	8,933	12.1%
Phillips and Perron (1988)	4,785	5,673	12.8%
Dickey and Fuller (1981)	4,676	5,368	10.8%
Perron (1989)	3,371	3,932	11.5%
KPSS (1992)	3,280	3,996	15.0%

Note: KPSS is Kwiakowski, Phillips, Schmidt and Shin (1992).

for example, there have been 1,300 more citations of the seminal Dickey and Fuller (1979) article in less than 18 months.

As in the case of Volume 1, appropriate prerequisites for this book include some knowledge of econometric theory at an intermediate or graduate level as, for example, in Davidson (2000), Davidson and Mackinnon (2004) or Greene (2011); also some exposure to the basic DF framework for testing for a unit root would be helpful, but this is now included in most introductory courses in econometrics (see, for example, Dougherty, 2011).

The results of a number of Monte Carlo studies are reported in various chapters, reflecting my view that simulation is a key tool in providing guidance on finite sample issues. Many more simulations were run than are reported in the various chapters, with the results typically illustrated for one or two sample sizes where they are representative of a wider range of sample sizes.

Organisation of the book

The contents are organised into topics as follows.

Chapter 1: Introduction

First, the opportunity is taken to offer a brief reminder of some key concepts that underlie the implicit language of unit root tests. In addition there is also an introduction to an econometric problem that occurs in several contexts in testing for a unit root and is referenced in later chapters. The problem is that of the presence of a 'nuisance' parameter which is only present under the alternative hypothesis, sometimes referred to as the 'Davies problem'. The problem and its solution occur in several other areas of econometrics and it originates, at least in its econometric interest, in application of the well-known Chow test (Chow, 1960) to the temporal stability of a regression model, but in the case that the temporal point of instability is unknown.

Chapter 2: Functional form and nonparametric tests for a unit root

There are two related issues covered in this chapter. The first is whether a unit root, if present, should be modelled in the original series or in the logarithm of the series, the difference being between random walk and an exponential random walk. Often the choice seems either arbitrary or a matter of convenience, whereas the choice can be vital to the outcome of a unit root test. The second issue is a contrast in the nature of the tests to be considered compared to those included in *UR, Vol. 1*. The tests in the latter were largely parametric in the sense that they were concerned with estimation in the context of an assumed parametric structure, usually in the form of an AR or ARMA model, the slight exception being in estimating the long-run variance based on a semi-parametric procedure. In this volume nonparametric tests are considered in greater detail, where such tests use information in the data, such as ranks, signs and runs, that do not require it to be structured by a model.

Chapter 3: Fractional integration

A considerable growth area in the development of techniques for nonstationary time series has been the extension to non-integer – or fractional – integration, a concept that is intimately tied to the idea of long memory. The basic idea has a long history, for example an influential paper by Hurst (1951) on the long-run capacity of reservoirs, introduced the Hurst exponent, which is related to the degree of fractional differencing necessary to ensure stationarity. As is the case with so many ideas in the development of econometrics, a paper by Granger, in this case co-authored with Joyeux (Granger and Joyeux, 1980), was seminal in bringing a number of essential concepts to the attention of econometricians. The basic idea is simple: consider an I(d) process that generates a series of observations on y_t, where d is the order of integration (the minimum number of differences necessary to ensure that y_t becomes stationary), then there is no need for d to be an integer. The standard DF test that assumes that d is either one (contains a single unit root) or zero (is stationary). There are two central aims in Chapter 3. The first is to lay out the I(d) framework, defining a process that is fractionally integrated and indicating how such processes may be generated; the second is to consider extensions of the unit root testing approach to tests for a fractional unit root.

Chapter 4: Semi-parametric estimation of the long-memory parameter

This chapter deals with the separate but related problem of estimating the long-memory parameter. The methods considered in this chapter are semi-parametric in the sense of not involving a complete parametric specification, in contrast to the extension of the autoregressive integrated moving average to the fractional case (ARFIMA models), which was considered in Chapter 3. Typically such methods either leave the short-run dynamic part of the model unspecified, or do not use all of the frequency range in estimating the long-memory parameter. Research on these methods has been a considerable growth area in the last ten years or so, leading to a potentially bewildering number of estimation methods.

Chapter 5: Smooth transition nonlinear models

The application of random walk models to some economic time series can be inappropriate when there are natural bounds to a series. For example, unemployment rates are bounded by zero and one, and negative nominal interest rates are not observed. There are a number of models that allow unit root-type behaviour, but take into account either the limits set by the data range or the economic mechanisms that are likely to prevent unbounded random walks in the data. These models have in common that they involve some form of nonlinearity. The models considered in this chapter involve a smooth form of nonlinearity, including smooth transition autoregressions based on the familiar AR model, allowing the AR coefficients to change as a function of the deviation of an 'activating' or 'exciting' variable from its target or threshold value. A second form of nonlinear model is the bounded random walk, referred to as a BRW, which allows random walk behaviour to be contained by bounds, or buffers, that 'bounce' the variable back if it is getting out of control.

Chapter 6: Threshold autoregressions

Another form of nonlinearity is when two or more linear AR models are discretely 'pieced' together, where each piece is referred to as a regime, and the indicator as to which regime is operating is based on the idea of a threshold parameter or threshold variable. This class of models is known as Threshold AR, or TAR, models. When the regime indicator is itself related to the dependent variable, TAR models are referred to as self-exciting, that is SETAR. These models are naturally extended to more than two regimes. In some ways these models share the kind of structural instability for which the well known Chow tests were designed; however, in contrast to the Chow case considered in calendar time, the regimes are not constrained to be composed of adjacent observations. The particular complication considered in this chapter allows a unit root in one or more of the regimes, so that not only may the component models be piecewise linear, they may be piecewise nonstationary.

Chapter 7: Structural breaks in AR models

Chapter 7 deals with a form of structural instability related to the passage of time. Perron's (1989) seminal article began an important reinterpretation of results from the unit root literature. What if, instead of a unit root process generating the data, there was a trend subject to a break due to 'exceptional' events? How would standard unit root tests perform, for example, what would be their power characteristics if the break was ignored in the alternative hypothesis?

The idea that regime change, rather than a unit root, was the cause of nonstationarity led to a fundamental re-evaluation of the simplicity of the dichotomy of 'opposing' the mechanisms of a unit root process on the one hand and a trend stationary process on the other. Examples of events momentous enough to be considered as structural breaks include the Great Depression (or Crash), the Second World War, the 1973 OPEC oil crisis and more recently '9/11', the financial crisis (the 'credit crunch') of 2008 and the government debt problem in the Euro area. Under the alternative hypothesis of stationarity, trended time series affected by such events could be modelled as stationary around a broken or split trend. Chapter 7 is primarily concerned with the development of an appropriate modelling framework for assessing the impact of a possible trend or mean break at a *known* break date, which is necessary to understand the more realistic cases dealt with in Chapter 8.

Chapter 8: Structural breaks with unknown break dates

In practice, although there are likely to be some contemporaneous and, later, historical indications of regime changes, there is almost inevitably likely to be uncertainty about not only the dating of such changes but also the nature of the changes. This poses another set of problems for econometric applications. If a break is presumed, when did it occur? Which model captures the nature of the break? If multiple breaks occurred, when did they occur? These are the concerns of Chapter 8.

Chapter 9: Conditional heteroscedasticity and unit root tests

Conditional heteroscedasticity is an important area of research and application in econometrics, especially since Engle's seminal article in 1982, which led to a voluminous literature that is especially relevant to modelling financial markets and where there are volatility 'clusters', such as in commodity prices. This chapter starts with an extension of the standard AR modelling framework that allows stochastic rather than deterministic coefficiencients; in context this means that a unit root, if it is present, may be stochastic rather than fixed (referred to as StUR). A model of this form gives rise to heteroscedasticity that is linked to the lagged dependent variable, which is a scale variable in the heteroscedasticity function. The chapter then covers the impact of ARCH/GARCH-type heteroscedasticity on unit root tests. The two types of conditional heteroscedasticity have different implications for conventional DF-type test statistics for a unit root.

The graphs in this book were prepared with MATLAB, www.mathworks.co.uk, which was also used, together with TSP (www.tspintl.com) and RATS (www.estima.com), for the programs written for this book. Martinez and Martinez (2002) again provided an invaluable guide to the use of MATLAB, and time and time again Hanselman and Littlefield (2005) was an essential and highly recommended companion to the use of MATLAB.

I would like to thank my publishers Palgrave Macmillan, and Taiba Batool and Ellie Shillito in particular, for their continued support not only for this book but in commissioning the series of which it is a part, namely 'Palgrave Texts in Econometrics', and more recently extending the concept in a practical way to a new series, 'Palgrave Advanced Texts in Econometrics', inaugurated by publication of *The Foundations of Time Series Analysis* (Mills, 2011). These two series have, at the time of writing, led to eight books, with more in press and, combined with the *Handbook of Econometrics* published in two volumes (Mills and Patterson, 2006, 2009), have made an important contribution to scholarly work and educational facilities in econometrics. Moreover, as noted above, the increasing ubiquity of modern econometric techniques means that the 'reach' of these contributions extends into non-economic areas, such as climatology, geography, meteorology, sociology, political science and hydrology.

In the earlier stages of preparation of the manuscript and figures for this book (as in the preparation of other books) I had the advantage of the willing and able help of Lorna Eames, my secretary in the School of Economics at the University of Reading, to whom I am grateful. Lorna Eames retired in April 2011. Those who have struggled with the many tasks involved in the preparation of a mansuscript, particularly one of some length, will know the importance of having a reliable and willing aide; Lorna was exceptionally that person.

If you have comments on any aspects of the book, please contact me at my email address given below.

k.d.patterson@reading.ac.uk

1
Some Common Themes

Introduction

The first four sections of this chapter review some basic concepts that, in part, serve to establish a common notation and define some underlying concepts for later chapters. In particular Section **1.1** considers the nature of a stochastic process, which is the conceptualisation underlying the generation of time series data. Section **1.2** considers stationarity in its strict and weak forms. As in Volume 1, the parametric models fitted to univariate time series are often variants of ARIMA models, so these are briefly reviewed in Section **1.3** and the concept of the long-run variance, which is often a key component of unit root tests where there is weak dependency in the errors, is outlined in Section **1.4**. Section **1.5** is more substantive and considers, by means of some simple illustrations, a problem and its solution that arises in Chapters 5, 6 and 8. The problem is that of designing a test statistic when there is a parameter that is not identified under the null hypothesis. The two cases considered here are variants of the Chow test (Chow, 1960), which tests for stability in a regression model. This section also considers how to devise a routine to obtain the bootstrap distribution of a test statistic and the bootstrap p-value of a sample test statistic.

1.1 Stochastic processes

An observed time series is a sample path or realisation of a sequence of random variables over an interval of time. Critical to understanding how such sample paths arise is the concept of a stochastic process, which involves a probability space defined over time. For simplicity we will consider stochastic processes defined in discrete time, so that the evolving index is t, $t = 1, \ldots, T$, a subset of the positive integers N^+. Denote the random variables that are the components of the stochastic process as $\varepsilon_t(\omega)$; Ω_t is the sample space of the random variable ε_t and $\omega \in \Omega_t$ indicates that ω is an element of Ω_t. (Note that ω is here used generically to denote an element of a sample space, the notation for a random variable is indicative, for example in particular applications it may be $y_t(\omega)$, $x_t(\omega)$ and so on and the sample space will be appropriately distinguished.)

In the case of a stochastic process, the sample space is not just the space of a single random variable but the sample space Ω of the sequence of random variables of length T. When there is no dependence between the random variables ε_t and ε_s for $t \neq s$ and all s, the sequence sample space Ω is particularly easy to determine in terms of the underlying sample spaces Ω_t, $t = 1, \ldots, T$ and an example is given below.

A discrete-time stochastic process, $\varepsilon(t)$, with $T \subseteq N^+$ may be summarised as:

$$\varepsilon(t) = (\varepsilon_t(\omega) : t \in T \subseteq N^+, \omega \in \Omega) \tag{1.1}$$

A particular realisation of $\varepsilon_t(\omega)$ is a sample 'observation', that is, a single number – a point on the overall sample path. Varying the element $\omega \in \Omega$, whilst keeping time constant, $t = s$, gives a distribution of outcomes at that point and so a 'slice' through $\varepsilon(t)$ at time $t = s$. On the other hand, when choosing a particular sequence from $\omega \in \Omega$, $\varepsilon_t(\omega)$ can then be regarded as a function of time, $t \in T$, and an 'outcome' is a complete sample path, $t = 1, \ldots, T$. By varying $\omega \in \Omega$, a different sample path is obtained, so that there are (potentially) different realisations for all $t \in T$.

Often the reference to $\omega \in \Omega$ is suppressed and a single random variable in the stochastic process is written ε_t, but the underlying dependence on the sample space should be recognised and means that different $\omega \in \Omega$ give rise to potentially different sample paths.

To illustrate the concepts consider a very simple example. In this case the random variable x arises from the single toss of a coin at time t, say x_t; this has the sample space $\Omega_1^x = \{H, T\}$ where italicised H and T refer to the coin landing heads or tails, respectively. Next suppose that T tosses of a coin take place sequentially in time, then the (sequence) sample space Ω^x is of dimension 2^T and ω, the generic element of Ω^x, refers to a T-dimensional *ordered* sequence. Assume that the coin tosses are independent, then the sample space Ω^x is just the product space of the individual sample spaces, $\Omega^x = \Omega_1^x \times \Omega_1^x \times \ldots \times \Omega_1^x = (\Omega_1^x)^T$ (the $\times$ symbol indicates the Cartesian product). By fixing $\omega \in \Omega^x$, where ω is a sequence, a whole path is determined, not just a single element at one point in time; thus as ω is varied, the whole sample path is varied, at least potentially. For example, suppose in the coin tossing experiment $T = 3$, then the sample space Ω^x for three consecutive tosses of a coin is $\Omega^x = \{(HHH), (HHT), (HTH), (HTT), (TTT), (TTH), (THT), (THH)\}$, which comprises $8 = 2^3$ elements, each one of which is a sample path.

Now map the outcomes of the coin toss into a (derived) random variable ε with outcomes $+1$ and -1; the sample space of $\varepsilon(t)$ for $T = 3$ is then $\Omega^\varepsilon = \{(1)(1)(1), (1)(1)(-1), (1)(-1)(1), (1)(-1)(-1), (-1)(-1)(-1), (-1)(-1)(1), (-1)(1)(-1),$

$(-1)(1)(1)\}$, again each of the 8 components is a sequence of $T = 3$ elements and each one of these ordered sequences is a sample path for $t = 1, 2, 3$; for example, the realisation $\{1, -1, 1\}$ is the path where the coin landed H, then T and finally H. Since by assumption H and T are equally likely the probability of each sample path is $1/2^3 = 1/8$.

Whilst this example is rather trivial, an interesting stochastic process can be derived from the coin-tossing game. Define the random variable $y_t = \sum_{j=1}^{t} \varepsilon_j = y_{t-1} + \varepsilon_t$, which further defines another stochastic process, say $Y(t)$ in terms of $\varepsilon(t)$. If $T = 3$, then the sample space of y_t, Ω^y, is simply related to Ω^ε, specifically $\Omega^y = \{(1)(2)(3), (1)(2)(1), (1)(0)(1), (1)(0)(-1), (-1)(-2)(-3)), (-1)(-2)(1), (-1)(0)(-1), (-1)(0)(1)\}$, and each ordered component of the sequence defines a sample path. The resulting stochastic process is a binomial random walk and in this case, as it has been assumed that the coin is fair, the random walk is symmetric. Whilst the original stochastic process for ε_t is stationary, that for y_t is not; the definition of stationarity is considered in the next section. A simple generalisation of this partial sum process is to generate the ε_t as draws from a normally distributed random variable, giving rise to a Gaussian random walk.

Stochastic processes arise in many different applications. For example, processes involving discrete non-negative numbers of arrivals or events are often modelled as a Poisson process, an illustrative example being the number of shoppers at a supermarket checkout in a given interval (for an illustration see A Primer, chapter 3). A base model for returns on financial assets is that they follow a random walk, possibly generated by a 'fat-tailed' distribution with a higher probability of extreme events compared to a random walk driven by Gaussian inputs. To illustrate, a partial sum process was defined as in the case of the binomial random walk with the exception that the 'inputs' ε_t were drawn from a Cauchy distribution (for which neither the mean nor the second moment exist), so that there was a greater chance of 'outliers' compared to draws from a normal distribution. The fat tails allow the simulations to mimic crisis events such as the events following 9/11 and the 2008 credit crunch. Figure 1.1 shows 4 of the resulting sample paths over $t = 1, \ldots, 200$. The impact of the fat tails is evident and, as large positive and large negative shocks can and do occur, the sample paths (or trajectories) show some substantial dips and climbs over the sample period.

1.2 Stationarity and some of its implications

Stationarity captures the idea that certain properties of a random or stochastic process generating the data are unchanging. If the process does not change at all over time, so that the underlying probability space (see A Primer, chapter 1) is constant, it does not matter which sample portion of observations are used to estimate the parameters of the process and we may as well, to improve the

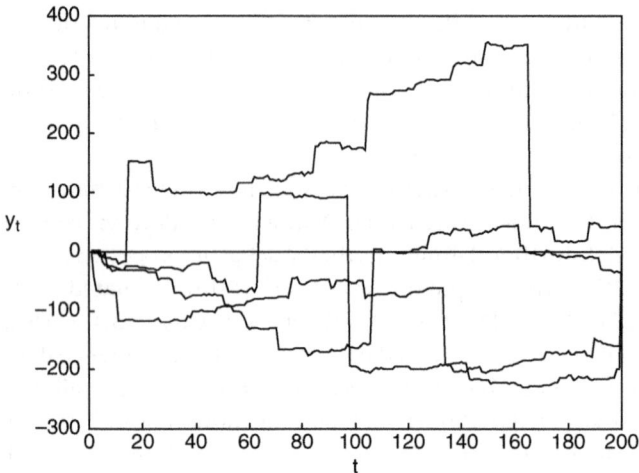

Figure 1.1   Illustrative fat-tailed random walks

efficiency of estimation and inference, use all available observations. However, stationarity comes in different 'strengths', starting with a strictly or completely stationary process. A stochastic process that is not stationary is said to be nonstationary.

1.2.1   A strictly stationary process

Let $\tau \neq s$ and T be arbitrary, if Y(t) is a strictly stationary, discrete-time stochastic process for a discrete random variable, y_t, then:

$$P(y_{\tau+1}, y_{\tau+2}, \ldots, y_{\tau+T}) = P(y_{s+1}, y_{s+2}, \ldots, y_{s+T}) \tag{1.2}$$

where P(.) is the joint probability mass function (pmf) for the random variables enclosed in (.). Strict stationarity requires that the joint pmf for the sequence of random variables of length T starting at time $\tau + 1$ is the same for any shift in the time index from τ to s and for any choice of T. This means that it does not matter which T-length portion of the sequence is observed, each has been generated by the same unchanging probability structure. A special case of this result in the discrete case is that for $T = 1$ where $P(y_\tau) = P(y_s)$, so that the marginal pmfs must also be the same for $\tau \neq s$ implying that $E(y_\tau) = E(y_s)$. These results imply that other moments, including joint moments such as the covariances, are invariant to arbitrary time shifts.

1.2.2   Stationarity up to order m

A weaker form of stationarity is where the moments up to order m are the same when comparing the (arbitrary) sequences $(y_{\tau+1}, y_{\tau+2}, \ldots, y_{\tau+T})$ and $(y_{s+1}, y_{s+2}, \ldots, y_{s+T})$. The following definition is adapted from Priestley (1981, Definition 3.3.2):

the process Y(t) is said to be stationary of order m if the joint moments up to order m of $(y_{\tau+1}, y_{\tau+2}, \ldots, y_{\tau+T})$ exist and are equal to the corresponding joint moments for $(y_{s+1}, y_{s+2}, \ldots, y_{s+T})$ for arbitrary τ, s and T.

This definition implies that if a process is stationary of order m then it is also stationary of order m − 1. The case most often used is m = 2, referred to as 2nd order, weak or covariance stationarity. It implies the following three conditions (the first of which is due to m = 1):
if a process is stationary of order 2 then for arbitrary τ and s, $\tau \neq s$:

$$E(y_\tau) = E(y_s) = \mu_y \Rightarrow \text{the mean of the process is constant}$$
$$\text{var}(y_\tau) = \text{var}(y_s) = \sigma_y^2 \Rightarrow \text{the variance of the process is constant}$$
$$\text{cov}(y_\tau, y_{\tau+k}) = \text{cov}(y_s, y_{s+k}) \Rightarrow \text{the autocovariances are invariant to}$$
$$\text{translation for all k}$$

Ross (2003) gives examples of processes that are weakly stationary but not strictly stationary; also, a process could be strictly stationary, but not weakly stationary by virtue of the non-existence of its moments. For example, a random process where the components have unchanging marginal and joint Cauchy distributions will be strictly stationary, but not weakly stationary because the moments do not exist.

The leading case of nonstationarity, at least in econometric terms, is that induced by a unit root in the AR polynomial of an ARMA model for y_t, which implies that the variance of y_t is not constant over time and that the k-th order autocovariance of y_t depends on t. A process that is trend stationary is one that is stationary after removal of the trend. A case that occurs in this context is where the process generates data as $y_t = \beta_0 + \beta_1 t + \varepsilon_t$, where ε_t is a zero mean weakly stationary process. This is not stationary of order one because $E(y_t) = \beta_0 + \beta_1 t$, which is not invariant to t. However, if the trend component is removed either as $E(y_t - \beta_1 t) = \beta_0$ or $E(y_t - \beta_0 - \beta_1 t) = 0$, the result is a stationary process.

Returning to the partial sum process $y_t = \sum_{j=1}^{t} \varepsilon_j$, with ε_t a binomial random variable, note that $E(\varepsilon_t) = 0$, $E(\varepsilon_t^2) = \sigma_\varepsilon^2 = 1$ and $E(\varepsilon_t \varepsilon_s) = 0$ for $t \neq s$. However, in the case of y_t, whilst $E(y_t) = \sum_{j=1}^{t} E(\varepsilon_j) = 0$, the variance of y_t is not constant as t varies and $E(y_t^2) = t\sigma_\varepsilon^2 = t$. Moreover, the autocovariances are neither equal to zero nor invariant to a translation in time; for example, consider the first-order autocovariances $\text{cov}(y_1, y_2)$ and $\text{cov}(y_2, y_3)$, which are one period apart, then:

$$\text{cov}(y_1, y_2) = \sigma_\varepsilon^2 + \text{cov}(\varepsilon_1, \varepsilon_2) \tag{1.3a}$$
$$= \sigma_\varepsilon^2$$

$$\text{cov}(y_2, y_3) = 2\sigma_\varepsilon^2 + 2\text{cov}(\varepsilon_1, \varepsilon_2) + \text{cov}(\varepsilon_1, \varepsilon_3) + \text{cov}(\varepsilon_2, \varepsilon_3)$$
$$= 2\sigma_\varepsilon^2 \tag{1.3b}$$

The general result is that $\text{cov}(y_t, y_{t+1}) = t\sigma_\varepsilon^2$, so the partial sum process generating y_t is not covariance stationary. The assumption that ε_t a binomial random variable is not material to this argument, which generalises to any process with $E(\varepsilon_t^2) = \sigma_\varepsilon^2$ and $E(\varepsilon_t \varepsilon_s) = 0$, such as $\varepsilon_t \sim N(0, \sigma_\varepsilon^2)$.

1.3 ARMA(p, q) models

ARMA (autoregressive, moving average) models form an important 'default' model in parametric models involving a unit root, and the notation of such models is established here for use in later chapters. An ARMA model of order p in the AR component and q in the MA component for the univariate process generating y_t, assuming that $E(y_t) = 0$, is written as follows:

$$\phi(L)y_t = \theta(L)\varepsilon_t \tag{1.4}$$

$$\phi(L) = (1 - \phi_1 L - \phi_2 L^2 - \ldots - \phi_p L^p) \qquad \text{AR(p) polynomial} \tag{1.5a}$$

$$\theta(L) = (1 + \theta_1 L + \theta_2 L^2 + \ldots + \theta_q L^q) \qquad \text{MA(q) polynomial} \tag{1.5b}$$

L is the lag operator, sometimes referred to as the backshift operator, usually denoted B, in the statistics literature, and defined as $L^j y_t \equiv y_{t-j}$, (see UR, Vol. 1, Appendix 2). The sequence $\{\varepsilon_t\}_{t=1}^T$ comprises the random variables ε_t, $t = 1, \ldots, T$, assumed to be independently and identically distributed for all t, with zero mean and constant variance, σ_ε^2, written as $\varepsilon_t \sim \text{iid}(0, \sigma_\varepsilon^2)$; if the (identical) distribution is normal, then $\varepsilon_t \sim \text{niid}(0, \sigma_\varepsilon^2)$.

Note that the respective sums of the coefficients in $\phi(L)$ and $\theta(L)$ are obtained by setting $L = 1$ in (1.5a) and (1.5b), respectively. That is:

$$\phi(L = 1) = \left(1 - \sum_{i=1}^{p} \phi_i\right) \text{ sum of AR(p) coefficients} \tag{1.6a}$$

$$\theta(L = 1) = \left(1 + \sum_{j=1}^{q} \theta_j\right) \text{ sum of MA(q) coefficients} \tag{1.6b}$$

For economy of notation it is conventional to use, for example, $\phi(1)$ rather than $\phi(L = 1)$. By taking the AR terms to the right-hand side, the ARMA(p, q) model is written as:

$$y_t = \phi_1 y_{t-1} + \ldots + \phi_p y_{t-p} + \varepsilon_t + \theta_1 \varepsilon_{t-1} + \ldots + \theta_q \varepsilon_{t-q} \tag{1.7}$$

Equation (1.4) assumed that $E(y_t) = 0$, if $E(y_t) = \mu_t \neq 0$, then y_t is replaced by $y_t - \mu_t$, and the ARMA model is written as:

$$\phi(L)(y_t - \mu_t) = \theta(L)\varepsilon_t \tag{1.8}$$

The term μ_t has the interpretation of a trend function, the simplest and most frequently occurring cases being where y_t has a constant mean, so that $\mu_t = \mu$, and y_t has a linear trend, so that $\mu_t = \beta_0 + \beta_1 t$.

The ARMA model can then be written in deviations form by first defining $\tilde{y}_t \equiv y_t - \mu_t$, with the interpretation that $\tilde{y}_t$ is the detrended (or demeaned) data, so that (1.8) becomes:

$$\phi(L)\bar{y}_t = \theta(L)\varepsilon_t \tag{1.9}$$

In practice, μ_t is unknown and a consistent estimator, say $\hat{\mu}_t$, replaces μ_t, so that the estimated detrended data is $\hat{\bar{y}}_t = y_t - \hat{\mu}_t$. Typically, in the constant mean case, the 'global' mean $\bar{y} = T^{-1}\sum_{t=1}^{T} y_t$ is used for $\hat{\mu}_t$, and in the linear trend case β_0 and β_1 are replaced by their LS estimators (denoted by $\hat{}$ over) from the prior regression of y_t on a constant and a time trend, so that $\hat{\mu}_t = \hat{\beta}_0 + \hat{\beta}_1 t$.

With $\hat{\mu}_t$ replacing μ_t, (1.9) becomes:

$$\phi(L)\hat{\bar{y}}_t = \theta(L)\varepsilon_t + \xi_t \tag{1.10}$$

where $\xi_t = \phi(L)(\mu_t - \hat{\mu}_t)$.

An alternative to writing the model as in (1.8) is to take the deterministic terms to the right-hand side, so that:

$$\phi(L)y_t = \mu_t^* + \theta(L)\varepsilon_t \tag{1.11}$$

where, for consistency with (1.8), $\mu_t^* = \phi(L)\mu_t$. For example, in the constant mean case and linear trend cases μ_t^* is given, respectively, by:

$$\mu_t^* = \mu^*$$

$$= \phi(1)\mu \tag{1.12a}$$

$$\mu_t^* = \phi(L)(\beta_0 + \beta_1 t) \tag{1.12b}$$

$$= \phi(1)\beta_0 + \beta_1\phi(L)t$$

$$= \beta_0^* + \beta_1^* t$$

where $\beta_0^* = \phi(1)\beta_0 + \beta_1\sum_{j=1}^{P} j\phi_j$, $\beta_1^* = \phi(1)\beta_1$ and $\phi(1) = 1 - \sum_{j=1}^{P}\phi_j$.

An equivalent representation of the ARMA(p, q) model is to factor out the dominant root, so that $\phi(L) = (1 - \rho L)\varphi(L)$, where $\varphi(L)$ is invertible, and the resulting model is specified as follows:

$$(1 - \rho L)y_t = z_t \tag{1.13a}$$

$$\varphi(L)z_t = \theta(L)\varepsilon_t \tag{1.13b}$$

$$z_t = \varphi(L)\theta(L)\varepsilon_t \tag{1.13c}$$

This error dynamics approach (see UR, Vol. 1, chapter 3) has the advantage of isolating all dynamics apart from that which might be associated with a unit root ($\rho = 1$ in this case) into the error z_t.

1.4 The long-run variance

An important concept in unit root tests is the long-run variance. It is one of three variances that can be defined when y_t is determined dynamically, depending on the extent of conditioning in the variance. This is best illustrated with a simple example and then generalised. Consider an AR(1) model:

$$y_t = \rho y_{t-1} + \varepsilon_t \qquad |\rho| < 1 \tag{1.14}$$

Then the three variances are distinguished as follows.

1. Conditional variance: $\text{var}(y_t|y_{t-1}) = \sigma_\varepsilon^2$, keeps everything but y_t constant.
2. Unconditional variance, $\sigma_y^2 \equiv \text{var}(y_t)$, allows variation in y_{t-1} to be taken into account:

$$\begin{aligned}
\sigma_y^2 &= \text{var}(\rho y_{t-1} + \varepsilon_t) \\
&= \rho^2 \text{var}(y_{t-1}) + \sigma_\varepsilon^2 + 2\rho\text{cov}(y_{t-1}\varepsilon_t) \\
&= \rho^2 \sigma_y^2 + \sigma_\varepsilon^2 \text{cov}(y_{t-1}\varepsilon_t) = 0 \\
&= (1 - \rho^2)^{-1}\sigma_\varepsilon^2
\end{aligned} \tag{1.15}$$

3. Long-run variance, $\sigma_{y,\text{lr}}^2$: allows variation in all lags of y_t to be taken into account:

$$\begin{aligned}
y_t &= \rho y_{t-1} + \varepsilon_t \\
&= (1 - \rho L)^{-1}\varepsilon_t \\
&= \sum_{i=0}^{\infty} \rho^i \varepsilon_{t-i} \quad \text{moving average form}
\end{aligned} \tag{1.16}$$

$$\begin{aligned}
\sigma_{y,\text{lr}}^2 &= \text{var}\left(\sum_{i=0}^{\infty} \rho^i \varepsilon_{t-i}\right) \\
&= \sum_{i=0}^{\infty} \rho^{2i}\sigma_\varepsilon^2 \quad \text{because} \quad \text{cov}(\varepsilon_{t-i}\varepsilon_{t-j}) = 0 \quad \text{for} \quad i \neq j \\
&= (1 - \rho)^{-2}\sigma_\varepsilon^2
\end{aligned} \tag{1.17}$$

The last line of (1.17) uses $\sum_{i=0}^{\infty} \rho^{2i} = \sum_{i=0}^{\infty} (\rho^2)^i = (1 - \rho)^{-2}$ for $|\rho| < 1$. Also note that it suggests an easy way of obtaining $\sigma_{y,\text{lr}}^2$: (1.16) is the MA representation of the AR(1) model, with lag polynomial $(1 - \rho L)^{-1}$ and $\sigma_{y,\text{lr}}^2$ is the square of this polynomial evaluated at $L = 1$, scaled by σ_ε^2. Hence, in generalising this result, it will be of importance to obtain the MA representation of the model, which we now consider.

A linear model is described as being causal if there exists an absolutely summable sequence of constants $\{w_j\}_0^{\infty}$, such that:

$$\begin{aligned}
y_t &= \sum_{j=0}^{\infty} w_j L^j \varepsilon_t \\
&= w(L)\varepsilon_t
\end{aligned} \tag{1.18}$$

If $E(y_t) = \mu_t \neq 0$, where μ_t is a deterministic function, then y_t is replaced by $y_t - \mu_t$ as in (1.8). The condition of absolute summability is $\sum_{j=0}^{\infty} |w_j| < \infty$. The lag polynomial $w(L)$ is the causal linear (MA) filter governing the response of $\{y_t\}$ to $\{\varepsilon_t\}$.

In an ARMA model, the MA polynomial is $w(L) = \sum_{j=0}^{\infty} w_j L^j = \phi(L)^{-1}\theta(L)$, with $\omega_0 = 1$; for this representation to exist the roots of $\phi(L)$ must lie outside the unit circle, so that $\phi(L)^{-1}$ is defined. The long-run variance of y_t, $\sigma_{y,lr}^2$, is then just the variance of $\omega(1)\varepsilon_t$; that is, $\sigma_{y,lr}^2 = \text{var}[w(1)\varepsilon_t] = w(1)^2 \sigma_\varepsilon^2$.

1.5   The problem of a nuisance parameter only identified under the alternative hypothesis

This section considers the general nature of a problem that occurs in different developments of models with unit roots, the solution to which has applications in later chapters, particularly Chapters 5, 6 and 8. The problem of interest is sometimes referred to as the Davies problem, after Davies (1977, 1987), who considered hypothesis testing when there is a nuisance parameter that is only present under the alternative hypothesis. An archetypal case of such a problem is where the null model includes a linear trend, which is hypothesised to change at some particular point in the sample. If that breakpoint was known, obtaining a test statistic would just require the application of standard principles, for example an F test based on fitting the model under the alternative hypothesis. However, in practice, the breakpoint is unknown and one possibility is to search over all possible breakpoints and calculate the test statistic for each possibility; however, the result of this enumeration over all possibilities is that, the F test for example, no longer has an F distribution.

To illustrate the problem and its solution, we consider a straightforward problem of testing for structural stability in a bivariate regression with a non-stochastic regressor and then in the context of a simple AR model. At this stage the exposition does not emphasise the particular problems that arise if a unit root is present (either under the null or the alternative). The aim here is to set out a simple framework that is easily modified in later chapters.

1.5.1   Structural stability

Two models of structural instability are considered here to illustrate how to deal with the problem that a parameter is only identified under the alternative hypothesis, H_A. These models have an important common element, that of specifying two regimes under H_A, but only one under H_0. They differ in the specification of the regime 'separator'; in one case the regimes are separated by the passage of time, the observations being generated by regime 1 up to a switching (or break) point in time and thereafter they are generated by regime 2. This is the case for which the well-known Chow test(s) were designed, (Chow, 1960). This kind of model may be considered for structural changes arising from discrete events, often institutional in nature, for example a change in the tax regime or in political institutions.

In the second case there are again two regimes, but they are separated by an endogenous indicator function, rather than the exogenous passage of time. The typical case here is where the data sequence $\{y_t\}_{t=1}^T$ is generated by an AR(p) model under H_0, whereas under H_A the coefficients of the AR(p) model change depending on a function of lagged y_t relative to a threshold κ. In both models the nature of the change is quite limited. According to H_A the basic structure of the model is not changed, but the coefficients are allowed to change. The common technical problem of interest here is that typically the structural breakpoint or threshold is not known, whereas standard tests assume that these are known.

Consider the standard problem of applying the Chow test for a structural break. A simple bivariate model, with a nonstochastic regressor, is sufficient to illustrate the problem.

$$y_t = \alpha_0 + \alpha_1 x_t + \varepsilon_t \quad t = 1,\dots,T \tag{1.19}$$

where $\varepsilon_t \sim \text{iid}(0, \sigma_\varepsilon^2)$, that is $\{\varepsilon_t\}_{t=1}^T$ is a sequence of independent and identically distributed 'shocks' with zero mean and constant variance, σ_ε^2; and, along the lines of an introductory approach, assume that x_t is fixed in repeated samples. The problem is to assess whether the regression model has been 'structurally stable' in the sense that the regression coefficients α_0 and α_1 have been constant over the period $t = 1, \dots T$, where for simplicity σ_ε^2 is assumed a constant. There are many possible schemes for inconstancy, but one that is particularly simple and has found favour is to split the overall period of T observations by introducing a breakpoint (in time) T_b, such that there are T_1 observations in period 1 followed by T_2 observations in period 2, thus $T = T_1 + T_2$ and $t = 1, \dots, T_1, T_1 + 1, \dots, T$; note that $T_b = T_1$. Such a structural break is associated with a discrete change related to external events.

According to this scheme the regression model may be written as:

$$y_t = \alpha_0 + \alpha_1 x_t + I(t > T_1)\delta_0 + I(t > T_1)\delta_1 x_t + \varepsilon_t \tag{1.20a}$$

$$y_t = \alpha_0 + \alpha_1 x_t + \varepsilon_t \qquad \text{if } 0 \leq t \leq T_1 \tag{1.20b}$$

$$y_t = \alpha_0 + \delta_0 + (\alpha_1 + \delta_1)x_t + \varepsilon_t \qquad \text{if } T_1 + 1 \leq t \leq T \tag{1.20c}$$

where $I(t > T_1) = 1$ if the condition in (.) is true and 0 otherwise. There will be several uses of this indicator function in later chapters. The null hypothesis of no change is H_0: $\delta_0 = \delta_1 = 0$ and the alternative hypothesis is that one or both of the regression coefficients has changed H_A: $\delta_0 \neq 0$ and/or $\delta_1 \neq 0$.

A regression for which H_0 is not rejected is said to be stable. A standard test statistic, assuming that, as we do here, there are sufficient observations in each regime to enable estimation, is an F test, which will be distributed as $F(2, T - 4)$ if we add the assumption that ε_t is normally distributed (so that it is niid) or T is sufficiently large to enable the central limit theorem to deliver normality. The F test statistic, first in general, is:

$$C = \frac{\hat{\varepsilon}_r'\hat{\varepsilon}_r - \hat{\varepsilon}_{ur}'\hat{\varepsilon}_{ur}}{\hat{\varepsilon}_{ur}'\hat{\varepsilon}_{ur}} \left(\frac{T-K}{g} \right) \sim F(g, T-K) \tag{1.21}$$

Specifically in this case $K = 4$ and $g = 2$, therefore:

$$C = \frac{\hat{\varepsilon}_r'\hat{\varepsilon}_r - \hat{\varepsilon}_{ur}'\hat{\varepsilon}_{ur}}{\hat{\varepsilon}_{ur}'\hat{\varepsilon}_{ur}} \left(\frac{T-4}{2} \right) \sim F(2, T-4) \tag{1.22}$$

where $\hat{\varepsilon}_r$ and $\hat{\varepsilon}_{ur}$ are the vectors of residuals from fitting (1.19) and (1.20a), respectively. At the moment the assumption that x_t is fixed in repeated samples removes complications due to possible stochastic regressors.

The F test is one form of the Chow test (Chow, 1960) and obviously generalises quite simply to more regressors. A large sample version of the Chow test uses the result that if a random variable, C, has the $F(g, T - K)$ distribution, then $gC \Rightarrow_D \chi^2(g)$, where $\Rightarrow_D$ indicates weak convergence or convergence in distribution (see A Primer, chapter 4). The test is sometimes presented in Wald form as:

$$W = T \left(\frac{\hat{\varepsilon}_r'\hat{\varepsilon}_r - \hat{\varepsilon}_{ur}'\hat{\varepsilon}_{ur}}{\hat{\varepsilon}_{ur}'\hat{\varepsilon}_{ur}} \right) \tag{1.23}$$

$$= g \left(\frac{T}{T-k} \right) C$$

$$\approx gC$$

$$W \Rightarrow_D \chi^2(g) \tag{1.24}$$

So that W is asymptotically distributed as χ^2 with g degrees of freedom.

The problem to be focussed on here is when, as is usually the case, the breakpoint is unknown rather than known. In such a case it may be tempting and, indeed, it may be regarded as 'good' econometrics to assess such a regression for stability by running the Chow test for all possible breakpoints. In this sense the test is being used as a regression diagnostic test. 'All possible' allows for some trimming to allow enough observations to empirically identify the breakpoint, if it is present (it is also essential for the distributional results to hold that λ_b is in the interior of the set $\Theta = (0, 1)$). A typical trimming would be 10% or 15% of the observations at either end of the sample.

The problem with checking for all possible breakpoints is that the Chow test no longer has the standard F distribution. To illustrate the problem consider the testing situation comprising (1.20b) and (1.20c), with $T = 100$ and define $\lambda_b = T_1/T$ as the break proportion, for simplicity ensuring that λ_b is an integer. Then simulate the model in (1.19) R times, with $\varepsilon_t \sim niid(0, 1)$, and note that σ_ε^2 may be set to unity without loss, each time computing C. Two variations are considered. In the first, the simplest situation, assume that λ_b is fixed for

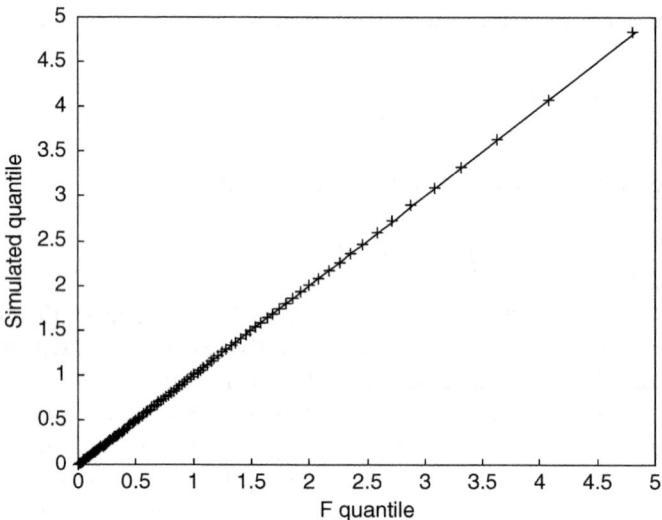

Figure 1.2   Quantile plot against F distribution: λ_b known

each set of simulations and, for simplicity $\lambda_b = 0.5$ is used for this illustration. Figure 1.2 shows a quantile-quantile plot taking the first quantile from $F(2, T - 4)$ and the matching quantile from the simulated distribution of C, based on R = 50,000 replications; as a reference point, Figure 1.2 also shows as the solid line the 45°line resulting from pairing the quantiles of $F(2, T - 4)$ against themselves. The conformity of the simulated and theoretical distributions is evident from the figure.

In the second case, λ_b is still taken as given for a particular set of R replications, but λ_b is then varied to generate further sets of R replications; in this case $\lambda_b \in \Theta = [0.10, 90]$ so that there is 10% trimming and the range of Θ is divided into a grid of G = 81 equally spaced points (so that each value of T = 100 is considered as a possible break point T_b). This enables an assessment of the variation, if any, in the simulated distributions as a function of λ_b (but still taking λ_b as given for each set of simulations). The 95% quantiles for each of the resulting distributions are shown in Figure 1.3, from which it is evident that although there are some varaiations these are practically invariant even for λ_b close to the beginning and end of the sample; (the maximum number of observations in each regime is at the centre of the sample period, so that the number of observations in one of the regimes declines as $\lambda_b \to 0.1$ or $\lambda_b \to 0.9$).

As noted, the more likely case in practice is that a structural break is suspected but the timing of it is uncertain, so that λ_b is unknown. The questions then arise as to what might be taken as a single test statistic in such a situation and

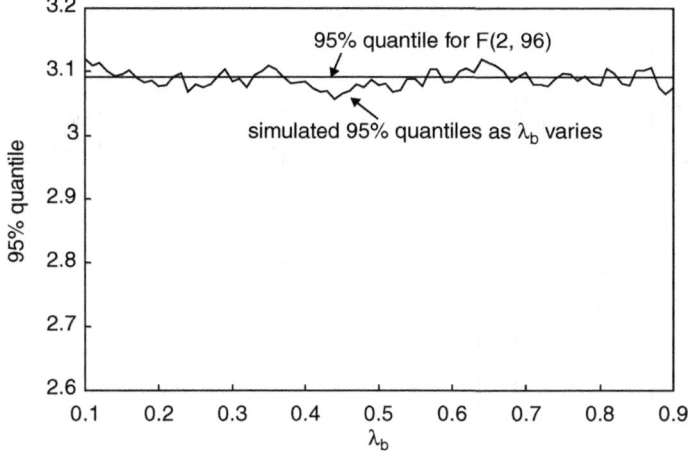

Figure 1.3   95% quantile for Chow test: λ_b known

what is the distribution of the chosen test statistic. One possibility is to look for the greatest evidence against the null hypothesis of structural stability in the set Θ, so that the chosen test statistic is the supremum of C, say Sup-C, over all possible breakpoints, with the breakpoint being estimated as that value of λ_b, say $\hat{\lambda} = \lambda_b^{max}$, that is associated with Sup-C. This idea is associated with the work of Andrews (1993) and Andrews and Ploberger (1994), who also suggested the arithmetic average and the exponential average of the C values over the set Θ. For reference these are as follows:

$$\text{Supremum}: \qquad \text{Sup-C}(\hat{\lambda}) = \sup_{\lambda_b \in \Theta} C(\lambda_b) \qquad (1.25)$$

$$\text{Mean}: \qquad C(\lambda_b)^{ave} = G^{-1} \sum_{\lambda_b \in \Theta} C(\lambda_b) \qquad (1.26)$$

$$\text{Exponential mean}: \qquad C(\lambda_b)^{exp} = \log\left(G^{-1} \sum_{\lambda_b \in \Theta}\right) \exp[C(\lambda_b)/2] \qquad (1.27)$$

Note that Andrews presents these test statistics in their Wald forms referring to, for example, Sup-W(.). The F form is used here as it is generally the more familiar form in which the Chow test is used; the translation to the Wald form was given in (1.23). Note that neither $C(\lambda_b)^{ave}$ nor $C(\lambda_b)^{exp}$ provide an estimate of the breakpoint.

Andrews (1993) obtained the asymptotic null distributions of Supremum versions of the Wald, LR and LM Chow tests and shows them to be standardised tied-down Bessel processes. Hansen (1997b) provides a method to obtain approximate asymptotic p-values for the Sup, Exp and Ave tests. Diebold and Chen (1996) consider structural change in an AR(1) model and suggested using a bootstrap procedure which allows for the dynamic nature of the regression

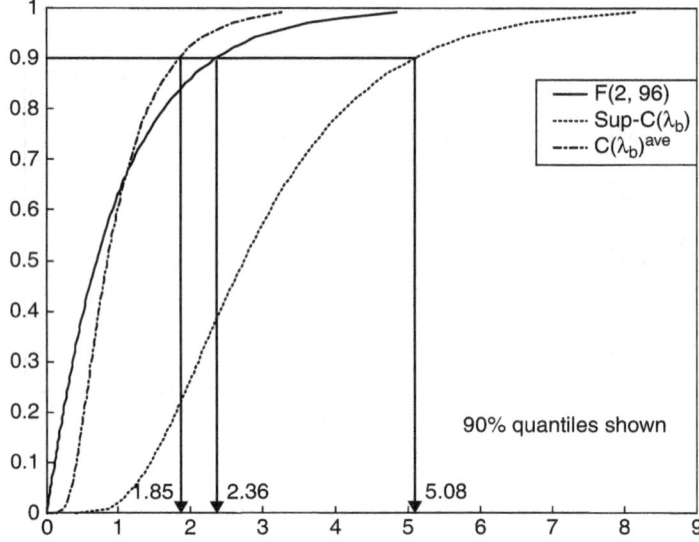

Figure 1.4   CDFs of F, Sup-C and C^{ave} tests

model and the practicality of a finite sample, for which asymptotic critical values are likely to be a poor guide.

Note that the distribution of Sup-C($\hat{\lambda}$) is not the same as the distribution of C(λ_b), where the latter is C evaluated at λ_b, so that quantiles of C(λ_b) should not be used for hypothesis testing where a search procedure has been used to estimate λ_b. The effect of the search procedure is that the quantiles of Sup-C($\hat{\lambda}$) are everywhere to the right of the quantiles of the corresponding F distribution. On the other hand the averaging procedure in C(λ_b)^{ave} tends to reduce the quantiles that are relevant for testing.

To illustrate, the CDFs for F(2, 96), Sup-C($\hat{\lambda}$) and C(λ_b)^{ave} are shown in Figure 1.4 for the procedure that searches over $\lambda_b \in \Theta = [0.10, 90]$, with G = 81 equally spaced points, for each simulation. For example, the 90% and 95% quantiles of the distributions of Sup-C($\hat{\lambda}$) are 5.08 and 6.02, and those of C(λ_b)^{ave} are 1.85 and 2.28, compared to 2.36 and 3.09 from F(2, 96).

1.5.2   Self-exciting threshold autoregression (SETAR)

A second example of a nuisance parameter that is only identified under the alternative hypothesis arises in a threshold autoregression. This can also be interpreted as model of the generation of y_t that is not structurally stable, but the form of the instability depends not on a fixed point in time, but on a threshold, κ, that is usually expressed in terms of a lagged value or values of y_t or the change in y_t. If the threshold is determined in this way such models are termed self-exciting, as the movement from one regime to another is generated within the model's own dynamics. This gives rise to the acronym SETAR for self-exciting threshold autoregression if the threshold depends on lagged y_t, and to MSETAR (sometimes shortened to MTAR) for momentum SETAR if the threshold depends on Δy_t.

The following illustrates a TAR of order one, TAR(1), which is 'self-exciting':

$$y_t = \varphi_0 + \varphi_1 y_{t-1} + \varepsilon_t \qquad \text{if } f(y_{t-1}) \leq \kappa, \text{ regime 1} \tag{1.28}$$

$$y_t = \phi_0 + \phi_1 y_{t-1} + \varepsilon_t \qquad \text{if } f(y_{t-1}) > \kappa, \text{ regime 2} \tag{1.29}$$

where $\varepsilon_t \sim \text{iid}(0, \sigma_\varepsilon^2)$, thus, for simplicity, the error variance is assumed constant across the two regimes; for expositional purposes we assume that $f(y_{t-1})$ is simply y_{t-1}, so that the regimes are determined with reference to a single lagged value of y_t. It is also assumed that $|\varphi_1| < 1$ and $|\phi_1| < 1$, so that there are no complications arising from unit roots (the unit root case is considered in chapter 6). Given this stability condition, the implied long-run value of y_t, y^*, depends upon which regime is generating the data, being $y_1^* = \varphi_0/(1 - \varphi_1)$ in Regime 1 and $y_2^* = \phi_0/(1 - \phi_1)$ in Regime 2.

The problem is to assess whether the regression model has been 'structurally stable', with stability defined as $H_0: \varphi_0 = \phi_0$ and $\varphi_1 = \phi_1$, whereas $H_A: \varphi_0 \neq \phi_0$ and/or $\varphi_1 \neq \phi_1$ is 'instability' in this context. As in the Chow case, the two regimes can be written as one generating model:

$$y_t = \phi_0 + \phi_1 y_{t-1} + I(y_{t-1} \leq \kappa)\gamma_0 + I(y_{t-1} \leq \kappa)\gamma_1 y_{t-1} + \varepsilon_t \tag{1.30}$$

where $\gamma_0 = \varphi_0 - \phi_0$ and $\gamma_1 = \varphi_1 - \phi_1$. (In Chapter 6, where TAR models are considered in greater detail, the indicator function is written more economically as I_{1t}.) The null and alternative hypotheses can then be formulated as $H_0: \gamma_0 = \gamma_1 = 0$ and $H_A: \gamma_0 \neq 0$ and/or $\gamma_1 \neq 0$, so that the overall model is analogous to (1.20a), with the following minor qualification. Specifically, conventions differ in specifying the indicator function; for the TAR model, Regime 2 is taken to be the base model, see Q1.1, whereas for the Chow test, Regime 1 was taken to be the base model. Note that the long-run value of y_t in Regime 1 is obtained as $y_1^* = (\phi_0 + \gamma_0)/[1 - (\phi_1 + \gamma_1)] = \varphi_0/(1 - \varphi_1)$.

The F test for H_0 against H_A, or its Wald, LM or LR form, parallels that for the Chow test and, with a suitable reinterpretation, the test statistic C of (1.23) here for *known* κ, provides a test statistic of the hypothesis of one regime against that of two regimes. However, the presence of stochastic regressors in the AR/TAR model means that C is not exactly $F(g, T - K)$ even in this case.

The parallel between the model of instability under the Chow formulation and the threshold specification should be clear and extends to the more likely situation of unknown κ; in both cases, the regime separator is only identified under H_A. The problems of estimating κ and, hence, the regime split, and testing for the existence of one regime against two regimes, are solved together and take the same form as the in the Chow test.

In respect of the first problem, specify a grid of N possible values for κ, say $\kappa_{(i)} \in$ **K** and then estimate (1.30) for each $\kappa_{(i)}$, taking as the estimate of κ the value of $\kappa_{(i)}$ that results in a minimum for the residual sum of squares over all possible values of $\kappa_{(i)}$. As in the Chow test, some consideration must be given to the range of the grid search. In a SETAR, κ is expected to be in the observed range of y_t, but this criterion must be coupled with the need to enable sufficient observations in each regime to empirically identify the TAR parameters. Thus, starting from the ordered values of y_t, say $y_{(1)} < y_{(2)} < \ldots < y_{(T)}$, some observations are trimmed from the beginning and end of the sequence $\{y_{(t)}\}_1^T$, with, typically, 10% or 15% of the observations trimmed out of the sample to define **K**.

Obtaining the minimum of the residual sum of squares from the grid search also leads to the Supremum F test: the residual sum of squares under H_0 is given for all values of $\kappa_{(i)}$, hence the F-test will be maximised where the residual sum of squares is minimised under H_A. The resulting estimator of κ is denoted $\hat{\kappa}$. The average and exponential average test statistics are defined as in (1.26) and (1.27), respectively. The test statistics are referred to as Sup-C($\hat{\kappa}$), $C(\kappa_{(i)})^{ave}$ and $C(\kappa_{(i)})^{exp}$ to reflect their dependence on $\hat{\kappa}$ and $\kappa_{(i)} \in$ **K**, respectively.

In the case of unknown κ (as well as known κ), the finite sample quantiles can be obtained by simulation or by a natural extension to a bootstrap procedure. To illustrate, the quantiles of the test statistics with unknown κ were simulated as in the case for Figure 1.4, but with data generated as follows: y_0 is a draw from $N(0, 1/(1 - \varphi_1^2))$, that is a draw from the unconditional distribution of y_t, which assumes stationarity (and normality); subsequent observations were generated by $y_t = \varphi_1 y_{t-1} + \varepsilon_t$, $\varepsilon_t \sim N(0, 1)$, with $\varphi_1 = 0.95$ and $T = 100$. The Chow-type tests allowed for a search over $\kappa_{(i)} \in$ **K** $= [y_{(0.1T)}, y_{(0.9T)}]$, $i = 1, \ldots, 81$. The CDFs for F(2, 96), Sup-C($\hat{\lambda}$) and $C(\lambda_b)^{ave}$ are shown in Figure 1.5, from which it is evident that the CDFs are very similar to those in the corresponding Chow case (see Figure 1.4). The 90% and 95% quantiles of the simulated distributions of Sup-C($\hat{\lambda}$) are 5.15 and 6.0, and those of $C(\lambda_b)^{ave}$ are 1.83 and 2.25, compared to 2.36 and 3.09 from F(2, 96).

1.5.3   Bootstrap the test statistics

In Chapters 5, 6 and 8 a bootstrap procedure is used to generate the bootstrap quantiles and the bootstrap p-value associated with a particular test statistic. The method is illustrated here for the two-regime model where the regime separation is unknown. A key difference in the bootstrap procedure compared to a standard simulation, in order to obtain quantiles of the null distribution, is that the bootstrap procedure samples with replacement from the empirical distribution function of the residuals from estimation of the model under H_0. In a standard

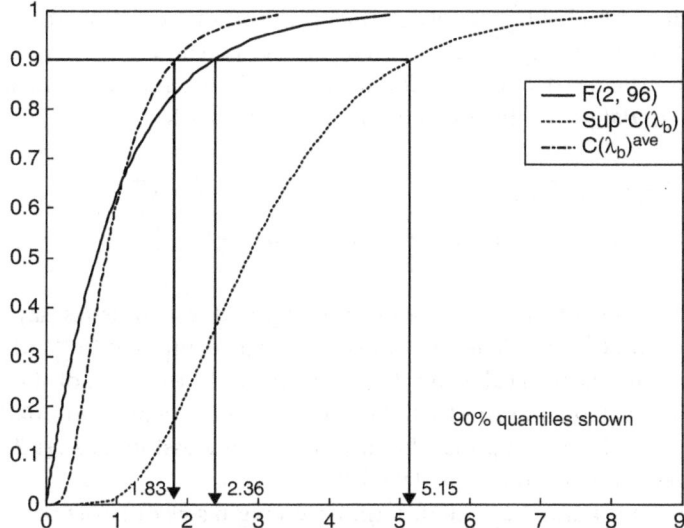

Figure 1.5 CDFs of F, Sup-C and C^{ave} tests for TAR

simulation the process shocks, ε_t, are drawn from a specified distribution, usually $N(0, \sigma_\varepsilon^2)$, and normally it is possible to set $\sigma_\varepsilon^2 = 1$ without loss. Thus, if the empirical distribution function is not close to the normal distribution function, differences are likely to arise in the distribution of the test statistic of interest.

The bootstrap steps are as follows.

1. Estimate the initial regression assuming no structural break, $y_t = \varphi_0 + \varphi_1 y_{t-1} + \varepsilon_t$, $t = 1, \ldots, T$, to obtain $y_t = \hat{\varphi}_0 + \hat{\varphi}_1 y_{t-1} + \hat{\varepsilon}_t$, and save the residual sequence $\{\hat{\varepsilon}_t\}_{t=1}^T$. Note that the inclusion of an intercept ensures that the mean of $\{\hat{\varepsilon}_t\}_{t=1}^T$ is zero so that the residuals are centred.

2. Estimate the maintained regression over the grid of values of $\lambda_b \in \Theta$ or $\kappa_{(i)} \in K$, with the appropriate indicator function $I(.)$ defined either by a temporal break or the self-exciting indicator:

$$y_t = \varphi_0 + \varphi_1 y_{t-1} + I(.)\gamma_0 + I(.)\gamma_1 y_{t-1} + \varepsilon_t$$

Calculate the required test statistics, for example Sup-C(.), C(.)^{ave} and C(.)^{exp}.

3. Generate the bootstrap data as:

$$y_t^b = \hat{\varphi}_0 + \hat{\varphi}_1 y_{t-1}^b + \hat{\varepsilon}_t^b \qquad t = 1, \ldots, T$$

The bootstrap innovations $\hat{\varepsilon}_t^b$ are drawn with replacement from the residuals $\{\hat{\varepsilon}_t\}_{t=1}^T$. The initial value in the bootstrap sequence of y_t^b is y_0^b, and there are (at least) two possibilities for this start up observation. One is to choose y_0^b

at random from $\{y_t\}_{t=1}^T$, another, the option taken here, is to choose $y_0^b = y_1$, so that the bootstrap is 'initialised' by an actual value.

4. Estimate the restricted (null) and unrestricted (alternative) regressions using bootstrap data over the grid, either $\lambda_b \in \Theta$ or $\kappa_{(i)} \in \mathbf{K}$:

$$\hat{y}_t^{b,r} = \hat{\varphi}_0^{b,r} + \hat{\varphi}_1^{b,r} y_{t-1}, \qquad\qquad e_t^{b,r} = y_t^b - \hat{y}_t^{b,r}$$

$$\hat{y}_t^b = \hat{\varphi}_0^b + \hat{\varphi}_1^b y_{t-1} + I(.)\hat{\gamma}_0^b + I(.)\hat{\gamma}_1^b y_{t-1}^b, \quad e_t^b = y_t^b - \hat{y}_t^b$$

Calculate the bootstrap versions of the required test statistics, say Sup-$C^b(.)$, $C^b(.)^{ave}$ and $C^b(.)^{exp}$. The test statistics may be computed in Wald or F form. It is usually easier to calculate the test statistics in the form of the difference in the restricted and unrestricted residual sums of squares; alternatively they can be calculated using just the unrestricted regression, see Q1.2, in which case there is no need to calculate $\hat{y}_t^{b,r}$.

5. Repeat steps 3 and 4 a total of B times, giving B sets of bootstrap data and B values of the test statistics.

6. From step 5, sort each of the B test statistics and thus obtain the bootstrap (cumulative) distribution.

7. The quantiles of the bootstrap distribution are now available from step 6; additionally the bootstrap p-value, p^{bs}, of a sample test statistic can be estimated by finding its position in the corresponding bootstrap distribution. For example, $p^{bs}[\text{Sup-}C(.)] = \#[\text{Sup-}C^b(.) > \text{Sup-}C(.)]/B$, where $\#$ is the counting function ($B + 1$ is sometimes used in the denominator, but B should at least be large enough for this not to make a material difference).

This bootstrap procedure assumes that there is not a unit root under the null hypothesis. If this is not the case, then the bootstrap data should be generated with a unit root (see Chapters 6 and 8 and UR, Vol. 1, chapter 8).

1.5.4   Illustration: Hog–corn price

By way of illustrating the techniques introduced here, they are applied to a time series on the ratio of the hog to corn price, with US monthly data for the period 1910m1 to 2008m12, an overall sample of 1,188 observations (see Figure 1.6 for the data in logarithms). The data is measured as the log ratio of the $ price of market hogs per cwt to the price of corn in $ per bushel and denoted y_t. Interest in the hog–corn price ratio derives from its status as a long-standing proxy for profitability in the hog industry as the largest cost in producing hogs is the cost of feed.

For simplicity of exposition, the basic model to be fitted is an AR(1); (for substantive modelling of the hog–corn price ratio, see, for example, Holt and

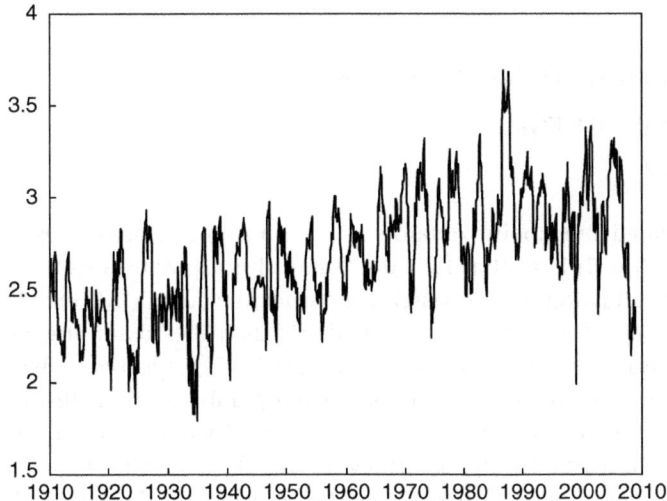

Figure 1.6   Hog–corn price ratio (log, monthly data)

Craig, 2006). The estimated model resulted in:

$$\hat{y}_t = 0.101 + 0.962 y_{t-1}$$

$$\hat{y}^* = 2.67$$

where $\hat{y}^*$ is the estimated steady state. Two variations are considered, the first where possible structural instability is related to an unknown breakpoint and the other to the regime indicator $y_{t-1} \leq \kappa$, where κ is unknown, so that the alternative is a two-regime TAR.

1.5.4.i   Chow test for temporal structural stability

The Chow test allows for a search with $\lambda_b \in \Theta = [0.10, 90]$; each observation included in the set is a possible breakpoint, so that $G = 952$. The resulting estimated two-regime model is:

$$\hat{y}_t = 0.189 + 0.923 y_{t-1} - 0.034 I(t > T_1) + I(t > T_1) 0.023 y_{t-1}$$

The models estimated for each regime, together with their implied steady states, are as follows.

Regime 1

$$\hat{y}_t = 0.189 + 0.923 y_{t-1} \qquad \text{if } 1 \leq t \leq 561$$

$$\hat{y}_1^* = 2.46$$

Regime 2

$$\hat{y}_t = 0.189 - 0.034 + (0.924 + 0.023)y_{t-1} \qquad \text{if } 562 \leq t \leq 1,187$$

$$= 0.155 + 0.947y_{t-1}$$

$$\hat{y}_2^* = 2.87$$

The estimated breakpoint obtained at the minimum of the residual sum of squares is $\hat{T}_1 = 561$, so that $\hat{\lambda}_b = 0.473$ and this estimate is well inside Θ. The Chow tests allowing for the search procedure are: Sup-$C(\hat{\lambda}_b) = 8.42$, $C(\lambda_b)^{\text{ave}} = 4.21$ and $C(\lambda_b)^{\text{exp}} = 2.69$. The simulated 95% quantiles (assuming an exogenous regressor) for $T = 1,187$ are 5.75, 2.22 and 1.37, respectively, which suggest rejection of the null hypothesis of no temporal structural break. There are two variations on obtaining appropriate critical values. First, by way of continuing the illustration, the quantiles should allow for the stochastic regressor y_{t-1}, so a second simulation was undertaken using an AR(1) generating process $y_t = \varphi_1 y_{t-1} + \varepsilon_t$, with $\varphi_1 = 0.95$ and $y_0 \sim N(0, 1/(1 - \varphi_1^2))$, $\varepsilon_t \sim N(0, 1)$. In this case, the 95% quantiles were 7.26, 2.49 and 1.55, each slightly larger than in the case with a nonstochastic regressor. Lastly, the quantiles and p-values were obtained by bootstrapping the test statistic to allow for the possibility that the empirical distribution function is generated from a dependent or non-normal process. The bootstrap 95% quantiles, with $B = 1,000$, for Sup-$C(\hat{\lambda}_b)$, $C(\lambda_b)^{\text{ave}}$ and $C(\lambda_b)^{\text{exp}}$ were 6.08, 2.17 and 1.36, with bootstrap p-values, p^{bs}, of 1%, 0.5% and 1%, respectively.

1.5.4.ii Threshold autoregression

In the second part of the illustration a two-regime TAR was fitted with $K = [y_{(0.05T)}, y_{(0.95T)}]$, so that the search allows for κ to lie between the 5% and 95% percentiles and $N = 1,100$ (an initial search between the 10% and 90% percentiles suggested that a slight extension to the limits was needed as the results below indicate). Minimising the residual sum of squares over this set resulted in $\hat{\kappa} = 2.21$, with the following estimates:

Two-regime TAR

$$\hat{y}_t = 0.080 + 0.970y_{t-1} + 0.772I(y_{t-1} \leq \hat{\kappa}) - 0.363I(y_{t-1} \leq \hat{\kappa})y_{t-1}$$

Regime 1: $y_{t-1} \leq \hat{\kappa} = 2.21$, $T_1 = 61$

$$\hat{y}_t = 0.080 + 0.772 + (0.970 - 0.363)y_{t-1}$$

$$= 0.852 + 0.607y_{t-1}$$

$$\hat{y}_1^* = 2.16$$

Regime 2: $y_{t-1} > \hat{\kappa} = 2.21$, $T_2 = 1,127$

$$\hat{y}_t = 0.080 + 0.970 y_{t-1}$$

$$\hat{y}_2^* = 2.64$$

The bulk of the observations are in Regime 2, which exhibits much more persistence than Regime 1. The estimates of the coefficient on y_{t-1} from the AR(1) model and in Regime 2 are both quite close to the unit root, so that would be an issue to investigate in a more substantive analysis. (Fitting an AR(2) model, though preferably to try and pick up some of the cyclical behaviour, resulted in very similar estimates of $\hat{\kappa}$ and $\hat{y}^*$, and does not add to the illustration.)

The Chow-type tests allowing for a search over $\kappa_{(i)} \in K$ were: Sup-$C(\hat{\kappa}) = 7.80$, $C(\kappa_{(i)})^{ave} = 3.01$ and $C(\kappa_{(i)})^{exp} = 2.07$. Two variations were considered to obtain the quantiles of the null distribution. As in the case of the Chow test for temporal stability, the generating process was taken to be AR(1), $y_t = \varphi_1 y_{t-1} + \varepsilon_t$, with $\varphi_1 = 0.95$ and $y_0 \sim N(0, 1/(1 - \varphi_1^2))$, $\varepsilon_t \sim N(0, 1)$. In this case, the 95% quantiles for Sup-$C(\hat{\kappa})$, $C(\kappa_{(i)})^{ave}$ and $C(\kappa_{(i)})^{exp}$ were 6.37, 2.2 and 1.35, so H_0 is rejected. The choice of $\varphi_1 = 0.95$ was motivated by the LS estimate of φ_1, which is fairly close to the unit root; but having started in this way a natural extension is to bootstrap the quantiles and p-values. The bootstrap 95% quantiles for Sup-$C(\hat{\lambda}_b)$, $C(\lambda_b)^{ave}$ and $C(\lambda_b)^{exp}$ were 6.82, 2.3 and 1.41, with bootstrap p-values, p^{bs}, of 2.2%, 0.2% and 0.6%, respectively.

1.6   The way ahead

A chapter summary was provided in the preface, so the intention here is not to repeat that but to focus briefly on the implications of model choice for forecasting, which implicitly encapsulates inherent differences, such as the degree of persistence, between different classes of models. Stock and Watson (1999) compared forecasts from linear and nonlinear models for 215 monthly macroeconomic time series for horizons of one, six and twelve months. The models considered included AR models in levels and first differences and LSTAR (logistic smooth transition autoregressive) models (see Chapter 5). They used a form of the ERS unit root test as a pretest (see Elliott, Rothenberg and Stock, 1996, and UR, Vol. 1, chapter 7). One of their conclusions was that "forecasts at all horizons are improved by unit root pretests. Severe forecast errors are made in nonlinear models in levels and linear models with time trends, and these errors are reduced substantially by choosing a differences or levels specification based on a preliminary test for a unit root" (Stock and Watson, 1999, p. 30). On the issue of the importance of stationarity and nonstationarity to forecasting see Hendry and Clements (1998, 1999).

The study by Stock and Watson was extensive, with consideration of many practical issues such as lag length selection, choice of deterministic components and forecast combination, but necessarily it had to impose some limitations. The class of forecasting models could be extended further in several directions. A unit root pretest suggests the simple dichotomy of modelling in levels vs differences, whereas a fractional unit root approach, as in Chapters 3 and 4, considers a continuum of possibilities; and Smith and Yadav (1994) show there are forecasting costs to first differencing if the series is generated by a fractionally integrated process. Second, even if taking account of a unit root seems appropriate there remains the question of whether the data should first be transformed, for example in logarithms, so that another possibility is a pretest that includes the choice of transformation, which itself could affect the outcome of a unit root test, (see Chapter 2).

The LSTAR models considered by Stock and Watson were one in a class of nonlinear models that allow smooth transition (ST) between AR regimes, others include: the ESTAR (exponential STAR) and BRW (bounded random walk), considered in Chapter 5; piecewise linear models either in the form of threshold models, usually autoregressive, TAR, as in Chapter 6; or temporally contiguous models that allow for structural breaks ordered by time, with known breakpoint or unknown breakpoint, as considered in Chapters 7 and 8, respectively.

Questions

Q1.1. Rewrite the TAR(1) model of Equation (1.30) so that the base model is Regime 1:

$$y_t = \varphi_0 + \varphi_1 y_{t-1} - I(y_{t-1} > \kappa)\gamma_0 - I(y_{t-1} > \kappa)\gamma_1 y_{t-1} + \varepsilon_t \tag{1.31}$$

A1.1. Note that indicator function is written so that it takes the value 1 if $y_{t-1} > \kappa$, whereas in Equation (1.30) it was written as $I(y_{t-1} \leq \kappa)$. Recall that $\gamma_0 = \varphi_0 - \phi_0$ and $\gamma_1 = \varphi_1 - \phi_1$, hence making the substitutions then, as before:

Regime 1

$$y_t = \varphi_0 + \varphi_1 y_{t-1} + \varepsilon_t$$

Regime 2

$$y_t = \varphi_0 + \varphi_1 y_{t-1} + (\phi_0 - \varphi_0) + (\phi_1 - \varphi_1)y_{t-1} + \varepsilon_t$$
$$= \phi_0 + \phi_1 y_{t-1} + \varepsilon_t$$

This is exactly as in (1.28) and (1.29), respectively.

Q1.2. The Wald test is often written as follows:

$$W = T(Rb)'[\hat{\sigma}_\varepsilon^2 R'(Z'Z)R]^{-1}Rb$$

Using the TAR(1) as an example, show that:

$$W = T\left(\frac{\hat{\varepsilon}_r'\hat{\varepsilon}_r - \hat{\varepsilon}_{ur}'\hat{\varepsilon}_{ur}}{\hat{\varepsilon}_{ur}'\hat{\varepsilon}_{ur}}\right)$$

$$= g\left(\frac{T}{T-k}\right)C$$

$$\approx gC$$

that is, approximately in finite samples W is g times the corresponding F-version of the test.

A1.2. First rewrite the TAR(1) in an appropriate matrix-vector form:

$$y_t = \alpha_0 + \alpha_1 x_t + I(t > T_1)\delta_0 + I(t > T_1)\delta_1 x_t + \varepsilon_t$$

$$y = Z\beta + \varepsilon$$

where:

$$z_t = (1, x_t, I(t > T_1), I(t > T_1)x_t)$$

$$\beta = (\alpha_0, \alpha_1, \delta_0, \delta_1)'$$

$$\hat{\sigma}_\varepsilon^2 = \hat{\varepsilon}_{ur}'\hat{\varepsilon}_{ur}/(T-k)$$

$$\tilde{\sigma}_\varepsilon^2 = \hat{\varepsilon}_{ur}'\hat{\varepsilon}_{ur}/T$$

$$\hat{\varepsilon}_{ur} = y - Zb, \text{ where b is the (unrestricted) LS estimator of } \beta.$$

The restricted model imposes $(\delta_0, \delta_1) = (0, 0)$. The restrictions can be expressed in the form $R\beta = s$, where

$$\begin{pmatrix} 0 & 0 & 1 & 0 \\ 0 & 0 & 0 & 1 \end{pmatrix} \begin{pmatrix} \beta_1 \\ \beta_2 \\ \beta_3 \\ \beta_4 \end{pmatrix} = \begin{pmatrix} 0 \\ 0 \end{pmatrix}$$

Although this could be reduced to

$$\begin{pmatrix} 1 & 0 \\ 0 & 1 \end{pmatrix} \begin{pmatrix} \beta_3 \\ \beta_4 \end{pmatrix} = \begin{pmatrix} 0 \\ 0 \end{pmatrix}$$

it is convenient to use the 'padded' version to maintain the dimension of R as conformal with β.

The LS estimator of β imposing $R\beta = s$ is denoted b_r and the resulting residual sum of squares is $\hat{\varepsilon}_r'\hat{\varepsilon}_r$, where $\hat{\varepsilon}_r = y - Zb_r$. Note that $Rb_r = 0$ or $Rb_r = s$ in general.

Next consider estimation of the regression model subject to the restrictions $R\beta = s$ (the general case is considered as the application here is just a special case). A natural way to proceed is to form the following Lagrangean function:

$$L = (y - Z\beta)'(y - Z\beta) - 2\lambda'(R\beta - s)$$

Where λ is a g x 1 vector of unknown Lagrangean multipliers. Minimising L with respect to β and λ results in:

$$\partial L /_{\partial \beta} = -2Z'y + 2Z'Z\beta - 2R'\lambda = 0$$

$$\partial L /_{\partial \lambda} = -2(R\beta - s) = 0$$

Let b_r and λ_r denote the solutions to these first order conditions, then:

$$b_r = (Z'Z)^{-1}Z'y + (Z'Z)^{-1}R'\lambda_r$$
$$= b + (Z'Z)^{-1}R'\lambda_r$$

λ_r is obtained on noting that b_r must satisfy $Rb_r = s$:

$$Rb_r = Rb + R(Z'Z)^{-1}R'\lambda_r \Rightarrow$$
$$\lambda_r = -[R(Z'Z)^{-1}R']^{-1}(Rb - s)$$

Now substitute λ_r into b_r, to obtain:

$$b_r = b - (Z'Z)^{-1}R'[R(Z'Z)^{-1}R']^{-1}(Rb - s)$$

The restricted residual vector and restricted residual sum of squares are then obtained as follows:

$$y - Zb_r = y - Zb + Z(Z'Z)^{-1}R'[R(Z'Z)^{-1}R']^{-1}(Rb - s)$$
$$= \hat{\varepsilon}_{ur} + A, A = Z(Z'Z)^{-1}R'[R(Z'Z)^{-1}R']^{-1}(Rb - s)$$
$$\hat{\varepsilon}'_r\hat{\varepsilon}_r = (y - Zb_r)'(y - Zb_r)$$
$$= \hat{\varepsilon}'_{ur}\hat{\varepsilon}_{ur} + A'A + 2A'\hat{\varepsilon}_{ur}$$
$$= \hat{\varepsilon}'_{ur}\hat{\varepsilon}_{ur} + A'A \text{ as } A'\hat{\varepsilon}_{ur} = 0 \text{ because } Z'\hat{\varepsilon}_{ur} = 0$$
$$\Rightarrow \hat{\varepsilon}'_r\hat{\varepsilon}_r - \hat{\varepsilon}'_{ur}\hat{\varepsilon}_{ur} = A'A$$

Considering $A'A$ further note that:

$$A'A = [Z(Z'Z)^{-1}R'[R(Z'Z)^{-1}R']^{-1}(Rb - s)]'[Z(Z'Z)^{-1}R'[R(Z'Z)^{-1}R']^{-1}(Rb - s)]$$
$$= (Rb - s)'[R(Z'Z)^{-1}R']^{-1}R(Z'Z)^{-1}Z'Z(Z'Z)^{-1}R'[R(Z'Z)^{-1}R']^{-1}(Rb - s)$$
$$= (Rb - s)'[R(Z'Z)^{-1}R']^{-1}R(Z'Z)^{-1}R'[R(Z'Z)^{-1}R']^{-1}(Rb - s)$$
$$= (Rb - s)'[R(Z'Z)^{-1}R']^{-1}(Rb - s)$$

Hence:

$$W = (Rb - s)'[\tilde{\sigma}^2_{\varepsilon,ur}R(Z'Z)^{-1}R']^{-1}(Rb - s)$$

$$= T\left(\frac{\hat{\varepsilon}'_r\hat{\varepsilon}_r - \hat{\varepsilon}'_{ur}\hat{\varepsilon}_{ur}}{\hat{\varepsilon}'_{ur}\hat{\varepsilon}_{ur}}\right)$$

$$= g\frac{T}{(T-k)}\left(\frac{\hat{\varepsilon}'_r\hat{\varepsilon}_r - \hat{\varepsilon}'_{ur}\hat{\varepsilon}_{ur}}{\hat{\varepsilon}'_{ur}\hat{\varepsilon}_{ur}}\right)\frac{(T-k)}{g}$$

$$= g\frac{T}{(T-k)}C, \quad C = \left(\frac{\hat{\varepsilon}'_r\hat{\varepsilon}_r - \hat{\varepsilon}'_{ur}\hat{\varepsilon}_{ur}}{\hat{\varepsilon}'_{ur}\hat{\varepsilon}_{ur}}\right)\frac{(T-k)}{g}$$

$$\approx gC$$

as required in the question.

Q1.3.i. Define the partial sum of y_t as $S_t = \sum_{i=1}^t y_i$, and assume for simplicity that $E(y_t) = 0$, then the long run variance of y_t is equivalently written as follows:

$$\sigma^2_{y,lr} = \lim_{T\to\infty} T^{-1}E(S_T^2)$$

Assuming that y_t is generated by a covariance stationary process, show that by multiplying out the terms in S_T, $\sigma^2_{y,lr}$ may be expressed in terms of the variance, $\gamma(0)$, and autocovariances, $\gamma(k)$, as follows:

$$\sigma^2_{y,lr} = \sum_{k=-\infty}^{\infty}\gamma(k)$$

$$= \gamma(0) + 2\sum_{k=1}^{\infty}\gamma(k)$$

$$= \sum_{k=-\infty}^{\infty}\gamma(k) \quad \text{assuming stationarity so that } \gamma(-k) = \gamma(k)$$

Q1.3.ii. Confirm that if $y_t = \rho y_{t-1} + \varepsilon_t$, $|\rho| < 1$, then $\sigma^2_{lr,y} = \sum_{j=-\infty}^{\infty}\gamma_j = (1-\rho)^{-2}\sigma^2_{\varepsilon}$; show that this is the same as $\omega(1)^2\sigma^2_{\varepsilon}$.

Q1.3.iii. Show that the spectral density function of y_t at frequency zero is proportional to $\sigma^2_{y,lr}$.

A1.3.i. Write $S_T = i'y$, where $i' = (1,\ldots,1)$ and $y = (y_1,\ldots,y_T)'$ and, therefore:

$$E(S_T^2) = E(i'y)(i'y) \quad \text{because } i'y \text{ is a scalar}$$

$$= E[i'y(y'i)] \quad i'y = y'i$$

$$= i'E(yy')i$$

$$= (1,\ldots,1)E\begin{bmatrix} y_1^2 & y_1y_2 & \cdots & y_1y_T \\ y_2y_1 & y_2^2 & \cdots & y_1y_T \\ \vdots & \vdots & \vdots & \vdots \\ y_Ty_1 & y_Ty_2 & \cdots & y_T^2 \end{bmatrix}\begin{pmatrix} 1 \\ 1 \\ \vdots \\ 1 \end{pmatrix}$$

$E(S_T^2)$ sums all the elements in the covariance matrix of y, of which there are T of the form $E(y_t^2)$, $2(T-1)$ of the form $E(y_t y_{t-1})$, $2(T-2)$ of the form $E(y_t y_{t-2})$ and so on until $2E(y_1 y_T)$, which is the last in the sequence. If the $\{y_t^2\}$ sequence is covariance stationary, then $E(y_t^2) = \gamma_0$ and $E(y_t y_{t-k}) = \gamma_k$, hence:

$$E(S_T^2) = T\gamma_0 + 2(T-1)\gamma_1 + 2(T-2)\gamma_2 + \cdots + 2\gamma_{T-1}$$

$$= T\gamma_0 + 2\sum_{j=1}^{T-1} (T-j)\gamma_j$$

$$T^{-1}E(S_T^2) = \gamma_0 + 2T^{-1}\sum_{j=1}^{T-1} (T-j)\gamma_j$$

$$= \gamma_0 + 2\sum_{j=1}^{T-1} (1-j/T)\gamma_j$$

Taking the limit as $T \to \infty$, we obtain:

$$\sigma_{y,lr}^2 \equiv \lim_{T\to\infty} T^{-1}E(S_T^2)$$

$$= \gamma_0 + 2\sum_{j=1}^{\infty} \gamma_j$$

In taking the limit it is legitimate to take j as fixed and let the ratio j/T tend to zero. Covariance stationarity implies $\gamma_k = \gamma_{-k}$, hence $\gamma_0 + 2\sum_{j=1}^{\infty} \gamma_j = \sum_{j=-\infty}^{\infty} \gamma_j$ and, therefore:

$$\sigma_{y,lr}^2 = \sum_{j=-\infty}^{\infty} \gamma_j$$

A1.3.ii. First back substitute to express y_t in terms of current and lagged ε_t and then obtain the variance and covariances:

$$y_t = \sum_{i=0}^{\infty} \rho^i \varepsilon_{t-i}$$

$$\gamma_0 \equiv \text{var}(y_t)$$

$$= \sum_{i=0}^{\infty} \rho^{2i} E(\varepsilon_{t-i}^2) \qquad\qquad \text{using cov}(\varepsilon_t \varepsilon_s) = 0 \text{ for } t \neq s$$

$$= (1-\rho^2)^{-1}\sigma_\varepsilon^2 \qquad\qquad \text{using } E(\varepsilon_{t-i}^2) = \sigma_\varepsilon^2$$

$$\gamma_k = \rho\gamma_{k-1} \qquad\qquad\qquad k = 1,\ldots$$

$$= \rho^k (1-\rho^2)^{-1}\sigma_\varepsilon^2$$

The long-run variance is then obtained:

$$\sigma_{y,lr}^2 = \sum_{k=-\infty}^{\infty} \gamma_k$$

$$= \gamma_0 + 2 \sum_{k=1}^{\infty} \gamma_k$$

$$= (1-\rho^2)^{-1} \sigma_\varepsilon^2 \left[1 + 2 \left((1-\rho)^{-1} - 1 \right) \right]$$

$$= \frac{1}{(1-\rho)(1+\rho)} \left[\frac{(1+\rho)}{(1-\rho)} \right] \sigma_\varepsilon^2$$

$$= (1-\rho)^{-2} \sigma_\varepsilon^2$$

Next note that $y_t = \rho y_{t-1} + \varepsilon_t$ may be written as $y_t = w(L)\varepsilon_t$ where $w(L) = (1 - \rho L)^{-1}$, hence:

$$var[w(1)\varepsilon_t] = w(1)^2 \sigma_\varepsilon^2$$

$$= (1-\rho)^{-2} \sigma_\varepsilon^2 \qquad \text{as } w(1)^2 = (1-\rho)^{-2}$$

$$= \sigma_{y,lr}^2$$

A1.3.iii. The sdf for a (covariance stationary) process generating data as $y_t = w(L)\varepsilon_t$ is:

$$f_y(\lambda_j) = |w(e^{-i\lambda_j})|^2 f_\varepsilon(\lambda_j) \quad \lambda_j \in [-\pi, \pi]$$

where $w(e^{-i\lambda_j})$ is $w(.)$ evaluated at $e^{-i\lambda_j}$, and $f_\varepsilon(\lambda_j) = (2\pi)^{-1}\sigma_\varepsilon^2$, for all λ_j, is the spectral density function of the white noise input ε_t. Thus, the sdf of y_t is the power transfer function of the filter, $|w(e^{-i\lambda_j})|^2$, multiplied by the sdf of white noise. If $\lambda_j = 0$, then:

$$f_y(0) = |w(e^0)|^2 f_\varepsilon(0)$$

$$= (2\pi)^{-1} w(1)^2 \sigma_\varepsilon^2$$

$$= (2\pi)^{-1} \sigma_{y,lr}^2$$

Hence, $f_y(0)$ is proportional to $\sigma_{y,lr}^2$.

The sdf may also be defined directly in terms of either the Fourier transform of the autocorrelation function, as in Fuller (1996, chapter 3), or of the auto-covariance function, as in Brockwell and Davis (2006, chapter 4), the resulting definition differing only by a term in $\gamma(0)$. We shall take the latter representation so that

$$f_y(\lambda_j) = (2\pi)^{-1} \sum_{k=-\infty}^{k=\infty} \gamma(k) e^{-ik\lambda_j}$$

Hence if $\lambda_j = 0$, then:

$$f_y(0) = (2\pi)^{-1} \sum_{k=-\infty}^{k=\infty} \gamma(k)$$

$$= (2\pi)^{-1} \sigma_{y,lr}^2$$

as before using the result in **A1.3.i.**

2
Functional Form and Nonparametric Tests for a Unit Root

Introduction

The chapter starts with an issue that is often not explicitly addressed in empirical work. It is usually not known whether a variable of interest, y_t, should be modelled as it is observed, referred to as the levels (or 'raw' data), or as some transformation of y_t, for example by taking the logarithm or reciprocal of the variable. Regression-based tests are sensitive to this distinction, for example modelling in levels when the random walk is in the logarithms, implying an exponential random walk, will affect the size and power of the test.

The second topic of this chapter is to consider nonparametric or semi-parametric tests for a unit root, which are in contrast to the parametric tests that have been considered so far. The DF tests, and their variants, are parametric tests in the sense that they are concerned with direct estimation of a regression model, with the test statistic based on the coefficient on the lagged dependent variable. In the DF framework, or its variants (for example the Shin and Fuller, 1998, ML approach), the parametric structure is that of an AR or ARMA model. Nonparametric tests use less structure in that no explicit parametric framework is required and inference is based on other information in the data, such as ranks, signs and runs. Semi-parametric tests use some structure, but it falls short of a complete parametric setting; an example here is the rank score-based test outlined in Section 2.3, which is based on ranks, but requires an estimator of the long-run variance to neutralise it against non-iid errors.

This chapter progresses as follows. The question of what happens when a DF-type test is applied to the wrong (monotonic) transformation is elaborated in Section 2.1. If the unit root is in the log of y_t, but the test is formulated in terms of y_t, there is considerable 'over-rejection' of the null hypothesis, for example, an 80% rejection rate for a test at a nominal size of 5%, which might be regarded as incorrect. However, an alternative view is that this is a correct decision, and indeed 100% rejection would be desirable as the unit root is not present in y_t (in the terminology of Section **2.2**, the generating process is a log-linear integrated process not a linear integrated process). There are two possible responses to the question of whether if a unit root is present it is in the levels or the logs. A parametric test, due to Kobayashi and McAleer (1999) is outlined in Section **2.2**; this procedure involves two tests reflecting the non-nested nature of the alternative specifications. This combines a test in which the linear model is estimated and tested for departures in the direction of log-linearity and then the roles are reversed, resulting in two test statistics. An alternative approach is to use a test for a unit root that is invariant to a monotonic transformation, such as the log transformation.

A test based on the ranks is necessarily invariant to monotonic transformations. Hence, a more general question can be posed: is there a unit root in the process generating y_t or some monotonic transformation of y_t? Thus, getting the transformation 'wrong', for example using levels when the data should be in logs, will not affect the outcome of such a test. Equally you don't get something for nothing, so the information from such a test, for example that the unit root is not rejected, does not tell you which transformation is appropriate, just that there is one that will result in H_0 not being rejected.

Some unit root tests based on ranks are considered in Section **2.3**. An obvious starting point is the DF test modified to apply to the ranks of the series. The testing framework is to first rank the observations in the sample, but then use an AR/ADF model on the ranked observations. It turns out that this simple extension is quite hard to beat in terms of size retention and power. Another possibility is to take the parametric score or LM unit root test (see Schmidt and Phillips, 1992), and apply it to the ranks. This test deals with the problem of weakly dependent errors in the manner of the PP tests by employing a semi-parametric estimator of the long-run variance, although an AR-based estimator could also be used.

The rank ADF test and the rank-score test take parametric tests and modify them for use with the ranked observations. A different approach is to construct a nonparametric test without reference to an existing parametric test. One example of this approach, outlined in Section **2.4**, is the range unit root test, which is based on assessing whether the marginal sample observation represents a new record in the sense of being larger than the existing maximum or smaller than the existing minimum, the argument being that stationary and nonstationary series add new records at different rates, which will serve to distinguish the two processes.

Finally, Section **2.5** considers a test based on the variance-ratio principle. In this case, the variance of y_t generated by a nonstationary process with a unit root grows at a different rate compared to when y_t is generated by a stationary process, and a test can be designed around this feature. The test described, due to Breitung (2002), which is particularly simple to construct, has the advantage that it does not require specification of the short-run dynamics, and in that sense it is 'model free'.

Some simulation results and illustrations of the tests in use are presented in Section 2.6. The emphasis in the simulation results is on size retention and relative power of the different tests; however, the advantages of nonparametric and semi-parametric tests are potentially much wider. In different degrees they are derived under weaker distributional assumptions than parametric tests and are invariant to some forms of structural breaks and robust to misspecification of generally unknown short-run dynamics.

2.1   Functional form: linear versus logs

Consider a simplified version of a question that often arises in empirical work. The data at hand are clearly trended in some form, either a stochastic trend or a deterministic trend, and a preliminary analysis concerns distinguishing between these two alternative explanations. However, it is not clear whether the (likely) trend is better modelled in the levels or logs of the series. Whilst there may be a presumption that some series are better modelled (in the sense of being motivated by economic theory) in logs, for example real or volume measures, because they are positive, and first differences then relate to growth rates, and some in levels, for example rate or share variables, or variables that are not inherently positive, there is naturally a lack of certainty as to the precise nature of the data generation process (DGP) – and intuition is not necessarily a good guide as to how to proceed.

Granger and Hallman (1991), hereafter GH, showed, by means of some Monte Carlo experiments, that the standard DF tests over-reject in the case that the random walk is generated in the logs, but when the test is applied to the levels of the data series. Kramer and Davies (2002), hereafter KD, showed, by theoretical means, that the GH result is not general, and the rejection probability includes both cases of over-rejection and under-rejection. We consider some of these issues in this section.

2.1.1   A random walk and an exponential random walk

To indicate the importance of the basic step of data transformation in the modelling procedure, assume that $w_t = f(y_t)$ is a transformation of the original data, such that w_t is generated by a first order autoregressive process, with a unit root under the null hypothesis. That is, respectively with and without drift:

$$w_t = w_{t-1} + \varepsilon_t \tag{2.1}$$

$$w_t = \beta_1 + w_{t-1} + \varepsilon_t \tag{2.2}$$

where $\varepsilon_t \sim \text{iid}(0, \sigma_\varepsilon^2)$, $\sigma_\varepsilon^2 < \infty$. The two cases considered are the no transformation case, $w_t = y_t$, and the log transformation case, $w_t = \ln y_t$. In the first case, y_t

follows a random walk (RW) in the natural units of y_t, and in the second case y_t follows an exponential random walk (ERW), which is a log-random walk; thus, in the second case, the generating process is, first without drift:

$$\ln y_t = \ln y_{t-1} + \varepsilon_t \tag{2.3}$$

In levels, this is:

$$y_t = y_{t-1} e^{\varepsilon_t} \Rightarrow \tag{2.4a}$$

$$y_t = y_0 e^{(\varepsilon_1 + \cdots + \varepsilon_t)} \qquad y_0 > 0 \tag{2.4b}$$

The generating process with drift is:

$$\ln y_t = \beta_1 + \ln y_{t-1} + \varepsilon_t \tag{2.5}$$

In levels, this is:

$$y_t = y_{t-1} e^{(\varepsilon_t + \beta_1)}$$

$$\Rightarrow$$

$$y_1 = y_0 e^{(\varepsilon_1 + \beta_1)}$$

$$y_1 = y_1 e^{(\varepsilon_2 + \beta_1)}$$

$$= y_0 e^{(\varepsilon_1 + \varepsilon_2)} e^{(\beta_1 + \beta_1)}$$

$$\vdots \quad \vdots \quad \vdots$$

$$y_t = y_0 e^{(\varepsilon_1 + \cdots + \varepsilon_t)} e^{\beta_1 t} \tag{2.6a}$$

$$\Rightarrow$$

$$\ln y_t = \ln y_0 + \beta_1 t + \sum_{i=1}^{t} \varepsilon_i \tag{2.6b}$$

Evidently for (2.6b) to be valid it is necessary that $y_0 > 0$ otherwise $\ln(y_0)$ is an invalid operation; however note that (2.6a) is still valid for $y_0 < 0$.

There are two situations to be considered as follows: Case 1, the random walk is in 'natural' units, $w_t = y_t$, but the testing takes place using $\ln y_t$, so that the maintained regression is of the form $\ln y_t = \mu_t^* + \rho \ln y_{t-1} + \upsilon_t$; whereas in Case 2, there is an exponential random walk, so the appropriate transformation is $w_t = \ln(y_t)$, but the testing takes place using y_t, and the maintained regression is of the form $y_t = \mu_t^* + \rho y_{t-1} + \upsilon_t$. For Case 1, in practice and in simulations, y_t has to be positive otherwise $\ln y_t$ cannot be computed; typical examples include macroeconomic time series such as aggregate consumption, which naturally have large positive magnitudes.

KD (op. cit.) consider the implications of misspecification for the DF test statistic $\hat{\delta} = T(\hat{\rho} - 1)$ when the maintained regression has neither a constant nor a trend. ($\hat{\rho}$ is the LS estimator of the coefficient on the lagged dependent variable in the misspecified maintained regression.) To interpret their results, the following terminology is used: let $\hat{\delta}_\alpha$ be the $\alpha\%$ critical value for the test statistic $\hat{\delta}$, then under-rejection occurs if $P(\hat{\delta} < \hat{\delta}_\alpha | H_0) < \alpha$, whereas over-rejection occurs if $P(\hat{\delta} < \hat{\delta}_\alpha | H_0) > \alpha$. In the former case if, for example, $\alpha = 0.05$, then the true rejection probability is less than 5%, but in the latter case the true rejection probability exceeds 5%.

In Case 1, the random walk is in the levels, so that the unit root generating process applies to y_t; however, the test is applied to the log of y_t. Assuming that $y_t > 0$ for all t in the sample, then KD (op. cit., Theorem 2) show that both as $\sigma_\varepsilon^2 \to 0$ and as $\sigma_\varepsilon^2 \to \infty$, then $\hat{\rho} \Rightarrow 1$, implying $P(\hat{\delta} < \hat{\delta}_\alpha | H_0) \to 0$; that is, the rejection probability, conditional on $y_t > 0$, tends to zero, which is the limit to under-rejection. The simulation results reported below confirm this result for $\hat{\delta}$, but indicate that it does not apply generally when there is a constant and/or a trend in the maintained regression.

In Case 2, that is the random walk is in logs, $w_t = \ln y_t$, but the test is applied to the levels, y_t, the rejection probabilities increase with σ_ε^2 and T; under-rejection occurs only for small sample sizes or for very small values of the innovation variance, for example $\sigma_\varepsilon^2 = 0.01$ (see KD, op. cit., theorem 1). Thus, generally, the problem is of over-rejection. These results are confirmed below and do carry across to the maintained regression with either a constant and/or a trend and to the pseudo-t tests.

2.1.2   Simulation results

In the simulations for Case 1, y_t has to be positive otherwise $w_t = \ln y_t$ cannot be computed, whereas in Case 2, given $y_0 > 0$, all y_t values are necessarily positive. There is more than one way to ensure positive y_t in Case 1, and the idea used here is to start the random walk from an initial value sufficiently large to ensure that the condition is met over the sample; of course, apart from the unit root, this imparts a deliberate nonstationarity into the series; see also GH (op. cit.). The notion of 'sufficiently large' is first calibrated by reference to a number of macroeconomic time series; for example, based on UK real GDP which, scaled to have a unit variance, has a starting value of 50, and to illustrate sensitivity, the variance is halved to 0.5.

Case 1

KD (op. cit.) reported that the rejection probabilities for $\hat{\delta}$ in Case 1 were zero for all design parameters tried (for example, variations in σ_ε^2, T and starting values, and introducing drift in the DGP), and this result was confirmed in the simulations reported in Table 2.1, which are based on 50,000 replications. However, some qualifications are required that limit the generality of this result. First, $\hat{\tau}$ was under-sized, but at around 2.6%–2.8% rather than 0%, for a nominal 5% test. The biggest difference arises when the maintained regression includes

Table 2.1 Levels or logs: incorrect specification of the maintained regression, impact on rejection rates of DF test statistics (at a 5% nominal level)

Case 1			unit root in y_t			
$\ln(y_t)$	$\hat{\delta}$	$\hat{\delta}_\mu$	$\hat{\delta}_\beta$	$\hat{\tau}$	$\hat{\tau}_\mu$	$\hat{\tau}_\beta$
T = 100						
$\sigma_\varepsilon^2 = 0.5$	0.0%	4.8%	4.5%	2.7%	5.2%	4.2%
$\sigma_\varepsilon^2 = 1.0$	0.0%	5.2%	4.8%	2.6%	5.3%	4.7%
T = 200						
$\sigma_\varepsilon^2 = 0.5$	0.0%	5.7%	5.6%	2.8%	5.3%	4.6%
$\sigma_\varepsilon^2 = 1.0$	0.0%	5.4%	5.2%	2.6%	5.1%	4.3%
Case 2			unit root in $\ln y_t$			
y_t	$\hat{\delta}$	$\hat{\delta}_\mu$	$\hat{\delta}_\beta$	$\hat{\tau}$	$\hat{\tau}_\mu$	$\hat{\tau}_\beta$
T = 100						
$\sigma_\varepsilon^2 = 0.5$	72.6%	60.0%	46.4%	73.8%	49.2%	41.5%
$\sigma_\varepsilon^2 = 1.0$	94.3%	92.5%	87.7%	93.1%	89.2%	84.1%
T = 200						
$\sigma_\varepsilon^2 = 0.5$	93.5%	90.3%	82.6%	91.5%	83.9%	74.5%
$\sigma_\varepsilon^2 = 1.0$	96.2%	95.9%	95.8%	95.4%	94.8%	94.4%

Notes: Case 1 is where the random walk is in 'natural' units, $w_t = y_t$, but the testing takes place using $\ln y_t$; Case 2 is where there is an exponential random walk, so the correct transformation is $w_t = \ln y_t$, but the testing takes place using y_t. The test statistics are of the general form $\hat{\delta} = T(\hat{\rho} - 1)$ and $\hat{\tau} = (\hat{\rho} - 1)/\hat{\sigma}(\hat{\rho})$, where $\hat{\sigma}(\hat{\rho})$ is the estimated standard error of $\hat{\rho}$; a subscript indicates the deterministic terms included in the maintained regression, with μ for a constant and β for a constant and linear trend. The maintained regressions are deliberately misspecified; for example in Case 1, the maintained regression is specified in terms of $\ln y_t$ and $\hat{\rho}$ is the estimated coefficient on $\ln y_{t-1}$ and in Case 2 the maintained regression is specified in terms of y_t and $\hat{\rho}$ is the estimated coefficient on y_{t-1}.

a constant and/or a trend. In these cases, both the n-bias ($\hat{\delta}$) and pseudo-t tests ($\hat{\tau}$) maintain their size.

Case 2

The results in KD (op. cit.) are confirmed for $\hat{\delta}$ in Case 2. Further, the over-rejection is found to be more general, applying to $\hat{\tau}$, and to the $\hat{\delta}$-type and $\hat{\tau}$-type tests when the maintained regression includes a constant and/or a trend; however, the over-rejection tends to decline as more (superfluous) deterministic terms are included in the maintained regression. For example, the rejection rates for $\sigma_\varepsilon^2 = 1.0$ and T = 100 are as follows: $\hat{\tau}$, 93.1%; $\hat{\tau}_\mu$, 89.2%; and $\hat{\tau}_\beta$, 84.1%. Also the rejection rates tend to increase as σ_ε^2 increases.

The over-rejection in Case 2 is sometimes looked on as an incorrect decision – there is a unit root so it might be thought that the test should not reject; however, it is the correct decision in the sense that the unit root is *not* in the levels of the series, it is in the logs. Failure to reject could be more worrying as modelling might then proceed incorrectly in the levels, whereas it should be in the logarithms.

2.2   A parametric test to discriminate between levels and logs

Kobayashi and McAleer (1999a), hereafter KM, have suggested a test to discriminate between an integrated process defined alternately on the levels or the logarithms of the data. In each testing situation, the asymptotic distribution of the test statistic depends on whether there is drift in the unit root process. Such tests are likely to be of particular use when a standard unit root test, such as one of the DF tests, indicates that the series both in the levels and in the logarithms is I(1). In this case, the tests are not sufficiently discriminating and the appropriate KM test is likely to be useful.

2.2.1   Linear and log-linear integrated processes

As in Section 2.1, the data are assumed either to be generated by a unit root process in the levels of the data, y_t, or in the logarithms of the levels so that the levels are generated by an exponential random walk or the generalisation of that model that follows when the errors are generated by an AR process. In either case, the data are generated by an integrated process, referred to as an IP. The test statistics are motivated as diagnostic tests; for example, if the wrong model is fitted then heteroscedasticity is induced in the conditional variance of the dependent variable, whereas if the correct model is fitted the conditional variance is homoscedastic.

The data-generating framework, covering both the linear and log-linear cases and which is more general than a simple random walk is:

$$w_t = \beta_1 + w_{t-1} + z_t \tag{2.7a}$$

$$\psi(L)z_t = \varepsilon_t \qquad \psi(L) = 1 - \sum_{i=1}^{p} \psi_i L^i \tag{2.7b}$$

$$\Rightarrow$$

$$\left(1 - \sum_{i=1}^{p} \psi_i L^i\right)(\Delta w_t - \beta_1) = \varepsilon_t \tag{2.8}$$

If the data are generated by a linear integrated process, LIP, with drift, then $w_t = y_t$; whereas if the data are generated by a log-linear integrated process, LLIP, with drift, then $w_t = \ln(y_t)$. In both cases KM assume that $\varepsilon_t \sim \text{iid}(0, \sigma_\varepsilon^2)$, $E(\varepsilon_t^3) = 0$, $E(\varepsilon_t^4) < \infty$, and the roots of $\psi(L)$ are outside the unit circle (see UR Vol.1, Appendix 2). The drift parameter β_1 is assumed positive, otherwise negative values of y_t are possible, thus precluding a log-linear transformation. ε_t and the

LS residuals are given by:

$$\varepsilon_t = \Delta w_t - \beta_1^* - \sum_{i=1}^{p} \psi_i \Delta w_{t-i} \qquad \text{where } \beta^* = \beta \, \psi(1) \tag{2.9a}$$

$$\hat{\varepsilon}_t = \Delta w_t - \hat{\beta}_1^* - \sum_{i=1}^{p} \hat{\psi}_i \Delta w_{t-i} \qquad \text{LS residuals} \tag{2.9b}$$

LS estimators of β_1 and σ_ε^2 are given by:

$$\hat{\beta}_1 = \frac{\hat{\beta}_1^*}{\hat{\psi}(1)} \tag{2.9c}$$

$$\hat{\sigma}_\varepsilon^2 = \frac{1}{T - (p+1)} \sum_{t=p+1}^{T} \hat{\varepsilon}_t^2 \tag{2.9d}$$

In the case of no drift, then $\beta_1 = 0$ in (2.7a), and an additional assumption is required to prevent negative values of y_t, which would preclude a log-transformation. (The assumption, applicable to both the LIP and LLIP cases, is that the limit of $T^{1+\zeta}\sigma_T^2$ as $T \to \infty$ is a constant for some $\zeta > 0$, where σ_T^2 is the variance of ε_T (see KM, op. cit, for more details).

The relevant quantities for constructing the test statistics, when $\beta_1 = 0$, are:

$$\hat{\varepsilon}_t = \Delta w_t - \sum_{i=1}^{p} \hat{\psi}_i \Delta w_{t-i} \qquad \text{LS residuals} \tag{2.10a}$$

$$\hat{\sigma}_\varepsilon^2 = \frac{1}{T - (p+1)} \sum_{t=p+1}^{T} \hat{\varepsilon}_t^2 \qquad \text{LS variance} \tag{2.10b}$$

Note that when testing the LIP against the LLIP, the estimator and residuals are defined in terms of $w_t = y_t$; whereas when testing the LLIP against the LIP, the estimator and residuals are defined in terms of $w_t = \ln y_t$. The following test statistics, derived by KM (op. cit.), make the appropriate substitutions.

2.2.2 Test statistics

2.2.2.i The KM test statistics

The test statistics are:

1.i V_1: LIP with drift is the null model: test for departure in favour of the LLIP;

1.ii U_1: LIP without drift is the null model: test for departure in favour of the LLIP;

2.i V_2: LLIP with drift is the null model: test for departure in favour of the LIP;

2.ii U_2: LLIP without drift is the null model: test for departure in favour of the LIP.

1.i The first case to consider is when the null model is the LIP with drift, then:

$$T^{-3/2} \sum_{t=p+1}^{T} y_{t-1}(\varepsilon_t^2 - \sigma_\varepsilon^2) \qquad \Rightarrow_D N(0, \omega^2) \tag{2.11}$$

$$\omega^2 = \left(\frac{\beta_1^2}{6} \sigma_\varepsilon^4 \right)$$

$\Rightarrow$ Test statistic, V_1:

$$V_1 \equiv T^{-3/2} \left(\frac{1}{\hat{\omega}} \sum_{t=p+1}^{T} y_{t-1} (\hat{\varepsilon}_t^2 - \hat{\sigma}_\varepsilon^2) \right) \Rightarrow_D N(0,1) \tag{2.12}$$

$$\hat{\omega} = \left(\frac{\hat{\beta}_1^2}{6} \hat{\sigma}_\varepsilon^4 \right)^{1/2}$$

Comment: when there is drift, then with an appropriate standardisation, the test statistic is asymptotically normally distributed with zero mean and unit variance, and the sample value is compared to a selected upper quantile of $N(0, 1)$, for example the 95% quantile; large values lead to rejection of the null of a LIP with drift.

1.ii The next case is when the null model is the LIP without drift, then:

$$T^{-1} \sum_{t=p+1}^{T} y_{t-1} (\varepsilon_t^2 - \sigma_\varepsilon^2) \Rightarrow_D \overline{\omega} \left[\int_0^1 B_1(r) dB_2(r) - \int_0^1 B_1(r) dr \int_0^1 dB_2(r) \right]$$

$$\overline{\omega} = \sqrt{2} \left(\frac{\sigma_\varepsilon^3}{\psi(1)} \right) \tag{2.13}$$

$\Rightarrow$ Test statistic, U_1:

$$U_1 = T^{-1} \left(\frac{1}{\hat{\overline{\omega}}} \sum y_{t-1} (\hat{\varepsilon}_t^2 - \hat{\sigma}_\varepsilon^2) \right) \Rightarrow_D \int_0^1 B_1(r) dB_2(r) - \int_0^1 B_1(r) dr \int_0^1 dB_2(r)$$

$$\hat{\overline{\omega}} = \sqrt{2} \left(\frac{\hat{\sigma}_\varepsilon^3}{\hat{\psi}(1)} \right) \tag{2.14}$$

where $B_1(r)$ and $B_2(r)$ are two standard Brownian motion processes with zero covariance.

Comment: when drift is absent, the test statistic has a non-standard distribution; the null hypothesis (of a LIP) is rejected when the sample value of U_1 exceeds the selected (upper) quantile; some quantiles are provided in Table 2.2 below.

2.i. The second case to consider is when the null model is the LLIP with drift, then:

$$T^{-3/2} \sum_{t=p+1}^{T} (-\ln y_{t-1}) (\varepsilon_t^2 - \sigma_\varepsilon^2) \Rightarrow_D N(0, \omega^2) \tag{2.15}$$

$$\omega^2 = \left(\frac{\beta_1^2}{6} \sigma_\varepsilon^4 \right)$$

⇒ Test statistic, V_2:

$$V_2 \equiv T^{-3/2} \left(\frac{1}{\hat{\omega}} \sum_{t=p+1}^{T} (-\ln y_{t-1})(\hat{\varepsilon}_t^2 - \hat{\sigma}_\varepsilon^2) \right) \Rightarrow_D N(0,1) \tag{2.16}$$

$$\hat{\omega} = \left(\frac{\hat{\beta}_1^2}{6} \hat{\sigma}_\varepsilon^4 \right)^{1/2}$$

Comment: as in the LIP case, the test statistic is asymptotically distributed as standard normal, and the sample value is compared with a selected upper quantile of $N(0, 1)$; large values lead to rejection of the null of a LLIP with drift.

2.ii. The next case is when the null model is the LLIP without drift:

$$T^{-1} \sum_{t=p+1}^{T} (-\ln y_{t-1})(\varepsilon_t^2 - \sigma_\varepsilon^2) \Rightarrow_D -\varpi \left[\int_0^1 B_1(r)dB_2(r) - \int_0^1 B_1(r)dr \int_0^1 dB_2(r) \right]$$

$$\varpi = \sqrt{2} \left(\frac{\sigma_\varphi^3}{\psi(1)} \right) \tag{2.17}$$

⇒ Test statistic, U_2:

$$U_2 = T^{-1} \left(\frac{1}{\hat{\varpi}} \sum y_{t-1}(\hat{\varepsilon}_t^2 - \hat{\sigma}_\varepsilon^2) \right) \Rightarrow_D - \left[\int_0^1 B_1(r)dB_2(r) - \int_0^1 B_1(r)dr \int_0^1 dB_2(r) \right]$$

$$\hat{\varpi} = \sqrt{2} \left(\frac{\hat{\sigma}_\varepsilon^3}{\hat{\psi}(1)} \right) \tag{2.18}$$

where $W_1(r)$ and $W_2(r)$ are two Brownian motion processes with zero covariance.

Comment: as in the LIP case, 'large' values of U_2 lead to rejection of the null, in this case that the DGP is a LLIP.

2.2.2.ii   Critical values

Some quantiles for the KM tests V_1 and V_2 for $\psi_1 = 0$ and an AR(1) process with $\psi_1 = 0.8$ are given in Table 2.2a, and similarly for U_1 and U_2 in Table 2.2b. The results are presented in the form of simulated quantiles with size in (.) parentheses, where the size is the empirical size if the quantiles from the asymptotic distribution are used. The asymptotic distributions of V_1 and V_2 are both standard normal.

Taking the finite sample results in Table 2.2a first, size is reasonably well maintained throughout for V_1 and V_2 using quantiles from the asymptotic distribution, which in this case is $N(0, 1)$; size retention is marginally better for $\psi_1 = 0.8$ compared to $\psi_1 = 0.0$ and, as one would expect, as T increases. The picture is similar for U_1 and U_2 (see in Table 2.2b); in this case size retention is marginally better for $\psi_1 = 0.0$ than for $\psi_1 = 0.8$, but (again) deviations are slight, for example, an actual size of 4.4% for T = 200 and $\psi_1 = 0.8$.

Table 2.2   Critical values for KM test statistics
Table 2.2a   Critical values for V_1 and V_2

T	90%		95%		99%	
	$\psi_1 = 0.0$	$\psi_1 = 0.8$	$\psi_1 = 0.0$	$\psi_1 = 0.8$	$\psi_1 = 0.0$	$\psi_1 = 0.8$
100	1.21	1.27	1.55	1.62	2.20	2.24
	(8.6)	(9.8)	(4.0)	(4.7)	(0.6)	(0.8)
200	1.24	1.26	1.58	1.61	2.29	2.28
	(9.2)	(9.6)	(4.5)	(4.6)	(0.8)	(0.9)
1,000	1.32	1.28	1.65	1.62	2.36	2.30
	(10.7)	(10.0)	(5.0)	(4.7)	(1.1)	(0.9)
∞	1.282	1.282	1.645	1.645	2.326	2.326
	(10.0)	(10.0)	(5.0)	(5.0)	(1.0)	(1.0)

Table 2.2b   Critical values for U_1 and U_2

T	90%		95%		99%	
	$\psi_1 = 0.0$	$\psi_1 = 0.8$	$\psi_1 = 0.0$	$\psi_1 = 0.8$	$\psi_1 = 0.0$	$\psi_1 = 0.8$
100	0.48	0.49	0.66	0.68	1.08	1.17
	(10.1)	(10.6)	(4.9)	(5.4)	(0.9)	(1.2)
200	0.47	0.47	0.66	0.64	1.13	1.01
	(9.8)	(9.4)	(4.9)	(4.4)	(1.1)	(0.7)
1,000	0.47	0.48	0.65	0.64	1.10	1.05
	(9.6)	(12.1)	(4.7)	(4.4)	(0.9)	(0.7)
∞	0.477	0.477	0.664	0.664	1.116	1.116
	(10.0)	(10.0)	(5.0)	(5.0)	(1.0)	(1.0)

Notes: Table entries are quantiles, with corresponding size in (.) brackets, where size is the empirical size using quantiles from $N(0, 1)$ for Table 2.2a and from the quantiles for $T = \infty$ for Table 2.2b; the latter are from KM (op. cit., table 1). Results are based on 20,000 replications. A realised value of the test statistic exceeding the $(1 - \alpha)\%$ quantile leads to rejection of the null model at the $\alpha\%$ significance level.

The likely numerical magnitude of the drift will depend on whether the integrated process is linear or loglinear and on the frequency of the data. In Table 2.2a, the critical values were presented for $\beta_1 = \sigma_\varepsilon$, which is 1 in units of the innovation standard error ($\beta_1/\sigma_\varepsilon = 1$). Table 2.3 considers the sensitivity of these results to the magnitude of the drift for the case $T = 100$ and $\kappa = \beta_1/\sigma_\varepsilon$; as in Table 2.2a the results are in the form of quantile with size in (.) parentheses, where the size is the empirical size if the quantiles from $N(0, 1)$ are used.

The results in Table 2.3 show that apart from small values of κ, for example $\kappa = 0.1$, the quantiles are robust to variations in κ, but there is a slight undersizing throughout for $\psi_1 = 0.0$; the actual size of a test at the nominal 5% level

Table 2.3 Sensitivity of critical values for V_1, V_2; $T = 100$, $\psi_1 = 0.0$, $\psi_1 = 0.8$

$\beta_1 = \kappa\sigma_\varepsilon$	90%		95%		99%	
	$\psi_1 = 0.0$	$\psi_1 = 0.8$	$\psi_1 = 0.0$	$\psi_1 = 0.8$	$\psi_1 = 0.0$	$\psi_1 = 0.8$
$\kappa = 0.1$	2.22	2.08	3.87	3.32	14.84	17.98
	(19.7)	(18.5)	(14.8)	(12.7)	(9.4)	(8.6)
$\kappa = 1.0$	1.21	1.27	1.55	1.62	2.20	2.24
	(8.6)	(9.8)	(4.0)	(4.7)	(0.6)	(0.8)
$\kappa = 2.0$	1.22	1.24	1.57	1.61	2.24	2.27
	(9.2)	(9.2)	(4.1)	(4.7)	(0.8)	(0.8)
$\kappa = 3.0$	1.21	1.29	1.57	1.62	2.19	2.21
	(8.9)	(10.3)	(4.2)	(4.8)	(0.8)	(0.6)
$\kappa = 4.0$	1.22	1.29	1.57	1.63	2.26	2.29
	(8.9)	(10.3)	(4.2)	(4.8)	(0.8)	(0.9)
$N(0, 1)$	1.282	1.282	1.645	1.645	2.326	2.326
	(10.0)	(10.0)	(5.0)	(5.0)	(1.0)	(1.0)

being between about 4% and 4.5%. However, the empirical size is closer to the nominal size for $\psi_1 = 0.8$. Overall, whilst there is some sensitivity to sample size and the magnitudes of ψ_1 and β_1, the quantiles are reasonably robust to likely variations.

2.2.2.iii Motivation

Whilst four test statistics were outlined in the previous section, they are motivated by similar considerations. In general, fitting the wrong model will induce heteroscedasticity, which is detected by assessing the (positive) correlation between the squared residuals from the fitted model and y_{t-1} in the case of the LIP and $-\log y_{t-1}$ in the case of the LLIP.

To illustrate, consider the case where the DGP is the exponential random walk, ERW, which is the simplest LLIP, but the analysis is pursued in levels; further, assume initially that $z_t = \varepsilon_t$. Then:

$$\Delta y_t = \left(\frac{y_t}{y_{t-1}} - 1\right) y_{t-1} \tag{2.19}$$

$$= \eta y_{t-1} \tag{2.20}$$

where $\eta = \left(\frac{y_t}{y_{t-1}} - 1\right) = (e^{(\varepsilon_t + \beta_1)} - 1)$.

Given that ε_t and y_{t-1} are uncorrelated, the conditional variance of Δy_t given y_{t-1} is proportional to y_{t-1}^2 when the DGP is the ERW, implying that Δy_t^2 and y_{t-1}^2 are positively correlated. However, when the DGP is a LIP, there is no correlation between Δy_t^2 and y_{t-1}^2; thus, the correlation coefficient between Δy_t^2 and y_{t-1}^2 provides a diagnostic test statistic when the linear model is fitted. KM (op. cit.) find that the asymptotic distribution is better approximated when the test statistic is taken as the correlation coefficient between Δy_t^2 and y_{t-1}, with large values leading to rejection of the linear model. When z_t follows an AR process, then $\Delta \varepsilon_t^2$, or $\Delta \hat{\varepsilon}_t^2$ in practice, replaces Δy_t^2, which should be uncorrelated with y_{t-1} under the null of the LIP.

2.2.2.iv   *Power considerations*

The power of the KM tests depends not only on T, but also on whether the test is based on the 'with drift' or 'without drift' model. Some of the results from the power simulations in KM (op. cit.) are summarised in Table 2.4. Evidently the test when there is drift is considerably more powerful than in the without drift case. Even with T = 100, power is 0.96 when the fitted model is linear, but the DGP is actually log-linear, and 0.81 when the fitted model is log-linear, but the DGP is actually linear; however, in the absence of drift, these values drop to 13% and 12% in the former case and only 3% and 4% in the latter case, respectively. This suggests caution in the use of the non-drifted tests, which require quite substantial sample sizes to achieve decent power. However, KM (op. cit.) found that their tests are more powerful, sometimes considerably so, than an alternative test for heteroscedasticity, in this case the well-known LM test for ARCH (see Kobayashi and McAleer, 1999b) for a power comparison of a number of tests.

2.2.3   Illustrations

As noted, the KM non-nested tests are likely to be of greatest interest when a standard unit root test is unable to discriminate between which of two integrated processes generated the data, finding that both the levels and the logarithms support non-rejection of the unit root. An important prior aspect of the use of the tests is whether the integrated process is with or without drift. Also, since the issue is whether the unit root is in either the levels or the logarithms, the data should be everywhere positive. The economic context is thus likely to be of

Table 2.4   Power of KM nonnested tests for 5% nominal size

T	Null: linear Alternative: log-linear		Null: log-linear Alternative: linear	
	V_1(drift)	U_1 (no drift)	V_2(drift)	U_2 (no drift)
50	0.29	0.07	0.23	0.08
100	0.96	0.13	0.81	0.12
200	1.00	0.25	1.00	0.26

Note: extracted from KM (1999a, table 4), based on 10,000 replications; in the with drift cases, $\beta_1 = 0.01$ and $\kappa = 1$.

importance, for example the default for aggregate expenditure variables, such as consumption, GDP, imports and so on, is that if generated by an integrated process, there is positive drift; whereas series such as unemployment rates are unlikely to be generated by a drifted IP.

2.2.3.i   Ratio of gold to silver prices

The first illustration is the ratio of the gold to silver price, obtained from the daily London Fix prices for the period 2nd January, 1985 to 31st March 2006; weekends and bank holidays are excluded, giving an overall total of 5,372 observations. This series, referred to as y_t is obviously everywhere positive, so there is an issue to be decided about whether it is generated by an LIP or an LLIP. In neither case is there evidence of drift (see Figure 2.1, left-hand panel for the 'levels' data and right-hand panel for the log data), and the appropriate KM tests are, therefore, U_1 and U_2, that is the driftless non-nested tests. The levels series has a maximum of 100.8 and a minimum of 38.3, and it crosses the sample mean of 68.5, 102 times, that is just 1.9% of the sample observations, which is suggestive of random walk-type behaviour. The size of the sample provides a safeguard against the low power indicated by the simulation results for the driftless tests.

For illustrative purposes, the standard ADF test $\hat{\tau}_\mu$ was used. The maximum lag length was set at 20, but in both the linear and loglinear versions of the ADF regressions BIC selected ADF(1); in any case, longer lags made virtually no difference to the value of $\hat{\tau}_\mu$, which was -2.433 for the LIP and -2.553 for the LLIP, both leading to non-rejection of the null hypothesis at conventional significance levels, for example, the 5% critical value is -2.86. For details of the regressions, see Table 2.5.

As an indicator of the likely result of the KM tests, note that the correlation coefficient between the squared residuals and y_{t-1} in the linear case is 0.004, indicating no misspecification, whereas for the log-linear case the correlation coefficient between the squared residuals and $-\ln y_{t-1}$ is 0.107. The latter correlation suggests that if a unit root is present it is in the linear model, which is confirmed by the sample value of $U_1 = 0.097$, which is not significant, whereas $U_2 = 4.57$, which is well to the right of the 99th percentile of 1.116, see Table 2.2b.

2.2.3.ii   World oil production

The second illustration is world oil production, using monthly data from January 1973 to February 2005, giving an overall sample of 386 observations. The data are graphed in Figure 2.2 in levels and logs; it is apparent that there is a positive trend and, hence, it is necessary to allow for drift if there is a unit root in the generating process. The appropriate KM tests are, therefore, V_1 and

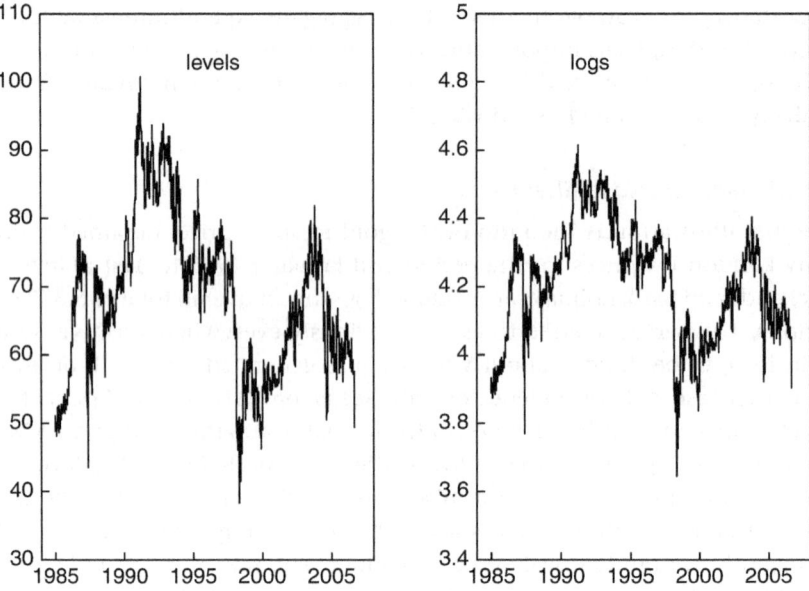

Figure 2.1   Gold–silver price ratio

Table 2.5   Regression details: gold–silver price ratio

	Constant	y_{t-1}	Δy_{t-1}	KM test (95% quantile)
Δy_t	0.181	–0.0026	–0.161	
('t')		(–2.43)	(–11.91)	
Δy_t	–	–	–0.162	$U_1 = 0.097$ (0.664)
('t')			(–12.01)	
	Constant	$\ln(y_{t-1})$	$\Delta \ln(y_{t-1})$	KM test (95% quantile)
$\Delta \ln(y_t)$	0.012	–0.0029	–0.150	
('t')		(–2.55)	(–11.08)	
$\Delta \ln(y_t)$	–	–	–0.151	$U_2 = 4.57$ (0.664)
('t')			(–11.19)	

V_2, which allow for drift; and the DF pseudo-t test is now $\hat{\tau}_\beta$, which includes a time trend in the maintained regression to allow the alternative hypothesis to generate a trended series.

Whilst, in this example, BIC suggested an ADF(1) regression, the marginal-t test criterion indicated that higher order lags were important and, as the sample is much smaller than in the first illustration, an ADF(5) was selected as the test model. As it happens, the KM test statistics were, however, virtually unchanged

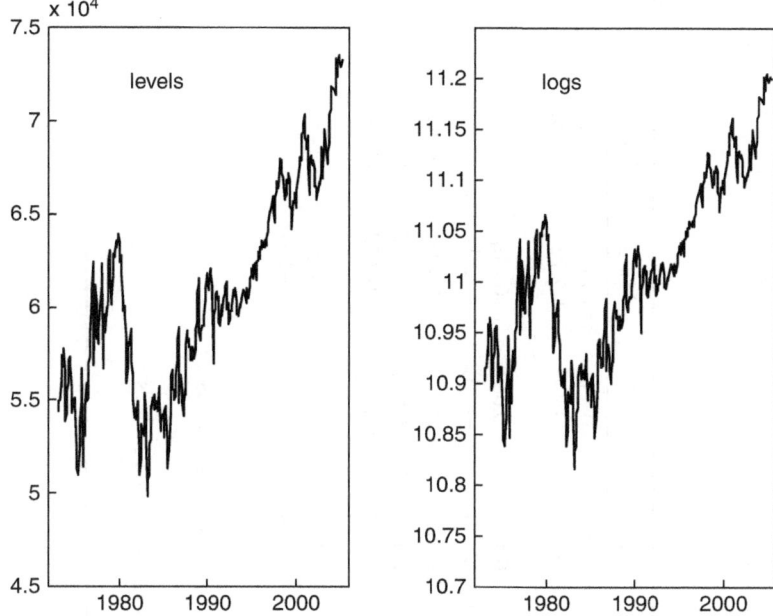

Figure 2.2   World oil production

between the two specifications and the conclusions were not materially different for the different lag lengths. As in the first illustration, the DF test is unable to discriminate between the linear and log specification, with $\hat{\tau}_\beta = -1.705$ and $\hat{\tau}_\beta = -1.867$, respectively, neither of which lead to rejection of the null hypothesis, with a 5% critical value of -3.41; see Table 2.6 for the regression details.

The correlation coefficient between the squared residuals and y_{t-1} in the linear case is -0.094, indicating no misspecification in the log-linear direction, whereas for the log-linear case the correlation coefficient between the squared residuals and $-\ln y_{t-1}$ is 0.132. The KM test values are $V_1 = -4.77$, which is wrong-signed for rejection of the null hypothesis, and $V_2 = 15.83$, which has a p-value of zero under the null. Hence, if there is an integrated process, then the conclusion is in favour of a linear model rather than a log-linear model.

2.3   Monotonic transformations and unit root tests based on ranks

A problem with applying standard parametric tests for a unit root is that the distribution of the test statistic under the null and alternative is not generally invariant to a monotonic transformation of the data. This was illustrated in Section **2.1** for the DF tests for the case of loglinear transformations, although

Table 2.6 Regression details: world oil production

Levels	C	y_{t-1}	Δy_{t-1}	Δy_{t-2}	Δy_{t-3}	Δy_{t-4}	Δy_{t-5}	Time
Δy_t	1,531	−0.029	−0.102	−0.093	−0.101	−0.058	−0.139	1.60
('t')		(−1.71)	(−1.93)	(−1.79)	(−1.95)	(−1.12)	(−2.72)	(1.94)
Δy_t	75.5	–	−0.120	−0.108	−0.113	−0.069	−0.148	–
('t')			(−2.32)	(−2.11)	(−2.22)	(−1.35)	(−2.91)	
Logs	C	$\ln y_{t-1}$	$\Delta \ln y_{t-1}$	$\Delta \ln y_{t-2}$	$\Delta \ln y_{t-3}$	$\Delta \ln y_{t-4}$	$\Delta \ln y_{t-5}$	Time
$\Delta \ln(y_t)$	0.36	−0.033	−0.097	−0.089	−0.104	−0.071	−0.139	0.000
('t')		(−1.87)	(−1.84)	(−1.71)	(−2.01)	(−1.39)	(−2.72)	(1.98)
$\Delta \ln(y_t)$	0.001	–	−0.120	−0.107	−0.119	−0.084	−0.149	–
('t')			(−2.29)	(−2.09)	(−2.33)	(−1.65)	(−2.94)	

KM tests

Null, linear;
Alternative, loglinear

$V_1 = -4.77$ 95% quantile = 1.645

Null, log-linear;
Alternative, linear

$V_2 = 15.83$ 95% quantile = 1.645

it holds more generally (see GH, 1991). Basing a unit root test on ranks avoids this problem.

The situation to be considered is that the correct form of the generating process is not known other than, say, it is a monotonic and, hence, order-preserving function of the data y_t; thus, as before, let y_t be the 'raw' data and let $w_t = f(y_t)$ be some transformation of the data, such that w_t is generated by an autoregressive process. The linear versus log-linear choice was a special case of the more general situation. In the (simplest) first order case, the generating process is:

$$w_t = \rho w_{t-1} + \varepsilon_t \qquad \varepsilon_t \sim \text{iid}(0, \sigma_\varepsilon^2), \sigma_\varepsilon^2 < \infty \qquad (2.21)$$

The parameter ρ is identified and, hence, we can form the null hypothesis, H_0: there exists a monotonic function $w_t = f(y_t)$ such that $\rho = 1$ in $w_t = \rho w_{t-1} + \varepsilon_t$.

A ranks-based test proceeds as follows. Consider the values of $Y = \{y_t\}_{t=1}^T$ ordered into a set (sequence) such that $Y_{(r)} = \{y_{(1)} < y_{(2)} \ldots < y_{(T)}\}$; then $y_{(1)}$ and $y_{(T)}$ are the minimum and maximum values in the complete set of T observations. To find the rank, r_t, of y_t, first obtain its place in the ordered set $Y_{(r)}$, then its rank is just the subscript value; the set of ranks is denoted $\mathbf{r}$. For example, if $\{y_t\}_{t=1}^T = \{y_1 = 3, y_2 = 2, y_3 = 4\}$, then $Y_{(r)} = \{y_{(1)} = y_2, y_{(2)} = y_1, y_{(3)} = y_3\}$, and the ranks are $\mathbf{r} = \{r_t\}_{t=1}^3 = \{2, 1, 3\}$. Most econometric software contains a routine to sort and rank a set of observations.

2.3.1 DF rank-based test

An obvious extension of the DF tests is to apply them to the ranks of the series, that is to $\mathbf{r}$, rather than the series, Y, itself. The time series of (simple) ranks can be centred by subtracting the mean of the ranks, say $R_t = r_t - \bar{r}$, where $\bar{r} = (T+1)/2$, assuming T observations in the series. A simple extension of the DF pseudo-t tests using the ranks, either the centred rank R_t with no intercept in the regression, or r_t, with an intercept in the regression, was suggested by GH (1991). In the former case the DF regression is:

$$R_t = \rho_r R_{t-1} + \nu_t \qquad (2.22)$$

$$\Delta R_t = \gamma_r R_{t-1} + \nu_t \qquad (2.23)$$

The standard DF tests are now defined with respect to the (centred) ranks:

$$\hat{\delta}_{r,\mu} = T(\hat{\rho}_r - 1) \qquad (2.24)$$

$$\hat{\tau}_{r,\mu} = \frac{\hat{\gamma}_r}{\hat{\sigma}(\hat{\gamma}_r)}$$

$$= \frac{(\hat{\rho}_r - 1)}{\hat{\sigma}(\hat{\rho}_r)} \qquad (2.25)$$

where $\hat{\rho}_r$ is the LS estimator based on the ranks and the testing procedure is otherwise as in the standard case.

The case of a deterministic mean under the alternative hypothesis causes no problems, as the ranks are not altered by the subtraction of a non-zero mean. However, for trended possibilities that occur if the random walk under the null is drifted or the stationary deviations under the alternative are around a deterministic trend, the ranks will tend to increase over time. A supposition in this context is that it is valid to derive test statistics analogous to the parametric case by first detrending the series and then constructing the ranks from the detrended series; the ranked DF, RDF, test statistics in this case are denoted $\hat{\delta}_{r,\beta}$ and $\hat{\tau}_{r,\beta}$.

In the event that the v_t are serially correlated, GH suggested a rank extension of the augmented DF test, say RADF, in which the maintained regression (here using the centred ranks) is:

$$\Delta R_t = \gamma_r R_{t-1} + \sum_{j=1}^{k-1} c_{r,j} \Delta R_{t-j} + v_t \qquad (2.26)$$

Although this version of a rank test has not been justified formally, the simulation results reported in Section **2.6**, show that the RADF test, with the simulated quantiles for the DF rank test, does perform well; see also Fotopoulus and Ahn (2003). By way of comparison, the rank-score test due to Breitung and Gouriéroux (1997), hereafter BG, considered in the next Section, **2.3.2**, does allow explicitly for serially correlated errors.

BG (op.cit.) show that the asymptotic distribution of $\hat{\tau}_{r,\mu}$ is not the same as the corresponding parametric test, the key difference being that $T^{-1}r_{[rT]}$ does not converge to a Brownian motion, where T is the sample size, $0 \leq r \leq 1$ and [.] represents the integer part. Fotopoulus and Ahn (2003) extend these results to obtain the relevant asymptotic distributions. If the standard DF quantiles are used, then the rank tests tend to be oversized. The extent of the over-sizing is illustrated in Table 2.7, which reports the results of a small simulation study where the innovations are drawn from one of $N(0, 1)$ and t(2). Although the wrong sizing is not great, for example 6.1% for a nominal 5% test with T = 200, the over-sizing does not decline with the sample size.

Table 2.7   Size of $\hat{\delta}_{r,\mu}$ and $\hat{\tau}_{r,\mu}$ using quantiles for $\hat{\delta}_\mu$ and $\hat{\tau}_\mu$

Size	T = 200				T = 500			
	$\hat{\delta}_\mu$	$\hat{\delta}_{r,\mu}$	$\hat{\tau}_\mu$	$\hat{\tau}_{r,\mu}$	$\hat{\delta}_\mu$	$\hat{\delta}_{r,\mu}$	$\hat{\tau}_\mu$	$\hat{\tau}_{r,\mu}$
N(0, 1)	5.1%	10.5%	5.0%	6.1%	4.9%	11.7%	4.9%	7.4%
t(2)	4.9%	10.5%	5.0%	6.8%	5.0%	11.8%	4.9%	7.6%

Source: results are for 50,000 replications of a driftless random walk; the $\hat{\delta}_\mu$ and $\hat{\tau}_\mu$ tests are from a DF regression with an intercept; and $\hat{\delta}_{r,\mu}$ and $\hat{\tau}_{r,\mu}$ use the mean-adjusted ranks.

Table 2.8   Critical values of the rank-based DF tests $\hat{\delta}_{r,\mu}$, $\hat{\tau}_{r,\mu}$, $\hat{\delta}_{r,\beta}$ and $\hat{\tau}_{r,\beta}$

T	$\hat{\delta}_{r,\mu}$			$\hat{\tau}_{r,\mu}$		
	100	200	500	100	200	500
1%	−23.04	−24.19	−25.39	−3.58	−3.58	−3.62
5%	−16.39	−17.15	−17.84	−2.98	−3.00	−3.03
10%	−12.40	−14.07	−14.65	−2.68	−2.71	−2.75
T	$\hat{\delta}_{r,\beta}$			$\hat{\tau}_{r,\beta}$		
	100	200	500	100	200	500
1%	−31.30	−32.67	−34.14	−4.20	−4.20	−4.20
5%	−23.17	−24.53	−25.68	−3.60	−3.60	−3.63
10%	−19.80	−20.85	−21.85	−3.29	−3.30	−3.35

Source: own calculations based on 50,000 replications.

Rather than use the percentiles for the standard DF tests, the percentiles can be simulated; for example, some critical values for $\hat{\delta}_{r,\mu}$, $\hat{\tau}_{r,\mu}$, $\hat{\delta}_{r,\beta}$ and $\hat{\tau}_{r,\beta}$ for T = 100, 200 and 500 are provided in Table 2.8. The maintained regression for these simulations used the mean-adjusted ranks, with 50,000 replications of the null of a driftless random walk. Other sources for percentiles are Fotopoulus and Ahn (2003), who do not use mean-adjusted ranks, and GH (1991), who include a constant in the maintained regression rather than use the mean-adjusted ranks.

As in the case of the standard DF tests, a sample value that is larger negative than the critical value leads to rejection of the null of a unit root in the data. More precisely, rejection of the null carries the implication that, subject to the limitations of the test, there is not an order preserving transformation of the data that provides support for the null hypothesis. Non-rejection implies that there is such a transformation (or trivially that no transformation is necessary) that is consistent with a unit root AR(1) process; but it does not indicate what is the transformation, which has to be the subject of further study. Other rank-based tests are considered in the next sections.

2.3.2   Rank-score test, Breitung and Gouriéroux (1997)

BG (op. cit.) developed a 'rank' extension of the parametric score test suggested by Schmidt and Phillips (1992). The test is designed for a DGP that under H_0 is a random walk with drift, whereas under H_A it is trend stationary, that is H_0 and H_A are, respectively:

$$H_0 : \Delta y_t = \beta_1 + \varepsilon_t \tag{2.27}$$

$$H_A : (1 - \rho L)(y_t - \beta_0 - \beta_1 t) = \varepsilon_t \qquad |\rho| < 1 \tag{2.28}$$

where $\varepsilon_t \sim$ iid and $E(\varepsilon_t) = 0$, with common cdf denoted $F(\varepsilon_t)$. The aim is to obtain a test statistic for H_0 that can be extended to the case where the sequence $\{\varepsilon_t\}$ comprises serially correlated elements. It is convenient here to adopt the notation that the sequence starts at $t = 0$, thus $\{\varepsilon_t\}_{t=0}^{t=T}$.

The test statistic is an application of the score, or LM, principle to ranks. In the parametric case, the test statistic for H_0 is:

$$S_\beta = \frac{\sum_{t=2}^{T} x_t S_{t-1}}{\sum_{t=2}^{T} S_{t-1}^2} \tag{2.29}$$

where $x_t = \Delta y_t - \hat{\beta}_1$, which is the demeaned data under the null, where $\hat{\beta}_1$ is the sample mean of Δy_t, that is $(y_T - y_0)/T$; and S_t is the partial sum of x_t, given by $S_t = \sum_{j=1}^{t} x_j$.

Analogously, the rank version of this test is:

$$S_{r,\beta} = \frac{\sum_{t=2}^{T} \tilde{r}_t S_{t-1}^R}{\sum_{t=2}^{T} (S_{t-1}^R)^2} \tag{2.30}$$

where $\tilde{r}_t$ is the (demeaned) rank of Δy_t among $(\Delta y_1, \ldots, \Delta y_T)$, that is, the ordering is now with respect to Δy_t, such that $\Delta y_{(1)} < \Delta y_{(2)} < \ldots < \Delta y_{(T)}$. The ordering is with respect to Δy_t rather than y_t, as under the LM principle the null of a unit root is imposed. A β subscript indicates that the null and alternative hypotheses allow trended behaviour in the time series either due to drift under H_0 or a trend under H_A.

S_t^R is the partial sum to time t of the (demeaned) ranks of Δy_t, that is $S_t^R = \sum_{j=1}^{t} \tilde{r}_j$. The ranks of the differences are demeaned by subtracting $(T+1)/2$, or $T/2$ if the convention is that $t = 1$ is the start of the process. That is, let $\tilde{R}(\Delta y_t)$ be the rank of Δy_t among $(\Delta y_1, \ldots, \Delta y_T)$, then define:

$$\tilde{r}_t \equiv \tilde{R}(\Delta y_t) - \left(\frac{T+1}{2}\right) \quad \text{demeaned ranks} \tag{2.31}$$

The normalised, demeaned ranks are then:

$$\tilde{r}_t^n \equiv \frac{\tilde{r}_t}{T} \tag{2.32}$$

BG (op. cit.) show that the numerator of $S_{r,\beta}$ is a function of T only, in which case a test statistic can be based just on the denominator, with their test statistic a function of λ, where:

$$\lambda = \sum_{t=1}^{T} (S_t^R)^2 \tag{2.33}$$

This expression just differs from the denominator of $S_{r,\beta}$ by the inclusion of $(S_T^R)^2$.

The BG rank-score test statistic is $\lambda_{r,\beta}$, defined as follows:

$$\lambda_{r,\beta} \equiv \frac{12}{T^4} \lambda$$

$$= \frac{12}{T^4} \sum_{t=1}^{T} \left(\sum_{j=1}^{t} \tilde{r}_j \right)^2$$

$$= \frac{12}{T^2} \sum_{t=1}^{T} \left(\sum_{j=1}^{t} \tilde{r}_j^n \right)^2 \tag{2.34}$$

where $\tilde{r}_t^n$ is given by (2.32). The distributional results are based on the following theorem (BG, op. cit., Theorem and Remark A):

$$\lambda_{r,\beta} \Rightarrow_D \int_0^1 (B(r) - rB(1))^2 dr \tag{2.35}$$

Finite sample critical values are given in BG (op. cit.), some of which are extracted into Table 2.9 below. The test is 'left-sided' in the sense that a sample value less than the $\alpha\%$ quantile implies rejection of the null hypothesis. The null distribution of $\lambda_{r,\beta}$ is independent of $F(\varepsilon_t)$, both asymptotically and in finite samples; and the limiting null distribution is invariant to a break in the drift during the sample, that is, if, under the null, β_1 shifts to $\beta_1 + \delta$ at some point T_b.

Referring back to H_0 of (2.27), a special case arises if $\beta_1 = 0$, corresponding to the random walk without drift, $\Delta y_t = \varepsilon_t$; however as $\lambda_{r,\beta}$ is a test statistic based on the ranks of Δy_t, it is invariant to monotonic transformations of Δy_t, a simple example being when a constant is added to each observation in the series, so the same test statistic would result under this special null.

Variations on the test statistic $\lambda_{r,\beta}$ are possible, which may improve its small sample performance. BG (op. cit.) suggest a transformation of the ranks using the inverse standard normal distribution function, which can improve power when the distribution of ε_t is not normal (see BG, op. cit., table 2). A nonlinear transformation is applied to the normalised ranks $\tilde{r}_t^n$ using the inverse of the standard normal distribution function $\Phi(.)$ and the revised ranks are:

$$\tilde{r}_t^{ins} = \Phi^{-1}(\tilde{r}_t^n + 0.5) \tag{2.36}$$

The test statistic, with this variation, using $\tilde{r}_t^{ins}$ rather than $\tilde{r}_t^n$, is referred to as $\lambda_{r,\beta}^{(ins)}$. Some quantiles for $\lambda_{r,\beta}$ and $\lambda_{r,\beta}^{(ins)}$ are given in Table 2.9.

2.3.2.i Serially correlated errors

In the event that the errors are serially correlated, then the asymptotic distribution of $\lambda_{r,\beta}$ is subject to a scale component, which is analogous to (but simpler than) the kind of adjustment applied to obtain the Phillips-Perron versions of the standard DF tests (see UR Vol. 1, chapter 6). The end result is a simple transformation of $\lambda_{r,\beta}$, using a consistent estimator of the long-run variance (see Chapter 1, Section **1.4**).

Table 2.9   Quantiles for the BG rank-score tests

	T = 100		T = 250		T = ∞	
	$\lambda_{r,\beta}$	$\lambda_{r,\beta}^{(ins)}$	$\lambda_{r,\beta}$	$\lambda_{r,\beta}^{(ins)}$	$\lambda_{r,\beta}$	$\lambda_{r,\beta}^{(ins)}$
1%	0.0257	0.0259	0.0252	0.0252	0.0250	0.0250
5%	0.0373	0.0376	0.0367	0.0370	0.0362	0.0368
10%	0.0466	0.0470	0.0461	0.0465	0.0453	0.0464

Source: extracted from BG (op. cit., table 6).

First, define the long-run variance in the usual way (see also UR Vol. 1, chapter 6), but applied to the ranks:

$$\sigma_{r,lr}^2 = \lim_{T\to\infty} E[T^{-1}(S_T^R)^2]$$

$$= \lim_{T\to\infty} E[T^{-1}\left(\sum_{j=1}^T \tilde{r}_j\right)^2] \tag{2.37}$$

Then asymptotically under H_0:

$$\frac{1}{\sigma_{r,lr}^2}\lambda_{r,\beta} \Rightarrow_D \int_0^1 (B(r) - rB(1))^2 dr \tag{2.38}$$

The long-run variance $\sigma_{r,lr}^2$ can be estimated by first regressing the ranks $\tilde{r}_t$ on S_{t-1}^R, to obtain the residuals denoted $\tilde{v}_t$, see (2.22). A consistent estimator of the long-run variance is given by:

$$\tilde{\sigma}_{r,lr}^2 = 12\left(\sum_{t=1}^T \tilde{v}_t^2/T + 2\sum_{k=1}^M \sum_{t=\kappa+1}^T \tilde{v}_{t-\kappa}\tilde{v}_t/T\right) \tag{2.39}$$

This is a semi-parametric form in keeping with the intention of the test not to involve a parametric assumption, such as would be involved in using an AR approximation to obtain $\tilde{\sigma}_{r,lr}^2$. Provided $M/T \to 0$ and $M \to \infty$ as $T \to \infty$, then $\tilde{\sigma}_{r,lr}^2$ is consistent for $\sigma_{r,lr}^2$. The scaling factor 12 outside the brackets in (2.39) arises because if the errors are iid, then the limiting value of the first term in the brackets is 12^{-1} as $T \to \infty$; see BG (op. cit., p.16). This reflects the normalisation that if the errors are iid, then $\sigma_{r,lr}^2 = 1$.

To ensure positivity of the estimator, a kernel function $\omega_M(\kappa)$ can be used, resulting in the revised estimator, $\tilde{\sigma}_{r,lr}^2$:

$$\tilde{\sigma}_{r,lr}^2(\kappa) = 12\left(\sum_{t=1}^T \tilde{v}_t^2/T + 2\sum_{k=1}^M \omega_M(\kappa) \sum_{t=\kappa+1}^T \tilde{v}_{t-\kappa}\tilde{v}_t/T\right) \tag{2.40}$$

The frequently applied Newey-West kernel function uses the Bartlett weights $\omega_M(\kappa) = 1 - \kappa/(M+1)$ for $\kappa = 1, ..., M$.

The revised test statistic, with the same asymptotic distribution as $\lambda_{r,\beta}$, is:

$$\lambda_{lr,\beta} = \frac{1}{\hat{\sigma}^2_{r,lr}} \lambda_{r,\beta} \tag{2.41}$$

An obvious extension gives rise to the version of the test that transforms the ranks using the inverse standard normal distribution and is referred to as $\lambda^{(ins)}_{lr,\beta}$. Some asymptotic critical values are given in the last column ($T = \infty$) of Table 2.7.

2.3.2.ii Simulation results

Some results from Monte-Carlo simulations in BG (op. cit.) are summarised here, taking $T = 100$ as illustrative of the general results. First, with $w_t = y_t$, so that the data are not transformed, and the ε_t are drawn alternately from $N(0, 1)$, $t(2)$ and the centred $\chi^2(2)$ distribution. Both the rank test statistic $\lambda_{r,\beta}$ and the DF test statistic $\hat{\tau}_\beta$ maintain their size well under the different distributions for ε_t, with $\hat{\tau}_\beta$ generally more powerful than $\lambda_{r,\beta}$, especially for non-normal innovations. The advantage of $\lambda_{r,\beta}$ is apparent when an additive outlier (AO) of 5σ contaminates the innovations at $T/2$. In this case $\lambda_{r,\beta}$ maintains its size, whereas $\hat{\tau}_\beta$ becomes over-sized, for example an empirical size of 17.4% for a nominal 5% size.

The AO situation apart, there is no consistent advantage in the rank-score test when it is applied to $w_t = y_t$, which is not surprising given its nonparametric nature. More promising is the case when the random walk is generated in a nonlinear transformation of the data, so that $w_t = f(y_t)$, with $f(.)$ alternately given by $f(y_t) = y_t^3$, $f(y_t) = y_t^{1/3}$, $f(y_t) = \ln y_t$ and $f(y_t) = \tan(y_t)$, and the random walk, without drift, is in w_t, that is the random walk DGP is in the transformed data. In these cases, $\lambda_{r,\beta}$ maintains its size, whereas $\hat{\tau}_\beta$ is seriously over-sized and the $\lambda^{(ins)}_{r,\beta}$ version of the test becomes moderately over-sized.

The case where $w_t = \ln(y_t)$, but the test uses y_t, was considered in Section 2.1, (referred to as Case 2), where it was noted that the rejection rate was typically of the order of 80–90% using $\hat{\tau}_\mu$ at a nominal 5% size. As to the rank-score tests, BG (op. cit.) report that $\lambda_{r,\beta}$ and $\lambda^{(ins)}_{r,\beta}$ have empirical sizes of 5.1% and 8.2%, respectively. The problematic case of MA(1) errors was also considered and it also remains a problem for the rank-score tests $\lambda_{r,\beta}$ and $\lambda^{(ins)}_{r,\beta}$, with serious under-sizing when the MA(1) coefficient is positive and serious over-sizing when the MA(1) coefficient is negative.

Illustrations of the rank-score tests and other tests introduced below are given in Section 2.6.3.

2.4 The range unit root test

Another basis on which to construct a nonparametric test, which is also invariant to monotonic transformations, is the range of a time series; such a test was suggested by Aparicio, Escribano and Sipols (2006), hereafter AES. Consider the sorted, or ordered, values of $\{y_t\}_{t=1}^{t=i}$, such that $y_{(1)} < y_{(2)} \ldots < y_{(i)}$; then $y_{(1)}$ and $y_{(i)}$ are the minimum and maximum values respectively, referred to as the extremes, of $y_{(t)}$ in a set of i observations. The range is simply the difference between the extremes, $R_i^F = y_{(i)} - y_{(1)}$; the i subscript indicates the length of the sequence of y_t, hence, within each sequence, t runs from 1 to i. The F superscript is redundant at this stage, but just indicates that this is the range when the observations are in their 'natural' time series order, running 'forward' in time; in a later section this order is reversed.

2.4.1   The range and new 'records'

Let the length of the sequence, i, vary from $i = 2, \ldots, T$, where T is the overall sample size. For $i = 2$, $y_{(1)} < y_{(2)}$ (for simplicity equality is ruled out); now consider adding a 3rd observation, y_3, then there are three possibilities: Case 1, $y_{(1)} < y_3 < y_{(2)}$; Case 2, $y_3 < y_{(1)} < y_{(2)}$; Case 3, $y_{(1)} < y_{(2)} < y_3$. In Case 1, $y_{(1)}$ is still the minimum value and $y_{(2)}$ is still the maximum value; thus, the extremes are unchanged and, therefore, the range remains unchanged; in Case 2, y_3 is the new minimum; whereas in Case 3, y_3 is the new maximum. In Cases 2 and 3, the range has changed and, therefore, $\Delta R_i^F > 0$. In the terminology associated with this test, in Cases 2 and 3 there is a new 'record'.

Next, define an indicator variable as a function of ΔR_i^F so that it counts the number of new records as the index i runs from 1 to T. The number of these new records in a sample of size T is: $\sum_{i=1}^{T} I(\Delta R_i^F > 0)$, where $I(\Delta R_i^F > 0) = 0$ if $\Delta R_i^F = 0$ and $I(\Delta R_i^F > 0) = 1$ if $\Delta R_i^F > 0$. (For simplicity, henceforth, when the nature of the condition is clear then the indicator variable is written as $I(\Delta R_i^F)$.) Thus, in the simple illustration above, in Case 1 $\Delta R_3^F = 0$, whereas in Cases 2 and 3 $\Delta R_3^F > 0$. The average number of new records is $\bar{R}^F = T^{-1} \sum_{i=1}^{T} I(\Delta R_i^F)$. The intuition behind forming a test statistic based on the range is that the number of new records declines faster for a stationary series (detrended if necessary) compared to a series that is nonstationary because of a unit root. The test described here, referred to as the range unit root, or RUR, test is due to AES (op.cit.).

It is intuitively obvious that a new record for y_t is also a new record for a monotonic transformation of y_t, including nonlinear transformations, and this is the case under the null and the stationary alternative. Simulations reported in AES (op. cit., especially table IV) show that the nominal size is very much better maintained for the RUR test compared to the DF t-type test for a wide range of monotonic nonlinear transformations, typically 3–5% for a nominal 5% test when $T \leq 250$, and is virtually the same for $T = 500$.

Thus far the implicit background has been of distinguishing a driftless random walk from a trendless stationary process. However, care has to be taken when dealing with DGPs that involve trending behaviour, since the number of new records will increase with the trend. The alternatives in this case are a series non-stationary by virtue of a unit root, but with drift, and a series stationary around a deterministic (linear) time trend: both will generate trending behaviour. The trending case is dealt with after the trendless case; see Section **2.4.4**.

2.4.1.i *The forward range unit root test*

This section describes the rationale for a unit root test statistic based on the range of a variable generically denoted y_t. For simplicity this can be thought of as the level of a series (the 'raw' data), but equally, as the test statistic is invariant (asymptotically and in finite samples) to monotonic transformations, this could be the logarithm, or some other monotonic transformation, of the original series. Otherwise the set-up is standard, with the null hypothesis (in this section) that y_t is generated by a driftless random walk, $y_t = y_{t-1} + z_t$, where $z_t = \varepsilon_t$, and $\{\varepsilon_t\}_{t=1}^T$ is an iid sequence of random variables with $E(\varepsilon_t) = 0$ and variance σ_ε^2.

As a preliminary to the test statistic, consider the properties of a function of the sample size, say $\Gamma(T)$, such that the following condition is (minimally) satisfied:

$$\Gamma(T) \sum_{i=1}^T I(\Delta R_i^F > 0) = O_p(1) \tag{2.42}$$

Then for stationary series, $\Gamma(T) = \ln(T)^{-1}$; for I(1) series without drift, $\Gamma(T) = 1/\sqrt{T}$; and for I(1) series with drift, $\Gamma(T) = 1/T$. (The first of these results requires a condition that the i-th autocovariance decreases to zero faster than $\ln(T)^{-1}$; see AES (op. cit.), known as the Berman condition, which is satisfied for Gaussian ARMA processes.)

Thus, a test statistic based on the range for an I(1) series without drift can be formed as:

$$\begin{aligned} R_{UR}^F &= T^{-1/2} \sum_{i=1}^T I(\Delta R_i^F) \\ &= T^{1/2} \bar{R}^F \end{aligned} \tag{2.43}$$

Given the rate of decrease of new records, R_{UR}^F will tend to zero in probability for a stationary time series and to a non-degenerate random variable for an I(1) series. Thus, relatively, R_{UR}^F will take *large* values for an I(1) series and *small* values for an I(0) series.

In order to formalise the test procedure let $R_{UR}^F(\alpha, T) > 0$ be the $\alpha\%$ quantile of the null distribution of R_{UR}^F based on T observations, and let $\tilde{R}_{UR}^F$ be a sample value of the range unit root test statistic R_{UR}^F, then the null hypothesis is rejected at the $\alpha\%$ significance level if $\tilde{R}_{UR}^F < R_{UR}^F(\alpha, T)$, otherwise the unit root null is not rejected. To be clear, in the case of trendless alternatives it is 'small' values of the test statistic that lead to rejection.

Theorem 1 of AES (op. cit.) establishes the asymptotic null distribution of R_{UR}^F, as follows. Let the DGP under the null be $y_t = y_{t-1} + \varepsilon_t$, where $\varepsilon_t \sim$

Table 2.10   Critical values of the forward range unit
root test, $R_{UR}^F(\alpha, T)$

T =	100	250	500	1,000	2,000	5,000
1%	0.9	0.939	1.012	1.044	1.118	1.146
5%	1.1	1.201	1.208	1.265	1.275	1.315
10%	1.3	1.328	1.386	1.423	1.453	1.451
90%	2.8	2.973	3.040	3.060	3.080	3.110
95%	3.1	3.289	3.354	3.352	3.444	3.470

Source: based on AES (op. cit., table I); 10,000 replications of
the null model with $\varepsilon_t \sim niid(0, 1)$.

$iid(0, \sigma_\varepsilon^2)$, $t \geq 1$, the random variable ε_t is continuous and y_0 has a bounded
pdf and finite variance, then the asymptotic distribution of R_{UR}^F is given by:

$$R_{UR}^F \Rightarrow_D \sqrt{\frac{2}{\pi}} (\xi + \eta)^2 e^{-\frac{1}{2}(\xi+\eta)^2} \tag{2.44}$$

where $\xi \rightarrow_p \bar{B}(1), \eta \rightarrow_p L_B(0, 1)$ where the latter is the local time at zero of a
Brownian motion in [0, 1]; see AES (op. cit., Definition 1 and Appendix A2).
Critically, AES (op. cit., Theorem 1, part (3)) show that if y_t is a stationary
Gaussian series (with covariance sequence satisfying the Berman condition),
then as $T \rightarrow \infty$:

$$R_{UR}^F \rightarrow_p 0$$

Thus, R_{UR}^F is consistent against a stationary Gaussian series and left-sided
quantiles of the distribution of R_{UR}^F lead to rejection against such stationary
(trendless) alternatives. The Berman condition is not particularly demanding;
as noted it is satisfied by stationary Gaussian ARMA processes and the simulation
results in AES, which cover fat-tailed as well as asymmetric error distributions,
suggest that the size and power of R_{UR}^F are robust to deviations from Gaussianity.

AES show that the finite sample distribution for $T = 1,000$, under the null $y_t = y_{t-1} + \varepsilon_t$, with $\varepsilon_t \sim niid(0, 1)$, $t \geq 1$, is very close to the asymptotic distribution
of their Theorem 1, especially in the left-tail. Some critical values are reported
in Table 2.10; this table also includes the quantiles for 90% and 95% as they are
required for the situation when the test is used to distinguish between trending
alternatives.

2.4.1.ii   Robustness of R_{UR}^F

AES (op. cit.) show that R_{UR}^F is robust to a number of important departures from
standard assumptions. In particular it is robust to the following problems.

i) Departures from niid errors (which is what the finite sample, simulated critical values are based upon); for example, draws from fat-tailed distributions

such as the 't' distribution and the Cauchy distribution, and asymmetric distributions (see AES, op. cit., table III).

ii) Unlike the parametric DF-based tests, R_{UR}^F is reasonably robust to structural breaks in the stationary alternative. The presence of such breaks may 'confuse' standard parametric tests into assigning these to permanent effects, whereas they are transitory, leading to spurious non-rejection of the unit root null hypothesis. The cases considered by AES were: the single level shift, that is where the alternative is of the form $y_t = \rho y_{t-1} + \delta_1 D_{t,1} + \varepsilon_t$, $|\rho| < 1$ and $D_{t,1} = 0$ for $t \le t_{b,1}$, $D_{t,1} = 1$ for $t > t_{b,1}$, where $t_{b,1} = T/2$; and a multiple shift, in this case where the alternative is of the form $y_t = \rho y_{t-1} + \sum_{i=1}^2 \delta_i D_{t,i} + \varepsilon_t$, where $|\rho| < 1$ and $D_{t,i} = 0$ for $t \le (T/4)i$, $D_{t,i} = 1$ for $(T/4)i < t_{b,i} \le (T/2)i$.

AES (op. cit.) found that: (a) R_{UR}^F outperformed the DF t-type test, which had no power in most break scenarios; (b) but for the power of R_{UR}^F to be maintained, compared to the no-break case, the sample size had to be quite large, for example $T = 500$; (c) the power of R_{UR}^F deteriorated in the two-break case and, generally, deteriorated as the break magnitude increased.

iii) Additive outliers in I(1) time series. If the true null model is a random walk 'corrupted' by a single additive outlier, AO, then the DGP is: $x_t = y_t + \delta A_{t_a}$ where $y_t = y_{t-1} + \varepsilon_t$, $A_{t_a} = 1$ for $t = t_a$ and 0 otherwise. An AO in standard parametric tests, for example, the DF t-type test, shifts the null distribution to the left, so that a rejection that occurs because the sample value is to the left of the standard (left-tailed) $\alpha\%$ critical value may be spurious given that the correct critical value is to the left of the standard value. Provided that t_a is not too early in the sample, generally later than the first quarter of the sample, R_{UR}^F improves upon the DF test. Protection against the problem of an early AO is achieved by the forward–backward version of R_{UR}^F.

2.4.2 The forward–backward range unit root test

Resampling usually takes place through bootstrapping, which is a form of random sampling based on the original sample (see UR, Vol. 1, chapter 8). AES (op. cit.) suggested a particularly simple form of resampling that improves the power of the range unit–root test and offers more protection against additive outliers. This is the forward–backward version of R_{UR}^F, denoted, R_{UR}^{FB}, which, as in the case of R_{UR}^F, runs the test forward in the sample counting the number of new records from the beginning to the end, it then continues the count but now starting from the end of the sample back to the beginning; thus, R_{UR}^F counts forward, whereas R_{UR}^B counts backwards. The revised test statistic, analogous to (2.43), is:

$$R_{UR}^{FB} = \frac{1}{\sqrt{2T}} \left(\sum_{i=1}^T I(\Delta R_i^F) + \sum_{t=1}^T I(\Delta R_i^B) \right) \tag{2.45}$$

Table 2.11   Critical values of the empirical distribution of $R_{UR}^{FB}(\alpha, T)$

T =	100	250	500
1%	1.556	1.565	1.613
5%	1.839	1.878	1.897
10%	1.980	2.057	2.087
90%	3.960	4.159	4.300
95%	4.384	4.651	4.743
99%	5.303	5.545	5.692

Source: own calculations; 10,000 replications of the null model with $\varepsilon_t \sim$ niid(0, 1).

where ΔR_i^B counts the new records on the time-reversed series $y_t^B = y_{T-t+1}$, $t = 1, 2, \ldots, T$. The revised test statistic R_{UR}^{FB} has better size fidelity than R_{UR}^{FB} for AOs in the null DGP early or late in the sample (but not both); it has better power compared to R_{UR}^F against stationary alternatives with a single structural break; and there are no significant differences for monotonic nonlinearities. Some critical values for R_{UR}^{FB} are given in Table 2.11.

2.4.3   Robustness of R_{UR}^F and R_{UR}^{FB}

Given the motivation for using a nonparametric test statistic, we consider the sensitivity of the finite sample critical values to different error distributions. By way of an illustration of robustness, consider the quantiles when $\{\varepsilon_t\}$ is an iid sequence and alternatively drawn from a normal distribution and the t distribution with 3 degrees of freedom, t(3). The latter has substantially fatter tails than the normal distribution. One question of practical interest is: what is the actual size of the test if the quantiles assume $\varepsilon_t \sim$ niid(0, 1) but, in fact, they are drawn from t(3)? In this case, and taking T = 100, 250, by way of example, the actual size is then compared with the nominal size. By way of benchmarking, the results are also reported for the DF t-type test statistics, $\hat{\tau}_\mu$ and $\hat{\tau}_\beta$. Some of the results are summarised in Table 2.12, with QQ plots in Figures 2.3 and 2.4, the former for the DF tests and the latter for the range unit root tests. (The QQ plots take the horizontal axis as the quantiles assuming $\varepsilon_t \sim$ niid(0, 1) and the vertical axis as the quantiles realised from $\varepsilon_t \sim$ t(3); differences from the 45° line indicate a departure in the pairs of quantiles.)

There are four points to note about the results in Table 2.12 and the illustrations in Figures 2.3 and 2.4.

i) The DF t-type tests are still quite accurate as far as nominal and actual size are concerned; where there is a departure, it tends to be at the extremes of the distribution (see Figure 2.3), which are not generally used for hypothesis testing.

Table 2.12   Actual and nominal size of DF and range unit root tests when $\varepsilon_t \sim t(3)$ and niid(0, 1) quantiles are used

	1%	5%	10%
T = 100			
$\hat{\tau}_\mu$	1.08%	5.06%	10.32%
$\hat{\tau}_\beta$	1.39%	5.03%	9.55%
R^F_{UR}	1.85%	5.35%	**10%**
R^{FB}_{UR}	1.22%	**5%**	**10%**
T = 250			
$\hat{\tau}_\mu$	1.39%	5.43%	10.27%
$\hat{\tau}_\beta$	1.07%	4.87%	9.32%
R^F_{ur}	1.57%	6.16%	10.13%
R^{FB}_{ur}	**1%**	**5%**	**10%**

Notes: based on 10,000 replications of the null model; embolden indicates exact size maintained.

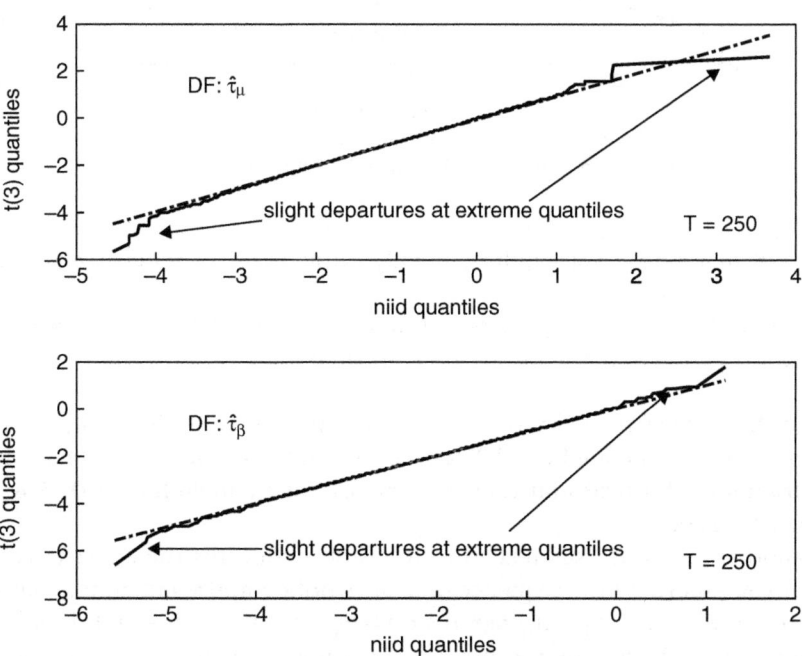

Figure 2.3   QQ plots of DF $\hat{\tau}$ tests, niid vs. t(3)

ii) In the case of R^F_{UR}, the 1% and 5% sized tests are slightly oversized, and there is some departure from the 45° degree line in the QQ plot of Figure 2.4 (upper panel), but this is confined to the upper quantiles.

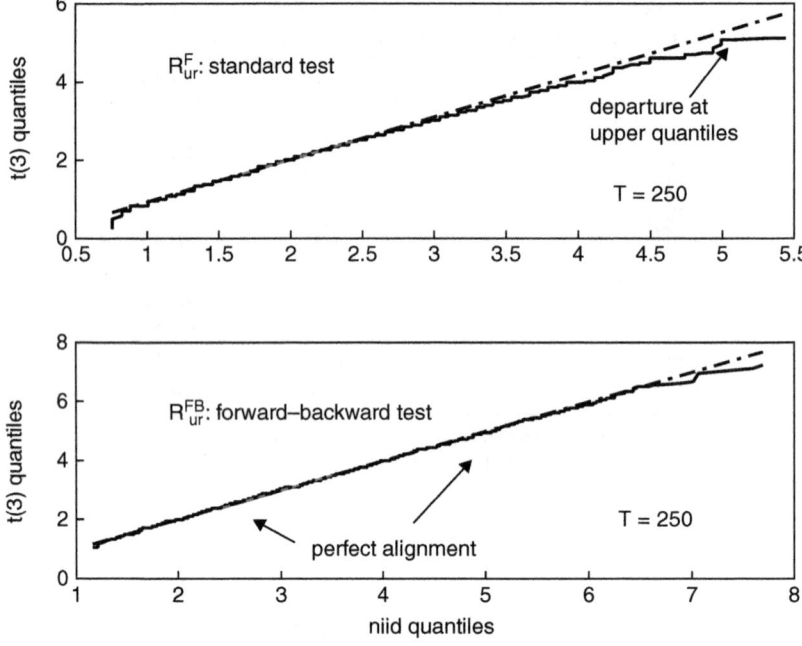

Figure 2.4   QQ plots of range unit-root tests, niid vs. t(3)

iii) R_{UR}^{FB} maintains its size quite generally throughout the range (see Figure 2.4, lower panel).

iv) Overall, although the differences among the test statistics are not marked, R_{UR}^{FB} is better at maintaining size compared to R_{UR}^{F}.

Further, reference to AES (op. cit., table III) indicates that there is a power gain in using R_{UR}^{F}, compared to a DF test, for a number of fat-tailed or asymmetric distributions, for near-unit root alternatives, for example for $\rho = 0.95$ to 0.99 when $T = 250$.

Another practical issue related to robustness concerns the characteristics of the RUR tests when the sequence of z_t does not comprise iid random variables (that is $z_t \neq \varepsilon_t$. AES (op. cit.) report simulation results for size and power when z_t is generated by an MA(1) process: $z_t = \varepsilon_t + \theta_1 \varepsilon_t$, with $\theta_1 \in (0.5, -0.8)$. AES find that, with a sample size of $T = 100$ and a 5% nominal size, both R_{UR}^{F} and R_{UR}^{FB} are (substantially) under-sized for $\theta_1 = 0.5$, for example, an actual size of 0.4% for R_{UR}^{F} and 0.6% for R_{UR}^{FB}. In the case of negative values of θ_1, which is often the case of greater interest, R_{UR}^{F} and R_{UR}^{FB} are over-sized, which is the usual effect on unit root tests in this case, but R_{UR}^{FB} performs better than R_{UR}^{F} and is only moderately over-sized; for example, when $\theta_1 = -0.6$, the empirical sizes of R_{UR}^{F}

and R_{UR}^{FB} are 20% and 9%, respectively. However, the DF t-type test, with lag selection by MAIC, due to Ng and Perron (2001) (see UR Vol. 1, chapter 9 for details of MAIC) outperforms both R_{UR}^{F} and R_{UR}^{FB} in terms of size fidelity.

2.4.4 The range unit root tests for trending alternatives

Many economic time series are trended, therefore to be of practical use a test based on the range must be able to distinguish a trend arising from a drifted random walk and a trend arising from a trend stationary process. This section describes the procedure suggested by AES (op. cit.) for such situations.

The stochastically trended and deterministically trended alternatives are, resepectively:

$$y_t = \beta_1 + y_{t-1} + \varepsilon_t \qquad\qquad \beta_1 \neq 0 \qquad\qquad (2.46)$$

$$(y_t - \beta_0 - \beta_1 t) = z_t \qquad\qquad z_t \sim I(0) \qquad\qquad (2.47)$$

In the first case, $y_t \sim I(1)$ with drift, which generates a direction or trend to the random walk, whereas in the second case, $y_t \sim I(0)$ about a deterministic trend. The key result as far as the RUR test is concerned is that R_{UR}^{F} and R_{UR}^{FB} are divergent under *both* processes, so that:

$$R_{UR}^{F} = O_p(T^{1/2}) \to \infty \text{ as } T \to \infty$$

under both (2.46) and (2.47). Thus, the right tail of the distribution of R_{UR}^{F} is now relevant for hypothesis testing, where the null hypothesis is of a driftless random walk against a trended alternative of the form (2.46) or (2.47); for appropriate upper quantiles, see Table 2.10 for R_{UR}^{F} and Table 2.11 for R_{UR}^{FB}.

As the divergence of R_{UR}^{F} follows for both the I(1) and trended I(0) alternative, it is necessary to distinguish these cases by a further test. The following procedure was suggested by AES (op. cit.). The basis of the supplementary test is that if $y_t \sim I(1)$ then $\Delta y_t \sim I(0)$, whereas if $y_t \sim I(0)$ then $\Delta y_t \sim I(-1)$, so the test rests on distinguishing the order of integration of Δy_t.

If a variable x_t is I(0), then its infinite lagged sum is I(1), so summation raises the order of integration by one each time the operator is applied. The same result holds if a series is defined as the infinite lagged sum of the even-order lags. The sequence $\tilde{x}_t^{(k)}$ is defined as follows:

$$\tilde{x}_t^{(k)} \equiv \sum_{j=0}^{\infty} L^j x_{t-j}^{(k-1)} \qquad\qquad (2.48)$$

where the identity operator is $\tilde{x}_t^{(0)} \equiv x_t$.

Some of the possibilities of interest for determining the integration order are:

A: if $x_t \sim I(0)$, then $\tilde{x}_t^{(1)} \equiv \sum_{j=0}^{\infty} L^j x_{t-j} \sim I(1)$.

B.i: if $x_t \sim I(-1)$, then $\tilde{x}_t^{(1)} \equiv \sum_{j=0}^{\infty} L^j x_{t-j} \sim I(0)$.

B.ii: if $\tilde{x}_t^{(1)} \sim I(0)$, then $\tilde{x}_t^{(2)} \equiv \sum_{j=0}^{\infty} L^j \tilde{x}_{t-j}^{(1)} \sim I(1)$.

These results may now be applied in the context that $x_t = \Delta y_t$, giving rise to two possibilities to enable a distinction between a drifted random walk from a trend stationary process:

Possibility A:
If $x_t \sim I(0)$, then $\tilde{x}_t^{(1)} \sim I(1)$, and using R_{UR}^F (or R_{UR}^{FB}) with the variable $\tilde{x}_t^{(1)}$, should result in *non-rejection* of the null hypothesis.

Possibility B:
If $x_t \sim I(-1)$, then $\tilde{x}_t^{(1)} \sim I(0)$, and using R_{UR}^F (or R_{UR}^{FB}) with the variable $\tilde{x}_t^{(1)}$, should result in *rejection* of the null hypothesis.
If $x_t \sim I(-1)$, then $\tilde{x}_t^{(2)} \sim I(1)$, and using R_{UR}^F (or R_{UR}^{FB}) with the variable $\tilde{x}_t^{(2)}$, should result in *non-rejection* of the null hypothesis.

Hence, given that in the case of the trending alternative, a rejection of the null hypothesis has first occurred using the upper quantiles of the null distribution, then the decision is to conclude in favour of a drifted random walk if possibility A results, but conclude in favour of a trend stationary process if possibility B results. Since the overall procedure will involve not less than two tests, each at, say, an $\alpha\%$ significance level, the overall significance level will (generally) exceed $\alpha\%$; for example in the case of two tests, the upper limit on the size of the tests is $\alpha_\Sigma = (1 - (1 - \alpha)^2)$. Note that the infinite sums implied in the computation of $\tilde{x}_t^{(j)}$ have to be truncated to start at the beginning of the sample. An example of the use of these tests is provided in Section **2.6.3.ii** as part of a comparison of a number of tests in Section **2.6**.

2.5   Variance ratio tests

Another basis for unit root tests that uses less than a full parametric structure is the different behaviour of the variance of the series under the I(1) and I(0) possibilities. Variance ratio tests are based on the characteristic of the partial sum (unit root) process with iid errors that the variance increases linearly with time. Consider the simplest case where $y_t = y_0 + \sum_{i=1}^{t} \varepsilon_i$, with $\varepsilon_t \sim$ iid$(0, \sigma_\varepsilon^2)$ and, for simplicity, assume that y_0 is nonstochastic, then var$(y_t) = t\sigma_\varepsilon^2$. Further, note that $\Delta_1 y_t \equiv y_t - y_{t-1} = \varepsilon_t$, where the subscript on Δ_1 emphasises that this is the first difference operator and higher-order differences are indicated by the subscript. Also var$(\Delta_1 y_t) = \sigma_\varepsilon^2$ and $\Delta_2 y_t \equiv y_t - y_{t-2} \equiv \Delta_1 y_t + \Delta_1 y_{t-1}$; hence, by the iid assumption, var$(\Delta_2 y_t) = 2\sigma_\varepsilon^2$. Generalising in an obvious way, var$(\Delta_q y_t) = q\sigma_\varepsilon^2$, where $\Delta_q y_t \equiv y_t - y_{t-q} \equiv \sum_{i=0}^{q-1} \Delta_1 y_{t-i}$. In words, the variance of the q-th order difference is q times the variance of the first-order difference.

Moreover, this result generalises to heteroscedastic variances provided $\{\varepsilon_t\}$ remains a serially uncorrelated sequence. For example, introduce a time subscript on σ_ε^2, say $\sigma_{\varepsilon,t}^2$, to indicate heteroscedasticity, then var$(\Delta_q y_t) = \sum_{i=0}^{q-1} \sigma_{\varepsilon,t-i}^2$, so that the variance of the q-th order difference is the sum of the variances of the q first-order differences.

For simplicity consider the homoscedastic case, then a test of whether y_t has been generated by the partial sum process with iid errors, can be based on a comparison of the ratio of $(1/q)$ times the variance of $\Delta_q y_t$ to the variance of $\Delta_1 y_t$. This variance ratio is:

$$VR(q) \equiv \frac{1}{q} \frac{var(\Delta_q y_t)}{var(\Delta_1 y_t)} \qquad (2.49)$$

The quantity $VR(q)$ can be calculated for different values of q and, in each case, it will be unity under the null hypothesis, H_0: $y_t = y_0 + \sum_{i=1}^{t} \varepsilon_i$, with $\varepsilon_t \sim$ iid$(0, \sigma_\varepsilon^2)$. Hence, the null incorporates, or is conditional on, a unit root, with the focus on the lack of serial correlation of the increments, ε_t. The nature of the alternative hypothesis, therefore, requires some consideration; for example, rejection of H_0 that $VR(q) = 1$ could occur because the ε_t are serially correlated, which is not a rejection of the unit root in H_0; however, rejection could also occur if there is not a unit root. Values of $VR(q)$ that are different from unity will indicate rejection of the null hypothesis of a serially uncorrelated random walk.

This procedure generalises if there are deterministic components in the generation of y_t. For example, assume that $(y_t - \mu_t) = u_t$, where $E(u_t) = 0$ and μ_t includes deterministic components, typically a constant and/or a (linear) trend; and, as usual, the detrended series is $\tilde{y}_t = y_t - \hat{\mu}_t$, where $\hat{\mu}_t$ is an estimate of the trend, then the variance ratio is based on $\tilde{y}_t$ rather than the original series, y_t.

2.5.1 A basic variance ratio test

Evidently a test can be based on constructing an empirical counterpart for $VR(q)$ and determining its (asymptotic) distribution. First, define an estimator of $VR(q)$, using the sample quantities as follows:

$$OVR(q) = \frac{1}{q} \left(\frac{\frac{1}{T}\sum_{t=q+1}^{T} (\Delta_q y_t - q\hat{\beta}_1)^2}{\frac{1}{T}\sum_{t=1}^{T} (\Delta_1 y_t - \hat{\beta}_1)^2} \right) \qquad (2.50)$$

The notation reflects the overlapping nature of the data uses in this estimator – this is discussed further below. The estimator $OVR(q)$ allows for drift in the null of a random walk as the variable in the numerator is the q-th difference, $\Delta_q y_t$, minus an estimator of its mean, $q\hat{\beta}_1$. An alternative notation is sometimes used: let $x_t \equiv \Delta_1 y_t$, then $\Delta_q y_t \equiv \sum_{j=0}^{q-1} x_{t-j}$ and this term is used in $OVR(q)$. Also, the quantity in (2.50) is sometimes defined using a divisor that makes a correction for degrees of freedom in estimating the variance; see Lo and MacKinlay (1989).

A test based on $OVR(q)$ uses overlapping data in the sense that successive elements have a common term or terms; for example, for $q = 2$, $\Delta_2 y_t = \Delta_1 y_t + \Delta_1 y_{t-1}$ and $\Delta_2 y_{t-1} = \Delta_1 y_{t-1} + \Delta_1 y_{t-2}$, so the 'overlapping' element is $\Delta_1 y_{t-1}$. To put this in context, suppose $\Delta_1 y_t$ are daily returns on a stock price, then assuming five working days, a weekly return can be defined as the sum of five consecutive daily returns; thus, for example, $\Delta_5 y_t = \sum_{j=0}^{4} \Delta_1 y_{t-j}$, with successive element $\Delta_5 y_{t+1} = \sum_{j=0}^{4} \Delta_1 y_{t+1-j}$, so that the overlapping elements are $\sum_{j=0}^{3} \Delta_1 y_{t-j}$. This differs from the weekly return defined as the sum of the five daily returns in each week, the time series for which would be non-overlapping. A non-overlapping version of VR(q), NOVR, is also possible, although the overlapping version is likely to have higher power; see Lo and MacKinlay (1989) and also Tian, Zhang and Huan (1999) for the exact finite-sample distribution of the NOVR version.

Lo and MacKinlay (1988) showed that:

$$\sqrt{T}(\text{OVR} - 1) \Rightarrow_D N\left(0, \frac{2q(q-1)(q-1)}{3q}\right) \tag{2.51}$$

The asymptotics here are large sample asymptotics, that is as $T \to \infty$, with q fixed; thus, $(q/T) \to 0$ as $T \to \infty$. On this basis, the following quantity has a limiting null distribution that is standard normal:

$$\sqrt{T}(\text{OVR} - 1)\left(\frac{3q}{2q(q-1)(q-1)}\right)^{1/2} \Rightarrow_D N(0, 1) \tag{2.52}$$

Remember that the null hypothesis is of a process generated by a unit root combined with uncorrelated increments. The test is two-sided in the usual sense that large positive or large negative values of the left-hand side of (2.52) relative to $N(0, 1)$ lead to rejection in favour of the random walk alternative.

There are four points to note about the use of the variance ratio test statistic (2.52).

The first point is that the result in (2.52) is asymptotic and the test statistic is biased downward in small samples. The statistic under-rejects the null in the lower tail, with the null distribution skewed to the right. There have been several suggestions to remove or ameliorate this bias. First, rather than use the $N(0, 1)$ distribution, Lo and MacKinlay (1989) provided some simulated critical values for T and q. Second, Tse, Ng and Zhang (2004) suggested a revised version of the OVR test statistic (the circulant OVR), derived its expectation, which is not equal to unity in small samples, and then adjusted their test statistic so that it does have a unit expectation. Whilst Tse et al. (2004) consider the expectation of the ratio, Bod et al. (2002) derived adjustment factors for the numerator and denominator of the variance ratio in its original OVR form; although the expectation of a ratio of random variables is not the ratio of expectations, for practical purposes the suggested adjustment delivers an unbiased version of OVR(q).

Finally, a small sample adjustment can be based on noting, as in UR, Vol.1, chapter 3, that if a scalar parameter ψ is estimated with bias, such that $E(\hat{\psi}) = \psi + b(\psi)$, then an unbiased estimator results after subtraction of the bias from the estimator; thus $E(\hat{\psi} - b(\psi)) = \psi$. To this end, OVR(q) can be written as a function of the sample autocorrelations $\hat{\rho}(q)$ (see Cochrane (1988)), that is:

$$OVR(q) = 1 + 2\sum_{j=1}^{q-1}\left(1 - \frac{j}{q}\right)\hat{\rho}(j) \tag{2.53}$$

The expected value of $\hat{\rho}(j)$ for a serially uncorrelated time series, ignoring terms of higher order than $O(T^{-1})$, is given by Kendall (1954):

$$E(\hat{\rho}(j)) = -\frac{1}{T-j} \tag{2.54}$$

This expectation should be zero for an unbiased estimator, hence (2.54) is also the first-order bias of $\hat{\rho}(j)$. The bias is negative in finite sample, but disappears as $T \to \infty$, for fixed j. Thus, $\hat{\rho}(j) + (T-j)^{-1}$, is a first order-unbiased estimator of $\hat{\rho}(j)$ under the null of serially uncorrelated increments.

Noting that as $E(OVR(q))$ is a linear function of $E\{\hat{\rho}(j)\}$, then the bias in estimating OVR(q) by (2.50) is:

$$B\{OVR(q)\} = E(OVR(q)) - 1 \tag{2.55}$$

where:

$$B(OVR(q)) = -\left(\frac{2}{q}\right)\sum_{j=1}^{q-1}\left(\frac{q-j}{T-j}\right) \tag{2.56}$$

See also Shiveley (2002). An estimator that is unbiased to order $O(T^{-1})$ under the null is obtained as ROVR(q) = OVR(q) – B(OVR(q)). However, the bias in (2.54) is only relevant under the null, so that a more general approach would correct for bias in estimating $\rho(j)$ whether under the null or the alternative hypotheses.

The second point is that if q is large relative to T, then the standard normal distribution does not approximate the finite sample distribution very well, with consequent problems for size and power; see Richardson and Stock (1989) and Deo and Richardson (2003). An alternative asymptotic analysis with $q \to \infty$ as $T \to \infty$, such that $(q/T) \to \lambda$, provides a better indication of the relevant distribution from which to obtain quantiles for testing H_0. Deo and Richardson (2003) showed that under this alternative framework, OVR(q) converges to the following non-normal limiting distribution:

$$OVR(q) \Rightarrow \frac{1}{(1-\lambda)^2\lambda}\int_{\lambda}^{1}(B(s) - B(s-\lambda) - \lambda B(1))^2 ds \tag{2.57}$$

for which critical values can be obtained by simulation; see Deo and Richardson (2003).

The third point is that variance ratio tests are often carried out for different values of q; this is because under the null the variance ratio is unity for all values of q. However, such a procedure involves multiple testing with implications for the cumulative (or overall) type-I error. For example, suppose n values of q are chosen and the null is not rejected if and only if none of the n variance ratio test statistics are not rejected at the α% level, then this multiple comparison implies an overall significance level that (generally) exceeds the individual significance level. Assuming independence of tests the overall type-I error is $\alpha_\Sigma = 1 - (1 - \alpha)^n$, but this reduces for dependent tests (as is the case here); further, Fong, Koh and Ouliaris (1997) show that this can be improved for joint applications of the variance ratio test and it is, therefore, possible to control for the overall size.

Fourth, the random walk hypothesis is often of interest in situations where homoscedasticity is unlikely; for example, there is a considerable body of evidence that financial time series are characterised by heteroscedasticity, including ARCH/GARCH-type conditional heteroscedasticity. The variance ratio test can be generalised for this and some other forms of heteroscedasticity, which again results in a test statistic that is distributed as standard normal under the null hypothesis; see Lo and MacKinlay (1989).

A problem with variance ratio-based tests is that they are difficult to generalise to non-iid errors; however, Breitung (2002) has suggested a test on a variance ratio principle, outlined in the next section, that is invariant to the short-run dynamics and is robust to a number of departures from normally distributed innovations.

2.5.2   Breitung variance ratio test

A test statistic for the null hypothesis that y_t is I(0), and in particular that it is white noise with zero mean, against the alternative that y_t is I(1), based on the variance ratio principle, was suggested by Tanaka (1990) and Kwiakowski et al. (1992) and is given by:

$$\upsilon_K = \frac{\sum_{t=1}^{T} Y_t^2 / T^2}{\sum_{t=1}^{T} y_t^2 / T} \tag{2.58}$$

where $Y_t \equiv \sum_{i=1}^{t} y_i$; see also UR, Vol. 1, chapter 11. The denominator is an estimator of the long-run variance assuming no serial correlation; if this is not the case, it is replaced by an estimator, either a semi-parametric estimator, with a kernel function, such as the Newey-West estimator, or a parametric version that uses a 'long' autoregression (see UR, Vol. 1, chapter 6).

The statistic υ_K is the ratio of the variance of the partial sum to the variance of y_t normalised by T. If the null hypothesis is of stationarity about a non-zero constant or linear trend, then the KPSS test statistic is defined in terms of $\tilde{y}_t$, so that the test statistic is:

$$
\nu_{rat} = \frac{\sum_{t=1}^{T} \tilde{Y}_t^2 / T^2}{\sum_{t=1}^{T} \tilde{y}_t^2 / T}
$$

$$
= T^{-1} \frac{\sum_{t=1}^{T} \tilde{Y}_t^2}{\sum_{t=1}^{T} \tilde{y}_t^2} \tag{2.59}
$$

where $\tilde{y}_t = (y_t - \hat{\mu}_t)$, $\hat{\mu}_t$ is an estimator of the deterministic components of y_t and $\tilde{Y}_t$ is the partial sum of $\tilde{y}_t$, that is $\tilde{Y}_t \equiv \sum_{i=1}^{t} \tilde{y}_i$.

When using ν_{rat} as a test of stationarity, the critical values come from the right tail of the null distribution of ν_{rat}. However, Breitung (2002) suggests using ν_{rat} as a unit root test, that is of the null hypothesis that y_t is I(1) against the alternative that y_t is I(0), so that the appropriate critical values for testing now come from the left tail of the null distribution.

Breitung (2002, Proposition 3) shows that ν_{rat} has the advantage that its asymptotic null distribution does not depend on the nuisance parameters, in this case those governing the short-run dynamics. The limiting null distribution when there are no deterministic components is:

$$
\nu_{rat} \Rightarrow \frac{\int_0^1 \left[\int_0^r B(s) \right]^2 dr}{\int_0^1 B(r)^2 dr} \tag{2.60}
$$

whereas when y_t is adjusted for deterministic components, this becomes:

$$
\nu_{rat,j} \Rightarrow \frac{\int_0^1 \left[\int_0^r B(s)_j \right]^2 dr}{\int_0^1 B(r)_j^2 dr} \tag{2.61}
$$

$B(s)_j$ is Brownian motion adjusted for deterministic components, with $j = \mu, \beta$ (see Breitung 2002, Proposition 3), defined as follows:

$$
B(r)_\mu = B(r) - \int_0^1 B(s) ds \tag{2.62}
$$

$$
B(r)_\beta = B(r) + (6r - 4) \int_0^1 B(s) ds - (12r - 6) \int_0^1 s B(s) ds \tag{2.63}
$$

See also UR, Vol. 1, chapter 6.

The advantage of this form of the test is that it generalises simple variance tests to the case of serially correlated errors, but without the need to specify and estimate the short-run dynamics. Some critical values for $\nu_{rat,j}$ are given in Table 2.13; as usual, a second subscript indicates whether the series has been adjusted for a mean and or a trend. The test is left-tailed, with a sample value less than the $\alpha\%$ critical value leading to rejection of H_0.

Table 2.13   Critical values of $\nu_{rat,j}$

	ν_{rat}			$\nu_{rat,\mu}$			$\nu_{rat,\beta}$		
	1%	5%	10%	1%	5%	10%	1%	5%	10%
100	0.0313	0.0215	0.0109	0.0144	0.0100	0.0055	0.0044	0.0034	0.0021
250	0.0293	0.0199	0.0097	0.0143	0.0100	0.0056	0.0044	0.0034	0.0022
500	0.0292	0.0199	0.0099	0.0147	0.0105	0.0054	0.0044	0.0036	0.0026

Source: Breitung (2002, table 5); a sample value less than the α% critical value leads to rejection of H_0.

The models for which $\nu_{rat,j}$ is relevant are of the following familiar form:

$$\tilde{y}_t = \rho\tilde{y}_{t-1} + z_t$$

$$z_t = \omega(L)\varepsilon_t$$

$$\tilde{y}_t = y_t - \mu_t$$

where $\omega(L) = 1 + \sum_{j=1}^{J} \omega_j L^j$, with $\varepsilon_t \sim iid(0, \sigma_\varepsilon^2)$, $\sigma_\varepsilon^2 < \infty$, $\sum_{j=1}^{J} j^2 \omega_j^2 < \infty$, and J may be infinite as in the case that the generation process of z_t includes an AR component. Error processes other than the standard linear one are allowed; for example, z_t could be generated by a fractional integrated process, or a non-linear process, provided that they admit a similar Beveridge-Nelson type of decomposition; see Breitung (2002) for details.

Breitung (2002) reports some simulation results for the problematic case with an MA(1) error generating process, so that $\Delta y_t = z_t$, with $z_t = (1 + \theta_1 L)\varepsilon_t$ and $T = 200$, where the test is based alternatively on the residuals from a regression of y_t on a constant or a constant and a trend. The results indicate a moderate size bias for $\nu_{rat,j}$, $j = \mu$, β, when $\theta_1 = -0.5$, which increases when $\theta_1 = -0.8$; whether this is better than the ADF for $\hat{\tau}_j$, depends on the lag length used in the latter, with ADF(4) suffering a worse size bias, but ADF(12) being better on size but worse on power.

2.6   Comparison and illustrations of the tests

In this section some consideration is given to a comparison of the test statistics described in the previous sections in terms of empirical size and power. Whilst this is necessarily limited (for example, the robustness of the tests to MA(1) errors is considered, but not to other forms of weak dependency) the comparison brings the assessments in the individual studies onto a comparable basis. Two empirical illustrations of the tests are given in Section **2.6.2**.

2.6.1 Comparison of size and power

The test statistics to be considered are first, by way of benchmarking the results, the standard DF test statistics $\hat{\delta}_\mu$ and $\hat{\tau}_\mu$, and then their rank-based counterparts $\hat{\delta}_{r,\mu}$ and $\hat{\tau}_{r,\mu}$, and then the range unit root tests in their forward and forward–backward versions, R_{UR}^F and R_{ur}^{FB}, and the variance ratio test, $\nu_{rat,\mu}$; the rank based score tests $\lambda_{r,\beta}$ and $\lambda_{lr,\beta}$ are also included. The simulations consider the robustness of the tests to MA(1) errors and to the form of the innovation distribution.

The simulation results are for the case where the DGP is: $y_t = \rho y_{t-1} + z_t$, $z_t = (1 + \theta_1 L)\varepsilon_t$, $\rho = (1, 0.95, 0.9, 0.85)$, $\theta_1 = (0, -0.5)$ and T = 200; size is evaluated when $\rho = 1$ for different values of θ_1, whereas power is evaluated for $\rho < 1$. The nominal size of the tests is 5%, and the alternative innovation distributions are $N(0, 1)$, t(3) and Cauchy. The rank score-based test $\lambda_{lr,\beta}$ requires estimation of the long-run variance when $\theta_1 = -0.5$ (see Section **2.3.2**, especially Equation (2.40)). A truncation parameter of M = 8 was used, as in BG (op. cit.) and, for comparability, a lag of 8 was used for the DF tests.

The results are summarised in Table 2.14 in two parts. The first set of results is for $\theta_1 = 0$ and the second set for $\theta_1 = -0.5$. In the case of $\rho < 1$, both 'raw' and size-adjusted power are reported, the latter is indicated by SA.

The first case to be considered is when the errors are normal and there is no MA component, so there is no distinction between errors and innovations. Size is generally maintained at its nominal level throughout. The standard DF tests might be expected to dominate in terms of power; however, this advantage is not uniform and the rank-based DF tests are comparable despite the fact that they only use part of the information in the sample; for example, the rank-based t-type test is more powerful than its parametric counterpart and is the most powerful of the nonparametric tests considered here. An explanation for this finding is that the mean of the ranks, $\bar{r}$, is known, whereas in the parametric test the mean has to be estimated, which uses up a degree of freedom; see Granger and Hallman (1991) for further details. Of the other nonparametric tests R_{UR}^{FB} is in second place. In the power evaluation, the rank-based score tests $\lambda_{r,\beta}$ and $\lambda_{lr,\beta}$ are likely to be at a disadvantage because in this simulation set-up, where H_0 is a random walk without drift, they over-parameterise the alternative, and this is confirmed in the simulation results. Their relative performance would improve against $\hat{\delta}_{r,\beta}$ and $\hat{\tau}_{r,\beta}$.

In the case of the innovations drawn from t(3), there are some slight size distortions. Otherwise the ranking in terms of power is generally as in the normal innovations case; of note is that $\nu_{rat,\mu}$ gains somewhat in power whereas $\lambda_{r,\mu}$ loses power. The Cauchy distribution is symmetric, but has no moments, and may be considered an extreme case: in this case the DF rank-based tests $\hat{\delta}_{r,\mu}$ and $\hat{\tau}_{r,\mu}$ become under-sized, whereas R_{UR}^F is over-sized and apart from R_{UR}^F,

Table 2.14 Summary of size and power for rank, range and variance ratio tests

$\theta_1 = 0$								
$\varepsilon_t \sim$ Normal	$\hat{\delta}_\mu$	$\hat{\tau}_\mu$	$\hat{\delta}_{r,\mu}$	$\hat{\tau}_{r,\mu}$	R^F_{UR}	R^{FB}_{UR}	$\lambda_{r,\beta}$	$\nu_{rat,\mu}$
$\rho = 1$	4.84	5.08	5.32	5.20	4.90	4.28	5.18	5.44
$\rho = 0.95$	18.68	14.20	19.62	18.26	15.34	16.84	11.22	15.62
$\rho = 0.90$	52.30	40.10	47.56	44.94	28.66	35.16	27.32	30.34
$\rho = 0.85$	**84.74**	73.52	**78.18**	75.32	42.96	55.42	46.28	43.28
$\varepsilon_t \sim$ t(3)	$\hat{\delta}_\mu$	$\hat{\tau}_\mu$	$\hat{\delta}_{r,\mu}$	$\hat{\tau}_{r,\mu}$	R^F_{UR}	R^{FB}_{UR}	$\lambda_{r,\beta}$	$\nu_{rat,\mu}$
$\rho = 1$	4.10	5.20	3.90	3.85	5.95	3.60	4.95	4.85
$\rho = 0.95$	22.35	12.90	19.45	19.50	17.75	20.85	10.25	18.10
$\rho = 0.90$	56.30	37.50	47.20	45.15	31.55	32.90	16.85	32.95
$\rho = 0.85$	**90.10**	74.55	**77.40**	76.15	44.45	54.20	24.70	47.45
$\varepsilon_t \sim$ Cauchy	$\hat{\delta}_\mu$	$\hat{\tau}_\mu$	$\hat{\delta}_{r,\mu}$	$\hat{\tau}_{r,\mu}$	R^F_{UR}	R^{FB}_{UR}	$\lambda_{r,\beta}$	$\nu_{rat,\mu}$
$\rho = 1$	5.68	7.34	1.58	1.68	11.30	4.86	5.52	4.42
$\rho = 0.95$	8.62	7.12	12.34	12.08	37.74	7.04	0.38	12.32
$\rho = 0.90$	34.92	14.40	30.78	27.94	43.94	8.64	0.26	27.20
$\rho = 0.85$	**89.40**	38.42	**56.64**	52.88	47.90	12.64	0.22	44.34
$\theta_1 = -0.5$								
$\varepsilon_t \sim$ Normal	$\hat{\delta}_\mu$	$\hat{\tau}_\mu$	$\hat{\delta}_{r,\mu}$	$\hat{\tau}_{r,\mu}$	R^F_{UR}	R^{FB}_{UR}	$\lambda_{lr,\beta}$	$\nu_{rat,\mu}$
$\rho = 1$	9.7	4.2	6.9	1.5	26.6	33.4	53.5	6.5
$\rho = 0.95$	59.8	24.8	48.2	16.9	76.2	91.8	86.4	40.1
SA	43.1	28.5	39.9	38.1	19.4	36.0	14.1	35.4
$\rho = 0.90$	93.3	61.1	84.5	51.2	92.0	99.5	93.8	69.1
SA	82.6	66.3	78.0	76.9	38.6	67.2	34.0	64.5
$\rho = 0.85$	99.1	84.7	95.7	78.0	95.8	1.0	93.6	85.8
SA	96.0	**88.5**	93.2	93.0	51.2	81.4	41.4	**83.2**
$\varepsilon_t \sim$ t(3)	$\hat{\delta}_\mu$	$\hat{\tau}_\mu$	$\hat{\delta}_{r,\mu}$	$\hat{\tau}_{r,\mu}$	R^F_{UR}	R^{FB}_{UR}	$\lambda_{lr,\beta}$	$\nu_{rat,\mu}$
$\rho = 1$	9.9	4.0	7.7	1.0	57.7	45.9	62.2	7.7
$\rho = 0.95$	59.2	19.8	44.0	14.6	92.0	91.5	70.3	36.4
SA	41.8	24.2	38.2	34.7	8.7	18.2	8.7	27.2
$\rho = 0.90$	91.5	60.1	80.6	47.6	96.3	99.1	77.1	69.9
SA	82.5	68.2	75.3	72.0	11.3	34.1	11.9	58.0
$\rho = 0.85$	98.3	84.0	94.4	74.4	97.1	99.9	77.3	85.4
SA	95.9	**88.0**	92.1	91.1	14.7	44.5	20.2	**77.8**
$\varepsilon_t \sim$ Cauchy	$\hat{\delta}_\mu$	$\hat{\tau}_\mu$	$\hat{\delta}_{r,\mu}$	$\hat{\tau}_{r,\mu}$	R^F_{UR}	R^{FB}_{UR}	$\lambda_{lr,\beta}$	$\nu_{rat,\mu}$
$\rho = 1$	9.1	7.1	6.0	1.0	97.4	57.0	67.3	6.2
$\rho = 0.95$	55.1	16.6	26.6	7.5	97.8	39.7	5.3	32.6
SA	24.5	11.7	24.5	24.5	1.0	1.7	0.0	28.5
$\rho = 0.90$	95.9	64.2	58.1	24.5	97.6	54.6	6.7	58.1
SA	87.5	37.2	56.1	55.1	1.0	3.0	0.0	55.7
$\rho = 0.85$	98.2	90.3	83.3	48.2	97.3	69.9	8.1	86.8
SA	95.7	81.3	**80.9**	79.4	0.3	4.9	0.0	**76.8**

Notes: T = 200; the first row of each entry is the 'raw' size or power, and the second row is the size-adjusted power when $\rho < 1$; best performers in bold.

power is clearly lost throughout. Overall, $\hat{\delta}_\mu$ remains the best test; R^F_{UR} has size-adjusted power that is better than $\hat{\delta}_\mu$ for part of the parameter space, but since it is over-sized the comparison is misleading; however, a bootstrapped version that potentially corrects the incorrect size may have something to offer, but is not pursued here.

A more problematic case for unit root testing is when there is an MA component to the serial correlation, for example, in the first-order case when the MA coefficient is negative, and $\theta_1 = -0.5$ is taken here as illustrative. This case is known to cause difficulties to the parametric tests, for which some protection is offered by extending the lag in an ADF model. A priori this may offer some relative advantage to v_{rat}, which is invariant to the serial correlation.

The first question, when $\rho = 1$ and $\theta_1 = -0.5$, relates to the retention of size, where the nominal size is 5% in these simulations. In the case of normal innovations, the best of the nonparametric tests is now $v_{rat,\mu}$, with R^F_{UR} and R^{FB}_{UR} suffering size distortions. The size distortion of $\lambda_{lr,\mu}$ is consistent with the results reported in BG (op. cit.). Of the ADF rank tests, $\hat{\delta}_{r,\mu}$ is slightly over-sized, whereas $\hat{\tau}_{r,\mu}$ is under-sized. The lag length is long enough for $\hat{\tau}_\mu$ to maintain its size reasonably well at 4.2%, but as observed in UR, Vol. 1, chapter 6, $\hat{\delta}_\mu$ is not as robust as $\hat{\tau}_\mu$. The comparisons of power will now only make sense for the tests that do not suffer (gross) size distortions. Both 'raw' and size-adjusted (SA) power is given for the cases where $\rho < 1$. In this case, it is now the simple nonparametric test $v_{rat,\mu}$ that has the advantage over the other tests.

The next question relates to the effect of non-normal innovations on the performance of the various tests. When the innovations are drawn from t(3), $\hat{\tau}_\mu$ again maintains its size reasonably well at 4.0%, and $\hat{\delta}_{r,\mu}$ and $v_{rat,\mu}$ are next best at 7.7%. The most powerful (in terms of SA power) of the tests is $\hat{\delta}_{r,\mu}$, with honours fairly even between $\hat{\tau}_\mu$ and $v_{rat,\mu}$ depending on the value of ρ. With Cauchy innovations, $\hat{\tau}_\mu$, $\hat{\delta}_{r,\mu}$ and $v_{rat,\mu}$ are comparable in terms of size retention, whereas $\hat{\delta}_{r,\mu}$ and $v_{rat,\mu}$ are best in terms of SA power.

Overall, there is no single test that is best in all the variations considered here. The rank-based tests are useful additions to the 'stock' of tests, but the test least affected by the variations considered here is Breitung's v_{rat}, which shows good size fidelity and reasonable power throughout, and is particularly simple to calculate.

2.6.2 Linear or exponential random walks?

The second issue on which the various tests are compared is their sensitivity to whether the unit root is in the levels or the logarithms of the variable of interest, as this choice is often the central one in empirical work. The DF rank tests and the RUR tests are invariant to the transformation, hence this comparison is concerned with the characteristics of $\lambda_{r,\mu}$ and $v_{rat,\mu}$, compared to the standard DF tests. The set-up is as in Section **2.1**, which considered two cases, denoted

Table 2.15 Levels or logs, additional tests: score and variance ratio tests (empirical size)

Case 1			Unit root in y_t			
$\ln(y_t)$	$\hat{\delta}_\mu$	$\hat{\delta}_\beta$	$\hat{\tau}_\mu$	$\hat{\tau}_\beta$	$\lambda_{r,\mu}$	$\nu_{rat,\mu}$
$\sigma^2 = 0.5$	5.7%	5.6%	5.3%	4.6%	5.9%	4.7%
$\sigma^2 = 1.0$	5.4%	5.2%	5.1%	4.3%	6.2%	5.0%

Case 2			Unit root in $\ln y_t$			
y_t	$\hat{\delta}_\mu$	$\hat{\delta}_\beta$	$\hat{\tau}_\mu$	$\hat{\tau}_\beta$	$\lambda_{r,\mu}$	$\nu_{rat,\mu}$
$\sigma^2 = 0.5$	90.3%	82.6%	83.9%	74.5%	6.2%	5.0%
$\sigma^2 = 1.0$	95.9%	95.8%	94.8%	94.4%	6.4%	4.7%

Notes: T = 200; other details are as in Section **2.1.2** and Table 2.2; 5% nominal size.

Case 1 where the unit root is in the levels of the series, but the tests are based on the logs, and Case 2 where the unit root is the logs, but the tests are based on the levels. The results, reported in Table 2.15, show that, unlike the DF tests, $\lambda_{r,\mu}$ and $\nu_{rat,\mu}$ maintain close to their nominal size of 5% in both cases.

2.6.3 Empirical illustrations

This chapter concludes by illustrating the use of the various tests introduced in this chapter with two time series.

2.6.3.i Ratio of gold–silver price (revisited)

The first illustration revisits the time series of the ratio of the gold to silver price used in Section **2.2.3**. This series has an overall total of 5,372 observations and is referred to in levels as y_t. The unit root tests are reported in Table 2.16. Apart from R_{UR}^{FB}, and possibly R_{UR}^{F}, the unit root tests point to clear non-rejection of the null for both y_t and $\ln y_t$ at a 5% significance level. (Note that the rank-based tests are invariant to the log transformation.)

Of the two RUR tests, R_{UR}^{FB} has better characteristics when the errors are serially correlated (see Section **2.4.3** and AES, op. cit., section 6), and here the sample value is virtually identical to the 5% critical value. Of course, the use of multiple tests implies that the overall significance level will generally exceed the nominal significance level used for each test.

The usual predicament, anticipated in Section **2.2**, therefore, arises in that the tests are unable to discriminate between whether the unit root is in the ratio or the log ratio, and a further test is required to assess this issue. This point was considered in Section **2.2.3**, where the KM test suggested that the unit root was in the ratio (levels) if the comparison is limited to linear integrated or log-linear integrated processes.

Table 2.16 Tests for a unit root: ratio of gold to silver prices

	$\hat{\delta}_\mu$	$\hat{\tau}_\mu$	$\hat{\delta}_{r,\mu}$	$\hat{\tau}_{r,\mu}$	R^F_{UR}	R^{FB}_{UR}	$\lambda_{r,\mu}$	$\lambda_{lr,\mu}$	$\nu_{rat,\mu}$
y_t	−12.00	−2.43	−12.74	−2.62	1.23	1.89	0.092	0.160	0.018
$\ln y_t$	−12.74	−2.53	−12.74	−2.62	1.23	1.89	0.078	0.130	0.016
5% cv	−14.02	−2.86	−17.84	−3.03	1.31	1.90	0.036	0.036	0.010

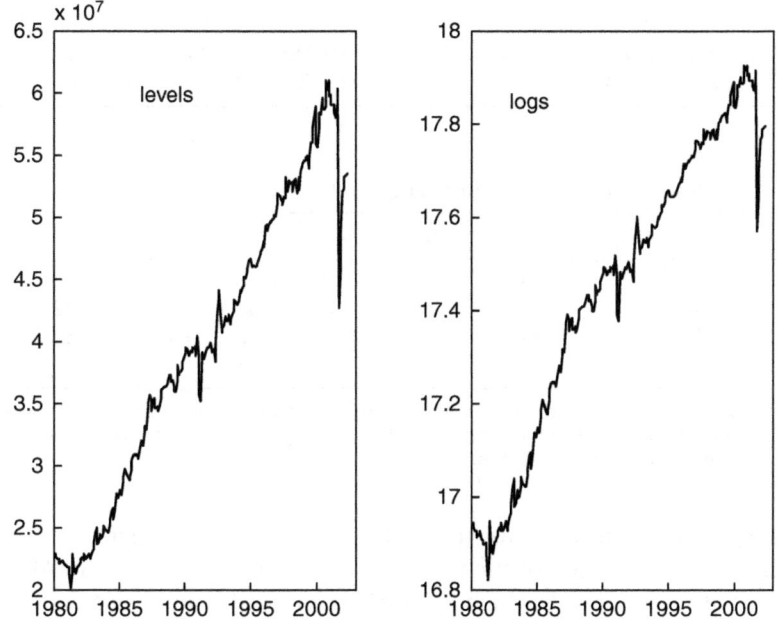

Figure 2.5   US air passenger miles

2.6.3.ii   *Air revenue passenger miles (US)*

The second illustration uses a smaller sample and a series with a trend. The data are for air revenue passenger miles in the US (one revenue passenger transported one mile; source: US Bureau of Transportation Studies; seasonally adjusted data). The time series are graphed in Figure 2.5 as levels and logs respectively. These data are monthly and the overall sample period is January 1980 to April 2002; however, there is an obvious break in the series in September 2001 (9/11) and the estimation sample ends in August 2001. Comparing the left-hand and right-hand panels of Figure 2.5, it is evident that the levels and log data share the same general characteristics making it quite difficult to assess whether, if there is a unit root, it is in the levels or the logs of the series.

Table 2.17   Regression details: air passenger miles, US

	C	y_{t-1}	Δy_{t-1}	Δy_{t-2}	Δy_{t-3}	Δy_{t-4}	Δy_{t-5}	Time
Δy_t	2.247	-0.112	-0.039	-0.214	-0.161	-0.49	-0.153	0.017
('t')	(3.14)	(-2.85)	(-0.57)	(-3.20)	(-2.42)	(-0.77)	(-2.40)	(2.91)
Δy_t	3.158	-0.165						0.026
('t')	(4.86)	(-4.73)						(4.74)
	C	$\ln y_{t-1}$	$\Delta \ln y_{t-1}$	$\Delta \ln y_{t-2}$	$\Delta \ln y_{t-3}$	$\Delta \ln y_{t-4}$	$\Delta \ln y_{t-5}$	Time
$\Delta \ln y_t$	0.240	-0.017	-0.090	-0.237	-0.162			0.0003
('t')		(-0.45)	(-1.42)	(-3.81)	(-2.54)			(1.98)
$\Delta \ln y_t$	0.294	-0.020	-0.113	-0.282	-0.213	-0.087	-0.185	-0.0004
('t')	(1.77)	(-0.53)	(-1.77)	(-4.42)	(-3.28)	(-1.36)	(-2.90)	(0.75)

Table 2.18   Tests for a unit root: air passenger miles, US

	$\hat{\delta}_\beta$	$\hat{\tau}_\beta$	$\hat{\delta}_{r,\beta}$	$\hat{\tau}_{r,\beta}$	R_{UR}^{F}	R_{UR}^{FB}	$\lambda_{r,\mu}$	$\lambda_{lr,\mu}$	$\nu_{rat,\beta}$
y_t	-20.79	-2.85	-14.77	-2.77	5.47	7.38	0.083	0.140	0.049
$\ln y_t$	-3.83	-0.45	-14.77	-2.77	5.47	7.38	0.051	0.622	0.014
5%	-21.23	-3.43	-24.55	-3.60	1.31	1.90	0.036	0.036	0.0034
95%					3.29	4.65			

Notes: a β subscript indicates that the test statistic is based on a linear trend adjusted series; appropriate critical values were simulated for the test statistics, based on 50,000 replications; Breitung's $\lambda_{r,\mu}$ statistic allows for drift, if present, under the null.

The parametric structure for the ADF tests is now somewhat more difficult to pin down. In the case of the levels data, y_t, both marginal-t and AIC select ADF(5), with $\hat{\tau}_\beta = -2.853$, which is greater than the 5% cv of -3.4, which leads to non-rejection of the null; however, BIC suggests that no augmentation is necessary and $\hat{\tau}_\beta = -4.729$, which, in contrast, leads to rejection of the null hypothesis. When the log of y_t is used, BIC selects ADF(3), with $\hat{\tau}_\beta = -0.45$, which strongly suggests, non-rejection; AIC and marginal-t both select ADF(5), with $\hat{\tau}_\beta = -0.53$, confirming this conclusion. At this stage, based on these tests, it appears that if there is a unit root then it is in the logarithm of the series; see Table 2.17 for a summary of the regression results and Table 2.18 for the unit root test statistics.

The next step is to compare the ADF test results with those for the nonparametric tests. The rank ADF tests now use (linearly) detrended data, and both $\hat{\delta}_{r,\beta} = -14.77$ and $\hat{\tau}_{r,\beta} = -2.77$ suggest non-rejection of the null. The RUR tests now compare the sample value with the 95% quantile (that is, they become right-sided tests), and both lead to rejection of a driftless random walk in favour of a trended alternative. Also the rank-score test, $\lambda_{lr,\mu}$, leads to non-rejection, as does the variance ratio test $\nu_{rat,\beta}$.

Table 2.19   KM tests for linearity or log-linearity: air passenger miles, US

KM tests		
Null, linear; alternative, loglinear	$V_1 = 2.84$ (5 lags) $V_1 = 2.83$ (3 lags)	95% quantile $= 1.645$
Null, log-linear; alternative, linear	$V_2 = -3.18$ (5 lags) $V_2 = -3.14$ (6 lags)	95% quantile $= 1.645$

There are two remaining issues. First, for the RUR tests, the question is whether the rejection arises from a drifted random walk or a trend stationary process. To assess this issue, the auxiliary procedure suggested in Section **2.4.4** is the next step. This procedure involves a second application of the RUR test, but this time on a variable that is based on a sum of the variable in the first test, referred to in Section **2.4.4** as $\tilde{x}_t^{(1)}$. The resulting sample values of the R_{UR}^F and R_{ur}^{FB} test statistics applied to $\tilde{x}_t^{(1)}$ are now 2.98 and 4.30 respectively; these values strongly indicate non-rejection of the null hypothesis of a unit root as the 5% quantiles are (approximately) 1.31 and 1.90 respectively; moreover, there is now no rejection in the upper tail, as the 95% quantiles are (approximately) 3.29 and 4.65 respectively. In terms of the decisions outlined in Section **2.4.4**, we find in favour of *possibility A*, that is, the drifted integrated process rather than the trend stationary process.

Second, non-rejection does not indicate the transformation of the series that is appropriate. If the choice is limited to either a linear integrated or a log-linear integrated process, the KM tests are relevant. In this case, the tests referred to as V_1 and V_2, which allow for a drifted integrated process, are reported in Table 2.19. The first of these leads to rejection of the null of linearity, whereas the second suggests non-rejection of the null of log-linearity.

2.7   Concluding remarks

A practical problem for empirical research often relates to the precise choice of the form of the variables to be used. This can be a key issue, even though quite frequently there is no clear rationale for a particular choice; a leading case, but of course by no means the only one, is whether a time series should be modelled in levels (the 'raw' data) or transformed into logarithms. There are very different implications for choosing a linear integrated process over a log-linear integrated process, since the former relates to behaviour of a random walk type whereas the other relates to an exponential random walk. A two-stage procedure was suggested in this chapter as a way of addressing this problem. In the first instance, it is necessary to establish that the null hypothesis of a unit root is not rejected for *some* transformation of the data; this stage may conclude that there is an integrated process generating the data, and the second stage decision is then to assess whether this is consistent with a linear or log-linear integrated process. Of course this just addresses two possible outcomes, linear or log-linear; others transformations may also be candidates, although these are the leading ones.

The second strand to this chapter was to extend the principle for the construction of unit root tests from variants of direct estimation of an AR model, as in the DF and ADF tests, to include tests based on the rank of an observation in a set and the range of a set of observations. Neither of these principles depended on a particular parametric structure. Another framework for constructing a unit root test without a complete parametric structure is based on the behaviour of the variance of a series with a unit root. Variance-ratio tests are familiar from the work of Lo and MacKinlay (1988), but the basic test is difficult to extend for non-iid errors; however, a simple and remarkably robust test due to Breitung (2002) overcomes this problem.

Other references of interest in this area include Delgado and Velasco (2005), who suggested a test based on signs that is applicable to testing for a unit root as well as other forms of nonstationarity. Hasan and Koenker (1997) suggested a family of rank tests of the unit root hypothesis. Charles and Darné (1999) provided an overview of variance ratio tests of the random walk. Wright (2000) devised variance-ratio tests using ranks and signs. Tse, Ng and Zhang (2004) have developed a non-overlapping version of the standard VR test. Nielsen (2009) develops a simple nonparametric test that includes Breitung's (2002) test as a special case and has higher asymptotic local power.

Questions

Q2.1. Define a monotonic transformation and give some examples.

A2.1. A monotonic transformation is one for which the derivative is everywhere positive. For example, consider $w_t = f(y_t) = \ln y_t$, then $\partial w_t / \partial y_t = 1/y_t$, which is positive provided that $y_t > 0$, otherwise the transformation is not defined. Thus, the following transformations, considered by Breitung and Gouriéroux (1997), are all monotonic: $f(y_t) = y_t^{1/3}$, $f(y_t) = y_t^3$, $f(y_t) = \tan(y_t)$.

Q2.2. Consider the ADF version of the rank-based test and explain how to obtain the equivalent DF-type test statistics.

A2.2. The suggested ADF regression is:

$$\Delta R_t = \gamma_r R_{t-1} + \sum_{j=1}^{k-1} c_{r,j} \Delta R_{t-j} + v_t \tag{A2.1}$$

where R_t are the centred ranks (otherwise a constant is included in the regression). The pseudo-t test based on the ranks is straightforward and is just:

$$\hat{\tau}_{R,\mu} = \frac{\hat{\gamma}_r}{\hat{\sigma}(\hat{\gamma}_r)} \tag{A2.2}$$

where $\hat{\tau}_{R,\mu}$ is the usual t-type test statistic; the extension of $\hat{\delta}_{R,\mu}$ to the rank-ADF case requires a correction, as in the usual ADF case, that is:

$$\hat{\delta}_{R,\mu} = T\frac{(\hat{\rho}_r - 1)}{(1 - \hat{c}_r(1))} \tag{A2.3}$$

where $\hat{c}_r(1) = \sum_{j=1}^{k-1}\hat{c}_{r,j}$. Note that, without formal justification, a detrended version of this test was used in the application to air revenue passenger miles in Section **2.6.3.ii**.

Q2.3. Develop a variance ratio test based on the signs; see for example, Wright (2000).

A2.3. The signs define a simple transformation of the original series. If x_t is a random variable, then the sign function for x_t is as follows:

$$S(x_t) = 2 * I(x_t > 0) - 1 \tag{A2.4}$$

Thus, if $x_t > 0$, then $I(x_t > 0) = 1$ and $S(x_t) = 1$, and if $x_t \leq 0$ then $I(x_t \leq 0) = 0$ and $S(x_t) = -1$. $S(x_t)$ has zero expectation and unit variance.

Wright (2000) suggested a sign-based version of a variance-ratio test. Assume that x_t has a zero mean, then $S(x_t)$ replaces x_t in the definition of the variance ratio, to give:

$$\text{OVR}_{\text{sign}} = \left(\frac{\frac{1}{Tq}\sum_{t=q+1}^{T}\left(\sum_{j=0}^{q-1}s_{t-j}\right)^2}{\frac{1}{T}\sum_{t=1}^{T}s_t^2} - 1 \right) \left(\frac{2(2q-1)(q-1)}{3qT} \right)^{-1/2} \tag{A2.5}$$

where $s_{t-j} \equiv S(x_{t-j})$. This form of this test arises because the identity used in the parametric form of the test does not hold with the sign function, that is $\Delta_q s_t \neq \sum_{j=0}^{q-1}s_{t-j}$.

This test could be applied to the case where $x_t \equiv \Delta y_t$ and the null model is $\Delta y_t = \sigma_t \varepsilon_t$, with assumptions that: (i) σ_t and ε_t are independent conditional on the information set, $I_t = \{y_t, y_{t-1}, \ldots\}$, and (ii) $E(\varepsilon_t|I_{t-1}) = 0$. The first assumption allows heteroscedasticity, for example of the ARCH/GARCH form or stochastic volatility type. Note that these assumptions do not require ε_t to be iid.

There are variations on these tests. For example, the sign-based test is easily extended if the null hypothesis is of a random walk with drift, and rank-based versions of these test can be obtained by replacing the signs by ranks (see Wright (op. cit.), who simulates the finite sample distributions and provides some critical values for different T and q).

3
Fractional Integration

Introduction

This chapter is the first of two that consider the case of fractional values of the integration parameter. That is suppose a process generates a time series that is integrated of order d, I(d), then the techniques in *UR, Vol. 1*, were wholly concerned with the case where d is an integer, the minimum number of differences necessary, when applied to the original series, to produce a series that is stationary. What happens if we relax the assumption that d is an integer? There has been much recent research on this topic, so that the approach in these two chapters must necessarily be selective. Like so many developments in the area of time series analysis one again finds influential original contributions from Granger; two of note in this case are Granger (1980) on the aggregation of 'micro' time series into an aggregate time series with fractional I(d) properties, and Granger and Joyeux (1980) on long-memory time series.

One central property of economic time series is that the effect of a shock is often persistent. Indeed in the simplest random walk model, with or without drift, a shock at time t is incorporated or *integrated* into the level of the series for all periods thereafter. One can say that the shock is always 'remembered' in the series or, alternatively, that the series has an infinite memory for the shock. This infinite memory property holds for more general I(1) processes, with AR or MA components. What will now be of interest is what happens when d is fractional, for example $d \in (0, 1.0)$. There are two general approaches to the analysis of fractional I(d) process, which may be analysed either in the time domain or in the frequency domain. This chapter is primarily concerned with the former, whereas the next chapter is concerned with the latter.

This chapter progresses by first spending some time on the definition of a fractionally integrated process, where it is shown to be operational through the binomial expansion of the fractional difference operator. The binomial expansion of $(1 - L)^d$ is an elementary but essential tool of analysis, as is its representation in terms of the gamma function. Having shown that a meaning can be assigned to the fractional differencing operator applied to y_t, it is important to note some of the properties of the resulting process; for example, what are the implied coefficients in the respective AR and MA representations of the process? The MA representation is particularly important because it is the usual tool for analysing the persistence of the process.

For $d \in (0, 0.5)$, the I(d) process is said to have 'long memory', defined more formally in the time domain as saying that the sum of the autocorrelations does not converge to a finite constant; in short, they are nonsummable. Of course, this property also applies to the autocovariances as $\rho(k) = \gamma(k)/\gamma(0)$. However, an I(d) process is still stationary for $d \in (0, 0.5)$, only becoming nonstationary for $d \geq 0.5$. Even though the $\rho(k)$ are not summable for $d \geq 0.5$, the coefficients in the MA representation of the process do tend to zero provided $d < 1$ and this is described as 'nonpersistence'. Thus, a process could be nonstationary but nonpersistent, and care has to be taken in partitioning the parameter space according to these different properties.

Having established that we can attach a meaning to $(1 - L)^d y_t$, a question to arise rather naturally is whether such processes are likely to occur in an economic context. Two justifications are presented here. The first relates to the error duration model, presented in the form due to Parke (1999), but with a longer history due, in part to the 'micropulse' literature, with important contributions by Mandelbrot and his co-authors (see, for example, Cioczek-Georges and Mandelbrot, 1995, 1996, and Mandelbrot and Van Ness, 1968). The other justification is due to Granger (1980), who presented a model of micro relationships which were not themselves fractionally integrated, but which on aggregation became fractionally integrated.

We also consider the estimation of d in the time domain and, particularly, hypothesis testing. One approach to hypothesis testing is to extend the DF testing procedure to the fractional d case. This results in an approach that is relatively easy to apply in a familiar framework.

The range of hypotheses of interest is rather wider than, but includes, the unit root null hypothesis. In a fractional d context, the alternative to the unit root null is not a stationary AR process, but a nonstationary fractional d process; because, in this case, the alternative hypothesis is still one of nonstationarity, another set-up of interest may be that of a stationary null against a nonstationary alternative.

The sections in this chapter are arranged as follows. Section 3.1 is concerned with the definition of a fractionally integrated process and some of its basic properties. Section 3.2 considers the ARFIMA(p, d, q), where d can be fractional, which is one of the leading long-memory models. Section 3.3 considers the kind of models that can generate fractional d. Section 3.4 considers the situation in which a Dickey-Fuller type test is applied when the process is fractionally integrated, whereas Section 3.5 is concerned with developments of Dickey-Fuller tests for fractionally integrated processes, and Section 3.6 considers how deterministic components are dealt with in such models. Section 3.7 outlines an LM test and a regression counterpart which is easy to calculate and has the same limiting distribution as the LM test. Power is considered in Section 3.8 and an illustrative example using time series data on US wheat production in included in Section 3.9. The final Section, 3.10, offers some concluding remarks.

3.1 A fractionally integrated process

In the integer differencing case, y_t is said to be integrated of order d, I(d), if the d-th difference of y_t is stationary (and d is the minimum number of differences necessary to achieve stationarity). In a convenient shorthand this is often written as $y_t \sim I(d)$. Thus, an I(d) process, with integer d, satisfies the following model:

$$(1 - L)^d y_t = u_t \qquad u_t \sim I(0) \tag{3.1}$$

$$\Rightarrow y_t \sim I(d)$$

One could add that $(1 - L)^{d-1} y_t = v_t$, $v_t \sim I(1)$, to indicate that d is the minimum number of differences necessary for the definition to hold. A special case of (3.1) is where $u_t = \varepsilon_t$ and $\{\varepsilon_t\}$ is a white noise sequence with zero mean and constant variance σ_ε^2.

In a sense defined in the next section it is possible to relax the assumption that d is an integer and interpret (3.1) accordingly either as an I(d) process for y_t or as a unit root process with fractional noise of order $I(d - 1)$.

3.1.1 A unit root process with fractionally integrated noise

To see that the generating equation can be viewed as a unit root process with fractional noise, rewrite Equation (3.1) separating out the unit root, so that:

$$(1 - L)^{d-1}(1 - L)y_t = u_t \tag{3.2}$$

$$\Rightarrow$$

$$(1 - L)y_t = \upsilon_t$$

$$= (1 - L)^{-\tilde{d}} u_t$$

$$\tilde{d} \equiv d - 1$$

In this representation y_t is generated with a unit root and possibly fractionally integrated noise $\upsilon_t \sim FI(\tilde{d})$. If $d = 1$, then $\tilde{d} = 0$, there is a single unit root and

$v_t = u_t$; if $d = 2$, then $\tilde{d} = 1$ and $(1 - L)v_t = u_t$; and if $d = 0$, then $\tilde{d} = -1$ and $v_t = (1 - L)u_t$. If it is thought that that y_t has a near-fractional unit root, with d close to but less than unity, then this implies that $\tilde{d}$ is close to but less than zero; for example, if the set $d \in (\frac{1}{2}, 1)$ is likely to be of interest, then that translates to the set $\tilde{d} \in (-\frac{1}{2}, 0)$ in terms of the noise of the process.

However, so far it has just been asserted that a meaning may be attached to $(1 - L)^d$ if d is not an integer but a number with a fractional part. In order to make progress on the precise definition of an I(d) process, it is first useful to consider the binomial expansion of the fractional differencing operator $(1 - L)^d$ and, in particular, the AR and MA coefficients associated with this operator.

3.1.2   Binomial expansion of $(1 - L)^d$

3.1.2.i   AR coefficients

In the fractional d case, the binomial expansion of $(1 - L)^d$ is given by:

$$(1 - L)^d = 1 + \sum_{r=1}^{\infty} {}_dC_r(-1)^r L^r \qquad (3.3)$$

$$= \sum_{r=0}^{\infty} A_r^{(d)} L^r$$

where:

$$A_0^{(d)} \equiv 1, A_r^{(d)} \equiv (-1)^r {}_dC_r \qquad (3.4)$$

$$_dC_r \equiv \frac{(d)_r}{r!} = \frac{d!}{r!(d-r)!} \qquad (3.5a)$$

$$_dC_0 \equiv 1, 0! \equiv 1$$

$$(d)_r = d(d-1)(d-2)\ldots(d-(r-1))$$

$$= d!/(d-r)! \qquad (3.5b)$$

In the integer case, the binomial coefficient $_dC_r$ is the number of ways of choosing r from d without regard to order. However, in the fractional case, d is not an integer and this interpretation is not sustained, although $_dC_r$ is defined in the same way.

Applying the operator $(1 - L)^d$ to y_t results in:

$$(1 - L)^d y_t = y_t + \sum_{r=1}^{\infty} {}_dC_r(-1)^r y_{t-r} \qquad (3.6)$$

$$= A^{(d)}(L)y_t$$

$$A^{(d)}(L) \equiv \sum_{r=0}^{\infty} A_r^{(d)} L^r \qquad (3.7)$$

$A^{(d)}(L)$ is the AR polynomial associated with $(1 - L)^d$. Using the AR polynomial, the model of Equation (3.1) with fractional d can be represented as the following infinite autoregression:

$$y_t = (-1)\sum_{r=1}^{\infty} {}_dC_r(-1)^r y_{t-r} + u_t \qquad (3.8a)$$

$$= (-1)\sum_{r=1}^{\infty} A_r^{(d)} y_{t-r} + u_t \qquad (3.8b)$$

$$= \sum_{r=1}^{\infty} \Lambda_r^{(d)} y_{t-r} + u_t \qquad \text{where } \Lambda_r^{(d)} \equiv -A_r^{(d)} \qquad (3.8c)$$

3.1.2.ii   MA coefficients

Also of interest is the MA representation of Equation (3.1), that is operating on the left-hand-side of (3.1) with $(1-L)^{-d}$, gives:

$$y_t = (1-L)^{-d} u_t \qquad (3.9)$$

This has an infinite moving average representation obtained by the binomial expansion of $(1-L)^{-d}$, that is:

$$(1-L)^{-d} = 1 + \sum_{r=1}^{\infty} {}_{-d}C_r(-1)^r L^r$$

$$= \sum_{r=0}^{\infty} B_r^{(d)} \qquad (3.10)$$

where:

$$B_0^{(d)} = 1, B_r^{(d)} \equiv (-1)^r {}_{-d}C_r \qquad (3.11)$$

Operating with $(1-L)^{-d}$ on u_t gives the MA representation of y_t:

$$y_t = [1 + \sum_{r=1}^{\infty} {}_{-d}C_r(-1)^r L^r] u_t \qquad (3.12)$$

$$= B^{(d)}(L) u_t$$

$$B^{(d)}(L) \equiv \sum_{r=0}^{\infty} B_r^{(d)} L^r \qquad (3.13)$$

There are recursions for both the AR and the MA coefficients (for details see Appendix 3.1 at the end of this chapter) respectively, as follows for $r \geq 1$:

$$A_r^{(d)} = [(r-d-1)/r] A_{r-1}^{(d)} \qquad (3.14)$$

$$B_r^{(d)} = [(r+d-1)/r] B_{r-1}^{(d)} \qquad (3.15)$$

3.1.2.iii   The fractionally integrated model in terms of the Gamma function

The coefficients of the fractionally integrated model are often written in terms of the Gamma function. The latter is defined by:

$$\Gamma(k) = \int_0^{\infty} x^{k-1} e^{-x} dx \qquad k > 0 \qquad (3.16)$$

The notation k is used for the general case and n for an integer. The Gamma function interpolates the factorials between integers, so that for integer $n \geq 0$ then $\Gamma(n+1) = n!$. The property $(n+1)! = (n+1)n!$, which holds for integer $n \geq 0$, also holds for the Gamma function without the integer restriction; this is written as $\Gamma(n+2) = (n+1)\Gamma(n+1)$.

In the notation of this chapter, the binomial coefficients defined in (3.5) are:

$$
\begin{aligned}
_dC_r &= \frac{(d)_r}{r!} \\
&= \frac{\Gamma(d+1)}{\Gamma(r+1)\Gamma(d-r+1)}
\end{aligned}
\tag{3.17}
$$

The autoregressive and moving average coefficients are given, respectively, by:

$$
A_r^{(d)} = \frac{\Gamma(r-d)}{\Gamma(r+1)\Gamma(-d)}
\tag{3.18}
$$

$$
B_r^{(d)} = \frac{\Gamma(r+d)}{\Gamma(r+1)\Gamma(d)}
\tag{3.19}
$$

Using Stirling's theorem, it can be shown that $\Gamma(r+h)/\Gamma(r+1) = O(r^{h-1})$ and applying this for $h = -d$ and $h = d$, respectively, then as $r \to \infty$, we have:

$$
A_r^{(d)} \sim \frac{1}{\Gamma(-d)} r^{-d-1} \qquad \text{AR coefficients}
\tag{3.20}
$$

$$
B_r^{(d)} \sim \frac{1}{\Gamma(d)} r^{d-1} \qquad \text{MA coefficients}
\tag{3.21}
$$

The AR coefficients will not (eventually) decline unless $d > -1$ and the MA coefficients will not (eventually) decline unless $d < 1$. Comparison of the speed of decline is interesting. For $0 < d < 1$, the AR coefficients (asymptotically) decline faster than the MA coefficients, whereas for $d < 0$, the opposite is the case. For given d, the relative rate of decline of $A_r^{(d)}/B_r^{(d)}$ is governed by r^{-2d}, which can be substantial for large r and even quite modest values of d. For example, for $d = 0.5$ the relative decline is of the order of r^{-1}, whereas for $d = -0.5$ it is the other way around. This observation explains the visual pattern in the figures reported below in Section **3.1.5**.

Notice that the MA coefficients $B_r^{(d)}$ for given d are the same as the AR coefficients $A_r^{(d)}$ for $-d$ and vice versa. To see this note that for $d = \bar{d}$, $A_r^{(\bar{d})} = (-1)^r {}_{\bar{d}}C_r$ and $B_r^{(\bar{d})} = (-1)^r {}_{-\bar{d}}C_r$ whereas for $d = -\bar{d}$, $A_r^{(-\bar{d})} = (-1)^r {}_{-\bar{d}}C_r$ and $B_r^{(-\bar{d})} = (-1)^r {}_{\bar{d}}C_r$.

The MA coefficients trace out the impulse response function for a unit shock in ε_t. A point to note is that even though the sequence $\{y_t\}$ is nonstationary for $d \in [\frac{1}{2}, 1)$, the MA coefficients eventually decline. This property is often referred to as mean reversion, but since the variance of y_t is not finite for $d \geq 0.5$, this term may be considered inappropriate. An alternative is to say that the process is not (infinitely) persistent or, bearing in mind the qualification, just non-persistent.

3.1.3   Two definitions of an I(d) process, d fractional

In this section we consider how to define an I(d) process such as:

$$(1 - L)^d (y_t - y_0) = u_t \tag{3.22}$$

where u_t is a weakly dependent stationary process and y_0 is a starting value. There are two ways of approaching the definition of a fractional I(d) process, referred to as partial summation and the direct approach (or definition), and these are considered in the next two sections.

3.1.3.i   Partial summation

One approach is to start with z_t, a stationary I(d − 1) process, and then sum z_t in order to obtain an I(d) process, $d \in [\frac{1}{2}, \frac{3}{2})$. (The unit root process can be considered a special case of this procedure as it sums an I(0) process.) This can be repeated in order to obtain higher-order processes.

Consider the following recursion typically associated with a unit root process:

$$y_t = y_{t-1} + z_t \qquad t \geq 1 \tag{3.23}$$

where z_t are termed increments. Then by back-substitution, and given the starting value y_0, we obtain:

$$y_t = y_0 + \sum_{j=1}^{t} z_j \qquad t \geq 1 \tag{3.24}$$

The difference from the standard unit root process is that the z_t are generated by an I(d − 1) process given by:

$$(1 - L)^{d-1} z_t = u_t$$

$$\Rightarrow$$

$$z_t = (1 - L)^{1-d} u_t$$

$$= \sum_{r=0}^{\infty} B_r^{(d-1)} u_{t-r} \qquad \text{for } d \in [\tfrac{1}{2}, \tfrac{3}{2}) \tag{3.25}$$

Substituting for z_t in (3.24) shows the dependence of y_t on lagged u_t, including 'pre-sample' values $(t < 1)$:

$$y_t = y_0 + \sum_{j=1}^{t} \sum_{r=0}^{\infty} B_r^{(d-1)} u_{j-r} \qquad t \geq 1 \tag{3.26}$$

The increments z_t are I(d − 1), with (d − 1) < ½, and the MA coefficients $B_r^{(d-1)}$ are defined in (3.19). Also notice that if $d \in (\frac{1}{2}, 1)$, then $d - 1 \in (-\frac{1}{2}, 0)$, so that z_t, on this definition, exhibits negative autocorrelation and is referred to as being anti-persistent.

Noting that $(1 - L)^{-1}$ is the summation operator and $z_t = 0$ for $t < 1$, then (3.26) can be rewritten as follows:

$$y_t = y_0 + (1 - L)^{-1}(1 - L)^{1-d}u_t \tag{3.27}$$

$\Rightarrow$

$$(1 - L)(y_t - y_0) = (1 - L)^{1-d}u_t \qquad \text{multiply through by } (1 - L)$$

$\Rightarrow$

$$(1 - L)^d(y_t - y_0) = u_t \qquad \text{multiply through by } (1 - L)^{d-1}, \quad t \geq 1 \tag{3.28}$$

Thus, given the definition of z_t as the moving average of a stationary input u_t with moving average coefficients given by the binomial expansion of $(1 - L)^{d-1}$, $(d - 1) < \frac{1}{2}$, then the process $y_t \sim I(d)$, $d \in [\frac{1}{2}, \frac{3}{2})$, follows by summation.

Next, consider $x_t \sim I(d)$ with $d \in [\frac{3}{2}, \frac{5}{2})$, then such a process is defined by the double summation of the primary input $z_t \sim I(d - 2)$, where $(d - 2) < \frac{1}{2}$, that is:

$$x_t = x_0 + \sum_{j=1}^{t} y_j \qquad t \geq 1$$

$$= x_0 + \sum_{i=1}^{t} \left(\sum_{j=1}^{i} z_j \right) \tag{3.29}$$

where:

$$(1 - L)^{d-2}z_t = u_t \qquad (d - 2) < \frac{1}{2} \tag{3.30}$$

$$z_t = \sum_{r=0}^{\infty} B_r^{(d-2)} u_{t-r}$$

By applying the summation operator twice,

$$(1 - L)^d(x_t - x_0) = u_t \qquad d \in [\frac{3}{2}, \frac{5}{2}), \qquad t \geq 1 \tag{3.31}$$

Note that the partial summation approach nests the unit root framework that is associated with integer values of d. For purposes of tying in this approach with the corresponding fractional Brownian motion (fBM) this will be referred to as a type I fractional process.

3.1.3.ii Direct definition

An alternative definition of an FI(d) process is based directly on the binomial expansion without the intermediate step of partial summation. Recall that the operator $(1 - L)^d$ has the AR and MA forms given by:

$$(1 - L)^d \equiv \sum_{r=0}^{\infty} A_r^{(d)} \qquad \text{and} \qquad (1 - L)^{-d} \equiv \sum_{r=0}^{\infty} B_r^{(d)}$$

Of particular interest for the direct definition of an I(d) process is the MA form *truncated* to reflect the start of the process $t = 1$, that is:

$$y_t - y_0 = (1 - L)^{-d} u_t \qquad t \geq 1$$

$$= \sum_{r=0}^{t-1} B_r^{(d)} u_{t-r} \qquad (3.32)$$

This is a well-defined expression for all values of d, not just $d \in (-\frac{1}{2}, \frac{1}{2})$.

The direct summation approach leads to what will be referred to as a type II fractional process. Where it is necessary to distinguish between y_t generated according to the different definitions they will be denoted $y_t^{(I)}$ and $y_t^{(II)}$, respectively.

3.1.4 The difference between type I and type II processes

Do the type I and type II definitions imply different processes? When $d < \frac{1}{2}$, $y_t^{(I)}$ and $y_t^{(II)}$ differ by terms of order $O_p(t^{(2d-1)})$, so that the difference vanishes asymptotically. However, this is not the case for $d \geq \frac{1}{2}$, with the different processes, appropriately normalised, leading to different types of fractional Brownian motion, known as type I and type II fBM respectively; see Marinucci and Robinson (1999), Robinson (2005) and Shimotsu and Phillips (2006).

It can be shown that the difference between $y_t^{(I)}$ and $y_t^{(II)}$ is essentially one of how the presample observations are treated. Note from the type I definition that $z_t = \sum_{r=0}^{\infty} B_r^{(d-1)} u_{t-r}$ involves an infinite summation of present u_t and u_{t-r}, $r > 0$, whereas on the type II definition, the u_t are assumed to be zero before $t = 1$. Indeed, as Shimotsu and Phillips (2000, 2006) observe, for $d \in [\frac{1}{2}, \frac{3}{2})$, y_t on the type I definition can be written as:

$$y_t^{(I)} = y_t^{(II)} + \xi_0(d) \qquad (3.33)$$

where the term $\xi_0(d)$ captures the presample values of u_t; this term is of the same stochastic order as $y_t^{(II)}$ for $d \in [\frac{1}{2}, \frac{3}{2})$, which means that it cannot be ignored asymptotically, so that the initialisation, or how the pre-sample values are treated, does matter.

3.1.5 Type I and type II fBM

As indicated, type I and type II definitions of an FI(d) process are associated with different types of fBM, which are defined as follows when considered as real continuous processes:

$$Y^{(I)} = \frac{1}{\Gamma(d+1)} \int_0^r (r-s)^d dW(s) + \frac{1}{\Gamma(d+1)} \int_{-\infty}^0 [(r-s)^d - (-s)^d] dW(s) \quad (3.34)$$

$$Y^{(II)} = \frac{1}{\Gamma(d+1)} \int_0^r (r-s)^d dW(s) \qquad (3.35)$$

where $r, s \in \mathfrak{R}, |d| < \frac{1}{2}$ and $W(s)$ is regular Brownian motion (BM) with variance σ^2_{lr}, which is a standard BM if $\sigma^2_{lr} = 1$, in which case it is usually referred to as $B(s)$.

The unit root case is worth recalling briefly. Consider the simple random walk $y_t = \sum_{j=1}^{t} z_j$, $t = 1, \ldots, T$, which is an example of a partial sum process based on the 'increments' $z_t \sim I(0)$, which may therefore exhibit weak dependence and, for simplicity, the initial value y_0 has been set to zero. This can be written equivalently by introducing the indexing parameter r:

$$y_T(r) = \sum_{t=1}^{[rT]} z_t \tag{3.36}$$

The notation $[rT]$ indicates the integer part of rT; thus, rT is exactly an integer for $r = j/T$, $j = 1, \ldots, T$ and, for these values, gives the random walk over the integers. However, $y_T(r)$ can be considered as continuous function of r, albeit it will be a step function, with the 'steps' becoming increasingly smaller as $T \to \infty$, so that, in the limit, $y_T(r)$ is a continuous function of r. To consider this limit, $y_T(r)$ is first normalised by $\sigma_{lr,z}\sqrt{T}$, where $\sigma^2_{lr,z} < \infty$ is the 'long-run' variance of z_t, so that:

$$Y_T(r) \equiv \frac{y_T(r)}{\sigma_{lr,z}\sqrt{T}} \tag{3.37}$$

(See *UR, Vol. 1*, chapters 1 and 6 for the definition of $\sigma^2_{lr,z}$). There then follows a form of the functional central limit theorem (FCLT) in which $Y_T(r)$ weakly converges to standard Brownian motion:

$$Y_T(r) \Rightarrow_D B(r) \tag{3.38}$$

where $B(r)$ is standard BM.

However, when $z_t \sim I(d)$, $d < |\frac{1}{2}|$, $d \neq 0$, and therefore $y_t \sim I(d+1)$, the result in (3.38) no longer holds. The limiting result in this case depends on whether y_t is generated by a type I or type II process. In the case of a type I process (and setting $y_0 = 0$),

$$y_t^{(I)}(r) = \sum_{t=1}^{[rT]} z_t \quad \text{where } z_t \sim I(d), d < |\frac{1}{2}|, d \neq 0$$

and $y_t^{(I)}(r)$ is then normalised as follows:

$$Y^{(I)}(r) = \frac{y_T^{(I)}(r)}{\sigma_{lr,z}T^{\frac{1}{2}+d}} \tag{3.39}$$

The FCLT leads to type I fBM, $Y^{(I)}$, that is:

$$Y^{(I)}(r) \Rightarrow_D Y^{(I)} \tag{3.40}$$

where $Y^{(I)}$ is defined in (3.34). An analogous result holds for a type II process:

$$Y^{(II)}(r) = \frac{y_T^{(II)}(r)}{\sigma_{lr,z} T^{1/2+d}}$$

$$\Rightarrow_D Y^{(II)} \qquad\qquad (3.41)$$

The impact of the different forms of fBM on some frequently occurring esti-mators is explored by Robinson (2005). (For example, if the properties of an estimator of, say, d are derived assuming type I fBM, what are its properties if type II fBM is assumed?)

3.1.6   Deterministic components and starting value(s)

3.1.6.i   *Deterministic components*

The DGP of (3.1) assumes that there are no deterministic components in the mean. In practice this is unlikely. As in the case of standard DF tests, the leading specification in this context is to allow the fractionally integrated pro-cess to apply to the deviations from a deterministic mean function, usually a polynomial function of time, such that:

$$(1 - L)^d [y_t - \mu_t] = u_t \qquad\qquad (3.42)$$

As in the standard unit root framework, the subsequent analysis is in terms of observations that are interpreted as deviations from the mean (or detrended observations). Defining $\tilde{y}_t$ as the deviation from the trend function, then the FI(d) process is written as:

$$(1 - L)^d \tilde{y}_t = u_t \qquad\qquad (3.43)$$

$$\tilde{y}_t \equiv y_t - \mu_t$$

The cases likely to be encountered in practice are $\mu_t = 0$, $\mu_t = y_0$ and $\mu_t = \beta_0 + \beta_1 t$; more generally one could allow for the trend to be α-order, that is $\mu_t = \beta t^\alpha$, where α is not necessarily an integer (this is a case considered by Phillips and Shimotsu, 2004). To simplify the notation, the expositional case will be written as in (3.1), however, there are occasions in this and the next chapter where the presence of a deterministic trend will be taken into account explicitly and in these cases (3.42) will be adopted. Also since most of the development does not make use of the more general property of u_t rather than ε_t, then the latter will be the default choice.

3.1.6.ii   *Starting value(s)*

There are two cases to be considered depending on how 'prehistorical' obser-vations are dealt with, and they correspond to the different types of fBM. The leading case is where $\varepsilon_t = 0$ for $t \leq 0$, so that the process is assumed to start in period $t = 1$. This is written as $\varepsilon_t 1_{(t>0)}$, so that:

$$(1 - L)^d \widetilde{y}_t = \varepsilon_t 1_{(t>0)} \qquad t = 1, \ldots, T \tag{3.44}$$

For obvious reasons, this is sometimes referred to as a truncated process. Alternatively, the process can be viewed as starting in the infinite past, that is:

$$(1 - L)^d \widetilde{y}_t = \varepsilon_t \qquad t = -\infty, \ldots, 0, 1, \ldots, T \tag{3.45}$$

As noted these two representations give rise to type I and type II fractional Brownian motion (fBM) where the latter is associated with the truncated process and the former with the non-truncated process; see Marinucci and Robinson (1999), and Davidson and Hashimzade (2009), and see Section **3.1.5**.

Recall that to write $t = 1$ is a matter of notational convenience; in practice, the sample starting date will be a particular calendar date, for example, 1964q1, which raises the question, for actual economic series, of whether it is reasonable to set all shocks before the particular start date at hand to zero. As Davidson and Hashimzade (2009) note, whilst in time series modelling such a truncation often does not matter, at least asymptotically, in this case it does. This issue is of importance in, for example, obtaining the critical values for a particular test statistic that is a functional of type I fBM, then it would be natural enough to start the process at some specific date in a simulation setting, but type II fBM may be a better approximation for 'real world' data. See Robinson (2005) for a discussion of some of the issues.

3.2   The ARFIMA(p, d, q) model

The value of d determines the long-run characteristics of y_t. Additional short-run dynamics are obtained by generalising the simple fractional model to include AR and MA components represented in the lag polynomials $\phi(L)$ and $\theta(L)$, respectively. This results in the fractional ARMA(p, d, q) model, usually referred to as an ARFIMA(p, d, q) model, given by:

$$\phi(L)(1 - L)^d y_t = \theta(L)\varepsilon_t \tag{3.46}$$

For simplicity it is assumed that $\mu_t = 0$. Alternatively, (3.46) can be written as:

$$(1 - L)y_t = u_t \tag{3.47a}$$

$$(1 - L)^{\widetilde{d}} u_t = w(L)\varepsilon_t \tag{3.47b}$$

where $\widetilde{d} \equiv d - 1$ and $w(L) = \phi(L)^{-1}\theta(L)$. This allows the interpretation of y_t as a unit process subject to fractionally integrated and serially correlated noise; see Section **3.1.1**.

Provided that the roots of $\phi(L)$ are outside the unit circle (see *UR, Vol. 1*, chapter 2), the ARFIMA(p, d, q) model can be inverted to obtain the moving average representation given by:

$$y_t = \phi(L)^{-1}\theta(L)(1-L)^{-d}\varepsilon_t \tag{3.48}$$

$$= w(L)B(L)\varepsilon_t$$

$$= \omega(L)\varepsilon_t$$

(The polynomials in (3.46) are scalar so the rearrangement of their order in (3.48) is valid.) The moving average polynomial is $\omega(L) = \phi(L)^{-1}\theta(L)(1-L)^{-d}$, which is now the convolution of three polynomials.

3.2.1 Autocovariances and autocorrelations of the ARFIMA(0, d, 0) process, d ≤ 0.5

3.2.1.i   Autocovariances

The lag k autocovariances (for a zero mean process), $E(y_t y_{t-k}) \equiv \gamma(k)$, of the fractionally integrated process are as follows:

$$\gamma(k) = \frac{(-1)^k(-2d)!}{(k-d)!(-k-d)!}\sigma_\varepsilon^2 \tag{3.49a}$$

$$= \frac{(-1)^k\Gamma(1-2d)}{\Gamma(k-d+1)\Gamma(1-k-d)}\sigma_\varepsilon^2 \tag{3.49b}$$

The variance is obtained by setting $k = 0$ in (3.49a) or (3.49b):

$$\gamma(0) = \frac{(-2d)!}{[(-d)!]^2}\sigma_\varepsilon^2 \tag{3.50}$$

Note that the variance, $\gamma(0)$, is not finite for $d \geq 0.5$. Throughout it has been assumed that $\mu_t = 0$, otherwise y_t should be replaced by $y_t - \mu_t$.

3.2.1.ii   Autocorrelations

The lag k autocorrelations, $\rho(k) \equiv \gamma(k)/\gamma(0)$, are given by:

$$\rho(k) = \frac{(-d)!(k+d-1)!}{(d-1)!(k-d)!} \qquad k = 0, \pm 1, \ldots \tag{3.51a}$$

$$= \frac{d(1+d)\ldots(k-1+d)}{(1-d)(2-d)\ldots(k-d)} \tag{3.51b}$$

$$= \frac{\Gamma(1-d)\Gamma(k+d)}{\Gamma(d)\Gamma(k+1-d)} \tag{3.51c}$$

As $k \to +\infty$, then $\frac{\Gamma(k+d)}{\Gamma(k+1-d)} \to k^{(d)-(1-d)} = k^{2d-1}$, therefore:

$$\rho(k) \to \frac{\Gamma(1-d)}{\Gamma(d)}k^{2d-1} \tag{3.52}$$

As $\gamma(k)$ and $\rho(k)$ are related by a constant (the variance), they have the same order, namely $O_p(k^{2d-1})$.

The first-order autocorrelation coefficient for $d < 0.5$ is obtained by setting $k = 1$ in (3.51a), that is:

$$\rho(1) = \frac{(-d)!(d!)}{(d-1)!(1-d)!}$$

$$= d/(1-d) \tag{3.53}$$

Note that $\rho(1) \to 1$ as $d \to 0.5$. Application of (3.53) for $d > 0.5$ results in $\rho(k) > 1$, which is clearly invalid for an autocorrelation.

From (3.52), the behaviour of the sum of the autocorrelations will be determined by the term k^{2d-1}. Clearly, such a sum will not be finite if $d \geq 0.5$ because the exponent on k will be greater than 0, but is $d < 0.5$ a sufficient condition? The answer is no. The condition for its convergence can be established by the p-series convergence test (a special case of the integral test); write $k^{2d-1} = 1/k^{-(2d-1)}$, then convergence requires that $-(2d-1) > 1$, that is $d < 0$. (The case $d = 0$ is covered trivially as all $\rho(k)$ apart from $\rho(0)$ are zero.) It is this nonsummability aspect that is sometimes taken as the defining feature of 'long memory'.

3.2.1.iii Inverse autocorrelations

The inverse autocorrelations, $\rho(k)_{inv}$, are given by:

$$\rho(k)_{inv} = \frac{(d)!(k-d-1)!}{(-d-1)!(k+d)!}$$

$$= \frac{-d(1-d)\dots(k-1-d)}{(1+d)(2+d)\dots(k+d)}$$

$$= \frac{\Gamma(d+1)\Gamma(k-d)}{\Gamma(-d)\Gamma(k+d+1)} \tag{3.54}$$

As $k \to \infty$, $\frac{\Gamma(k-d)}{\Gamma(k+d+1)} \to k^{(-d)-(d+1)} = k^{-2d-1}$; thus,

$$\rho(k)_{inv} \to \frac{\Gamma(d+1)}{\Gamma(-d)} |k|^{-2d-1} \tag{3.55}$$

Note that the inverse autocorrelations are obtained by replacing d by $-d$ in the standard autocorrelations (hence $\rho(k|d)_{inv} = \rho(k|-d)$).

3.2.2 Graphical properties of some simple ARFIMA models

Figures 3.1a to 3.3b illustrate some of the properties of the fractional d process. The autocorrelations and sum of the autocorrelations are shown in Figures 3.1a and 3.1c for $d = 0.3$, 0.45, and in Figures 3.1b and 3.1d for $d = -0.3$, -0.45. Evidently, it is positive values of d that exhibit a slow decline in the autocorrelation function, $\rho(k)$, which are likely to characterise economic time series.

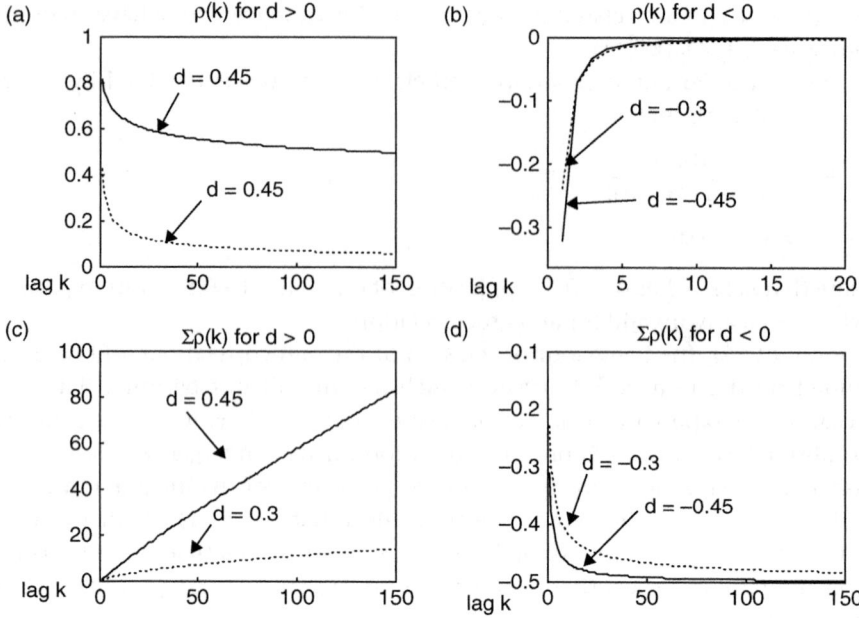

Figure 3.1   Autocorrelations for FI(d) processes

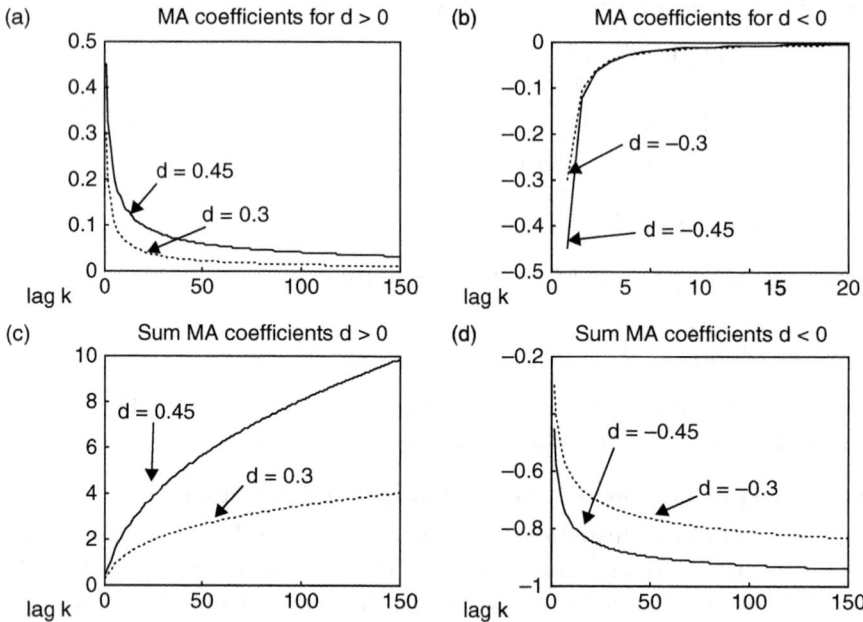

Figure 3.2   MA coefficients for FI(d) processes

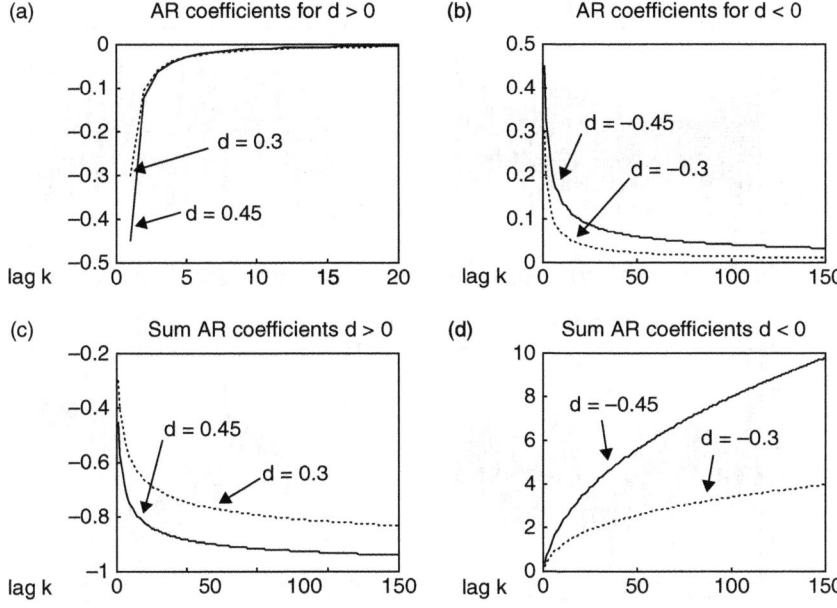

Figure 3.3   AR coefficients for FI(d) processes

The slow decline and non-summability is more evident as d moves toward the nonstationary boundary of 0.5. The autocorrelations for negative d are slight, decline quickly and are summable.

The corresponding MA and AR coefficients and their respective sums are shown in Figures 3.2a–3.2d, and Figures 3.3a–3.3d, respectively. These are plots of the $B_r^{(d)}$ coefficients from (3.15) and the $A_r^{(d)}$ coefficients from (3.14), respectively. Again it is evident from the positive and declining MA coefficients in Figure 3.2a that positive values of d are more likely to characterize economic times series than negative values of d, which, as Figure 3.2b shows, generate negative MA coefficients. Note also that the AR coefficients, and their respective sums, in Figures 3.3a to 3.3d are simply the negative of the corresponding MA coefficients in Figures 3.2a to 3.2d.

To give a feel of what time series generated by fractional d look like, simulated time series plots are shown in Figures 3.4a to 3.4d. In these illustrations $d = -0.3$ (to illustrate 'anti-persistence'), then $d = 0.3$ and $d = 0.45$ each illustrates some persistence, and $d = 0.9$ illustrates the nonstationary case. In addition, to mimic economic time series somewhat further, some serial correlation is introduced into the generating process; in particular, the model is now ARFIMA(1, d = 0.9, 0) with $\phi_1 = 0.3, 0.5, 0.7$ and 0.9 in Figures 3.5a to 3.5d respectively. As the degree of serial correlation increase, the series becomes 'smoother' and characterises well the typical features of economic time series.

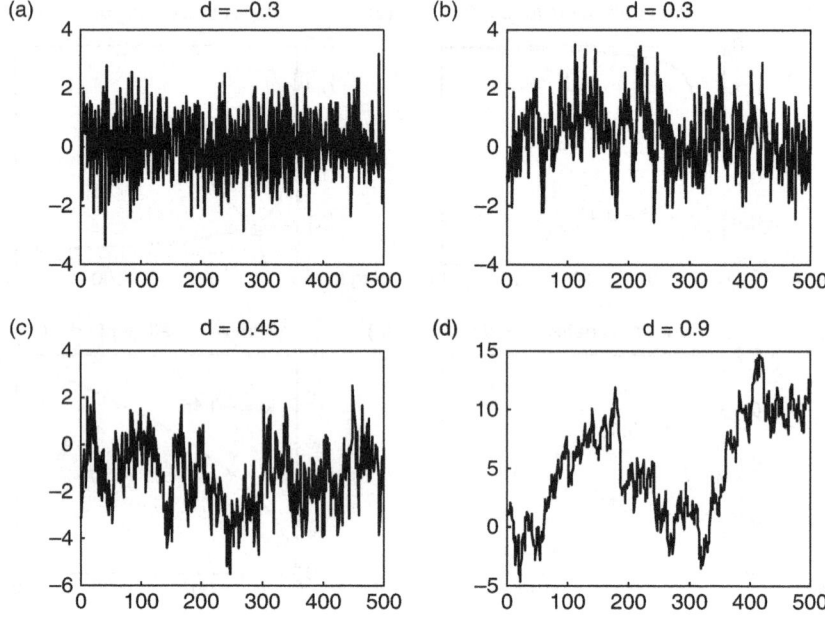

Figure 3.4   Simulated data for fractional d

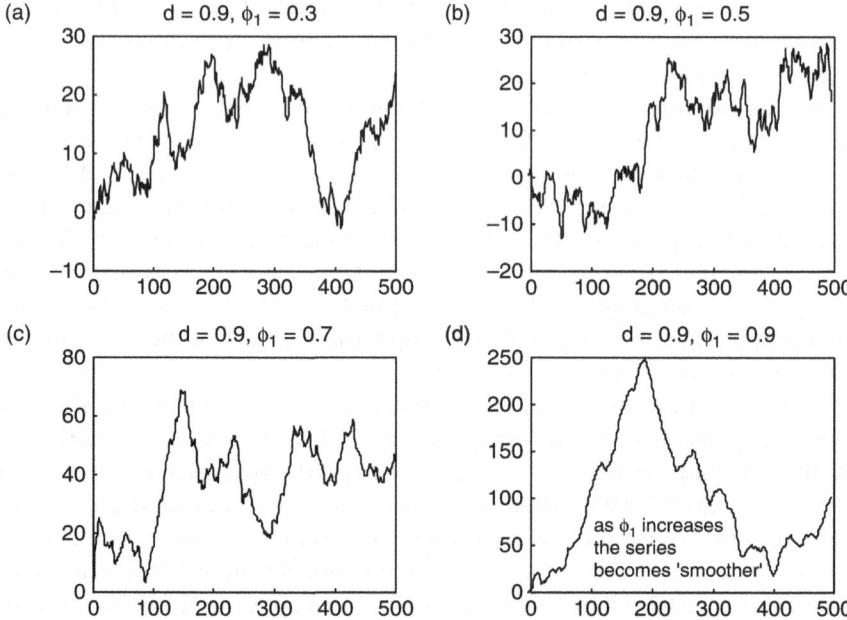

Figure 3.5   Simulated data for fractional d and serial correlation

3.3 What kind of models generate fractional d?

We outline two justifications for allowing d to be fractional; the first is due to Parke (1999) and relates to the error duration model and the second to Granger (1980), which relates to the aggregation of micro processes that are individually ARMA.

3.3.1 The error duration model (Parke, 1999)

The essence of the error duration model, EDM, is that there is a process generating stochastic shocks in a particular economic activity that have a lifetime, or duration, which is itself stochastic, and that the generic time series variable y_t is the sum of those shocks surviving to period t. The probability of a shock originating in period s surviving to period t, referred to as the survival probability, is critical in determining whether the process is one with inherently long memory (defined as having nonsummable autocovariances). Parke (op. cit.) gives two examples of the EDM. In the first, shocks to aggregate employment originate from the birth and death of firms; if a small proportion of firms have a very long life, then long memory is generated in aggregate employment. The second example concerns financial asset positions: once an asset position is taken, how long will it last? This is another example of a survival probability; if some positions, once taken, are held on a long term basis, there is the potential for a process, for example that of generating volatility of returns, to generate long memory. This line of thought generates a conceptualisation in which events generate shocks, for example productivity shocks, oil price shocks, stock market crashes, which in turn affect and potentially impart long memory into economic activity variables, such as output and employment. Equally, one can regard innovations as 'shocks' in this sense; for example, mobile phone technology, plasma television and computer screens, and then consider whether such shocks have a lasting effect on consumer expenditure patterns.

3.3.1.i The model

Let $\{\xi_t\}_1^\infty$ be a series of iid shocks with mean zero and finite variance σ^2. Each shock has a survival period; it originates in period s and survives until period $s + n_s$. The duration or survival period of the shock $n_s - (s - 1)$ is a random variable. Define a function $g_{s,t}$ that indicates whether a shock originating in period s survives to period t; thus, $g_{s,t} = 0$ for $t > s + n_s$, and $g_{s,t} = 1$ for $s \leq t < s + n_s$; the function is, therefore, a form of indicator function. The shock and its probability of survival are assumed to be independent. The probability that the shock survives until period s +k is $p_k = p(g_{s,s+k} = 1)$, which defines a sequence of probabilities $\{p_k\}$ for $k \geq 0$. For $k = 0$, corresponding to $g_{s,s} = 1$, assume $p_0 = 1$, so that a shock lasts its period of inception; also, assume that the sequence of probabilities $\{p_k\}$ is monotone non-increasing. The $\{p_k\}$ are usually referred to as the survival probabilities. A limiting case is when the process has no memory, so that $p_k = p(g_{s,s+k} = 1) = 0$ for $k \geq 1$. Otherwise, the process has a memory characterised by the sequence $\{p_k\}$, and the interesting question is the persistence of this memory, and does it qualify, in the sense of the McLeod-Hippel (1978) definition, to be described as long memory.

Finally, the realisation of the process at time is y_t, which is the sum of all the errors that survive until period t; that is:

$$y_t = \sum_{s=-\infty}^{t} g_{s,t} \xi_s \tag{3.56}$$

3.3.1.ii   *Motivation: the survival rate of firms*

A brief motivation, to be expanded later, may help this framework. The stock of firms in existence at any one time comprises firms of very different ages; some are relatively recent (new firms are 'born'), others range from the moderately lived through to those that have been in business a long time. Let each of these firms have an impact ξ_s (for simplicity assumed constant here) on total employment, y_t. At time t, total employment is the sum of present and past employment 'shocks' that have survived to time t. For example, $g_{10,t}\xi_{10}$ comprises the employment impact or shock that occurred at time s = 10, ξ_{10}, multiplied by the indicator function, $g_{10,t}$, which indicates whether that impact survived until period t. Now, whether total employment is best modelled as a short-or long-memory process depends upon the survival probabilities. If the survival probabilities decline slowly, to be defined precisely below, total employment, y_t, will have long memory in the sense used earlier in this chapter.

3.3.1.iii   *Autocovariances and survival probabilities*

The k-th autocovariance of y_t is $\gamma(k) = \sigma^2 \sum_{j=k}^{\infty} p_j$. Recall that a process is said to have long memory if the sequence of its autocovariances is not summable; that is, $\lim \sum_{k=-n}^{k=n} \gamma(k) \to \infty$ as $n \to \infty$. Normalising σ^2 to 1 without loss, the autocovariances, $\gamma(k)$, $k = 1, \ldots, n$, are given by a sequence of partial sums:

$$\gamma(1) = p_1 + p_2 + p_3 + \cdots$$
$$\gamma(2) = p_2 + p_3 + p_4 + \cdots$$
$$\vdots$$
$$\gamma(k) = p_k + p_{k+1} + p_{k+1} + \cdots \tag{3.57}$$

The pattern is easy to see from these expressions, and the sum of the autoco-variances, for $k = 1, \ldots, n$, is:

$$\sum_{k=1}^{n} \gamma(k) = p_1 + 2p_2 + 3p_3 \cdots$$

$$= \sum_{k=1}^{n} k p_k \tag{3.58}$$

Hence, the question of long memory concerns whether $\sum_{k=1}^{n} k p_k$ is summable. (Omitting $\gamma(0)$ from the sum does not affect the condition for convergence.) The survival probabilities are assumed not to increase; let them be characterised as:

$$p_k = c k^{\alpha} \tag{3.59}$$

where c is positive constant and $\alpha < 0$, such that $0 \leq c k^{\alpha} \leq 1$; $\alpha > 0$ is ruled out as otherwise $p_k > 1$ is possible for k large enough. (All that is actually required is that the expression for $p_k(k)$ in (3.59) holds in the limit for $n \to \infty$.)

The individual terms in $\sum_{j=1}^{n} k p_k$ are given by $k p_k = k(c k^{\alpha}) = c k^{(1+\alpha)}$. The series to be checked for convergence is, therefore:

$$\sum_{k=1}^{n} \gamma(k) = c + 2c(2^{\alpha}) + 3c(3^{\alpha}) + k c k^{\alpha} + \cdots$$

$$= c[1 + 2^{\alpha+1} + 3^{\alpha+1} + k^{\alpha+1} + \cdots]$$

$$= c[1 + 1/2^{-(1+\alpha)} + 1/3^{-(1+\alpha)} + 1/k^{-(1+\alpha)} + \cdots] \tag{3.60}$$

The series in [.] parentheses is well known (it is sometimes referred to as a p-series) and the condition for its convergence can again be established by the p-series convergence test, which requires that the exponent on k (as written in the last line of (3.60)) is greater than 1, that is $-(1 + \alpha) > 1$, which implies that $\alpha < -2$. The series is divergent, that is 'nonsummable', if $\alpha \geq -2$. We can write $\alpha = -2(1 - d)$, in which case the condition for convergence in terms of d is $d < 0$, otherwise the series is divergent; in particular, if $0 < d \leq 1$, then the series is not convergent; the right-hand side of this limit can also be extended, since the series continues to be nonsummable as d becomes increasingly positive. This implies that if the survival probabilities approach zero at or more slowly than k^{-2}, then long memory is generated. For example, if $d = 0.25$, then $\alpha = -1.5$, and the survival probabilities are $p_k(k) = c/k^{1.5}$, which generates long memory.

The first order autocorrelation of the error duration process is $\rho(1) = \lambda/(1 + \lambda)$ where $\lambda = \sum_{k=1}^{\infty} p_k$. Higher-order autocorrelations are also a function of λ, which, therefore, determine whether the autocorrelations are well defined. For $p_k = c k^{\alpha}$, the integral test can again be used to assess convergence; this time the exponent is $-\alpha$ and the condition for convergence is $\alpha < -1$, which implies $d < 0.5$ (from $\alpha = -2(1 - d)$). That is, the survival probabilities must approach zero at least as fast as k^{-1} for the autocorrelations to be defined. If $d \in [0.5, \infty)$, then λ is not finite and, therefore, y_t has long memory and is nonstationary.

This outline shows that the error duration model can generate a series y_t that is integrated of order d, that is I(d). Hence, applying the d-th difference operator $(1 - L)^d$ to y_t reduces the series to stationarity; that is $(1 - L)^d y_t = \varepsilon_t$, where ε_t is iid.

3.3.1.iv   *The survival probabilities in a long-memory process*

The survival probabilities for the I(d) error duration model with $0 < d \leq 1$ are:

$$p_k = \frac{\Gamma(k+d)\Gamma(2-d)}{\Gamma(k+2-d)\Gamma(d)} \tag{3.61}$$

with the following recursion starting from $p_0 = 1$:

$$p_{k+1} = \frac{k+d}{k+2-d} p_k \tag{3.62}$$

Note that the p_k in (3.61) satisfy (3.59) as a limiting property. The recursion shows that the ratio of successive probabilities p_{k+1}/p_k, referred to as the conditional survival probabilities, tends to one as k increases; thus, there is a tendency for shocks that survive a long time to continue to survive. Figure 3.6 plots the survival probabilities and the conditional survival probabilities for $d = 0.15$ and $d = 0.45$. The relatively slow decline in the survival probabilities is more evident in the latter case; in both cases $p_{k+1}/p_k \to 1$ as $k \to \infty$.

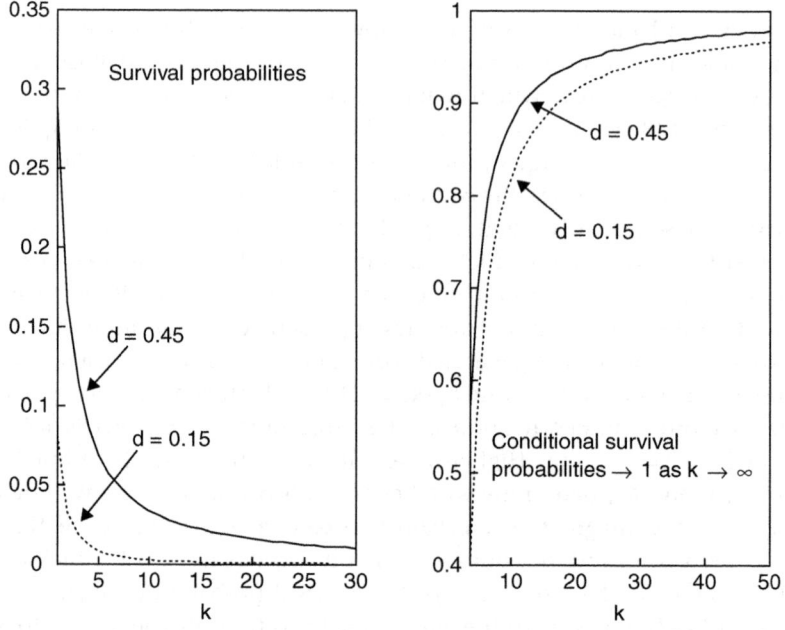

Figure 3.6   Survival and conditional survival probabilities

Table 3.1   Empirical survival rates S_k and conditional survival rates S_k/S_{k-1} for US firms

Year k $\Rightarrow$	1	2	3	4	5	6	7	8	9	10
S_k	0.812	0.652	0.538	0.461	0.401	0.357	0.322	0.292	0.266	0.246
S_k/S_{k-1}	0.812	0.803	0.826	0.857	0.868	0.891	0.902	0.908	0.911	0.923

Source: Parke (1999); based on Nucci's (1999) data, which is from 5,727,985 active businesses in the 1987 US Bureau of the Census Statistical Establishment List.

3.3.1.v   *The survival probabilities in a short-memory, AR(1), process*

It is useful to highlight the difference between the survival probabilities for a long-memory process and those for a short-memory process, taking the AR(1) model as an example of the latter. Consider the AR(1) model $(1 - \phi_1 L)y_t = \eta_t$ where η_t is iid and $0 \le \phi_1 < 1$. The autocorrelation function is $\rho(k) = \phi_1^k$, with recursion $\rho(k+1) = \phi_1 \rho(k)$. This matches the case where the conditional survival probabilities are constant, that is $p_{k+1}/p_k = \phi_1$ starting with $p_0 = 1$; thus, $p_2 = \phi_1 p_1$, $p_3 = \phi_1 p_2$ and so on. Hence, the survival probability p_k is just the k-th autocorrelation, $\rho(k)$. The AR(1) model with $\phi_1 = 0.5$ and the I(d) model with $d = 1/3$ have the same first order autocorrelation of 0.5, but the higher-order autocorrelations decline much more slowly for the I(d) model.

3.3.2   An example: the survival rate for US firms

Parke (1999) gives the following example of a long memory model. The empirical survival rate S_k for US business establishments is the fraction of start-ups in year 0 that survive to year k; S_k is the empirical counterpart of p_k. Table 3.1 gives these survival rates for nearly 6 million active businesses. For example, 81.2% of businesses last one year after starting, 65.2% last two years through to 24.6% of businesses lasting ten years after starting. Note from the row in the table headed S_k/S_{k-1} that the conditional survival rates increase, making the long-memory model a candidate for the survival process. The AR(1) model is not a candidate. The latter implies $p_{2k} = (p_k)^2$ but, taking S_k as an estimate of p_k, this is clearly not the case, especially as k increases; for example, $S_5 = 0.401$ would predict $S_{10} = 0.401^2 = 0.161$ compared with the actual $S_{10} = 0.246$.

To estimate d requires a specification of how the survival probabilities are generated. Consider the model $p_k = ck^\alpha$ as in (3.59), with $\alpha = -2(1 - d)$, either as an exact representation or as an approximation for a long-memory process; then the conditional survival probabilities j periods apart are:

$$p_k/p_{k+j} = ck^\alpha/c(k+j)^\alpha \tag{3.63}$$
$$= k^\alpha/(k+j)^\alpha$$

Hence, solving for α and then d by, taking logs and rearranging, we obtain:

$$\ln(p_k/p_{k+j}) = \ln\left(\frac{k}{(k+j)}\right)^{\alpha} \tag{3.64}$$

$$\Rightarrow \quad \alpha = \frac{\ln(p_k/p_{k+j})}{\ln[k/(k+j)]} \tag{3.65}$$

$$\Rightarrow \quad d = 1 + 0.5\frac{\ln(p_k/p_{k+j})}{\ln[k/(k+j)]} \tag{3.66}$$

Given data, as in Table 3.1, an estimate, $\hat{d}$, can be obtained by choosing k and j; for example, $k = 5$ and $j = 5$ gives $\hat{d} = 0.65$, indicating long memory and non-stationarity. With an estimate of d, an estimate of the scaling factor c can be obtained from $c = k^{\alpha}/p_k$ for a choice of k. For $\hat{d} = 0.65$ and $k = 10$, Parke obtains $\tilde{c} = 1.25$ by ensuring that $\tilde{p}_{10} = S_{10}$, so the estimated model is:

$$\tilde{p}_k = 1.25k^{-2+2(0.65)} \tag{3.67}$$

This estimated function is graphed in Figure 3.7, and fits the actual survival rates particularly well from S_5 onward, but not quite as well earlier in the sequence; and also note that $\tilde{p}_1 = 1.25(1)^{-2+2(0.65)} = 1.25 > 0$.

An alternative to using just one choice of k and j, which might reduce the problem of the estimated probabilities above one, is to average over all possible

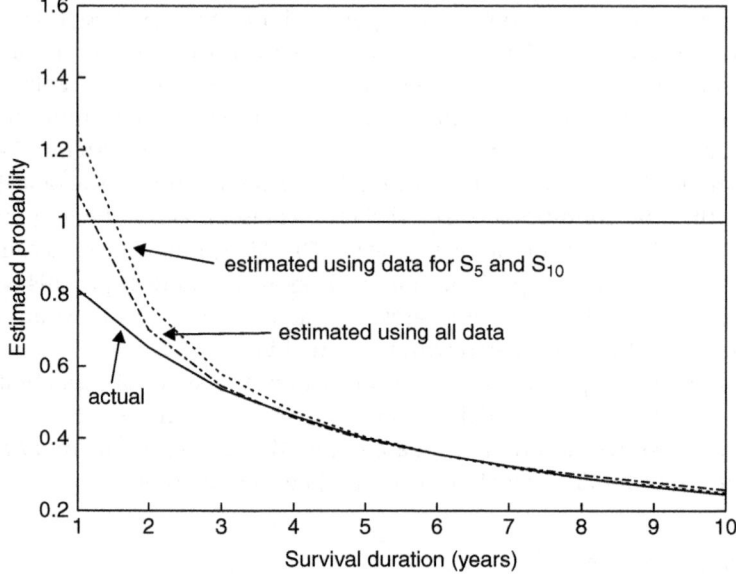

Figure 3.7   Survival rates of US firms: actual and fitted

choices. There are 9 combinations for $j = 1$, that is the set comprising the pairs $(1, 2)$, $(2, 3)$ ... $(9, 10)$; 8 combinations for $j = 2$ through to 1 combination for $j = 9$, that is the pair $(1, 10)$. In all there are $\sum_{j=1}^{9} j = 45$ combinations; evaluating these and averaging, we obtain $\hat{d} = 0.69$ and $\tilde{c} = 1.08$, so the fitted probability function is:

$$\tilde{p}_k = 1.08 k^{-2+2(0.69)} \tag{3.68}$$

This function fits the actual survival rates particularly well for S_3 onward; the problem with the starting probabilities has been reduced but not completely removed. The survival rates and fitted values from the revised estimate are also shown in Figure 3.7.

3.3.3 Error duration and micropulses

The generating method for a fractionally integrated process described by Parke is related to the generation of stochastic processes that are obtained as sums of micropulses (ε), a literature due to the ideas of Mandelbrot (for example Cioczek-Georges and Mandelbrot 1995, 1996). The micropulses, which we can interpret from an economic perspective as infinitesimal shocks, are generated by one of several stochastic processes; then, once generated, they have a random lifetime.

The use of the term micropulses reflects the eventual asymptotic nature of the argument, with $\varepsilon \rightarrow 0$ asymptotically, whereas the number of pulses $\rightarrow \infty$. In the simplest version the pulses are just 'up-and-down' or 'down-and-up' shocks, that is a shock of size $\varepsilon(-\varepsilon)$ at time t has a 'cancelling echo' of $-\varepsilon(\varepsilon)$ at $t + s$, where s is a random variable. The shocks appear at uniformly distributed points in time. The sum of these micropulses generates fractional Brownian motion. The generating process for the micropulses can be generalised to be more flexible than this starting model, for example, allowing the micropulses to have varying amplitude over their lifetime. These models seem quite promising for the prices of actively traded stocks, where the micropulses can be interpreted as increments of news that potentially have an impact on the stock price.

3.3.4 Aggregation

Granger (1980) showed that an I(d) process, for d fractional, could result from aggregating micro relationships that were not themselves I(d). Chambers (1998) has extended and qualified these results (and see also Granger, 1990). We deal with the simplest case here to provide some general motivation for the generation of I(d) processes; the reader is referred to Granger (1980, 1990) for detailed derivations and other cases involving dependent micro processes.

There are many practical cases where the time series considered in economics are aggregates of component series; for example, total unemployment aggregates both male unemployment and female unemployment, and each of these, in turn aggregates different age groups of the unemployed or unemployment by duration of the unemployment spell. Aggregate consumption expenditure is the result of summing expenditures by millions of households. Granger (1980) provided a model in which the AR(1) coefficients of the (stationary) micro relationships were drawn from a beta distribution and the resulting aggregate had the property of long memory even though none of the micro relationships had this property.

Note that this section involves some of the explanatory material on frequency domain analysis, especially spectral density functions, included in the next chapter; and the reader unfamiliar with frequency domain concepts will need to first review Sections **4.1.1** to **4.1.4**.

3.3.4.i   Aggregation of 'micro' relationships

First, consider two 'micro' time series comprising the sequences $\{y_{kt}\}_{t=1}^T$, $k = 1, 2$, each generated by a separate, stationary AR(1) process:

$$y_{kt} = \phi_{k1} y_{kt-1} + \varepsilon_{kt} \tag{3.69}$$

where the ε_{kt} are zero-mean, independent white noise 'shocks' (across the components and time). The aggregate variable, which in this case is just the sum of these two series, is denoted y_t, so that:

$$y_t = y_{1t} + y_{2t} \tag{3.70}$$

This model follows an AR(2, 1) process, with AR lag polynomial given by $(1 - \phi_{11}L)(1 - \phi_{21}L)$; see, for example, Granger and Newbold (1977). In general, the aggregation of N independent series of the kind (3.69) leads to an AR(N, N − 1) model; cancellation of the roots of the AR and MA polynomials can occur, which then reduces the order of the ARMA model.

In the case that the components y_{it} of the aggregate y_t are independent, the spectral density function (referred to by the shorthand sdf) of y_t is just the sum of the sdfs for each component. For the individual component, the sdf for an AR(1) process in the time domain is:

$$f_{y_k}(\lambda_j) = \frac{1}{|(1 - e^{-i\lambda_j})|^2} f_{\varepsilon_k}(\lambda_j) \tag{3.71a}$$

$$= \frac{1}{|(1 - e^{-i\lambda_j})|^2} \frac{\sigma_{\varepsilon_k}^2}{2\pi} \tag{3.71b}$$

where $f_{\varepsilon_k}(\lambda_j)$ is the sdf of the white noise input, ε_{kt}. Thus, the sdf for the aggregate variable is:

$$f_y(\lambda_j) = \sum_{k=1}^N f_{y_k}(\lambda_j) \tag{3.72}$$

In Granger's model, the AR(1) coefficients ϕ_{k1} are assumed to be the outcomes of a draw from a population with distribution function $F(\phi_1)$. To illustrate the argument, Granger (1980) uses the beta distribution, and whilst other distributions would also be candidates, the link between the beta distribution and the gamma distribution is important in this context.

3.3.4.ii   *The AR(1) coefficients are 'draws' from the beta distribution*

The beta distribution used in Granger (1980) is:

$$dF(\phi_1) = \frac{2}{B(u,v)} \phi_1^{2u-1}(1-\phi_1^2)^{v-1} d\phi_1 \tag{3.73}$$

where $0 \le \phi_1 \le 1$, u, $v > 0$, and $B(u, v)$ is the beta function with parameters u and v. The coefficients ϕ_{k1}, $k = 1$, 2, are assumed to be drawn from this distribution. The beta function is $B(u, v) = \int_0^1 \phi_1^{u-1}(1-\phi_1)^{v-1} d\phi_1 = 2\int_0^1 \phi_1^{2u-1}(1-\phi_1^2)^{v-1} d\phi_1$, which operates here to scale the distribution function so that its integral is unity over the range of ϕ_1. In this case, the k-th autocovariance, $\gamma(k)$, with $v > 1$, is given by:

$$\gamma(k) = \frac{B(u+k/2, v-1)}{B(u,v)} \tag{3.74a}$$

$$= \frac{\Gamma(v-1)}{B(u,v)} \frac{\Gamma(k/2+u)}{\Gamma(k/2+u+v-1)} \tag{3.74b}$$

Applying Sterling's theorem, the ratio $\Gamma(k/2+u)/\Gamma(k/2+(u+v-1))$ is $O_p(k^{1-v})$; note that the first term in each of these $\Gamma(.)$ functions is $(1/2)k$, so $(1/2)$ can be regarded as a constant in determining $O_p(.)$. Then:

$$\gamma(k) \to Ck^{1-v} \tag{3.75}$$

where C is a constant.

On comparison with (3.52), $1 - v = 2d - 1$, therefore $d = 1 - (1/2)v$; thus, the aggregate y_t is integrated of order $1 - (1/2)v$; for example, if $v = 1.5$, then $d = 0.25$. Typically, the range of interest for d in the case of aggregate time series is the long-memory region, $d \in (0, 1]$, which corresponds to $v \in (2,0]$, with perhaps particular attention in the range $v \in [1,0]$, corresponding to $d \in [0.5,1]$. Note that the parameter u does not appear in the order of integration, and it is v that is the critical parameter, especially for ϕ_1 close to unity, (see Granger 1980).

3.3.4.iii   *Qualifications*

Granger (1990) considers a number of variations on the basic model including changing the range of ϕ_1; introducing correlation in the ε_{it} sequences, and allowing the component series to be dependent; generally these do not change the essential results. However, Granger (1990) notes that if there is an upper bound to ϕ_1, say $\overset{\leftrightarrow}{\phi}_1 < 1$, so that $p(\phi_1 \ge \overset{\leftrightarrow}{\phi}_1) = 0$, then the fractional d result no longer holds. Chambers (1998) shows that linear aggregation, for example in the form of simple sums or weighted sums, does not by itself lead to an aggregate series with long memory. As Chambers notes (op.cit.), long memory in the aggregate requires the existence of long memory in at least one of the components, and the value of d for the aggregate will be the maximum value of d for the components. In this set-up, whilst the aggregation is linear, the values of ϕ_{k1} are not draws from a beta distribution.

3.4 Dickey-Fuller tests when the process is fractionally integrated

Given the widespread use of Dickey-Fuller tests, a natural question to ask is: what are their characteristics when the time series is generated by a fractional integrated process? It is appropriate to start with the LS estimator $\hat{\rho}$ in the following simple model:

$$y_t = \rho y_{t-1} + \varepsilon_t \tag{3.76}$$

However, in contrast, the generating process is a simple random walk with fractionally integrated noise, that is:

$$(1 - L)(y_t - y_0) = u_t \tag{3.77a}$$

$$(1 - L)^{\tilde{d}} u_t = \varepsilon_t \tag{3.77b}$$

with $y_0 = 0, -0.5 < \tilde{d} < 0.5 \Rightarrow \tilde{d} = d - 1$. The sequence $\{\varepsilon_t\}$ is iid with a zero mean and $E|\varepsilon_t|^k$, for $k \geq [4, -8d/(1 + 2d)]$; see Sowell (1990, theorem 2), who shows that appropriately normalised, the LS estimator $\hat{\rho}$ converges to 1. However, the normalisation to ensure a nondegenerate limiting distribution depends on the value of $\tilde{d}$.

Sowell (1990, theorem 3) shows that:

$$T(\hat{\rho} - 1) \Rightarrow_D \frac{\frac{1}{2}[B^{(1)}_{\tilde{d}}]^2}{\int_0^1 B^{(1)}_{\tilde{d}}(r)^2 dr} \qquad \text{for } \tilde{d} \in [0, 0.5) \tag{3.78}$$

$$T^{(1+2\tilde{d})}(\hat{\rho} - 1) \Rightarrow_D -\frac{(\tilde{d} + \frac{1}{2})}{\int_0^1 B^{(1)}_{\tilde{d}}(r)^2 dr} \frac{\Gamma(1 + \tilde{d})}{\Gamma(1 - \tilde{d})} \qquad \text{for } \tilde{d} \in (-0.5, 0) \tag{3.79}$$

where $B^{(1)}_{\tilde{d}}(r)$ is a standard (type I) fBM. (This statement of Sowell's results embodies the amendment suggested by Marinucci and Robinson (1999), which replaces type II fBM with type I fBM.)

Note that the normalisation to ensure the convergence of $\hat{\rho}$ depends on $\tilde{d}$, and in general one can refer to the limiting distribution of $T^{\min(1, 1+2\tilde{d})}(\hat{\rho} - 1)$. In the case that $\tilde{d} = 0$, one recovers the standard unit root case in the first of these expressions (3.78).

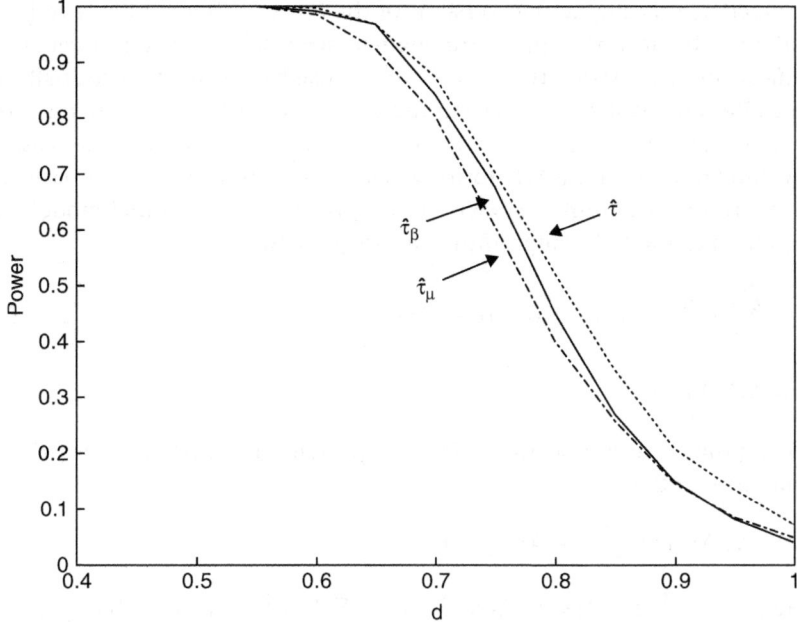

Figure 3.8   Power of DF tests against an FI(d) process

For the case $\tilde{d} \in (-0.5, 0.5)$, Sowell (op. cit., theorem 4) shows that the t statistic associated with the LS estimator $\hat{\rho}$, which in DF notation is referred to as $\hat{\tau}$, diverges in probability to ∞ if $\tilde{d} > 0$ and to $-\infty$ if $\tilde{d} < 0$. This result shows the consistency of the (simple) DF tests against fractionally integrated alternatives. (Dolado, Gonzalo and Mayoral, 2002 generalise this result to cover the asymptotically stationary and nonstationary cases in one theorem.) However, the rate of convergence of $\hat{\rho}$ is T for $\tilde{d} \in [0, 0.5)$, but $T^{(1+2\tilde{d})}$ for $\tilde{d} \in (-0.5, 0]$, and note that as $\tilde{d} \to -0.5$ from above, convergence is very slow. Hence, there are parts of the parameter space for which the power of the (simple) DF test $\hat{\tau}$ is likely to be small, a conjecture that is supported by the Monte Carlo results in Diebold and Rudebusch (1991).

The small sample power of the DF tests $\hat{\tau}$, $\hat{\tau}_\mu$ and $\hat{\tau}_\beta$ against fractional alternatives is illustrated in Figure 3.8 for $T = 200$, with 5,000 replications and a nominal size of 5% for each of the tests (using the DF critical values for $T = 200$). As anticipated, power increases as d declines from 1 toward zero, being 100% at about $d = 0.6$. Interestingly, and in contrast to the standard situation, including superfluous deterministic components does not uniformly reduce the power of the tests. (Power is not size-adjusted and $\hat{\tau}$ is slightly oversized, whereas $\hat{\tau}_\beta$ is slightly undersized.)

In practice, it is augmented versions of the DF tests that are likely to be used and the results from the simple (unaugmented) test do not carry over without qualification. The ADF $\hat{\tau}$-type test is not consistent against fractionally integrated alternatives if the lag length increases 'too fast' with T. This is a result due to Hassler and Wolters (1994), amended by Krämer (1998). The reason for likely difficulties with the ADF form of $\hat{\tau}$ can be seen from the AR representation of an FI(d) process. From (3.8c) with $u_t = \varepsilon_t$, the simple fractional model can be represented as the following infinite autoregression:

$$y_t = \sum_{r=1}^{\infty} \Lambda_r^{(d)} y_{t-r} + \varepsilon_t \qquad \text{where } \Lambda_r^{(d)} \equiv -A_r^{(d)} \tag{3.80}$$

$$= \Lambda_r^{(d)}(L)y_t + \varepsilon_t$$

Applying the DF decomposition of the lag polynomial to $\Lambda_r^{(d)}(L)$ (see *UR, Vol. 1*, chapter 3) we obtain:

$$\Delta y_t = \gamma_\infty y_{t-1} + \sum_{j=1}^{\infty} c_j \Delta y_{t-j} + \varepsilon_t \tag{3.81}$$

where $\gamma_\infty = [\Lambda_r^{(d)}(1) - 1] = \sum_{r=0}^{\infty} \Lambda_r^{(d)} - 1 = -\sum_{r=0}^{\infty} A_r^{(d)} - 1$. Note that $\sum_{r=1}^{\infty} A_r^{(d)} = -1$ for $d > 0$, (Hassler and Wolters, 1994), hence $\gamma_\infty = 0$ in the ADF(∞) representation; also note that one can interpret d as $d = 1 + \tilde{d}$.

In practice the ADF(∞) model is truncated to a finite number of lags, so that:

$$\Delta y_t = \gamma_k y_{t-1} + \sum_{j=1}^{k-1} c_j \Delta y_{t-j} + \varepsilon_{t,k} \tag{3.82}$$

$$\varepsilon_{t,k} = \left(\sum_{r=k+1}^{\infty} \Lambda_r^{(d)}\right) y_{t-1} + \sum_{j=k}^{\infty} c_j \Delta y_{t-j} + \varepsilon_t$$

$$= \sum_{r=k+1}^{\infty} \Lambda_r^{(d)} y_{t-r} + \varepsilon_t \tag{3.83}$$

$$\gamma_k = \sum_{r=1}^{k} \Lambda_r^{(d)} - 1$$

$$= \left(1 - \sum_{r=k+1}^{\infty} \Lambda_r^{(d)}\right) - 1 \tag{3.84}$$

(Note that (3.83) follows on comparison of (3.82) with (3.81).) Hence, γ_k will not, generally, be zero but will tend to zero as $k \to \infty$, since $\sum_{r=k+1}^{\infty} \Lambda_r^{(d)} \to 0$. This means that despite the generating process not being the one assumed under the null hypothesis, as k increases it would be increasingly difficult to discover this from the ADF regression. It was this result, supported by Monte Carlo simulations, that led Hassler and Wolters (1994) to the conclusion that the probability of rejection of the unit root null in favour of a fractional alternative decreases for the $\hat{\tau}$ type ADF test as k increases. An extract from their simulation results is shown in Table 3.2 below. For example, if $d = 0.6$ ($\tilde{d} = -0.4$) then the probability of rejection, whilst increasing with T, decreases with k from 86.6% for ADF(2) to

Table 3.2 Power of ADF $\hat{\tau}_\mu$ test for fractionally integrated series

	d = 1.0		d = 0.9		d = 0.6		d = 0.3	
	T = 100	T = 250	T = 100	T = 250	T = 100	T = 250	T = 100	T = 250
ADF(0)	5.1	4.6	13.8	19.9	92.7	99.9	100	100
ADF(2)	5.1	4.7	8.5	12.1	48.4	86.6	98.2	100
ADF(10)	4.6	4.5	4.8	6.2	9.4	28.8	22.9	85.2
ADF(12)	4.5	4.5	3.8	6.0	6.9	22.8	16.2	73.0

Source: Extracted from Hassler and Wolters, 1994, table 1.
Notes: The generating process is $(1 - L)^d y_t = \varepsilon_t$; the maintained regression is $\Delta y_t = \mu^* + \gamma_k y_{t-1} + \sum_{j=1}^{k-1} c_j \Delta y_{t-j} + \varepsilon_{t,k}$, with 5,000 replications.

28.8% for ADF(10) when T = 250. Hassler and Wolters (1994) note that the PP test, which is designed around the unaugmented DF test, does not suffer from a loss of power as the truncation parameter in the semi-parametric estimator of the variance increases.

Krämer (1998) points out that consistency can be retrieved if k, whilst increasing with T, does not do so at a rate that is too fast. A sufficient condition, where $\tilde{d} \in (-0.5, 0.5)$, to ensure divergence of $\hat{\tau}$ is $k = o(T^{1/2+\tilde{d}})$, in which case $\hat{\tau} \to_p -\infty$ for $\tilde{d} < 0$ and $\hat{\tau} \to_p \infty$ for $\tilde{d} > 0$. For further details and development see Krämer (1998). Of course, from a practical point of view, the value of $\tilde{d}$ is not known but it may be possible in some circumstances to narrow the range of likely values of $\tilde{d}$ in order to check that this condition is satisfied.

Given difficulties with the standard ADF test, if fractional integration is suspected a more sensible route is to use a test specifically designed to test for this feature. There are a number of such tests; for example one possibility is simply to estimate d and test it against a specific alternative. This kind of procedure is considered in the next chapter. In this chapter we first consider test procedures based on the general DF approach, but applied to the fractional case.

3.5 A fractional Dickey-Fuller (FDF) test for unit roots

This section introduces an extension of the standard Dickey-Fuller (DF) test for a unit root to the case of fractionally integrated series. This is a time series approach, whereas frequency domain approaches are considered in the next chapter. The developments outlined here are due to Dolado, Gonzalo and Mayoral (2002), hereafter DGM as modified by Lobato and Velasco (2007), hereafter LV. The fractional DF tests are referred to as the FDF tests and the efficient FDF, EFDF, tests.

3.5.1   FDF test for fractional integration

This section outlines the original fractional DF test. As both the FDF test and the EFDF test are based on the DF test, a brief reminder of the rationale of the latter will be useful. In the simplest DF case the maintained regression is:

$$\Delta y_t = \gamma y_{t-1} + z_t \tag{3.85}$$

$$z_t = \varepsilon_t$$

where $\varepsilon_t \sim iid(0, \sigma_\varepsilon^2)$ and $\gamma = (\rho - 1)$. If y_t is I(1), then the regression (3.85) is unbalanced in the sense that the orders of integration of the regressand and regressor are different, being I(0) and I(1) respectively. The LS (or indeed any consistent) estimator of γ should be zero asymptotically to reflect this lack of balance. The regression (3.85) is relevant to testing the null hypothesis of a unit root against the alternative of stationarity, that is, in the context of the notation of FI processes, $H_0 : d = 1$ against $H_A : d = 0$.

The essence of the FDF test is that the regression and testing principle can be applied even if y_t is fractionally integrated. Consider the maintained regression for the case in which H_0: $d = d_0$ against the simple (point) alternative H_A: $d = d_1 < d_0$; then, by analogy, the following regression model could be formulated:

$$\Delta^{d_0} y_t = \gamma \Delta^{d_1} y_{t-1} + z_t \tag{3.86}$$

The superscripts on Δ refer to the value of d under the null and the (simple) alternative hypothesis(es) or under specific non-unit root alternatives, respectively, and the properties of z_t have yet to be specified. In the standard DF case, the null hypothesis is $d_0 = 1$ and the (simple) alternative is $d_A = 0$; thus, under the null $\Delta^{d_0} y_t = \Delta^1 y_t = \Delta y_t$ and $\Delta^{d_1} y_{t-1} = y_{t-1}$, resulting in the DF regression $\Delta y_t = \gamma y_{t-1} + \varepsilon_t$ (so that in this special case $z_t = \varepsilon_t$). However, in the FDF case, neither value of d is restricted to an integer.

DGM suggest using Equation (3.86), or generalisations of it, to test $H_0 : d = d_0$ against the simple (point) alternative H_A : $d = d_1 < d_0$ or the general alternative H_A : $d < d_0$. The DF-type test statistics are constructed in the usual way as, for example, with an appropriately normalised version of $\hat{\gamma} = (\hat{\rho} - 1)$ or the t statistic on $\hat{\gamma}$, for the null that $\gamma = 0$, denoted $\hat{t}_y$. In the case that a simple alternative is specified, the value of $d = d_1$ under the alternative is known and can be used as in (3.86); however, the more general (and likely) case is $H_A : 0 \leq d < d_0$, or some other composite hypothesis, and a (single) input value of d, say d_1 for simplicity of notation, is required to operationalise the test procedure. In either case LS estimation of (3.86) gives the following:

$$\hat{\gamma}(d_1) = \frac{\sum_{t=2}^{T} \Delta^{d_0} y_t \Delta^{d_1} y_{t-1}}{\sum_{t=2}^{T} (\Delta^{d_1} y_{t-1})^2} \tag{3.87}$$

$$t_\gamma(d_1) = \frac{\hat{\gamma}}{\tilde{\sigma}_{d_1}(\hat{\gamma})} \tag{3.88}$$

$$\tilde{\sigma}_{d_1}(\hat{\gamma}) = \left(\tilde{\sigma}_{d_1}^2 / \sum_{t=2}^{T} (\Delta^{d_1} y_{t-1})^2\right)^{1/2}$$

$$\tilde{\sigma}_{d_1}^2 = \sum_{t=2}^{T} \tilde{\varepsilon}_t^2 / T$$

$$\tilde{\varepsilon}_t = \Delta^{d_0} y_t - \hat{\gamma}\Delta^{d_1} y_{t-1}$$

In the special case where $d_0 = 1$, then $\hat{\gamma}(d_1)$ is given by:

$$\hat{\gamma}(d_1) = \frac{\sum_{t=2}^{T} \Delta y_t \Delta^{d_1} y_{t-1}}{\sum_{t=2}^{T} (\Delta^{d_1} y_{t-1})^2} \tag{3.89}$$

d_1 is usually an estimator that is consistent under H_A and is considered in greater detail below.

3.5.2   A feasible FDF test

To make the FDF test feasible, the series under the alternative defined by $\Delta^{d_1} y_t$ has to be generated. The variable $\Delta^{d_1} y_t$ is computed using the coefficients in the expansion of $(1 - L)^{d_1}$, as in Equation (3.3). (For the Present the binomial expansion is assumed to be infinite; finite sample considerations are taken up below.) That is:

$$\Delta^{d_1} y_t = \sum_{r=0}^{\infty} A_r^{(d_1)} y_{t-r} \tag{3.90}$$

where $A_0^{(d_1)} = 1$, $A_r^{(d_1)} \equiv (-1)^r {}_{d_1}C_r$. (The case of interest is assumed to relate to H_0: $d = d_0 = 1$, otherwise if $d_0 \neq 1$, then $\Delta^{d_0} y_t$ is generated in an analogous way.)

In practice, the infinite sum in the expansion of Δ^{d_1} is truncated and the 'observed' series is generated as:

$$(1 - L)^{d_1} y_t = \sum_{r=0}^{t-s-1} A_r^{(d_1)} y_{t-r} \tag{3.91}$$

where s is the integer part of $(d_1 + \frac{1}{2})$, which reflects the assumption that pre-sample values of y_t are zero. (DGM, op. cit., report that this initialisation, compared to the alternative that the sequence $\{y_t\}_{-\infty}^{t=0}$ exists, has only minor effects on the power of the test; see also Marinucci and Robinson, 1999, for discussion of the general point.)

3.5.3   Limiting null distributions

The distributions of $\hat{\gamma}(d)$ and $t_\gamma(d)$ depend on the value of $d = d_1$ and whether d_1 is known, or estimated as in the composite alternative $d < d_0$. Since the more usual case is the composite alternative, a summary of the distributions in the case of known d_1 is contained in Appendix 3.2.

For the case of interest here, there are two key results as follows that inform subsequent sections. Let the data be generated by a random walk with iid errors:

$$\Delta y_t = \varepsilon_t \tag{3.92}$$

$$\varepsilon_t \sim \text{iid}(0, \sigma_\varepsilon^2) \qquad E|\varepsilon_t^4| < \infty$$

1. In the first case, note that for $d_1 \in [0.5, 1.0]$:

$$t_\gamma(d_1) \Rightarrow_D N(0, 1) \tag{3.93}$$

   See DGM, op. cit., theorem 2, (25). Testing $H_0: d = d_0 = 1$ against $H_A: d < d_0$ with $t_\gamma(d_1)$ uses left-sided critical values and rejection follows from 'large' negative values of $t_\gamma(d_1)$.
2. The second case concerns the typical case in practice, where the 'input' value $d_1 = \hat{d}_T$, and $\hat{d}_T$ is a consistent estimator of d, with convergence rate T^κ, $\kappa > 0$; if $\hat{d}_T \geq 1$, then it is trimmed to $1 - c$ where $c \approx 0$, but satisfies $c > 0$, for example DGM select $c = 0.02$ in their simulation experiments. Then, it is also the case that the limiting distribution is standard normal:

$$t_\gamma(\hat{d}_T) \Rightarrow_D N(0, 1) \tag{3.94}$$

   see DGM, op. cit., theorem 5. Thus provided that the range of $\hat{d}_T$ is restricted (although not unduly so), standard $N(0, 1)$ testing procedures apply.

3.5.4 Serially correlated errors: an augmented FDF, AFDF

For the case of serially correlated errors, DGM suggested an augmented FDF regression along the lines of an ADF regression. First, consider a development analogous to the simple case, but here for $y_t \sim \text{FI}(d)$ with an AR(p) error, that is:

$$\varphi(L)\Delta^d y_t = \varepsilon_t \qquad \varphi(L) = 1 - \sum_{j=1}^{p} \varphi_j L^j$$

$$\Rightarrow$$

$$\varphi(L)\Delta y_t = \Delta^{1-d}\varepsilon_t$$

$$= \varepsilon_t + (d-1)\varepsilon_{t-1} + 0.5d(d-1)\varepsilon_{t-2} + \cdots$$

$$= (d-1)\varphi(L)\Delta^d y_{t-1} + z_t \tag{3.95}$$

$$z_t = \varepsilon_t + 0.5d(d-1)\varepsilon_{t-2} + \cdots \tag{3.96}$$

The second line of (3.95) uses the binomial expansion of $\Delta^{1-d} \equiv (1-L)^{1-d}$; see Equations (3.3) to (3.5). The third line of (3.95) uses $\varphi(L)\Delta^d y_{t-1} = \varepsilon_{t-1}$. Substituting for $\varphi(L)$ in (3.95) and using the BN decomposition

$\varphi(L) = \varphi(1) + \varphi^*(L)\Delta$ results in:

$$\varphi(L)\Delta y_t = (d-1)[\varphi(1) + \varphi^*(L)\Delta]\Delta^d y_{t-1} + z_t$$

$$\Rightarrow$$

$$\Delta y_t = (d-1)\varphi(1)\Delta^d y_{t-1} + [1 - \varphi(L)]\Delta y_t + z_t^* \tag{3.97a}$$

$$= \gamma\Delta^d y_{t-1} + [1 - \varphi(L)]\Delta y_t + z_t^* \tag{3.97b}$$

$$= \gamma\Delta^d y_{t-1} + \sum_{j=1}^{p} \varphi_j \Delta y_{t-j} + z_t^* \tag{3.97c}$$

$$\gamma = (d-1)\varphi(1) \tag{3.97d}$$

$$z_t^* = z_t + (d-1)\varphi^*(L)\Delta^{1+d} y_{t-1} \tag{3.97e}$$

Note that (3.97a) follows on adding $[1 - \varphi(L)]\Delta y_t$ to both sides of the previous equation, where $1 - \varphi(L) = \sum_{j=1}^{p} \varphi_j L^j$. Also note that if $d = 1$, then the terms in $(d-1)$ disappear and, therefore, $\varphi(L)\Delta y_t = \varepsilon_t$; if $d = 0$, then $\varphi(L)y_t = \varepsilon_t$ (see Q3.5); also if $\varphi(1) > 0$, then γ becomes more negative as d increases. Note also that the case with $\varphi(1) = 1 - \sum_{j=1}^{p} \varphi_j = 0$ is ruled out because it implies a unit root (or roots) in $\varphi(L)$, and $\varphi(1) > 0$ implies the stability condition $\sum_{j=1}^{p} \varphi_j < 1$. This, as in the simple case, motivates the estimation of (3.97c) and use of the t-statistic on $\hat{\gamma}$. (However, as in the simple case, LS estimation is inefficient because of the serially correlated property of z_t^*; see the EFDF test below.)

In the DGM procedure, d is (generally) replaced by d_1, and the regression to be estimated, the AFDF(p) regression, is:

$$\Delta y_t = \gamma\Delta^{d_1} y_{t-1} + \sum_{j=1}^{p} \varphi_j \Delta y_{t-j} + z_t^*(d_1) \tag{3.98}$$

The suggested test statistic is the t statistic on γ and, in practice, $d_1 = \hat{d}_T$ in (3.98). The critical values are left-sided (which assumes that $\varphi(1) > 0$). Under the null hypothesis that y_t is generated by an ARIMA(p, 1, 0) process, the limiting distributions of the AFDF t-type statistics, $t_\gamma(d_1)$ and $t_\gamma(\hat{d}_T)$, are the same as their counterparts in the iid case, (DGM, op. cit., theorem 6); see (3.93) and (3.94) respectively, and see also Appendix 3.2, Table A3.1 for known $d = d_1$.

So far p has been assumed known, whereas the more likely case is that the correct lag length is unknown and k, rather than p, is the selected lag length. In the context of the standard ADF(k) $\hat{\tau}$-type test, a key result due to Said and Dickey (1984), states that if the errors follow a stationary invertible ARMA process, then a sufficient condition to obtain the same limiting null distribution and the consistency of the pseudo t test is that $k \to \infty$ as $T \to \infty$, such that $k^3/T \to 0$. DGM (op. cit., theorem 7) show that this result also applies to the AFDF(k) regression. For example, as to consistency, let the DGP be: $\Delta^d y_t = u_t 1_{(t>0)}$, $d \in [0, 1)$ and $u_t = \phi(L)^{-1}\theta(L)\varepsilon_t$ is a stationary process; if $k \to \infty$ as $T \to \infty$, such that $k^3/T \to 0$, then the t-type test on $\hat{\gamma}$, $t_\gamma(d_1) \to -\infty$ and is, therefore, consistent.

Replacement of d_1 with $\hat{d}_T$, where $\hat{d}_T$ is a consistent estimator, does not alter the consistency of the test. We also discuss later (see section **3.8.3**) an alternative formulation of the augmented FDF test, the pre-whitened AFDF, due to Lobato and Velasco (2006), which considers the optimal choice of d_1.

3.5.5   An efficient FDF test

LV have suggested a simple modification of the DGM framework, which leads to a more efficient test. To understand this development start with the simple FI(d) process:

$$\Delta^d y_t = \varepsilon_t 1_{(t>0)} \tag{3.99}$$

The start-up condition $\varepsilon_t 1_{(t>0)}$ indicates that the process is initialized at the start of the sample; (which, incidentally, identifies the resulting fractional Brownian motion as type II). Where the context is clear, this initialization will be left implicit. The usual assumptions on ε_t are that it has finite fourth moment and the sequence $\{\varepsilon_t\}$ is iid.

Now note that by a slight rearrangement (3.99) may be written as follows:

$$\Delta^{d-1}\Delta y_t = \varepsilon_t \Rightarrow$$

$$\Delta y_t = \Delta^{1-d}\varepsilon_t$$

$$= \varepsilon_t + (d-1)\varepsilon_{t-1} + 0.5d(d-1)\varepsilon_{t-2} + \cdots$$

$$= (d-1)\Delta^d y_{t-1} + z_t \tag{3.100}$$

The third line uses $\varepsilon_{t-1} = \Delta^d y_{t-1}$ by one lag of Equation (3.99) and note that:

$$z_t = \varepsilon_t + 0.5d(d-1)\varepsilon_{t-2} + \cdots \tag{3.101}$$

In effect, Equation (3.100) is just a more explicit statement of Equation (3.86) due to DGM, with $d_0 = 1$, $d = d_1$ and z_t spelled out by Equation (3.101); also note that $\gamma = (d-1) = (d_1 - 1)$ so that $\gamma < 0$ for $d_1 < 1$, justifying the use of the t statistic $t_\gamma(d_1)$, or $\hat{t}_\gamma(\hat{d}_T)$, with 'large' negative values leading to rejection of H_0: $d = d_0 = 1$ against H_A: $d = d_1 \in [0.5, 1)$ or $d_1 \in [0, 1)$. However, whilst z_t is orthogonal to $\Delta^d y_{t-1}$ it is serially correlated and so LS in the context of (3.100) is a consistent but not an efficient estimation method.

LV suggested the following way around this problem. Adding and subtracting Δy_t to Equation (3.99), and rearranging, gives:

$$\Delta y_t \equiv \Delta y_t - \Delta^d y_t + \varepsilon_t$$

$$\equiv (1 - \Delta^{d-1})\Delta y_t + \varepsilon_t$$

$$\equiv -(\Delta^{d-1} - 1)\Delta y_t + \varepsilon_t \tag{3.102}$$

This is, so far, just a restatement of an FI(d) process, and though in that sense trivial it maintains ε_t as the 'error', in contrast to the serially correlated z_t in Equation (3.100). It may be interpreted as a testing framework in the following way. First, introduce the coefficient π on the variable $(\Delta^{d-1} - 1)\Delta y_t$, which then enables a continuum of cases to be distinguished, including $d = 0$ and $d = 1$. That is:

$$\Delta y_t = \pi(\Delta^{d-1} - 1)\Delta y_t + \varepsilon_t \tag{3.103}$$

If $\pi = 0$, then $\Delta y_t \equiv \varepsilon_t$ so that $d = 1$, whereas if $\pi = -1$ then $\Delta^d y_t \equiv \varepsilon_t$ for $d \neq 1$.

Although the variable $(\Delta^{d-1} - 1)\Delta y_t$ appears to include Δy_t, this is not the case. Specifically from an application of Equation (3.3) note that $\Delta^{d-1}\Delta y_t = \sum_{r=0}^{t-1} A_r^{(d-1)}\Delta y_{t-r}$, where $A_0^{(d-1)} \equiv 1$, $A_r^{(d-1)} \equiv (-1)^r{}_{d-1}C_r$, and, therefore:

$$(1 - \Delta^{d-1})\Delta y_t = \Delta y_t - [1 - (d-1)L + A_2^{(d-1)}L^2 + \cdots]\Delta y_t$$

$$= (d-1)\Delta y_{t-1} - \sum_{r=2}^{t-1} A_r^{(d-1)}L^r \Delta y_t \tag{3.104}$$

The development leading to (3.103), suggests that it would be possible to nest the generating process, assuming that d is not known, in a suitable generalisation of (3.103). Consider replacing d by $d_1 \neq 1$, and let $d^+ = d_1 - d$; then multiply both sides of (3.99) through by Δ^{d^+}, so that:

$$\Delta^{d_1} y_t = \Delta^{d^+} \varepsilon_t \tag{3.105}$$

Then following the steps leading to (3.103) results in:

$$\Delta y_t = \pi(\Delta^{d_1-1} - 1)\Delta y_t + v_t \tag{3.106}$$

$$v_t = \Delta^{d^+} \varepsilon_t$$

If $d_1 = d$, then $v_t = \varepsilon_t$, otherwise $v_t \neq \varepsilon_t$ and is serially correlated. This differs from (3.95) and (3.96) because z_t is serially correlated even for the correct value of d, whereas this is not the case for v_t.

A test of $H_0 : d = d_0 = 1$ against $H_A : d < 1$ could be obtained by first constructing $(\Delta^{d_1-1} - 1)\Delta y_t$ using the value of d_1 in (3.106), with a suitable truncation to the beginning of the sample, and then regressing Δy_t on $(\Delta^{d_1-1} - 1)\Delta y_t$ and using a left-sided t test. In this formulation, the use of $d_1 = 1$ is ruled out because $(\Delta^0 - 1)\Delta y_t = 0$; and the coefficients in (3.106) using d_1, for example, $(d_1 - 1)$, change sign for $d_1 > 1$, so that such a left-sided test is not appropriate for $d_1 > 1$. To avoid the discontinuity at $d_1 = 1$, LV suggest scaling the regressor by $(1 - d_1)$, so that the regression model is as follows:

$$\Delta y_t = \eta z_{t-1}(d_1) + v_t \tag{3.107}$$

$$z_{t-1}(d_1) \equiv \frac{(\Delta^{d_1-1} - 1)}{(1 - d_1)} \Delta y_t \tag{3.108}$$

$$\eta = \pi(1 - d_1) \tag{3.109}$$

$$v_t = \Delta^{d^+} \varepsilon_t \tag{3.110}$$

This set-up, therefore, differs from that of DGM because the suggested regressor is $z_{t-1}(d_1)$ and not $\Delta^{d_1} y_{t-1}$. The problem as $d_1 \to 1$ is solved because applying L'Hôpital's rule results in $(\Delta^{d_1-1} - 1)/(1 - d_1) \to -\ln(1 - L) = \sum_{j=1}^{\infty} j^{-1} L^j$. Considering this limiting case, the regression model becomes:

$$\Delta y_t = \eta \sum_{j=1}^{\infty} j^{-1} \Delta y_{t-j} + v_t \tag{3.111}$$

In this form, the resulting t test based on η is asymptotically equivalent to Robinson's (1991) LM test formulated in the time domain (see also Tanaka, 1999).

Considering (3.107), if $d_1 = d$, then $\eta = \pi(1 - d) = (d - 1)$. Thus, a test of a unit root null against a fractionally integrated alternative can be based on the departure of η from 0, becoming more negative as d differs from 1, which is the value under the unit root null. If $d = 1$, as under H_0, then $\eta = 0$ whatever the value d_1 and, hence, $\Delta y_t = \varepsilon_t$; whereas $\eta = (d - 1) < 0$ under H_A and $\Delta^d y_t = \varepsilon_t$, thus $y_t \sim FI(d)$.

The proposed test statistic is analogous to the FDF set-up, the t test on η, denoted $t_\eta(d_1)$ for the simple alternative or $t_\eta(\hat{d}_T)$ for the composite alternative with input $d_1 = \hat{d}_T$. Note that in practice there is no constraint to choose d_1 in the EFDF test to be the same as that in the FDF test, although that is one possibility. Large negative values of the test statistic lead to rejection of $H_0 : d = d_0 = 1$ against the left-sided alternative (either simple or composite). This test will be referred to as the efficient FDF (EFDF) test.

3.5.6 EFDF: limiting null distribution

The limiting null distribution for $t_\eta(\hat{d}_T)$ is as follows. Let the data be generated as in (3.99), under the null hypothesis $d = 1$, then:

$$t_\eta(\hat{d}_T) \Rightarrow_D N(0, 1) \tag{3.112}$$

$$\hat{d}_T = d_1 + o_p(T^{-\kappa}), \ \hat{d}_T > 0.5, \ d_1 > 0.5, \ \kappa > 0, \ \hat{d}_T \neq 1 \tag{3.113}$$

See LV, theorem 1, who show that the test statistic(s) is, again, normally distributed in large samples.

3.5.7 Serially correlated errors: an augmented EFDF, AEFDF

The more realistic DGP rather than a simple FI(d) process is a form of the ARFIMA model. LV consider the ARFIMA(p, d, 0) case, in which the model is as follows:

$$\Delta^d y_t = z_t$$

$$\varphi(L)z_t = \varepsilon_t 1_{(t>0)}$$

$$\Rightarrow$$

$$\varphi(L)\Delta^d y_t = \varepsilon_t 1_{(t>0)} \tag{3.114}$$

where $\varphi(L) = 1 - \varphi_1 L - \cdots - \varphi_p L^p$, with all roots outside the unit circle. Proceeding as in the simple case, but here adding and subtracting $\varphi(L)\Delta^{d_0} y_t$ and rearranging, results in:

$$\Delta^{d_0} y_t = \varphi(L)(1 - \Delta^{d-d_0})\Delta^{d_0} y_t + [1 - \varphi(L)]\Delta^{d_0} y_t + \varepsilon_t \tag{3.115}$$

If $d_0 = 1$, then this becomes:

$$\Delta y_t = \varphi(L)(1 - \Delta^{d-1})\Delta y_t + [1 - \varphi(L)]\Delta y_t + \varepsilon_t \tag{3.116}$$

As in the simple case, introduce the coefficient π such that if $\pi = 0$, then $\varphi(L)\Delta y_t = \varepsilon_t$, and if $\pi = -1$, then $\varphi(L)\Delta^d y_t = \varepsilon_t$, so that:

$$\Delta y_t = \pi[\varphi(L)(\Delta^{d-1} - 1)]\Delta y_t + [1 - \varphi(L)]\Delta y_t + \varepsilon_t \tag{3.117}$$

Next proceed along lines identical to Equation (3.107), but using the scaled regressors $\varphi(L)z_{t-1}(d_1)$ rather than $\varphi(L)(\Delta^{d_1-1} - 1)\Delta y_t$, so that:

$$\Delta y_t = \eta[\varphi(L)z_{t-1}(d_1)] + [1 - \varphi(L)]\Delta y_t + v_t \tag{3.118}$$

As in the simple case, the proposed test statistic is the t test associated with η, $t_\eta(d_1)$, or using an estimated value $\hat{d}_T$, $t_\eta(\hat{d}_T)$, which will be referred to as the efficient augmented FDF (EAFDF) test.

The regression (3.118) is nonlinear in the coefficients because of the multiplicative form $\eta\varphi(L)$, with the $\varphi(L)$ coefficients also entering through the second set of terms, although in the first set they are associated with $d = d_1$ and in the second set with $d = d_0$. There are several ways to deal with this. First, it is fairly straightforward to estimate the nonlinear form; for example suppose that $p = 1$, so that $\varphi(L) = 1 - \varphi_1 L$ and, therefore, $1 - \varphi(L) = \varphi_1 L$; and using a consistent estimator $\hat{d}_T$ of d, then:

$$\Delta y_t = \eta(1 - \varphi_1 L)z_{t-1}(\hat{d}_T) + \varphi_1 L\Delta y_t + \varepsilon_t$$

$$= \eta z_{t-1}(\hat{d}) - \eta\varphi_1 z_{t-2}(\hat{d}_T) + \varphi_1 \Delta y_{t-1} + \varepsilon_t$$

$$= \alpha_1 z_{t-1}(\hat{d}_T) + \alpha_2 z_{t-2}(\hat{d}_T) + \alpha_3 \Delta y_{t-1} + \varepsilon_t \tag{3.119}$$

where $\alpha_1 = -\alpha_2/\alpha_3$.

Alternatively, LV suggest first estimating the coefficients in $1 - \varphi(L) = \sum_{j=1}^{p} \varphi_j L^j$ using $\hat{d}_T$, that is from the p-th order autoregression in $\Delta^{\hat{d}_T} y_t$:

$$\Delta^{\hat{d}_T} y_t = \sum_{j=1}^{p} \varphi_j \Delta^{\hat{d}_T} y_{t-j} + \xi_t \qquad \text{first step} \qquad (3.120)$$

Then use the estimates $\hat{\varphi}_j$ of φ_j in the second stage to form $\eta[\hat{\varphi}(L) z_{t-1}(\hat{d}_T)]$:

$$\Delta y_t = \eta[\hat{\varphi}(L) z_{t-1}(\hat{d}_T)] + \sum_{j=1}^{p} \varphi_j \Delta y_{t-j} + \hat{v}_t \qquad \text{second step} \qquad (3.121)$$

$$z_{t-1}(\hat{d}_T) = \frac{(\Delta^{\hat{d}_T - 1} - 1)}{(1 - \hat{d}_T)} \Delta y_t \qquad (3.122)$$

Notice that the coefficients $\hat{\varphi}_j$ are not used in the second set of terms as, in this case, they relate to the null value $d_0 = 1$, with $\Delta^{d_0} y_{t-j} = \Delta y_{t-j}$. As before, the test statistic is the t test, $t_\eta(\hat{d}_T)$, on the η coefficient. Also DGM suggest a one-step procedure, for details see DGM (2008, section 3.2.2 and 2009, section 12.6).

3.5.8   Limiting null distribution of $t_\eta(\hat{d}_T)$

Let the data be generated by $\varphi(L) \Delta y_t = \varepsilon_t$ with ε_t as specified in (3.99), where $\varphi(L)$ is a p-th order lag polynomial with all roots outside the unit circle, $\hat{\varphi}(L)$ is estimated as in (3.120) (or DGM's alternative one-step procedure or nonlinear estimation) and $\hat{d}_T$ is specified as in (3.113), then (under the same conditions as for the simple case):

$$t_\eta(\hat{d}_T) \Rightarrow_D N(0, 1)$$

see LV (op. cit., theorem 2).

3.6   FDF and EFDF tests: deterministic components

Note that the DGP so far considered has assumed that there are no deterministic components in the mean. In practice this is unlikely. As in the case of standard DF tests, the leading specification in this context is to allow the fractionally integrated process to apply to the deviations from a deterministic mean function, usually a polynomial function of time, as in Equation (3.42). For example, the ARFIMA(p, d, q) model with mean or trend function μ_t is:

$$\Delta^d[y_t - \mu_t)] = \phi(L)^{-1} \theta(L) \varepsilon_t \qquad (3.123)$$

$$\theta(L)^{-1} \phi(L) \Delta^d \tilde{y}_t = \varepsilon_t \qquad (3.124)$$

$$\tilde{y}_t \equiv [y_t - \mu_t)] \qquad (3.125)$$

As a special case the ARFIMA(0, d, 0) model is:

$$\Delta^d \tilde{y}_t = \varepsilon_t \qquad (3.126)$$

Thus, as in the standard unit root framework, the subsequent analysis is in terms of observations that are interpreted as deviations from the mean (or detrended observations), $\tilde{y}_t$. As usual, the cases likely to be encountered in practice are $\mu(t) = 0$, $\mu(t) = y_0$ and $\mu_t = \beta_0 + \beta_1 t$. The presence of a mean function has implications for the generation of data for the regressor(s) in the FDF and EFDF tests. It is now necessary to generate a regressor variable of the form $\Delta^{d_1}\tilde{y}_{t-1}$, where $\tilde{y}_t = y_t - \mu_t$, and application will require a consistent estimator $\hat{\mu}_t$ of μ_t and, hence, $\hat{\tilde{y}}_t = y_t - \hat{\mu}_t$.

The simplest case to consider is $\mu_t = y_0$. Then the generating model is

$$y_t = y_0 + w(L)\Delta^{-d}\varepsilon_t \tag{3.127}$$

Notice that $E(y_t) = y_0$, given $E(\varepsilon_t) = 0$, so the initial condition can be viewed as the mean of the process. Let Δ_s^d be the truncated (finite sample) operator (defined below), then it does not follow that $\Delta_s^d\tilde{y}_t = \Delta_s^d y_t$ except asymptotically and given a condition on d. The implication is that adjusting the data may be desirable to protect against finite sample effects (and is anyway essential in the case of a trend function). The notation in this section will explicitly acknowledge the finite sample nature of the Δ^d operator, a convention that was previously left implicit.

The argument is as follows. First define the truncated binomial operator Δ_s^d, which is appropriate for a finite sample, that is:

$$\Delta_s^d = \sum_{r=0}^{s} A_r^{(d)} L^r \tag{3.128}$$

$$\Rightarrow$$

$$\Delta_s^d y_t = \sum_{r=0}^{s} A_r^{(d)} y_{t-r} = y_t + A_1^{(d)} y_{t-1} + \cdots + A_s^{(d)} y_{t-s} \tag{3.129}$$

and $\Delta_s^d(1) = \sum_{r=0}^{s} A_r^{(d)}$, that is Δ_s^d evaluated at $L = 1$.

Next consider the definition $\tilde{y}_t \equiv y_t - y_0$, (the argument is relevant for y_0 known), so that:

$$\Delta_s^d\tilde{y}_t = \Delta_s^d(y_t - y_0)$$
$$= \Delta_s^d y_t - \Delta_s^d y_0 \tag{3.130}$$

Hence, $\Delta_s^d\tilde{y}_t$ and $\Delta_s^d y_t$ differ by $-\Delta_s^d y_0 = -y_0\Delta_s^d(1)$, which is not generally 0 as it is for d = 1, (that is the standard first difference operator for which $\Delta y_0 = 0$). Using a result derived in Q3.5.a, $\Delta_s^d(1) = A_s^{(d-1)}$ and, therefore,

$$\Delta_s^d\tilde{y}_t - \Delta_s^d y_t = -A_s^{(d-1)} y_0 \tag{3.131}$$

Thus, $\Delta_s^d\tilde{y}_t \neq \Delta_s^d y_t$, differing by a term that depends on y_0; as $s \to \infty$, $\Delta_s^{d-1}\Delta\tilde{y}_t \to \Delta_s^d y_t$, provided that $A_s^{(d-1)} y_0 \to 0$. In the case of a zero initial value, $y_0 = 0$, then $\Delta_s^d\tilde{y}_t = \Delta_s^d y_t$, as expected. If y_0 is fixed or is a bounded random variable then

$\Delta_s^d(1) = A_s^{(d-1)} \to 0$ is sufficient for $A_s^{(d-1)}y_0 \to 0$ as $s \to \infty$. Referring to Equation (3.20), but with $d-1$, then $A_s^{(d-1)} \to 0$ if $d > 0$. If $s = t-1$, as is usually the case, then $\Delta_{t-1}^d \tilde{y}_t - \Delta_{t-1}^d y_t = -A_{t-1}^{(d-1)}y_0$.

The result in (3.131) implies the following for $s = t-1$ and the simple FI(d) process $\Delta_s^d \tilde{y}_t = \varepsilon_t$:

$$\Delta_{t-1}^d y_t = \Delta_{t-1}^d y_0 + \varepsilon_t$$
$$= A_{t-1}^{(d-1)}y_0 + \varepsilon_t$$

and for $t = 2, \ldots, T$, this gives rise to the sequence:

$$\Delta_1^d y_2 = A_1^{(d-1)}y_0 + \varepsilon_2, \Delta_2^d y_3 = A_2^{(d-1)}y_0 + \varepsilon_3, \ldots, \Delta_{T-1}^d y_T = A_{T-1}^{(d-1)}y_0 + \varepsilon_t$$

Notice that even though y_0 is a constant, terms of the form $A_s^{(d-1)}y_0$ are not, as they vary with s. Under the null hypothesis $d = 1$, all of these terms disappear, but otherwise do not vanish for $d \in [0, 1)$, becoming less important as $d \to 1$ and as $s \to \infty$. The implication of this argument is that demeaning y_t using an estimator of y_0 that is consistent under the alternative should, in general, lead to a test with better finite sample power properties.

The linear trend case is now simple to deal with. In this case $\mu_t = \beta_0 + \beta_1 t$ and $\Delta_s^d \tilde{y}_t = \Delta_s^d[y_t - (\beta_0 + \beta_1 t)] = \Delta_s^d y_t - \Delta_s^d(1)\beta_0 - \beta_1 \Delta_s^d t = \Delta_s^d y_t - A_s^{(d-1)}\beta_0 - \beta_1 \Delta_s^d t$, and it is evident that $\Delta_s^d \tilde{y}_t \neq \Delta_s^d y_t$. Even if the finite sample effect of $A_s^{(d-1)}\beta_0$ is ignored, the term due to the trend remains.

Shimotsu (2010) has suggested a method of demeaning or detrending that depends on the value of d. The context of his method is semi-parametric estimation of d, but it also has relevance in the present context. In the case that y_0 is unknown, two possible estimators of y_0 are the sample mean $\bar{y} = T^{-1}\sum_{t=1}^T y_t$ and the first observation, y_1. $\bar{y}$ is a good estimator of y_0 when $d < |0.5|$, whereas y_1 is a good estimator of y_0 when $d > 0.75$, and both are good estimators for $d \in (0.5, 0.75)$; for further details see Chapter 4, Section **4.9.2.iv**. Shimotsu (2010) suggests the estimator $\tilde{y}_0$ that weights $\bar{y}$ and y_1, using the weighting function $\kappa(d)$:

$$\tilde{y}_0 = \kappa(d)\bar{y} + [1 - \kappa(d)]y_1 \tag{3.132}$$

The weighting function $\kappa(d)$ should be twice differentiable, such that:

$$\kappa(d) = 1 \text{ for } d \leq \tfrac{1}{2}, \kappa(d) \in (0, 1) \text{ for } d \in (\tfrac{1}{2}, \tfrac{3}{4}); \kappa(d) = 0 \text{ for } d \geq \tfrac{3}{4}.$$

A possibility for $\kappa(d)$ when $d \in (\tfrac{1}{2}, \tfrac{3}{4})$, is $\kappa(d) = 0.5(1 + \cos[4\pi(d - 0.5)])$, which weights $\tilde{y}_0$ toward y_1 as d increases. In practice, d is unknown and is replaced by an estimator that is consistent for $d \in (0, 1)$. For example, replacing d by $\hat{d}_T$ results in an estimated value of $\tilde{y}_0$ given by:

$$\hat{\tilde{y}}_0 = \kappa(\hat{d}_T)\bar{y} + [1 - \kappa(\hat{d}_T)]y_1 \tag{3.133}$$

The demeaned data $\tilde{y}_t \equiv y_t - \hat{\tilde{y}}_0$ are then used in the FDF and EFDF tests and $\hat{d}_T$ is the value of d required to operationalise the test (or d_1 where an estimator is not used).

Next consider the linear trend case, so that $\mu_t = \beta_0 + \beta_1 t$ (note that $\beta_0 \equiv y_0$). The LS residuals are:

$$\hat{y}_t = y_t - (\hat{\beta}_0 + \hat{\beta}_1 t) \tag{3.134}$$

where ^ above indicates the coefficient from a LS regression of y_t on (1,t). As $\hat{\bar{y}} = 0$ from the properties of LS regression, the weighting function applied to the residuals $\hat{y}_t$ simplifies relative to $\kappa(\hat{d}_T)$ of Equation (3.113); thus, let

$$\hat{\tilde{y}}_0 = [1 - \kappa(\hat{d}_T)]\hat{y}_1 \tag{3.135}$$

The data is then transformed as $\tilde{y}_t = \hat{y}_t - \hat{\tilde{y}}_0$.

DGM (2008) suggest a slightly different approach, which provides an estimator of the parameters of $\Delta\mu_t$, which is consistent under H_0: d = 1 and H_A: d < 1 at different rates (see also Breitung and Hassler, 2002). Imposing the unit root, then:

$$\Delta y_t = \Delta\mu_t + \varepsilon_t \tag{3.136}$$

which is the regression of Δy_t on $\Delta\mu_t$, where the order of μ_t is assumed to be known. For example, suppose that $\mu_t = \beta_0 + \beta_1 t$, then:

$$\Delta y_t = \beta_1 + \varepsilon_t \tag{3.137}$$

and the LS estimator of β_1 is $\hat{\beta}_1 = (T-1)^{-1} \sum_{t=2}^{T} \Delta y_t$. Under H_0, $\hat{\beta}_1$ is consistent at rate $T^{1/2}$ and under H_A it is consistent at rate $T^{(3/2)-d}$ (see DGM, 2008). The adjusted observations are obtained as:

$$\Delta\hat{\tilde{y}}_t = \Delta y_t - \hat{\beta}_1 \tag{3.138}$$

The revised regressions for the FDF and EFDF tests (in the simplest cases) using adjusted data are as follows, where $\hat{\tilde{y}}_t$ generically indicates estimated mean or trend adjusted data as the case requires:

FDF

$$\Delta\hat{\tilde{y}}_t = \gamma\Delta_{t-1}^{\hat{d}_T}\hat{\tilde{y}}_{t-1} + z_t \tag{3.139}$$

EFDF

$$\Delta\hat{\tilde{y}}_t = \eta z_{t-1}(\hat{d}_T) + v_t \tag{3.140}$$

$$z_{t-1}(\hat{d}_T) = \frac{(\Delta_{t-1}^{\hat{d}_T-1} - 1)}{(1 - \hat{d}_T)}\Delta\hat{\tilde{y}}_t \tag{3.141}$$

As before the test statistics are the t-type statistics on γ and η, respectively, considered as functions of $\hat{d}_T$.

In the case of the DF tests, the limiting null distributions depend on μ_t, leading to, for example, the test statistics $\hat{\tau}$, $\hat{\tau}_\mu$ and $\hat{\tau}_\beta$; however, that is not the case for the DGM, LV and LM tests considered here. Provided that the trend function is replaced by a consistent estimator, say $\hat{\mu}_t$, then the limiting distributions remain the same as the case with $\mu_t = 0$.

3.7   Locally best invariant (LBI) tests

This section outlines a locally best invariant (LBI) test due to Tanaka (1999) and Robinson (1991, 1994a), the former in the time domain and the latter in the frequency domain. This test is also asymptotically uniformly most powerful invariant, UMPI. The test, referred to generically as an LM test, is particularly simple to construct in the case of no short-run dependence in the error; and whilst it is somewhat more complex in the presence of short-run dynamics, a development due to Agiakloglou and Newbold (1994) and Breitung and Hassler (2002) leads to a test statistic that can be readily computed from a regression that is analogous to the ADF regression in the standard case.

An advantage of an LM-type test is that it only requires estimation under the null hypothesis, which is particularly attractive in this case as the null hypothesis leads to the simplification that the estimated model uses first differenced data. This contrasts with test statistics that are based on the Wald principle, such as the DGM and LV versions of the FDF test, which also require specification and, practically, estimation of the value of d under the alternative hypothesis. Tanaka's results hold for any d; specifically, for typical economic time series where $d \geq 0.5$, and usually $d \in [0.5, 1.5)$, so that d is in the nonstationary region. The asymptotic results do not require Gaussianity.

3.7.1   An LM-type test

3.7.1.i   No short-run dynamics

Under an alternative local to d, the data is generated by:

$$(1 - L)^{(d+c)}y_t = \varepsilon_t \tag{3.142}$$

The case emphasized so far has been $d = d_0 = 1$, that is, there is a unit root, and a test of the null hypothesis $H_0 : c = 0$ against the alternative $H_A : c < 0$ is, in effect, a test of the unit root null against the left-sided alternative, with local alternatives of the form $c = \delta/\sqrt{T}$. However, other cases can be treated within this framework, for example $H_0 : d = d_0 = \frac{1}{2}$, so that the process is borderline nonstationary; against $H_A : d_0 + c$, with $c < 0$, implying that the generating process is stationary; and again with $d_0 = 1$ but $H_A : d_0 + c, c > 0$, so that the alternative is explosive.

There is a simple way to compute an asymptotic equivalent to the LM test, which does not require estimation of the alternative; however, to understand the basis of these tests it is helpful to consider its underlying principle. First, allow for the data to be generated by a FI(d + c) process, that is:

$$(1 - L)^{(d+c)}y_t = \varepsilon_t 1_{t>0} \qquad \varepsilon_t \sim N(0, \sigma_\varepsilon^2 < \infty) \tag{3.143}$$

Hereafter, the initialization $\varepsilon_t 1_{t>0}$ will be left implicit; and note that normality is not required for the asymptotic results to hold and can be replaced with an iid assumption. The (conditional) log-likelihood function is:

$$LL(\{y_t\}; c, \sigma_\varepsilon^2) = -\frac{T}{2} \log(2\pi\sigma_\varepsilon^2) - \frac{1}{2\sigma_\varepsilon^2} \left(\sum_{t=1}^{T} [(1 - L)^{d+c} y_t]^2 \right) \tag{3.144}$$

Let $x_t \equiv (1 - L)^d y_t$, then the first derivative of LL(.) with respect to c (the score), evaluated at c = 0, is:

$$S_{T1} \equiv \left(\frac{\partial LL(\{y_t\}; c, \sigma_\varepsilon^2)}{\partial c} | H_0 : c = 0, \sigma_\varepsilon^2 = \hat{\sigma}_\varepsilon^2 \right)$$

$$= \frac{1}{\hat{\sigma}_\varepsilon^2} \sum_{t=2}^{T} x_t \left(\sum_{j=1}^{t-1} j^{-1} x_{t-j} \right)$$

$$= T \sum_{j=1}^{T-1} j^{-1} \hat{\rho}(j) \tag{3.145}$$

$$\hat{\rho}(j) = \frac{\sum_{t=j+1}^{T} x_t x_{t-j}}{\sum_{t=1}^{T} x_t^2} \tag{3.146}$$

$$\hat{\sigma}_\varepsilon^2 = T^{-1} \sum_{t=2}^{T} \Delta^d y_t^2$$

$$= T^{-1} \sum_{t=2}^{T} x_t^2$$

$\hat{\rho}(j)$ is the j-th autocorrelation coefficient of $\{x_t\}$; more generally $\hat{\rho}(j)$ is the j-th autocorrelation coefficient of the residuals (obtained under H_0) and $\hat{\sigma}_\varepsilon^2$ is an estimator of σ_ε^2.

In the case of the unit root null, H_0: $d = d_0 = 1$, hence $x_t = \Delta y_t$ and, therefore, $\hat{\rho}(j)$ simplifies to:

$$\hat{\rho}(j) = \frac{\sum_{t=j+1}^{T} \Delta y_t \Delta y_{t-j}}{\sum_{t=1}^{T} \Delta y_t^2} \tag{3.147}$$

Thus, $\hat{\rho}(j)$ is the j-th autocorrelation coefficient of Δy_t.

At this stage it is convenient to define the following:

$$x^*_{t-1} \equiv \sum_{j=1}^{t-1} j^{-1} x_{t-j} \tag{3.148}$$

and write (3.145) as

$$S_{T1} = \frac{1}{\hat{\sigma}^2_\varepsilon} \sum_{t=2}^{T} x_t x^*_{t-1} \tag{3.149}$$

(See Q3.7) Apart from the scale factor $\hat{\sigma}^{-2}_\varepsilon$, this is the numerator of the least squares estimator of the coefficient in the regression of x_t on x^*_{t-1}, an interpretation that will be useful below.

A suitably scaled version of S_{T1} provides the LM-type test statistic. First note that:

$$\frac{S_{T1}}{\sqrt{T}} = \sqrt{T} \sum_{j=1}^{T-1} j^{-1} \hat{\rho}(j)$$

$$\Rightarrow_D N((\pi^2/6)\delta, (\pi^2/6)) \tag{3.150}$$

where $c = \delta/\sqrt{T}$ for δ fixed (see Tanaka, 1999, theorem 3.1). Hence, dividing $S_{T1}/\sqrt{T}$ by $\sqrt{(\pi^2/6)}$ gives an LM-type statistic with a limiting standard normal distribution:

$$LM_0 \equiv \frac{S_{T1}}{\sqrt{T}\sqrt{(\pi^2/6)}} = \frac{\sqrt{T}}{\sqrt{(\pi^2/6)}} \sum_{j=1}^{T-1} j^{-1} \hat{\rho}(j)$$

$$\Rightarrow_D N\left(\delta\sqrt{\pi^2/6}, 1\right) \tag{3.151}$$

Hence, under the null $\delta = 0$ and $LM_0 \Rightarrow_D N(0, 1)$, leading to the standard decision rule that, with asymptotic size α, reject $H_0 : c = 0$ against $H_A : c < 0$ if $LM_0 < z_\alpha$, and reject H_0 against $H_A : c > 0$ if $LM_0 > z_{1-\alpha}$, where z_α and $z_{1-\alpha}$ are the lower and upper quantiles of the standard normal distribution. Alternatively for a two-sided alternative take the square of LM_0 and compare this with the critical values from $\chi^2(1)$, rejecting for 'large' values of the test statistic.

The LM_0 test is locally best invariant, LBI, and asymptotically uniformly most powerful invariant, UMPI. For a detailed explanation of these terms, see Hatanaka (1996, especially section 3.4.1). In summary:

locally: the alternatives are local to d_0, that is $d_0 + c$, where $c > 0$ ($c < 0$), but $c \approx 0$; the 'local' parameterization here is $c = \delta/\sqrt{T}$.

best: considering the power function of the test statistic at d_0 and $d_0 + c$, the critical region for a size α test is chosen such that the slope of the power function is steepest.

invariant: the test is invariant to linear transformations of $\{y_t\}$, that is to changes in location (translation) and scale.

asymptotically: as $T \to \infty$.

uniformly most powerful: consider testing H_0: d_0 against the composite alternative H_A: $d_0 + c$, $c > 0$ $(c < 0)$; a test that maximizes power (for a given size) for all (permissible) values of c is said to be uniformly most powerful.

Tanaka (1999, theorem 3.2 and corollary 3.2) shows that asymptotically the power of LM_0 coincides with the power envelope of the locally best invariant test.

3.7.1.ii *Deterministic components*

We now consider the general case and show how it specialises to other cases of interest. Let y be determined by Z, a T x K matrix of nonstochastic variables, and u a T x 1 vector of errors generated as FI($-d$), such that:

$$y_t = Z_t'\beta + u_t \qquad t = 1, \ldots, T \tag{3.152}$$

$$y = Z\beta + u \tag{3.153}$$

where $y = (y_1, \ldots, y_T)'$, $Z = (z_1, \ldots, z_K)$, $z_i = (z_{i1}, \ldots, z_{iT})'$, $Z_t' = (z_{1t}, \ldots, z_{Kt})$, $u = \Delta^{-d}\varepsilon$, $u = (u_1, \ldots, u_T)'$, $\varepsilon = (\varepsilon_1, \ldots, \varepsilon_T)'$, where $\{\varepsilon_t\}$ is a white noise sequence. On multiplying through by Δ^d, the model can be written as:

$$x = \tilde{Z}\beta + \varepsilon \tag{3.154}$$

$x \equiv \Delta^d y$ (as before), $\tilde{Z} \equiv \Delta^d Z$, on the understanding that the operator Δ^d is applied to every element in the matrix or vector that it precedes. The conditional ML (and LS) estimator of β is $\hat{\beta} = (\tilde{Z}'\tilde{Z})^{-1}\tilde{Z}'x$ and $\hat{\sigma}_\varepsilon^2 = \hat{\varepsilon}'\hat{\varepsilon}/T$, where $\hat{\varepsilon} = x - \tilde{Z}\hat{\beta} = \Delta^d(y - Z\hat{\beta})$ and $\hat{\varepsilon}_t = x_t - \tilde{Z}_t'\hat{\beta}$.

The likelihood and score are:

$$LL(\{y_t\}; c, \sigma_\varepsilon^2) = -\frac{T}{2}\log(2\pi\sigma_\varepsilon^2) - \frac{1}{2\sigma_\varepsilon^2}\left(\sum_{t=1}^{T}[(1 - L)^{d+c}(y_t - Z_t'\beta)]^2\right)$$

$$S_{T1} \equiv \left(\frac{\partial LL(\{y_t\}; c, \sigma_\varepsilon^2)}{\partial c} \Big| H_0 : c = 0, \sigma_\varepsilon^2 = \hat{\sigma}_\varepsilon^2\right)$$

$$= -\frac{1}{\hat{\sigma}_\varepsilon^2}\sum_{t=1}^{T}[\ln(1 - L)(1 - L)^d(y_t - Z_t'\hat{\beta})](1 - L)^d(y_t - Z_t'\hat{\beta})$$

$$= -\frac{1}{\hat{\sigma}_\varepsilon^2}\sum_{t=1}^{T}[\ln(1 - L)\hat{\varepsilon}_t]\hat{\varepsilon}_t$$

$$= T\sum_{j=1}^{T-1} j^{-1}\hat{\rho}(j) \tag{3.155}$$

where $\hat{\rho}(j) = \sum_{t=j+1}^{T} \hat{\varepsilon}_t \hat{\varepsilon}_{t-j} / \sum_{t=1}^{T} \hat{\varepsilon}_t^2$ is the j-th autocorrelation coefficient of the residuals $\hat{\varepsilon}$; (the derivation of the last line is the subject of an end of chapter question). H_0: $c = 0$ is rejected against $c > 0$ for large values of S_{T1} and rejected against $c < 0$ for small values of S_{T1}. As noted above, a suitably scaled version of S_{T1} has a standard normal distribution, which then provides the appropriate quantiles.

We can now consider some special cases. The first case of no deterministics has already been considered, in that case Z is empty and y is generated as $\Delta^d y = \varepsilon$. In this case, the value of d is given under the null and no parameters are estimated. The test statistic is based on $x = \Delta^d y$, so that for $d = 1$, $x = \Delta y = (y_2 - y_1, \ldots, y_2 - y_{T-1})$. It is also important to consider how to deal with the presence of a linear trend (the test statistics are invariant to $y_0 \neq 0$). Suppose the generating process is:

$$y_t = \beta_0 + \beta_1 t + \Delta^{-d} \varepsilon_t \tag{3.156}$$
$$= Z_t' \beta + \Delta^{-d} \varepsilon_t$$

where $Z_t' = (1, t)$ and $\beta = (\beta_0, \beta_1)'$. $\hat{\beta}$ is obtained as $\hat{\beta} = (\tilde{Z}'\tilde{Z})^{-1} \tilde{Z}' x$ where, under the unit root null $d = 1$, $x = \Delta y$ and $\tilde{Z} = \Delta Z = (\Delta z_1, \Delta z_2)$, with Δz_1 a column of 0s and Δz_2 a column of 1s; the first element of $\hat{\beta}$ is, therefore, annihilated and $\hat{\beta}_1$ is the LS coefficient from the regression of Δy_t on a constant, equivalently the sample mean of Δy_t. The residuals on which the LM test is based are, therefore, $\hat{\varepsilon}_t = x_t - \tilde{Z}_t' \hat{\beta} = \Delta y_t - \hat{\beta}_1$ and the adjustment is, therefore, as described in Section **3.6**.

3.7.1.iii   Short-run dynamics

In the more general case the coefficients of the AR and MA parts of an ARFIMA model have to be estimated. We consider this extension briefly in view of previous developments connected with the FDF test. In this case, data has been generated by an ARFIMA $(p, d_0 + c, q)$ process, that is:

$$\varphi(L)(1 - L)^{(d_0+c)} y_t = \theta(L) \varepsilon_t \tag{3.157}$$

The (conditional) log-likelihood function is:

$$LL(\{y_t\}; c, \psi, \sigma_\varepsilon^2) = -\frac{T}{2} \log(2\pi\sigma_\varepsilon^2) - \frac{1}{2\sigma_\varepsilon^2} \left(\sum_{t=1}^{T} [\theta(L)^{-1} \varphi(L)(1 - L)^{d_0+c} y_t]^2 \right) \tag{3.158}$$

where $\psi = (\varphi', \theta')$ collects the coefficients in $\varphi(L)$ and $\theta(L)$. The score with respect to c is as before, but with the residuals now defined to reflect estimation of the coefficients in $\varphi(L)$ and $\theta(L)$:

$$S_{T2} \equiv \left(\frac{\partial LL(\{y_t\}; c, \psi, \sigma_\varepsilon^2)}{\partial c} | H_0 : c = 0, \sigma_\varepsilon^2 = \hat{\sigma}_\varepsilon^2, \psi = \hat{\psi} \right)$$

$$= T \sum_{j=1}^{T-1} j^{-1} \hat{\rho}(j) \tag{3.159}$$

where $\hat{\rho}(j)$ is the j-th order autocorrelation coefficient of the residuals given by:

$$\hat{\rho}(j) = \frac{\sum_{t=j+1}^{T} \hat{\varepsilon}_t \hat{\varepsilon}_{t-j}}{\sum_{t=1}^{T} \hat{\varepsilon}_t^2} \tag{3.160}$$

$$\hat{\varepsilon}_t = \hat{\theta}(L)^{-1} \hat{\varphi}(L)(1 - L)^{d_0} y_t \tag{3.161}$$

Notice that the residuals are estimated using the null value d_0, hence d itself is not a parameter that is estimated. Then, see Tanaka (op. cit., theorem 3.3):

$$\frac{S_{T2}}{\sqrt{T}} = \sqrt{T} \sum_{j=1}^{T-1} j^{-1} \hat{\rho}(j) \tag{3.162}$$

$$\Rightarrow_D N(\delta\omega^2, \omega^2) \qquad c = \delta/\sqrt{T}$$

Hence, the LM statistic, LM_0, is now:

$$LM_0 = \frac{\sqrt{T}}{\omega} \sum_{j=1}^{T-1} j^{-1} \hat{\rho}(j) \tag{3.163}$$

$$\Rightarrow_D N(\delta\omega, 1)$$

and if $\delta = 0$, then

$$LM_0 \Rightarrow_D N(0, 1)$$

The (long-run) variance, ω^2, in (3.162) is given by:

$$\omega^2 = \frac{\pi^2}{6} - (\kappa_1, \ldots, \kappa_p, \lambda_1, \ldots, \lambda_q) \Phi^{-1} (\kappa_1, \ldots, \kappa_p, \lambda_1, \ldots, \lambda_q)' \tag{3.164}$$

The sequences $\{\kappa_i\}$ and $\{\lambda_i\}$ are given by:

$$\kappa_i = \sum_{j=1}^{\infty} j^{-1} g_{j-i} \qquad \lambda_i = -\sum_{j=1}^{\infty} j^{-1} h_{j-i} \tag{3.165}$$

where g_j and h_j are the coefficients of L^j in the expansions of $\varphi(L)^{-1}$ and $\theta(L)^{-1}$ respectively, and Φ is the Fisher information matrix for $\psi = (\varphi', \theta')$.

To operationalise (3.163) following the LM principle, a consistent estimate of ω under the null denoted $\hat{\omega}_0$ is substituted for ω, resulting in the feasible LM test statistic, $LM_{0,0}$:

$$LM_{0,0} = \frac{\sqrt{T}}{\hat{\omega}_0} \sum_{j=1}^{T-1} j^{-1} \hat{\rho}(j) \tag{3.166}$$

$$\Rightarrow_D N(\delta\omega, 1)$$

A consistent estimator of ω can be obtained from (3.164) using the estimates of ψ.

3.7.2   LM test: a regression approach

3.7.2.i   No short-run dynamics

An easy way of computing a test statistic that is asymptotically equivalent to the LM test statistic and using the same principle is to adopt the regression approach due to Agiakloglou and Newbold (1994), as developed by Breitung and Hassler (2002). First consider the simple case in which the data are generated as $(1 - L)^{d+c}y_t = \varepsilon_t$ and note that the structure of S_{T1} suggested the least squares regression of x_t on x_{t-1}^*, that is:

$$x_t = \alpha x_{t-1}^* + e_t \tag{3.167}$$

$$x_t \equiv \Delta^d y_t \tag{3.168}$$

$$x_{t-1}^* \equiv \sum_{j=1}^{t-1} j^{-1} x_{t-j} \tag{3.169}$$

The least squares estimator and associated test statistic for H_0: $c = 0$ against H_A: $c \neq 0$ translates to H_0: $\alpha = 0$ and H_A:$\alpha \neq 0$. The LS estimator of α and the square of the t statistic for $\hat{\alpha}$, are, respectively:

$$\hat{\alpha} = \frac{\sum_{t=2}^T x_t x_{t-1}^*}{\sum_{t=2}^T (x_{t-1}^*)^2} \tag{3.170}$$

$$t_\alpha^2 = \frac{\left(\sum_{t=2}^T x_t x_{t-1}^*\right)^2}{\hat{\sigma}_e^2 \sum_{t=2}^T (x_{t-1}^*)^2} \tag{3.171}$$

where $\hat{\sigma}_e^2 = T^{-1} \sum_{t=2}^T \hat{e}_t^2$ and $\hat{e}_t = x_t - \hat{\alpha} x_{t-1}^*$.

Breitung and Hassler (2002, theorem 1) show that for data generated as $(1 - L)^{(d+c)}y_t = \varepsilon_t 1_{t>0}$, with $\varepsilon_t \sim (0, \sigma_\varepsilon^2 < \infty)$, then:

$$t_\alpha^2 = (LM_0)^2 + o_p(1) \tag{3.172}$$

$$\Rightarrow$$

$$t_\alpha^2 \Rightarrow_D \chi^2(1) \tag{3.173}$$

So that t_α^2 is asymptotically equivalent to $(LM_0)^2$. Recall that if $d_0 = 1$, then $x_t = \Delta y_t$ and, therefore, the regression (3.167) takes the following form:

$$\Delta y_t = \delta y_{t-1}^* + \upsilon_t \tag{3.174}$$

$$y_{t-1}^* = (\Delta y_{t-1} + \tfrac{1}{2}\Delta y_{t-2} + \tfrac{1}{3}\Delta y_{t-3} + \cdots + \tfrac{1}{(t-1)}(y_1 - y_0)) \tag{3.175}$$

In contrast the corresponding DF regression is:

$$\Delta y_t = \gamma(\Delta y_{t-1} + \Delta y_{t-2} + \cdots + y_1 - y_0) + v_t \tag{3.176}$$

$$= \gamma y_{t-1} + v_t$$

and the last line assumes the initialization $y_0 = 0$. Thus, the difference between the two approaches is that the regressor y_{t-1}^* weights the lagged first differences, with weights that decline in the pattern $(1, \frac{1}{2}, \frac{1}{3}, \ldots, \frac{1}{t-1})$.

Note that the test statistic is here presented in the form t_α^2 as $H_A: c \neq 0$, however, the more likely case is $H_A: c < 0$, so the test statistic t_α can be used with left-sided critical values from $N(0, 1)$.

3.7.2.ii Short-run dynamics

In the more likely case there will be short-run dependence introduced through either or both the (non-redundant) lag polynomials $\varphi(L)$ and $\theta(L)$. In this case, the approach suggested by Agiakloglou and Newbold (1994) and Breitung and Hassler (2002) follows the familiar route of augmenting the basic regression. First, start with short-run dynamics captured by an AR(p) component, so that data is generated by an ARFIMA(p, d, 0) model:

$$\varphi(L)(1-L)^d y_t = \varepsilon_t \tag{3.177}$$

$$\varphi(L)x_t = \varepsilon_t \tag{3.178}$$

$$x_t \equiv (1-L)^d y_t$$

Then under the null $d = d_0$ and $x_t \equiv (1-L)^{d_0} y_t$ can be formed. The resulting p-th order autoregression in x_t is:

$$x_t = \varphi_1 x_t + \cdots + \varphi_p x_{t-p} + \varepsilon_t \tag{3.179}$$

Estimation results in the residuals and regressor given by:

$$\hat{\varepsilon}_t = x_t - (\hat{\varphi}_1 x_t + \cdots + \hat{\varphi}_p x_{t-p}) \tag{3.180}$$

$$\hat{\varepsilon}_{t-1}^* = \sum_{j=1}^{t-(p+1)} j^{-1} \hat{\varepsilon}_{t-j} \tag{3.181}$$

The augmented test regression is then:

$$\hat{\varepsilon}_t = \alpha \hat{\varepsilon}_{t-1}^* + \sum_{i=1}^{p} \psi_i x_{t-i} + \upsilon_t \tag{3.182}$$

Notice that the regressors x_{t-i} are also included in the LM regression. (See Wooldridge, 2011, for a discussion of this point for stochastic regressors.) The suggested test statistics are t_α or t_α^2, which are asymptotically distributed as $N(0, 1)$ and $\chi^2(1)$ respectively.

In the case of the invertible ARFIMA (p, d, q) model, the generating process is:

$$\theta(L)^{-1}\varphi(L)(1-L)^d y_t = \varepsilon_t \tag{3.183}$$

$$A(L)x_t = \varepsilon_t \qquad A(L) = \theta(L)^{-1}\varphi(L) \tag{3.184}$$

$$x_t \equiv (1-L)^d y_t$$

Then, as before, under the null $d = d_0$ and $x_t \equiv (1-L)^{d_0} y_t$. In this case A(L) is an infinite order lag polynomial and, practically, is approximated by truncating the lag to a finite order, say k. A reasonable conjecture based on existing results is that provided the order of the approximating lag polynomial expands at an appropriate rate, then the limiting null distributions of t_α and t_α^2 are maintained as in the no serial correlation case. The suggested rate for the ADF(k) test is k $\rightarrow \infty$ as T $\rightarrow \infty$ such that $k/T^{1/3} \rightarrow 0$ (see also Ng and Perron (2001) and *UR Vol. 1*, chapter 9).

3.7.2.iii Deterministic terms

By extension of the results for the standard LM test (see Section **3.7.1**) the test statistic, t_α, under the unit root null uses differenced data and is, therefore, invariant to a non-zero mean, $\mu_t = y_0 \neq 0$. In the case of a linear trend, the adjusted data for the test is $\Delta y_t - \hat{\beta}_1$; see Breitung and Hassler (2002), and Section **3.7.1.ii.**

3.7.3 A Wald test based on ML estimation

Noting that ω^{-1} is the standard error of the limiting distribution of $\sqrt{T}(\hat{d} - d)$, where $\hat{d}$ is the ML estimator of d, then a further possibility is to construct a Wald-type test based on estimation under the alternative (Tanaka, 1999, section 4). Specifically we have the following limiting distribution under the null:

$$\sqrt{T}(\hat{d} - d_0) \Rightarrow_D N(0, \omega^{-2}) \Rightarrow \tag{3.185}$$

$$t_{\hat{d}} = \left(\frac{(\hat{d} - d_0)}{(\omega^{-1}/\sqrt{T})} \right) \tag{3.186}$$

$$= \sqrt{T}\omega(\hat{d} - d_0) \tag{3.187}$$

$$\Rightarrow_D N(0,1)$$

The first statement, (3.185), says that $\sqrt{T}(\hat{d} - d)$ is asymptotically normally distributed with zero mean and variance ω^{-2}. The 'asymptotic variance' of $\sqrt{T}\hat{d}$ is ω^{-2} and the 'asymptotic standard error' is $\sqrt{T}\omega^{-1}$, therefore standardising $(\hat{d} - d)$ by $\omega^{-1}/\sqrt{T}$ gives the t-type statistic, denoted $t_{\hat{d}}$ in (3.186). This results in a quantity that is asymptotically distributed as standard unit normal. In practice, one could use the estimator of ω^{-1} obtained from ML estimation to obtain a feasible (finite sample) version of $t_{\hat{d}}$.

In the simplest case $\omega^2 = (\pi^2/6)$, therefore $\omega^{-2} = (\pi^2/6)^{-1} = 6/\pi^2$, $\sqrt{T}(\hat{d} - d) \Rightarrow_D N(0, 6/\pi^2)$ and the test statistic is $t_{\hat{d}} = 1.2825\sqrt{T}(\hat{d} - d_0) \Rightarrow_D N(0, 1)$. The more general case is considered in a question (see Q3.8).

Tanaka (op. cit.) reports some simulation results to evaluate the Wald and LM tests (in the forms LM_0 and $LM_{0,0}$) and compare them with the limiting power envelope. The data-generating process is an ARFIMA(1, d, 0) model, with niid input, $T = 100$, and 1,000 replications with a nominal size of 5%. The results most likely to be of interest in an economic context are those around the stationary/nonstationary boundary, with $H_0: d_0 = 0.5$ against $H_A: d_A > 0.5$ and, alternately, $H_0: d_0 = 1$ against $H_A: d_A < 1$.

With a simple fractional integrated process, the simulated size is 5.7% for LM_0 and 3.9% for $t_{\hat{d}}$ using $\omega^2 = (\pi^2/6)$, and 3.9% and 3.5%, respectively, when ω^2 is estimated. In terms of size-unadjusted power, the LM test is better than the Wald test, for example when $d = 0.6$, 36.2% and 30.7% respectively using $\omega^2 = (\pi^2/6)$, and 30% and 29.3% when ω^2 is estimated. Given the difference in nominal size, it may well be the case that there is little to choose between the tests in terms of size-adjusted power.

In the case of the unit root test against a fractional alternative, ARFIMA(1, d, 0), Tanaka (op. cit.) reports the results with $\varphi_1 = 0.6$ and, alternately, $\varphi_1 = -0.8$; we concentrate on the former as that corresponds to the more likely case of positive serial correlation in the short-run dynamics. The simulation results suggest a sample of $T = 100$ is not large enough to get close to the asymptotic results. For example, the nominal sizes of LM_0 and $LM_{0,0}$ were 1.4% and 2.2%, whereas the corresponding figures for the Wald test were 9.8% and 16.1%; these differences in size suggest that a power comparison could be misleading.

3.8 Power

3.8.1 Power against fixed alternatives

The LM, FDF and EFDF tests are consistent against fixed alternatives, where the versions of the FDF test assume an estimator d_1 that is consistent at rate T^κ, with $\kappa > 0$. DGM (2008, theorem 2) derive the non-centrality parameter associated with each test under the fixed alternative $d \in (0, 1)$ and $\Delta^d y_t = \varepsilon_t$ and they suggest an analysis of asymptotic power in terms of Bahadur's asymptotic relative efficiency, ARE. Consider two tests, say, $\hat{t}_1$ and $\hat{t}_2$, with corresponding non-centrality parameters nc_1 and nc_2, then the ARE of the two tests as a function of d is:

$$\text{ARE}(\hat{t}_1, \hat{t}_2, d) = (nc_1/nc_2)^2 \tag{3.188}$$

where $\text{ARE}(\hat{t}_1, \hat{t}_2, d) > 1$ implies that test $\hat{t}_1$ is asymptotically more powerful than test $\hat{t}_2$. ARE is a function of the ratio of the non-centrality parameters, so that $\hat{t}_1$ is asymptotically more powerful than $\hat{t}_2$ as $|nc_1| > |nc_2|$ (because of the nature of H_A, large negative values of the test statistic lead to rejection, thus $nc_i < 0$). DGM (2008, figures 1 and 2) show that the non-centrality parameters of $t_\gamma(d_1)$, $t_\eta(d_1)$ and LM_0, say nc_γ, nc_η and nc_0, are similar for $d \in (0.9, 1)$, but for $d \in (0.3, 0.9)$ approximately, $|nc_\eta| > |nc_\gamma| > |nc_0|$, whereas for $d \in (0.0, 0.3)$, $|nc_\eta| \approx |nc_\gamma| > |nc_0|$. In turn these imply that in terms of asymptotic power $t_\eta(d_1) \approx t_\gamma(d_1) \approx LM_0$ for $d \in (0.9, 1)$; $t_\eta(d_1) > t_\gamma(d_1) > LM_0$ for $d \in (0.3, 0.9)$; and $t_\eta(d_1) \approx t_\gamma(d_1) > LM_0$ for $d \in (0.0, 0.3)$. Perhaps of note is that there is little to choose (asymptotically) between the tests close to the unit root for fixed alternatives.

3.8.2   Power against local alternatives

When the null hypothesis is that of a (single) unit root, so that $\Delta y_t = \varepsilon_t$, then the local alternative is:

$$\Delta^d y_t = \varepsilon_t \tag{3.189}$$

$$d = 1 + \delta/\sqrt{T} \qquad \delta < 0 \tag{3.190}$$

where ε_t is iid with finite fourth moment. This is as in the LM test, with $d_0 = 1$. As $T \to \infty$ then $d \to 1$, which is the value under H_0, so that there is a sequence of alternative distributions that get closer and closer to that under H_0 as $T \to \infty$. The test statistic $t_\eta(\hat{d}_T)$ is asymptotically equivalent to the Robinson/Tanaka LM test which is, as noted above, UMPI under a sequence of local alternatives. To see this equivalence it is helpful to consider the distributions of the test statistics under the local alternatives.

Let the DGP be $\Delta^d y_t = \varepsilon_t 1_{(t>0)}$, then under the sequence of local alternatives $d = 1 + \delta/\sqrt{T}$, $\delta < 0$, $d_1 \geq 0.5$, the limiting distributions are as follows:

LM

$$LM_0 \Rightarrow_D N(\delta h_0, 1)$$

$$h_0 = \sqrt{\sum_{j=1}^{\infty} j^{-2}} = \sqrt{\pi^2/6} = 1.2825 \tag{3.191}$$

FDF

$$t_\gamma(d) \Rightarrow_D N(\lambda_1, 1) \qquad d = 1 + \delta/\sqrt{T}$$

$$\lambda_1 = \delta h_1, h_1 = 1 \tag{3.192}$$

$$t_\gamma(d_1) \Rightarrow_D N(\lambda_2(d_1), 1)$$

$$\lambda_2(d_1) = \delta h_2(d_1)$$

$$h_2(d_1) = \frac{\Gamma(d_1)}{d_1 \sqrt{\Gamma(2d_1 - 1)}} \qquad d_1 > 0.5 \tag{3.193}$$

EFDF

$$t_\eta(\hat{d}_T) \Rightarrow_D N(\lambda_3(d_2), 1)$$

$$\lambda_3(d_2) = \delta h_3(d_2)$$

$$h_3(d_2) = \frac{\sum_{j=1}^\infty j^{-1} A_j^{(d_2-1)}}{\sqrt{\sum_{j=1}^\infty (A_j^{(d_2-1)})^2}} \qquad d_2 > 0.5, d_2 \neq 1 \qquad (3.194)$$

$$\hat{d}_T = d_2 + o_p(T^{-\kappa}), \kappa > 0, \hat{d}_T > 0.5$$

The $A_j^{(d)}$ coefficients are defined in Equations (3.4) and (3.5); $t_\gamma(d)$ is computed from the regression of Δy_t on $\Delta^d y_{t-1}$, $d = 1 + \delta/\sqrt{T}$; whereas $t_\gamma(d_1)$ is computed from the regression of Δy_t on $\Delta^{d_1} y_{t-1}$; $t_\eta(\hat{d}_T)$ is computed from the regression of Δy_t on $z_{t-1}(\hat{d}_T)$ and LM_0 is given by (3.151). The input values d_1 and d_2 are distinguished as they need not be the same in the two tests.

The effect of the local alternative on the distributions of the test statistics is to shift each distribution by a non-centrality parameter λ_i, which depends on δ and a drift function denoted, respectively, as h_0, h_1, $h_2(d_1)$ and $h_3(d_2)$; note that the first two are just constants. In each case, the larger the shift in $\lambda_i(d_j)$ relative to the null value, $\delta = 0$, the higher the power of the test.

The drift functions are plotted in Figure 3.9. Note that $h_2(d_1)$ and $h_3(d_2)$ are everywhere below h_0, but that $h_3(d_2) \rightarrow h_0$ as $d \rightarrow 1$, confirming the asymptotic local equivalence of $t_\eta(\hat{d}_T)$ and LM_0, given a consistent estimator of d for the former. The drift functions of $t_\gamma(d_1)$ and $t_\eta(\hat{d}_T)$ cross over at $d = 0.77$, indicating that $t_\eta(\hat{d}_T)$ is not everywhere more powerful than $t_\gamma(d_1)$, with the former more powerful for $d \in (0.5, 0.77)$, but $t_\eta(\hat{d}_T)$ more powerful closer to the unit root.

Note also that there is a unique maximum value of d_1 for $h_2(d_1)$, say d_1^*, which occurs at $d_1^* \approx 0.69145$ and, correspondingly, $h_2(d_1^*) = 1.2456$, at which the power ratio of $t_\gamma(d_1)$ relative to LM_0 is 97%. What is also clear is that $d_1 = 0.5$ is a particularly poor choice as an input for $\hat{t}_\gamma(d_1)$ as $\lambda_1 = 0$. Comparing $t_\gamma(d)$ and $t_\gamma(d_1^*)$, the squared ratio of the relative noncentrality parameters, which gives the asymptotic relative efficiency is $(1/1.2456)^2 = 0.6445$; note that the power of $t_\gamma(d_1)$ is greater than that of $t_\gamma(d)$ for $d_1 \in (0.56, 1.0)$.

3.8.3 The optimal (power maximising) choice of d_1 in the FDF test(s)

Lobato and Velasco (2006), hereafter LV (2006), consider choosing d_1 optimally, where optimality refers to maximising the power of $t_\gamma(d_1)$. In the case of local alternatives, it has already been observed that there is a unique maximum to the drift function $h_2(d_1)$, where $d_1^* \approx 0.69145$ (in the white noise case).

In the case of *fixed* alternatives $d \in (0.5, 1.0)$, the optimal value of d_1, the input value of the test, follows from maximising the squared population correlation between Δy_t on $\Delta^{d_1} y_{t-1}$. In general, the value that maximises this correlation,

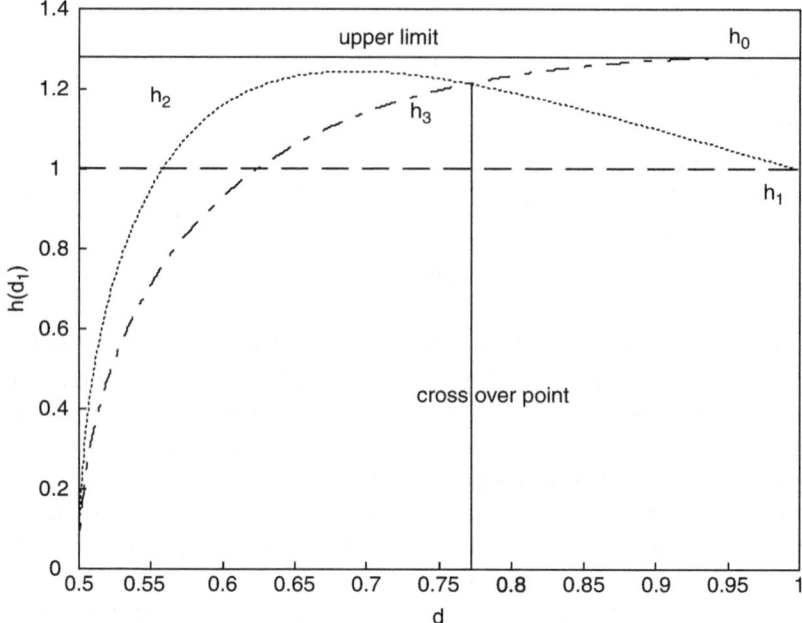

Figure 3.9　The drift functions for t_γ, t_η and LM_0

d_1^*, is a function of d and is written as $d_1^*(d)$ to emphasise this point. LV (2006) find that this relationship is in effect linear over the range of interest $(d \geq 0.5)$ and can be fitted by $\hat{d}_1^*(d) = -0.031 + 0.719d$, thus for a specific alternative, $d = d_1$, the 'best' input value for the FDF regressor is $d_1^* = -0.031 + 0.719d_1$. In the composite alternative case, d_1 is estimated by a T^κ-consistent estimator $\hat{d}_T$, with $\kappa > 0$, and the optimal selection of the input value is:

$$\hat{d}_1^*(\hat{d}_T) = -0.031 + 0.719\hat{d}_T \tag{3.195}$$

To emphasise, note that in this procedure it is not the consistent estimator $\hat{d}_T$ that is used to construct the regressor, (nor is it $d = d_1$ in the simple alternative), but a simple linear transform of that value, which will be less than $\hat{d}_T$ (or d_1). Using this value does not alter the limiting standard normal distribution (LV, 2006, lemma 1).

In the more general case with $\varphi(L)\Delta^d y_t = \varepsilon_t 1_{(t>0)}$, the optimal d_1 depends on the order of $\varphi(L)$ for local alternatives and the values of the φ_j coefficients for fixed alternatives. LV (2006) suggest a prewhitening approach, in which the first step is to estimate $\varphi(L)$ from a p-th order autoregression using Δy_t, that is under the null $d = 1$, $\varphi(L)\Delta y_t = \varepsilon_t \Rightarrow \Delta y_t = \sum_{j=1}^{p} \hat{\varphi}_j \Delta y_{t-j} + \hat{\varepsilon}_t$, where ^ over denotes a LS estimator; the second step is to form $\breve{y}_t = y_t - \hat{y}_t$, where $\hat{y}_t = \sum_{j=1}^{p} \hat{\varphi}_j y_{t-j}$; $\breve{y}_t$ is

Table 3.3   Optimal d_1, d_1^*, for use in PAFDF $\hat{t}_\gamma(d_1)$ tests (for local alternatives)

p	0	1	2	3	4	5
d_1^*	0.691	0.846	0.901	0.927	0.942	0.951

Source: LV (2006, table 1).

referred to as the prewhitened data. The final step is to run an AFDF regression, but with $\breve{y}_t$ rather than y_t, that is:

$$\Delta\breve{y}_t = \gamma\Delta^{d_1}\breve{y}_{t-1} + \sum_{j=1}^{p} \varphi_j \Delta y_{t-j} + u_t^*(d_1) \tag{3.196}$$

(Note that this differs from DGM's, 2002, ADF in that it uses $\breve{y}_t$ rather than y_t; also note that y_t may be subject to a prior adjustment for deterministic components.) As before, the test statistic is the t-type test associated with γ, $t_\gamma(d_1)$, and the inclusion of lags of Δy_t control, the size of the test. The revised AFDF regression is referred to as the prewhitened AFDF, PAFDF.

In the case of a sequence of local alternatives, the optimal d_1 depends on the order p of $\varphi(L)$, but not the value of φ_j coefficients themselves, so that $d_1^* \approx 0.69145$ can be seen to correspond to the special case where $p = 0$, as shown in Table 3.3, for $p \leq 5$.

In the case of fixed alternatives, there is no single function such as (3.195) but a function that varies with p and the values of φ_j. LV (2006, op. cit., section 4), to which the reader is referred for details, suggest an algorithm that leads to the automatic selection d_1^* ; they note that the method is computationally intensive, but can offer power improvements in finite samples.

3.8.4   Illustrative simulation results

This section is concerned with a brief evaluation of the power of the various competing tests outlined earlier against fixed alternatives. It is not intended to be comprehensive in the sense of a full evaluation for all possible DGPs, rather to illustrate some of the issues that may guide the application of such tests. The feasible versions of the FDF and EFDF tests require either a single input value d_1 or an input based on a consistent estimator $\hat{d}_T$ of d; in the case of the FDF test, with no weak dependence in the errors, there are three possibilities, namely $d_1^* = 0.69145$, $\hat{d}_1^*(\hat{d}_T) = -0.031 + 0.719\hat{d}_T$ and $\hat{d}_T$ without adjustment, whereas the EFDF test uses $\hat{d}_T$. As discussed extensively in Chapter 4, the performance of 'local' semi-parametric frequency estimators is dependent on the number of included frequencies designated m (for bandwidth), and the simulations use $m = T^{0.65}$.

Two estimators, $\hat{d}_T$, were considered. These are the widely used semi-parametric estimator due to Robinson (1995b), based on the local Whittle

(LW) contrast, and the exact LW estimator of Shimotsu and Phillips (2005) and Shimotsu (2010), referred to as $\hat{d}_{LW}$ and $\hat{d}_{ELW}$, respectively (see Chapter 4). In the case of $\hat{d}_{LW}$, the data are differenced and one is added back to the estimator, whereas this is not necessary for $\hat{d}_{ELW}$, which is consistent for d in the nonstationary region. The FDF tests were relatively insensitive to the choice of $\hat{d}_T$, so a comparison is only made for the EFDF tests.

The tests to be considered are:

FDF: $t_\gamma(d_1^*)$ where $d_1^* = 0.69145$; $t_\gamma(\hat{d}_1^*)$ based on $\hat{d}_{LW}$; and $t_\gamma(\hat{d}_{LW})$.

EFDF: $t_\eta(\hat{d}_{LW})$ and $t_\eta(\hat{d}_{ELW})$.

LM: Tanaka's LM_0 and the Breitung-Hassler regression version t_α.

The DGPs are variations of the following:

$$[y_t - \mu_t] = u_t \tag{3.197}$$

$$\Delta^d u_t = \varepsilon_t 1_{(t>0)} \qquad d \in (0.5, 1] \tag{3.198}$$

$$\hat{\tilde{y}}_t \equiv y_t - \hat{\mu}_t \tag{3.199}$$

The first case to be considered is when $\mu_t = y_0$ and the estimators are as described in Section **3.6**. The simulations are with one exception invariant to the value of y_0. The exception arises when using the test based on $\hat{d}_{ELW}$. To see why, compare this method with that based on $\hat{d}_{LW}$; in the latter case, the data are first differenced and then 1 is added back to the estimator, thus, like the LM tests, invariance to a non-zero initial value is achieved by differencing. In the case of $\hat{d}_{ELW}$, the estimator is asymptotically invariant to a non-zero initial value for $d \geq 0.5$, but Shimotsu (2010) recommends demeaning by subtracting y_1 from y_t to protect against finite sample effects. Thus, if it is known that $y_0 = 0$, there would be no need to use the Shimotsu demeaned data; however, in general this is not the case (and it is not the case here), so the data is demeaned both in the estimation of d and in the construction of the tests. The results, based on 5,000 replications, are given in Table 3.4a for $T = 200$ and Table 3.4b for $T = 500$, and illustrated in Figures 3.10a and 3.10b.

The central case is $T = 200$. The empirical sizes of $t_\gamma(d_1^*)$, $t_\gamma(\hat{d}_{LW})$ and t_α are close to the nominal size of 5%, whereas $t_\gamma(\hat{d}_1^*)$, $t_\eta(\hat{d}_{LW})$ and $t_\eta(\hat{d}_{ELW})$ are slightly oversized, whilst LM_0 is undersized. Relative power differs across the parameter space. In terms of size-adjusted power (given the differences in nominal size), for d close to the unit root, say $d \in (0.9, 1.0)$, $t_\gamma(d_1^*)$ is best, but it loses a clear advantage as d decreases, with a marginal advantage to $t_\eta(\hat{d}_{LW})$ and t_α. The least powerful of the tests is $t_\gamma(\hat{d}_{LW})$. Figure 3.10a shows how, with one exception, the power of the tests clusters together, whereas the region $d \in (0.9, 1.0)$ is shown in Figure 3.10b.

The results for $T = 500$ (see Table 3.4b for a summary), show that the tests, with the exception of $t_\gamma(\hat{d}_{LW})$, which has the lowest power, are now quite difficult to

Table 3.4a   Power of various tests for fractional d, (demeaned data), T = 200

d	$t_\gamma(d_1^*)$	$t_\gamma(\hat{d}_{LW})$	$t_\gamma(\hat{d}_1^*)$	$t_\eta(\hat{d}_{LW})$	$t_\eta(\hat{d}_{ELW})$	LM_0	t_α
0.60	1.000	1.000	1.000	1.000	1.000	1.000	1.000
0.65	1.000	1.000	1.000	1.000	1.000	0.999	1.000
0.70	0.999	0.998	1.000	1.000	1.000	0.999	0.999
0.75	0.991	0.979	0.999	0.999	0.999	0.991	0.999
0.80	0.934	0.896	0.971	0.971	0.970	0.920	0.960
0.85	0.777	0.698	0.856	0.846	0.849	0.728	0.820
0.90	0.491	0.424	0.587	0.546	0.548	0.403	0.511
0.95	0.210	0.195	0.268	0.228	0.234	0.148	0.212
1.00	0.049	0.053	0.079	0.067	0.069	0.030	0.054
size-adjusted power							
0.60	1.000	1.000	1.000	1.000	1.000	1.000	1.000
0.65	1.000	1.000	1.000	1.000	1.000	1.000	1.000
0.70	0.999	0.998	1.000	1.000	1.000	0.999	0.997
0.75	0.992	0.977	0.994	0.998	0.999	0.996	0.996
0.80	0.937	0.893	0.945	0.959	0.956	0.950	0.954
0.85	0.779	0.691	0.773	0.812	0.805	0.796	0.805
0.90	0.494	0.413	0.467	0.496	0.481	0.483	0.490
0.95	0.214	0.186	0.181	0.195	0.193	0.196	0.199
1.00	0.050	0.050	0.050	0.050	0.050	0.050	0.050

Table 3.4b   Power of various tests for fractional d, (demeaned data), T = 500

d	$t_\gamma(d_1^*)$	$t_\gamma(\hat{d}_{LW})$	$t_\gamma(\hat{d}_1^*)$	$t_\eta(\hat{d}_{LW})$	$t_\eta(\hat{d}_{ELW})$	LM_0	t_α
0.60	1.000	1.000	1.000	1.000	1.000	1.000	1.000
0.65	1.000	1.000	1.000	1.000	1.000	1.000	1.000
0.70	1.000	1.000	1.000	1.000	1.000	1.000	1.000
0.75	1.000	1.000	1.000	1.000	1.000	1.000	1.000
0.80	1.000	0.999	1.000	1.000	1.000	1.000	1.000
0.85	0.990	0.967	0.996	0.997	0.997	0.989	0.996
0.90	0.848	0.731	0.889	0.892	0.892	0.831	0.877
0.95	0.404	0.322	0.456	0.426	0.424	0.351	0.414
1.00	0.057	0.058	0.076	0.063	0.063	0.042	0.056
size-adjusted power							
0.60	1.000	1.000	1.000	1.000	1.000	1.000	1.000
0.65	1.000	1.000	1.000	1.000	1.000	1.000	1.000
0.70	1.000	1.000	1.000	1.000	1.000	1.000	1.000
0.75	1.000	1.000	1.000	1.000	1.000	1.000	1.000
0.80	1.000	0.999	1.000	1.000	1.000	1.000	1.000
0.85	0.989	0.963	0.994	0.995	0.995	0.994	0.995
0.90	0.823	0.711	0.841	0.862	0.859	0.849	0.851
0.95	0.373	0.303	0.368	0.380	0.378	0.378	0.374
1.00	0.050	0.050	0.050	0.050	0.050	0.050	0.050

Note: $d_1^* = 0.69145, \hat{d}_{LW}$ is the local Whittle estimator, $\hat{d}_{ELW}$ is the exact Whittle estimator.

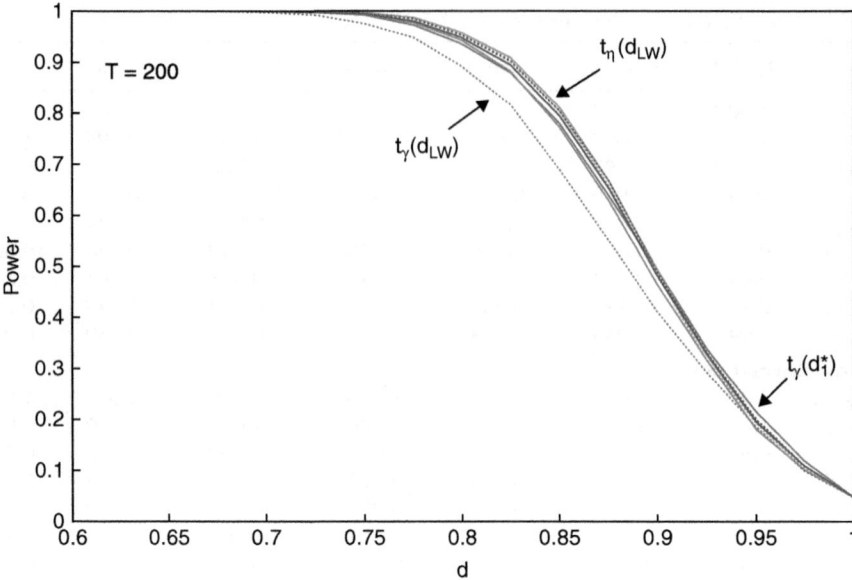

Figure 3.10a   Size-adjusted power (data demeaned)

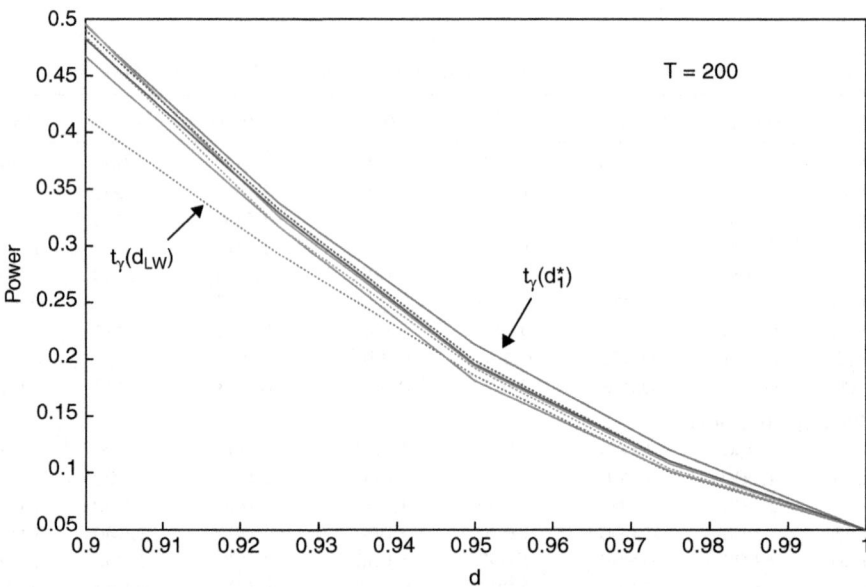

Figure 3.10b   Size-adjusted power (data demeaned), near unit root

distinguish across the range $d \in (0.6, 1.0)$, although $t_\gamma(\hat{d}_1^*)$, $t_\eta(\hat{d}_{LW})$ and $t_\eta(\hat{d}_{ELW})$ are still slightly oversized.

The other case likely to arise in practice is that of a possible non-zero deterministic trend, which is illustrated with detrended data. The LM tests now also require detrending, which is achieved by basing the test on $\hat{\varepsilon}_t = \Delta y_t - \hat{\beta}_1$, where $\hat{\beta}_1$ is the sample mean of Δy_t. The FDF-type tests are based on (Shimotsu) detrended data obtained as $\hat{\tilde{y}}_t = \tilde{y}_t - \tilde{y}_1$, where $\tilde{y}_t = y_t - (\hat{\beta}_0 + \hat{\beta}_1 t)$; see Section 3.6, Shimtosu (2010) and Chapter 4. The results are summarised in Tables 3.5a and 3.5b; and power and s.a power are shown in Figures 3.11a and 3.11b for $T = 200$.

Starting with $T = 200$, all of the tests are oversized, but the LM_0 test is only marginally so at 5.5% followed by $t_\gamma(\hat{d}_{LW})$ at 6.6% (see the upper panel of Table 3.5a); the other tests have broadly twice their nominal size. The oversizing indicates that the finite sample 5% critical value is to the left of the nominal critical value. One way of overcoming the inaccuracy of the nominal critical values is to bootstrap the test statistics (See *UR, Vol. 1*, chapter 8). An approximate estimate of the finite sample 5% cv can be obtained by finding the normal distribution that delivers a 10% rejection rate with a critical value of -1.645; such a distribution has a mean of -0.363 (rather than 0) and a (left-sided) 5% cv of -2.0. Overall, in terms of s.a power, LM_0 is the best of the tests although the differences between the EFDF tests and the LM tests are relatively slight and $t_\gamma(\hat{d}_1^*)$ is also competitive (see Figure 3.11a); when the alternative is close to the unit root, $t_\gamma(d_1^*)$ and $t_\gamma(\hat{d}_{LW})$ also become competitive tests (see Figure 3.11b); whilst away from the unit root $t_\gamma(\hat{d}_{LW})$ is least powerful. Thus, LM_0 is recommended in this case as it is close to its nominal size and the best overall in terms of (s.a) power.

The general oversizing results for $T = 200$ suggest that it would be of interest to increase the sample size for a better picture of what is happening to the empirical size of the test, and Table 3.5b presents a summary of the results for $T = 1,000$. In this case, $t_\gamma(d_1)$, $t_\gamma(\hat{d}_{LW})$, and t_α are now much closer to their nominal size and LM_0 maintains its relative fidelity; there are improvements in the empirical size of $t_\gamma(\hat{d}_1^*)$, $t_\eta(\hat{d}_{LW})$ and $t_\eta(\hat{d}_{ELW})$, but these remain oversized. The s.a power confirms that the LM tests still have a marginal advantage in terms of power, whereas $t_\gamma(\hat{d}_{LW})$ is clearly the least powerful. Overall, the advantage does again seem to lie with the LM type tests, LM_0 and t_α.

3.9 Example: US wheat production

The various tests for a fractional root are illustrated with a time series on US wheat production; the data is in natural logarithms and is annual for the period 1866 to 2008, giving a sample of 143 observations. The data, y_t, and the data detrended by a regression on a constant and a linear trend, $\tilde{y}_t$, are graphed

Table 3.5a   Power of various tests for fractional d, (detrended data), T = 200

d	$t_\gamma(d_1^*)$	$t_\gamma(\hat{d}_{LW})$	$t_\gamma(\hat{d}_1^*)$	$t_\eta(\hat{d}_{LW})$	$t_\eta(\hat{d}_{ELW})$	LM_0	t_α
0.60	1.000	1.000	1.000	1.000	1.000	1.000	1.000
0.65	1.000	1.000	1.000	1.000	1.000	1.000	1.000
0.70	1.000	0.998	1.000	1.000	1.000	0.999	1.000
0.75	0.994	0.982	0.999	0.999	0.999	0.994	0.999
0.80	0.948	0.907	0.981	0.980	0.980	0.934	0.972
0.85	0.806	0.715	0.889	0.877	0.877	0.756	0.842
0.90	0.541	0.449	0.650	0.630	0.630	0.453	0.580
0.95	0.270	0.219	0.345	0.311	0.311	0.187	0.274
1.00	0.082	0.066	0.124	0.108	0.115	0.055	0.093
size-adjusted power							
0.60	1.000	1.000	1.000	1.000	1.000	1.000	1.000
0.65	1.000	1.000	1.000	1.000	1.000	1.000	1.000
0.70	0.999	0.998	1.000	1.000	1.000	0.999	1.000
0.75	0.977	0.974	0.990	0.993	0.991	0.992	0.993
0.80	0.899	0.877	0.928	0.930	0.929	0.925	0.927
0.85	0.698	0.655	0.735	0.735	0.733	0.740	0.742
0.90	0.422	0.381	0.430	0.425	0.424	0.437	0.427
0.95	0.176	0.167	0.170	0.165	0.166	0.178	0.174
1.00	0.050	0.050	0.050	0.050	0.050	0.050	0.050

Table 3.5b   Power of various tests for fractional d, (detrended data), T = 1,000

d	$t_\gamma(d_1^*)$	$t_\gamma(\hat{d}_{LW})$	$t_\gamma(\hat{d}_1^*)$	$t_\eta(\hat{d}_{LW})$	$t_\eta(\hat{d}_{ELW})$	LM_0	t_α
0.825	1.000	1.000	1.000	1.000	1.000	1.000	1.000
0.850	1.000	0.999	1.000	1.000	1.000	1.000	1.000
0.875	1.000	0.991	1.000	1.000	1.000	1.000	1.000
0.900	0.986	0.942	0.999	0.991	0.991	0.989	0.992
0.925	0.909	0.774	0.931	0.932	0.935	0.901	0.923
0.950	0.660	0.509	0.707	0.707	0.698	0.632	0.681
0.975	0.300	0.209	0.336	0.337	0.299	0.244	0.274
1.00	0.059	0.045	0.079	0.079	0.069	0.044	0.059
size-adjusted power							
0.825	1.000	1.000	1.000	1.000	1.000	1.000	1.000
0.850	1.000	0.999	1.000	1.000	1.000	1.000	1.000
0.875	1.000	0.992	1.000	1.000	1.000	1.000	1.000
0.900	0.985	0.949	0.982	0.989	0.989	0.990	0.989
0.925	0.890	0.788	0.900	0.911	0.912	0.911	0.910
0.950	0.635	0.525	0.624	0.647	0.649	0.657	0.659
0.975	0.267	0.222	0.250	0.258	0.258	0.256	0.256
1.00	0.050	0.050	0.050	0.050	0.050	0.050	0.050

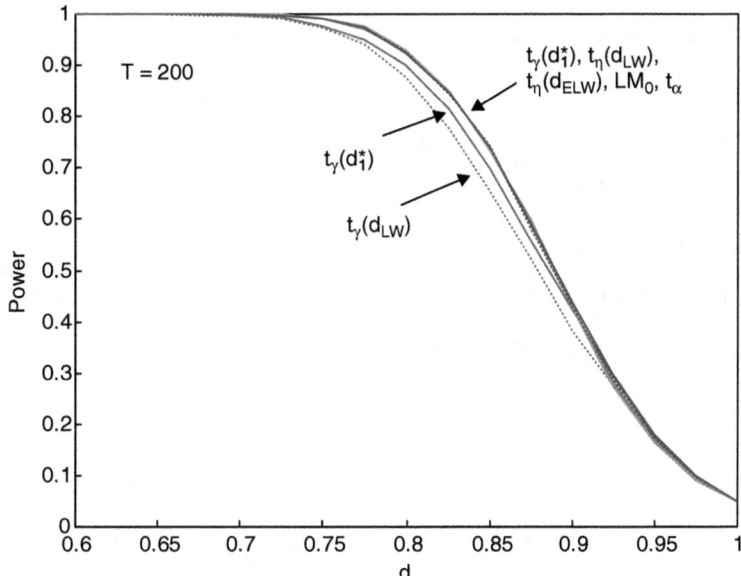

Figure 3.11a   Size-adjusted power (data detrended)

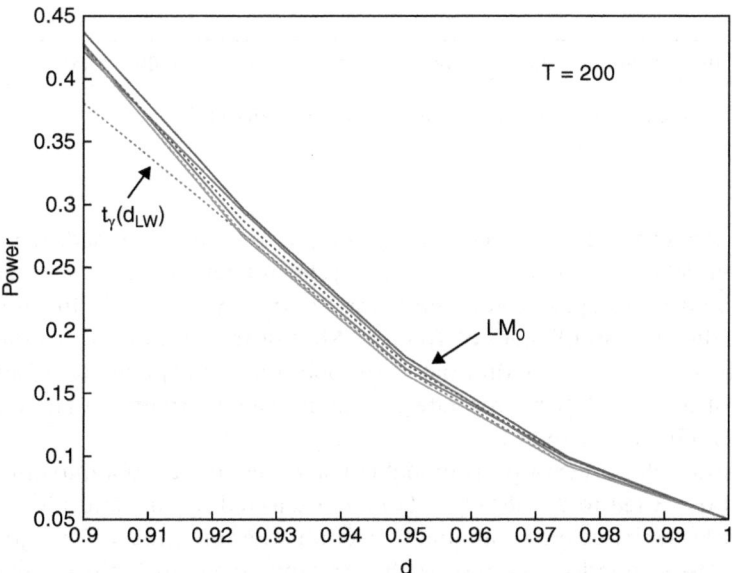

Figure 3.11b   Size-adjusted power (data detrended), near unit root

(a)

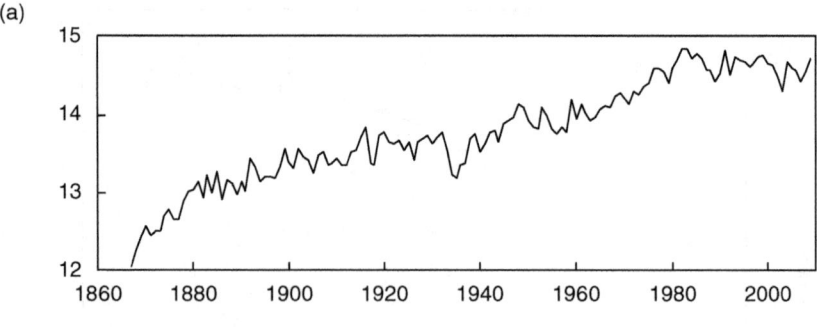

Figure 3.12a   US wheat production (logs)

(b)

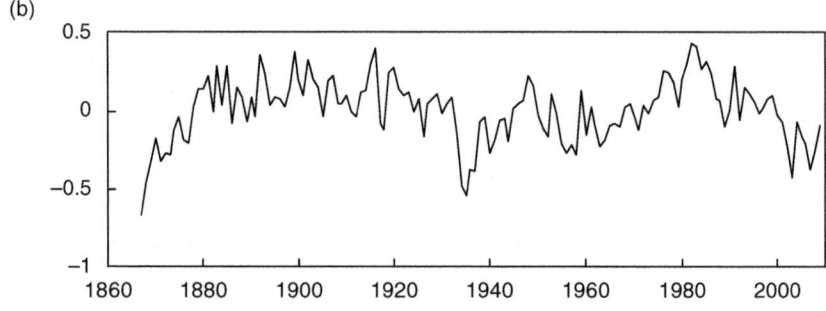

Figure 3.12b   US wheat production (logs, detrended)

in Figures 3.12a and 3.12b, respectively; the data is obviously trended, whilst the detrended data shows evidence of a long cycle and persistence. A standard approach would be to apply a unit root test to either y_t or $\tilde{y}_t$. To illustrate we calculate the Shin and Fuller (1998) exact ML test for a unit root, denoted $LR_{UC,\beta}$ and based on the unconditional likelihood function, applied to $\tilde{y}_t$, and the version of the ADF test appropriate to a detrended alternative, that is $\hat{\tau}_\beta$, (see *UR, Vol. 1*, chapters 3 and 6).

ML estimation of an ARMA(p, q) model considering all lags to a maximum of p = 3 and q = 3 led to an ARMA(1, 1) model selected by AIC and BIC. An ADF(1) was selected based on a marginal t selection criterion using a 10% significance level. The estimation and test results are summarized in Table 3.6, with 't' statistics in parentheses.

The results show that whilst the ML test does not reject the null hypothesis of a unit root, the ADF test leads to the opposite conclusion and both are reasonably robust in that respect to variations in the significance level.

Table 3.6 Estimation of models for
US wheat production

Exact ML estimation of ARMA(1, 1)
$$\tilde{y}_t = 0.824\tilde{y}_{t-1} + \hat{\varepsilon}_t - 0.326\hat{\varepsilon}_{t-1}$$
$$(10.04)(-2.20)$$

$LR_{UC,\beta} = 2.20, 5\%\ cv = 2.80$

ADF(1) estimation
$$\Delta\tilde{y}_t = -0.317\tilde{y}_{t-1} - 0.142\Delta\tilde{y}_{t-1}$$
$$(-4.72)(-1.71)$$

$\hat{\tau}_\beta = -4.52, 5\%\ cv = -3.44$

Note: (For critical values see *UR Vol. 1*,
Chapters 3 and 6.)

Table 3.7   Tests with a fractionally integrated alternative; $p = 0$ and $p = 2$ for augmented tests

	$t_\gamma(d_1^*)$	$t_\gamma(\hat{d}_{LW})$	$t_\gamma(\hat{d}_1^*)$	$t_\eta(\hat{d}_{LW})$	$t_\eta(\hat{d}_{ELW})$	LM_0	t_α
$p = 0$							
test value	−4.43	−4.14	−5.17	−4.55	−5.73	−3.19	−4.65
$p = 2$	$t_\gamma(d_1^*)$	$t_\gamma(\hat{d}_{LW})$	$t_\gamma(\hat{d}_{ELW})$	$t_\eta(\hat{d}_{LW})$	$t_\eta(\hat{d}_{ELW})$	$LM_{0,0}$	t_α
test value	−2.54	−2.43	−3.23	−1.93	−3.19	−4.88	−3.00

Notes: $d_1^* = 0.691$ if no serial dependence allowed ($p = 0$); $d_1^* = 0.901$ for $p = 2$; throughout the limiting 5% cv is −1.645, from the standard normal distribution; two-step estimation procedure used for $t_\eta(\hat{d}_{LW})$ and $t_\eta(\hat{d}_{ELW})$ when $p = 2$ (see Section **3.5.7**); similar results were obtained from nonlinear estimation.

The tests in this chapter allow for the possibility that the alternative is a fractionally integrated, but still nonstationary, process. The test results are summarised in Table 3.7. The ML and ADF models suggest that the tests have to take into account the possibility of serially dependent errors, but for comparison the tests are first presented in their simple forms.

As a brief recap in the case of weakly dependent errors, the test statistics are obtained as follows. In the case of the (LV version of the) FDF test, the suggested augmented regression (see Section **3.8.3**) is:

$$\Delta\breve{y}_t = \gamma\Delta^{d_1}\breve{y}_{t-1} + \sum_{j=1}^{p} \varphi_j \Delta\tilde{y}_{t-j} + u_t^*(d_1) \tag{3.200}$$

This is referred to as the prewhitened ADF, PADF, where $\breve{y}_t = \tilde{y}_t - \sum_{j=1}^{p} \hat{\varphi}_j\tilde{y}_{t-j}$, and $\hat{\varphi}(L)$, a p-th order polynomial, is obtained from fitting an AR(p) model to $\Delta\tilde{y}_t$, where $\tilde{y}_t$ is the detrended data. The test statistic is the t-type test associated with γ, $t_\gamma(d_1)$, and the optimal d_1 depends on the order of $\varphi(L)$ for local alternatives (and the values of the φ_j coefficients for fixed alternatives); see Table 3.3. We find $p = 2$ is sufficient to 'whiten' y_t. Reference to Table 3.3 indicates that $d_1^* = 0.901$ for the choice of $p = 2$.

The augmented EFDF test regression takes the following form:

$$\Delta\tilde{y}_t = \eta[\varphi(L)z_{t-1}(\hat{d}_T)] + [1 - \varphi(L)]\Delta\tilde{y}_t + v_t \tag{3.201}$$

The test statistic is the t test associated with η, $t_\eta(\hat{d}_T)$. This regression is non-linear in the coefficients because of the multiplicative form $\eta\varphi(L)$, and the $\varphi(L)$ coefficients also enter through the second set of terms. As noted above, there are several ways to deal with this and the illustration uses the two stage procedure (see Section **3.5.7**); see Q3.11 for the nonlinear estimation method. For $p = 2$, the first stage comprises estimation of the AR(2) coefficients obtained from:

$$\Delta^{\hat{d}_T}\tilde{y}_t = \hat{\varphi}_1\Delta^{\hat{d}_T}\tilde{y}_{t-1} + \hat{\varphi}_2\Delta^{\hat{d}_T}\tilde{y}_{t-2} + \hat{\xi}_t$$

The regression in the second stage is

$$\Delta\tilde{y}_t = \eta w_{t-1} + \varphi_1\Delta\tilde{y}_{t-1} + \varphi_2\Delta\tilde{y}_{t-2} + \hat{v}_t$$

where $w_{t-1} = (1 - \hat{\varphi}_1 L - \hat{\varphi}_2 L^2)z_{t-1}(\hat{d}_T)$, with $\hat{\varphi}_1$ and $\hat{\varphi}_2$ obtained from the first stage.

The relevant version of the LM test is $LM_{0,0}$, given by:

$$LM_{0,0} = \frac{\sqrt{T}}{\hat{\omega}_0}\sum_{j=1}^{T-1} j^{-1}\hat{\rho}(j) \Rightarrow_D N(0, 1) \text{ under the null}$$

$$\hat{\rho}(j) = \frac{\sum_{t=j+1}^{T}\hat{\varepsilon}_t\hat{\varepsilon}_{t-j}}{\sum_{t=1}^{T}\hat{\varepsilon}_t^2}$$

$$\hat{\varepsilon}_t = \hat{\theta}(L)^{-1}\hat{\varphi}(L)(1 - L)\tilde{y}_t$$

In this case, exact ML estimation suggested an ARMA(1, 1) model, so that if a unit root is imposed then this becomes an ARIMA(0, 1, 1) specification. Imposing the unit root we find that $\hat{\theta}(L) = (1 - 0.46L)$; the residuals can then be obtained from the recursion $\hat{\varepsilon}_t = \Delta\tilde{y}_t + 0.46\hat{\varepsilon}_{t-1}$, $t = 2, \ldots, T$. If the MA(1) coefficient is zero, then this reduces to $\hat{\varepsilon}_t = \Delta\tilde{y}_t$. To complete the test statistic an estimate $\hat{\omega}_0$ of ω (under the null) is required; this is obtained by specializing (3.164) and (3.165) to the first-order case (see also Q3.10).

The BH version of the LM test is constructed by first estimating an AR(p) model in $x_t \equiv \Delta y_t$, saving the residuals $\hat{\varepsilon}_t$, then constructing the lagged weighted sum of the $\hat{\varepsilon}_t$, denoted $\hat{\varepsilon}_{t-1}^*$ and running the (augmented) test regression as follows:

$$\hat{\varepsilon}_t = \alpha\hat{\varepsilon}_{t-1}^* + \sum_{i=1}^{p}\psi_i x_{t-i} + v_t$$

The various test results are illustrated with $p = 2$ and summarised in Table 3.7. The results suggest that the unit root null should be rejected at usual significance levels; moreover the semi-parametric estimates of d, which are $\hat{d}_{LW} = 0.765$ and $\hat{d}_{ELW} = 0.594$, with $m = T^{0.65}$, tend to confirm this view. (Note that as $\hat{d}_{ELW}$ is further from the unit root than $\hat{d}_{LW}$, the test statistics are more negative when $\hat{d}_{ELW}$ is used.) The simulation results reported in Table 3.5 suggested that LM_0 maintained its size (at least at 5%), whereas the other tests had an actual size about twice the nominal size, that is the finite sample critical value was to the left of the nominal critical value. The test results are generally robust to likely variations due to the oversizing found using critical values from the standard normal distribution.

It is possible to set up different null hypotheses of interest; one in particular is H_0: $d = 0.5$ against H_A: $d \in (0.5, 1]$, which is a test of borderline nonstationarity, and it is simple enough to apply the various tests to this situation, for details see Tanaka (1999) and Breitung and Hassler (2002) for the LM tests and DGM (2009) for the extension of the EFDF test.

3.10 Concluding remarks

This chapter has introduced some key concepts to extend the range of integrated processes to include the case of fractional d. These are of interest in themselves and for their use in developments of the I(d) literature to fractionally cointegrated series. The first question to be addressed is: can a meaning be attached to the fractional difference operator applied to a series of observations? The answer is yes and relies upon an expansion of $(1 - L)^d$ using the binomial theorem either directly or through a gamma function representation, whilst the treatment of the initial condition gives rise to two types of fractionally integrated processes. Corresponding developments have been made in the mathematical analysis of fractionally integrated processes in continuous time; see for example Oldham and Spanier (1974), Kalia (1993) and Miller and Ross (1993).

Most of the analysis of this chapter was in the time domain, with the frequency domain approach to be considered in the following chapter. Once fractional d is allowed, the range of hypotheses of interest is extended quite naturally. The set-up is no longer simply that of nonstationarity, $d = 1$, to be tested against stationarity, $d = 0$; one obvious generalisation is $d = 1$ against $d \in [0, 1)$, but others may be of interest in context, for example $d = 0$ against $d \in (0, 1]$ or $d = 0.5$ (borderline nonstationary) against $d \in (0.5, 1]$. Processes with $d > 0$ have long memory in the sense that the autocorrelations are not summable, declining hyperbolically for $0 < d < 1$, whilst a process with $d \geq 0.5$ is also nonstationary.

One way to approach the testing problem is to extend the familiar framework of Dickey and Fuller, as suggested by Dolado, Gonzalo and Mayoral (2002) and further extended by Lobato and Velasco (2006, 2007) and Dolado, Gonzalo and Mayoral (2008, 2009). Tanaka (1999) has suggested an LM test in the time domain which just requires estimation under the null hypothesis, which is particularly simple in the case of the unit root null hypothesis (see also Robinson 1994a). More directly, the long-memory coefficient d can be estimated in the time domain, using for example the results of Sowell (1992a, b) and hypotheses formed and tested directly using standard Wald, LR or LM principles. An analysis in terms of the frequency domain, an area of growing interest, is considered in the next chapter.

The FDF-type tests have been extended by Kew and Harris (2009) to the case of heteroscedasticity, covering both the unconditional and conditional cases. They show that existing tests can be put within a nesting framework and can be 'robustified' effectively by using White's (1980) standard errors, otherwise there are asymptotic size distortions in the presence of heteroscedasticity. They illustrate the the benefit of their method in a model with a variance shifts, respectively a single and a double shift, which are particular kinds of structural break just affecting the unconditional variance, an ARCH model and a stochastic volatility model.

Questions

Q3.1 Obtain the first 5 terms in the binomial expansion of $(1-L)^d$ and calculate their values for $d = 0.4$.

A3.1 This is covered in detail in Appendix 3.1 (see especially the development around Equation A3.2). Using the definition of $(1-L)^d$, that is:

$$(1-L)^d = (1 + {}_dC_1(-1)L + {}_dC_2(-1)^2L^2 + {}_dC_3(-1)^3L^3 + {}_dC_4(-1)^4L^4 \ldots)$$

which, using the definition of ${}_dC_r$, gives:

$$(1-L)^d = 1 - dL + \tfrac{1}{2}d(d-1)L^2 - \tfrac{1}{6}d(d-1)(d-2)L^3 + \tfrac{1}{24}d(d-1)(d-2)(d-3)L^4 \ldots$$

hence for $d = 0.4$, the expansion is:

$$(1-L)^{0.4} = 1 - 0.4L + \tfrac{1}{2}0.4(-0.6)L^2 - \tfrac{1}{6}0.4(-0.6)(-1.6)L^3$$
$$+ \tfrac{1}{24}0.4(-0.6)(-1.6)(-2.6)L^4 \ldots$$
$$= 1 - 0.4L - 0.12L^2 - 0.064L^3 - 0.0416L^4 \ldots$$

Q3.2.i What asymptotically governs the rate of decline of the $A_r^{(d)}$ coefficients? Does the sum $\sum_{r=0}^{\infty} A_r^{(d)}$ have a finite limit (that is, are the $A_r^{(d)}$ summable)?

Q3.2.ii What asymptotically governs the rate of decline of the $B_r^{(d)}$ coefficients? Does the sum $\sum_{r=0}^{\infty} B_r^{(d)}$ have a finite limit (that is, are the $B_r^{(d)}$ summable)?

A3.2.i This is most easily answered using the gamma function representation of $A_r^{(d)} = \frac{\Gamma(r-d)}{\Gamma(-d)\Gamma(r+1)}$, see (3.18); then using Sterling's theorem, the order of decline will be governed by r^δ where $\delta = r - d - (r+1) = -1 - d$. The individual $A_r^{(d)}$

coefficients $\to 0$ if $d > -1$; however, this is not sufficient for $\sum_{r=0}^{\infty} A_r^{(d)}$ to be summable. The latter requires $(1 + d) > 1$, that is $d > 0$ ($d = 0$ trivially satisfies summability), by the p-series convergence test. These points were illustrated in Figure 3.3, where summability of the AR coefficients is shown for $d = 0.3$ and 0.45, but not for -0.3 and -0.45.

A3.2.ii Following the same principle as the previous answer note that:

$$B_r^{(d)} = \frac{\Gamma(r+d)}{\Gamma(d)\Gamma(r+1)} \sim \frac{1}{\Gamma(d)} r^{d-1}$$

Asymptotically, the rate of decline is governed by r^{δ} where $\delta = r + d - (r+1) = d - 1$; and the $B_r^{(d)}$ coefficients $\to 0$ if $d < 1$, and are summable if $d < 0$, the sum diverging for $d \geq 0$ (by the p-series convergence test). These points were illustrated in Figure 3.2.

Q3.3 Consider the AR(1) model $(1 - \phi_1 L)y_t = \varepsilon_t$; show that this does not have long memory and that the autocorrelations are summable.

A3.3 The autocorrelations in the AR(1) model are particularly simple: $\rho(k) = \phi_1^k$, which clearly decay if $|\phi_1| < 1$; the sum of the autocorrelations is $\sum_{k=0}^{\infty} \phi_1^k$, which is a geometric series, the condition for convergence is $|\phi_1| < 1$ (see for example Sydsaeter and Hammond, 2008).

Q3.4 What merit, if any, is there in using a simple DF-type test against a fractionally integrated alternative, even though the implied AR process is infinite?

A3.4 The simple versions of the DF tests, with no augmentation, have better properties at least as far as consistency against the fractional alternative is concerned; alternatively use a test with a fixed k (lag length), since for fixed k power will increase with T.

The maintained regression for the simple DF tests, when the data is generated by an FI(d) process, is:

$$\Delta y_t = \gamma_0 y_{t-1} + \varepsilon_{t,0}$$

where

$$\gamma_0 = \Lambda_1^{(d)} - 1 = -A_1^{(d)} - 1$$
$$= d - 1 \qquad \text{as } A_1^{(d)} = -d$$

Hence, such a test will have power that increases as d tends to zero, so that γ_0 approaches -1, which is the case for which the DF tests were designed. Moreover, the rate of decline of the $A_r^{(d)}$ coefficients is a function of d, specifically it is governed by $1/r^{(1+d)}$ (see Equation (3.18) and **Q3.2.i**), such that for d in the likely region $d \in [0, 1]$, then $\sum_{r=0}^{k} \Lambda_r^{(d)} - 1$ converges to 0 more slowly for small d than large d; for example, if $d = 0.75$ then $\sum_{r=0}^{k=100} \Lambda_r^{(0.75)} - 1 = -0.0087$, whereas if $d = 0.1$ then $\sum_{r=0}^{k=100} \Lambda_r^{(0.1)} - 1 = -0.2578$. Thus, even if additional lags are included in the ADF representation, power will increase as d declines.

Q3.5.a Show the following:

(i) $\Delta_{s-1}^{d-1} \Delta = \Delta_s^d - A_s^{(d-1)} L^s$.

(ii) $\Delta_{s-1}^{d-1} \Delta y_t = \Delta_s^d y_t - A_s^{(d-1)} L^s y_t$

(iii) $\Delta_t^d y_t = \Delta_{t-1}^{d-1} \Delta y_t + A_t^{(d-1)} y_0$

(iv) If $d = 0$, then $y_t = \Delta_t^{-1} \Delta y_t + y_0$.

(v) $\Delta_s^d \tilde{y}_t - \Delta_s^d y_t = -A_s^{(d-1)} y_0$

Q3.5.b Consider the argument that $\Delta_s^d y_0 = \Delta_{s-1}^{d-1} \Delta y_0 = 0$, because $\Delta y_0 = 0$; (the argument is based on the infinite operators where $\Delta^d y_0 = \Delta^{d-1} \Delta y_0$); therefore $\Delta_s^d \tilde{y}_t = \Delta_s^d y_t$. Is this argument correct?

A3.5.a

(i) Show: $\Delta_{s-1}^{d-1} \Delta = \Delta_s^d - A_s^{(d-1)} L^s$

Step 1:

$$\Delta_{s-1}^{d-1} \Delta = (1 + A_1^{(d-1)} L + \cdots + A_{s-1}^{(d-1)} L^{s-1})(1 - L)$$
$$= (1 + A_1^{(d-1)} L + A_2^{(d-1)} L^2 + \cdots + A_{s-1}^{(d-1)} L^{s-1}$$
$$\quad - L - A_1^{(d-1)} L^2 - \cdots - A_{s-2}^{(d-1)} L^{s-1} - A_{s-1}^{(d-1)} L^s)$$
$$= 1 + A_1^{(d)} L + \cdots + A_{s-1}^{(d)} L^{s-1} - A_{s-1}^{(d-1)} L^s$$
$$= \Delta_{s-1}^d - A_{s-1}^{(d-1)} L^s$$

Step 2:

$$\Delta_s^d = 1 + A_1^{(d-1)} L + \cdots + A_s^{(d)} L^s$$
$$= 1 + A_1^{(d-1)} L + \cdots + A_{s-1}^{(d)} L^{s-1} + A_s^{(d)} L^s$$
$$= \Delta_{s-1}^d + A_s^{(d)} L^s$$

Step 3:

$$\Delta_{s-1}^{d-1} \Delta = \Delta_s^d - (A_s^{(d)} + A_{s-1}^{(d-1)}) L^s$$
$$= \Delta_s^d - (A_s^{(d)} + A_{s-1}^{(d-1)}) L^s$$
$$= \Delta_s^d - A_s^{(d-1)} L^s$$

In Step 1 use is made of $A_r^{(d-1)} - A_{r-1}^{(d-1)} = A_r^{(d)} \Rightarrow A_s^{(d-1)} = A_s^{(d)} + A_{s-1}^{(d-1)}$.

(ii) $\Delta_{s-1}^{d-1}\Delta y_t = \Delta_s^d y_t - A_s^{(d-1)} L^s y_t$. Apply the operator in the previous question to y_t.

(iii) $\Delta_t^d y_t = \Delta_{t-1}^{d-1}\Delta y_t + A_t^{(d-1)} y_0$. Set $s = t$ and substitute into the expression in (ii) and rearrange, noting that $L^t y_t = y_0$.

(iv) If $d = 0$, then $y_t = \Delta_{t-1}^{-1}\Delta y_t + y_0$. In this case $A_s^{(d-1)}$ is given by (setting $d = 0$ and $s = t$):

$$A_t^{(-1)} = (-1)^t \frac{(-1)(-2)\dots(-t)}{t!} = (-1)^t(-1)^t \frac{t!}{t!} = (-1)^{2t} = 1$$

and applying the general result, gives:

$$\Delta_t^0 y_t = \Delta_{t-1}^{-1}\Delta y_t + A_t^{(-1)} y_0$$

$$\Rightarrow$$

$$y_t = \Delta_{t-1}^{-1}\Delta y_t + y_0 \qquad \text{where } \Delta_t^0 y_t = \sum_{r=0}^t A_r^{(0)} y_{t-r} \equiv y_t$$

That is, in finite samples applying the summation operator, in this case Δ_{t-1}^{-1} for a finite sample, does not recover y_t; rather the 'starting' value y_0 has to be added in (analogous to the constant of integration in the case of continuous time).

(v) Show $\Delta_s^d \tilde{y}_t - \Delta_s^d y_t = -A_s^{(d-1)} y_0$. Applying the result in (ii) to y_t and $\tilde{y}_t$, respectively, after a slight rearrangement, gives:

$$\Delta_s^d y_t = \Delta_{s-1}^{d-1}\Delta y_t + A_s^{(d-1)} L^{s+1} y_t$$

$$\Delta_s^d \tilde{y}_t = \Delta_{s-1}^{d-1}\Delta \tilde{y}_t + A_s^{(d-1)} L^{s+1} \tilde{y}_t$$

Hence:

$$\Delta_s^d \tilde{y}_t - \Delta_s^d y_t = \Delta_{s-1}^{d-1}\Delta(\tilde{y}_t - y_t) + A_s^{(d-1)} L^s(\tilde{y}_t - y_t)$$

$$= -\Delta_{s-1}^{d-1}\Delta y_0 + A_s^{(d-1)} L^s[(y_t - y_0) - y_t] \qquad \text{using } \tilde{y}_t \equiv y_t - y_0$$

$$= -A_s^{(d-1)} y_0 \qquad \Delta_{s-1}^{d-1}\Delta y_0 = 0, L^s y_0 = y_0$$

A3.5.b Consider the first part of the argument. Note from (i) that $\Delta_s^d y_0 = \Delta_{s-1}^{d-1}\Delta y_0 + A_s^{(d-1)} y_0 = A_s^{(d-1)} y_0$ as $\Delta_{s-1}^{d-1}\Delta y_0 = 0$; thus although $\Delta_{s-1}^{d-1}\Delta y_0 = 0$, generally $\Delta_s^d y_0 \neq \Delta_{s-1}^{d-1}\Delta y_0$ and, therefore, $\Delta_s^d \tilde{y}_t = \Delta_s^d y_t - \Delta_s^d y_0 \neq \Delta_s^d y_t$. Also note that as $\Delta_s^d \tilde{y}_t - \Delta_s^d y_t = -\Delta_s^d y_0$ then (see the answer to (v)), $\Delta_s^d(1) = A_s^{(d-1)}$, where $\Delta_s^d(1) = A_s^{(d-1)}$, is the sum of the (AR) coefficients in the binomial operator Δ_s^d.

Q3.6 In the development of (3.97),

$$\Delta y_t = \gamma \Delta^d y_{t-1} + [1 - \varphi(L)]\Delta y_t + z_t^*$$

$$z_t^* = z_t + (d-1)\varphi^*(L)\Delta^{1+d} y_{t-1}$$

show that $d = 0$ implies $\varphi(L) y_t = \varepsilon_t$.

A3.6 From the derivation in the text we know that:

$$\gamma = (d-1)\varphi(1)$$

$$z_t^* = z_t + (d-1)\varphi^*(L)\Delta^{1+d}y_{t-1}$$

Setting $d = 0$, it follows that:

$$\gamma = -\varphi(1)$$

$$z_t^* = \varepsilon_t - \varphi^*(L)\Delta y_{t-1}$$

Hence, on substitution and collecting terms:

$$\varphi(L)\Delta y_t = -[\varphi(1) + \varphi^*(L)\Delta]y_{t-1} + \varepsilon_t$$

$$= -\varphi(L)y_{t-1} + \varepsilon_t \qquad \text{as } \varphi(L) = \varphi(1) + \varphi^*(L)\Delta$$

$$\Rightarrow$$

$$\varphi(L)[\Delta y_t + y_{t-1}] = \varepsilon_t$$

$$\varphi(L)y_t = \varepsilon_t \text{ as required.}$$

Q3.7 As required in the Breitung-Hassler version of the LM test, show that:

$$T\sum_{j=1}^{T-1} j^{-1}\hat{\rho}(j) = \frac{1}{\hat{\sigma}_\varepsilon^2} \sum_{t=2}^{T} x_t \left(\sum_{j=1}^{t-1} j^{-1}x_{t-j} \right)$$

A3.7 Expand the terms of the right-hand side:

$$\frac{1}{\hat{\sigma}_\varepsilon^2} \sum_{t=2}^{T} x_t \left(\sum_{j=1}^{t-1} j^{-1}x_{t-j} \right)$$

$$= \frac{1}{\hat{\sigma}_\varepsilon^2} \left[\sum_{t=2}^{T} x_t x_{t-1} + \frac{1}{2}\sum_{t=3}^{T} x_t x_{t-2} + \frac{1}{3}\sum_{t=4}^{T} x_t x_{t-3} + \cdots \right.$$

$$+ \frac{1}{j}\sum_{t=j+1}^{T} x_t x_{t-j} + \cdots$$

$$\left. + \frac{1}{T-1}\sum_{t=T}^{T} x_t x_{t-(T-1)} \right]$$

$$= \frac{1}{\hat{\sigma}_\varepsilon^2} \sum_{j=1}^{T-1} j^{-1} \sum_{t=j+1}^{T} x_t x_{t-j}$$

$$= T\sum_{j=1}^{T-1} j^{-1} \left(\frac{\sum_{t=j+1}^{T} x_t x_{t-j}}{\sum_{t=1}^{T} x_t^2} \right) \qquad \text{using } \hat{\sigma}_\varepsilon^2 = T^{-1}\sum_{t=1}^{T} x_t^2$$

$$= T\sum_{j=1}^{T-1} j^{-1}\rho_j(x) \qquad \text{where } \rho_j(x) = \frac{\sum_{t=j+1}^{T} x_t x_{t-j}}{\sum_{t=1}^{T} x_t^2}$$

where $\rho(j)$ is the j-th order autocorrelation coefficient of the residuals, as required.

Q3.8 You are given the following:

$$
\left(\begin{array}{c} \sqrt{T}(\hat{d}-d) \\ \sqrt{T}(\hat{\psi}-\psi) \end{array} \right) \Rightarrow_D N\left[\left(\begin{array}{c} 0 \\ 0 \end{array} \right), \Omega^{-1} \right] \qquad \text{where } \Omega = \left(\begin{array}{cc} \pi^2/6 & \kappa' \\ \kappa & \Phi \end{array} \right)
$$

Obtain the variance of $\sqrt{T}(\hat{d}-d)$.

A3.8 Partition the inverse of Ω as follows:

$$
\left[\begin{array}{cc} A & B \\ C & D \end{array} \right] = \left[\begin{array}{cc} E & -EBD^{-1} \\ -D^{-1}CE & D^{-1}+D^{-1}CEBD^{-1} \end{array} \right] \qquad \text{where } E = (A-BD^{-1}C)^{-1}
$$

The element of interest is the $(1,1)$ element of Ω, designated as E in the partition; therefore, using the definitions of the elements of the matrix Ω, obtain: $E = (\pi^2/6 - \kappa'\Phi^{-1}\kappa)^{-1}$; for reference see (3.137). Recall that $\kappa = (\kappa_1,\ldots,\kappa_p,\lambda_1,\ldots,\lambda_q)'$ and, thus, the required variance is:

$$
\omega^{-2} = \left(\frac{\pi^2}{6} - (\kappa_1,\ldots,\kappa_p,\lambda_1,\ldots,\lambda_q)\Phi^{-1}(\kappa_1,\ldots,\kappa_p,\lambda_1,\ldots,\lambda_q)' \right)^{-1}
$$

Q3.9.i Consider the case where the AR and MA components of the ARFIMA model are redundant, so that $\omega^2 = \pi^2/6$. Obtain the asymptotic Wald (t-type) statistic, $t_{\hat{d}}$, for testing the null that $d = d_0$.

Q3.9.ii Suppose $\hat{d} = 0.9$; illustrate the calculation of the test statistic $t_{\hat{d}}$ for $T = 100$ and $T = 500$, and carry out the test for $d = d_0 = 0.5$ against $d > 0.5$, and for $d = d_0 = 1$ against $d < 1$.

A3.9.i In this case, $\omega^{-1} = \sqrt{6/\pi^2}$, hence $t_{\hat{d}}$ is given by:

$$
\begin{aligned}
t_{\hat{d}} &= \frac{(\hat{d}-d_0)}{(\sqrt{(6/\pi^2)}/\sqrt{T}} \\
&= \sqrt{T}\sqrt{(\pi^2/6)}(\hat{d}-d_0)
\end{aligned}
$$

where $t_{\hat{d},asym}$ is asymptotically distributed as $N(0,1)$ under the null.

A3.9.ii The asymptotic standard error of $\hat{d}$ is $\sqrt{(6/\pi^2)}/\sqrt{T} = 0.078$ for $T = 100$ and 0.035 for $T = 500$. Hence, for $T = 100$, to test the null hypothesis, H_0: $d = d_0$, against the alternative, H_A: $d < d_0$, form the test statistic $t_{\hat{d}} = \sqrt{T}\sqrt{(\pi^2/6)}$ $(\hat{d}-d_0) = (\hat{d}-d_0)/0.078$ for $T = 100$ and $(\hat{d}-d_0)/0.035$ for $T = 500$, and reject the null for values $t_{\hat{d}}$ more negative than the $\alpha\%$ quantile of $N(0,1)$ at the (asymptotic) significance level α.

In the case of $d_0 = 0.5$ against $d > 0.5$, the resulting test statistics are $(0.9 - 0.5)/0.078 = 5.128$ for $T = 100$ and $(0.9 - 0.5)/0.035 = 11.429$ for $T = 500$; each asymptotic p-value is 0%, so H_0 is rejected. In the case of the unit root null, $H_0 : d_0 = 1$, whereas H_A: $d < 1$, and the resulting test statistic is $(0.9 - 1)/0.078 = -1.28$ for $T = 100$ and $(0.9 - 1)/0.035 = -2.86$ for $T = 500$; the asymptotic p-value of the former is 9.99%, whereas that of the latter is 0.21%. Such tests could also be based on the ML estimator of the standard error rather than the asymptotic standard error $\sqrt{6/T\pi^2}$; the resulting t-type test is also asymptotically standard normal.

Q3.10 Obtain ω^2 for the ARFIMA(1, d, 0) or ARFIMA(0, d, 1) models.

A3.10 If the AR(1) component is present let $\lambda = \varphi_1$ and if the MA(1) component is present let $\lambda = -\theta_1$. (The negative sign on the latter arises from our notational convention that signs in the MA polynomial are opposite to those in the AR polynomial.)

In either the AR(1) or MA(1) case, $g_j = \lambda^j$ (see 3.165) and, thus:

$$\kappa_1 = \sum_{j=1}^{\infty} \frac{1}{j} g_{j-1}$$

$$= \sum_{j=1}^{\infty} \frac{1}{j} \lambda^{j-1}$$

$$= -\frac{1}{\lambda} \log(1 - \lambda)$$

also $\Phi^{-1} = 1 - \lambda^2$. Putting the pieces together:

$$\omega^2 = \frac{\pi^2}{6} - \kappa_1 \Phi^{-1} \kappa_1$$

$$= \frac{\pi^2}{6} - \frac{(1 - \lambda^2)}{\lambda^2} [\log(1 - \lambda)]^2$$

See Tanaka, 1999, p. 564, expression (52).

Q3.11 Consider the augmented EFDF test for the wheat example, which is a particular case of the following:

$$\Delta\tilde{y}_t = \eta[\varphi(L)z_{t-1}(\hat{d}_T)] + [1 - \varphi(L)]\Delta\tilde{y}_t + v_t \tag{3.201}$$

Obtain the regression for nonlinear estimation in the case that $p = 2$.

A3.11 If $p = 2$, then $\varphi(L) = 1 - \varphi_1 L - \varphi_2 L^2$ and $1 - \varphi(L) = \varphi_1 L + \varphi_2 L^2$, so that the augmented regression is:

$$\Delta\tilde{y}_t = \eta(1 - \varphi_1 L - \varphi_2 L^2)z_{t-1}(\hat{d}_T) + (\varphi_1\Delta\tilde{y}_{t-1} + \varphi_2\Delta\tilde{y}_{t-2}) + v_t$$

$$= \eta z_{t-1}(\hat{d}_T) - \eta\varphi_1 z_{t-2}(\hat{d}_T) - \eta\varphi_2 z_{t-3}(\hat{d}_T) + \varphi_1\Delta\tilde{y}_{t-1} + \varphi_2\Delta\tilde{y}_{t-2} + v_t$$

$$= \alpha_1 z_{t-1}(\hat{d}_T) + \alpha_2 z_{t-2}(\hat{d}_T) + \alpha_3 z_{t-3}(\hat{d}_T) + \alpha_4\Delta\tilde{y}_{t-1} + \alpha_5\Delta\tilde{y}_{t-2} + v_t$$

There are 3 coefficients η, φ_1 and φ_2 and 5 variables, so there are 2 nonlinear constraints: $\alpha_1 = -\alpha_2/\alpha_4 = -\alpha_3/\alpha_5$.

Appendix 3.1 Factorial expansions for integer and non-integer d

This appendix gives greater consideration to some of the 'five-fingered' exercises in deriving and using various results connected to the binomial expansion, and provides justification for some of the results in Chapter 3.

A3.1 What is the meaning of $(1-L)^d$ for fractional d?

The focus of this section is to attach a meaning to $(1-L)^d$ for fractional d. This is approached in two steps. In step 1, we consider the standard binomial case, expanding $(1+s)^d$ for $|s| < 1$. Then from Newton's Binomial formula, for d a positive integer:

$$(1+s)^d = (1 + {_d}C_1 s + {_d}C_2 s^2 + {_d}C_3 s^3 \cdots + {_d}C_d s^d) \tag{A3.1}$$

$$= \sum_{r=0}^{d} {_d}C_r s^r$$

$$= 1 + \sum_{r=1}^{d} {_d}C_r s^r$$

This formula is a straightforward application of the binomial theorem, where:

$$_d C_r \equiv \frac{d(d-1)(d-2)\ldots(d-(r-1))}{r!}$$

$$\equiv \frac{(d)_r}{r!}$$

$$(d)_r \equiv d(d-1)(d-2)\ldots(d-r+1))$$

where $_d C_0 \equiv 1$ and $0! \equiv 1$. If r is a positive integer, $(d)_r$ is well-defined for all real d even if d is not an integer. Note that for $r \geq 1$, the $_d C_r$ binomial coefficients can be calculated recursively using:

$$_d C_r = {_d}C_{r-1}\left(\frac{d-r+1}{r}\right)$$

The coefficients on the powers of s are coefficients from the n-th row of Pascal's triangle, starting and ending with coefficients equal to 1 if d is an integer because $_d C_0 \equiv 1$ and $_d C_d \equiv 1$.

Example: integer d
$(d)_r$ and $_dC_r$ are defined for $r = 0, 1, 2, 3, 4, 5$ as follows:

$_5C_0 = 1;$

$(5)_1 = 5,$	$_5C_1 = 5/1 = 5;$
$(5)_2 = 5(4) = 20,$	$_5C_2 = 20/2 = 10;$
$(5)_3 = 5(4)(3) = 60,$	$_5C_3 = 60/6 = 10;$
$(5)_4 = 5(4)(3)(2) = 120,$	$_5C_4 = 120/24 = 5;$
$(5)_5 = 5(4)(3)(2)(1) = 120,$	$_5C_5 = 120/120 = 1.$

If d is an integer, the right hand side of (A3.1) has a finite number of terms. However, the expansion is also valid if d is not a positive integer, with the difference that the right hand side of (A10.1) is now an infinite series. Thus, for d not an integer:

$$(1 + s)^d = 1 + \sum_{r=1}^{\infty} {_dC_r} s^r$$

Example: non-integer d
$d = 1.5;$ $(d)_r$ and $_dC_r$ are defined for all positive integers r; the first five terms, for $r = 0, 1, 2, 3, 4, 5$, are as follows:

$_{1.5}C_0 = 1;$

$(1.5)_1 = 1.5, {}_{1.5}C_1 = 1.5/1 = 1.5;$

$(1.5)_2 = 1.5(0.5) = 0.75, {}_{1.5}C_2 = 0.75/2 = 0.375;$

$(5)_3 = 1.5(0.5)(-0.5) = -0.375, {}_5C_3 = -0.375/6 = -0.0625;$

$(5)_4 = 1.5(0.5)(-0.5)(-1.5) = 0.5625, {}_5C_4 = 0.5625/24 = 0.0234;$

$(5)_5 = 1.5(0.5)(-0.5)(-1.5)(-2.5) = -1.40625, {}_5C_5 = -1.40625/120 = -0.01172.$

A3.2   Applying the binomial expansion to the fractional difference operator

We now use the expansion for $(1 - L)$ rather than $(1 + s)$ by using a change of variable $s = -L$. First separate out the sign so that $s^r = +(-1)^r L^r$, then for *non-integer* d:

$$(1 - L)^d = (1 + {_dC_1}(-1)L + {_dC_2}(-1)^2 L^2 + {_dC_3}(-1)^3 L^3 \ldots) \tag{A3.2}$$

$$= 1 + \sum_{r=1}^{\infty} {_dC_r}(-1)^r L^r$$

$$= 1 - dL + \tfrac{1}{2}d(d-1)L^2 - \tfrac{1}{6}d(d-1)(d-2)L^3 \ldots$$

Now apply the operator $(1 - L)^d$ to the series y_t:

$$(1 - L)^d y_t = [1 + \sum_{r=1}^{\infty} {}_d C_r (-1)^r L^r] y_t$$

$$= y_t + \sum_{r=1}^{\infty} {}_d C_r (-1)^r y_{t-r}$$

$$= \sum_{r=0}^{\infty} A_r^{(d)} y_{t-r}$$

Thus, for fractional d, $(1 - L)^d y_t$ is an infinite autoregression with coefficients $A_0 = 1, A_r^{(d)} = (-1)^r {}_d C_r$. Given the recursion for ${}_d C_r$ we can easily obtain a recursion for the $A_r^{(d)}$ coefficients. That is:

$$A_r^{(d)} = (-1)^r {}_d C_r$$

$$= (-1)(-1)^{r-1} {}_d C_{r-1}[(d - r + 1)/r]$$

$$= (-1)[(d - r + 1)/r] A_{r-1}^{(d)} \quad \Rightarrow$$

$$A_r^{(d)} = [(r - d - 1)/r] A_{r-1}^{(d)} \quad \text{for } r \geq 1.$$

The process $(1 - L)^d y_t = \varepsilon_t$ also has an infinite moving average representation obtained by multiplying through by the inverse operator $(1 - L)^{-d}$, that is:

$$y_t = (1 - L)^{-d} \varepsilon_t$$

The moving average coefficients are again obtained from a binomial expansion but now for the binomial with $-d$ rather than d.

$$y_t = \varepsilon_t + \sum_{r=1}^{\infty} {}_{-d} C_r (-1)^r \varepsilon_{t-r}$$

$$= \varepsilon_t + \sum_{r=1}^{\infty} B_r^{(-d)} \varepsilon_{t-r}$$

A recursion for the $B_r^{(-d)}$ is available, as in the case of the $A_r^{(d)}$ coefficients, but allowing for $-d$ rather than d:

$$B_r^{(-d)} = (-1)^r {}_{-d} C_r$$

$$= (-1)(-1)^{r-1} {}_{-d} C_{r-1}[\{(-d) - r + 1)\}/r]$$

$$= (-1)[-d - r + 1)/r] B_{r-1}^{(-d)} \Rightarrow$$

$$B_r^{(-d)} = [r + d - 1)\}/r] B_{r-1}^{(-d)} \quad \text{for } r \geq 1 \qquad B_0^{(-d)} = 1$$

A3.3   The MA coefficients in terms of the gamma function

We use the following binomial relationship for any $d > 0$:

$${}_{-d} C_r = (-1)^r {}_{(d+r-1)} C_r$$

$$= (-1)^r \left(\frac{(d + r - 1)!}{r!(d - 1)!} \right)$$

Hence, in terms of Gamma functions:

$$_{-d}C_r = (-1)^r \frac{\Gamma(d+r)}{\Gamma(r+1)\Gamma(d)}$$

We can now apply this result to the MA coefficients (the reader should be able to obtain equivalent expressions for the AR coefficients, which is left as an exercise). We know that $B_r^{(-d)} \equiv (-1)^r {_{-d}C_r}$, hence:

$$B_r^{(-d)} = (-1)^r \left((-1)^r \frac{\Gamma(d+r)}{\Gamma(r+1)\Gamma(d)} \right) \quad \text{as } (-1)^{2r} \text{ is always positive.}$$

$$= \frac{\Gamma(d+r)}{\Gamma(r+1)\Gamma(d)}$$

Appendix 3.2   FDF test: assuming known d_1

This appendix summarises the asymptotic distributions for the FDF test assuming known d_1. Under H_O: $d_O = 1$, the LS estimator $\hat{\gamma}$ is a consistent estimator of γ with convergence rate(s) and asymptotic distribution that depends upon the range of d_1, assuming d_1 to be *known*. These properties are summarised in Tables A3.1 and A3.2 for $\hat{\gamma}$ and $t_\gamma(d_1)$ respectively, and are due to DGM (2002, theorems 1 and 2).

$B(r)$ is standard Brownian motion, and B_d is standard fractional Brownian motion with parameter d (corresponding to the definition of 'type II' fBM as defined in Marinucci and Robinson, 1999).

These results hold for the maintained regression (3.86) with $d_O = 1$, and data either generated by $\Delta^{d_1} y_t = \varepsilon_t$ or the truncated process with $y_t = 0$ for $t \leq 0$ and $\Delta^{d_1} y_t = \varepsilon_t$ for $t > 0$. The respective sequences $\{\varepsilon_t\}_{-\infty}^t$ and $\{\varepsilon_t\}_1^t$ are assumed to be iid random variables with finite variance and $E|\varepsilon_t^4| < \infty$.

Note that these results generalise the standard DF set-up, where $d_O = 1, d_1 = 0$ and $\hat{\gamma}$ is T-consistent. When $d_1 = 0$, the asymptotic distributions are just those of the DF case, that is functions of standard (not fractional) Brownian motion. The appropriate normalisation of $\hat{\gamma}$ is, therefore, T. When $d_1 \in (0, 0.5)$, the relevant distributions are based on fractional Brownian motion (fBM); for example, the

Table A3.1   Convergence and asymptotic distribution of LS estimator $\hat{\gamma}$ (d_1 known)

Range of d_1	Convergence rate	Normalised quantity	Asymptotic distribution
$d_1 \in [0, 0.5)$	T^{1-d_A}	$T^{1-d_A}\hat{\gamma}$	$\Rightarrow_D \dfrac{\int_0^1 B_{-d_1}(r)dB(r)}{\int_0^1 B_{-d_1}^2(r)d(r)}$
$d_1 = 0.5$	$(T\log T)^{1/2}$	$(T\log T)^{1/2}\hat{\gamma}$	$\Rightarrow_D N(0, \pi)$
$d_1 \in (0.5, 1.0)$	$T^{1/2}$	$T^{1/2}\hat{\gamma}$	$\Rightarrow_D N\left(0, \dfrac{(\Gamma(d_1))^2}{\Gamma(2d_1 - 1)}\right)$

Table A3.2   Asymptotic distribution of LS t statistic $\hat{t}_\gamma(d_1)$, d_1 known

Range of d_1	Asymptotic distribution
$d_1 \in [0, 0.5)$	$\hat{t}_\gamma(d) \Rightarrow_D \dfrac{\int_0^1 B_{-d_1}(r)dB(r)}{\left(\int_0^1 B^2_{-d_1}(r)d(r)\right)^{1/2}}$
$d_1 \in [0.5, 1.0)$	$\hat{t}_\gamma(d) \Rightarrow_D N(0,1)$

Source: DGM (2002, theorems 1 and 2).

5% critical values of $t_\gamma(d_1)$, which depend upon d_1, become less negative as $d \to 0.5$. In the case likely to be of particular practical interest, that is $d_1 \in (0.5, 1.0)$, so that the generating process has long memory and is nonstationary, the normalisation is $T^{1/2}$ and the asymptotic distribution of the t statistic is standard normal.

The FDF test based on $t_\gamma(d_1)$ is consistent (asymptotically diverges to infinity), but is not asymptotically uniformly most powerful invariant (AUMPI) with Gaussian errors. However, the FDF test retains its size well compared to three other tests considered by DGM and was only outperformed in terms of finite sample power for very local alternatives (d_1 very close to the unit root).

The critical values for the nonstandard distributions, that is when $d_1 \in [0, 0.5)$, and d_1 is known, also depend, as in the DF case, on whether a constant and/or trend are included in the deterministics, in which case fractional Brownian motion (fBM) is replaced by demeaned (or detrended) fBM in the asymptotic distributions. Critical values are available in DGM (2002).

4
Semi-Parametric Estimation of the Long-memory Parameter

Introduction

There is a large body of research on different methods of estimating the long-memory parameter, d, using semi-parametric methods. These are methods that fall short of a complete parametric specification of the model. An example of a parametric model is the ARFIMA model of the previous chapter, which fully specifies the long-memory parameter d, and the dynamics governing the short-run adjustment of the process. Typical semi-parametric models leave the short-run dynamic part of the model unspecified, apart from some key characteristics, or, while specifying the form of the short-run dynamics, they do not use all of the frequency range in estimating the long-memory parameter.

For example, neither the popular GPH (Geweke and Porter-Hudak, 1983) estimator nor the GSE (Gaussian semi-parametric estimator, Robinson, 1995b) make an explicit assumption about the structure of the short-run dynamics, except to assume that they may be approximated by a constant in the frequency domain when frequencies are in the neighbourhood of zero. This description also points to an important aspect of the methods considered here in that they are based on frequency domain rather than time domain methods; of these it is the latter that tend to be more familiar in economics. Hence, the first section of this chapter gives a brief outline of some of the basic concepts in frequency domain analysis. While this is no substitute for a more complete analysis, it is hoped that this chapter will be reasonably self-contained for those unfamiliar with the key concepts of frequency and period.

A class of frequency domain estimators that have proved popular, largely due to their simplicity, are based on least squares estimation of a log-periodogram (LP) regression using frequencies that are in the 'shrinking neighbourhood of zero', a term to be defined more formally later in this chapter, but these are, essentially, those frequencies associated with 'long' periods and the resulting methods are known as 'local' (to the zero frequency). These are referred to in this chapter as LP regressions. The most influential and simplest is the GPH estimator. There are several developments of this estimator available; for example, Andrews and Guggenberger (2003), hereafter AG, suggest a development to reduce the finite sample bias of the GPH estimator; and Kim and Phillips (2000) suggest a redefinition of the discrete Fourier transform (DFT) on which the LP regression is based, which offers the desirable properties of an LP estimator of d across a wider range of values of d.

Another popular class of estimators is known as Gaussian semi-parametric estimators, usually referred to as GSE (or Local Whittle); in fact, this description is somewhat misleading as these estimators are not dependent on an assumption of Gaussianity of the data. Probably the most popular form of this estimator is due to Robinson (1995b); see also Künsch (1987). These estimators are based on minimising the Whittle contrast, which, in essence, is a distance function measuring the 'nearness' of the true spectral density and an estimator of that quantity. Again, there are several important developments of the class of Gaussian semi-parametric estimators to enable consistent estimation in the nonstationary region of d, without first taking any differences of a time series.

There is another group of semi-parametric estimators, probably less familiar in application, that uses the full range of frequencies, and are known as global (or broad-band) frequency methods. However, these methods are still semi-parametric in nature in that they fall short of a full parametric specification.

Whether local or global semi-parametric methods are used, some choice has to be exercised over certain key parameters. For example, in the case of local estimators, it is necessary to make a judgment about just what is the 'shrinking neighbourhood of zero'; in essence how many frequencies, working away from the positive side of zero (referred to as 0^+), should be included in the GPH and Gaussian semi-parametric estimators. This choice becomes critical in the usual case that short-run dynamics impart serial correlation, and hence persistence, into the time series, and 'muddy the waters' as to what is long-range dependence and what is short-run dependence. Similarly, in the case of global methods, a decision has to be made about the appropriate order of truncation of the cosine expansion.

The chapter starts in Section **4.1** with an application of the Fourier transform to linear filters, which is an essential first step to obtain the spectral density function (SDF) for ARFIMA models. The log-periodogram and the discrete Fourier transform (DFT) are introduced in Section **4.2** as a precursor to their use in the familiar GPH estimator. The GPH estimator is introduced in Section **4.3**, together with a consideration of its properties and the problem of choosing the number of frequencies (the bandwidth) to be included in the log-periodogram. Variants of the log periodogram involving the pooling of adjacent frequencies and tapering frequencies are considered in Section **4.4** and these developments are applied to the GPH estimator in Section **4.5**. Some finite sample results are reported in Section **4.6**, together with some observations on practical considerations as to the bandwidth choice based on the infeasible optimal selection; this section includes an outline of the developments due to AG that can reduce the asymptotic bias in the GPH estimator.

A modification of the DFT due to Phillips (1999a, b) is outlined in Section **4.7**, which leads to an estimator that can be applied to some nonstationary processes without the need for differencing; in particular, the implications for unit root testing are considered in this section. Another popular semi-parametric estimator is based on the Whittle contrast, which is considered in Section **4.8**, with modifications and extensions considered in Section **4.9**. Some illustrative simulation results are reported in Section **4.10**.

The estimators so far considered are what are known as 'narrowband' or 'local' semi-parametric estimators, whereas another class of estimators, considered in Section **4.11**, are 'broadband' in the sense that whilst all frequencies are kept, the SDF of the short-run, or dynamic, components is expanded and then truncated. Section **4.12** considers two empirical illustrations, the first for the rate on the US three-month Treasury Bill and the second the price of gold in US$. A brief appendix outlines the Taylor series expansion of a logarithmic function, which is used on a number of occasions in the text.

4.1 Linear filters

4.1.1 A general result

In this section we consider the case where the Fourier transform is applied directly to the sequence $\{y_t\}$ or some linear transformation of this sequence. An important case of interest arises when white noise is filtered through a linear filter, such as arises when the stochastic process is either an MA, an invertible AR or a mixed ARMA, process. These specific cases are considered in the next section. Here we state the general result.

Consider the time series representation where y_t is a linear filter of the input ε_t:

$$y_t = w(L)\varepsilon_t \tag{4.1a}$$

$$= \sum_{j=0}^{\infty} w_j \varepsilon_{t-j} \tag{4.1b}$$

where $\{\varepsilon_t\}$ is white noise sequence, $w(L)$ is a linear filter in the lag operator L as indicated by (4.1b). Then (see, for example, Anderson, 1971, theorem 7.5.1 and Fuller, 1996, theorem 4.3.1), the spectral density function

(SDF), $f_y(\lambda_j)$, of $\{y_t\}$ is:

$$f_y(\lambda_j) = |w(e^{-i\lambda_j})|^2 f_\varepsilon(\lambda_j) \tag{4.2a}$$

$$= |w(e^{-i\lambda_j})|^2 \frac{1}{2\pi}\sigma_\varepsilon^2 \tag{4.2b}$$

$$= |\sum_{r\in R} w_r e^{-i\lambda_j r}|^2 \frac{1}{2\pi}\sigma_\varepsilon^2 \tag{4.2c}$$

Note that some authors do not use the minus sign in the exponential in (4.2c). The notation $r \in R$ for the index indicates that the process could be, for example, an infinite two-sided filter, provided that the (square summability) convergence condition $\sum_{r\in R} w_r^2 < \infty$ is met; equally, other filters, for example positive one-sided filters, are allowed.

Also, notice the adoption of the notation $f_\varepsilon(\lambda_j) = (2\pi)^{-1}\sigma_\varepsilon^2$ in (4.2b), which is the spectral density function of white noise. The result in (4.2) states that the spectral density $f_y(\lambda_j)$ of y_t is the spectral density $f_\varepsilon(\lambda_j)$ of the input ε_t multiplied by $|w(e^{-i\lambda_j})|^2$, referred to as the power transfer function of the filter. An example of this filtering and its effect on the spectral density is given by an ARMA process, which is considered next.

4.1.2 The spectral density of an ARMA process

Consider a stationary ARMA(p, q) process, then:

$$\phi(L)y_t = \theta(L)\varepsilon_t \tag{4.3}$$

$$y_t = \phi(L)^{-1}\theta(L)\varepsilon_t \tag{4.4}$$

$$= w(L)\varepsilon_t$$

where $w(L) = \phi(L)^{-1}\theta(L)$, $\phi(L) = (1 - \phi_1 L - \phi_2 L^2 - \cdots - \phi_p L^p)$ and $\theta(L) = (1 + \theta_1 L + \theta_2 L^2 + \cdots + \theta_q L^q)$; ε_t is white noise $\sim \text{iid}(0, \sigma_\varepsilon^2)$; and y_t is assumed to have a zero mean (for example, by subtracting the mean or trend of the original series).

The SDF of y_t is:

$$f_y(\lambda_j) = |w(e^{-i\lambda_j})|^2 f_\varepsilon(\lambda_j)$$

$$= \frac{|\theta(e^{-i\lambda_j})|^2}{|\phi(e^{-i\lambda_j})|^2} f_\varepsilon(\lambda_j) \tag{4.5}$$

Note that $f_y(\lambda_j)$ is an even function, that is, $f_y(-\lambda_j) = -f_y(\lambda_j)$. The notation $\theta(e^{-i\lambda_j})$ means that the polynomial $\theta(L)$ is evaluated at $e^{-i\lambda_j}$ and, analogously, for the polynomials $\phi(L)$ and $w(L)$. For example, if $\phi(L) = (1 - \phi_1 L - \phi_2 L^2)$ and $\theta(L) = (1 + \theta_1 L)$, then $\theta(e^{-i\lambda_j}) = (1 + \theta_1 e^{-i\lambda_j})$ and $\phi(e^{-i\lambda_j}) = (1 - \phi_1 e^{-i\lambda_j} - \phi_2 e^{-i\lambda_j})$. Also, in this case:

$$|\theta(e^{-i\lambda_j})|^2 = (1 + \theta_1 e^{-i\lambda_j})(1 + \theta_1 e^{i\lambda_j}) \tag{4.6}$$

$$|\phi(e^{-i\lambda_j})|^2 = (1 - \phi_1 e^{-i\lambda_j} - \phi_2 e^{-i2\lambda_j})(1 - \phi_1 e^{i\lambda_j} - \phi_2 e^{i2\lambda_j}) \tag{4.7}$$

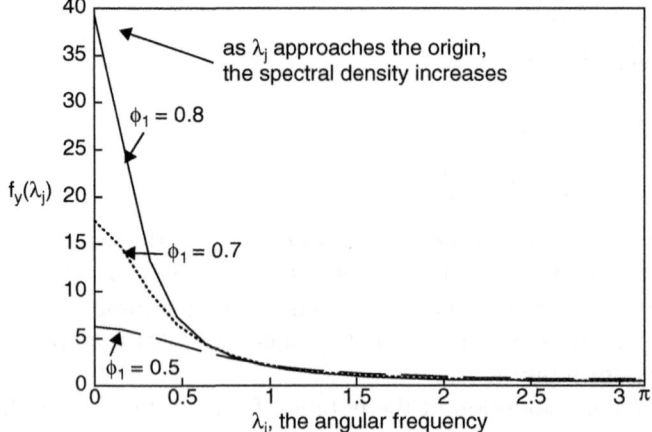

Figure 4.1   The spectral density of y_t for an AR(1) process

The ARMA model can be viewed as the output y_t of transformed or filtered white noise ε_t, where the filter is the linear moving average filter given by $w(L) = \phi(L)^{-1}\theta(L)$, which, in this case, is a one-sided filter.

4.1.3   Examples of spectral densities

This section illustrates the previous concepts by means of some frequently occurring models.

AR(1)

The spectral density of the AR(1) model is given by:

$$f_y(\lambda_j) = \frac{1}{(1 + \phi_1^2 - 2\phi_1\cos\lambda_j)}\frac{\sigma_\varepsilon^2}{2\pi} \tag{4.8}$$

Noting that as $\cos\lambda = 1$ for $\lambda = 0$ then $f_y(0)$, that is, $f_y(\lambda_j)$ evaluated at $\lambda_j = 0$, is:

$$f_y(0) = \frac{1}{(1 + \phi_1^2 - 2\phi_1)}\frac{\sigma_\varepsilon^2}{2\pi}$$

$$= \frac{1}{(1 - \phi_1)^2}\frac{\sigma_\varepsilon^2}{2\pi} \to \infty \text{ as } \phi_1 \to 1$$

Thus, as $\phi_1 \to 1$, the spectral density evaluated at 0, $f_y(0)$, approaches infinity. To show the influence of ϕ_1 on the spectral density, $f_y(\lambda_j)$ for the AR(1) process is shown in Figure 4.1 for $\phi_1 = 0.5$, $\phi_1 = 0.7$ and $\phi_1 = 0.8$ (the plots assume that $\sigma_\varepsilon^2 = 1$); notice how the spectral density becomes increasingly peaked around λ_j close to zero as ϕ_1 increases.

ARMA(2, 1)

The spectral density of the ARMA(2, 1) model with white noise inputs is:

$$f_y(\lambda_j) = \frac{(1 + \theta_1 e^{-i\lambda_j})(1 + \theta_1 e^{i\lambda_j})}{(1 - \phi_1 e^{-i\lambda_j} - \phi_2 e^{-2i\lambda_j})(1 - \phi_1 e^{i\lambda_j} - \phi_2 e^{2i\lambda_j})} \frac{\sigma_\varepsilon^2}{2\pi} \tag{4.9}$$

or expanding the parentheses:

$$f_y(\lambda_j) = \frac{(1 + 2\theta_1 \cos \lambda_j + \theta_1^2)}{(1 + \phi_1^2 + \phi_2^2 + 2(\phi_2\phi_1 - \phi_1)\cos \lambda_j - 2\phi_2 \cos 2\lambda_j)} \frac{\sigma_\varepsilon^2}{2\pi} \tag{4.10}$$

(Where $\cos^2 \lambda \equiv (\cos \lambda)^2$ and repeated use has been made of $e^{i\lambda} = \cos \lambda + i \sin \lambda$, $e^{-i\lambda} = \cos \lambda - i \sin \lambda$, so that $e^{i\lambda} + e^{-i\lambda} = 2\cos \lambda$, and note $e^{i\lambda}e^{-i\lambda} = e^{i\lambda - i\lambda} = e^0 \equiv 1$.)

Now consider $f_y(0)$, for which it is easier to use (4.9),

$$f_y(0) = \frac{(1 + \theta_1)(1 + \theta_1)}{[1 - (\phi_1 + \phi_2)][1 - (\phi_1 + \phi_2)]} \frac{\sigma_\varepsilon^2}{2\pi} \tag{4.11}$$

It is evident from the denominator of (4.11) that as $\phi_1 + \phi_2 \to 1$, then $f_y(0) \to \infty$; thus, as the unit root is approached the spectral density increases in the 'neighbourhood' of the zero frequency.

Pure random walk

The simple random walk is the AR(1) model with $\phi_1 = 1$; in this case (4.8) becomes:

$$\begin{aligned}
f_y(\lambda_j) &= \frac{1}{(2 - 2\cos \lambda_j)} \frac{\sigma_\varepsilon^2}{2\pi} \\
&= \frac{1}{2(1 - \cos \lambda_j)} \frac{\sigma_\varepsilon^2}{2\pi} \\
&= |(1 - e^{-i\lambda_j})|^{-2} \frac{\sigma_\varepsilon^2}{2\pi}
\end{aligned} \tag{4.12}$$

As a random walk does not have a finite variance, (4.12) is sometimes referred to as a pseudo-spectrum. Another way of interpreting (4.12) is that $(2(1 - \cos \lambda_j))^{-1} = |(1 - e^{-i\lambda_j})|^{-2}$ is the power transfer function of the first difference filter $(1 - L)$, which is considered next.

4.1.4 The difference filter

The filter with $w_0 = 1$ and $w_1 = -1$, with all other w_j equal to zero, is the first difference filter, $w(L) = 1 - L$. In this case, the power transfer function $|w(e^{-i\lambda_j})|^2$ is:

$$|w(e^{-i\lambda_j})|^2 = |(1 - e^{-i\lambda_j})|^2 \tag{4.13a}$$

$$= 2(1 - \cos \lambda_j) \tag{4.13b}$$

See Anderson (op. cit., especially corollary 5.3.1). The first difference filter is often used to define a new series, say $z_t = (1 - L)y_t$; then, see (4.2), the SDF of z_t is:

$$f_z(\lambda_j) = |1 - e^{-i\lambda_j}|^2 f_y(\lambda_j) \tag{4.14}$$

The extension of this to the d-th difference, results in:

$$f_z(\lambda_j) = |1 - e^{-i\lambda_j}|^{2d} f_y(\lambda_j) \tag{4.15}$$

If $(1 - L)^d y_t = \varepsilon_t$, then $y_t = (1 - L)^{-d} \varepsilon_t$, where ε_t is white noise; and the SDF of y_t is:

$$f_y(\lambda_j) = |(1 - e^{-i\lambda_j})|^{-2d} \frac{\sigma_\varepsilon^2}{2\pi} \tag{4.16}$$

The case of $d = 1$ has already been given in (4.14). The extension of this result to fractional d is considered in the next section.

4.1.5   Processes with fractional long-memory parameter

4.1.5.i   ARFIMA processes

This section considers the case where d is allowed to be fractional. The ARFIMA model of Chapter 3 is:

$$(1 - L)^d y_t = u_t \tag{4.17a}$$

$$u_t = w(L)\varepsilon_t \tag{4.17b}$$

$$y_t = (1 - L)^{-d} w(L)\varepsilon_t \tag{4.17c}$$

where $w(L) = \phi(L)^{-1}\theta(L)$ and $\varepsilon_t \sim$ iid$(0, \sigma_\varepsilon^2)$. The SDF of the ARFIMA process, (4.17c), is:

$$f_y(\lambda_j) = |(1 - e^{-i\lambda_j})|^{-2d} f_u(\lambda_j) \tag{4.18a}$$

$$f_u(\lambda_j) = \frac{|\theta(e^{-i\lambda_j})|^2}{|\phi(e^{-i\lambda_j})|^2} \frac{\sigma_\varepsilon^2}{2\pi} \tag{4.18b}$$

Although (4.18a) refers specifically to an ARMA filter, with white noise input ε_t, it is valid more generally for time-invariant linear filters with absolutely summable coefficients. Also, to ensure stationarity of the process, we assume $d < 0.5$; other, possibly more relevant cases, with $d \geq 0.5$, are considered below.

The expression in (4.18a) sometimes appears using the following approximation and assumption. To a first order approximation $|1 - e^{-i\lambda_j}|^{-2d} \approx |\lambda_j|^{-2d}$; also assume that $f_u(\lambda_j)$ is constant in the neighbourhood of 0, specifically if $\lambda_j \to 0^+$, then $f_u(\lambda_j) \to G$, a constant. Thus, on this basis, (4.18a) is approximated as:

$$f_y(\lambda_j) \approx |\lambda_j|^{-2d} G \tag{4.19}$$

where $\approx$ indicates approximation under the stated conditions. This form is sometimes used in estimation, as we shall see later in this chapter.

4.1.5.ii Fractional noise (FN) processes

We start with the idea of a self-similar process, which is one that, in a geometrical sense, looks similar whether you look at a piece of a shape or the whole of a shape (see, for example, Mandelbrot, 1982). Translating this idea to a set comprising T time series observations, the idea is that a, indeed any, subset, or window, of the T observations should, with appropriate scaling, look similar to the complete set of observations. Given the sequence $\{y_t\}_1^T$, with individual observation y_t, to obtain a new time series of observations scale the t index such that $t^* = t/c$, $c > 0$, and scale the y index such that $y^* = c^H y$; taken together the new relationship is $y_t^* = c^H y_t^*$; then if $y_t =^D y_t^*$, where $=^D$ indicates equality in distribution, the series is said to be self-similar with self-similarity parameter H. The parameter H is sometimes referred to as the Hurst exponent following Hurst (1951).

In practice, equality in distribution is weakened in two ways by looking at the moments of the distributions. First, an exactly second-order, self-similar process is one for which the k-th order autocorrelations are the same for y_t and y_t^*, and an asymptotically self-similar process is one for which the k-th order autocorrelations are asymptotically the same. Self-similar processes have the property of long-range dependence, usually characterised as a slower than exponential rate of decline of the autocorrelations and a non-summable auto-correlation function. This contrasts with, for example, stationary AR models for which the autocorrelations decline exponentially fast and are summable.

An implication of the definition of self-similarity, related to the temporal aggregation of y_t, can be taken advantage of to gain an indication of whether self-similarity holds. First, take an average y_t over non-overlapping blocks (NOB) of (window) length ω; for example, if the initial data is quarterly then taking a NOB of $\omega = 4$ simply gets the annual data, which are then averaged so that the resulting data represent the average quarterly figure for a year. This is just one choice of ω from $\omega = 1, 2, 3, ..., W$, where $W = \text{int}[T/2]$, although in practice W is usually not taken to exceed 15 or so, unless the number of observations is very large (for example, 20,000). The property of self-similarity implies that autocorrelation functions for these averaged series are unchanged by the aggregation.

An example of a self-similar process is provided by fractional Gaussian noise (fGn). In this case, y_t is normally distributed and generated by a self-similar process with index H and stationary increments, referred to as H-SSSI (see, for example, Davies and Harte, 1987). Fractional Gaussian noise is a special case of a fractional noise process. Although much of the emphasis in economics is on ARFIMA processes, fractional noise or fGn processes are important in modelling financial times series and fGn is the dominant model in several other areas, for example hydrology and network modelling. It shares the characteristic of ARFIMA processes that, in some cases, it has 'long memory' as judged by the relative slow decay of the autocorrelation function and the non-summability of the autocovariances.

The fGn process has an autocovariance sequence given by:

$$\gamma_k = 0.5\sigma_\varepsilon^2[(k+1)^{2H} + (k-1)^{2H} - (2k)^{2H}] \tag{4.20}$$

where σ_ε^2 is the innovation variance; if $k = 0$, then $\gamma_0 = \sigma_\varepsilon^2$. If $H = \frac{1}{2}$, then $\gamma_k = 0$ for $k > 0$. An important result, as far as long-range dependence is concerned, is the slow decline of the k-th order autocorrelation as $k \to \infty$. That is:

$$\gamma_k \sim \sigma_\varepsilon^2 H(2H-1)k^{2(H-1)} \qquad \text{as } k \to \infty, k > 0 \tag{4.21}$$

The case of particular interest is where $\frac{1}{2} < H < 1$, which corresponds to long-range dependence.

The spectral density of fGn is:

$$f_y^{fGn}(\lambda_j) = C_H(\lambda_j)2(1 - \cos\lambda_j)\sum_{j=-\infty}^{\infty} |2\pi j + \lambda_j|^{-2H-1} \quad \lambda_j \in (-\pi, \pi), H \neq \frac{1}{2} \tag{4.22}$$

where C_H is a constant. The first-order approximation to the right-hand side of (4.22) is:

$$f_y^{fGn}(\lambda_j) \sim |\lambda_j|^{1-2H} C_H \tag{4.23}$$

Comparison of (4.19) and (4.23) shows that $d = H - \frac{1}{2}$; and, therefore, to a first-order approximation, the spectral densities of ARFIMA and fractional Gaussian noise processes are not distinguishable.

4.2   Estimating d

4.2.1   Prior transformation of the data

In the context of economic series it is likely that $d \notin (-0.5, 0.5)$, in particular $d > 0.5$ seems probable for many macroeconomic time series, so that the process is nonstationary. Whilst some of the methods described below apply to some parts of the parameter space for which $d > 0.5$, a common practice is to integer difference the series so that d for the transformed series is in the region $(-0.5, 0.5)$.

Let y_t denote a series of interest that becomes stationary after a minimum of h integer differences, then $\Delta^h y_t \equiv (1 - L)^h y_t$ is stationary and is a fractional process of order that lies in the interval $\in (h - 0.5, h + 0.5)$. Prior first differencing will also remove a deterministic polynomial trend of order $h - 1$. Typically $h = 1$ with economic data, so that estimation is based on the first difference Δy_t; correspondingly define $d_y \equiv d_{\Delta y} + 1$, which is the d value for the original series. Some of the estimation methods considered here still deliver a consistent estimate of d if $d_y \in (-0.5, 2]$ (see, for example, Velasco, 1999a, b, and Shimotsu,

2010), and if tapering (a weighting scheme on the frequencies) is introduced (see Hurvich and Chen, 2000), d can be $\in (-0.5, 1.5)$; however, the cost of tapering is an increase in the variance, which can have a substantial effect on inference. This case will be considered later.

4.2.2 The discrete Fourier transform, the periodogram and the log-periodogram 'regression', pooling and tapering

4.2.2.i Reminders

Two essential tools to analyse time series in the frequency domain are the discrete Fourier transform, DFT, and the periodogram, which are as follows for the sequence with typical observation denoted y_t.

$$\omega_y(\lambda_j) = \left(\frac{1}{\sqrt{2\pi}}\right)\sum_{t=0}^{T-1} y_t e^{-i\lambda_j t} \qquad \text{DFT} \qquad (4.24)$$

$$I_y(\lambda_j) = \frac{|\omega_y(\lambda_j)|^2}{T} \qquad \text{Periodogram} \qquad (4.25)$$

$$= \left(\frac{1}{2\pi T}\right)|\sum_{t=0}^{T-1} y_t e^{-i\lambda_j t}|^2 \qquad (4.26)$$

The DFT and periodogram are practical tools for observations (signals) in discrete time; typically, they are evaluated for (all or some of) the harmonic or Fourier frequencies $\lambda_j = 2\pi j/T$, $j = l, ..., m$, where $m \leq \text{int}[T/2]$, and $l \geq 0$ is a trimming parameter. In later sections, we shall also need the DFT and periodogram for the error input sequence u_t; these are defined analogously to (4.24) and (4.25) respectively, with u_t replacing y_t, with the notation $\omega_u(\lambda_j)$ and $I_u(\lambda_j)$ respectively.

4.2.2.ii Deterministic components

The prior removal of deterministic components, for example a mean and linear or higher-order trend, is generally important for semi-parametric estimation methods. (However, there are some exceptions that will be detailed later in this chapter.) Suppose that the data generating mechanism includes a linear (deterministic) trend, with data generated by:

$$x_t = \beta t + y_t$$

The DFT of x_t is:

$$\omega_x(\lambda_j) = \beta \frac{e^{i\lambda_j}}{(e^{i\lambda_j} - 1)}\left(\frac{T}{\sqrt{2\pi}}\right) + \omega_y(\lambda_j) \qquad (4.27)$$

The expression (4.27) assumes that $\omega_y(\lambda_j)$ is defined as in (4.27) *without* the divisor $\sqrt{T}$; if the divisor is $\sqrt{2\pi T}$ rather than $\sqrt{2\pi}$, then T in the first term of (4.27) is replaced by $\sqrt{T}$.

Most semi-parametric methods assume that the first term on the right-hand side of (4.27) is absent. If a deterministic trend is suspected, then one of two methods is usually adopted for this purpose. As noted above, first differencing the data removes a linear trend; however, there is a danger that over-differencing can occur, for example if d < 0.5 for the original data, then $d_{\Delta y} < -0.5$ in the transformed data with adverse consequences if the over-differenced series is used for parameter estimation (see Hurvich and Ray, 1995). Alternatively, x_t is first regressed on a linear time trend, and the residuals are used in semi-parametric estimation. For an important variation of this detrending method see Shimotsu (2010) and Phillips (1999b); both developments are considered later.

4.3   Estimation methods

4.3.1   A log-periodogram estimator (GPH)

4.3.1.i   Setting up the log-periodogram (LP) regression framework

A frequently used estimation procedure, either of direct interest or to provide a starting value into another routine, is the GPH estimator. Consider the SDF in (4.18a), restated here for convenience:

$$f_y(\lambda_j) = |(1 - e^{-i\lambda_j})|^{-2d} f_u(\lambda_j) \tag{4.28}$$

Then take logs of both sides of:

$$\ln[f_y(\lambda_j)] = -2d\ln(|1 - e^{-i\lambda_j}|) + \ln[f_u(\lambda_j)] \tag{4.29}$$

Now add and subtract $\ln[f_u(0)]$ to the right-hand side then, with some minor rearrangement, obtain:

$$\ln[f_y(\lambda_j)] = \ln[f_u(0)] - 2d\ln(|1 - e^{-i\lambda_j}|) + \ln[f_u(\lambda_j)/f_u(0)] \tag{4.30}$$

Finally, add $\ln[I_y(\lambda_j)]$ to both sides of (4.30) and rearrange to obtain:

$$\ln[I_y(\lambda_j)] = \ln[f_u(0)] - 2d\ln[|1 - e^{-i\lambda_j}|] + \ln[f_u(\lambda_j)/f_u(0)] + \ln[I_y(\lambda_j)/f_y(\lambda_j)] \tag{4.31}$$

An equivalent expression uses $|1 - e^{-i\lambda_j}|^{-2d} = [4\sin^2(\lambda_j/2)]^{-d}$, in which case (4.31) is written as follows (this is the form in the original GPH log-periodogram regression):

$$\ln[I_y(\lambda_j)] = \ln[f_u(0)] - d\ln[4\sin^2(\lambda_j/2)] + \ln[f_u(\lambda_j)/f_u(0)] + \ln[I_y(\lambda_j)/f_y(\lambda_j)] \tag{4.32}$$

Evaluating (4.31), or equivalently (4.32), at the Fourier frequencies $\lambda_j = 2\pi j/T$, $j = 1, \ldots, m$ where $m \leq \text{int}[T/2]$, leads to the terminology that (4.31) is a pseudo-regression with m observations. In this pseudo-regression, the 'dependent' variable is $\ln[I_y(\lambda_j)]$; the 'explanatory' variable is $\ln[|(1 - e^{-i\lambda_j})|]$; the intercept is $\ln[f_u(0)]$ plus the mean of $\ln[I_y(\lambda_j)/f_y(\lambda_j)]$; and the 'disturbance' is $\ln[f_u(\lambda_j)/f_u(0)] + \ln[I_y(\lambda_j)/f_y(\lambda_j)]$, where $f_u(\lambda_j)/f_u(0) \to 1$ as $\lambda_j \to 0^+$ and, hence, $\ln[f_u(\lambda_j)/f_u(0)] \to 0^+$ leaving $\ln[I_y(\lambda_j)/f_y(\lambda_j)]$.

In Geweke and Porter-Hudak (1983), the asymptotic (as $T \to \infty$) mean of $\ln[I_y(\lambda_j)/f_y(\lambda_j)]$ was taken to be $-C$, where $C = 0.577216...$ is Euler's constant; hence on this basis, $\ln[I_y(\lambda_j)/f_y(\lambda_j)] + C$ will, asymptotically, have a zero mean and for the moment we proceed on that basis. (The terms $I_y(\lambda_j)/f_y(\lambda_j)$, $j = 1, ...,$ m, are referred to as the normalised periodogram ordinates, the correct mean of which is given below.)

Adding and subtracting C on the right-hand side of (4.32), and for simplicity taking the opportunity to define a number of terms, (4.32) can be rewritten so that it resembles a linear regression model:

$$Y_j = \beta_0 + \beta_1 X_j + \varepsilon_j \qquad j = 1, ..., m \tag{4.33}$$

where: $Y_j = \ln[I_y(\lambda_j)]$; $X_j = \ln[4\sin^2(\lambda_j/2)]$; $\beta_0 = \ln[f_u(0)] - C$; $\beta_1 = -d$; and $\varepsilon_j = \ln[f_u(\lambda_j)/f_u(0)] + \ln[I_y(\lambda_j)/f_y(\lambda_j)] + C$.

The choice of notation deliberately emphasises the regression-like nature of (4.33); and apart from the term $\ln[f_u(\lambda_j)/f_u(0)]$, equation (4.33) now looks like a regression model with m < T 'observations'. The term $\ln[f_u(\lambda_j)/f_u(0)]$, which is part of ε_j, only involves contributions from the short-run dynamics, and hence will be negligible for the harmonic frequencies λ_j close to zero; for emphasis, these are sometimes referred to as 'harmonic frequencies in the shrinking neighbourhood of zero'.

The unfinished element of this outline relates to the asymptotic mean of $\ln[I_y(\lambda_j)/f_y(\lambda_j)]$ originally taken to be $-C$ and constant for all d; however, Hurvich and Beltrao (1993) show that whilst this result holds for $d = 0$, it does not hold for $d \neq 0$; the asymptotic mean of $\ln[I_y(\lambda_j)/f(\lambda_j)]$ is typically greater than $-C$, but decreases with j. This problem implies a positive bias to the estimator based on (4.33), being worse for $d \in (-0.5, 0)$ compared to $d \in (0, 0.5)$; however, since a typical procedure for macroeconomic time series is to take the first difference of the series, the former region is relevant for empirical purposes. Methods of estimation that address the question of bias reduction are considered later in this chapter. The next section outlines the natural LS estimator following from the way that the pseudo-regression has been written in (4.33).

4.3.1.ii *A simple LS estimator based on the log-periodogram*

The GPH estimator based on the pseudo-regression is simple to calculate; it is just the estimator obtained by applying least squares to (4.33). The resulting estimator of d is denoted $\hat{d}_{GPH}$, and is the negative of the OLS estimator of the slope, β_1, from the regression of Y_j on X_j in (4.33). That is applying standard

LS formula, and depending on whether it is (4.31) or (4.32) that underlies the pseudo-regression (4.33), the GPH estimator is one of the following:

$$X_j \equiv \ln[4\sin^2(\lambda_{j,T}/2)]$$

$$\hat{d}_{GPH} = -\left[\frac{\sum_{j=1}^{m}(X_j - \bar{X})Y_j}{\sum_{j=1}^{m}(X_j - \bar{X})^2} \right] \tag{4.34}$$

$$\check{X}_j \equiv \ln[|1 - e^{-i\lambda_j}|]$$

$$\check{d}_{GPH} = -0.5\left[\frac{\sum_{j=1}^{m}(\check{X}_j - \bar{\check{X}})Y_j}{\sum_{j=1}^{m}(\check{X}_j - \bar{\check{X}})^2} \right] \tag{4.35}$$

where $\bar{X} = \sum\limits_{j=1}^{m} X_j/m$ and $\bar{\check{X}} = \sum\limits_{j=1}^{m} \check{X}_j/m$.

An estimator asymptotically equivalent to those in either (4.34) or (4.35) is given on noting that:

$$|1 - \exp(-i\lambda_j)|^{-2d} = \lambda_j^{-2d}(1 + o(1)) \quad \text{as } \lambda_j \to 0^+ \tag{4.36}$$

Hence, replacing $-2d\ln[|1 - e^{-i\lambda_j}|]$ with $-2d\ln[(\lambda_j)]$ results in the pseudo-regression:

$$Y_j = \beta_0 + \hat{\beta}_1 \hat{X}_j + \hat{\varepsilon}_j \qquad j = 1,...,m \tag{4.37}$$

where $\hat{\varepsilon}_j$ differs from ε_j by asymptotically negligible terms. The estimator is:

$$\hat{d}_{GPH} = -0.5\left[\frac{\sum_{j=1}^{m}(\hat{X}_j - \bar{\hat{X}})Y_j}{\sum_{j=1}^{m}(\hat{X}_j - \bar{\hat{X}})^2} \right] \text{ where } \hat{X}_j \equiv \ln[(\lambda_j)] \tag{4.38}$$

In finite samples (that is finite for m, the number of ordinates) $\hat{d}_{GPH}$ and $\hat{d}_{GPH}$ may differ, but there is no difference asymptotically (in m).

4.3.1.iii   *Properties of the GPH estimator(s)*

In this section the properties of $\hat{d}_{GPH}$ (equivalently $\check{d}_{GPH}$, and $\hat{d}_{GPH}$) are considered when $d \in (-0.5, 0.5)$; the nonstationary case is considered in Section **4.7**. There was some discussion in the early literature on the GPH estimator on whether it would be advisable to truncate the frequencies near zero in calculating $\hat{d}_{GPH}$, referred to as trimming. AG derived a number of properties of $\hat{d}_{GPH}$ without any truncation of the frequencies near zero; see also an important earlier work by Hurvich, Deo and Brodsky (1998). In particular, they proved that $\hat{d}_{GPH}$ is consistent and asymptotically normal without trimming. They assumed that y_t is a stationary Gaussian time series process and made two further assumptions to establish consistency (asymptotic normality is considered separately below):

A1: $m = m(T) \to \infty$ and $m/T \to 0$ as $m \to \infty$ and $T \to \infty$.

A2: $f_u(\lambda_j)$ is an even function on $\lambda_j \in [-\pi, \pi]$; $f_u(\lambda_j)$ is a smooth function of order s for some $s \geq 1$; $0 < f_u(0) < \infty$; $d \in (-0.5, 0.5)$; and $\int_{-\pi}^{\pi} |\lambda_j|^{-2d} f_u(\lambda_j) d\lambda_j < \infty$; $f_u''(0)$ and $f_u'''(0)$ are finite constants for λ_j in a neighbourhood of zero.

The shorthand and notation are as follows: generally, the shorthand for the k-th derivative of $f_u(\lambda_j)$ with respect to λ_j is $f_u^{(k)}(\lambda_j)$, but if k is small, for example k = 1, 2, this is written as $f_u'(\lambda_j)$ or $f_u''(\lambda_j)$ respectively. Evaluating the k-th derivative at zero is written $f_u^{(k)}(0)$. Smoothness of order s at zero for the function $f_u(\lambda_j)$ implies that it is continuously differentiable s times in some neighbourhood of the zero frequency.

Other assumptions have been used in proving consistency, see Robinson (1995a) and Velasco (1999a), but A1 and A2 appear to be the least restrictive for the stationary case, especially in not requiring trimming.

The asymptotic bias, asymptotic variance, and asymptotic mean square error of $\hat{d}_{GPH}$ are, respectively:

GPH (i): asymptotic bias

$$AE(\hat{d}_{GPH}) - d = \left[\frac{-2\pi^2}{9} \right] \left[\frac{f_u''(0)}{f_u(0)} \right] \left(\frac{m^2}{T^2} \right) \tag{4.39}$$

GPH (ii): asymptotic variance

$$\sigma_{GPH}^{2,asym}(\hat{d}_{GPH}) = \left(\frac{\pi^2}{6} \right) \frac{1}{4m} = \frac{\pi^2}{24m} \tag{4.40}$$

GPH(iii) : asymptotic mean square error

$$AMSE(\hat{d}_{GPH}) = \left[\frac{-2\pi^2}{9} \right]^2 \left| \frac{f_u''(0)}{f_u(0)} \right|^2 \left[\frac{m^2}{T^2} \right]^2 + \frac{\pi^2}{24m} \tag{4.41}$$

Hence, from (4.39), $\hat{d}_{GPH}$ is biased in finite samples, but the bias tends to zero as $T \to \infty$, therefore, $\hat{d}_{GPH}$ is consistent.

The expression in (4.40) is the asymptotic variance. In practice, there are three options found in applications for the estimated variance of the GPH estimator. The least squares estimator is of the standard form for a linear regression, that is:

$$\hat{\sigma}_{GPH}^{2,OLS} = \hat{\sigma}_\varepsilon^2 \left(\sum_{j=1}^{m} (X_j - \overline{X})^2 \right)^{-1} \tag{4.42}$$

where $\hat{\sigma}_\varepsilon^2 = \hat{\varepsilon}'\hat{\varepsilon}/(m-2)$ and $\hat{\varepsilon}$ is the m vector of OLS residuals from the GPH regression, (4.33). Alternatively, in what is usually referred to as the asymptotic variance, $\hat{\sigma}_\varepsilon^2$ is replaced by the asymptotic variance of $\hat{\varepsilon}_j$, which is $\pi^2/6$, and this is the variance usually programmed; the resulting estimator is denoted $\hat{\sigma}_{GPH}^{2,asy}$. However, the true asymptotic variance of the GPH estimator is given by Equation

(4.40) above, that is, $\sigma_{GPH}^{2,asym} = \pi^2/24m$. There is some simulation evidence (AG, 2003) in a related context that the least squares estimator with $\hat{\sigma}_\varepsilon^2 = \pi^2/6$ is more accurate in finite samples than the asymptotic variance. However, $\sigma_{GPH}^{2,asym}$ is easy to calculate and should not be too dissimilar from $\hat{\sigma}_{GPH}^{2,asy}$.

Under certain assumptions, considered further below, the standard GPH estimator $\hat{d}_{GPH}$ is asymptotically normal. Surprisingly, proof of this result has not been straightforward. It was not much more than a conjecture in GPH (1983) and various attempts before Robinson (1995a) were less than complete. The proof in Robinson (1995a) requires a positive trimming parameter (as does the extension to nonstationary time series in Velasco, 1999a); since such an estimator is referred to below it is denoted $\hat{d}_{GPH}(\ell)$, where ℓ is the trimming parameter that removes the first ℓ frequencies from the sample. However, a proof without trimming was provided by AG as a special case of a wider class of log-periodogram estimators. AG provided a bias-reduced version of the GPH estimator, which we consider in Section **4.8.1**. The relevance of their results here is that $\hat{d}_{GPH}$ is a special case of the class of estimators they propose, which have the property of asymptotic normality, with a simple convergence condition on m and without trimming.

Assumption A1 is replaced by assumption A3, given here by:

A3: $m = o(T^{2\kappa/(2\kappa+1)})$ where $\kappa = \min(s, 2)$, $s \geq 1$

Generally, we expect $s \geq 2$; for example, $f_u(\lambda_j)$ generated by an ARMA process is infinitely differentiable, hence we can take $\kappa = 2$ and $m = o(T^{4/5})$. Then, with assumptions 2 and 3 and $d \in (-0.5, 0.5)$, $\hat{d}_{GPH}$ is asymptotically normally distributed, summarized as follows:

$$\sqrt{m}(\hat{d}_{GPH} - d) \Rightarrow_D N(0, \pi^2/24) \tag{4.43}$$

If trimming is applied, then the assumptions and results of Robinson (1995a) are relevant, with consistency for the trimmed GPH estimator, $\hat{d}_{GPH}(\ell)$, so that:

$$\sqrt{m}(\hat{d}_{GPH}(\ell) - d) \Rightarrow_D N(0, \pi^2/24) \tag{4.44}$$

4.3.1.iv   An (approximately) mean squared error optimal choice for m

To terms of smaller order than m^{-1}, the variance of $\hat{d}_{GPH}$ is $\pi^2/24m$. Notice that the bias increases with m, but the variance decreases with m; see (4.39) and (4.40). In the former case, inclusion of the higher frequencies moves the neighbourhood of estimation away from zero, with the effect that the higher frequencies 'contaminate' the regression. On the other hand, increasing m is good for the variance. Hence, the first term in the asymptotic mean squared error is due to the squared bias and the second to the variance. As $f_u''(0)/f_u(0)$ is a constant for a particular process, and assuming that $f_u''(0) \neq 0$, there is a mean squared error trade-off between squared bias and variance. (Generally, given an estimator $\hat{d}$ of d the mse is $E(\hat{d} - d)^2 = E([\hat{d} - E(\hat{d})] + [E(\hat{d}) - d])^2 = E[\hat{d} - E(\hat{d})]^2 + [E(\hat{d}) - d]^2 = var(\hat{d}) + bias(\hat{d})^2$, the cross-product in the expansion being zero.)

Minimising the $AMSE(\hat{d}_{GPH})$ of (4.41) with respect to m, (that is, the two leading terms in the $MSE(\hat{d}_{GPH})$) results in m^{opt}_{GPH} given by:

$$\text{GPH (iv)}: m^{opt}_{GPH} = \left[\frac{27}{128\pi^2}\right]^{1/5} \left|\left(\frac{f_u(0)}{f''_u(0)}\right)\right|^{2/5} T^{4/5} \tag{4.45}$$

$$= (K_{GPH}) T^{4/5}$$

The estimator of $\hat{d}_{GPH}$ based on m^{opt}_{GPH} is denoted $\hat{d}^{MSE}_{GPH}$. It is evident that m^{opt}_{GPH} is of order $O(T^{-4/5})$, the other terms being constant for a particular application.

The mean square error criterion for selection of $\hat{d}_{GPH}$ can be simply illustrated assuming that the short-run dynamics are generated by an AR(1) process. In that case (see problem Q4.2.ii), m^{opt}_{GPH} is:

$$m^{opt}_{GPH} = \left[\frac{27}{128\pi^2}\right]^{1/5} \left|\left(\frac{(1-\phi_1)^2}{2\phi_1}\right)\right|^{2/5} T^{4/5} \tag{4.46}$$

Whilst this is clearly $O(T^{-4/5})$, the scaling constant, K_{GPH}, can also have an important impact on m^{opt}_{GPH}. Intuitively, fewer frequencies should be included as $\phi_1 \to 1$ to avoid contamination from the lower frequency component of the short-run dynamics. For example, if $\phi_1 = 0.5$ and $T = 512$, then $K_{GPH} = 0.266$ and $m^{opt}_{GPH} = 39$; if $\phi_1 = 0.7$, then $K_{GPH} = 0.155$ and $m^{opt}_{GPH} = 22$. The m^{opt}_{GPH} function when the short-run dynamics are generated by an AR(1) process, and $T = 512$, is shown in Figure 4.2.

When m is chosen by a criterion based on minimising the (asymptotic) mean squared error, it must also be the case that $m^{opt}_{GPH} \leq T/2$, otherwise $\lambda_j > \pi$ for $j > T/2$, which is not a permissible value. A condition has to be placed on m^{opt}_{GPH} to ensure that $m^{opt}_{GPH} \leq T/2$. Inspection of (4.45) shows that the following is required:

$$(2K_{GPH})^5 < T \tag{4.47}$$

This condition is generally satisfied, but boundary situations and very small samples can cause difficulties. For example, for AR(1) short-run dynamics with $\phi_1 = 0.01$, and $T = 512$, then $m^{opt}_{GPH} = 323$ according to (4.46), but the condition in (4.47) is not satisfied as $(2K_{GPH})^5 = 1,642$; however, if $\phi_1 = 0.02$, then the condition is satisfied and $m^{opt}_{GPH} = 242$. The lack of short-run dynamics indicates that the limiting boundary to m^{opt}_{GPH} is int[T/2].

Although the motivation for the choice of m is explicit in (4.45), m^{opt}_{GPH} is not an operational concept as it depends upon what is deliberately not being assumed known in the semi-parametric approach; that is, some key properties

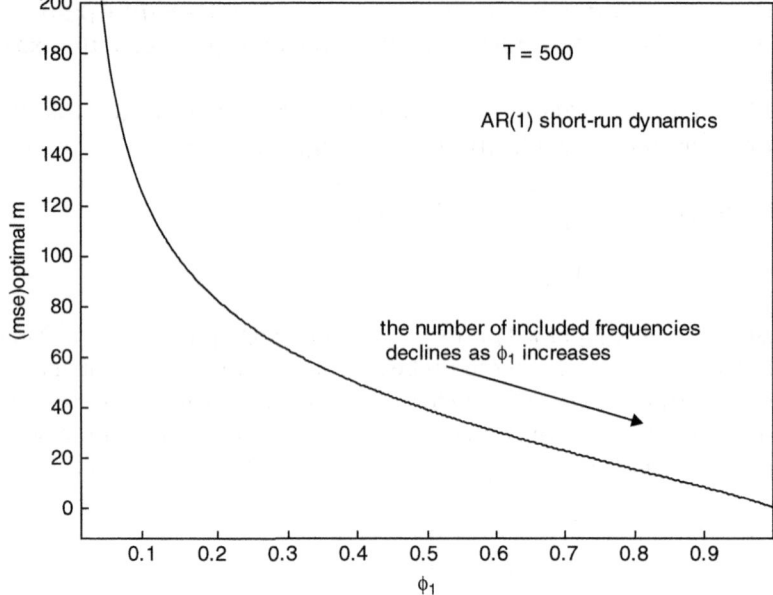

Figure 4.2   Optimal m for GPH estimator

of the short-run dynamics. Notwithstanding this present omission, the expression for m_{GPH}^{opt} shows that a simple rule for choosing m, for example of the form $m = AT^{\alpha}$, but with $A = 1$ and $\alpha = \frac{1}{2}$ irrespective of the dynamic properties, is unlikely to work well generally. Solutions to this problem are considered below; however, given the simplicity of the estimation procedure, it is not computationally demanding to calculate $\hat{d}_{GPH}$ for several different values of m and then assess the sensitivity of the results to different choices of m.

4.3.1.v   *Extensions to nonstationary time series*

The proofs of consistency and asymptotic normality in Robinson (1995a) are for $d \in (-0.5, 0.5)$. Velasco (1999a) extended these results (using type 1 fBM) to show that if y_t is Gaussian, then the GPH estimator is still asymptotically normal for $d \in (-0.5, 0.75)$ and consistent for $d \in (-0.5, 1)$. A typical procedure if it is suspected that $d \geq 0.5$ is to difference the data, estimate d and then add 1 to that estimated value to obtain an estimate of d for the series in levels. However, an alternative is to estimate d directly using the original data (which may be in logs). The question now relevant is: what are the properties of $\hat{d}_{GPH}$ when $d \geq 0.5$?

Kim and Phillips (2000, 2006, theorem 3.1) establish that $\hat{d}_{GPH}$ is not consistent for $d \in (1, 2)$; moreover, the probability limit of $\hat{d}_{GPH}$ in this range is unity. However, $\hat{d}_{GPH}$ is consistent for $d \in (0.5, 1)$; see Kim and Phillips, 2000, theorem 3.2 for details. Earlier related work by Velasco (1999a, 2000) has shown that a trimmed GPH estimator, referred to above as $\hat{d}_{GPH}(\ell)$, is consistent for $d < 1$ and asymptotic normality holds for the trimmed estimator for $d < 0.75$ (with a strengthening of the convergence condition on m/T relative to the consistency proof).

The important case of $d = 1$ has been considered by Phillips (1999a), who showed that $\hat{d}_{GPH}$ is consistent for $d = 1$, with a limit distribution that is mixed normal, *MN*, (rather than simply normal). Specifically, Phillips (1999a) obtains:

$$\sqrt{m}(\hat{d}_{GPH} - 1) \Rightarrow_D MN(0, \sigma^2_{GPH}(d = 1)) \tag{4.48}$$

where $\sigma^2_{GPH}(d = 1) = (1/4)\sigma^2(V)$, with V distributed as $\chi^2(1)$ (see Theorem 3.1 of Phillips, 1999a). The variance of the limiting distribution, $\sigma^2_{GPH}(d = 1)$, is:

$$\sigma^2_{GPH}(d = 1) = \frac{\pi^2}{24} - 0.0164 \tag{4.49}$$

$$= 0.3948$$

In contrast, when $d \in (-0.5, 0.5)$, the variance of the limiting distribution is just the first term on the right-hand side $\pi^2/24$, as noted in (4.40); hence, when $d = 1$, the limiting distribution shows greater concentration around d than the stationary case. The asymptotic variance of $\hat{d}_{GPH}$ when $d = 1$ is, therefore, 0.3948m, which could be used for hypothesis testing, an issue that is considered next.

4.3.1.vi   Testing the unit root hypothesis using the GPH estimator

On the basis of the distributional result in (4.49), the unit root hypothesis H_0: $d = 1$ could be tested by using the following t-type statistic:

$$t_{GPH} = \frac{(\hat{d}_{GPH} - 1)}{\sqrt{0.3948m}}$$

where critical values are obtained by simulation based on (4.49), with large negative values leading to rejection. However, this would only be valid for the left-sided alternative H_A: $d < 1$, because $\hat{d}_{GPH}$ is inconsistent under the alternative H_A: $d > 1$. An alternative which is consistent for the two-sided alternative is described in Section 4.7.4.

4.4   Pooling and tapering

Pooling the periodogram components and tapering the inputs to the DFT are important developments that are used in variants of GPH estimation (and other methods to be described). The concepts are now briefly described to make this discussion self-contained. Section **4.5** then takes up their use in a GPH-type estimation context.

4.4.1   Pooling adjacent frequencies

The basic idea of pooling is to separate the total number of available frequencies into K non-overlapping blocks each of size J, where $K = (m - l)/J$ is an integer. As before l is the trimming parameter if some of the frequencies are removed, usually at the beginning of the sample of frequencies. If there is no trimming $l = 0$ and the blocks are 1 to J, J + 1 to 2J, and so on; with trimming the blocks are l to $J + l$, $J + l + 1$ to $2J + l$ and so on. The periodogram is then defined for each block just over the frequencies in that block; for example, assuming no trimming then $I_{y,1}^{P,J}(\lambda_j)$ is the periodogram defined over frequencies 1 to J, $I_{y,2}^{P,J}(\lambda_j)$ is the periodogram over frequencies $j = J + 1$, ..., 2J, and so on until finally $I_{y,K}^{P,J}(\lambda_j)$ is the periodogram over frequencies $j = (K - 1)J + 1$, ..., m.

The notation is somewhat complicated, but serves to remind the reader that the pooled blocks are of length J. For example, the first and k-th pooled periodograms, assuming no trimming, $I_{y,1}^{P,J}(\lambda_j)$, are:

$$I_{y,1}^{P,J}(\lambda_j) = \frac{1}{2\pi T} \sum_{j=1}^{J} |\omega_y(\lambda_j)|^2$$

$$= \sum_{j=1}^{J} I_y(\lambda_j) \tag{4.50}$$

$$I_{y,k}^{P,J}(\lambda_j) = \sum_{j=(k-1)J+1}^{kJ} I_y(\lambda_j) \quad k = 1, ..., K \tag{4.51}$$

The (equally weighted) pooled periodogram is the sum over the K blocks of size J:

$$I_y^{P,J}(\lambda_j) = \sum_{k=1}^{K} I_{y,k}^{P,J}(\lambda_j) \tag{4.52}$$

Robinson (1995a) suggests using the log of the pooled periodograms for each of the K blocks in the GPH estimator; that is, use $\ln[I_{y,k}^{P,J}(\lambda_j)]$ in place of the usual periodogram ordinates. We consider this further below.

4.4.2   Tapering

In effect, each of the frequencies in the DFT has an equal weight in its overall contribution. Tapering is a scheme that introduces a set of weights, h_t, into the DFT and, combined with pooling, defines a 'window' of frequencies that are to be considered as a group with, potentially, unequal weights within the group. First, we consider just the operation of tapering.

Consider the following scheme for a weighted, or tapered, DFT:

$$\omega_y^{tap}(\lambda_j) = \frac{1}{(2\pi \sum_{t=1}^{T} |h_t|^2)^{1/2}} \sum_{t=1}^{T} h_t y_t e^{-i\lambda_j t} \tag{4.53}$$

If each contribution in the sum has an equal weight, then $h_t = 1$ for all t, and the 'standard' DFT results, but note that this is the version of the DFT with $(2\pi T)^{1/2}$ as the divisor. The presence of $(\sum_{t=1}^{T} |h_t|^2)^{1/2}$ can be viewed as 'normalising' the sum of the weights; that is, effectively, each weight is actually $h_t/(\sum_{t=1}^{T} |h_t|^2)^{1/2}$. Otherwise, $0 \le h_t \le 1$, and the weighting scheme is designed to emphasise the contribution of observations in the central part of the observed sequence and give relatively smaller weights to observations away from the centre, although not necessarily symmetrically; see Velasco (2000).

Examples of tapers that have been used in estimating d are the cosine bell taper (Tukey, 1967), the complex valued taper (Hurvich and Chen, 2000) and the rectangular window. The weights for these tapers are as follows:

$$h_t = 0.5\left\{1 - \cos\left(\frac{2\pi(t - 1/2)}{T}\right)\right\}; \quad \sum_{t=1}^{T} h_t^2 = 3/8 T \quad \text{cosine bell weights}$$

$$h_t = 0.5(1 - e^{i2(t-1/2)\pi/T}); \quad \sum_{t=1}^{T} |h_t|^2 = 1/2 T \quad \text{complex valued taper (HC)}$$

$$h_t = 1; \quad \sum_{t=1}^{T} h_t^2 = T \quad \text{rectangular window (RW)}$$

(Note that some authors use t rather than $(t - 1/2)$ in the cosine bell taper.) For other tapers see Velasco (1999b).

The tapered periodogram, denoted $I_y^{tap}(\lambda_j)$, follows naturally from the tapered DFT, except that, given the definition of $\omega_y^{tap}(\lambda_j)$, there is no need to use a divisor in the resulting periodogram:

$$I_y^{tap}(\lambda_j) = |\omega_y^{tap}(\lambda_j)|^2 \tag{4.54a}$$

$$= \frac{1}{2\pi \sum_{t=1}^{T} |h_t|^2} \left|\sum_{t=1}^{T} h_t y_t e^{-i\lambda_j t}\right|^2 \tag{4.54b}$$

Tapering can be helpful in reducing the bias of the log-periodogram estimator (see for example, Hurvich and Ray, 1995, and Hurvich and Chen, op. cit.)

but the reduction in the bias is bought at the expense of an increase in the variance.

The tapered periodogram could also be pooled as in the previous section. That is the inputs to $I_y^{P,k}(\lambda_j)$ in expression (4.52) would be the tapered periodograms given by $I_y^{tap}(\lambda_j)$ of (4.54). As this variation is used below, the following notation indicates a pooled and tapered periodogram: $I_{y,k}^{P,J,tap}(\lambda_j)$.

4.5   Variants of GPH estimation using pooling and tapering

4.5.1   Pooled and pooled and tapered GPH

We can now return to the central aim of this section, which is to consider the properties of the GPH estimator and variants thereof in the case of non-Gaussian generating processes. A variant of the standard GPH was suggested by Robinson (1995a) and analysed further by Velasco (2000). In this variant, the pooled periodogram replaces the standard periodogram. The resulting pseudo-regression is given by:

$$\ln[I_{y,k}^{P,J}(\lambda_j)] = \beta_0 - d[\ln(4\sin^2(\lambda_k/2))] + \xi_k \qquad k = 1,...,K \qquad (4.55)$$

In effect, the log of the k-th pooled periodogram is regressed on a constant and a function of the k-th (angular) frequency; the estimator of d is, as in the standard case, the negative of the LS slope coefficient. Another GPH variant is to use the pooled and tapered periodogram ordinates, in which case $\ln[I_{y,k}^{P,J,tap}(\lambda_j)]$ replaces $\ln[I_{y,k}^{P,J}(\lambda_j)]$ as the 'dependent' variable in the pseudo-regression, (4.55). To distinguish these estimators they will be referred to as $\hat{d}_{GPH}^P$ and $\hat{d}_{GPH}^{P,tap}$ for the pooled, and pooled and tapered, GPH estimators respectively.

Both variants of the GPH estimator, pooled and pooled and tapered, were considered by Velasco (2000) who relaxed the assumption of Gaussian inputs in favour of a fourth-order stationary linear process; that is, the random inputs in the moving average representation of y_t are iid with a finite fourth moment, and the MA coefficients are square summable. In the case of the $\hat{d}_{GPH}^P$ estimator, with $J \geq 3$, he was able to show that it was consistent for $d \in [0, \frac{1}{2})$, with a conjecture that consistency would also hold for $J \leq 2$; see Velasco (2000, theorem 1) for the full set of assumptions. To establish that the asymptotic distribution would remain the same without Gaussianity, for example that asymptotic normality would hold, Velasco (op. cit.) introduced pooling and tapering (and strengthened the moment condition to a finite third moment). The particular tapering scheme was the cosine bell, but other smooth tapers would deliver the same result; see Velasco (1999). The resulting asymptotic distribution is summarised in the following result (see

Velasco, 2000, theorem 3):

$$\sqrt{m}(\hat{d}_{GPH}^{P,tap} - d) \Rightarrow_D N\left(0, \frac{3J}{4}\psi(J)\right) \tag{4.56}$$

where $\psi(J)$ is the digamma function.

Velasco (op. cit.) reports some simulation results with Gaussian, Uniform, Exponential and t_5 (no moments beyond the fourth) distributions using an ARFIMA(0, d, 0) model (that is, no short-run dynamics). The unmodified GPH estimator remained robust in terms of bias, but the pooled estimator with $J = 2$ or 3 had a better mean squared error, with $J = 2$ generally slightly better unless m was large (for example, m = 90 when T = 512). Tapering increased the mean squared error with or without pooling and thus, on this criterion, pooling but not tapering is preferred. However, one benefit of tapering is that the simulated coverage of the 90%, 95% and 99% confidence intervals was much closer to the nominal value with tapering than without.

4.5.2 A bias-adjusted log-periodogram estimator

In practice, the number of frequencies, m, has to be chosen for the GPH estimator, with a trade-off between choosing m 'small' and m 'large'. Despite its widespread use in the empirical literature, the 'square root' rule, that is $m = T^{0.5}$, is not justified on mean squared error grounds, although it may be useful for illustrative purposes. In essence, whilst small m concentrates the 'frequencies in the shrinking neighbourhood of zero', this is at the cost of a larger variance for $\hat{d}_{GPH}$ than if m is chosen to be large. However, in the latter case the regression involves frequencies away from zero and this increases the bias in the estimator $\hat{d}_{GPH}$. One presently infeasible solution is to use a mean squared error criterion to derive an 'optimal' value for m, which is then, in principle, used in calculating $\hat{d}_{GPH}$; see Section **4.3.1.iv**.

A second problem is that simulation evidence strongly suggests that the GPH estimator is biased when the ARFIMA model includes short-run dynamics, and that this bias worsens as the 'persistence' in the short-run dynamics increases. The class of log-periodogram estimators suggested by AG is aimed at reducing this bias, and may also be used to provide a feasible data-driven choice of the number of frequencies to include in the LP regression.

4.5.2.i The AG bias-reduced estimator

4.5.2.i.a The estimator AG have suggested an estimator aimed at reducing the bias in $\hat{d}_{GPH}$. They note that the dominant term in the bias arises from the presence of the term $\ln[f_u(\lambda_j)/f_u(0)]$ in the GPH regression (if that term was absent, there would not be the term $f_u''(0)/f_u(0)$ in the bias expression). Hence, eliminating this term will reduce the bias; recall that $f_u(\lambda_j)$ is the contribution to the SDF of y_t, arising from the short-run dynamics.

To understand the motivation for the AG method, first expand $\ln[f_u(\lambda_j)]$ in a Taylor series about $\lambda_j = 0$, so that:

$$\ln[f_u(\lambda_j)] = \ln[f_u(0)] + \sum_{k=1}^{[s]} \frac{b_k}{k!} \lambda_j^k + O(\lambda_j^s) \qquad (4.57)$$

$$b_k = \frac{\partial^k}{\partial \lambda_j^k} \ln[f_u(\lambda_j)]|_{\lambda_j=0}$$

The notation [s] indicates the integer part of s and also shows that $f_u(\lambda_j)$ is [s] times continuously differentiable in the neighbourhood of zero (said to be 'smooth of order s'). Next, note that $\ln[f_u(\lambda_j)]$ is an even function with odd order derivatives equal to zero at $\lambda_j = 0$; thus, $b_k = 0$ for all odd integers $\leq$ [s]. As a result all odd terms in (4.57) disappear to leave (after a minor rearrangement to take $\ln[f_u(0)]$ to the left-hand side):

$$\ln[f_u(\lambda_j)/f_u(0)] = \sum_{k=1}^{[s/2]} \frac{b_{2k}}{(2k)!} \lambda_j^{2k} + O(\lambda_j^s) \qquad (4.58)$$

For example, the second order expansion (k = 1) is:

$$\ln[f_u(\lambda_j)/f_u(0)] = \left[\frac{b_2}{2}\right] \lambda_j^2 + O(\lambda_j^3) \qquad (4.59)$$

$$b_2 = \frac{f_u''(0)}{f_u(0)}$$

Thus, rather than omit the term $\ln[f_u(\lambda_j)/f_u(0)]$ from the GPH regression, the suggestion is to include λ_j^2 as a regressor, which should enable the bias to be reduced. The principle can be extended to include higher-order even powers of λ_j, but there are implications for the variance of the resulting estimator of d.

For pedagogical purposes, assume that the highest power of λ_j in (4.58) is 2r, and note that this must necessarily be even. The estimator is then based on the following extension of the GPH pseudo-regression:

$$\ln[I_y(\lambda_j)] = \beta_0 - 2d\ln[\lambda_j] + \sum_{k=1}^{r} \varphi_k \lambda_j^{2k} + \breve{\varepsilon}_j \quad r \text{ (integer)} \geq 0, j = 1, \ldots, m \quad (4.60)$$

This can be viewed as the regression of $\ln[I_y(\lambda_j)]$ on the regressors $x_j = (1, -2\ln(\lambda_j), \lambda_j^2, \ldots, \lambda_j^{2r})$, $j = 1, \ldots, m$, with coefficient vector $\beta = (\beta_0, d, \varphi_1, \ldots, \varphi_r)'$, and OLS estimator $\hat{\beta}$; the regression 'error' $\breve{\varepsilon}_j$ is revised relative to the GPH regression error $\widehat{\varepsilon}_j$ of (4.37) by moving the part of the latter due to $\ln[f_u(\lambda_j)/f_u(0)]$ and approximated by (4.59) directly into the regression. It will be useful to note that $\varphi_k = b_k/(2k)!$, where b_k are the derivatives defined following (4.57). Also note that AG use the regressor $\ln(\lambda_j)$, but asymptotically equivalent results follow throughout from use of the other forms of the regressor given in Section **4.3.1.ii.**

The estimator resulting from OLS estimation of d in (4.60) is denoted $\hat{d}_{AG}$, which is just the second element in the estimated coefficient vector $\hat{\beta}$. If emphasis on the order of the polynomial is needed $\hat{d}_{AG}$ is written as $\hat{d}_{AG}(r)$. Given smoothness conditions on $f_u(\lambda_j)$, the asymptotic bias of $\hat{d}_{AG}(r)$ is of order m^{2+2r}/T^{2+2r} compared to m^2/T^2 for $\hat{d}_{GPH}$. To illustrate, for $m = T^{0.5}$ and $r = 1$, these orders are $1/T^2$ and $1/T$ respectively.

In essence, the basic GPH pseudo-regression is augmented by even powers of λ_j, where r is typically small, for example $r = 1$ or $r = 2$. This results, respectively, in the following LP-type regressions:

$$\ln[I_y(\lambda_j)] = \beta_0 - 2d\ln(\lambda_j) + \varphi_1\lambda_j^2 + \breve{\varepsilon}_j \qquad (4.61a)$$

$$\ln[I_y(\lambda_j)] = \beta_0 - 2d\ln(\lambda_j) + \varphi_1\lambda_j^2 + \varphi_2\lambda_j^4 + \breve{\varepsilon}_j \qquad (4.61b)$$

and, in each case, $\hat{d}_{AG}(r)$ is $-\frac{1}{2}$ times the coefficient of the regressor $\ln(\lambda_j)$. Note that estimates of the b_k coefficients can also be obtained from these regressions; for example, $\hat{b}_2 = 2\hat{\varphi}_1$ and $\hat{b}_4 = (4!)\hat{\varphi}_2$. Thus, referring back to Section **4.3.1.iv** and the determination of m_{GPH}^{opt} in (4.45), a feasible version of this quantity, say $\hat{m}_{GPH}^{opt}$, can be obtained by estimating (4.61a), and an estimate of $f_u(0)/f_u''(0)$ is provided by $(\hat{b}_2)^{-1} = 1/(2\hat{\varphi}_1)$. This is part of a general principle considered further below.

Finally, defining $y = (Y_1, Y_2, \ldots, Y_m)'$, where $Y_j = \ln[I_y(\lambda_j)]$ and X as the m × (2 + r) matrix with rows given by X_j, $j = 1, \ldots, m$, then, as usual, $\hat{\beta} = (X'X)^{-1}X'y$. It is also convenient for later purposes to separate the regressors so that the model is written as:

$$y = \beta_0 + X_{(2)}d + Z\varphi + \breve{\varepsilon} \qquad (4.62a)$$

$$= \beta_0^* + X_{(2)}^*d + Z^*\varphi + \breve{\varepsilon} \qquad (4.62b)$$

Where $X_{(2)}$ is the second column of X, that is, it comprises the elements $[-2\ln(\lambda_j)]$, $j = 1, \ldots, m$; Z is the m × r matrix such that each row is $(\lambda_j^2, \ldots, \lambda_j^{2r})$; and $\breve{\varepsilon}$ is the m × 1 vector of revised errors. In the second line, the model is written in deviations from mean form, thus $X_{(2)}^* = X_{(2)} - 1_m\bar{X}_{(2)}$, where 1_m is the m × 1 vector with 1 in each element and $\bar{X}_{(2)} = -2\sum_{j=1}^{m}\ln(\lambda_j)/m$; similarly, $Z^* = Z - \bar{Z}$, where $\bar{Z}$ contains the column means of Z. Using the deviations from mean form, the AG estimator of d, $\hat{d}_{AG}$, and its variance are:

$$\hat{d}_{AG} = (X^{*'}M_{Z^*}X^*)^{-1}X^{*'}M_{Z^*}y \qquad (4.63)$$

$$\hat{\sigma}_{AG}^2(\hat{d}_{AG}) = \sigma_\varepsilon^2(X^{*'}M_{Z^*}X^*)^{-1} \qquad (4.64)$$

where $\sigma_\varepsilon^2 = \pi^2/6$ and M_{Z^*} is the projection matrix given by:

$$M_{Z^*} = (I_m - Z^*(Z^{*'}Z^*)^{-1}Z^{*'}) \qquad (4.65)$$

4.5.2.i.b   Asymptotic properties of the AG estimator   Asymptotic properties of $\hat{d}_{AG}$ for $d \in (-0.5, 0.5)$ and $s \geq 2 + 2r$ are established by AG (op. cit.), given assumptions A1 and A2 as stated in Section **4.3.1.iii**. Throughout it is assumed that y_t is a Gaussian stationary long-memory process.

With assumptions A1 and A2, AG (op. cit.) establish the following (asymptotic) properties of $\hat{d}_{AG}$:

$$AG(i) : AE(\hat{d}_{AG}) - d = \kappa_r b_{2+2r} \frac{m^{2+2r}}{T^{2+2r}} \tag{4.66}$$

$$AG(ii) : \sigma_{AG}^{2,asym}(\hat{d}_{AG}) = \left(\frac{\pi^2}{6}\right)\frac{c_r}{4m}$$

$$= \frac{\pi^2}{24m}c_r \tag{4.67}$$

$$AG(iii) : AMSE(\hat{d}_{AG}) = \kappa_r^2 b_{2+2r}^2 \frac{m^{4+4r}}{T^{4+4r}} + \frac{\pi^2}{24m}c_r \tag{4.68}$$

$$AG(iv) : m_{AG}^{opt}(r) = \left(\frac{\pi^2 c_r}{24(4+4r)\kappa_r^2 b_{2+2r}^2}\right)^{1/(5+4r)} T^{(4+4r)/(5+4r)} \tag{4.69a}$$

$$= K(r)T^{(4+4r)/(5+4r)} \tag{4.69b}$$

In place of assumption A1, AG (op. cit.) introduce assumption A3, which was stated in Section **4.3.1.iii**. Then, with assumptions 2 and 3, the following distributional result holds:

$$AG(v) : \sqrt{m}(\hat{d}_{AG} - d) \Rightarrow_D N(0, c_r(\pi^2/24)) \tag{4.70}$$

The constants are c_r: $c_0 = 1$, $c_1 = 2.25$, $c_2 = 3.52$, $c_3 = 4.79$, $c_4 = 6.06$, $c_5 = 7.33$; κ_r: $\kappa_0 = -2.19$, $\kappa_1 = 2.23$, $\kappa_2 = -0.793$, $\kappa_3 = 0.146$, $\kappa_4 = -0.0164$, $\kappa_5 = 0.00125$.

Of particular interest is the asymptotic normality of $\hat{d}_{AG}$; by setting $r = 0$, the corresponding properties of the GPH estimator, $\hat{d}_{GPH}$, are obtained on noting that $\kappa_0 = -2\pi^2/9 = -2.19$ and $b_2 = f_u''(0)/f_u(0)$. If $r = 0$, then the asymptotic properties of $\hat{d}_{GPH}$ and $\hat{d}_{AG}$ are the same. However, for $r \geq 1$ there are two differences as follows.

First, the rate of convergence of the bias is governed by (m^2/T^2) for $\hat{d}_{GPH}$ and by (m^{2+2r}/T^{2+2r}) for $\hat{d}_{AG}(r)$, so the bias converges to zero faster for $\hat{d}_{AG}(r)$, $r \geq 1$. Let $m = T^\gamma$, where $1 > \gamma > (2 + 2r)/(3 + 2r)$, then the required rate γ is satisfied by m_{GPH}^{opt} (that is, $r = 0$) and $m_{AG}^{opt}(r)$, $r \geq 1$ (see below). For $m = T^\gamma$, then $(m^2/T^2) = T^{2(\alpha-1)}$ and $(m^{2+2r}/T^{2+2r}) = T^{2(\gamma-1)}T^{2r(\gamma-1)}$, and the faster convergence is provided by the term $T^{2r(\gamma-1)}$.

Second, $\sigma_{AG}^{2,asym}(\hat{d}_{AG}) = \sigma_{GPH}^{2,asym}(\hat{d}_{GPH})$ for $r = 0$; and, for given m and $r \geq 1$, $\sigma_{AG}^{2,asym}(\hat{d}_{AG}) > \sigma_{GPH}^{2,asym}(\hat{d}_{GPH})$. Note that $\sigma_{AG}^{2,asym}(\hat{d}_{AG})$ is $\sigma_{GPH}^{2,asym}(\hat{d}_{GPH})$ multiplied by a constant c_r, which depends upon r and $c_r > 1$ for $r \geq 1$. For example, if $r = 1$, and given m, then $\sigma_{AG}^{2,asym}(\hat{d}_{AG}) = 2.25\sigma_{GPH}^{2,asym}(\hat{d}_{GPH})$, and so the standard error will be 1.5 times larger for $\hat{d}_{AG}$ compared with $\hat{d}_{GPH}$. At first sight this seems a substantial cost to reducing the bias; however, the faster rate of convergence of the bias to zero enables a larger value of m in the MSE trade-off, so m will not be fixed in the practical comparison.

Note that $c_i > c_1 = 1$ for $i > 1$, so the variance of $\hat{d}_{AG}(r)$ is inflated for $r > 0$ relative to $r = 0$; for example, by a factor of 2.25 comparing $\hat{d}_{AG}(1)$ and $\hat{d}_{AG}(0)$. Hence, whilst bias reduction is an important aim of the AG procedure, a reduction in rmse needs to come from allowing $m_{AG}^{opt}(1)$ to be sufficiently large relative to $m_{AG}^{opt}(0)$ so that the variance is driven down.

As in the case of $\hat{d}_{GPH}$, there are three options for the variance of $\hat{d}_{AG}$ in finite samples. The first option is the least squares estimator of the variance of the second element of $\hat{\beta}$ estimated from (4.62); that is, the second diagonal element in $\hat{\sigma}_\varepsilon^2 (X'X)^{-1}$ using the usual LS estimator of the residual variance $\hat{\sigma}_\varepsilon^2 = \hat{\varepsilon}'\hat{\varepsilon}/(m - (2 + r))$, where $\hat{\varepsilon}$ is the residual vector from least squares estimation of (4.62). This is as in (4.64) but with σ_ε^2 replaced by $\hat{\sigma}_\varepsilon^2$. The second option uses $\sigma_\varepsilon^2 = \pi^2/6$, with the resulting estimator denoted $\hat{\sigma}_{AG}^{2,asy}$. The third option is the asymptotic variance $\sigma_{AG}^{2,asym}$ of (4.67). In simulations of a comparison of $\hat{\sigma}_{AG}^{2,asy}$ and $\sigma_{AG}^{2,asym}$, AG (op. cit.) suggested the former was more accurate in finite samples.

4.5.2.i.c Bias, rmse and m^{opt} comparisons; comparisons of asymptotics The asymptotic bias and asymptotic rmse are calculated for the ARFIMA(1, d, 0) model with $\phi_1 = 0.3$, 0.6, 0.9, $T = 500$ and 2,500, and $r = 0$, 1, 2 3; these are shown in Figures 4.3a, b and 4.4a, b, respectively, for bias and rmse. The values of ϕ_1 generate short-run positive autocorrelations and are the relevant case for typical economic time series. Throughout each sub-figure of Figure 4.3, the legend identifying the value of r is shown in the first case illustrated, $\phi_1 = 0.9$, with other sub-figures the same; note that $r = 0$ in $\hat{d}_{AG}(r)$ corresponds to a version of $\hat{d}_{GPH}$.

The asymptotic bias is shown in Figures 4.3a and 4.3b for $T = 500$ and $T = 2,500$, respectively. The general pattern is the same for the three values of ϕ_1: comparing $\hat{d}_{AG}(r)$ with $\hat{d}_{AG}(r-1)$, the former initially shows a reduction in bias; however, this is not uniform across m, and as m increases there is a point at which $\hat{d}_{AG}(0)$ has the smallest bias. Perhaps the point of particular interest is the evaluation of bias at the 'optimal' value of m, $m_{AG}^{opt}(r)$; in fact, as is shown below, $\hat{d}_{AG}(1)$ has a smaller asymptotic bias at $m = m_{AG}^{opt}(1)$ than $\hat{d}_{AG}(0)$ with $m = m_{AG}^{opt}(0)$. (The more general case is considered in a question at the end of the chapter.)

The asymptotic rmse for $\hat{d}_{AG}(r)$ in the ARFIMA(1, d, 0) model is shown in Figures 4.4a and 4.4b for $T = 500$ and $T = 2,500$ respectively. Note that these

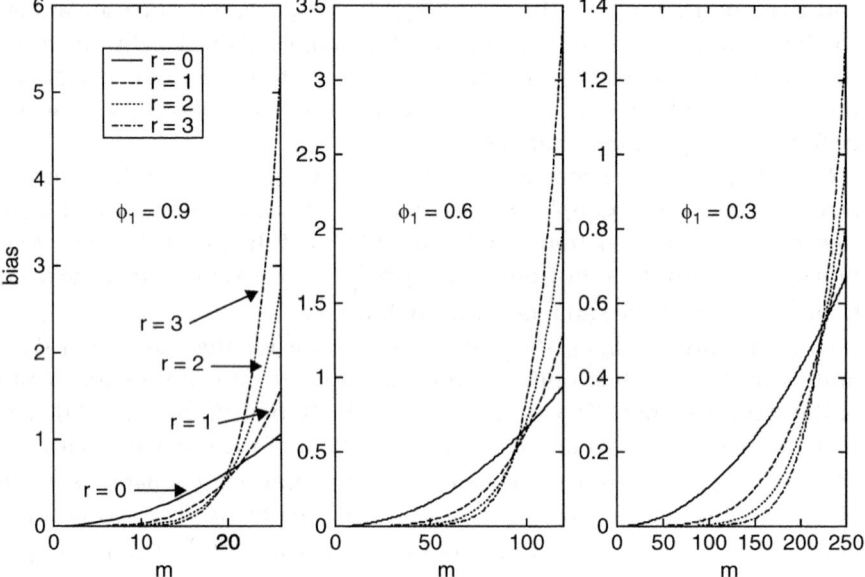

Figure 4.3a   Asymptotic bias, $\hat{d}_{(AG)}(r)$, T = 500

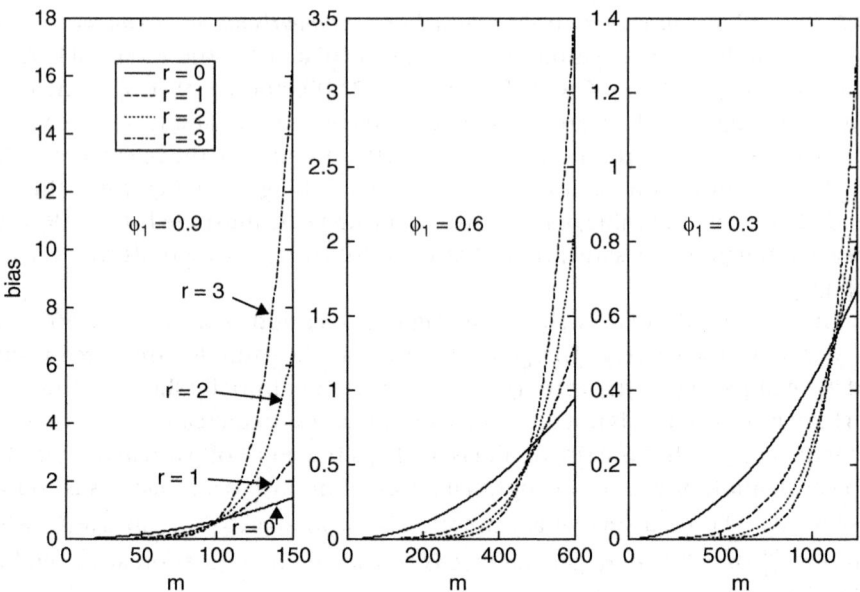

Figure 4.3b   Asymptotic bias, $\hat{d}_{(AG)}(r)$, T = 2,500

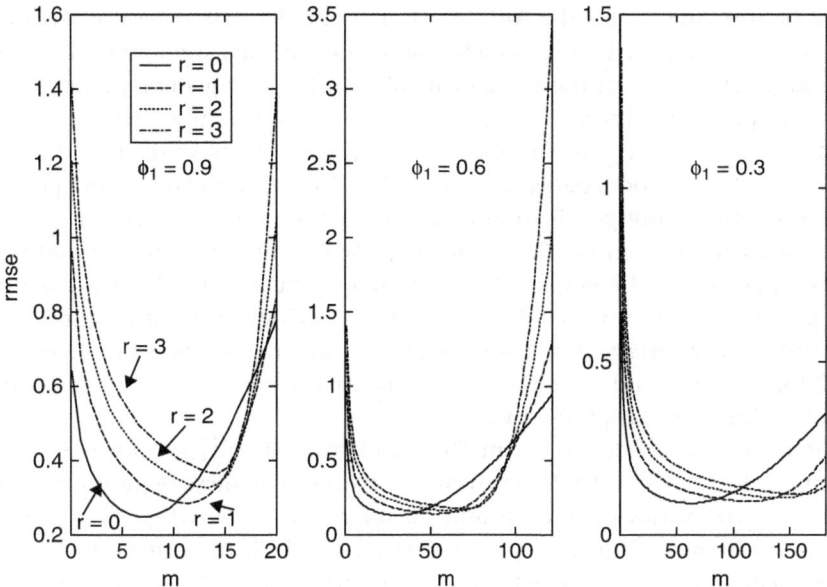

Figure 4.4a   Asymptotic rmse, $\hat{d}_{(AG)}(r)$, T = 500

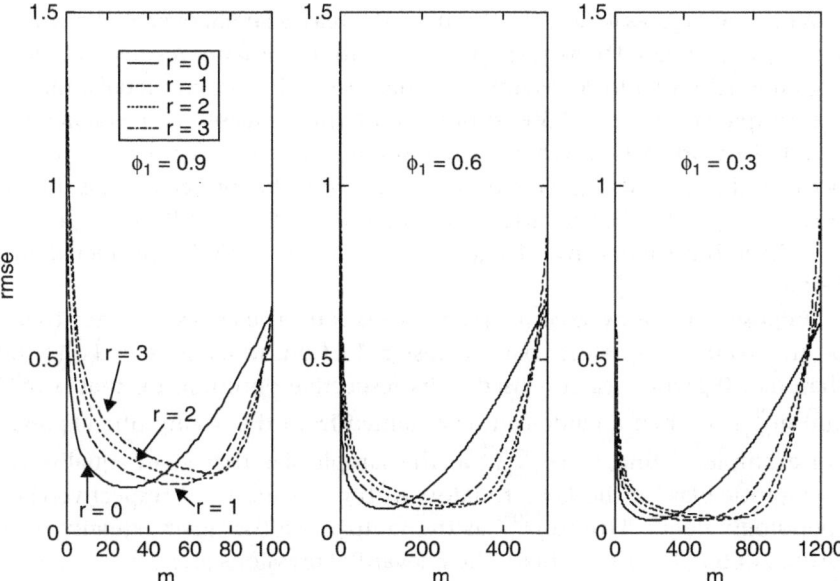

Figure 4.4b   Asymptotic rmse, $\hat{d}_{(AG)}(r)$, T = 2,500

will be the same as for the ARFIMA(0, d, 1) model with $\theta_1 = -\phi_1$. As in the previous figures, the legend, which is shown in the first sub-figure, is the same throughout. Although the numerical values differ, the general picture for the two sample sizes is the same and they can be considered together.

When $\phi_1 = 0.9$, $r = 0$ corresponds to the smallest rmse; in the other cases there is little to choose between $r = 0, 1, 2, 3$. Note that the minimum point for each function corresponds to $m = m_{AG}^{opt}(r)$. A feature of the cases for $\phi_1 = 0.3$, 0.6, other than when $r = 0$, is that the rmse function is flatter, so that if this was replicated in the empirical rmse function there would be less sensitivity to getting the 'right' value of m. What is also evident from the figures is that adopting an arbitrary rule, for example the square root rule for choosing m (22 for $T = 500$, and 50 for $T = 2,500$) is usually very misleading in terms of minimising the asymptotic rmse.

Some additional details for $m_{AG}^{opt}(r)$, and rmse at $m_{AG}^{opt}(r)$ and $m = \sqrt{T}$, are given in Tables 4.1 and 4.2. When the short-run dynamics are prominent ($\phi_1 = 0.9$), $m_{AG}^{opt}(r)$ varies between 8 and 16 for $r = 0, ..., 4$ when $T = 500$, and between 30 and 77 for $r = 0, ..., 4$ when $T = 2,500$. In both cases, as noted, the smallest rmse is for $r = 0$, that is, $\hat{d}_{AG}(0) = \hat{d}_{GPH}$. A 'wrong' choice can carry a heavy penalty; for example, when $\phi_1 = 0.9$, choosing $m = \sqrt{T} = 22$ for $T = 500$ results in an rmse of approximately three to four times that with $m = m_{AG}^{opt}(r)$.

Whilst, as Figures 4.3a and 4.3b show, the bias is larger, over a relevant range, for $\hat{d}_{AG}(0)$ compared to $\hat{d}_{AG}(1)$, the increase in the variance means that the latter does not achieve a lower asymptotic rmse; see Table 4.2 for details. Increasing the sample size to $T = 2,500$ should allow the increased convergence rate of $\hat{d}_{AG}(1)$ to work better. When $\phi_1 = 0.6$ and 0.3, there is now less to choose between $\hat{d}_{AG}(0)$ and $\hat{d}_{AG}(1)$ in terms of rmse, but the former still has the lowest rmse for $\phi_1 = 0.9$; in the latter case, the sample size is still not large enough to deliver the superiority of $\hat{d}_{AG}(1)$ over $\hat{d}_{AG}(0)$. This is considered further below.

Suppose, on a pairwise comparison, one is concerned to assess when there will be an advantage, in either bias or rmse reduction, to using, say, $\hat{d}_{AG}(1)$ rather than $\hat{d}_{AG}(0)$, where each is used at its respective optimum m, that is $m_{AG}^{opt}(1)$ and $m_{AG}^{opt}(0)$. Then a guide can be obtained from the asymptotic expressions. For example, define $T(bias)_{0=1}^{equal}$ as the sample size that gives equality of the asymptotic bias for the AG estimators with $r = 0$ and $r = 1$ respectively; correspondingly define $T(rmse)_{0=1}^{equal}$ as the sample size that gives equality of rmse. Using (4.66), (4.67) and (4.69), the relevant expressions are:

$$T(bias)_{0=1}^{equal} = (\kappa_1/\kappa_0)^{45/2} \qquad (4.71)$$

$$T(rmse)_{0=1}^{equal} = (\delta_1/\delta_0)^{45/4} \qquad (4.72)$$

Table 4.1   ARFIMA(1, d, 0): theoretical m_{AG}^{opt} for $\hat{d}_{AG}(r)$

T = 500	m_{AG}^{opt} for different values of r				
	r = 0 (GPH)	r = 1	r = 2	r = 3	r = 4
$\phi_1 = 0.9$	8	12	14	15	16
$\phi_1 = 0.6$	30	53	64	70	73
$\phi_1 = 0.3$	62	112	139	155	164
T = 2,500					
$\phi_1 = 0.9$	30	54	65	71	77
$\phi_1 = 0.6$	108	218	279	314	336
$\phi_1 = 0.3$	223	468	616	704	761

Note: Table entries are calculated using the expression for m_{AG}^{opt} in (4.69).

where:

$$\kappa_0 = \kappa_0 b_2 K(0)^2, \kappa_1 = \kappa_1 b_4 K(1)^4;$$
$$\delta_0 = \kappa_0^2 b_2^2 K(0)^4 + \pi^2/(24K(0)); \delta_1 = \kappa_1^2 b_4^2 K(1)^8 + c_1 \pi^2/(24K(1));$$

and K(r), r = 0, 1, are defined in (4.69b).

To give an idea of the magnitudes of $T(bias)_{0=1}^{equal}$ and $T(mse)_{0=1}^{equal}$, (4.71) and (4.72) were evaluated for $f_u(\lambda_j)$ generated by an ARMA(1, 0) process with $\phi_1 = 0.3, 0.6$ and 0.9; see Table 4.3. The (asymptotic) dominance of $\hat{d}_{AG}(1)$ over $\hat{d}_{AG}(0)$ in terms of bias is easily achieved; for example, $T(bias)_{0=1}^{equal} = 15$ for $\phi_1 = 0.9$, which means that if $\phi_1 = 0.9$, then (as a guide) the bias of $\hat{d}_{AG}(1)$ at $m_{AG}^{opt}(1)$ is less than the bias of $\hat{d}_{AG}(0)$ at $m_{AG}^{opt}(0)$ for all sample sizes ≥ 16.

However, the picture is more problematical for equality in rmse. It can be shown that the function relating $T(rmse)_{0=1}^{equal}$ to ϕ_1 has a minimum at $\phi_1 = 0.31$ with $T(rmse)_{0=1}^{equal} = 1,705$, so that $rmse(\hat{d}_{AG}(0)) < rmse(\hat{d}_{AG}(1))$ for T < 1,705, and $rmse(\hat{d}_{AG}(0)) > rmse(\hat{d}_{AG}(1))$ for T > 1,705. This means that whatever the value of ϕ_1, the smallest value of T at which $rmse(\hat{d}_{AG}(0)) > rmse(\hat{d}_{AG}(1))$ is 1,705; indeed, from Table 4.3, it can be observed that T has generally to be much larger, for example 2,580 for $\phi_1 = 0.6$. (Of course, these calculations are a guide given the use of asymptotic expressions for finite sample calculations; the latter are considered in a simulation context in the next section.)

4.6   Finite sample considerations and feasible estimation

There are two considerations that suggest the comparison so far, in terms of asymptotic bias and rmse, may need to be extended. First, finite sample effects may have an important contribution to make to the bias and rmse; and, second,

Table 4.2 ARFIMA(1, d, 0): theoretical rmse for $\hat{d}_{AG}(r)$

rmse at $m_{AG}^{opt}(r)$ and $m = \sqrt{T}$

T = 500	r = 0		r = 1		r = 2		r = 3		r = 4	
	m_{AG}^{opt}	$m = \sqrt{T}$	m_{AG}^{opt}	$m = \sqrt{T}$	m_{AG}^{opt}	$m = \sqrt{T}$	m_{AG}^{opt}	$m = \sqrt{T}$	m_{AG}^{opt}	$m = \sqrt{T}$
$\phi_1 = 0.9$	0.248	0.775	0.285	0.839	0.325	1.041	0.367	1.393	0.403	1.950
$\phi_1 = 0.6$	0.130	0.140	0.141	0.205	0.156	0.256	0.172	0.299	0.188	0.370
$\phi_1 = 0.3$	0.091	0.136	0.096	0.205	0.106	0.256	0.116	0.296	0.126	0.336
T = 2,500										
$\phi_1 = 0.9$	0.131	0.182	0.139	0.140	0.155	0.170	0.171	0.198	0.187	0.223
$\phi_1 = 0.6$	0.069	0.091	0.069	0.136	0.074	0.170	0.081	0.198	0.088	0.223
$\phi_1 = 0.3$	0.048	0.091	0.047	0.136	0.050	0.170	0.054	0.198	0.056	0.223

Note: Table entries are calculated using the expression for the square root of $\mathrm{AMSE}(\hat{d}_{AG})$ in (4.68).

Table 4.3 $T(bias)_{0=1}^{equal}$ and $T(mse)_{0=1}^{equal}$ for $f_u(\lambda_j) = ARMA(1, 0)$ process

	$\phi_1 = 0.3$	$\phi_1 = 0.6$	$\phi_1 = 0.9$
$T(bias)_{0=1}^{equal}$	2	3	15
$T(mse)_{0=1}^{equal}$	1,707	2,580	11,392

the estimator $\hat{d}_{AG}(r)$ is infeasible without some means of choosing m. These issues are considered in the following sections.

4.6.1 Finite sample considerations

The mean squared error that is minimised to obtain optimal m is an approximation based on neglecting terms of lower order of probability than m^{2+2r}/T^{2+2r}. Although negligible (when m grows as T^γ, $(2+2r)/(3+2r) < \gamma < 1$), such terms may be important for relatively small m, for example when ϕ_1 is close to unity in the ARFIMA(1, d, 0) model.

Hurvich et al. (1998) consider this issue for m_{GPH}^{opt} $(= m_{AG}^{opt}(0))$ by simulating an ARFIMA(1, d, 0) model for d between 0.1 and 0.49 and $\phi_1 = 0.1$, 0.3 and 0.9; sample sizes were varied from $T = 200$ to 2,000. As expected, the agreement between m_{GPH}^{opt} and the simulation minimum was very good for large m (which results from $\phi_1 = 0.1$, 0.3), even with the smallest sample size. The mse 'cost' of the differences, that is $(mse^{opt}/mse_{sim}^{opt})$, where mse_{sim}^{opt} is the simulation mse, was not large, for example with $d = 0.3$ and $\phi_1 = 0.3$, then $(mse^{opt}/mse_{sim}^{opt}) = 1.06$ for $T = 200$ and $(mse^{opt}/mse_{sim}^{opt}) = 1.02$ for $T = 2,000$. When ϕ_1 was increased to 0.9, then $(mse^{opt}/mse_{sim}^{opt}) = 2.45$ for $T = 200$; however, when the sample size was increased the 'cost' declined, with $(mse^{opt}/mse_{sim}^{opt}) = 1.07$ for $T = 700$ and 1.03 for $T = 2,000$.

AG (op. cit.) report simulations with a range of ARFIMA(1, d, 1) models and sample sizes 128, 512 and 2,048 (these are all powers of 2, which enable the benefits of the Fast Fourier Transform in the simulations). These simulations consider, for example, bias and rmse in finite samples for $\hat{d}_{AG}(r)$, $r = 0$, 1, 2, as m varies. The simulations results are somewhat more favourable to $\hat{d}_{AG}(r)$, with $r = 1$, 2, over $\hat{d}_{AG}(0)$, than the asymptotic results have indicated. The bias improvement, with $r = 1$, 2, is maintained over a wider range of m; indeed, in the cases of $\phi_1 = 0.9$, 0.6, with $T = 512$, the bias of $\hat{d}_{AG}(0)$ is uniformly larger than the bias of $\hat{d}_{AG}(r)$ with $r = 1$, 2 over $m = 1$, ..., 256. As a result, in this and other cases, the smallest rmse tends to be for one of the bias-adjusted estimators, either $\hat{d}_{AG}(1)$ or $\hat{d}_{AG}(2)$ depending on the particular value of ϕ_1 or θ_1.

To illustrate what is happening in the finite sample case we take an example from AG (op. cit., Table I); for $T = 512$ and $\phi_1 = 0.9$, the minimum of the simulation rmse is 0.327 for $\hat{d}_{AG}(2)$, with rmse $= 0.328$ for $\hat{d}_{AG}(1)$ and 0.337 for $\hat{d}_{AG}(0)$. The corresponding asymptotic rmse are: 0.323 for $r = 2$; 0.282 for $r = 1$; and 0.246 for $r = 0$. This suggests that, in practice, the asymptotic rmse tends to underestimate the actual rmse for $r = 0$.

4.6.2   Feasible estimation (iteration and 'plug-in' methods)

Although the concept of an optimal m based on minimizing the mse is a useful one, it does not result in a feasible estimator because it relies on knowledge of the true spectral density. The question then arises as to how to make the procedure feasible. We describe two essentially similar methods, both using a Taylor series approximation to the spectral density of the short-run dynamics; the first is due to Henry (2001) and the second to AG (op. cit.).

In the first method, assume a starting value for m, say $m^{(0)}$, and then obtain the corresponding starting value for d, say $\hat{d}^{(0)}$, from LS estimation of the extended log-periodogram regression of (4.62) with $r = 1$. Then, as in AG (op. cit.), expand $f_u(\lambda_j)$ in a second-order Taylor series with remainder. This results in $f_u(\lambda_j) = f_u(0) + f_u'(0)\lambda_j + f_u''(0)\lambda_j^2/2 + R_2$. Since $f_u(\lambda_j)$ is an even function, $f_u'(\lambda_j) = 0$; then, ignoring the remainder R_2, to an approximation, the SDF for y_t is:

$$f_y(\lambda_j) = |(1 - \exp(-i\lambda_j))|^{-2\hat{d}^{(0)}}[f_u(0) + f_u''(0)\lambda_j^2/2] \qquad (4.73)$$

Hence, an OLS regression of the periodogram, $I_y(\lambda_j)$, on the regressors $|(1 - \exp(-i\lambda))|^{-2\hat{d}^{(0)}}$ and $(|(1 - \exp(-i\lambda))|^{-2\hat{d}^{(0)}}\lambda_j^2/2)$ for frequencies $j = 1, ..., m^{(0)}$, provides estimators of $f_u(\lambda_j)$ and $f_u''(\lambda_j)$ respectively. Denote these $f_u(0)^{(1)}$ and $f_u''(0)^{(1)}$, and then use these to estimate m from m_{GPH}^{opt} as in (4.45), or (4.69) with $r = 0$; this results in, say, $m^{(1)}$; in the next step update $\hat{d}^{(0)}$ to $\hat{d}^{(1)}$ using $m^{(1)}$. Further rounds in the iteration are possible, but Henry (2001) found that these did not improve the results in his simulation experiments.

The second approach, which is a variation on the first but does not involve iteration, exploits the unity offered by the AG approach. In the case of $r = 1$, the AG regression is:

$$Y_j = \beta_0 - 2d\ln(\lambda_j) + \varphi_1\lambda_j^2 + \breve{\varepsilon}_j$$

where $\varphi_1 = f_u''(0)/2f_u(0) = b_2/2$. Hence, the AG regression for $r = 1$, with $j = 1, ..., m^{(0)}$, provides an estimate of $b_2 = 2\varphi_1$, which can be used to obtain an estimate of $m_{AG}^{opt}(0)$ and, thus, becomes feasible. For any choice of r, an estimator using m_{GPH}^{opt} is not feasible unless an estimate of b_{2+2r} is available. The general principle is that the AG regression of one higher order than the value of r in $\hat{d}_{AG}(r)$ can be used to estimate the unknown coefficient b_{2+2r} in $m_{AG}^{opt}(r)$. Thus, if $r = 0$, then b_2 is estimated; if $r = 1$, then b_4 is estimated; and so on. This method uses the AG regression for $r + 1$ to estimate b_{2+2r} by:

$$\hat{b}_{2+2r} = (2 + 2r)!\hat{\varphi}_{(1+r)}$$

As this is not the primary regression it will be referred to as the AG auxiliary regression. The estimated value of b_{2+2r} is then used to obtain an estimate of m_{GPH}^{opt} for r, denoted $\hat{m}_{AG}^{opt}(r)$, which is as follows:

$$\hat{m}_{AG}^{opt}(r) = \left(\frac{\pi^2 c_r}{24(4+4r)\kappa_r^2 \hat{b}_{2+2r}^2} \right)^{1/(5+4r)} T^{(4+4r)/(5+4r)} \tag{4.74}$$

The estimate $\hat{m}_{AG}^{opt}(r)$ is then used in the primary AG regression.

The remaining practical problem is to decide how many frequencies, say v, to use in the AG auxiliary regression; however, this seems to have set up an insoluble problem in which the regression to obtain b_{2+2r} requires $b_{2+2(r+1)}$, which in turn requires $b_{2+2(r+2)}$ and so on (an 'infinite-regression' problem). AG (op. cit.) show that $\hat{b}_{(2+2r)}$ is consistent if $v = HT^\alpha$, for a constant H and $\alpha \in$ A where $A = [(4 + 4r)/(5 + 4r), 1]$; v is 'rate-optimal' (minimises the asymptotic mse) if $\alpha = (8 + 4r)/(9 + 4r)$. For example, if $r = 0$, the GPH case, then estimate b_2 using the AG auxiliary regression with $v = HT^{8/9}$; if $r = 1$, estimate b_4 using the AG auxiliary regression with $v = KT^{12/13}$. However, it is now K that is not known because of the 'infinite-regression' problem; based on some calibrating simulation results, AG (op. cit.) suggest setting $K = 0.3$ or 0.4. Apart from the need to set the constant K, this method is 'plug-in' in the sense that it is data-dependent according to a well-defined rule, and the estimator can be obtained with two applications of the AG method (auxiliary plus primary regressions). The plug-in version of $\hat{d}_{AG}(r)$ will be denoted $\hat{d}_{PAG}(r)$.

The simulation results in AG (op. cit.) suggest that the cost of making $\hat{d}_{AG}(r)$ feasible is quite large in rmse terms; for example, the rmse increases by not less than 20% and by 50% in some cases relative to the infeasible estimator $\hat{d}_{AG}(r)$; but since this is so for all values of r, including $r = 0$, the GPH case, what matters is the relative cost as r varies.

Some simulation results are extracted into Table 4.4 for $T = 512$ and $K = 0.3$, 0.4. These results use a slight variation of $m_{AG}^{opt}(r)$, in particular the second term in (4.68) due to the variance is replaced by $\hat{\sigma}_{AG}^2(\hat{d}_{AG})$ of (4.64); AG (op. cit.) found that this produced slightly better results than when $\hat{\sigma}_{AG}^{2,asym}$ of (4.67) is used (this substitution does not affect the asymptotic properties). The asymptotic values of bias, rmse and $m_{AG}^{opt}(r)$ for the infeasible estimator are shown in the first three columns; thereafter, the results relate to the feasible estimator with $K = 0.3$ and 0.4, respectively.

If $\phi_1 = 0.3$, then bias reduction is effective for $r = 1, 2$; this reduction is not sufficient to reduce the rmse if $K = 0.3$, but it is just sufficient if $K = 0.4$. Note that as ϕ_1 increases, the bias increases for each value of r, but bias reduction remains effective and is sufficient to reduce rmse relative to the estimator with $r = 0$. Generally, there is an increase in rmse in the finite sample relative to the asymptotic results. Overall, AG recommend $\hat{d}_{AG}(1)$ combined with $K = 0.4$.

Table 4.4　Asymptotic results for $\hat{d}_{AG}(r)$ and simulation results for the plug-in version of the AG estimator, $\hat{d}_{PAG}(r)$; d = 0, T = 512

	Asymptotic results			K = 0.3			K = 0.4		
	abias	armse	$m_{AG}^{opt}(r)$	bias	rmse	m̄	bias	rmse	m̄
$\phi_1 = 0.3$									
r = 0	0.040	0.090	62	0.045	0.119	77	0.059	0.117	94
r = 1	0.032	0.095	114	0.011	0.125	99	0.021	0.110	129
r = 2	0.029	0.105	142	0.003	0.146	110	0.008	0.125	145
$\phi_1 = 0.6$									
r = 0	0.058	0.130	30	0.129	0.186	58	0.159	0.198	65
r = 1	0.047	0.140	53	0.077	0.156	95	0.113	0.168	119
r = 2	0.042	0.156	64	0.041	0.155	109	0.072	0.149	143
$\phi_1 = 0.9$									
r = 0	0.110	0.246	8	0.551	0.568	53	0.614	0.625	69
r = 1	0.094	0.282	13	0.486	0.507	86	0.554	0.568	111
r = 2	0.093	0.322	15	0.420	0.450	104	0.494	0.513	135

Note: Abias and armse are the asymptotic bias and asymptotic rmse at the optimum m, $m_{AG}^{opt}(r)$; m̄ is the simulation average of m.
Source: AG (op. cit., table II).

However, note that in general m is chosen too large with both choices of K, a result that is particularly noticeable as ϕ_1 increases and which results in rmse being away from its (asymptotic) minimum; it may be possible, therefore, to develop the method further so as to 'tune' K closer to a value that results in a closer agreement with the asymptotic results. On this latter point see Guggenberger and Sun (2006) who have developed a GPH-based estimator, averaged over different bandwidths, which addresses the question of variance inflation in the AG estimator.

4.7　A modified LP estimator

The discussion of the GPH and AG versions of the LP estimator assumed d ∈ (−0.5, 0.5) which, if necessary, can be achieved by prior differencing of the level of a series to obtain the minimum integer difference that places d in the stationary region. The unit root hypothesis then relates to d = 0 for the differenced series. However, methods that are based on the LP idea, but which do not difference the data, may be preferable given that in some cases there is little knowledge about the value of d. LP regression in the form used in the GPH estimator is not invariant to first differencing the data; that is adding 1 to the estimate of d using differenced data does not generally equal the estimated value of d from the original series (Hurvich and Ray, 1995).

It is important to ask whether progress can be made for LP-type regressions if d is outside the stationary range. The answer to this question is yes. First, note that Kim and Phillips (2000, 2006) have shown that $\hat{d}_{GPH}$ is consistent for d $\in$ (0.5, 1) without requiring normality of the innovations. By extension, this property applies to $\hat{d}_{AG}(r)$. These LP estimators are, however, inconsistent for d > 1 and tend to 1 in probability if d > 1; this result explains the simulation results in Hurvich and Ray (1995), where $\hat{d}_{GPH}$ is close to unity even when, for example, d = 1.4. The boundary case is the unit root, d = 1, and Phillips (1999b) and Kim and Phillips (2000, 2006) have obtained some important results in this case and for d > 1, which will be considered in detail below.

Second, for d $\notin$ (–0.5, 0.5), such that d < 1.5, Hurvich and Ray (1995) have shown that the GPH type estimator resulting from omitting the first periodogram ordinate and using a cosine bell data taper reduces the bias and is more nearly invariant to first-differencing, but at the cost of an increase in the variance relative to the standard $\hat{d}_{GPH}$ estimator.

4.7.1 The modified discrete Fourier transform and log-periodogram

The generating mechanism for y_t is taken to be the following:

$$(1 - L)^d (y_t - \mu_t) = u_t \tag{4.75a}$$

$$\mu_t = y_0 \tag{4.75b}$$

$$\Rightarrow$$

$$y_t = y_0 + z_t \tag{4.75c}$$

$$z_t = (1 - L)^{-d} u_t \tag{4.75d}$$

It is assumed that u_t is stationary for t > 0, with zero mean and continuous spectrum; whereas for t $\leq$ 0 it is assumed that $u_t = 0$, referred to as the absence of 'prehistorical' influences and, therefore, relates to type II fBM. The initial value y_0 is assumed be a random variable with a fixed distribution (an alternative to this assumption is that y_0 is a nonrandom finite constant interpreted as the mean of y_t; see Shimotsu, 2010.) The definition of the process with fractional d follows the development in Chapter 3, where the operators $(1 - L)^d$ and $(1 - L)^{-d}$ were defined by expansions using the binomial coefficients (or equivalently in terms of the gamma distribution).

Common to the various cases for different values of d are the following two conditions:

(i) For all $t > 0$, u_t has the Wold representation:

$$u_t = c(L)\varepsilon_t$$

$$= \sum_{j=0}^{\infty} c_j \varepsilon_{t-j} \qquad \text{where } \sum_{j=0}^{\infty} j|c_j| < \infty, \ c(1) \neq 0$$

(ii) $\varepsilon_t \sim \text{iid}(0, \sigma_\varepsilon^2)$, with finite fourth moment, $E|\varepsilon_t|^4 < \infty$

Assumptions (i) and (ii), which will be maintained throughout this section, are relatively unrestrictive and allow a wide range of processes generating the short-run dynamics; they are, for example, satisfied by stable ARMA processes where $u_t = \phi(L)^{-1}\theta(L)\varepsilon_t$ and ε_t is white-noise input. Gaussianity is not assumed unless specifically noted, although it clearly satisfies the moment condition.

4.7.2   The modified DFT and modified log-periodogram

In the usual approach, the DFT, $\omega_y(\lambda_j)$, is defined by (4.24), with corresponding periodogram $I_y(\lambda_j)$. Phillips (1999b) introduced the modified DFT, denoted here $v_y(\lambda_j)$ and, correspondingly, the modified periodogram $I_v(\lambda_j)$ as follows:

$$v_y(\lambda_j) = \omega_y(\lambda_j) + \frac{e^{i\lambda_j}}{1 - e^{i\lambda_j}} \frac{(y_T - y_0)}{\sqrt{2\pi}} \qquad j = 1, ..., m \qquad (4.76)$$

$$I_v(\lambda_j) = \frac{|v_y(\lambda_j)|^2}{T} \qquad j = 1, ..., m \qquad (4.77)$$

(Note that Phillips (1999b) defines the DFT with the divisor $\sqrt{2\pi T}$ and, hence, on that definition, the periodogram is then without the divisor T; here we adopt a definition of $v_y(\lambda_j)$ that is consistent with the earlier definition of $\omega_y(\lambda_j)$. The periodogram is, of course, the same in both cases.)

The motivation for the modified DFT is relatively simple. The usual DFT is an approximation, rather than an exact representation of the DFT, for a fractionally integrated process generating y_t; see Phillips (1999b). The modified representation adds in two terms, one of which is zero assuming $y_t = 0$ for $t = 0$; the other contributes the second term on the right-hand side of (4.76), which can be viewed as a correction term based on the last sample observation y_T.

Note that the second term is proportional to the DFT of a linear time trend (see (4.27)), hence its inclusion will implicitly 'protect' the resulting estimator of d against unknown linear trends in the generating process for y_t, and it is not necessary to first 'detrend' the data; however, in practice, as the results in Kim and Phillips (2000, 2006) show, the estimates resulting from 'raw' and detrended data can differ quite substantially. This is not the case for methods based on $\omega_y(\lambda_j)$ alone, for example, $\hat{d}_{GPH}$ and $\hat{d}_{AG}$, for which a linear trend should have first been removed. Typically this is dealt with by first differencing the data or regressing y_t on a linear time trend and then working with the residuals from this regression.

4.7.3 The modified estimator

The development of the revised LP pseudo-regression and estimator follows as in the GPH case, but using the modified DFT and the revised periodogram. The LP pseudo-regression is now:

$$\ln[I_v(\lambda_j)] = \ln[f_u(0)] - 2d\ln[|1 - e^{-i\lambda_j}|] + \ln[I_u(\lambda_j)/f_u(0)] \qquad j = 1,...,m \quad (4.78)$$

The practical effect of using the modified DFT is that the dependent variable in the pseudo-regression is the log of the modified periodogram $I_v(\lambda_j)$ given by (4.77) rather than the log of $I_y(\lambda_j)$. This is easy to implement: all that is required is to add the second term in (4.76) to the standard DFT and then compute the periodogram. The resulting LS estimator of d, denoted $\hat{d}_{MGPH}$, is straightforward since it just uses the revised dependent variable:

$$\hat{d}_{MGPH} = -0.5 \left[\frac{\sum_{j=1}^{m} (\check{X}_j - \bar{\check{X}}) \ln[I_v(\lambda_j)]}{\sum_{j=1}^{m} (\check{X}_j - \bar{\check{X}})^2} \right] \tag{4.79}$$

where $\check{X}_j \equiv \ln[|1 - e^{-i\lambda_j}|]$. The Phillips estimator $\hat{d}_{MGPH}$ of d, is $-\frac{1}{2}$ times the least squares estimator of the slope coefficient in the regression of $I_v(\lambda_j)$ on $\ln[|1 - e^{-i\lambda_j}|]$; alternatively the form using (4.34) can be used. A trimmed version of $\hat{d}_{MGPH}$ is also possible, but is not needed for the central results; it is defined by excluding the first l frequencies, that is, (4.78) is only defined for $j = l + 1$, ..., m and hence the summations in (4.78) start from $j = l + 1$, which results in $\hat{d}_{MGPH}(l)$, say.

4.7.4 Properties of the modified estimator

This section summarises the results in Kim and Phillips (2000, 2006), hereafter KP, and Phillips (1999a, 2007), who consider the case with $y_0 = 0$.

4.7.4.i Consistency

The estimator $\hat{d}_{MGPH}$ is consistent for a significantly larger part of the parameter space compared to the GPH estimator (and the AG versions); in particular, it avoids the problem that $\hat{d}_{GPH}$ is not consistent for $1 \leq d < 2$. The range of d is distinguished as follows.

For $0.5 < d < 2$, consistency of $\hat{d}_{MGPH}$ requires the weak condition $(m/T) \to 0$ as $T \to \infty$ and $m \to \infty$ and some weak conditions on ε_t. Of note is that neither Gaussianity nor trimming are required for consistency.

In the case of $d \in (0, 0.5]$, there is a slight variation in the convergence condition relating to m/T. For details see KP (2000), who provide the proofs of consistency of $\hat{d}_{MGPH}$ for the cases: $d \in (1, 2)$, KP (theorem 3.1); $d = 1$, KP (equation (28)); $d \in (0.5, 1)$, KP (theorem 3.4); and $d \in (0, 0.5)$, KP (theorem 3.6).

4.7.4.ii   Asymptotic normality for $d \in [0.5, 2]$ if the initial condition is zero or known

It is known that in the stationary case $\sqrt{m}(\hat{d}_{GPH} - d) \Rightarrow_D N(0, \pi^2/24)$, implying that the asymptotic variance of $\hat{d}_{GPH}$ is $\pi^2/24m$. In the case of the modified GPH estimator, the general result is that the limit distribution of $\hat{d}_{MGPH}$ is the same as the limit distribution of $\hat{d}_{GPH}$ but, as in the case of the consistency results, for a wider range of d values.

KP show that, in addition to the usual situation of asymptotic normality in the stationary region, $\hat{d}_{MGPH}$ is asymptotically normal, with distribution as for $\hat{d}_{GPH}$ if $d \in (0.5, 2)$, although the precise conditions required for this result distinguish between the three cases of $d \in (0.5, 1)$, $d = 1$ and $d \in (1, 2)$. A brief summary follows, and more detail can be found in KP (2000).

The general result is:

$$\sqrt{m}(\hat{d}_{MGPH} - d) \Rightarrow_D N(0, \pi^2/24) \text{ for } d \in (0.5, 1), d = 1, d \in (1, 2) \qquad (4.80)$$

In the case of $d \in (0.5, 1)$, to achieve asymptotic normality there is a trade-off in the assumptions required on the properties of ε_t and the convergence condition relating to the rates of growth of m and T. Broadly, if ε_t is not Gaussian, then the m/T-type condition has to be strengthened; see KP (2000, theorem 3.5) for details.

In the unit root case, that is, $d = 1$, no Gaussianity assumption is required for the asymptotic normality result (there is a slight variation on the m/T condition relative to the first case); see Phillips (1999b, theorem 4.1, 2007) for details. Finally, in the case of $d \in (1, 2)$, if ε_t is not Gaussian, then the m/T-type condition has to be strengthened; see KP (2000, theorem 3.2) for details. As in the stationary case, the asymptotic variance of $\hat{d}_{MGPH}$, $\sigma^{2,asym}(\hat{d}_{PGPH})$, is $\pi^2/24m$, which can be used for hypothesis testing and constructing confidence intervals.

4.7.4.iii   Reported simulation results

KP (2000) report a number of experimental designs of ARFIMA form to assess the small sample properties of $\hat{d}_{MGPH}$, for T = 256, 512, and 1,000 replications, and $y_0 = 0$. The generating process is $y_t = (1 - L)^{-d} u_t 1_{(t>0)}$, $u_t = \phi_1 u_{t-1} + \varepsilon_t$, with $\varepsilon_t \sim$ niid(0, 1). The theory has not yet developed to choose an asymptotically optimal value of m (in principle this is straightforward enough, all that is lacking is an expression for the asymptotic bias of $\hat{d}_{MGPH}$), so the results are presented for $m = T^\gamma$, with $\gamma = 0.4$, 0.5. 0.6 and 0.7. A brief summary of the results for the ARFIMA(1, d, 0) case is presented in Table 4.5.

The entries in Table 4.5 indicate a number of key features. The result that $\hat{d}_{GPH} \to_p 1$ is evident for $d > 1$; $\hat{d}_{MGPH}$ has less bias than $\hat{d}_{GPH}$ even for the case

Table 4.5 Simulation results for $\hat{d}_{GPH}$ and $\hat{d}_{MGPH}$ with AR(1) short-run dynamics, $\phi_1 = 0.3$, $T = 512$

d	$m = T^{0.5}$		$m = T^{0.6}$		$m = T^{0.7}$	
	$\hat{d}_{GPH}$	$\hat{d}_{MGPH}$	$\hat{d}_{GPH}$	$\hat{d}_{MGPH}$	$\hat{d}_{GPH}$	$\hat{d}_{MGPH}$
0.8	0.8280	0.8050	0.8440	0.8227	0.8646	0.8503
	(0.1658)	(0.1685)	(0.1196)	(0.0117)	(0.0833)	(0.0785)
1.25	1.0727	1.2778	1.0619	1.2812	1.0486	1.3073
	(0.1449)	(0.1652)	(0.1231)	(0.1178)	(0.1093)	(0.0812)
1.45	1.0348	1.4907	1.0275	1.4991	1.0028	1.5119
	(0.1160)	(0.1769)	(0.0098)	(0.1211)	(0.0604)	(0.0841)
1.8	1.0876	1.8353	1.0556	1.8452	1.0307	1.8592
	(0.1756)	(0.1711)	(0.1322)	(0.1212)	(0.1169)	(0.0840)

Source: Kim and Phillips (2000, tables 1-(ii) and 2-(ii)).
Notes: $T = 512$; table entries are the means over 1,000 replications; $int[T^{0.5}] = 22$, $int[T^{0.6}] = 42$, $int[T^{0.7}] = 78$. The standard error of the estimate is shown in (.); note that $\sigma^{2,asym}(\hat{d}_{MGPH})$ is $\pi^2/24m = 0.135$, 0.099, 0.072 respectively for the three values of m.

where the latter is valid; in the case of $\hat{d}_{MGPH}$ there is a positive bias that increases with T but, overall, the finite sample results are in line with the theoretical results. The results for an MA(1) process for the dynamics are also in line with the theoretical results, although there are some minor differences in detail compared to the AR(1) case; for example, biases tend to be negative and decline with m.

4.7.5 Additional simulations

The aim in this section is to compare the basic and the modified GPH estimators for the cases where $\mu_t = y_0$ and $\mu_t = y_0 + \beta_1 t$. The range of d of interest for $\hat{d}_{GPH}$ is $d \in (-\frac{1}{2}, 1]$, whereas for $\hat{d}_{MGPH}$ it is $d \in (\frac{1}{2}, 2)$. The reported simulations are for y_t generated by

$$y_t = \mu_t + (1 - L)^{-d} \varepsilon_t 1_{(t > 0)}$$

with $\varepsilon_t \sim niid(0, 1)$, $T = 512$, $m = T^{0.65}$ and 5,000 replications. Whilst more general error processes are clearly possible, they do not introduce any additional points of interest in this exercise (similarly, while different values of T alter the numerical results they did not add anything of substance).

The estimators will be subscripted depending on what allowance is made for deterministic terms in the estimation procedure. In the case of the GPH estimator, the standard practice is to detrend the data by a prior regression on a constant and time and that procedure is followed here, with estimator denoted $\hat{d}_{GPH,\beta}$; note that $\hat{d}_{GPH}$ is invariant to mean adjustment, but not to trend adjustment. A slightly different procedure, based on Shimotsu (2010), was found to be necessary for the modified GPH estimator to work. Shimotsu (2010) addresses the problem of demeaning and detrending in the context of local Whittle estimation and notes two possibilities when y_0 is a nonrandom constant; see Section **4.9.2.iv** and Chapter 3, Section **3.6**. For $d \in (-\frac{1}{2}, \frac{1}{2}]$, Shimotsu (2010) suggests estimating y_0 by the sample mean, $\bar{y} = T^{-1} \sum_{t=1}^{T} y_t$; however, the error in the estimation of y_0 by $\bar{y}$ is $O_p(T^{d-1/2})$, which will therefore increase as d increases for $d > \frac{1}{2}$. In the latter case, Shimotsu (2010) suggests estimating y_0 by y_1 and using the demeaned data given by $\tilde{y}_t \equiv y_t - y_1$, and this suggestion is followed here as the merit in $\hat{d}_{MGPH}$ is for d outside the stationary region; the resulting modified GPH estimator is denoted $\hat{d}_{MGPH,\mu}$. In the linear trend case, the data for the modified GPH estimator was first linearly detrended (as in the GPH case) and then the first detrended observation was subtracted, that is, $\tilde{y}_t \equiv \hat{y}_t - \hat{y}_1$, where $\hat{y}_t = y_t - (\hat{\beta}_0 + \hat{\beta}_1 t)$, which we describe as 'Shimotsu detrending'. The modified GPH estimator is referred to as $\hat{d}_{MGPH,\beta}$.

The estimators considered in these simulations are, therefore, $\hat{d}_{GPH}$ and $\hat{d}_{GPH,\beta}$, and $\hat{d}_{MGPH}$, $\hat{d}_{MGPH,\mu}$ and $\hat{d}_{MGPH,\beta}$; and to simplify the graphical presentation, the results are shown for the simulation bias in the first figure of each pair and the simulation mean squared error in the second figure.

As a benchmark, the simplest case with $y_0 = 0$ is considered first. The GPH estimators $\hat{d}_{GPH}$ and $\hat{d}_{GPH,\beta}$ are invariant to y_0. The simulations show that the small sample bias for $\hat{d}_{GPH}$ is somewhat less than that for $\hat{d}_{GPH,\beta}$ for $d \in [0, 1]$, but there is little to choose between their mean squared errors, and the results confirm that neither should be used outside this range. The simulations show that for $d \in [\frac{1}{2}, 2)$, there is little to choose between $\hat{d}_{MGPH}$, $\hat{d}_{MGPH,\mu}$ and $\hat{d}_{MGPH,\beta}$ in terms of bias and mse (note that this range differs slightly from KP (2000, theorem 3.6) in not including $d \in [0, \frac{1}{2})$). The bias and mse of selected GPH and MGPH variants are shown in Figures 4.5 and 4.6; over the common range $(\frac{1}{2}, 1)$ there is a slightly smaller bias for $\hat{d}_{MGPH,\mu}$ (and $\hat{d}_{MGPH,\beta}$), but little to choose between the estimators in terms of mse – see Figure 4.6.

The next case to be considered is with a non-zero initial value, $y_0 = 100$. As to the basic GPH estimators, $\hat{d}_{GPH}$ and $\hat{d}_{GPH,\beta}$ are invariant to y_0. Of the modified GPH estimators, $\hat{d}_{MGPH}$ is no longer a sensible estimator and mean adjustment is essential for the modified GPH estimator, that is $\hat{d}_{MGPH,\mu}$, to work over the range $d \in [\frac{1}{2}, 2)$, otherwise $\hat{d}_{MGPH} \approx 1$ for most of the range of d; $\hat{d}_{MGPH,\mu}$ and $\hat{d}_{MGPH,\beta}$ are not distinguishable, so there is no cost in allowing for a linear trend. Of the selected estimators there is little to choose between $\hat{d}_{GPH,\beta}$, $\hat{d}_{MGPH,\mu}$ and $\hat{d}_{MGPH,\beta}$ over the range for which these are sensible estimators; see Figure 4.7 for bias and Figure 4.8 for mse.

In the last case a linear trend is introduced into the DGP, with $y_0 = 100$ and $\beta_1 = 1$. Of the simple GPH estimators, it is vital to detrend the data as $\hat{d}_{GPH} \approx 1$ throughout the range of d, and only $\hat{d}_{GPH,\beta}$ is a sensible estimator. In the case of trended data, $\hat{d}_{MGPH}$ is not a sensible estimator, whilst $\hat{d}_{MGPH,\mu}$ works for $(\frac{1}{2}, 2)$, if the data are (as we shall see) not too highly trended. $\hat{d}_{MGPH,\beta}$ works

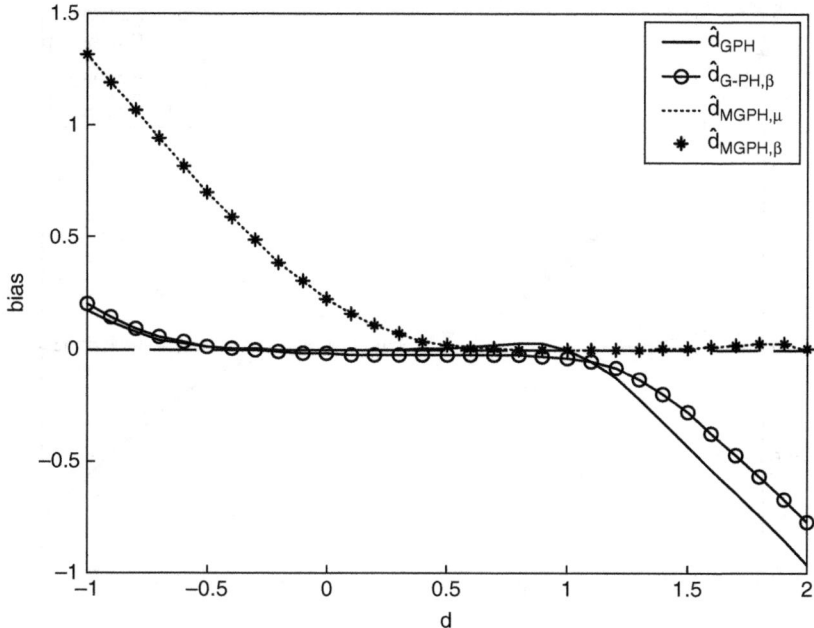

Figure 4.5   Bias of selected GPH and MGPH variants, zero initial value

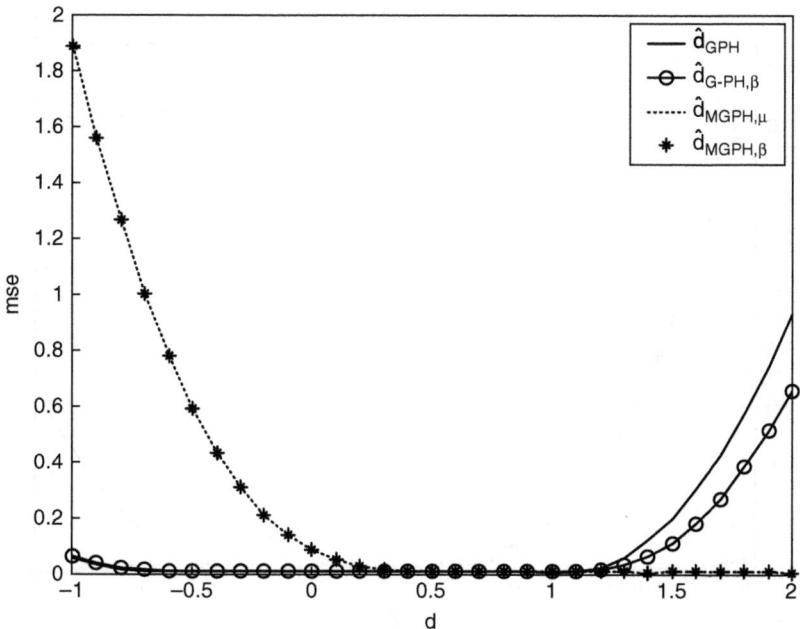

Figure 4.6   Mse of selected GPH and MGPH variants, zero initial value

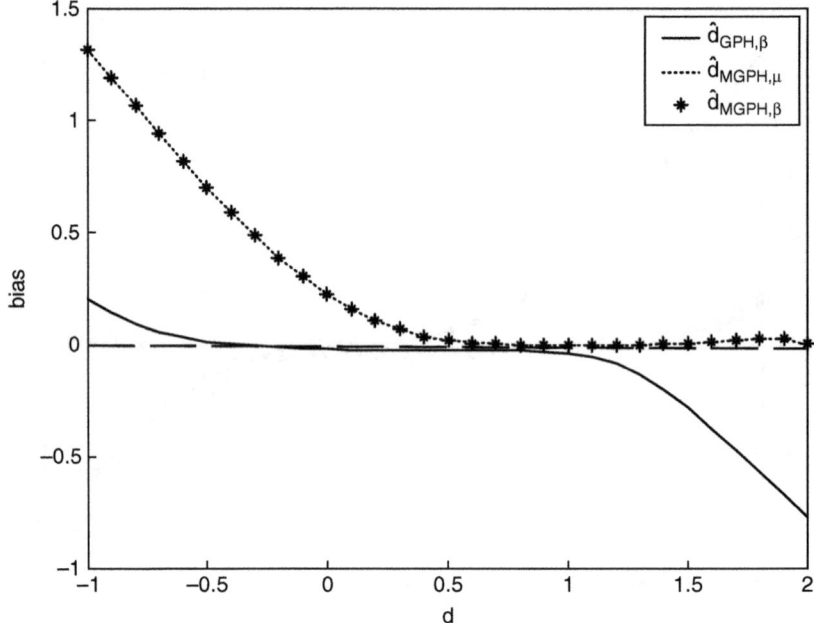

Figure 4.7   Bias of selected GPH and MGPH variants, non-zero initial value

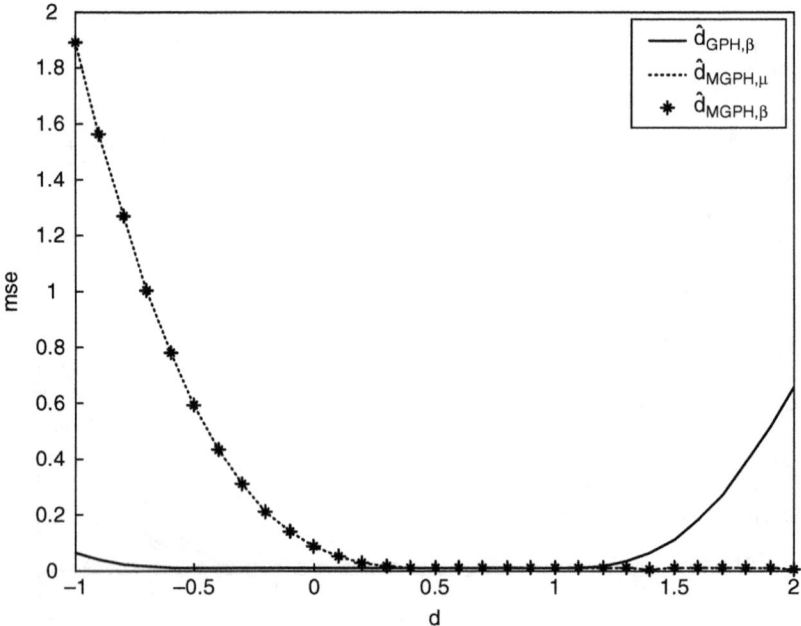

Figure 4.8   Mse of selected GPH and MGPH variants, non-zero initial value

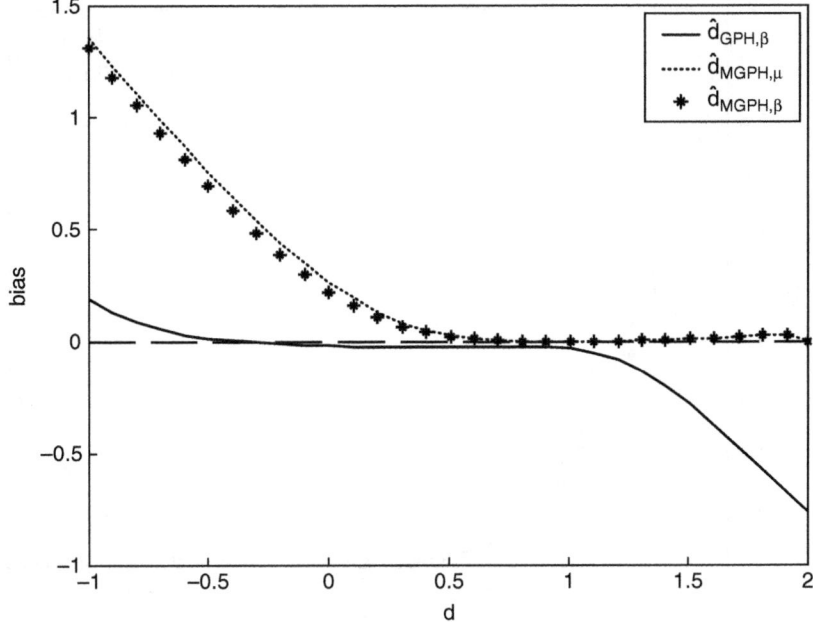

Figure 4.9   Bias of selected GPH and MGPH variants, trended data, $\beta_1 = 1$

over the range ($\frac{1}{2}$, 2), provided the data is 'Shimotsu' detrended. Figures 4.9 and 4.10 show the bias and mse for the competitive estimators, the differences being quite slight over the common range of d.

Notice that the slope of the trend, β_1, is effectively scaled in terms of units of $\sigma_\varepsilon = 1$ in this case; the estimators $\hat{d}_{GPH,\beta}$ and $\hat{d}_{MGPH,\beta}$ are not affected by this scaling, however, $\hat{d}_{MGPH,\mu}$ is affected. To illustrate briefly, β_1 was increased to 10 and it is evident that $\hat{d}_{MGPH,\mu}$ is no longer a competitive estimator (see Figures 4.11a and 4.11b). Thus, when there is uncertainty over whether there is a trend $\hat{d}_{MGPH,\beta}$ is preferred to $\hat{d}_{MGPH,\mu}$ as there is no mean squared error cost to $\hat{d}_{MGPH,\beta}$ when the trend is absent (see Figures 4.7 and 4.8).

4.7.6   Testing the unit root hypothesis

It was noted in Section **4.3.1.vi** that whilst $\hat{d}_{GPH}$ is consistent for $d = 1$, inference on the unit root issue is complicated because $\hat{d}_{GPH}$ is not consistent for $d > 1$, hence two-sided alternatives are ruled out; also even if attention is confined to the left-sided alternative, the limit distribution is not normal. These problems are avoided if the appropriate version of $\hat{d}_{MGPH}$ is used, with the t-type test

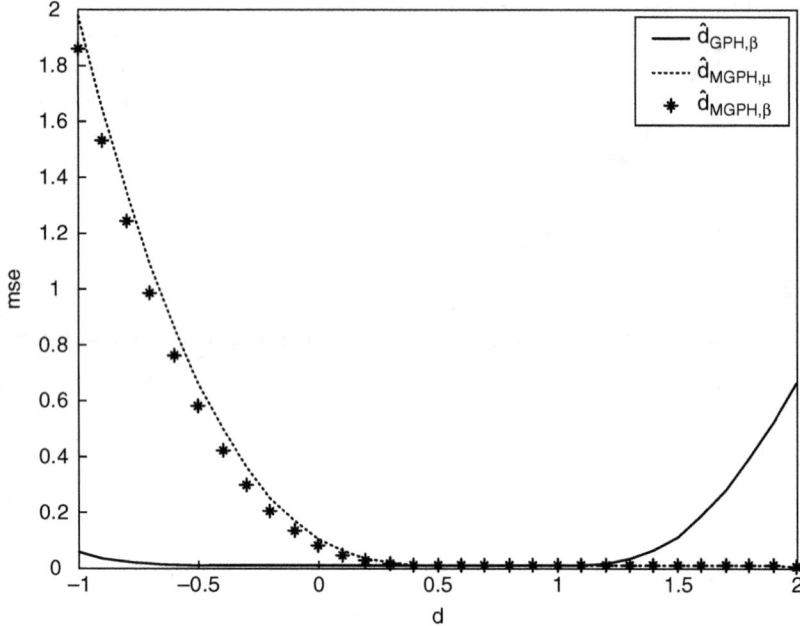

Figure 4.10   Mse of selected GPH and MGPH variants, trended data, $\beta_1 = 1$

statistic denoted Z_d defined as follows:

$$Z_d = \frac{(\hat{d}_{MGPH} - 1)}{\sqrt{\pi^2/24m}} \Rightarrow_D N(0, 1) \tag{4.81}$$

Thus, Z_d is easier to use than the test statistic based on $\hat{d}_{GPH}$ because, asymptotically, it has a standard normal distribution.

KP (2000) applied their method to the extended Nelson-Plosser data set comprising 14 US macroeconomic time series. They found most point estimates close to and slightly below 1; as a result, with relatively large standard errors, for example of the order of 0.1 to about 0.18, the respective 95% confidence intervals only exclude the unit root on two occasions; with the exception of the unemployment series, the stationary boundary of 0.5 is always to the left of the left-hand limit of the two-sided confidence interval, that is, stationarity is rejected.

In the case of the unemployment series using demeaned data, KP (2000) obtain $\hat{d}_{MGPH,\bar{y}} = 0.392$ (0.161) with $m = T^{0.7}$ and $\hat{d}_{MGPH,\bar{y}} = 0.554$ (0.114) with $m = T^{0.8}$ (estimated standard errors in parentheses, and $\hat{d}_{MGPH,\bar{y}}$ refers to the MGPH estimator using conventional demeaning rather than 'Shimotsu' demeaning). The results indicate some sensitivity of the point estimates to the

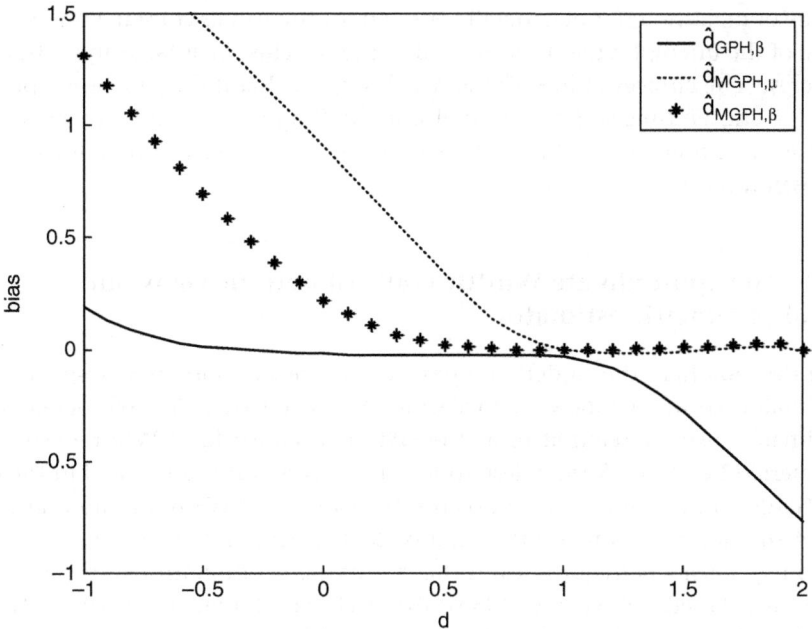

Figure 4.11a Bias of selected GPH and MGPH variants, trended data, $\beta_1 = 10$

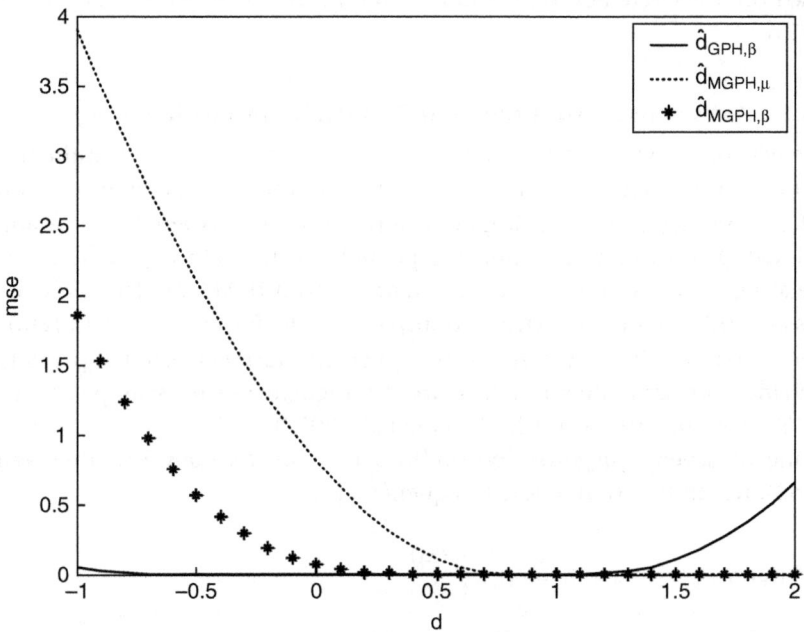

Figure 4.11b Mse of selected GPH and MGPH variants, trended data, $\beta_1 = 10$

value of m, although not generally enough for the values chosen, except in the case of the unemployment series, to alter the conclusion as far as hypotheses of interest. The choices of $m = T^{0.7}$ and $T^{0.8}$ suggest that if the generating process for the chosen time series is an ARFIMA(1, d, 0) model or similar process, then the AR component should not be very important otherwise m is likely to be chosen too large.

4.8   The approximate Whittle contrast and the Gaussian semi-parametric estimator

We describe here the underlying basis of another important semi-parametric estimator known (somewhat misleadingly) as the Gaussian semi-parametric estimator, usually referred to as the GSE or LWE for local Whittle estimator (preferred here); the former description suggests that it is only valid in the case of Gaussian time series, but this is not the case. The LWE is an important narrowband or local estimator; that is, it is like the GPH and AG estimators in that it (only) considers frequencies in the 'shrinking neighbourhood' of zero. This section is based on Whittle (1953), Beran (1994), Taniguchi (1979, 1981 and 1987), Künsch (1987) and Robinson (1995b). There are two related approaches to the LWE: the first considered here (although the second historically) is motivated by the idea of minimizing a distance or contrast function; the second is based on a discrete approximation to the likelihood function due to Whittle (1953).

4.8.1   A discrete contrast function (Whittle's contrast function)

One way of deriving the LW estimator is to view the problem of estimation as that of minimizing the distance between $f_y(\lambda_j)$ and a nonparametric estimator of $f_y(\lambda_j)$, say $\hat{f}_y(\lambda_j)$. $f_y(\lambda_j)$ is implicitly a function of a parameter or parameters denoted Θ, for example, d and the possibly vector valued parameters ϕ and θ relating to the AR and MA components of an ARFIMA model. (On occasion this dependence may be written transparently as, for example, $f_y(\lambda_j; \Theta)$.) One then needs to define a 'distance' function or 'contrast' between $f_y(\lambda_j)$ and $\hat{f}_y(\lambda_j)$. Such ideas are pre-eminent in the work of Taniguchi; see, for example, Taniguchi (1979, 1981 and 1987) and Taniguchi et al. (2003).

One of several suggested possibilities is Whittle's contrast, after Whittle (1953), which for an individual frequency λ_j is:

$$W[f_y(\lambda_j); I_y(\lambda_j)|\Theta] = \ln f_y(\lambda_j) + \frac{I_y(\lambda_j)}{f_y(\lambda_j)} \qquad (4.82)$$

In this case, the 'raw' periodogram $I_y(\lambda_j)$ has been taken as the estimator, $\hat{f}_y(\lambda_j)$, of $f_y(\lambda_j)$ (on the properties of the periodogram in this context see, for example, Priestley, 1981).

In principle, $\lambda_j \in \lambda = [-\pi, \pi]$ and one could proceed by minimizing the integral of (4.82) over λ with respect to an estimator of Θ; however, in practice only discrete elements of λ are chosen, usually those relating to the harmonic frequencies. The resulting (averaged) contrast function over m values of λ_j, denoted $W_m[f_y(\lambda_j); I_y(\lambda_j)|\Theta]$, or more simply, $W_m[.]$ is:

$$W_m[.] = m^{-1} \sum_{j=1}^{m} \left(\ln f_y(\lambda_j) + \frac{I_y(\lambda_j)}{f_y(\lambda_j)} \right) \quad \lambda_j = \frac{2\pi}{T} j, \ j = 1,\ldots,m \qquad (4.83)$$

This is the (approximate) Whittle contrast evaluated over m harmonic frequencies. For developments of this idea, see Künsch (1987) and Robinson (1995b). When $m < \text{int}[T/2]$ the estimator of d resulting from minimising $W_m[.]$ is, as in the case of the GPH and other log-periodogram estimators, referred to as a local (to the zero frequency) or narrowband estimator.

4.8.2 The Whittle contrast as a discrete approximation to the likelihood function

The Whittle contrast function may be derived in a different manner, relating it directly to the likelihood function and, hence, the estimator based upon it to a maximum likelihood estimator, MLE. As this derivation is somewhat more complex, only a brief outline is given here; the interested reader is referred to the excellent and comprehensive treatment in Beran (1994), on which this section is based.

Assume that $\{y\}_{t=1}^{T}$ is generated by a Gaussian process, so that the form of the likelihood function is known. As to notation, Ω_y is the $T \times T$ covariance matrix of $y = (y_1, y_2, \ldots, y_T)'$ and $|\Omega_y|$ is the determinant of Ω_y. The likelihood function to be considered is:

$$L(\Theta|y) = -\frac{T}{2} \ln 2\pi - \frac{1}{2} \ln |\Omega_y| - \frac{1}{2} y' \Omega_y^{-1} y \qquad (4.84)$$

The ML estimator $\hat{\Theta}$ is the solution to $\partial L(y|\hat{\Theta})/\partial \Theta = 0$, which is a system of equations with dimension equal to that of Θ, and where realisations of y replace y for a particular sample.

The component of $L(\Theta|y)$ that depends on Θ can be written in terms of the frequency domain as follows:

$$\lim_{T \to \infty} \frac{1}{T} \ln |\Omega_y| = \frac{1}{2\pi} \int_{-\pi}^{\pi} \ln f_y(\lambda; \Theta) d\lambda \qquad (4.85)$$

and define $A(\Theta)$ as follows:

$$A(\Theta) \equiv \frac{1}{(2\pi)^2} \int_{-\pi}^{\pi} \frac{1}{f_y(\lambda; \Theta)} e^{i(h-k)\lambda} d\lambda \qquad \text{for } h, k = 1, \ldots, T \qquad (4.86)$$

Asymptotically $A(\Theta)$ is Ω_y^{-1}; for details see Beran (1994). To obtain the Whittle function: (i) divide $L(\Theta|y)$ by T, noting that the constant can be ignored in maximisation; (ii) use (4.85) and (4.86) respectively to substitute for $T^{-1}\ln|\Omega_y|$ and Ω_y^{-1}; (iii) multiply by 2 and change the sign, so that the object is now to minimise the resulting function. This results in:

$$L_W(\Theta|y) = \frac{1}{2\pi}\int_{-\pi}^{\pi}\ln f_y(\lambda;\Theta)d\lambda + \frac{1}{T}y'A(\Theta)y \tag{4.87}$$

The next step is to obtain a version that does not involve the integrals in (4.87). This is achieved by using the following discrete approximations (where $\approx$ means is approximately equal to in the discrete approximation).

$$\frac{1}{2\pi}\int_{-\pi}^{\pi}\ln f_y(\lambda;\Theta)d\lambda \approx \frac{1}{2\pi}2\sum_{j=1}^{T}\ln f_y(\lambda_j)\frac{2\pi}{T} \tag{4.88}$$

$$\frac{1}{T}y'A_y(\Theta)y = \frac{1}{2\pi}\int_{-\pi}^{\pi}\frac{I_y(\lambda)}{f_y(\lambda)}d\lambda \tag{4.89a}$$

$$\approx \frac{1}{2\pi}2\sum_{j=1}^{T}\frac{I_y(\lambda_j)}{f_y(\lambda_j)}\frac{2\pi}{T} \tag{4.89b}$$

Substituting these terms into (4.87) and restricting the number of frequencies to m results in:

$$L_{W,m}(\Theta|y) = \frac{1}{2\pi}2\left(\sum_{j=1}^{m}\ln f_y(\lambda_j)\frac{2\pi}{T} + \sum_{j=1}^{m}\frac{I_y(\lambda_j)}{f_y(\lambda_j)}\frac{2\pi}{T}\right) \tag{4.90a}$$

$$= \frac{2}{T}\sum_{j=1}^{m}\left(\ln f_y(\lambda_j) + \frac{I_y(\lambda_j)}{f_y(\lambda_j)}\right) \tag{4.90b}$$

Minimisation of $L_{W,m}(.)$ results in the same estimator as that obtained from minimisation of $W_m[.]$ in (4.83), as they only differ by a scalar, which is constant for given m and T. Although the derivation through to $L_{W,m}(.)$ has assumed Gaussianity, its absence does not prevent the general approach with the resulting estimator referred to as quasi-ML or the minimum distance estimator.

4.8.3   Estimation details

4.8.3.i   *The estimation set-up*

The estimator of d using $W_m[.]$ is developed, for some choice of m, by Robinson (1995b) with the following approximations (which are familiar from the development around (4.36) for the local LP-based pseudo-regressions):

R.i. $f_y(\lambda_j) \approx \lambda_j^{-2d}f_u(\lambda_j)$ \qquad (4.91)

(Note that this approximation is not essential to the method, and $f_y(\lambda_j) = |(1-e^{-i\lambda_j}|^{-2d}f_u(\lambda_j)$.)

R.ii. As $\lambda_j \to 0^+$ it is assumed that $f_u(\lambda_j)$ tends to a constant G, that is

$$f_u(\lambda_j) \to G \in (0,\infty) \tag{4.92}$$

Assumption R.ii is essential to the LW estimator, but other approximations are possible and would still be faithful to the idea of minimising a contrast function. Together assumptions R.i and R.ii result in:

$$f_y(\lambda_j) \approx \lambda_j^{-2d}G \qquad \text{as } \lambda_j \to 0^+ \tag{4.93}$$

The second assumption contracts (reduces) the parameter vector from Θ to $\Phi = (d, G)$. The resulting specification of $W_m[.]$ is:

$$W_m[d,G] = m^{-1}\sum_{j=1}^m \left(\ln(\lambda_j^{-2d}G) + \frac{I_y(\lambda_j)}{\lambda_j^{-2d}G} \right) \qquad \text{for } m < \tfrac{1}{2}T \tag{4.94}$$

$$= m^{-1}\sum_{j=1}^m \left(\ln G - 2d\ln(\lambda_j) + \frac{I_y(\lambda_j)\lambda_j^{2d}}{G} \right) \tag{4.95}$$

Notice that because of the assumption on $f_u(\lambda_j)$ as $\lambda_j \to 0^+$, $W_m[d,G]$ is written as a function of just d and G, rather than the full parameter set, which will generally involve the parameters of the short-run dynamics as well as d, and the estimation problem is much simpler as a result. Thus, whatever the short-run dynamics their influence is assumed well approximated by a constant, provided that λ_j is 'close' to 0^+.

The solution to minimising $W_m[d,G]$ does not exist in closed form, but the estimators of d and G exist for $d \in (-0.5, 0.5)$. Examples of closed form estimators are given by the GPH and AG estimators, which can be written in advance of estimation as known functions of the data.

Initially take the closed interval of admissible estimates for d as $\Theta = [\Delta_1, \Delta_2]$ such that $-0.5 < \Delta_1 < \Delta_2 < +0.5$. This interval and the range of d are extended below. The criterion function (4.95) can be simplified by replacing G by an estimator that is conditional on d as outlined in the next two sub-sections.

4.8.3.ii Concentrating out G

From $f_y(\lambda_j) \approx \lambda_j^{-2d}G$, note that for a single frequency $G(\lambda_j|d) \approx f_y(\lambda_j)\lambda_j^{2d}$, where the notation indicates the dependence of G on d. Using the result that the periodogram is an estimator of the spectral density (for $d \in (-0.5, 0.5)$) and averaging over the included frequencies results in the following estimator for G:

$$\hat{G}(\lambda_j|d) = m^{-1} \sum_{j=1}^{m} I_y(\lambda_j)\lambda_j^{2d} \tag{4.96}$$

4.8.3.iii   Estimating d

Now on substituting $\hat{G}(\lambda_j|d)$ in (4.96) for G in (4.95), the function to be minimised is:

$$W_m[d|\hat{G}(\lambda_j|d)]$$

$$= m^{-1} \sum_{j=1}^{m} \left(\ln[\hat{G}(\lambda_j|d)] - 2d\ln(\lambda_j) \right) + \frac{1}{\hat{G}(\lambda_j|d)} \left(m^{-1} \sum_{j=1}^{m} I_y(\lambda_j)\lambda_j^{2d} \right)$$

$$= m^{-1} \sum_{j=1}^{m} \left(\ln[\hat{G}(\lambda_j|d)] - 2d \sum_{j=1}^{m} \ln(\lambda_j) \right) + 1 \quad \text{using}(4.96)$$

$$= m^{-1} m[\ln(\hat{G}(\lambda_j|d))] - m^{-1} 2d \sum_{j=1}^{m} \ln(\lambda_j) + 1$$

$$= R(d) + 1 \tag{4.97}$$

where $R(d) = \ln[\hat{G}(\lambda_j|d)] - m^{-1} 2d \sum_{j=1}^{m} \ln(\lambda_j)$.

Minimising $W_m[d|\hat{G}(\lambda_j|d)]$ with respect to d follows from minimising $R(d)$ because these functions only differ by a constant. The resulting estimator of d is denoted $\hat{d}_{LW}$. Additionally $\hat{G}(\lambda_j|d)$ is given from (4.96) by:

$$\hat{G}(\lambda_j|d) = m^{-1} \sum_{j=1}^{m} I_y(\lambda_j)\lambda_j^{2\hat{d}_{LW}}$$

Finally, note that following Robinson's (1995b) key developments $\hat{d}_{LW}$ is usually based on using the approximation λ_j^{-2d} of (4.93) in (4.94), which can be viewed in one of two ways. First, as an approximation to the spectral density of the fractionally differencing operator for the time series process $y_t = (1 - L)^{-d}\varepsilon_t$; thus, in keeping with modelling the behaviour of the spectrum for frequencies in the neighbourhood of zero, $|(1 - e^{-i\lambda_j}|^{-2d}$ is approximated by λ_j^{-2d}. However, note that an estimator with the same asymptotic properties results from using $|1 - e^{-i\lambda_j}|^{-2d}$ rather than the approximation. Second, it may also be viewed as an approximation to fractional Gaussian noise – see (4.23) – which has the same first-order approximation, so the approximation could be regarded as 'neutral' as far as the mechanism generating the long-range dependence of the process is concerned.

4.8.4 Properties of $\hat{d}_{LW}$ and mean squared error optimal m for the LW estimator

Some key (asymptotic) properties of $\hat{d}_{LW}$, with regard to bias, variance and mean square error, as follows:

$$\text{LW(i). AE}(\hat{d}_{LW} - d) = \left[\frac{-2\pi^2}{9} \right] \left[\frac{f_u''(0)}{f_u(0)} \right] \left[\frac{m^2}{T^2} \right] \tag{4.98}$$

$$\text{LW(ii). } \sigma_{LW}^{2,\text{asym}}(\hat{d}_{LW}) = \frac{1}{4m} \tag{4.99}$$

$$\text{LW(iii). MSE}(\hat{d}_{LW}) = \left[\frac{-2\pi^2}{9} \right]^2 \left| \frac{f_u''(0)}{f_u(0)} \right|^2 \left[\frac{m^2}{T^2} \right]^2 + \frac{1}{4m} \tag{4.100}$$

Correcting earlier results, Andrews and Sun (2004, remark 1) show that $\hat{d}_{GPH}$ and $\hat{d}_{LW}$ have the same asymptotic bias; thus, the difference in MSE results from the difference in asymptotic variances. The asymptotic variance is $\sigma_{LW}^{2,\text{asym}} = \frac{1}{4}m < \sigma_{GPH}^{2,\text{asym}} = \pi^2/24m$ and $\sigma_{GPH}^{2,\text{asym}}/\sigma_{LW}^{2,\text{asym}} = \pi^2/6 = 1.644 > 1$ so, for given m, $\hat{d}_{LW}$ is more efficient than $\hat{d}_{GPH}$.

The asymptotic variance of $\hat{d}_{LW}$ is $\sigma_{LW}^{2,\text{asym}} = \frac{1}{4}m$. An alternative expression for finite sample applications has been suggested by Hurvich and Chen (2000), who use $\hat{\sigma}_{LW}^{2,HC} = 1/[4\sum_{j=1}^{M}(z_j - \bar{z})^2]$, where $z_j = \log[2\sin(\lambda_j/2)]$, $\bar{z} = m^{-1}\sum_{j=1}^{m} z_j$ with $\lambda_j = 2\pi j/T$, which converges to $\sigma_{LW}^{2,\text{asym}}$, but was found to be a better approximation in finite samples for some experimental designs with ARFIMA processes.

4.8.5 An (approximately) mean squared error optimal choice for m

As in the case of the GPH estimator, it is possible to derive an estimator of m from minimising the asymptotic mean squared error. Minimising (4.100) with respect to m results in:

$$m_{LW}^{opt} = \left[\frac{9}{8\pi^2} \right]^{2/5} \left[\frac{f_u(0)}{f_u''(0)} \right]^{2/5} T^{4/5} \tag{4.101}$$

Note that m_{LW}^{opt} is $O(T^{4/5})$, but with a different scaling constant compared to m_{GPH}^{opt}. For a given $f_u(\lambda_j)$, optimal m is different for GPH and GSE, compare (4.45) and (4.101); the ratio $m_{GPH}^{opt}/m_{LW}^{opt}$ is $(\pi^2/6)^{1/5} = 1.105$, so m_{GPH}^{opt} is chosen larger than m_{LW}^{opt}, although both are $O(T^{4/5})$. Because the asymptotic bias is the same for $\hat{d}_{GPH}$ and $\hat{d}_{LW}$, the difference in optimal m is driven by the difference in variance which, for given m, is $(\pi^2/6)$, being the inverse of the relative efficiency of the two estimators.

Although the optimal choices for m for both $\hat{d}_{GPH}$ and $\hat{d}_{LW}$ depend upon $f_u(\lambda_j)$ and, hence, it cannot be known a priori what the variances of $\hat{d}_{GPH}$ and $\hat{d}_{LW}$

will be, it is simple to see that the variance ratio is a constant at the respective optimal m:

$$\frac{\sigma_{GPH}^{2,asym}(m_{GPH}^{opt})}{\sigma_{LW}^{2,asym}(m_{LW}^{opt})} = \left(\frac{\pi^2}{6}\right)^{4/5} = 1.489 \tag{4.102}$$

which implies a standard error ratio of 1.22. Although it would be possible to equalise the two variances by choosing $m_{GPH} = (\pi^2/6)m_{LW}$, this is too costly for $\hat{d}_{GPH}$ in terms of the increase in squared bias. Also, note that the expressions for asymptotically optimal m require knowledge of $g(\lambda_j)$, so feasible estimators must solve the problem that this will, generally, be unknown.

Using m_{LW}^{opt} and m_{GPH}^{opt} and basing inference on the resulting t statistics has to take account of one subtle consideration. Although the estimators $\hat{d}_{GPH}(m_{GPH}^{opt})$ and $\hat{d}_{GSE}(m_{GSE}^{opt})$ are (at least) asymptotically normally distributed at the mse optimum, which balances squared bias and variance, the bias is of the same stochastic order of magnitude as the standard error. This implies that the standard t ratios will not necessarily have a zero mean at the mse optimum. One way to assess the robustness of the results is to deliberately choose m to be smaller than at the mse optimum so that the squared bias is dominated by the variance.

4.8.6 The presence of a trend

The presence of a polynomial deterministic trend has important implications for the estimation of d using untransformed data. Let the process generating the data be given by:

$$(1 - L)^d[y_t - \mu_t] = u_t \tag{4.103}$$

$$\mu_t = y_0 + \beta_1 t^\alpha \tag{4.104}$$

So that under a type II initialisation, the generating process is:

$$y_t = y_0 + \beta_1 t^\alpha + (1 - L)^{-d} u_t \qquad t > 0; \beta_1 \neq 0, \alpha > 0 \tag{4.105}$$

The case $\alpha = 1$ corresponds to a linear trend.

Phillips and Shimotsu (2004, Theorem 5.1) show that with y_t generated by (4.105), and $d \in (\tfrac{1}{2}, K]$, $1 < K < \infty$ and $\alpha > \tfrac{1}{2}$, then:

$$\hat{d}_{LW} \rightarrow_p 1$$

The assumptions underlying this result (Phillips and Shimotsu, 2004, assumptions 1–4) are standard in this context; for example u_t is a linear process in ε_t, that is $u_t = C(L)\varepsilon_t = \sum_{j=0}^{\infty} c_j \varepsilon_{t-j}$, $\sum_{j=0}^{\infty} c_j^2 < \infty$, so that ARMA processes are included, and $(m^{-1} + T^{-1}m) \rightarrow 0$ as $T \rightarrow \infty$. Thus, data generated by an FI(d) process, with a value of d in the nonstationary range, plus a deterministic trend with $\alpha > \tfrac{1}{2}$, imply that the resulting estimator $\hat{d}_{LW}$ will be inconsistent and biased toward unity, except for the case d = 1. It is important, therefore, to remove a linear trend in the data before obtaining $\hat{d}_{LW}$, the resulting estimator being denoted $\hat{d}_{LW,\beta}$.

Phillips and Shimotsu (2004) report estimates of d using $\hat{d}_{LW}$ for a number of US macroeconomic times series in the Nelson and Plosser (1982) data set. They compare two versions of the estimator, the first using the raw data, $\hat{d}_{LW}$, and the second from differencing the data to remove a linear trend and then adding unity back to the estimate. They find that the estimates from these two methods differ quite markedly for data that is evidently trended, for example, GNP and employment, with the estimate for the raw data being much closer to unity than that from the differenced data.

4.8.7 Consistency and the asymptotic distribution of $\hat{d}_{LW}$

With some fairly standard regularity conditions, satisfied by ARFIMA and fGN processes, Robinson (1995b, Theorems 1 and 2) shows that the LWE is consistent and asymptotically normal for d ∈ (–0.5, 0.5). Crucially, consistency requires finite moments only up to the second and the condition $(m^{-1} + T^{-1}m) \to 0$ as $T \to \infty$.

Further work has shown that the LWE is asymptotically normal for d ∈ (–0.5, 0.75), with the same asymptotic variance as in the stationary situation, that is, $\sigma_{LW}^{2,asym}(\hat{d}_{LW}) = \frac{1}{4}m$; finite moments only up to the fourth are required and Gaussianity is not required for either proof. Velasco (1999b) extended these results, showing that the LWE is consistent for d ∈ (–0.5, 1) continuing, as in Robinson (1995b), under the weak condition of finite second moments (Velasco (op. cit., theorem 2)). Velasco (1999b) shows that to achieve consistency for d ≥ 1 requires suitable data tapers, but, as a result, the variance of the limit distribution increases.

4.8.8 More on the LW estimator for d in the nonstationary range

Phillips and Shimotsu (2004) consider d ∈ (0.5, 2) for the untapered case and prove, inter alia, that the LWE is consistent for d ∈ (0.5, 1], and converges to unity in probability for d ∈ (1, 2); a way of dealing with this latter case is considered further below. They also show that the LWE has a non-normal distribution for d ∈ (0.75, 1), but a mixed normal distribution for d = 1. (Velasco (1999b) and Phillips and Shimotsu (2004) differ in using type I and type II fractional processes respectively but the latter (op. cit. remarks 3.3 and 4.3b) conjecture that their results on consistency, inconsistency and asymptotic distributions continue to hold under the different initialization.)

As with the GPH estimator, the asymptotic variance of the LWE when $d = 1$ is smaller than for $d \in (-0.5, 0.75)$; specifically it is $0.2028/m$ compared to $\frac{1}{4}m$. Thus, some care has to be taken if d in the levels of a time series of interest is thought to be close to the unit root. If d is above unity, the GPH estimator and the LWE bias the estimator toward unity; neither have a normal limit distribution for $d \in (0.75, 1)$; and at $d = 1$ the limit distribution, whilst well defined, is not the same as for $d \in (-0.5, 0.75)$.

There is an important implication of Velasco's extensions of Robinson's results. Consider consistency: if $d \in (1, 2)$, conditional on this range for d, then a consistent estimator follows from first differencing the series and estimating d by the LWE to obtain $\hat{d}_{LW,\Delta y}$, and then adding 1 back so obtaining an estimator of d for the original series, say $\hat{d}_{LW,y} = \hat{d}_{LW,\Delta y} + 1$. Since this approach of initially taking the first difference also applies for $d \in (0.5, 1)$, $\hat{d}_{LW,y}$ is consistent for $d \in (0.5, 2)$ by this method; a similar argument shows that $\hat{d}_{LW,y}$ is asymptotically normal for $d \in (0.5, 1.75)$, again by this method. If $d \in (1, 2)$ and estimation uses the untransformed data, the resulting estimator will not be consistent. However, this differencing method assumes that the range of d is known; otherwise suppose that $d \in (0, 1)$ then differencing the series implies a value of d for the differenced series, say $d_{\Delta y}$, such that $d_{\Delta y} \in (-1, 0)$, which does not produce a consistent estimator over the complete range. Hence, unless the range of d is known, the LW estimator outlined in Section **4.9.2** below, which is consistent over a wider range than is the case for $\hat{d}_{LW}$, is preferred.

A summary of the known results on the properties of the LW estimator, $\hat{d}_{LW}$, when estimation proceeds using untransformed data is provided in Table 4.6.

Table 4.6   Properties of $\hat{d}_{LW}$ for stationary and nonstationary cases

	$d \in (-0.5, 0.5)$	$d \in [0.5, 1)$		$d = 1$	$d > 1$
		$d \in (0.5, 0.75)$	$d \in [0.75, 1)$		
$\rightarrow_p$	consistent	consistent	consistent (but at a slower rate)	consistent	not consistent, converges to 1
$\Rightarrow_D$	asymptotically normal	asymptotically normal	non-normal	mixed normal, variance smaller than when $d < 1$	not normal

Note: Reminder $d \in [a, b)$, means $a \leq d < b$. See Phillips and Shimotsu (2004, sections 3 and 4) for the underlying assumptions.

4.9 Modifications and extensions of the local Whittle estimator

4.9.1 The modified local Whittle estimator

In work related to an extension of the log-periodogram estimator to the nonstationary case, Shimotsu and Phillips (2000) modified the objective function of the local Whittle estimator. Recall that Section **4.7** described a modification to the spectral density and periodogram suggested by Phillips (1999b), which led to the revised expressions given by (4.76) and (4.77) respectively. In a similar vein, Shimotsu and Phillips (2000) suggested replacing $I_y(\lambda_j)$ in (4.94) by $I_v(\lambda_j)$, the modified periodogram, which leads to a modified version of the Whittle contrast function given by:

$$
MW_m[d, G] = m^{-1} \sum_{j=1}^{m} \left[\ln(\lambda_j^{-2d} G) + \frac{I_v(\lambda_j)}{\lambda_j^{-2d} G} \right] \tag{4.106a}
$$

$$
= m^{-1} \sum_{j=1}^{m} \left[\ln(\lambda_j^{-2d} G) + \frac{\lambda_j^{2d}}{G} \left| \frac{\omega_y(\lambda_j)}{\sqrt{T}} + \frac{e^{i\lambda_j}}{1 - e^{i\lambda_j}} \frac{(y_T - y_0)}{\sqrt{2\pi T}} \right|^2 \right] \tag{4.106b}
$$

The resulting estimator obtained by minimizing $MW_m[d, G]$ is denoted $\hat{d}_{MLW}$, where the subscript indicates that it is the modified local Whittle estimator. (Note that the divisor $\sqrt{T}$ appears in the moduli expression of (4.106b) compared to Shimotsu and Phillips (2000), because our definition of $\omega_y(\lambda_j)$ uses the divisor $\sqrt{2\pi}$ not $\sqrt{2\pi T}$; see (4.76) and (4.77).)

The estimator $\hat{d}_{MLW}$ is consistent for $d \in (0.5, 2)$, but the condition required for the expansion of m differs depending on the region for d (see Shimotsu and Phillips, 2000, theorems 3.1 to 3.3), as follows:

SP1. $d \in [\Delta_1, 0.5]$, $\Delta_1 > 0$:
if $(1/m) + (m/T) + T^{(1-2\Delta_1)}(\log T)(\log m)/m \to 0$ as $T \to \infty$ then $\hat{d}_{MLW}$ is consistent: $\hat{d}_{MLW} \to_p d$.

SP2. $d \in [0.5, 1.5)$:
if $(1/m) + (m/T) \to 0$ as $T \to \infty$ then $\hat{d}_{MLW}$ is consistent: $\hat{d}_{MLW} \to_p d$.

SP3. $d \in [1.5, 2)$:
if $(1/m) + (m/T) + (T^{\psi}/m) + (m^{2(d-1)}/T) \to 0$ as $T \to \infty$, for some $\psi > 0$, *then* $\hat{d}_{MLW}$ is consistent: $\hat{d}_{MLW} \to_p d$.

The convergence condition is at its strongest in case SP1 and Shimotsu and Phillips (2000) note it will be difficult to satisfy when Δ_1 is small. For case SP2, the convergence condition is weak and easily satisfied; and for case SP3, it is still relatively weak, for instance, it is satisfied by $m = o(T^{0.5})$.

For d $\in$ (0.5, 1.75), and with strengthening of the convergence conditions for m, $\hat{d}_{MLW}$ is asymptotically normal distributed with the same asymptotic variance as in the standard GSE/LWE case, that is $\frac{1}{4}$ m (see Shimotsu and Phillips, 2000, theorems 4.1 and 4.2); this result is of some importance as it obviously includes the unit root case, d $= 1$.

There is a connection between the results of Velasco (1999b) and Shimotsu and Phillips (2000), which is prompted by the correspondence between the ranges of d for consistency and asymptotic normality. In the standard local Whittle approach, consistency is extended to the region d $\in$ [0.5, 1) and asymptotic normality to d $\in$ [0.5, 0.75), and in the modified local Whittle approach to d $\in$ [0.5, 2) and d $\in$ [0.5, 1.75) respectively; that is, 1 is added to the right of the range. Shimotsu and Phillips (2000) note that the modified local Whittle function is asymptotically equivalent to the local Whittle function when used with first differenced data. Therefore, if d $\in$ (0.5, 2), then $\hat{d}_{MLW}$ based on the original series and $\hat{d}_{LW,y} = \hat{d}_{LW,\Delta y} + 1$, where $\hat{d}_{LW,\Delta y}$ is obtained from the first differenced data, will be asymptotically equivalent; nevertheless, some experience with these two estimators suggests that some quite large differences can occur using macroeconomic time series with fairly limited numbers of observations.

4.9.2   The 'exact' local Whittle estimator

Another variation of the local Whittle (LW) estimator due to Shimotsu and Phillips (2005) (see also Shimotsu, 2010) is referred to as the exact LW estimator, $\hat{d}_{ELW}$. Unlike the LW estimator, $\hat{d}_{LW}$, and the modified LW estimator, $\hat{d}_{MLW}$, $\hat{d}_{ELW}$ is valid for all values of d (however, note that if a constant is estimated the range for consistency and asymptotic normality is limited; see Section **4.9.2.iii**).

4.9.2.i   A problem

The availability of an estimator not limited to a particular range of values of d solves a number of problems. For example, whilst one explanation of a trended series is that it has been generated by an FI(d) process with d $\geq \frac{1}{2}$, an alternative explanation is that the trend is generated by a deterministic linear trend with deviations about the trend that are FI(d), with d $\in$ [0, $\frac{1}{2}$), that is the generating process is trend stationary, with persistent, but not infinitely so, deviations.

Differencing such a trend stationary series with a view to using the 'differencing + add-one-back' procedure puts d for the first differenced series into the region d $\in$ [-1, -$\frac{1}{2}$), which is outside the range usually assumed for d. (One could amend the notation here to say $d_{\Delta y}$ to make the context clear, but this will over-burden the notation below; thus bear in mind that in this section, the LWE of d for the levels series is the LWE of d for the first differenced series plus one.)

Shimotsu and Phillips (2006) show that $\hat{d}_{LW}$ may be consistent for d $\in$ [-1, -$\frac{1}{2}$), the 'may' depending on the number of frequencies m used in the estimation. The following results are summarised from Shimotsu and Phillips (2006, theorem 3.2 and remark 3.3):

$$C1 : \hat{d}_{LW} \rightarrow_p d \text{ if } r_1 \equiv \frac{m^{2d-2\Delta_1-1}}{T^{2d+1}} = m^{-2(\Delta_1+1)} \left(\frac{m}{T}\right)^{(2d+1)}$$

$$\rightarrow 0 \text{ for } d \in [-1 < \Delta_1, -\tfrac{1}{2}) \tag{4.107}$$

$$C2 : \hat{d}_{LW} \rightarrow_p 0 \text{ if } r_2 \equiv \frac{m^{2d}}{T^{2d+1}} = m^{-1} \left(\frac{m}{T}\right)^{(2d+1)}$$

$$\rightarrow \infty \text{ for } d \in [-1, -\tfrac{1}{2}) \tag{4.108}$$

(With respect to C2, there is a further technical condition that bounds deviations from y_0; see Shimotsu and Phillips, 2006, theorem 3.2.)

Whether $\hat{d}_{LW}$ is consistent depends upon the rate of expansion of m relative to T and also upon d and Δ_1, the lower limit to the set of allowable d. A key term in r_1 and r_2 is $(T/m)^v$, $v = -(2d + 1)$; thus, as $m < T$ and $v > 0$, this term tends to infinity with T, so for r_1 to be satisfied m must grow 'fast'; if it grows too slowly satisfying r_2, then $\hat{d}_{LW} \rightarrow_p 0$. Note that it is possible for $r_1 \rightarrow \infty$ and $r_2 \rightarrow 0$, so m is neither expanding fast enough to satisfy C1 nor slowly enough to satisfy C2, and Phillips and Shimotsu (2006) conjecture a bimodal distribution for $\hat{d}_{LW}$ in such a case. (Questions 4.7 and 4.8 further explore C1 and C2.)

The dependence of the properties of the LW estimator for the region $d \in [-1, -\tfrac{1}{2})$ on unknown parameters suggests that it is not generally desirable to use $\hat{d}_{LW}$ in this case. The alternative, that is, the exact LWE described in the next section, suggested by Phillips and Shimotsu (2005), in a sense encompasses both $\hat{d}_{LW}$ and $\hat{d}_{MLW}$.

4.9.2.ii The exact LW estimator, with known initial value: $\hat{d}_{ELW}$

The negative of the Whittle likelihood function of u_t based on frequencies λ_j, $j = 1, ..., m$, is, apart from a scalar multiple:

$$W_m[u] = m^{-1} \sum_{j=1}^{m} \left[\ln f_u(\lambda_j) + \sum_{j=1}^{m} \frac{I_u(\lambda_j)}{f_u(\lambda_j)} \right] \tag{4.109}$$

In the stationary case $I_u(\lambda_j)$ is approximated by $\lambda_j^{2d} I_y(\lambda_j)$, (that is use (4.91) applied to the periodogram); use the 'local' approximation $f_u(\lambda_j) \rightarrow G$ as $\lambda_j \rightarrow 0^+$ (see (4.92)); and transform from u to y by adding the Jacobian $\sum_{j=1}^{m} \ln(\lambda_j^{-2d})$. These steps result in the local Whittle likelihood function in terms of y (see (4.94)), that is:

$$W_m[y] = m^{-1} \sum_{j=1}^{m} \left[\ln(\lambda_j^{-2d} G) + \frac{I_y(\lambda_j)}{G\lambda_j^{-2d}} \right] \tag{4.110}$$

The reason for starting with the local likelihood function in terms of u is that it becomes clearer that it may be possible to improve upon the approximation

$I_u(\lambda_j) \approx \lambda_j^{2d} I_y(\lambda_j)$, which works well for $\bar{d} < \frac{1}{2}$, but not for other values of d, for example $\bar{d} > 1$.

To obtain the ELW estimator, first recall from Chapter 3 that the binomial expansion of $(1 - L)^d$, truncated to recognize the beginning of the process, leads to:

$$(1 - L)^d y_t \equiv \sum_{r=0}^{t} A_r^{(d)} y_{t-r}$$

where $A_r^{(d)} = (-1)^r \frac{(d)_r}{r!}$, $dC_r = \frac{(d)_r}{r!} = \frac{d!}{r!(d-r)!}$; the process has been assumed to start at $t = 0$. The DFT of y is given by:

$$\omega_y(\lambda_j) = (2\pi)^{-1/2} \sum_{t=0}^{T-1} y_t e^{-i\lambda_j t}$$

Then the periodogram of $u_t = \Delta^d y_t$ is given by:

$$I_u(\lambda_j) = I\Delta_y^d(\lambda_j)$$
$$= T^{-1} |D_T(\lambda_j)|^2 |v_y(\lambda_j)|^2 \tag{4.111}$$

(Note that the convention of introducing the divisor T in the periodogram rather than in the DFT is maintained here.)

For reference, the components of $I_u(\lambda_j) = I\Delta_y^d(\lambda_j)$ are as follows:

$$D_T(\lambda_j) = \sum_{k=0}^{T} A_k^{(d)} e^{ik\lambda_j} \tag{4.112}$$

$$v_y(\lambda_j) = \omega_y(\lambda_j) - \frac{1}{\sqrt{2\pi}} \frac{y_T^*(\lambda_j)}{D_T(\lambda_j)} \tag{4.113}$$

$$y_T^*(\lambda_j) = \sum_{k=0}^{T-1} \tilde{A}_p^{(d)} e^{-ip\lambda_j} y_{T-p} \tag{4.114}$$

$$\tilde{A}_p = \sum_{k=p+1}^{T} A_k^{(d)} e^{-ik\lambda_j} \tag{4.115}$$

See Phillips and Shimotsu (2005, especially section 2.1 and lemma 5.1). To simplify the notation, the dependence of the functions in (4.112)–(4.115) on the value of d has been left implicit.

The exact local Whittle contrast follows by:

(i) Using $I_u(\lambda_j) = I\Delta_y^d(\lambda_j)$ of (4.111) in the LW likelihood given by (4.109).
(ii) Maintaining the local approximation $f_u(\lambda_j) \approx G$.
(iii) Adding the Jacobean $\sum_{j=1}^{m} \ln|D_T(\lambda_j)|^{-2}$ with local approximation $D_T(\lambda_j)$ $\approx \lambda_j^{2d}$.

The resulting objective function, the 'exact' local Whittle contrast, is:

$$EW_m[d, G] = m^{-1} \sum_{j=1}^{m} \left[\ln(\lambda_j^{-2d} G) + \frac{1}{G} I\Delta_y^d(\lambda_j) \right] \tag{4.116}$$

The exact LW estimator $\hat{d}_{ELW}$ is obtained as the solution to:

$$(\hat{d}_{ELW}, \hat{G}) = \arg\min EW_m[d, G] \qquad (4.117)$$

where, as for the other Whittle estimators, minimization is over $G \in (0, \infty)$ and $d \in [\Delta_1, \Delta_2]$; however, in this case $-\infty < \Delta_1 < \Delta_2 < \infty$, rather than, for example, being limited to the stationary boundaries. Also, analogous to LW estimation – see Equations (4.96) and (4.97) – $EW_m[d, G]$ can be concentrated with respect to G and then $\hat{d}_{ELW}$ is obtained as the solution to:

$$\hat{d}_{ELW} = \arg\min R(d) \quad \text{for } d \in [\Delta_1, \Delta_2] \qquad (4.118)$$

$$R_{ELW}(d) = \ln[\hat{G}(\lambda_j|d)] - m^{-1}2d\sum_{j=1}^{m}\ln(\lambda_j) \qquad (4.119)$$

$$\hat{G}(\lambda_j|d) = m^{-1}\sum_{j=1}^{m}I_{\Delta^d y}(\lambda_j) \qquad (4.120)$$

On a computational note, estimation focuses on the expression in (4.116); all that is needed is to specify that one takes the periodogram of the fractional differenced variable of interest, that is $(1 - L)^d y_t$; the quantities in the expressions (4.112)–(4.115) are the underlying theoretical relationships, they are *not* needed explicitly in the computation. (Shimotsu's MATLAB code, which has also been converted to Gauss, for the calculation of $\hat{d}_{ELW}$ can be downloaded from: http://www.econ.queensu.ca/faculty/shimotsu/.)

4.9.2.iii Properties of the exact LWE

Phillips and Shimotsu (2005) establish the properties of $\hat{d}_{ELW}$ under the assumptions used in LW estimation, but differing in two respects. First, a condition on the expansion of m:

$$\frac{1}{m} + \frac{m(\ln m)^{1/2}}{T}\frac{\ln T}{m^\gamma} \to 0 \text{ for any } \gamma > 0$$

and, second, on the searchable boundary limits:

$$\Delta_2 - \Delta_1 \leq \tfrac{9}{2}$$

(These are assumptions 4 and 5 of Phillips and Shimotsu, 2005.) Whilst the latter assumption limits the range of searchable values of the estimator, the range is wide and the simulations in Phillips and Shimotsu suggest that this range can be extended without loss.

Phillips and Shimotsu (2005, theorems 2.1 and 2.2) show that $\hat{d}_{ELW}$ is consistent and asymptotically normally distributed. First, as to consistency: suppose y_t is generated by $y_t = y_0 + (1 - L)^{-d}u_t 1_{(t>0)}$, with y_0 known, for example, $y_0 = 0$, and under assumptions 1 to 5 detailed in Phillips and Shimotsu (2005), with

$d \in [\Delta_1, \Delta_2]$, then $\hat{d}_{ELW} \to_p d$. As to asymptotic normality: under the same generating process for y_t, and under assumptions $1'$ to $5'$ detailed in Phillips and Shimotsu (2005), with $d \in (\Delta_2, \Delta_1)$, then: $\hat{d}_{ELW} \Rightarrow_D N(d, \tfrac{1}{4}m)$

4.9.2.iv   *The exact LW estimator with unknown initial value:* $\hat{d}_{FELW,\mu}$

In practice it is likely that y_0 is unknown and this feature has to be included in the overall procedure and in considering its impact on the range of d for which consistency and asymptotic normality (CAN) hold. In brief, the need to estimate y_0 restricts the range of d for which $\hat{d}_{ELW}$ is CAN. The cases of an unknown initial value and a deterministic trend are considered in sequence.

Two possible estimators of y_0 are the sample mean $\bar{y} = T^{-1} \sum_{t=1}^{T} y_t$ and the first observation y_1; see Chapter 3, Section **3.6** and Shimotsu (2010). In the first case, if $\bar{y}$ replaces y_0, then:

$$\hat{d}_{ELW} \to_p d \text{ for } d \in (-\tfrac{1}{2}, 1)$$

$$\sqrt{m}(\hat{d}_{ELW} - d) \Rightarrow_D N(0, \tfrac{1}{4}) \text{ for } d \in (-\tfrac{1}{2}, \tfrac{3}{4}).$$

In the second case, if y_1 replaces y_0, then:

$$\hat{d}_{ELW} \to_p d \text{ for } d \in [0, \infty)$$

$$\sqrt{m}(\hat{d}_{ELW} - d) \Rightarrow_D N(0, \tfrac{1}{4}) \text{ for } d \in (0, 2);$$

In the case where y_1 is used it is assumed that $\Delta_2 - \Delta_1 \leq \tfrac{9}{2}$ and $d \in [\Delta_1, \min(\Delta_2, 2)]$; as to the latter condition, given the likely range of d, and the permissible range of search, the result has been stated for $\min(\Delta_2, 2) = 2$. As noted by Phillips and Shimotsu (2005), simulation evidence suggests that $\hat{d}_{ELW}$ is inconsistent for $d \leq 0$.

Thus, $\bar{y}$ is a good estimator for $d < |0.5|$, whereas y_1 is a good estimator for $d > 0.75$ and both are good estimators for $d \in (0.5, 0.75)$. A weighted estimator of y_0 exploits the different parts of the parameter space for these two results, that is:

$$\tilde{y}_0 = \kappa(d)\bar{y} + [1 - \kappa(d)]y_1 \tag{4.121}$$

where $\kappa(d)$ is a twice differentiable weighting function such that:

$$\kappa(d) = 1 \text{ for } d \leq \tfrac{1}{2}, \kappa(d) \in (0, 1) \text{ for } d \in (\tfrac{1}{2}, \tfrac{3}{4}), \kappa(d) = 0 \text{ for } d \geq \tfrac{3}{4}.$$

A possibility for $\kappa(d)$ when $d \in (\tfrac{1}{2}, \tfrac{3}{4})$, is $\kappa(d) = 0.5(1 + \cos[4\pi(d - 0.5)])$, which is virtually linear, with weights that vary between 0.5 and 0.1464, that is $\kappa(d) \approx 0.5$ for $d \approx 0.5$, but weighting toward y_1 as d increases (for example, for $d = 0.7$, $\kappa(d) = 0.0955$, so $1 - \kappa(d) = 0.9045$).

The next step is to use the demeaned data $\tilde{y}_t \equiv y_t - \tilde{y}_0$ in the periodogram, so that the minimisation problem is:

$$\hat{d}_{FELW,\mu} = \arg\min R(d) \quad \text{for } d \in [\Delta_1, \Delta_2] \tag{4.122}$$

$$R_F(d) = \ln[\hat{G}_{\tilde{y}}(\lambda_j|d)] - m^{-1}2d\sum_{j=1}^{m}\ln(\lambda_j) \tag{4.123}$$

$$\hat{G}_{\tilde{y}}(\lambda_j|d) = m^{-1}\sum_{j=1}^{m}I_{\Delta^d\tilde{y}}(\lambda_j) \tag{4.124}$$

Thus, $\hat{d}_{FELW,\mu}$, referred to as the feasible ELW estimator, with the subscript indicating estimation of y_0, is distinguished from $\hat{d}_{ELW}$, where the latter assumes that y_0 is known; the polynomial trend case is considered below.

4.9.2.v Properties of $\hat{d}_{FELW,\mu}$

Let y_t be generated as $y_t = y_0 + (1 - L)^{-d}u_t 1_{(t>0)}$, with y_0 unknown, and the assumptions detailed in Shimotsu (2010), then $\hat{d}_{FELW,\mu}$ is generally CAN, with the usual limiting distribution $N(d, \frac{1}{4}m)$, for $d \in (-\frac{1}{2}, 2)$. The qualification 'generally' is because there are two gaps in the range of d, the first around 0 and the second around 1, that is, $0 \pm \nu$ and $1 \pm \nu$, for ν arbitrarily small, where this result does not hold. These gaps can be plugged by the use of a two-stage estimation procedure that starts with a $\sqrt{m}$-consistent estimator (not the ELW estimator), which is then updated; see Shimotsu (2010) for details.

4.9.2.vi Allowing for a deterministic polynomial trend

A variation on $\hat{d}_{FELW,\mu}$ relates to the data generating process with a deterministic trend; thus suppose that $\{y_t\}_{t=1}^T$ is generated by:

$$y_t = \beta_0 + \sum_{j=1}^{k}\beta_j t^k + (1 - L)^{-d}u_t 1_{(t>0)}$$

The residuals are given by:

$$\tilde{y}_t = y_t - (\hat{\beta}_0 + \sum_{j=1}^{k}\hat{\beta}_j t^k)$$

where $\hat{\ }$ above indicates the coefficient from an LS regression of y_t on $(1, t, \ldots, t^k)$. As $\tilde{\bar{y}} = 0$ from the properties of LS regression, the weighting function applied to the residuals $\tilde{y}_t$ simplifies relative to $\kappa(d)$ of Equation (4.121); thus, let

$$\tilde{y}_0 = [1 - \kappa(d)]\hat{y}_1 \tag{4.125}$$

The data is then transformed as $\tilde{\tilde{y}}_t = \hat{y}_t - \tilde{y}_0$ (see also Chapter 3, Section 3.6), and the DFT and periodogram are defined in terms of $(1 - L)^d\tilde{\tilde{y}}_t$. An estimator analogous to $\hat{d}_{FELW,\mu}$, with prior removal of deterministic components, then follows the same general procedure. The case most likely to be considered is where there is a linear trend, so that $k = 1$, and in this case the estimator is denoted $\hat{d}_{FELW,\beta}$. Then (see Shimotsu, 2010, theorem 4):

$$\hat{d}_{FELW,\beta} \Rightarrow_D N(d, \tfrac{1}{4}m) \text{ for } d \in (\Delta_1, \Delta_2) \text{ and } -\tfrac{1}{2} < \Delta_1 < \Delta_2 < \tfrac{7}{4}$$

This result also holds for higher order linear trends. Note that the upper limit to d is now $7/4$. (As for $\hat{d}_{FELW,\mu}$, there are two arbitrary small 'gaps' where the CAN result does not hold for $\hat{d}_{FELW,\beta}$, but can be avoided with a two-stage estimation procedure.)

4.10   Some simulation results for LW estimators

Some of the key issues are illustrated here from simulation of the FI(d) generating process given by:

$$y_t = y_0 + (1 - L)^{-d} u_t 1_{(t>0)}$$

where $u_t = \varepsilon_t \sim N(0, 1)$. (Other cases were also considered where $u_t \neq \varepsilon_t$, for example, $u_t = \rho u_{t-1} + \varepsilon_t$; however the results were of the same general pattern and are not reported separately.) To simplify the issues, at first the initial value is assumed to be zero and then consideration is given to a non-zero initial value and a linear trend.

The results in this section are for $T = 512$ (again, other values of T were used, but the general pattern was the same). To assess the results to the sensitivity of choosing m, whilst the primary case is for $m = T^{0.65}$, a variation using the square root rule, $m = T^{0.5}$, was also considered. The evaluation is in terms of the simulation mean and the simulation mse of the estimators. There are, of course, other criteria for assessment; for example the actual coverage level of nominal $(1 - \alpha)\%$ confidence intervals, and the small sample distributions of the various estimators as for example in the extent to which asymptotic normality holds it also holds in finite samples.

The eight variants of the estimators considered in the simulations are shown in Table 4.7, with their section references.

Each estimator is first considered by group, and there is a graph for each of the cases $y_0 = 0$, $y_0 \neq 0$, and $y_0 \neq 0$ and $\beta_1 \neq 0$. The presentation of the

Table 4.7   LW estimators considered in the simulations

	LW		Modified LW			Exact LW		
Estimator	$\hat{d}_{LW}$	$\hat{d}_{LW,\beta}$	$\hat{d}_{MLW}$	$\hat{d}_{MLW,\mu}$	$\hat{d}_{MLW,\beta}$	$\hat{d}_{ELW}$	$\hat{d}_{FELW,\mu}$	$\hat{d}_{FELW,\beta}$
Section	4.8		4.9.1			4.9.2		

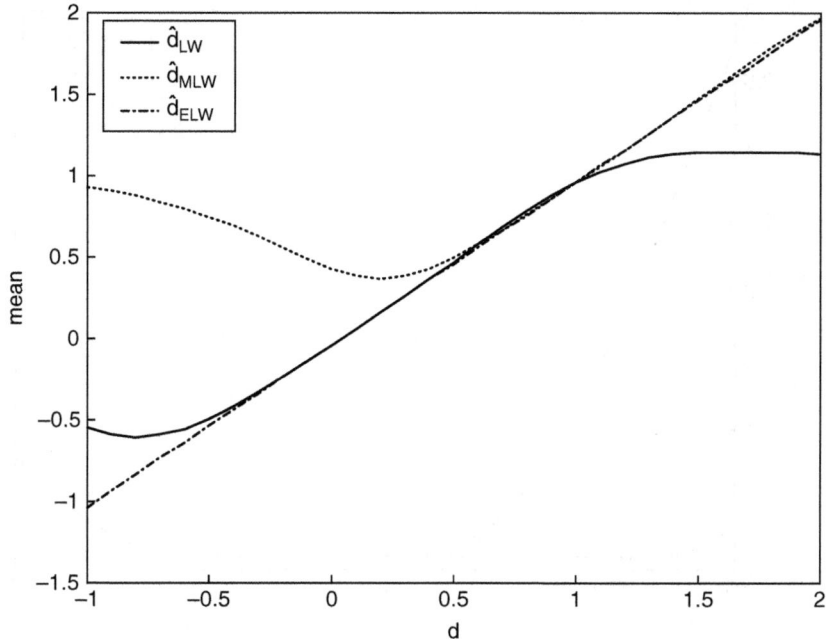

Figure 4.12   Mean of best of LW estimators, zero initial value

graphical results differs slightly from Section **4.7.5**, in part for the reader to judge which better summarises the results. The simulation means for each sub-group of estimators are presented first; the reference point for zero bias is, therefore, a 45° line (which is shown where it does not confuse interpretation of the graph). The simulation mse is then presented in the same order.

The results are first summarised briefly for each method and then the best estimator for each method are compared graphically.

i) $y_0 = 0$

LW: $\hat{d}_{LW}$ has a smaller bias and a smaller mse compared to $\hat{d}_{LW,\beta}$, but both should only be used for $d \in (-\frac{1}{2}, 1)$; neither are appropriate for $d \in (-1, -\frac{1}{2})$ and $d \in (1, 2)$.

MLW: $\hat{d}_{MLW}$, $\hat{d}_{MLW,\mu}$ and $\hat{d}_{MLW,\beta}$ are not appropriate for $d \in (-\frac{1}{2}, \frac{1}{2})$; for $d \in (\frac{1}{2}, 2)$, the differences between $\hat{d}_{MLW}$, $\hat{d}_{MLW,\mu}$ and $\hat{d}_{MLW,\beta}$ are marginal.

ELW: $\hat{d}_{ELW}$ is appropriate for $d \in (-1, -\frac{1}{2})$, whereas $\hat{d}_{ELW,\mu}$ and $\hat{d}_{ELW,\beta}$ are not appropriate for this range; in terms of a lack of bias $\hat{d}_{ELW}$ is best for

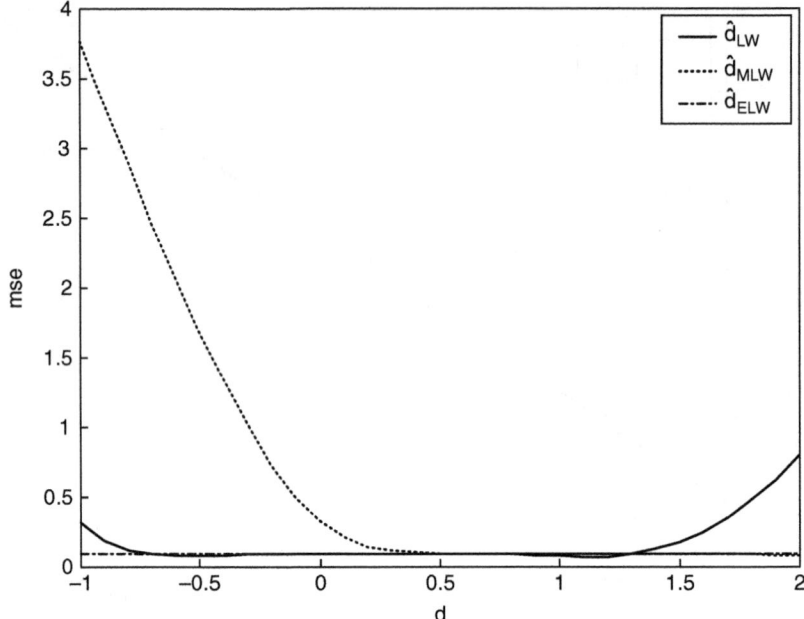

Figure 4.13   Mse of best of LW estimators, zero initial value

$d \in (-1, 2)$; superfluous detrending has a cost in terms of an increase in mse, being dominated by $\hat{d}_{ELW}$ and $\hat{d}_{ELW,\mu}$ for $d \in (-\frac{1}{2}, 2)$.

Overall, Figures 4.12, 4.13: in a comparison between estimators (taking the best in each group), $\hat{d}_{ELW}$ dominates for $d \in (-1, 2)$, whilst there is little to choose in mse terms between $\hat{d}_{LW}$ and $\hat{d}_{ELW}$ for $d \in (-\frac{1}{2}, 1)$ and between $\hat{d}_{MLW}$ and $\hat{d}_{ELW}$ for $d \in (1, 2)$.

ii) $y_0 \neq 0$

LW: $\hat{d}_{LW}$ is invariant to the non-zero value for y_0; as in the zero mean case, $\hat{d}_{LW}$ is less biased than $\hat{d}_{LW,\beta}$ and has a smaller mse, and it is again clear that neither are appropriate for $d \in (-\frac{1}{2}, \frac{1}{2})$ or $d \in (1, 2)$.

MLW: $\hat{d}_{MLW} \approx 1$ for $d \in (-1, 1)$, showing how important it is to adjust for the non-zero mean; when this is done $\hat{d}_{MLW,\mu}$ and $\hat{d}_{MLW,\beta}$ are not practically distinguishable.

ELW: $\hat{d}_{ELW}$ is not now an appropriate estimator. It assumes knowledge of $y_0 \neq 0$, which is not correct in this case and the estimator is not invariant to y_0, and

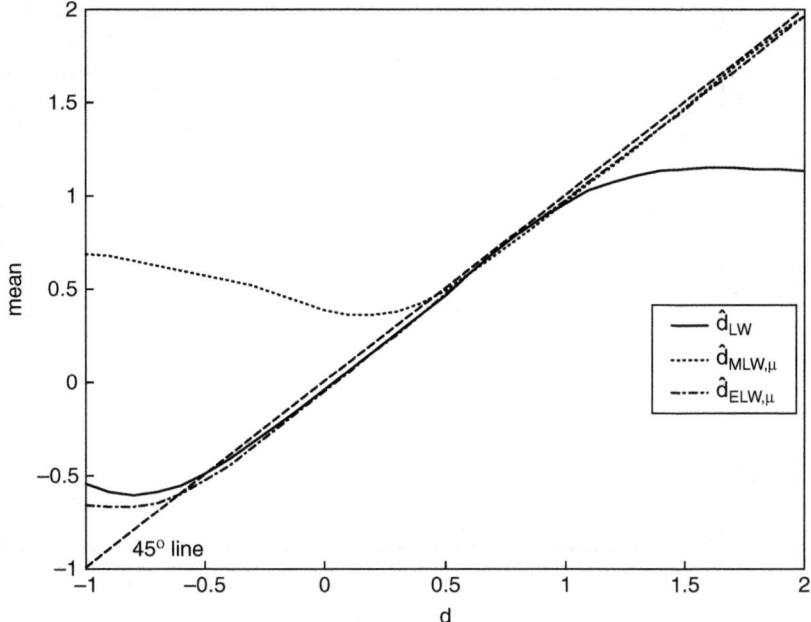

Figure 4.14   Mean of best of LW estimators, non-zero initial value

note that $\hat{d}_{ELW} \approx 0$ for $d \in (-1, 1)$; $\hat{d}_{ELW,\mu}$ is generally better than $\hat{d}_{ELW,\beta}$ in terms of bias and mse for $d \in (-\tfrac{1}{2}, 2)$, but note that both are biased for $d \in (-1, -\tfrac{1}{2})$.

Overall, Figures 4.14, 4.15: in a comparison between the best of these estimators, $\hat{d}_{ELW,\mu}$ is best for $d \in (-\tfrac{1}{2}, 2)$, but for $d \in (-\tfrac{1}{2}, 1)$, $\hat{d}_{LW}$ is a close competitor in bias and mse; and for $d \in (1, 2)$, $\hat{d}_{MLW,\mu}$ is then close competitor.

iii) $y_0 \neq 0$ and $\beta_1 \neq 0$
    This is likely to be the most frequently occurring case in practice.

LW: $\hat{d}_{LW}$ is not an appropriate estimator. The simulations show $\hat{d}_{LW} \approx 1$ for most of its range; however, $\hat{d}_{LW,\beta}$ is not so affected and is a reasonable estimator over the range $d \in (-\tfrac{1}{2}, 1)$, but with a tendency to underestimate the true value of d.

MLW: as in the case $y_0 \neq 0$, $\hat{d}_{MLW} \approx 1$ for $d \in (-1, 1)$ and is not elsewhere a good estimator; $\hat{d}_{MLW,\mu}$ is biased for $d < 1$, but acceptable for $d \in (1, 2)$; only $\hat{d}_{MLW,\beta}$ is acceptable for $d \in (\tfrac{1}{2}, 2)$.

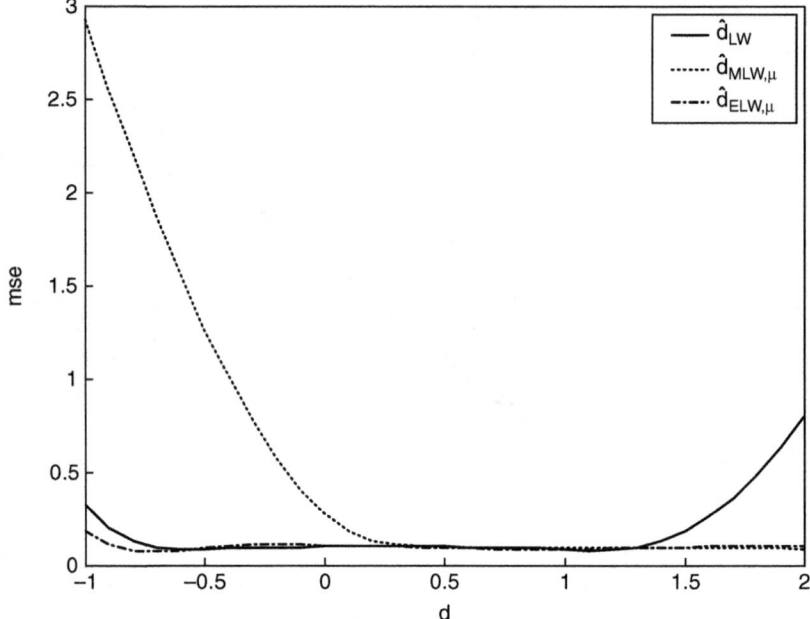

Figure 4.15   Mse of best of LW estimators, non-zero initial value

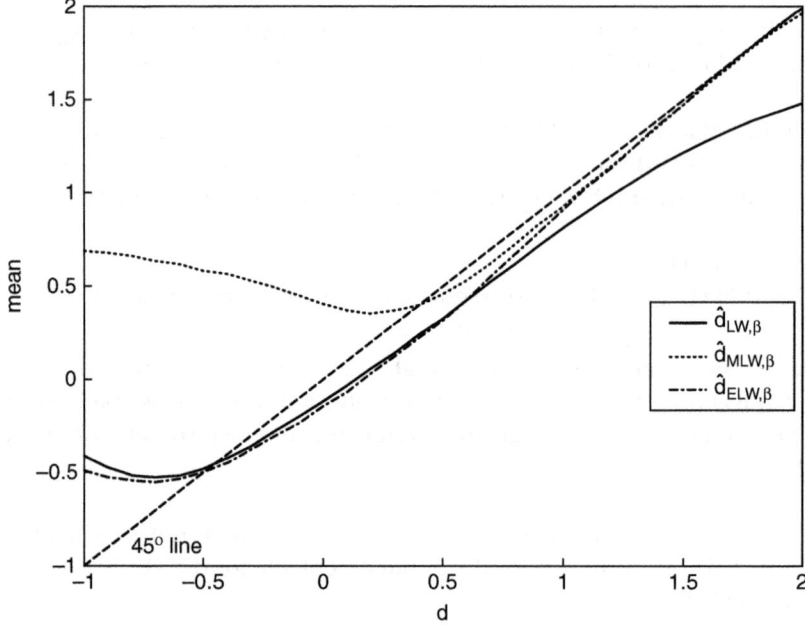

Figure 4.16   Mean of LW estimators, trended data

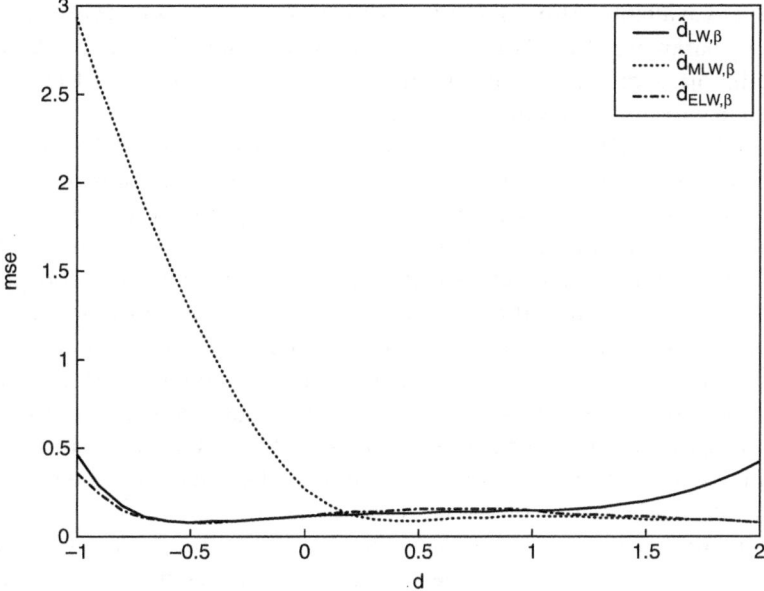

Figure 4.17   Mse of best of LW estimators, trended data

ELW: as anticipated from the theoretical results neither $\hat{d}_{ELW}$ nor $\hat{d}_{ELW,\mu}$ are satisfactory estimators, whereas the trend correction works for $\hat{d}_{ELW,\beta}$ over the range $d \in (-\frac{1}{2}, 2)$, although with a tendency for under-estimation of d for $d \in (-\frac{1}{2}, 1)$.

Overall, Figures 4.16, 4.17: the estimator that is best in mse terms depends on the range of d; for example, there is a slight advantage to $\hat{d}_{MLW,\beta}$ over $\hat{d}_{ELW,\beta}$ for $d \in (\frac{1}{2}, 2)$, an advantage that is lost for $d \in (-\frac{1}{2}, \frac{1}{2})$ because $\hat{d}_{MLW,\beta}$ is not consistent in that range; while there is little to choose between $\hat{d}_{LW,\beta}$ and $\hat{d}_{ELW,\beta}$ for $d \in (-\frac{1}{2}, \frac{1}{2})$. Thus, as Shimotsu and Phillips (2005) note, $\hat{d}_{ELW,\beta}$ is a good 'general purpose' estimator when there is doubt over the range of d.

Of course, this is just a small simulation study to provide some graphical illustrations of the theoretical properties of the various estimators, but it does indicate the importance of choosing correctly the estimator appropriate to the specification of the deterministic terms.

4.11   Broadband – or global – estimation methods

Although much has been covered in the previous sections, there is another class of frequency domain estimators known as broadband, or global, estimators, which are available. Although distinguished from local estimators of the long-memory parameter, there is a line of continuity in terms of recent developments, as both forms of estimation address concerns about how to deal with the persistence in time series induced by short-run dynamics.

The GPH and LWE approaches assume that if the included frequencies, m, can be limited to be close, in some sense, to 0^+, then the contribution of the short run can be treated as a constant. However, a problem with this approach is that, because the variance of the estimator of d is inversely proportional to m, the resulting standard errors for small m can be too large for reasonably practical purposes. The AG contribution can be viewed in this light as introducing a better approximation to $\ln[f_u(\lambda_j)/f_u(0)]$, rather than assuming that it is constant. This development enables m to be larger than in the standard GPH case, which, ceteris paribus, drives the variance down; but the gain is bought at the cost of increasing the variance, for a given m, relative to the GPH case; and whether, overall, the net effect is to reduce the variance depends upon the sample size and the particular nature of the short-run dynamics.

Broadband estimators also consider the question of how to improve the way that the short-run dynamics are dealt with in the frequency domain, but, as the description suggests, they use all rather than a limited number of frequencies. In the FEXP version of a broadband estimator, the logarithm of $f_u(\lambda_j)$ is expanded on the cosine basis. That is:

$$\ln[f_u(\lambda_j)] = \sum_{j=0}^{\infty} \eta_j c_j(\lambda_j) \tag{4.126}$$

where $c_0(\lambda_0) = 1/\sqrt{2\pi}$ and $c_j(\lambda_j) = \cos(j\lambda_j)/\sqrt{\pi}$, $\lambda_j \in [-\pi, \pi]$, $j \geq 1$. Provided that $f_u(\lambda_j)$ is smooth, then the method proceeds by truncating the infinite upper sum to p coefficients, η_j for $j = 0, ..., p - 1$. For example, the ARMA case is smooth in the sense used here and the coefficients η_j go to zero exponentially fast, suggesting that p need not be very large in order to capture the short-run spectral density. The broadband estimator again involves a choice on truncation, but in this case on the number of terms, p, in the cosine expansion of $\ln[f_u(\lambda_j)]$.

The cosine expansion can be incorporated into the LP regression or the Whittle contrast in much the same way as the AG development includes an expansion based on the even powers of λ_j, but with applicability over a different range of frequencies. In the case of the LP regression, in the simplest case (that is without tapering or pooling of the periodogram), the regression to be estimated is:

$$\ln[I_y(\lambda_j)] = \beta_0 + dX_j + \sum_{j=0}^{p-1} \eta_j c_j(\lambda_j) + \breve{\varepsilon}_j \qquad j = 1, ..., K \tag{4.127}$$

where $X_j = -2\ln[|1 - e^{-i\lambda_j}|]$ and $K = \text{int}[T/2]$. An estimator of d follows by straightforward application of LS; however, a key issue is the selection of the truncation parameter in the cosine expansion. Moulines and Soulier (1999) have considered a Mallows criterion for order selection, which has desirable properties but under some restrictive assumptions. Iouditsky, Moulines and Soulier (2001) have suggested an adaptive method based on a minimax criterion that involves trading off the bias and variance.

For development of the original FEXP idea see Beran (1994) and Robinson (1994b); and for further developments, see Hurvich and Brodsky (2001), Moulines and Soulier (1999), Hurvich, Moulines and Soulier (2002) and Iouditsky, Moulines and Soulier (2001). In particular, Hurvich, Moulines and Soulier (2001) show how the FEXP model is extended to cover the cases of pooling and tapering the periodogram, considered earlier in Section **4.4** for the GPH estimator, and they also give more detail on the variance and rate of convergence of the broadband estimator.

4.12   Illustrations

This section considers some practical illustrations of the estimation methods described in earlier sections. The five estimation methods reported here are summarised in Table 4.8, with equation references in parentheses and, for convenience of reference, their respective asymptotic standard errors.

Each of the resulting estimators is potentially subject to variation depending on m, the number of included frequencies. An often used default for m, which we use here for illustrative purposes, is the 'square root' rule for m, that is $m = \text{int}[\sqrt{T}]$. As noted at various points in this chapter, a theoretical optimum value for m, referred to as m^{opt}, can be derived, but it, in turn, depends on unknown

Table 4.8   Semi-parametric estimators considered in this section

Equation reference	Notation	Asymptotic variance
Log-periodogram-based estimators		
GPH Equations (4.34/35)	$\hat{d}_{\text{GPH}}$	$\frac{\pi^2}{24m}$
AG (Andrews and Guggenberger) Equation (4.63)	$\hat{d}_{\text{AG}}(r)$	$\frac{\pi^2}{24m}c_r; r = 1 \Rightarrow \frac{\pi^2}{24m}2.25$
Modified GPH (Kim and Phillips) Equation (4.79)	$\hat{d}_{\text{MGPH}}$	$\frac{\pi^2}{24m}$
Whittle contrast function-based estimators		
Local Whittle, LW Equation (4.97)	$\hat{d}_{\text{LW}}$	$\frac{1}{4m}$
Modified LW (Shimotsu and Phillips) Equation (4.106)	$\hat{d}_{\text{MLW}}$	$\frac{1}{4m}$
Exact LW (Phillips and Shimotsu) Equation (4.118)	$\hat{d}_{\text{ELW}}$	$\frac{1}{4m}$

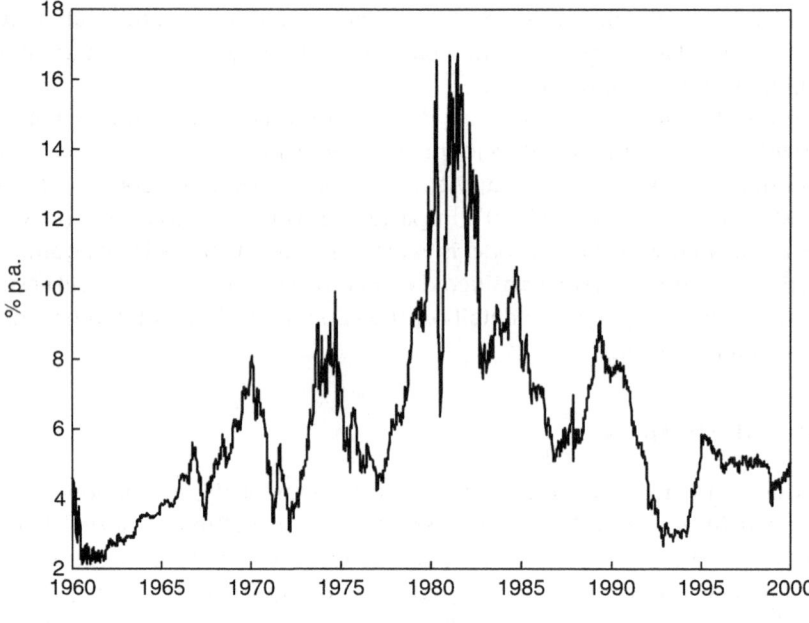

Figure 4.18   US T-bill rate

coefficients; however, these latter can be estimated and a feasible or 'plug-in' version of m^{opt} can be used, which is asymptotically optimal.

4.12.1   The US three-month T-bill

Two illustrations are used here to indicate some of the practical considerations that arise in estimating the long-memory parameter. The first illustration uses weekly (w) data on the US three-month T-bill rate over the period 1960w1 to 1999w52; there are 2,080 observations. This number of observations should deliver sample properties that are close to the asymptotic results.

The time series is graphed in Figure 4.18 and initial impressions are of a series that exhibits persistence. Estimates of d are obtained from the levels of the series for $\hat{d}_{MGPH}$, $\hat{d}_{MLW}$ and $\hat{d}_{ELW,\mu}$, and its first difference for the remaining methods; the results are reported in Table 4.9 and are broadly similar for each estimation method.

Using the 'square-root' rule for the number of included frequencies results in estimates based on the LP regressions that are generally no more than one asymptotic standard error away from the unit root; on the same basis, the estimates based on Whittle's contrast function are similar, but with a smaller asymptotic standard error they cast greater doubt on the unit root hypothesis.

Next, we consider approximations to an 'optimal' choice of the number of included frequencies. The GPH estimator, $\hat{d}_{GPH}$, has a theoretical MSE minimising optimum value for m given by Equation (4.45), m_{GPH}^{opt}, which, to be feasible,

Table 4.9 Estimates of d for the US three-month T-bill rate

m rule	$\hat{d}_{GPH}$		$\hat{d}_{AG}(1)$		$\hat{d}_{MGPH,\mu}$		$\hat{d}_{LW}$		$\hat{d}_{MLW,\mu}$		$\hat{d}_{ELW,\mu}$	
	$\sqrt{T}$	$\hat{m}^{opt}_{GPH}$	$\sqrt{T}$	$\hat{m}^{opt}_{AG}(1)$	$\sqrt{T}$	$\hat{m}^{opt}_{GPH}$	$\sqrt{T}$	$\hat{m}^{opt}_{LW}$	$\sqrt{T}$	$\hat{m}^{opt}_{LW}$	$\sqrt{T}$	$\hat{m}^{opt}_{LW}$
$\hat{m}$	0.919	0.913	1.119	0.897	0.917	0.899	0.875	0.920	0.875	0.917	0.880	0.924
	45	126	45	242	45	126	45	179	45	179	45	179
$\sigma^{asym}(.)$	0.096	0.057	0.137	0.062	0.096	0.057	0.075	0.037	0.075	0.037	0.075	0.037
't'	−0.84	−1.53	0.87	−1.66	−0.86	−1.77	−1.66	−2.16	−1.66	−2.24	−1.60	−2.05

Notes: The estimates $\hat{d}_{GPH}$, $\hat{d}_{AG}(1)$ and $\hat{d}_{LW}$ are obtained from first differenced data, with 1 added to the original estimate; 't' is the value of the asymptotic t statistic against the null H_0: d = 1.

requires an estimate of $f_u(0)/f_u''(0)$; as noted in Section **4.5.2.i.a**, one method is to obtain this from an AG(1) regression using $(\hat{b}_2)^{-1} = 1/(2\hat{\varphi}_1)$ as an estimator of $f_u(0)/f_u''(0)$. In this case, the AG(1) regression is regarded as an auxiliary, rather than the primary, regression, with v included frequencies where $v = HT^{8/9}$, and we take $H = 0.3$; the auxiliary regression results in $v = 266$, and hence, $\hat{m}_{GPH}^{opt} = 126$ in the primary regression. This is much larger than $int[\sqrt{T}] = 45$, and the asymptotic variance is therefore substantially reduced; as a result, the asymptotic p-value for the unit root null is also reduced.

The AG estimator with $r = 1$, $\hat{d}_{AG}(1)$, has a theoretical MSE minimising optimum value for m given by (4.69); in practice, the feasible version $\hat{m}_{AG}^{opt}(1)$ requires $b_{2+2r} = b_4$, which is estimated from the auxiliary AG regression for $r = 2$, with v included frequencies where here $v = HT^{12/13}$ and, as before, $H = 0.3$. In this case $v = 346$, which, in turn, results in $\hat{m}_{AG}^{opt}(1) = 242$. The increase in the number of included frequencies drives the variance down and, as a result, the asymptotic standard error is halved compared to the square root choice. Note that, as anticipated, the number of included frequencies at the approximate MSE optimum, is (substantially) larger for the AG estimator compared to the GPH estimator; however, even so, this is not sufficient to reduce the AG asymptotic variance below that for the GPH estimator.

The estimator $\hat{d}_{MGPH,\mu}$ for the choice of m given by $\hat{m}_{GPH}^{opt} = 126$, is included for illustrative purposes, given that there is not, as yet, an equivalent optimising quantity. As a result of including more frequencies in computing this estimator, the standard error declines, and $\hat{d}_{MGPH,\mu} = 0.899$ and an asymptotic t-statistic of -1.77, which leads to rejection of the unit root null hypothesis against the one-sided alternative.

The final three columns relate to local Whittle estimation. The estimates are similar to those for the GPH methods, that is broadly the estimates of d are about 0.9, with those using $m = \sqrt{T}$ slightly lower and those using $\hat{m}_{LW}^{opt} = 179$ slightly higher. The estimates for $\hat{d}_{MLW,\mu}$ and $\hat{d}_{ELW,\mu}$ with $m = \hat{m}_{LW}^{opt} = 179$ are included for illustrative purposes. As anticipated, the standard error declines; for example, $\hat{d}_{MLW,\mu} = 0.917$ with an asymptotic t-statistic of -2.24.

Overall, it is clear that the hypothesis of stationarity $d < 0.5$ is rejected, by whatever method, but the unit root null is sensitive to which estimator is used and with which choice of frequencies.

An informative way of presenting the sensitivity of the estimates of the long-memory parameter is to 'loop' the estimating method through an increasing number of frequencies. This enables a picture to be drawn of the sensitivity of the estimates to the number of included frequencies. The results from this procedure are shown in Figures 4.19a and 4.19b, first for the three LW estimators and then for $\hat{d}_{ELW,\mu}$, together with the respective one-sided 95% confidence interval for the unit root, assuming asymptotic normality, which is $\hat{d} + 1.645\sigma^{asym}$. It is clear from Figure 4.19a that the estimates are generally below unity as the number

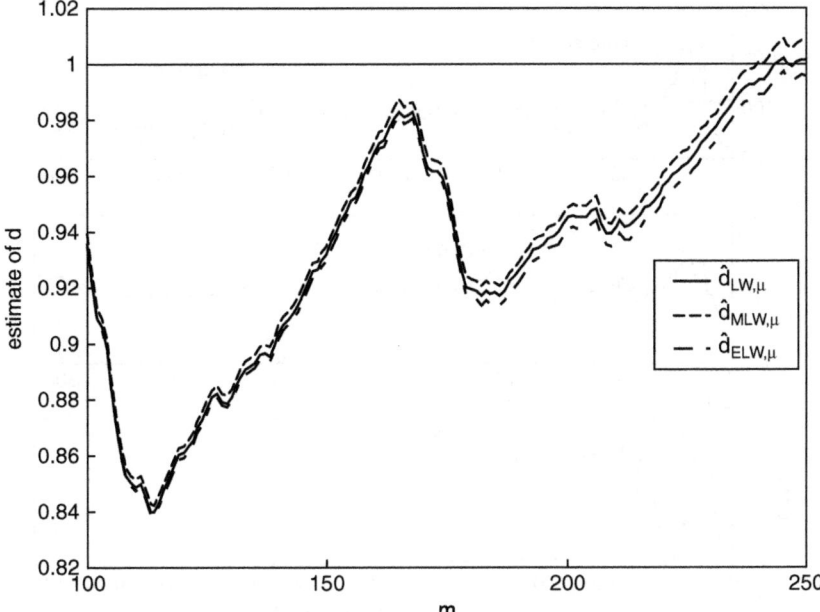

Figure 4.19a   Estimates of d as bandwidth varies, US T-bill rate

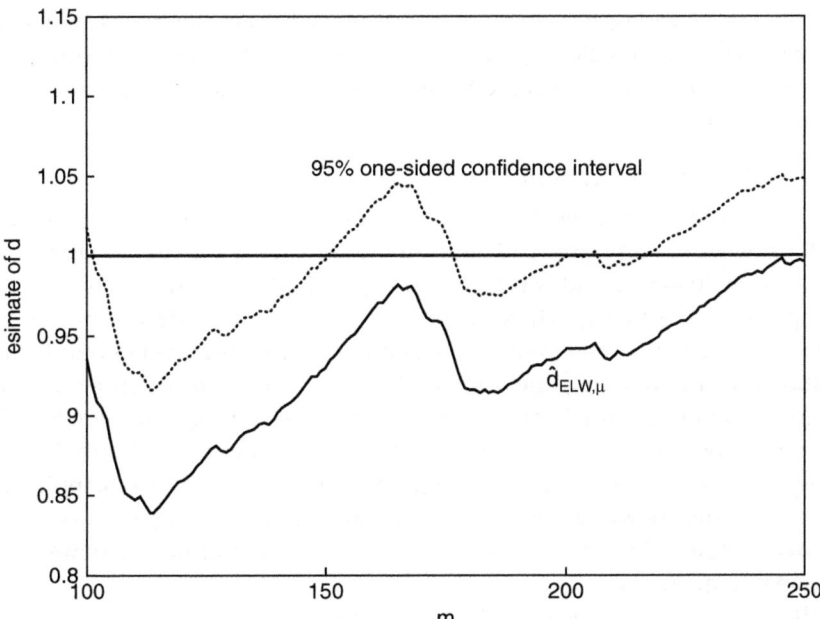

Figure 4.19b $\hat{d}_{ELW,\mu}$ as bandwidth varies, US T-bill rate

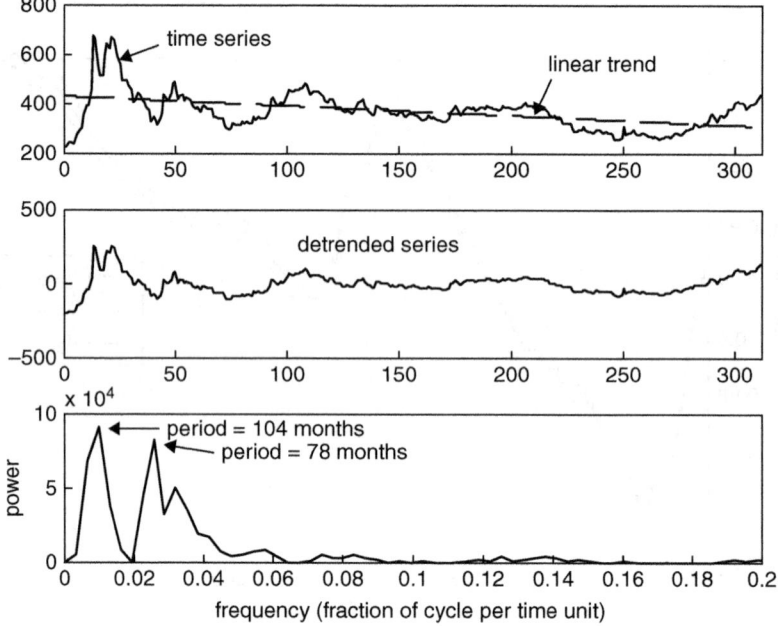

Figure 4.20   Gold price US$ (end month) 1974m1 to 2004m12

of frequencies varies, but they do exceed unity as the number of frequencies increases above about 250. Figure 4.19b indicates that hypothesis testing against a unit root null will be sensitive to the number of frequencies included in the estimation method.

4.12.2   The price of gold in US$

The second illustration uses the time series on the price of gold (per ounce in US$); the series comprises 312 monthly observations for the 26-year period, 1979m1 to 2004m12 and is graphed in Figure 4.20. The series is shown in the top panel of the figure, where there is a suggestion of a decline in the gold price and, perhaps, three 'saucer'-shaped cycles; the (linearly) detrended data is shown in the middle panel, where the trend decline in the price has been removed; and in the final panel, there is a plot of power against frequency. The maximum power is associated with a period of 104 months (slightly less than nine years), which seems to be picking up the three cycles in the sample; there is some further power at 76 months and some power for a period or periods between about 30 to 40 months. There is no suggestion of any power at the seasonal frequencies.

The estimates are reported in Table 4.10. In this case the estimators allow for a deterministic trend as there is some suggestion that, at least for the sample

Table 4.10 Estimates of d for the price of gold

m rule	$\hat{d}_{GPH,\beta}$		$\hat{d}_{AG,\beta}(1)$		$\hat{d}_{MGPH,\beta}$		$\hat{d}_{LW,\beta}$		$\hat{d}_{MLW,\beta}$		$\hat{d}_{ELW,\beta}$	
	$\sqrt{T}$	$\hat{m}^{opt}_{GPH}$	$\sqrt{T}$	$\hat{m}^{opt}_{AG}(1)$	$\sqrt{T}$	$\hat{m}^{opt}_{GPH}$	$\sqrt{T}$	$\hat{m}^{opt}_{LW}$	$\sqrt{T}$	$\hat{m}^{opt}_{LW}$	$\sqrt{T}$	$\hat{m}^{opt}_{LW}$
Estimate	1.357	1.201	0.600	1.255	1.357	1.195	1.227	1.126	1.217	1.116	1.073	1.110
m	(17)	(29)	(17)	(50)	(17)	(29)	(17)	(37)	(17)	(37)	(17)	(37)
$\sigma^{asym}(.)$	0.156	0.119	0.233	0.136	0.156	0.120	0.121	0.082	0.121	0.082	0.121	0.082
't'	2.29	1.69	-1.72	1.88	2.89	1.63	1.88	1.54	1.79	1.41	0.60	1.34

Notes: As for Table 4.9.

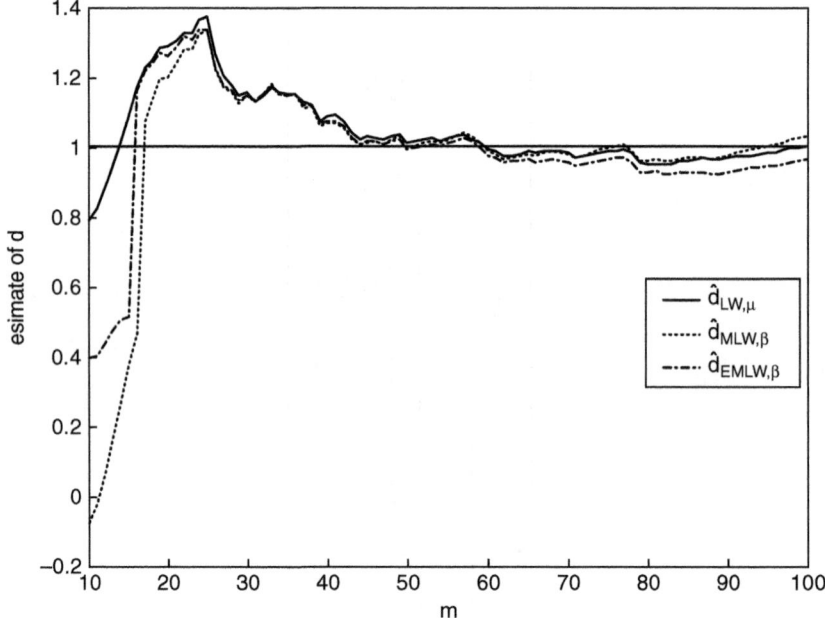

Figure 4.21a   Estimates of d as bandwidth varies, price of gold

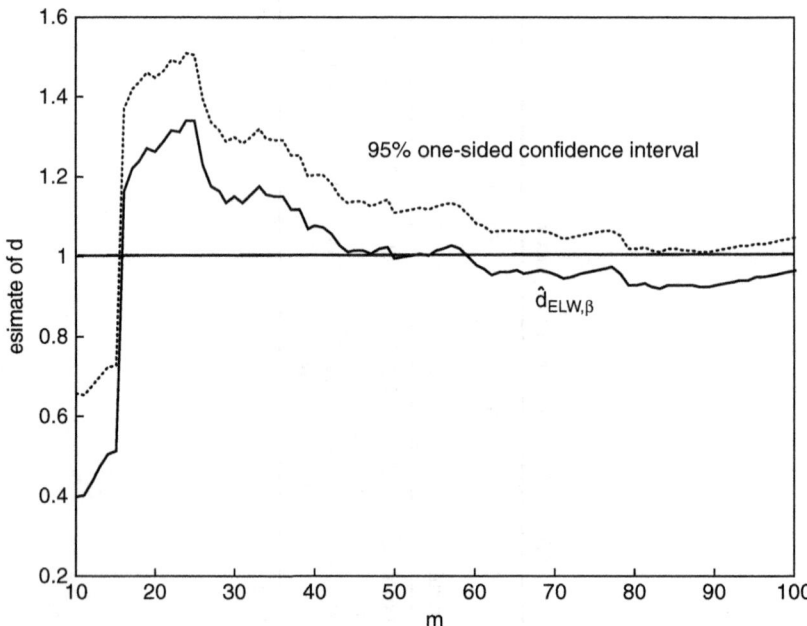

Figure 4.21b $\hat{d}_{ELW,\mu}$ as bandwidth varies, price of gold

period considered here, there is a 'local' linear trend (however, the presence of a 'global' trend might imply rather uncomfortable implications for the price of gold).

The most consistent picture emerges if attention is concentrated on the approximate mse minimizing choice, which is larger than $\sqrt{T}$ for the GPH and LW estimators (note that as in the previous example, the mse optimum for GPH is applied to MGPH, and the mse optimum for LW is applied to MLW and ELW). In this case, the estimates are broadly between 1.1 and 1.2, and, interpreted literally, this would imply an explosive series, but there is not overwhelming evidence that these values are significantly different from unity. As in the previous example, a graph of the estimates as the number of included frequencies varies provides an overall picture (see Figure 4.21a for the LW estimates, where these are either above or close to unity for m > 15). Figure 4.21b shows the ELW estimator, regarded as a good 'general purpose' estimator and one that produces asymptotically valid confidence intervals for d in the 'explosive' range (see Phillips and Shimotsu, 2005); and the asymptotic 95% one-sided confidence interval includes unity for m > 15.

4.13 Concluding remarks

The primary aim of this chapter has been to extend the analysis of fractionally integrated processes of the previous chapter to include a semi-parametric approach that enables estimation of the long-memory parameter, d, and tests of the hypothesis that d = 1. In so doing a lot of ground has been covered, but there is much more that could have been included. This reflects the extent of research into new methods of estimation and the properties of, and improvements to, existing methods. However, relative to the extent of the theoretical developments, empirical results are rather thin on the ground; there are many methods of estimation, with no consensus yet emerging, through the weight of applications, as to which method or methods to use or, at least, which to consider the baseline method. In part this is due to the vibrancy of this research area; previous practical emphasis on the original GPH estimator, with a selection of the number of included frequencies by the square root rule seems to have been, ex post at least, misplaced.

The exact local Whittle (ELW) estimator due to Phillips and Shimotsu (2005) is one of the most promising developments as it is not restricted in its consistency and asymptotic normality by an unduly limited range of d values, although there remains the problem of obtaining the 'optimum' number of frequencies to be included. Another method, not considered here, that differs from both the LP and LW narrowband methods, is based on the scaling function (see, for example, Calvet and Fisher, 2002, and Fillol and Tripier, 2003); Fillol (2007) reports some simulation evidence which shows that the scaling function method can improve the Andrews and Guggenberger bias-reduced GPH estimator for processes with fat-tailed distributions.

Part of the recent research emphasis has been to provide practical 'plug-in' solutions to the problem of selecting the number of frequencies to be included in 'local' estimators. However, these methods are not entirely automatic; their optimality depends, in part, on knowing the structure of the short-run dynamics and, since this is not known, a user-set constant is required to operationalise the 'plug-in' method. Although it is useful to know what the point estimate of the long-memory parameter, d, is for a given method of selecting m, a practical guide can be helped by estimating d for a selection of m values; at the least, this should give an idea of the sensitivity of a particular point estimate to variation in the number of included frequencies. Of course, this will not necessarily be conclusive, but it will allow the data to indirectly indicate whether short-run dynamics are an influential factor.

Of course, from a practical viewpoint, there is a cost associated with semi-parametric methods just as there is a cost of misspecifying a fully parametric model. A researcher may well be interested in the short-run dynamics as well as the long-run, where the latter are governed by long-memory parameter; hence, estimating d is not sufficient. A short-to-medium-term forecast will require more than an estimate of d; see, for example Barkoulas and Baum (2006) for forecasting US monetary indices based on semi-parametric estimation.

Questions

Q4.1 What are the period and the fundamental angular frequency of the following function?

$$y_t = \cos(2\pi t/4)$$

A.4.1 This is a periodic function with $\lambda = 2\pi/4 = \pi/2$ and period $P = 2\pi/(\pi/2) = 4$ time units; thus, the time series pattern shows an exact repeat every 4 periods: $y_t = y_{t+4} = y_{t+8} = \ldots$. The fundamental period is 4 and the fundamental angular frequency is $2\pi/4$.

Q4.2 Consider the SDF of the ARFIMA(1, d, 0) model, that is:

$$f_y(\lambda_j) = |(1 - e^{-i\lambda_j})|^{-2d} f_u(\lambda_j)$$

$$f_u(\lambda_j) = \frac{1}{(1 + \phi_1^2 - 2\phi_1 \cos\lambda_j)} \frac{\sigma_\varepsilon^2}{2\pi} \quad \text{see (4.8)}$$

Q4.2.i Obtain: $f_u'(\lambda)$, $f_u''(\lambda)$ and $f_u''(0)/f_u(0) = b_2$. (The notation can be simplified without loss by omitting the j subscript.)

Q4.2.ii Obtain the asymptotic root mean squared error at $m_{AG}^{opt}(0)$ of $\hat{d}_{AG}(0)$ for $T = 500$ and $\phi_1 = 0.9, 0.6, 0.3$; do the same for $\hat{d}_{AG}(1)$, given $b_4 = 97{,}380$.

A4.2.i The answer will use $\cos(\lambda = 0) = 1$ and

$$f_u(0) = (1 + \phi_1^2 - 2\phi_1)^{-1}(\sigma_\varepsilon^2/2\pi)$$
$$= (1 - \phi_1^2)^{-2}(\sigma_\varepsilon^2/2\pi)$$

The derivatives of $h(\lambda) = \cos\lambda$ repeat themselves in a pattern of four, as follows:

$$h'(\lambda) = -\sin\lambda, h'(0) = -\sin 0 = 0; h''(\lambda) = -\cos\lambda, h''(0) = -\cos 0 = -1;$$
$$h'''(\lambda) = \sin\lambda, h'''(0) = \sin 0 = 0; h^{(4)}(\lambda) = \cos\lambda, h^{(4)}(0) = \cos 0 = 1.$$

These rules use the result that if $f(\lambda) = \sin\lambda$, then $f'(\lambda) = h(\lambda) = \cos\lambda$. Now define $u(\lambda) \equiv (1 + \phi_1^2 - 2\phi_1\cos\lambda)$, then:

$$f_u'(\lambda) \equiv \frac{\partial f_u(\lambda)}{\partial\lambda}$$

$$= -\frac{\partial u(\lambda)}{\partial\lambda} u(\lambda)^{-2}\frac{\sigma_\varepsilon^2}{2\pi}$$

$$= -(2\phi_1\sin\lambda)u(\lambda)^{-2}\frac{\sigma_\varepsilon^2}{2\pi}$$

$f_u'(0) = 0$ because $\sin 0 = 0$

$$f_u''(\lambda) = \frac{\partial\left(-(2\phi_1\sin\lambda)u(\lambda)^{-2}\dfrac{\sigma^2}{2\pi}\right)}{\partial\lambda}$$

$$= -(2\phi_1\cos\lambda)u(\lambda)^{-2}\frac{\sigma^2}{2\pi} + (-2)(-2\phi_1\sin\lambda)u(\lambda)^{-3}(2\phi_1\sin\lambda)\frac{\sigma^2}{2\pi}$$

$$= -(2\phi_1\cos\lambda)u(\lambda)^{-2}\frac{\sigma^2}{2\pi} + 8\phi_1^2\sin^2\lambda\, u(\lambda)^{-3}\frac{\sigma^2}{2\pi}$$

$f_u''(0) = -2\phi_1 u(\lambda)^{-2}\dfrac{\sigma^2}{2\pi}$ note that $u(\lambda)^{-2} = (1 - \phi_1^2)^{-4}$

$$\frac{f_u''(0)}{f_u(0)} = -2\phi_1\frac{(1 - \phi_1)^{-4}}{(1 - \phi_1)^{-2}} = -\frac{2\phi_1}{(1 - \phi_1)^2}$$

For example, if $\phi_1 = 0.9$, then $\frac{f_u''(0)}{f_u(0)} = -\frac{2\phi_1}{(1-\phi_1)^2} = -\frac{1.8}{(1-0.9)^2} = -180$; if $\phi_1 = 0.6$, then $\frac{f_u''(0)}{f_u(0)} = -7.5$; if $\phi_1 = 0.3$, then $\frac{f_u''(0)}{f_u(0)} = -1.2245$. Note that in the notation of Section 4.5.2, $\frac{f_u''(0)}{f_u(0)} = b_2$.

A4.2.ii The asymptotic mean square error of the GPH estimator is:

$$\text{AMSE}(\hat{d}_{GPH}) = \left[\frac{-2\pi^2}{9}\right]^2\left|\frac{f_u''(0)}{f_u(0)}\right|^2\left[\frac{m^2}{T^2}\right]^2 + \frac{\pi^2}{24m}$$

Table A4.1 m_{GPH}^{opt} and ARMSE at m_{GPH}^{opt} for $\hat{d}_{AG}(r)$, r = 0, 1

ϕ_1	0.3	0.6	0.9
r = 0			
m_{GPH}^{opt}	61	29	8
ARMSE	0.0913	0.1312	0.2482
r = 1			
m_{GPH}^{opt}	12	52	112
ARMSE	0.2846	0.1411	0.1060

Hence, in this case:

$$AMSE(\hat{d}_{GPH}) = \left[\frac{2\pi^2}{9}\right]^2 \left|\left(\frac{2\phi_1}{(1-\phi_1)^2}\right)\right|^2 \left[\frac{m^2}{T^2}\right]^2 + \frac{\pi^2}{24m}$$

The question asks for the rmse at $m = m_{GPH}^{opt}$; hence, first obtain m_{GPH}^{opt} as follows:

$$m_{GPH}^{opt} = int\left[\frac{27}{128\pi^2}\right]^{1/5} \left|\left(\frac{(1-\phi_1)^2}{2\phi_1}\right)\right|^{2/5} T^{4/5}$$

$$= 8$$

Now use m_{GPH}^{opt} in $AMSE(\hat{d}_{GPH})$ and take the square root to obtain: $ARMSE(\hat{d}_{GPH}) = 0.248$.

Q4.3 Consider the spectral density of the ARFIMA(0, d, 1) model, that is:

$$f_y(\lambda_j) = |(1 - e^{-i\lambda_j})|^{-2d} f_u(\lambda_j)$$

$$f_u(\lambda_j) = (1 + \theta_1^2 + 2\theta_1 \cos \lambda_j)\frac{\sigma_\varepsilon^2}{2\pi}$$

Show that this model results in the same mse and m_{GPH}^{opt} for $\theta_1 = -\phi_1$.

A4.3 First obtain $f_u'(\lambda)$, $f_u''(\lambda)$ and $f_u''(0)/f_u(0)$. This is much simpler than the ARFIMA(1, d, 0) case.

$$f_u'(\lambda) = -2\theta_1 \sin \lambda(\sigma^2/2\pi); f_u'(0) = -2\theta_1 \sin 0(\sigma^2/2\pi) = 0;$$

$$f_u''(\lambda) = -2\theta_1 \cos \lambda(\sigma^2/2\pi); f_u''(0) = -2\theta_1(\sigma^2/2\pi);$$

$$f_u(\lambda) = (1 + \theta_1^2 + 2\theta_1 \cos \lambda)\frac{\sigma^2}{2\pi}; f_u(0) = (1 + \theta_1^2 + 2\theta_1)\frac{\sigma^2}{2\pi} = (1 + \theta_1)^2\frac{\sigma^2}{2\pi};$$

$$f_u''(0)/f_u(0) = -\frac{2\theta_1}{(1 + \theta_1)^2}$$

Table A4.2   Ratio of asymptotic bias for $\hat{d}_{AG}(0)$ and $\hat{d}_{AG}(1)$, $\phi_1 = 0.9$

M	1	5	10	15	20	21	22	23
AB ratio	454.5	18.17	4.55	2.02	1.14	1.03	0.94	0.86

Notes: T = 500, ARFIMA(1, d, 0) model, $\phi_1 = 0$.; AB ratio is $AB[\hat{d}_{AG}(0)]/AB[\hat{d}_{AG}(1)]$.

In the case of $\theta_1 = -\phi_1$:

$$f_u''(0)/f_u(0) = \frac{2\phi_1}{(1 - \phi_1)^2}.$$

The asymptotic bias for $\hat{d}_{GPH}$ is $\left[\frac{-2\pi^2}{9}\right]\left[\frac{f_u''(0)}{f_u(0)}\right]\left[\frac{m^2}{T^2}\right]$, which is opposite in sign for the MA component with $\theta_1 = -\phi_1$ compared to the AR component with ϕ_1. However, the difference in sign is not relevant for the AMSE, which squares the bias. Hence, AMSE and m_{GPH}^{opt} will be the same for ARFIMA(1, d, 0) and ARFIMA(0, d, 1) when $\theta_1 = -\phi_1$.

Q4.4 Consider the asymptotic bias given by AG (i), Equation (4.66). Is it possible that $AB[\hat{d}_{AG}(0)]] < AB[\hat{d}_{AG}(1)]$, where AB indicates the asymptotic bias? To illustrate your answer take T = 500, $\phi_1 = 0.9$, and note that in this case $b_2 = -180$ and $b_4 = 97,380$. Evaluate the asymptotic bias and asymptotic rmse at the (asymptotic) values for $m_{AG}^{opt}(r)$.

A4.4 First note that the asymptotic bias, AB(r), is given by $\kappa_r b_{2+2r} \frac{m^{2+2r}}{T^{2+2r}}$ (see AG (i), (4.66)), with $\kappa_0 = -2.19$, $\kappa_1 = 2.23$; hence, set r = 0, 1, to obtain the relevant biases and hypothesise the inequality as follows:

$$\kappa_0 b_2 \frac{m^2}{T^2} < \kappa_1 b_4 \frac{m^4}{T^4}$$

$$\Rightarrow -2.19 b_2 \frac{m^2}{T^2} < 2.23 b_4 \frac{m^4}{T^4}$$

$$\Rightarrow -\frac{2.19}{2.23} \frac{b_2}{b_4} T^2 < m^2$$

$$\Rightarrow -0.983 \frac{(-180)}{97380} (500)^2 < m^2$$

The integer part of the left-hand side of the last inequality is 21; hence, the condition for T = 500 is m ≥ 22, for which $AB[\hat{d}_{AG}(0)] < AB[\hat{d}_{AG}(1)]$ is satisfied. To illustrate, the ratio of the asymptotic bias for a grid of m values is given in Table A4.2.

Finally, compare the asymptotic bias at $m_{AG}^{opt}(r)$. First note that $int[m_{AG}^{opt}(0)] = 8$, and $m_{AG}^{opt}(1)$ is obtained from:

$$m_{AG}^{opt}(1) = \left(\frac{\pi^2 c_1}{24(4+4)\kappa_1^2 b_{2+2}^2} \right)^{1/(5+4)} T^{(4+4)/(5+4)}$$

where $c_1 = 2.25$ and $\kappa_1 = 2.23$. Elementary calculations give $m_{AG}^{opt}(1) = 12.85$, so that $int[m_{AG}^{opt}(1)] = 12$. Evaluating $AB[\hat{d}_{AG}(r)]$ for these values gives: $AB[\hat{d}_{AG}(0)] = 0.101$ and $AB[\hat{d}_{AG}(1)] = 0.072$, confirming that the (asymptotic) bias is reduced in this example.

Evaluating the asymptotic rmse at $m_{AG}^{opt}(r)$ gives: $ARMSE[\hat{d}_{AG}(0)] = 0.248$ and $ARMSE[\hat{d}_{AG}(1)] = 0.284$; thus, although asymptotic bias is reduced, this does not compensate sufficiently for the increase in the variance.

Q4.5 Given $f_y(\lambda_j) = |(1 - e^{-i\lambda_j}|^{-2d} f_u(\lambda_j)$, and taking logs of both sides it follows that $\ln[f_y(\lambda_j)] = -2d \ln[|(1 - e^{-i\lambda_j})|] + \ln[f_u(\lambda_j)]$; hence, obtain the pseudo log-periodogram regression.

A4.5 Start with $\ln[f_y(\lambda_j)] = -2d \ln[|(1 - \exp(-i\lambda_j))|] + \ln[f_u(\lambda_j)]$ and add $\ln[I_y(\lambda_j)]$ to both sides, add and subtract $\ln[f_u(0)]$ to the right-hand side and then rearrange so that:

$$\ln[I_y(\lambda_j)] = \ln[f_u(0)] - 2d \ln[|1 - e^{-i\lambda_j}|] + \ln[f_u(\lambda_j)/f_u(0)] + \ln[I_y(\lambda_j)/f_y(\lambda_j)]$$

This is then evaluated at the Fourier frequencies and interpreted as a regression equation.

Another way of interpreting these manipulations is to see $\ln[f_y(\lambda_j)]$ as being replaced by (estimated by) $\ln[I_y(\lambda_j)]$, which induces an error of $\ln[I_y(\lambda_j)] - \ln[f_y(\lambda_j)] = \ln[I_y(\lambda_j)/f_y(\lambda_j)]$.

Q4.6 Consider how to define the spectral density of a nonstationary series.

A4.6 The standard definition of the SDF does not hold for $d \geq 0.5$. For $d \in (0.5, 1.5)$ the usual approach is to consider y_t as the first difference of the original series Z_t; that is, $y_t = \Delta Z_t$ so that $Z_t \equiv \Delta^{-1} y_t + Z_0 = \sum_{i=1}^{t} y_i + Z_0$, where Z_0 is a random variable not depending on t. The SDF of y_t is now standard, because $d \in (-0.5, 0.5)$, and given by:

$$f_y(\lambda_j) = |1 - e^{-i\lambda_j})|^{-2(d-1)} f_u(\lambda_j)$$

Note that the exponent is positive because $d \in (-0.5, 0.5)$.

The analysis then proceeds as before but with the function $f_Z(\lambda_j)$ defined as:

$$f_Z(\lambda_j) = |1 - e^{-i\lambda_j})|^{-2} f_y(\lambda_j)$$
$$= |1 - e^{-i\lambda_j}|^{-2d} f_u(\lambda_j)$$

Then modify (4.93) for $d \in (0.5, 1.5)$, so that $f_y(\lambda_j) \sim G\lambda_j^{-2(d-1)}$ as $\lambda_j \to 0^+$, which implies $f_Z(\lambda_j) \sim G\lambda_j^{-2d}$ as $\lambda_j \to 0^+$. See Phillips (2000) for more detailed consideration and development of the arguments; in particular, note that the modification suggested by Phillips (see (4.76)), can be extended to the case considered in this question.

Q4.7.i Note that the rate r_1 that implies consistency of $\hat{d}_{LW}$, condition C1 of (4.107), depends on Δ_1, the lower limit to the set of allowable d and d itself. By way of example let $\Delta_1 = -0.9$, $d = -0.75$ and adopt the rule that $m = T^\alpha$: at what rate should m expand; is the square root rule, $m = T^{1/2}$, adequate for this purpose?

Q4.7.ii Next, continuing with $d = -0.75$, consider r_2 and derive the condition on m that avoids C2 of (4.108).

A4.7.i Note that $r_1 = 1$ for $m = T^{5/7}$, $r_1 \to 0$ for $m > T^{5/7}$ and $r_1 \to \infty$ for $m < T^{5/7}$. Thus, for consistency, m should expand faster than $T^{5/7}$, that is $\alpha > 5/7$. This is typically a faster rate than is imposed by some rules of thumb, such as $m = T^{1/2}$.

A4.7.ii Next consider r_2 continuing with $d = -0.75$, then $r_2 = 1$ for $m = T^{1/3}$, $r_2 \to \infty$ for $m < T^{1/3}$ and $r_2 \to 0$ for $m > T^{1/3}$; thus on this basis m should expand faster than $T^{1/3}$ to avoid an estimator biased to zero. Others things being equal, the closer d is to the -0.5 boundary, the more likely that $\hat{d}_{LW}$ is consistent and the closer d is to -1 the less likely $\hat{d}_{LW}$ is to be consistent.

Q4.8 Consider the consistency of $\hat{d}_{LW}$ for $d = -0.95$ and $d = -0.6$: is $\hat{d}_{LW}$ consistent for the rule $m = T^{1/2}$ and $\Delta_1 = -0.99$?

A4.8 Take $d = -0.95$ first and check r_2:

$$r_2 \equiv m^{-1}\left(\frac{m}{n}\right)^{(2d+1)}$$
$$= m^{-1}\left(\frac{m}{n}\right)^{-0.9}$$
$$= \frac{n^{0.9}}{m^{1.9}} \to \infty \quad \text{if} \quad m^{1.9} < n^{0.9} \Rightarrow m < n^{9/19}$$

Thus, if m expands at a rate α less than 9/19, then $r_2 \to \infty$ and $\hat{d}_{LW} \to_p 0$; the square root rule $m = T^{1/2}$ is faster than this, but is it fast enough to ensure consistency? That depends also on Δ_1. To illustrate, take $\Delta_1 = -0.99$.

$$r_1 = m^{-2(\Delta_1+1)}\left(\frac{m}{n}\right)^{(2d+1)}$$

$$= m^{-0.02}\left(\frac{n}{m}\right)^{0.9}$$

$$= \frac{n^{0.9}}{m^{0.92}} \quad \to 0 \quad \text{if } m^{0.92} > n^{0.9} \Rightarrow m > n^{90/92}$$

Consistency requires that m expands at the fast rate of $\alpha = 90/92$, so that $\alpha = \frac{1}{2}$ is not fast enough to ensure consistency.

Next consider $d = -0.6$, $\Delta_1 = -0.99$ and evaluate r_1 :

$$r_1 = \frac{n^{0.2}}{m^{0.22}} \quad \to 0 \quad \text{if } m^{0.22} > n^{0.2} \Rightarrow m > n^{10/11}$$

This is again a fast rate but not as fast as $\alpha = 90/92$.

Evaluate r_2:

$$r_2 = m^{-1}\left(\frac{m}{n}\right)^{(2d+1)}$$

$$= m^{-1}\left(\frac{m}{n}\right)^{-0.2}$$

$$= \frac{n^{0.2}}{m^{1.2}} \quad \to \infty \quad \text{if } m^{1.2} < n^{0.2} \Rightarrow m < n^{1/6}$$

The rule $m = T^{1/2}$ does not, therefore, imply $\hat{d}_{LW} \to_p 0$.

Appendix: The Taylor series expansion of a logarithmic function

The Taylor series expansion is used at several points in this chapter. First, consider the general function $f(\lambda)$, then the Taylor series expansion of $f(\lambda)$ at $\lambda = \lambda_0$ is:

$$f(\lambda) = f(0) + \frac{f'(\lambda_0)}{1!}(\lambda - \lambda_0) + \frac{f''(\lambda_0)}{2!}(\lambda - \lambda_0)^2 + \ldots \frac{f^{(s)}(\lambda_0)}{s!}(\lambda - \lambda_0)^s + R_s \quad \text{(A4.1)}$$

The remainder R_s comprises terms of higher order than λ^s.

For $f(\lambda) = \ln[g(\lambda)]$, the first two derivatives are: $f'(\lambda) = g'(\lambda)/g(\lambda)$; and $f''(\lambda) = -(g'(\lambda)/g(\lambda))^2 + g''(\lambda)/g(\lambda)$. So, for example, the second-order expansion about $\lambda_0 = 0$, with remainder, R_2, is:

$$\ln[g(\lambda)] = \ln[g(0)] + \frac{g'(0)}{g(0)}\lambda + \frac{1}{2}\left[\frac{g''(0)}{g(0)} - \frac{g'(0)^2}{g(0)^2}\right]\lambda^2 + R_2 \quad \text{(A4.2)}$$

AG (2003) use the result that a continuous even function, $f(\lambda)$, has odd-order derivatives that are continuous odd functions, and these equal zero when evaluated at $\lambda = 0$. An even function has the property $f(x) = f(-x)$ and an odd function

$f(-x) = -f(x)$. The function $g(\lambda)$ is such a case, hence all odd order derivatives up to the order s of differentiability of the function can be set to zero. Thus, (A4.2) becomes:

$$\ln[g(\lambda)] = \ln[g(0)] + \frac{g'(0)}{g(0)}\lambda\frac{1}{2}\left[\frac{g''(0)}{g(0)}\right]\lambda^2 + R_2 \tag{A4.3}$$

More generally:

$$\ln[g(\lambda)/g(0)] = \sum_{k=1}^{[s/2]}\left[\frac{f^{(2k)}(0)}{(2k)!}\right]\lambda^{2k} + R_s \tag{A4.4}$$

In terms of the notation of this chapter, let $g(\lambda_j) = f_u(\lambda_j)$.

5

Smooth Transition Nonlinear Models

Introduction

The application of random walk models to some economic time series can be inappropriate, as when there are natural bounds or limits to the values that the series can take. For example, there are a number of studies that have tested for unit root nonstationarities in unemployment rates and (nominal) interest rate data; however, the former are bounded by zero and one, and negative nominal interest rates are not observed. In practice, the boundedness of these variables may be recognised by referring to 'local' nonstationarity in the observed series. A way of modelling this is to allow unit root behaviour, for example persistence and the absence of mean reversion, over a range of possible values, but reversion at other values.

There are a number of models that allow unit root type behaviour but take into account either the limits set by the data range or the economic mechanisms that are likely to prevent unbounded random walks in the data. These models have in common that they involve some form of nonlinearity. Perhaps the simplest form of nonlinearity arises from piecewise linearity: that is, an overall model comprises two or more linear models for sub-periods where the component models differ not in their form, for example all are AR(p), but in their parameters. The sub-periods could be defined in chronological time, in which case they are usually considered under the heading of structural break models. See, for example, the literature on structural change deriving from Perron's seminal (1989) paper on breaks arising from the Great Crash of 1929 and the oil price crisis of the early 1970s. These models, where the index of change is time, are considered in Chapter 7.

Where the index of change is not time, the piecewise linearity is generated by an indicator variable which may or may not be contained within the model. A popular piecewise linear model is the threshold autoregression, or TAR, which, in its self-exciting form, switches between or among regimes as a function of the variable that is being modelled. This type of nonlinear model is considered in Chapter 6.

The models considered in this chapter involve a smooth rather than a discrete form of nonlinearity. The chapter starts by introducing smooth transition autoregressions, referred to as STAR models. These are based on the familiar AR model, but allow the AR coefficients to change as a function of the deviation of an 'activating' or 'exciting' variable from its target or threshold value. In the simplest case the exciting variable is a lag of the variable being modelled, for example y_{t-1}, and the target is a constant, say c; then the impulse to change the AR coefficients is a function of $(y_{t-1} - c)$.

There are two frequent choices of activating function. In the ESTAR model, the function is based on an exponential transformation of the quadratic distance $(y_{t-1} - c)^2$; in the LSTAR model, the function is based on a logistic transformation of the distance $(y_{t-1} - c)$. Although STAR models involve smooth transition between regimes, they are capable of looking like piecewise threshold autoregressions if the distance $y_{t-1} - c$ is large and the activating function exhibits substantial sensitivity to this distance; in such cases, there can appear to be just two regimes, as in a two-regime TAR model. The implied behaviour of the two popular nonlinear forms is quite different: for example, the simple ESTAR model has a quadratic response to deviations from threshold, whereas the LSTAR does not have this feature. Nevertheless, there may be circumstances in which both models are potential candidates to fit the data, and a number of criteria have been suggested to discriminate between the two forms. We outline the tests due to Teräsvirta (1994), their suggested modification due to Escribano and Jordá (2001) and, as an alternative or complement, give an application of the pseudo-score encompassing test of Chen (2003) and Chen and Kuan (2002).

A more recent form of nonlinear model also considered in this chapter is the bounded random walk, referred to as a BRW. Whilst this model allows random walk behaviour, such behaviour is contained by bounds, or buffers, that will 'bounce' the variable back if it is getting out of control. The sensitivity of the bound is governed, as in the ESTAR case, by an exponential function of some measure of distance from a threshold.

The organisation of this chapter is as follows. The opening Section (5.1) briefly introduces the idea of an autoregression with two regimes and smooth transition between these regimes. Section 5.2 considers one of two leading smooth transition models, that is, the exponential smooth transition autoregression (ESTAR), while Section 5.3 deals with the other leading model, that is the logistic STAR (LSTAR) model. Multi-regime extensions of these STAR models are considered in Section 5.4. A newer nonlinear model, the bounded random walk (BRW) model, follows in Section 5.5. Section 5.6 is concerned with complications arising from testing for nonstationarity caused by a unit root when the underlying model is nonlinear. Section 5.7 outlines a set of misspecification tests designed to indicate whether nonlinearity is present and an alternative encompassing type test is considered in Section **5.8**. Testing for serial correlation in STAR models is considered in Section **5.9** and the estimation of nonlinear models is considered in Sections **5.10** and **5.11**.

5.1   Smooth transition autoregressions

A 'smooth' transition threshold autoregression is a good place to start for nonlinear models. Consider the following extension of an AR(1) model:

$$y_t = \phi_1 y_{t-1} + \gamma_1 y_{t-1} e(y_{t-d}) + \varepsilon_t \tag{5.1a}$$

$$= [(\phi_1 + \gamma_1 e(y_{t-d})] y_{t-1} + \varepsilon_t \tag{5.1b}$$

For simplicity it is assumed that $\varepsilon_t \sim \text{iid}(0, \sigma_\varepsilon^2)$. The first part of the model is just an AR(1) model; however, the second part adds a modifying term 'driven' by the nonlinear component interpreted as a transition (or adjustment) function, $e(y_{t-d})$. This is specified as a function of the d-th lag of y_t, although, in principle, it could be specified as a function of variables outside the model. The transition function will be 'smooth' in the sense of being differentiable with respect to y_{t-d} and any parameters of the function. (Notation: the use of d as the lag parameter originates with the idea of a 'delay'; it should not be confused with the previous use of d as the long-memory parameter.)

By collecting terms associated with y_{t-1} as in (5.1b), it is evident that the overall AR coefficient is $\phi_1 + \gamma_1 e(y_{t-d})$. If the range of $e(y_{t-d})$ is [0, 1], then the overall coefficient will have a range of $[\phi_1, \phi_1 + \gamma_1]$. In this case, an important distinction, which we develop below, is that there are two limiting regimes, an 'inner' regime with $e(y_{t-d}) = 0$ and, hence, an AR coefficient of ϕ_1, and an 'outer' regime where $e(y_{t-d}) = 1$ and, hence, an AR coefficient of $\phi_1 + \gamma_1$. If $\phi_1 = 1$, then there is a unit root in the inner regime.

5.2   The exponential smooth transition function: ESTAR

5.2.1   Specification of the transition function

An example of a smooth transition function with range [0, 1] is the exponential smooth threshold autoregression, ESTAR, with transition function given by $e(y_t) = \Theta(y_{t-d})$, where:

ESTAR transition function

$$0 \leq \Theta(y_{t-d}) = (1 - e^{-\theta(y_{t-d}-c)^2}) \leq 1 \tag{5.2}$$

Combining (5.2) with (5.1a) gives an ESTAR(1) model:

$$y_t = \phi_1 y_{t-1} + \gamma_1 y_{t-1}(1 - e^{-\theta(y_{t-1}-c)^2}) + \varepsilon_t \tag{5.3}$$

In the ESTAR specification $\theta > 0$, and c, the threshold parameter, can take either sign. As a practical matter, provided that the process is sampled sufficiently from the potential range of y_t, c is expected to lie within observable experience of y_t. The transition function depends upon the delay parameter d, which is usually chosen to be one, and we use that value for illustration here. However, Teräsvirta and Anderson (1992), report a number of examples with d ranging from 1 to 5.

Important properties of the exponential transition function, $\Theta(y_{t-1})$, are $\Theta(0) = 0$ and $\Theta(\infty) = \Theta(-\infty) = 1$, the idea being that there is little or no adjustment for y_{t-1} close to c, but adjustment then increases nonlinearly as y_{t-1} increasingly departs from c in either direction. The triggering point – or threshold – is the parameter c. As $0 \leq \Theta(y_{t-1}) \leq 1$, the values generated by particular deviations from the threshold, $y_{t-1} - c$, can be interpreted as transition weights ranging from zero to one (which are multiplied by $\gamma_1 y_{t-1}$ to get an overall effect). In the exponential transition function, the deviation is squared so there is no distinction as to sign of the deviation. Note that as $\theta \to \infty$, the transition function approaches a discrete indicator function, such that for $y_{t-1} - c > 0$, $\Theta(y_{t-1}) \to 1$, and the 'switch' between regimes tends to instantaneous. Variations on the theme, capturing asymmetry, are considered later in this chapter.

5.2.2   An 'inner' and an 'outer' regime

Although the ESTAR model is continuous in $y_{t-1} - c$, it can be thought of as generating two regimes at the extremes for the (symmetric) ESTAR. If $y_{t-1} = c$, then $\Theta(0) = 0$ and $y_t = \phi_1 y_{t-1} + \varepsilon_t$, with $\phi_1 = 1$ corresponding to the unit root case; as noted above this is the 'inner' regime. In the symmetric ESTAR, as $(y_{t-1} - c) \to \pm\infty$, $\Theta(y_{t-1} - c) \to 1$ and $y_t = (\phi_1 + \gamma_1)y_{t-1} + \varepsilon_t$; this is the 'outer' regime and here it is the same whichever the sign of the deviation. The case $\phi_1 = 1$ and $\gamma_1 = -1$ corresponds to $y_t = \varepsilon_t$, that is, white noise, in the outer regime. This is an example where shocks are less persistent in the outer regime than in the inner regime.

Some of the key properties of the ESTAR model can be represented graphically. First, from Figure 5.1, note that the transition function, $\Theta(y_{t-1})$, is symmetric and U-shaped about c. For simplicity we set $c = 0$, so the 'centre' of the symmetry in Figure 5.1 is 0. The graph shows the transition function for two values of θ, $\theta = 0.02$ and $\theta = 2$. In the former case, other things equal, there is only a slight modification to the linear AR model, whereas in the latter case there is a much stronger nonlinear response. The effect of the adjustment function on potential random walk behaviour is illustrated in Figure 5.2 for $\theta = 0.02$ and Figure 5.3 for $\theta = 2$. In these figures an ESTAR(1) model is simulated for $t = 1, \ldots, 1000$, with $\phi_1 = 1$ and $\gamma = -1$; these values correspond to a unit root in the inner regime and white noise in the outer regime. For comparison, the figure also shows a random walk, that is $\phi_1 = 1$ and $e(y_{t-d}) = 0$, with the same sequence of errors

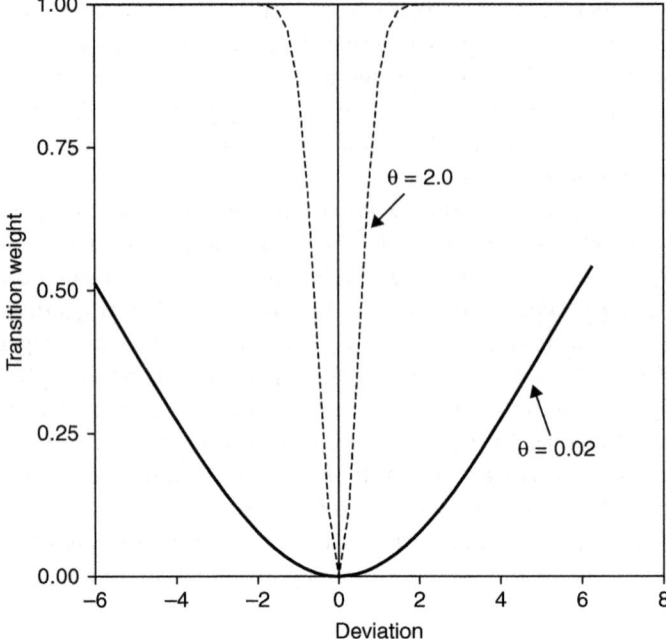

Figure 5.1   ESTAR transition function

$\{\varepsilon_t\}_1^{1000}$, where $\varepsilon_t \sim \text{niid}(0, 0.16)$. These simulated values are shown in panel a of each figure; whereas panel b of each figure shows the corresponding transition weights. Note from Figures 5.2a and 5.3a how the nonstationarity implicit in the inner regime is kept under control, the more so with a larger value of θ. The transition weights shown in Figures 5.2b and 5.3b are, in a sense, doing the work of keeping the process away from the random walk inner regime. Note the difference between Figures 5.2b and 5.3b, where the transition weights in the latter are generally much larger.

5.2.3 An asymmetric ESTAR, AESTAR

The ESTAR model can be modified in two ways to capture asymmetric adjustment, the essence being to allow the effect of positive and negative deviations to differ in some form. First, define the functions $(y_{t-1} - c)^+ = (y_{t-1} - c)$ if $(y_{t-1} - c) < 0$ and 0 otherwise, and $(y_{t-1} - c)^- = (y_{t-1} - c)$ if $(y_{t-1} - c) < 0$ and 0 otherwise. Then $\Theta(y_{t-d}) = (1 - e^{-\theta^+[(y_{t-d}-c)^+]^2} e^{-\theta^-[(y_{t-d}-c)^-]^2})$, which allows asymmetric responses if $\theta^+ \neq \theta^-$ depending on the sign of the deviation. A possible asymmetric ESTAR is then:

$$y_t = \phi_1 y_{t-1} + \gamma_1 y_{t-1}(1 - e^{-\theta^+[(y_{t-1}-c)^+]^2} e^{-\theta^-[(y_{t-1}-c)^-]^2}) + \varepsilon_t \qquad (5.4)$$

(a)

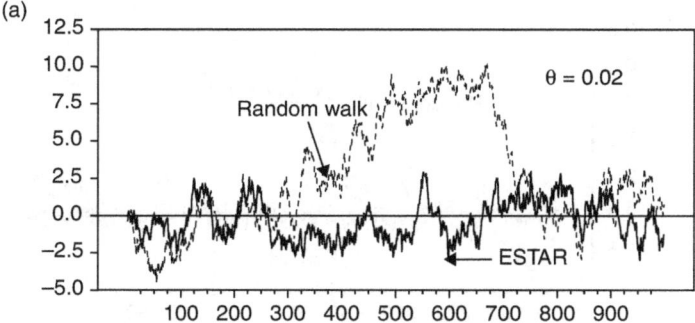

Figure 5.2a   ESTAR and random walk, $\theta = 0.02$

(b)

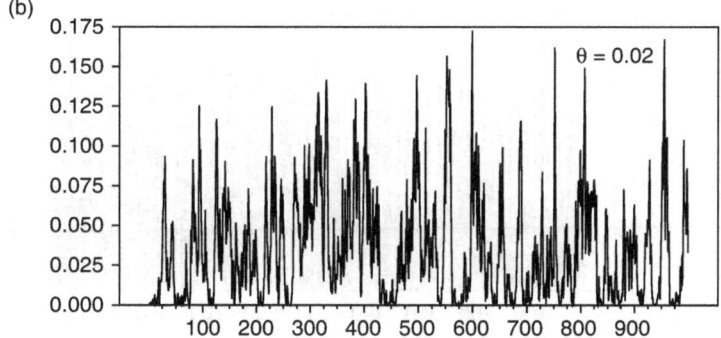

Figure 5.2b   ESTAR transition function, $\theta = 0.02$

In this version the transition effects differ as $\theta^+ \neq \theta^-$ but the 'outer' regime is the same for positive and negative deviations since both imply $\Theta(y_{t-1} - c) \to 1$, so that $y_t = (\phi_1 + \gamma_1)y_{t-1} + \varepsilon_t$. An alternative is to allow the outer regimes to be distinguished, so that the AESTAR is:

$$y_t = \phi_1 y_{t-1} + \gamma_1^+ y_{t-1}(1 - e^{-\theta_1 [(y_{t-1}-c)^+]^2}) + \gamma_1^- y_{t-1}(1 - e^{-\theta_2 [(y_{t-1}-c)^-]^2}) + \varepsilon_t \quad (5.5)$$

The positive outer regime is $y_t = (\phi_1 + \gamma_1^+)y_{t-1} + \varepsilon_t$ and the negative outer regime is $y_t = (\phi_1 + \gamma_1^-)y_{t-1} + \varepsilon_t$. Such a model may be attractive where there are thought to be different degrees of persistence and adjustment speeds for positive and negative deviations.

Simulated paths for two asymmetric ESTAR models are illustrated in Figures 5.4 and 5.5. In Figure 5.4, the generating model is the two-regime AESTAR with $\theta^+ = 0.02$, $\theta^- = 2.0$ and $\gamma = -1$. This pairing implies that whilst having the same outer regime, the nonlinear effect is far more pronounced for negative

(a)
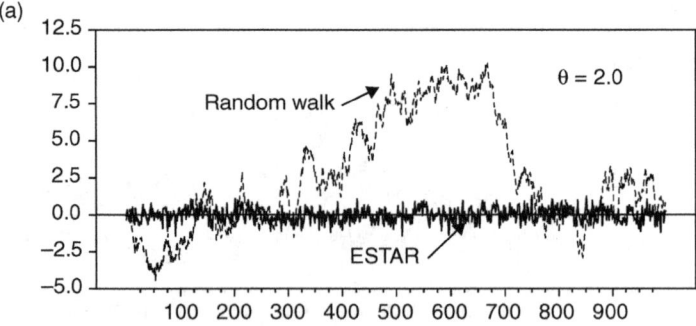

Figure 5.3a   ESTAR and random walk, $\theta = 2.0$

(b)
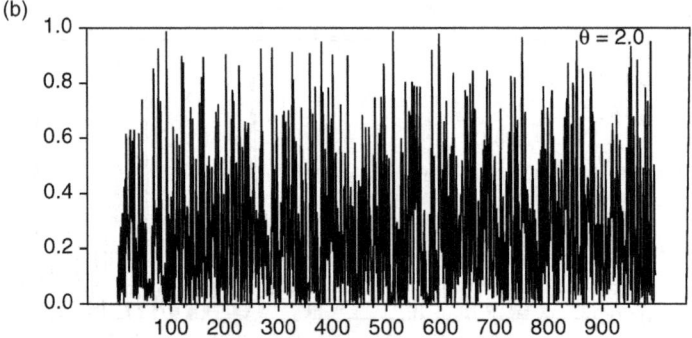

Figure 5.3b   ESTAR transition function, $\theta = 2.0$

compared to positive deviations. (The sequence $\{\varepsilon_t\}$ is the same as used in generating Figures 5.2 and 5.3.) Again there are two parts to each figure, with panel a showing the time series and panel b the transition function. The asymmetric adjustment is particularly apparent in Figure 5.4b, which shows periods of relative inactivity in the transition function. In Figure 5.5, the generating model is the two-regime AESTAR with $\theta^+ = \theta^- = 2.0$ and the asymmetry comes through $\gamma^+ = -1$ and $\gamma^- = -0.1$. Now the outer regimes differ, being $y_t = \varepsilon_t$ for $\gamma^+ = -1$ but relatively persistent at $y_t = 0.9y_{t-1} + z_t$ for $\gamma^- = -0.1$. Another aspect of what is happening in these ESTAR models relates to viewing the overall AR coefficient $\phi_1 + \gamma_1\Theta(y_{t-1})$ as variable because the transition function is not constant; this aspect in now considered in further detail.

5.2.4   The ESTAR as a variable coefficient model

As noted, the exponential function of the deviation $y_{t-1} - c$ can be viewed as generating a transition variable $\Theta(y_{t-1})$, say Θ_t for economy of notation, which is applied to the coefficient, γ_1, on y_{t-1}. Thus, with $\gamma_1 \leq 0$ and $\Theta_t \geq 0$

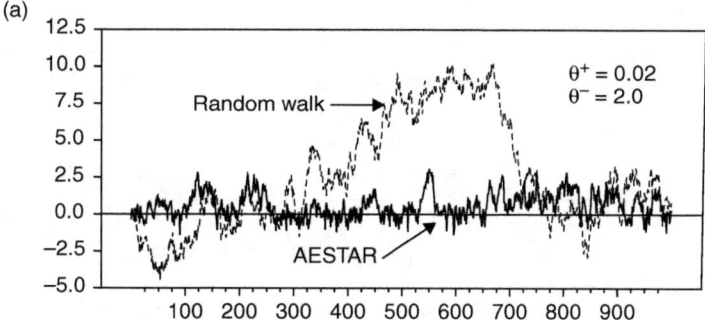

Figure 5.4a   AESTAR and random walk, $\theta^+ = 0.02, \theta^- = 2.0$

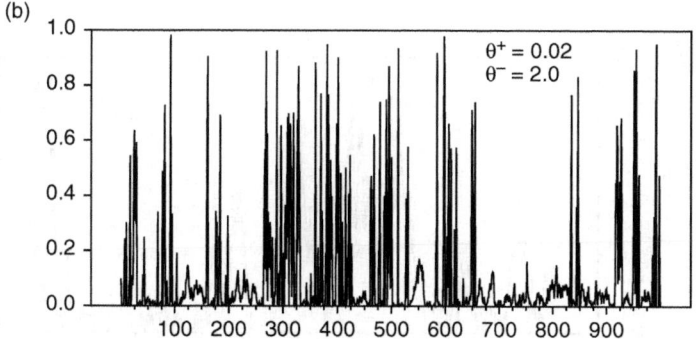

Figure 5.4b   AESTAR transition function, $\theta^+ = 0.02, \theta^- = 2.0$

then $\Theta_t \gamma_1 \le 0$, and the overall coefficient on y_{t-1} is $(\phi_1 + \Theta_t \gamma_1)$. At time t, the ESTAR(1) representation, with $d = 1$, emphasising this aspect is:

$$y_t = \phi_1 y_{t-1} + \gamma_1 y_{t-1}(1 - e^{-\theta(y_{t-1}-c)^2}) + \varepsilon_t \tag{5.6a}$$

$$= \phi_1 y_{t-1} + \gamma_1 \Theta_t y_{t-1} + \varepsilon_t \qquad \text{where } \Theta_t = (1 - e^{-\theta(y_{t-1}-c)^2}) \tag{5.6b}$$

$$= (\phi_1 + \gamma_1 \Theta_t) y_{t-1} + \varepsilon_t \tag{5.6c}$$

$$= \beta_t y_{t-1} + \varepsilon_t \qquad \text{where } \beta_t = \phi_1 + \gamma_1 \Theta_t \tag{5.6d}$$

$$\Delta y_t = \alpha_t y_{t-1} + \varepsilon_t \qquad \text{where } \alpha_t = (\phi_1 - 1) + \gamma_1 \Theta_t \tag{5.7}$$

The transition variable Θ_t is a continuous function of deviations from the inner regime (defined by $y_{t-1} = c$) bounded between 0 and 1; its role is to impose stationarity as $\Theta_t \to 1$. As Θ_t varies between 0 and 1, the coefficient $\beta_t = \phi_1 + \Theta_t \gamma_1$ ranges from the inner regime, where $\beta_t = \phi_1$, to the outer regime, where $\beta_t = \phi_1 + \gamma_1$.

(a)

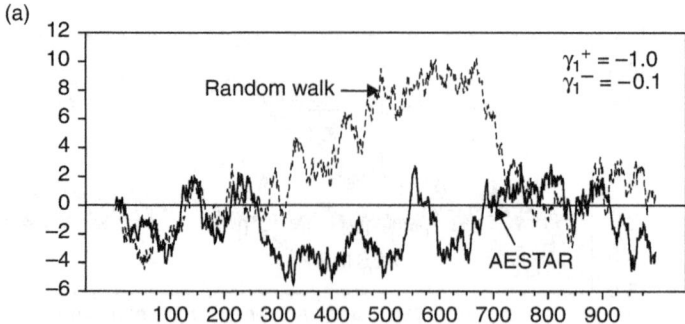

Figure 5.5a　AESTAR and random walk, $\gamma_1^+ = -1.0, \gamma_1^- = -0.1$

(b)

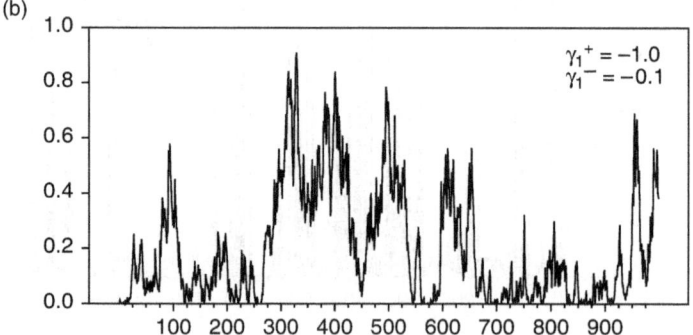

Figure 5.5b　AESTAR transition function, $\gamma_1^+ = -1.0, \gamma_1^- = -0.1$

By setting $\phi_1 = 1$, the ESTAR(1) model can allow for unit root behaviour in, or close to, the inner regime, that is, if y_{t-1} is in the vicinity of c. In this case $\Theta_t \approx 0$ and so the coefficient on y_{t-1} in (5.7), that is α_t, is (approximately) zero. At the other extreme, a substantial deviation in y_{t-1} leads to $\Theta_t \approx 1$, so that $\beta_t \approx \phi_1 + \gamma_1 = 1 + \gamma_1$ when $\phi_1 = 1$, and $\alpha_t \approx \gamma_1$, so that stationarity then requires $\gamma_1 < 0$. The inner regime could even generate explosive behaviour with $\phi_1 > 1$, provided $\phi_1 + \gamma_1 < 1$.

Figures 5.6a and 5.6b show the variable AR coefficient β_t corresponding to Figures 5.2 and 5.3, respectively. In the random walk case there is no variation in β_t, which is unity throughout. Comparing Figure 5.6a with Figure 5.6b, note how β_t is much more 'active' as θ increases. The (conditional) AR coefficient, β_t, shows, in a sense, how much work is being done to ensure mean reversion, especially so for $\theta = 2.0$, where the effect is to induce a 'corridor' of observations.

(a)

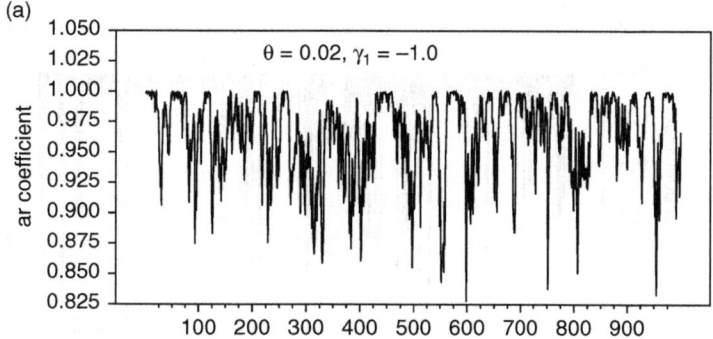

Figure 5.6a   ESTAR variable ar coefficient, $\beta_t, \theta = 0.02, \gamma_1 = -1.0$

(b)

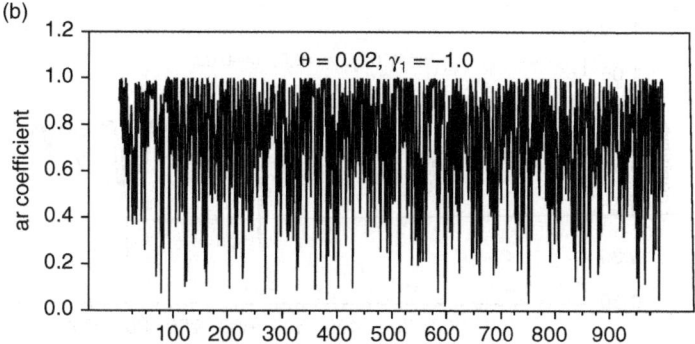

Figure 5.6b   ESTAR variable ar coefficient, $\beta_t, \theta = 2.0, \gamma_1 = -1.0$

Figures 5.7a and 5.7b show the variable AR coefficient β_t corresponding to Figures 5.4 and 5.5 respectively, which relate to the asymmetric ESTAR models. As a reminder, in Figure 5.7a, $\theta^+ = 0.02$, $\theta^- = 2.0$, so that nonlinear responses to negative deviations are much stronger than nonlinear responses to positive deviations; thus when the outer regime is reached this is primarily due to negative deviations. In Figure 5.7b, the asymmetry comes through different outer regimes.

5.2.5   An ESTAR(2) model

There are a number of possible generalisations of the ESTAR(1) model. The simplest is to an ESTAR(2) as follows:

$$y_t = \phi_1 y_{t-1} + \phi_2 y_{t-2} + (\gamma_1 y_{t-1} + \gamma_2 y_{t-2})(1 - e^{-\theta(y_{t-1}-c)^2}) + \varepsilon_t \qquad (5.8)$$

(a)

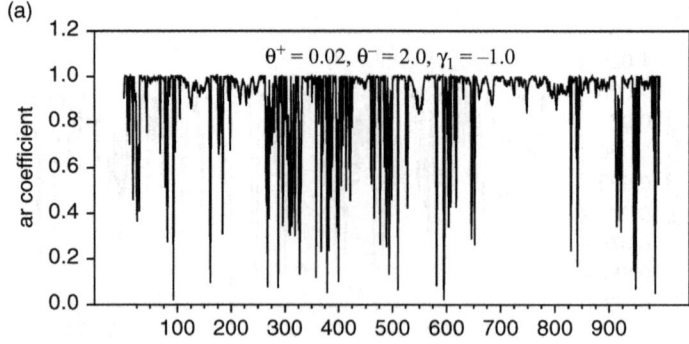

Figure 5.7a ESTAR variable ar coefficient, $\beta_t, \theta^+ = 0.02, \theta^- = 2.0, \gamma_1 = -1.0$

(b)

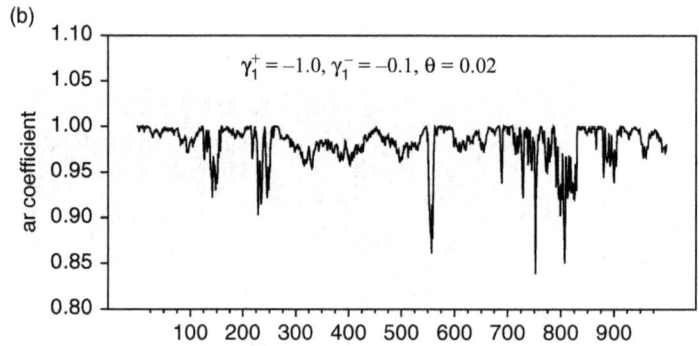

Figure 5.7b ESTAR variable ar coefficient, $\beta_t, \gamma_1^+ = -1.0, \gamma_1^- = -0.1, \theta = 0.02$

It is sometimes convenient to put this in ADF form as follows:

$$y_t = \phi_1 y_{t-1} + \phi_2 y_{t-2} + (\gamma_1 y_{t-1} + \gamma_2 y_{t-2})\Theta_t + \varepsilon_t \tag{5.9a}$$

where $\Theta_t = (1 - e^{-\theta(y_{t-1}-c)^2})$

$$= (\phi_1 + \phi_2)y_{t-1} + \delta_1 \Delta y_{t-1} + (\gamma_1 + \gamma_2)\Theta_t y_{t-1} + \eta_1 \Theta_t \Delta y_{t-1} + \varepsilon_t \tag{5.9b}$$

where $\delta_1 = -\phi_2$ and $\eta_1 = -\gamma_2$

$$= \rho y_{t-1} + \delta_1 \Delta y_{t-1} + \rho_t y_{t-1} + \eta_1 \Theta_t \Delta y_{t-1} + \varepsilon_t \tag{5.9c}$$

where $\rho = \phi_1 + \phi_2$ and $\rho_t = (\gamma_1 + \gamma_2)\Theta_t$

$$= \beta_t y_{t-1} + \delta_t \Delta y_{t-1} + \varepsilon_t \tag{5.9d}$$

where $\beta_t = \phi_1 + \phi_2 + (\gamma_1 + \gamma_2)\Theta_t$ and $\delta_t = -(\phi_2 + \gamma_2 \Theta_t)$

$\Rightarrow$

$$\Delta y_t = \alpha_t y_{t-1} + \delta_t \Delta y_{t-1} + \varepsilon_t \quad \text{where } \alpha_t = \beta_t - 1 \tag{5.10}$$

Now in the case of a unit root $\phi_1 + \phi_2 = 1$ and $\beta_t = 1 + (\gamma_1 + \gamma_2)\Theta_t$. If the outer regime, that is when $\Theta_t = 1$, is to 'contain' the nonstationarity in the inner regime, then we require $(\gamma_1 + \gamma_2) < 0$, so that $(\beta_t | \Theta_t = 1) < 1$.

5.2.6 The ESTAR(p) model

The further generalisation to an ESTAR(p) model is now straightforward and the steps are shown below:

$$y_t = \sum_{i=1}^{p} \phi_i y_{t-i} + \sum_{i=1}^{p} \gamma_i y_{t-i} \Theta_t + \varepsilon_t \tag{5.11a}$$

$$= \left(\sum_{i=1}^{p} \phi_i \right) y_{t-1} + \left(\sum_{i=1}^{p} \gamma_i \Theta_t \right) y_{t-1}$$

$$+ \sum_{j=1}^{p-1} \delta_j \Delta y_{t-j} + \sum_{j=1}^{p-1} \eta_j \Theta_t \Delta y_{t-j} + \varepsilon_t \tag{5.11b}$$

where $\delta_j = -\sum_{i=j+1}^{p} \phi_i$ and $\eta_j = -\sum_{i=j+1}^{p} \gamma_i$

$$= \left[\sum_{i=1}^{p} (\phi_i + \gamma_i \Theta_t) \right] y_{t-1} + \sum_{j=1}^{p-1} (\delta_j + \eta_j \Theta_t) \Delta y_{t-j} + \varepsilon_t \tag{5.11c}$$

$$= \beta_t y_{t-1} + \sum_{j=1}^{p-1} \delta_{tj} \Delta y_{t-j} + \varepsilon_t \tag{5.11d}$$

where $\beta_t = \sum_{i=1}^{p} (\phi_i + \gamma_i \Theta_t)$ and $\delta_{tj} = (\delta_j + \eta_j \Theta_t)$

$$\Delta y_t = \alpha_t y_{t-1} + \sum_{j=1}^{p-1} \delta_{tj} \Delta y_{t-j} + \varepsilon_t \quad \text{where } \alpha_t = \beta_t - 1 \tag{5.12}$$

Evidently, the expression in (5.12) is the ADF(p – 1) form of the ESTAR(p) model. Now in the case of a unit root $\sum_{i=1}^{p} \phi_i = 1$ and $\beta_t = 1 + \Theta_t \sum_{i=1}^{p} \gamma_i$. If the outer regime, that is when $\Theta_t = 1$, is to 'contain' the nonstationarity in the inner regime, then we require $-2 < \sum_{i=1}^{p} \gamma_i < 0$, so that $(|\beta_t; \Theta_t = 1)| < 1$. If there is a unit root in the inner regime and the outer regime restrictions enforce $y_t = \varepsilon_t$, then $\beta_t = e^{-\theta(y_{t-1}-c)^2}$.

Using a theorem of Tweedie (1975), Ozaki (1985) shows that the ESTAR model is an ergodic Markov chain if the roots in the outer regime lie outside the unit circle; this ergodicity follows even if the roots of the inner regime lie *inside* the unit circle. Intuitively, the stationarity in the outer regime ensures the 'shift back to centre' (or origin) property when $(y_{t-1}, y_{t-2}, \ldots, y_{t-p})$ are outside some finite region. Ergodicity implies that the ESTAR model defines a stationary process. The roots contain useful information about the ESTAR process and we consider them further below.

5.2.7   The ESTAR specification including a constant

The ESTAR model can be extended to include a constant that is also subject to the transition function. For example, the ESTAR(1) with constant is:

$$y_t = \phi_0 + \phi_1 y_{t-1} + (\gamma_0 + \gamma_1 y_{t-1})(1 - e^{-\theta(y_{t-1}-c)^2}) + \varepsilon_t \tag{5.13}$$

The inner and outer regimes are then:

$$y_t = \phi_0 + \phi_1 y_{t-1} + \varepsilon_t \tag{5.14}$$

$$y_t = (\phi_0 + \gamma_0) + (\phi_1 + \gamma_1) y_{t-1} + \varepsilon_t \tag{5.15}$$

Assuming stationarity in the outer regime, then $|\phi_1 + \gamma_1| < 1$, and if the outer regime is sustained then, apart from noise, $y_t = (\phi_0 + \gamma_0)/(1 - \phi_1 - \gamma_1)$.

If there is a unit root in the inner regime, $\phi_1 = 1$, then the presence of ϕ_0 also implies drift in the random walk. This can be avoided by specifying the constant such that $\phi_0 \to 0$ as $\phi_1 \to 1$. For example, an alternative specification builds on the idea that the ESTAR is operating on the deviation from threshold, that is $y_{t-1} - c$, so that:

$$(y_t - c) = \phi_1(y_{t-1} - c) + \gamma_1(y_{t-1} - c)(1 - e^{-\theta(y_{t-1}-c)^2}) + \varepsilon_t \tag{5.16}$$

The inner and outer regimes are now:

$$(y_t - c) = \phi_1(y_{t-1} - c) + \varepsilon_t \qquad \text{inner regime} \tag{5.17}$$

$$(y_t - c) = \phi_1(y_{t-1} - c) + \gamma_1(y_{t-1} - c) + \varepsilon_t$$

$$= (\phi_1 + \gamma_1)(y_{t-1} - c) + \varepsilon_t \qquad \text{outer regime} \tag{5.18}$$

This specification follows from equation (5.13) if $\phi_0 = (1 - \phi_1)c$ and $\gamma_0 = -\gamma_1 c$; then as $\phi_1 \to 1$, $\phi_0 \to 0$. It is of relevance to note that the deviation from threshold specification imposes two restrictions on the general specification.

5.2.8   Trended data

The specification of the 'driving' force in STAR models, and ESTAR models in particular, as $y_{t-1} - c$, that is, the deviation from a threshold modelled as a constant, suggests that y_t is not trended or has been detrended, otherwise it will be continually moving away from any particular constant. Two approaches based on what is done in linear models have also been applied to nonlinear models, even though the analogy is not perfect. One approach is to detrend the data by prior regression on a constant and a deterministic time trend, thus using the residuals from such a regression in the ESTAR model. Also, since in linear models a deterministic polynomial trend of order s is reduced by one on differencing once, a linear trend can be removed by first (or, for example, fourth) differencing of the data. An alternative is to incorporate the trend directly in the ESTAR specification.

5.2.9 Standardisation of the transition variable in the ESTAR model

To assist an interpretation of the relative magnitude of changes (deviations), it can be helpful to standardise the variable in the exponent of the ESTAR model. The change is to put the 'driving' variable in units of the standard deviation of y_t. Thus, the basic variable becomes $(y_{t-d} - c)/s_y$, where s_y is the sample standard deviation of y_t; as the deviation variable in an ESTAR model is squared, so that sign does not matter, this becomes $((y_{t-d} - c)/s_y)^2$. The idea is that variables that might have quite different scales can be compared in terms of units of standard deviation. The ESTAR transition function is now written as:

$$\Theta(y_{t-d}) = (1 - e^{-\theta^*[(y_{t-d}-c)/s_y]^2}) \tag{5.19}$$

where, to maintain consistency with the original interpretation in (5.2), $\theta^* = \theta s_y^2$, since typically with economic time series (especially where they are in logs) $s_y^2 < 1$ so that $\theta^* < \theta$.

5.2.10 Root(s) and ergodicity of the ESTAR model

In the case of an AR(p) model, it is a simple enough matter to obtain the characteristic polynomial and, hence, the roots of the process. In the case of an ESTAR model, the situation is more complex and to understand it we start with an ESTAR(1) model. At the simplest level it is usually informative to at least obtain the roots in the inner and outer regimes, which are denoted λ_0 and λ_∞ respectively.

5.2.10.i ESTAR(1)

In an ESTAR(1) model the roots are simply $\lambda_0 = \phi_1$ in the inner regime and $\lambda_\infty = \phi_1 + \gamma_1$ in the outer regime; however, there is more information in the ESTAR process. Viewing (5.6d) as an AR(1) process with variable AR coefficient, the root, conditional on y_{t-1}, will also be variable; this is of interest in itself and as an indicator of persistence of the process, with the process becoming more persistent as the root approaches unity.

To obtain the root conditional on the realisation of y_{t-1}, use is made of $\beta_t = \phi_1 + \Theta_t \gamma$ in (5.6d), which is the 'quasi' AR(1) way of writing the ESTAR(1) model. First, write:

$$\Theta_t = (1 - e^{-\theta(y_{t-1}-c)^2})$$
$$= (1 - \lambda_t)$$

so that $\lambda_t = e^{-\theta(y_{t-1}-c)^2}$; then rewriting (5.6d) to make the lag polynomial explicit, we have: $[1 - (\phi_1 + (1 - \lambda_t)\gamma_1)L]y_t = \varepsilon_t$, and, thus, the AR(1) lag polynomial is $[1 - (\phi_1 + (1 - \lambda_t)\gamma_1)L]$, say $f(L)$. The polynomial in this form is the reverse characteristic polynomial, and the root is the value of z, say z_1, such that $f(z_1) = 0$; in this case $z_1 = (\phi_1 + (1 - \lambda_t)\gamma_1)^{-1}$. Stability requires that

$|z_1| > 1$. Alternatively, but equivalently, we can focus on the characteristic polynomial $\lambda - (\phi_1 + (1 - \lambda_t)\gamma_1)$, with root $\lambda_1 = (\phi_1 + (1 - \lambda_t)\gamma_1)$ and stability condition $|\lambda_1| < 1$. (See *UR, Vol. 1* for more on the roots of a lag polynomial.)

5.2.10.ii   The conditional root in the unit root plus white noise case

The case of a unit root in the inner regime and white noise in the outer regime is often of interest. This corresponds to $\phi_1 = 1$ and $\gamma_1 = -1$, so that $y_t = \varepsilon_t$ in the outer regime. The root of the lag polynomial in this case is $z_1 = (\phi_1 + (1 - \lambda_t)\gamma_1)^{-1} = (\lambda_t)^{-1} = (e^{-\theta(y_{t-1}-c)^2})^{-1} = e^{\theta(y_{t-1}-c)^2} > 1$ for $(y_{t-1} - c) \neq 0$ and $z_1 = 1$ for $(y_{t-1} - c) = 0$. In terms of the root of the characteristic polynomial, this is $\lambda_1 = \lambda_t = e^{-\theta(y_{t-1}-c)^2}$ and, thus, $\lambda_t < 1$ for $(y_{t-1} - c) \neq 0$. The root is a function of θ and $(y_{t-1} - c)^2$, with less persistence for a given deviation, $(y_{t-1} - c)$, as θ increases.

5.2.10.iii   ESTAR(2)

To obtain the lag polynomial for the ESTAR(2) model, following on from (5.8), we have:

$$y_t = \phi_1 y_{t-1} + \phi_2 y_{t-2} + (\gamma_1 y_{t-1} + \gamma_2 y_{t-2})\Theta_t + \varepsilon_t \tag{5.20a}$$

$$= (\phi_1 + \gamma_1 \Theta_t)y_{t-1} + (\phi_2 + \gamma_2 \Theta_t)y_{t-2} + \varepsilon_t \tag{5.20b}$$

$$y_t[1 - (\phi_1 + \gamma_1 \Theta_t)L - (\phi_2 + \gamma_2 \Theta_t)L^2] = \varepsilon_t \tag{5.21}$$

The characteristic polynomial for the term in square brackets is $\lambda^2 - (\phi_1 + \gamma_1 \Theta_t)\lambda - (\phi_2 + \gamma_2 \Theta_t)$, and the dominant root provides useful information on the behaviour of the ESTAR model away from the inner and outer regimes. In some special cases the characteristic polynomial simplifies, but otherwise it would have to be solved for each given realisation of Θ_t.

The lag polynomials in the inner and outer regimes give rise to the following characteristic polynomials to solve for the roots (which may now be complex). The inner regime roots are denoted $\lambda_0, \bar{\lambda}_0$ and the outer regime roots are denoted $\lambda_\infty, \bar{\lambda}_\infty$.

	Inner regime	Outer regime
Model:	$y_t = \phi_1 y_{t-1} + \phi_2 y_{t-2}$	$y_t = (\phi_1 + \gamma_1)y_{t-1} + (\phi_2 + \gamma_2)y_{t-2}$
Lag polynomial:	$(1 - \phi_1 L - \phi_2 L^2)$	$(1 - (\phi_1 + \gamma_1)L - (\phi_2 + \gamma_2)L^2)$
Characteristic Polynomial:	$(\lambda^2 - \phi_1 \lambda - \phi_2) = 0$	$(\lambda^2 - \varphi_1 \lambda - \varphi_2) = 0$
		$\varphi_1 = \phi_1 + \gamma_1, \varphi_2 = \phi_2 + \gamma_2$
Roots:	$\lambda_0, \bar{\lambda}_0$	$\lambda_\infty, \bar{\lambda}_\infty$
Ergodic if:		$|\lambda_\infty| < 1, |\bar{\lambda}_\infty| < 1$

5.2.11 Existence and stability of stationary points

5.2.11.i Equilibrium of the STAR process

Another aspect of interest is to find the stable stationary point of the process, if it exists – see Ozaki (1985). The ESTAR model is a particular example of a nonlinear autoregressive model that can be written generally as $y_t = G(y_{t-1}, \ldots, y_{t-p}; \Lambda) + \varepsilon_t$, where Λ is the set of parameters. The deterministic part of the model is $G(y_{t-1}, \ldots, y_{t-p}; \Lambda)$, and an equilibrium is $y^* = G(y^*; \Lambda)$. The interpretation of equilibrium is that if y^* is reached by the process, then subsequent values of y will also be y^*; thus, y^* is self-replicating.

Consider an ESTAR(2) model with $c = 0$, that is:

$$y_t = \phi_1 y_{t-1} + \phi_2 y_{t-2} + (\gamma_1 y_{t-1} + \gamma_2 y_{t-2})(1 - e^{-\theta y_{t-1}^2}) + \varepsilon_t \tag{5.22a}$$

$$= (\varphi_1 + \pi_1 e^{-\theta y_{t-1}^2}) y_{t-1} + (\varphi_2 + \pi_2 e^{-\theta y_{t-1}^2}) y_{t-2} + \varepsilon_t \tag{5.22b}$$

where $\varphi_1 = \phi_1 + \gamma_1$, $\varphi_2 = \phi_2 + \gamma_2$, $\pi_1 = -\gamma_1$ and $\pi_2 = -\gamma_2$. Disregarding ε_t, the definition of equilibrium implies that if y^* exists it must satisfy the following equation:

$$y^* = (\varphi_1 + \pi_1 e^{-\theta(y^*)^2}) y^* + (\varphi_2 + \pi_2 e^{-\theta(y^*)^2}) y^* \tag{5.23}$$

This definition implies that y^* is self-replicating. Clearly trivially $y^* = 0$ satisfies (5.23), so what we are referring to here is whether there exists a solution of (5.23) such that $y^* \neq 0$. This gives rise first to a condition for the existence of a solution and second to the solution, if it exists.

The existence conditions for the ESTAR(2) model and ESTAR(p) models, respectively, are:

$$0 < \frac{(1 - \varphi_1 - \varphi_2)}{\pi_1 + \pi_2} < 1 \tag{5.24}$$

$$0 < \frac{(1 - \sum_{i=1}^{p} \varphi_i)}{\sum_{i=1}^{p} \pi_i} < 1 \qquad \text{in general} \tag{5.25}$$

The lower limit 0 arises because it is not possible to take the log of a negative number in the solution below, and the upper limit arises because θ times $(y^*)^2$ is positive, hence $-\theta$ times $(y^*)^2$ is negative and the log of a number less than 1 is negative.

If the existence condition is satisfied then the solution is given by:

$$y^* = \pm \theta^{-1/2} \left[-\ln \left(\frac{(1 - \varphi_1 - \varphi_2)}{\pi_1 + \pi_2} \right) \right]^{1/2} \tag{5.26}$$

$$y^* = \pm \theta^{-1/2} \left[\ln \left(\frac{(1 - \sum_{i=1}^{p} \varphi_i)}{\sum_{i=1}^{p} \pi_i} \right) \right]^{1/2} \qquad \text{in general} \tag{5.27}$$

5.2.11.ii   *Stability of singular points and limit cycles*

A singular point, or equilibrium in these terms, may not be stable. The singular point y^* is defined as stable if it is approached as $t \to \infty$. An example of an unstable equilibrium is when a limit cycle is generated even though y_t is initially near y^* but receives a shock. It is possible that neither stable singular points nor a stable limit cycle exists and, in this case, the model is sometimes referred to as generating *chaos*.

A limit cycle of period r involves a repetition of different values of y_i, $i = 1,\ldots,p$, at an interval of r time units. For example, if $p = 2$, and y_1 and y_2 are in a cycle of period r, then $y_1 = y_{1+r}$ and $y_2 = y_{2+r}$. By extension to the general case, the cycle implies that $(y_{1+jr}, y_{2+jr}, \ldots, y_{p+jr}) = (y_1, y_2, \ldots, y_p)$ for j an integer. A limit cycle generates exact repetitions of the sequence $\{y_i\}_{j=1}^{p}$. By extension, a limit cycle can be stable or unstable. It is possible to derive analytical conditions to assess these stability issues; see Ozaki (1985, especially propositions 2.1 and 2.2, but for the special case with $\theta = 1$). It is also practically useful to simulate the models to obtain a 'picture' of what is happening. Some different characteristics are illustrated next with two examples.

Example 1: The ESTAR(2) model is given by:

$$y_t = 1.5y_{t-1} - 0.4y_{t-2} - (0.8y_{t-1} - 0.1y_{t-2})(1 - e^{-0.5y_{t-1}^2}) + \varepsilon_t$$

$$= (0.7 + 0.8e^{-0.5y_{t-1}^2})y_{t-1} - (0.3 + 0.1e^{-0.5y_{t-1}^2})y_{t-2} + \varepsilon_t \qquad (5.28)$$

Inner regime:	$y_t = 1.5y_{t-1} - 0.4y_{t-2} + \varepsilon_t$
Outer regime:	$y_t = 0.7y_{t-1} - 0.3y_{t-2} + \varepsilon_t$
Roots:	$\lambda_0 = 1.153, \bar{\lambda}_0 = 0.347;$
	$\lambda_\infty = 0.35 + 0.421i, \bar{\lambda}_\infty = 0.35 - 0.421i$
	$|\lambda_\infty| = |\bar{\lambda}_\infty| = 0.5477 < 1$

Note that the dominant root in the inner regime is 1.153, which is, therefore, not stationary, but the roots (a complex pair) in the outer regime have modulus less than one; thus the ESTAR(2) process is ergodic and stationary. It is of interest to know whether this model generates a singular point and whether that singular point is stable.

First, we verify that the existence condition is satisfied. The relevant coefficients are $\varphi_1 = \phi_1 + \gamma_1 = 1.5 - 0.8 = 0.7$, $\varphi_2 = \phi_2 + \gamma_2 = -0.4 + 0.1 = -0.3$ and $\pi_1 = 0.8$, $\pi_2 = -0.1$. Hence, the following are obtained:

$$\frac{(1 - \varphi_1 - \varphi_2)}{\pi_1 + \pi_2} = \frac{(1 - 0.7 - (-0.3))}{0.8 - 0.1} = \frac{0.6}{0.7} < 1$$

with singular points given by:

$$y^* = \pm\theta^{-1/2}\left[-\ln\left(\frac{(1-\varphi_1-\varphi_2)}{\pi_1+\pi_2}\right)\right]^{1/2}$$

$$= \pm\left[-\ln\left(\frac{(1-0.7-(-0.3))}{0.8-0.1}\right)\Big/0.5\right]^{1/2}$$

$$= \pm0.555$$

Economic circumstances will usually determine the appropriate sign of y^* and here, for illustration, we take the positive solution. To check that this is indeed a solution, set $\varepsilon_t = 0$ in (5.28) and substitute $y^* = 0.555$ for y_t and its lags on the right-hand side of the equation. This value is self-replicating, as required:

$$y^* = (0.7+0.8e^{-0.5(0.555)^2})0.555 - (0.3+0.1e^{-0.5(0.555)^2})0.555 = 0.555 \quad (5.29)$$

The model is dynamically simulated and two cases are considered. In each case, the initial equilibrium is shocked; in case A, y_t is initially shocked to 2, whereas in case B the shock takes y_t to 2.4. These two cases show how it is possible to reach different equilibria depending on the shock. In case A, the equilibrium is reached is $y^* = 0.555$, whereas in case B it is $y^* = -0.555$; see Figure 5.8.

Example 2 This example, which is a variation on Ozaki (1985, equation 2.39), generates a stable limit cycle. Compared to the first example, this model has a

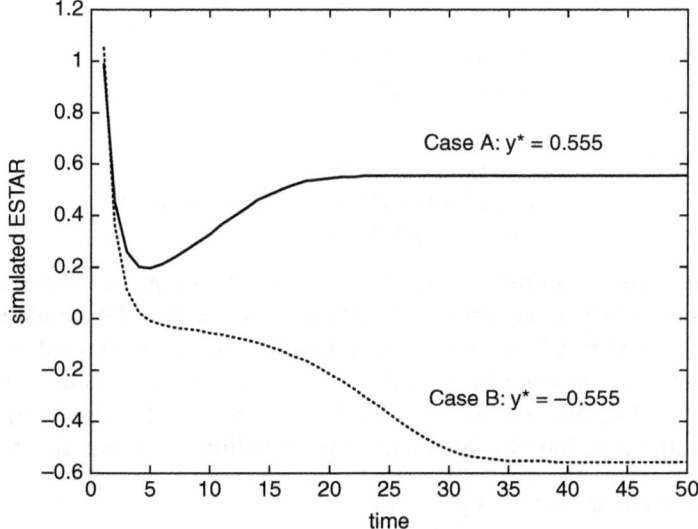

Figure 5.8   ESTAR(2) simulation, multiple equilibria

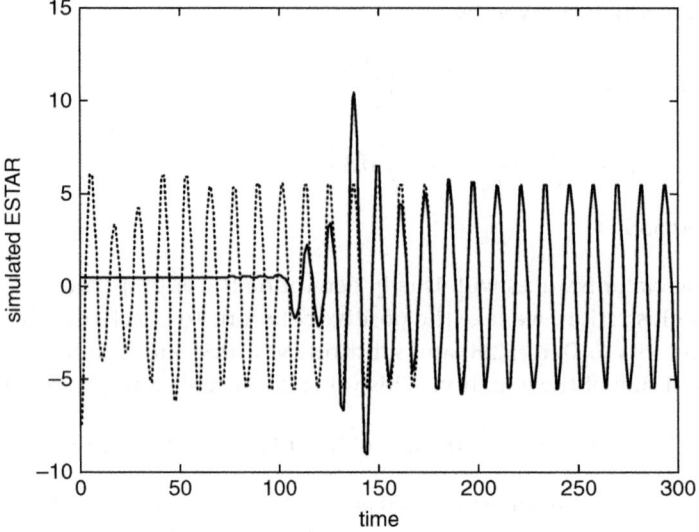

Figure 5.9   ESTAR(2) simulation, stable limit cycle

much sharper adjustment to deviations from the implied threshold of 0. The model is:

$$y_t = 4.57y_{t-1} - 2.83y_{t-2} - (3.0y_{t-1} - 2.0y_{t-2})(1 - e^{-5y_{t-1}^2}) + \varepsilon_t$$

$$= (1.57 + 3.0e^{-5y_{t-1}^2})y_{t-1} - (0.83 + 2.0e^{-5y_{t-1}^2})y_{t-2} + \varepsilon_t \qquad (5.30)$$

Inner regime: $y_t = 4.57y_{t-1} - 2.83y_{t-2} + \varepsilon_t$
Outer regime: $y_t = 1.57y_{t-1} - 0.83y_{t-2} + \varepsilon_t$

$y^* = 0.519$

Roots: $\lambda_0 = 3.831, \bar{\lambda}_0 = 0.739;$
$\lambda_\infty = 0.785 + 0.462i, \bar{\lambda}_\infty = 0.785 - 0.4624i$
$|\lambda_\infty| = |\bar{\lambda}_\infty| = 0.911 < 1$

The inner regime is nonstationary, whereas the outer regime is stationary. This model generates a stable limit cycle with period $p = 12$. The simulated model is shown in Figure 5.9, where two simulation paths are generated; in one the model is initially in equilibrium with $y_t = y^* = 0.519$, but a stable limit cycle is eventually self-generated; in the second, there is an initial shock ($y_1 = y_2 = -10$) and, after a period of adjustment, the stable limit cycle is generated.

5.2.11.iii   *Some special cases of interest*

It is often quite difficult to empirically identify an unrestricted ESTAR(p) model. Problems include those of convergence and coefficient estimates that are outside reasonably anticipated bounds. It is, therefore, quite useful to see if some special but interesting cases are as acceptable as the unrestricted model in terms of no loss in fit, but result in more readily interpretable coefficient estimates. Two interesting cases relate, respectively, to the outer regime and the inner regime; the combination of both cases leads to a third special case.

Case 1 $y_t = \varepsilon_t$ in the outer regime.
The restrictions that ensure this are: $\gamma_i = -\phi_i$, which implies $\varphi_i = 0$, for all i, so $y_t = \varepsilon_t$ in the outer regime; the existence condition is then $\sum_{i=1}^{p} \pi_i > 1$; but, under the restrictions $\gamma_i = -\phi_i$ this implies

$$\sum_{i=1}^{p} \pi_i = -\sum_{i=1}^{p} \gamma_i = \sum_{i=1}^{p} \phi_i > 1$$

that is, an explosive inner regime for $y^* \neq 0$. If $\sum_{i=1}^{p} \phi_i < 1$, then the solution of the ESTAR(p) model, as in (5.11), is $y^* = 0$. If the ESTAR model has been formulated in terms of the deviations variable, $y_t - c$, then the zero solution corresponds to $y^* = c$.

Case 2 A unit root in the inner regime.
A unit root in the inner regime implies that $\sum_{i=1}^{p} \phi_i = 1$; hence,

$$1 - \sum_{i=1}^{p} \phi_i = 1 - \sum_{i=1}^{p} (\phi_i + \gamma_i) = -\sum_{i=1}^{p} \gamma_i$$

but $\sum_{i=1}^{p} \pi_i = -\sum_{i=1}^{p} \gamma_i$, therefore, $(1 - \sum_{i=1}^{p} \varphi_i)(\sum_{i=1}^{p} \pi_i)^{-1} = 1$. The existence condition for a nonzero solution is not satisfied, hence only $y^* = 0$ satisfies (5.23) or its p-th order equivalent. This is not as restrictive as it seems. Again, if the ESTAR model has been formulated in terms of the variable $y_t - c$, then the zero solution corresponds to $y^* = c$.

Case 3 $y_t = \varepsilon_t$ in the outer regime and a unit root in the inner regime.
It is sometimes useful to consider combining the two previous cases; this is partly because estimation of ESTAR models sometimes encounters difficulties empirically identifying the γ_i coefficients, the objective function being quite 'flat' for variations in the estimated values, $\hat{\gamma}_i$, of γ_i. A set of restrictions which may serve to empirically identify these coefficients is motivated by the view that the outer regime is characterised by white noise, whilst the inner regime has a unit root. To this end, set $\gamma_i = -\phi_i$ and combine this with the unit root case, $\sum_{i=1}^{p} \phi_i = 1$; hence, the restrictions imply $\sum_{i=1}^{p} \gamma_i = -1$. Stationarity in the outer regime requires $\sum_{i=1}^{p} \gamma_i < 0$, but this is clearly satisfied because $\sum_{i=1}^{p} \gamma_i = -\sum_{i=1}^{p} \phi_i = -1 < 0$.

5.3   LSTAR transition function

The ESTAR transition function has been a popular choice in recent work on modelling possible nonlinearities in the real exchange rate. Since deviations from c, the threshold parameter, are squared, the response in the standard ESTAR, which is captured by the transition weight, is the same for equally-sized positive and negative deviations. An alternative choice that does not impose such equality of response is the logistic transition function given by:

LSTAR transition function

$$\Psi(y_{t-d}) = \frac{1}{1 + e^{-\Psi(y_{t-d}-c)}} \quad \text{or} \tag{5.31}$$

$$\Psi_C(y_{t-d}) = \frac{1}{1 + e^{-\Psi(y_{t-d}-c)}} - 0.5 \tag{5.32}$$

where $\psi > 0$. The centred logistic transition function is $\Psi_C(y_{t-d})$ in (5.32). The reason for subtracting 0.5 is that $\Psi(y_{t-d})$ is centred on 0.5, that is $\Psi(y_{t-d}) = 0.5$ for $y_{t-d} = c$, whereas the interpretation of the transition function as generating transition weights is aided by having $\Psi_C(y_{t-d}) = 0$ for $y_{t-d} = c$. Some relevant properties of $\Psi(y_{t-d})$ are $\Psi(-\infty) = 0$, $\Psi(0) = 0.5$, $\Psi(\infty) = 1$; and for $\Psi_C(y_{t-d})$ these imply $\Psi_C(-\infty) = -0.5$, $\Psi_C(0) = 0$, $\Psi_C(\infty) = 0.5$. Which interpretation is chosen is a matter of convenience not substance.

In terms of $\Psi(y_{t-d})$, the weights increase from 0 to 1 and are centred on 0.5, whereas the transition weights generated by $\Psi_C(y_{t-d})$ are monotonically increasing from –0.5 to 0.5, being centred on 0; note that there is a symmetry of a kind in the transition weights as $\Psi_C(\delta) = -\Psi_C(-\delta)$. The transition function $\Psi_C(y_{t-d})$ is illustrated for two cases, $\psi = 0.75$ and $\psi = 2$, in Figure 5.10. As ψ increases the transition function soon approaches a switching function, with quite small deviations inducing a move to one limit or the other of the logistic function.

5.3.1 Standardisation of the transition variable in the LSTAR model

As in the ESTAR model, it can be helpful to standardise the deviations by the sample standard deviation, s_y, of y_t. That is, use $(y_{t-d} - c)/s_y$ rather than $(y_{t-d} - c)$ in the exponent of the logistic function and, to maintain consistency with the original specification, the logistic coefficient is then $\psi^* = \psi s_y$. In this revised form, the transition functions are:

$$\Psi(y_{t-d}) = \frac{1}{1 + e^{-\psi^*(y_{t-d}-c)/s_y}} \tag{5.33}$$

and

$$\Psi_C(y_{t-d}) = \frac{1}{1 + e^{-\psi^*(y_{t-d}-c)/s_y}} - 0.5 \tag{5.34}$$

In the case of economic time series data s_y is typically quite small, to illustrate, say $s_y = 1/20$ and $\psi = 40$, then $\psi^* = 40/20 = 2$ and $[1 + e^{-2(y_{t-d}-c)/(1/20)}] = [1 + e^{-2*20(y_{t-d}-c)}]$.

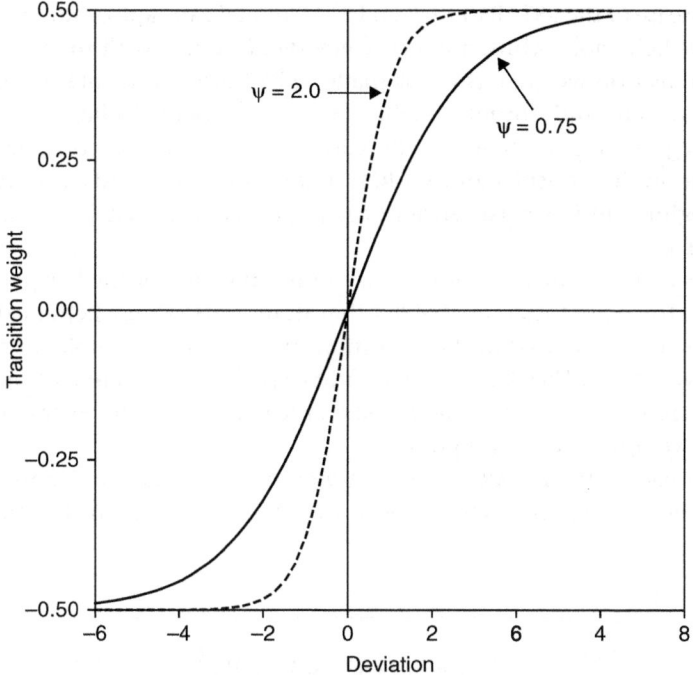

Figure 5.10   LSTAR transition function

5.3.2   The LSTAR model

Substitution of the logistic transition function, $\Psi(y_{t-d})$, into the STAR model gives rise to the LSTAR model; for example, the LSTAR(1) is:

$$y_t = \phi_1 y_{t-1} + \gamma_1 y_{t-1}[(1 + e^{-\Psi(y_{t-d}-c)})^{-1}] + \varepsilon_t \tag{5.35}$$

Unlike the ESTAR model there are three regimes in the LSTAR model: the lower regime for $y_{t-d} - c \rightarrow -\infty$, the middle regime for $y_{t-d} = c$ and the upper regime for $y_{t-d} - c \rightarrow +\infty$. These are as follows:

Regime

$$\text{Lower}: \ y_t = \phi_1 y_{t-1} + \varepsilon_t \qquad\qquad y_{t-d} - c \rightarrow -\infty; \Psi(-\infty) = 0 \tag{5.36}$$

$$\text{Middle}: \ y_t = \phi_1 y_{t-1} + 0.5\gamma_1\, y_{t-1} + \varepsilon_t \ \ y_{t-d} = c; \Psi(0) = 0.5$$

$$= (\phi_1 + 0.5\gamma_1) y_{t-1} + \varepsilon_t \tag{5.37}$$

$$\text{Upper}: \ y_t = \phi_1 y_{t-1} + \gamma_1 y_{t-1} + \varepsilon_t \qquad y_{t-d} - c \rightarrow +\infty; \Psi(\infty) = 1$$

$$= (\phi_1 + \gamma_1) y_{t-1} + \varepsilon_t \tag{5.38}$$

As in the case of the ESTAR model, an LSTAR model can capture unit root or even explosive behaviour whilst still being ergodic. Coupled with the monotonicity of the transition weights, this has made LSTAR models popular for modelling unemployment and output variables where an essential characteristic is different degrees of persistence in the contraction and expansion phases of the economic cycle. Consider an LSTAR(1) model with $d = 1$ and a unit root in the lower regime but less persistence in the upper regime, then $\phi_1 = 1$ and $\gamma_1 < 0$, so that $1 + \gamma_1 < 1$.

In this context, the deviation $(y_{t-d} - c)$ is a measure of the 'deepness' of the cycle. To the extent that a variable other than y_t itself may better capture this deepness, it can be used in the transition function. For example, in modelling the growth rate of US GDP, Beaudry and Koop (1993) and van Dijk and Franses (1999) define deepness measures based on the maximum difference of past levels of log GNP from a reference point.

The impact of the LSTAR transition function is illustrated in Figures 5.11 and 5.12, where each figure has two panels, a and b. The DGP for the LSTAR(1) model

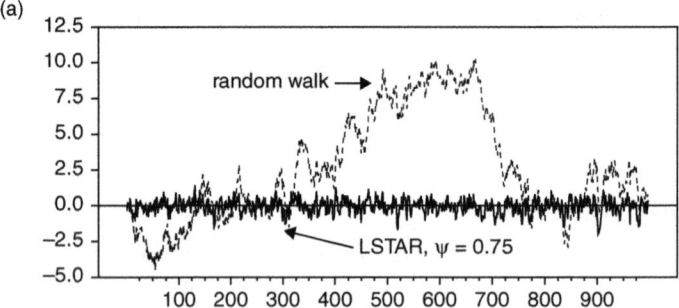

Figure 5.11a   LSTAR and random walk, $\psi = 0.75$

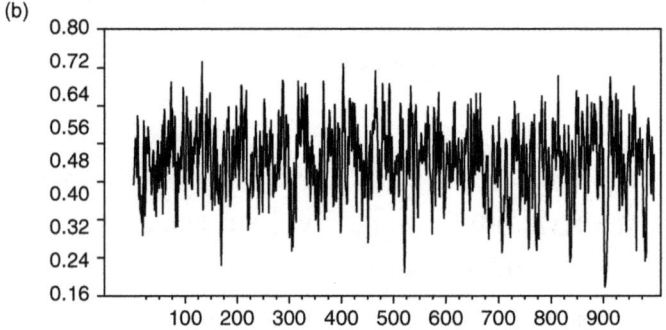

Figure 5.11b   LSTAR transition function, $\psi = 0.75$

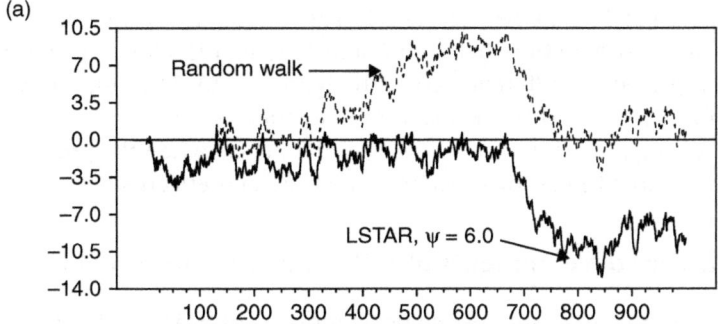

Figure 5.12a   LSTAR and random walk, $\psi = 6.0$

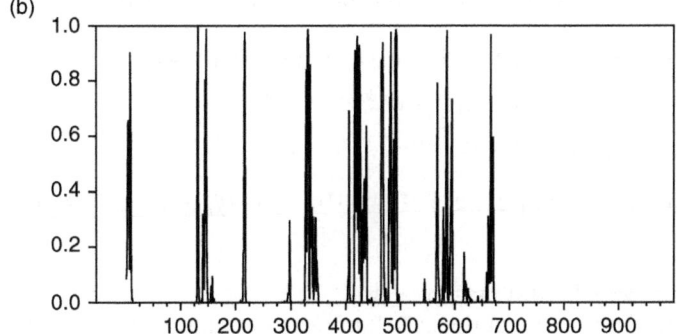

Figure 5.12b   LSTAR transition function, $\psi = 6.0$

is as in (5.35) with $\phi_1 = 1$, $\gamma_1 = -1$ and $\psi = 0.75$ in Figure 5.11 and $\psi = 6.0$ in Figure 5.12. (The innovation sequence $\{\varepsilon_t\}$, where $\varepsilon_t \sim$ niid(0, 0.16), is kept the same for each case.) In Figures 5.11a and 5.12a, there are graphs of two time series, one generated by a random walk and the other by an LSTAR(1) model, whereas Figures 5.11b and 5.12b show the corresponding values of the LSTAR transition function, $\Psi(y_{t-1})$.

Considering the case of $\psi = 0.75$, illustrated in Figures 5.11a and 5.11b, the first point to note is that the transition function is very effective at avoiding the random walk: why? Recall that $\Psi(0) = 0.5$, in which case the LSTAR adjustment, with $\gamma_1 = -1$, is $0.5\gamma_1 y_{t-1} = -0.5y_{t-1}$ and hence the LSTAR data is generated by $y_t = y_{t-1} - 0.5y_{t-1} + \varepsilon_t = 0.5y_{t-1} + \varepsilon_t$, thus the data is generated by a clearly stationary process. The second panel is informative, indicating that the transition function is quite 'active' in the centre around $\Psi(y_{t-1}) = 0.4$ to 0.6, which, in turn, holds back the random walk.

In the case of $\psi = 6.0$ (see Figures 5.12a and 5.12b) a larger value of ψ induces what is close to switching between 0 and 1, which is shown in the transition function in Figure 5.12b (contrast with Figure 5.11b). The random walk is sustained for $\Psi(-\infty) = 0$ but white noise is generated for $\Psi(\infty) = 1$. There are relatively long periods toward the end of the period when the process is clearly random walking in the lower regime and $\Psi(y_{t-1})$ is effectively zero.

5.4   Further developments of STAR models: multi-regime models

There are a number of interesting developments of STAR models that have not yet been considered. We briefly review some of these. So far the STAR models only allow two regimes, referred to as an inner and an outer regime, but the general idea is more flexible than this. Consider the following 3-regime LSTAR(1) model, with two logistic transition functions denoted $\Psi_1(y_{t-1})$ and $\Psi_2(y_{t-1})$:

$$y_t = \phi_1 y_{t-1} + (\phi_2 - \phi_1)y_{t-1}\Psi_1(y_{t-1}) + (\phi_3 - \phi_2)y_{t-1}\Psi_2(y_{t-1}) + \varepsilon_t \qquad (5.39a)$$

$$= \phi_1 y_{t-1} + (\phi_2 - \phi_1)y_{t-1}(1 + e^{-\psi_1(y_{t-1} - c_1)})^{-1}$$

$$+ (\phi_3 - \phi_2)y_{t-1}(1 + e^{-\psi_2(y_{t-1} - c_2)})^{-1} + \varepsilon_t \qquad (5.39b)$$

$$= \phi_1 y_{t-1} + \gamma_{11} y_{t-1}(1 + e^{-\psi_1(y_{t-1} - c_1)})^{-1} + \gamma_{12} y_{t-1}(1 + e^{-\psi_2(y_{t-1} - c_2)^2})^{-1} + \varepsilon_t \qquad (5.39c)$$

where $\gamma_{11} = \phi_2 - \phi_1$ and $\gamma_{12} = \phi_3 - \phi_2$.

Recall that in the case of the LSTAR transition function $\Psi(-\infty) = 0$, $\Psi(0) = 0.5$ and $\Psi(\infty) = 1$. Now for simplicity of interpretation assume $c_1 < c_2$, and consider a time path for y_{t-1} such that initially $y_{t-1} < c_1$ and then y_{t-1} increases moving through c_1 such that $y_{t-1} > c_1$. Then the nonlinear adjustment is (primarily) driven by $\Psi_1(y_{t-1})$ and, ceteris paribus, the AR(1) coefficient changes from ϕ_1 to $\phi_1 + \gamma_{11} = \phi_2$. As y_{t-1} increases, approaching c_2 then moving through c_2 such that $y_{t-1} > c_2$, the nonlinear adjustment is now (primarily) driven by $\Psi_2(y_{t-1})$ and, ceteris paribus, the AR(1) coefficient changes from ϕ_2 to $\phi_1 + \gamma_{11} + \gamma_{12} = \phi_1 + (\phi_2 - \phi_1) + (\phi_3 - \phi_2) = \phi_3$. Thus, three regimes are defined by this model, corresponding to AR(1) coefficients of ϕ_1, ϕ_2 and ϕ_3. The number of regimes can be extended further, for details see van Dijk and Franses (1999) and van Dijk et al. (2002).

The degree of separation of the two transition effects, that is from ϕ_1 to ϕ_2, and then from ϕ_2 to ϕ_3, will depend on the (numerical) separation of the thresholds c_1 and c_2, and the impact of the deviations through the scaling given by ψ_1 and ψ_2. As ψ_1 and ψ_2 (both are > 0) increase, the switch from 0 to 1 becomes faster and $\Psi_1(y_{t-1})$ and $\Psi_2(y_{t-1})$ approach 0/1 indicator functions; the resulting model is effectively a three-regime threshold AR(1) model (TAR), which is considered in the next chapter.

Other developments of interest include allowing time variation in the autoregressive coefficients, as in the time-varying STAR (TVSTAR) of Lundbergh et al. (2003); allowing for autoregressive conditional heteroscedasticity effects, ARCHSTAR, as in Lundbergh and Teräsvirta (1999); combining with long memory in Kılıç (2011); and extending the univariate nonlinear framework to a multivariate basis, (see van Dijk et al. 2002, for development and references).

Whilst STAR models in various forms have been popular, another model of interest that contains or 'bounds' the random walk is the bounded random walk considered in the next section.

5.5   Bounded random walk (BRW)

5.5.1   The basic model

A framework similar in some respects to the STAR family is the bounded random walk, BRW, suggested by Nicolau (2002). This model also allows nonstationarity for some regions, but stationarity elsewhere. Consider:

$$y_t = y_{t-1} + a(y_t) + \varepsilon_t \tag{5.40}$$

where $a(y_t)$ is an adjustment function such that if $a(y_t) = 0$ then the data is generated by a random walk. As in the STAR family, the term $a(y_t)$ serves to provide a corrective check on just how far the random walk is allowed to 'walk' without being brought back to some interval.

The adjustment function works like an error correction mechanism so that if $E(\Delta y_t | y_{t-1}) > 0$, then $a(y_t) < 0$, and if $E(\Delta y_t | y_{t-1}) < 0$, then $a(y_t) > 0$. Thus, if y_t is expected to increase given y_{t-1}, the adjustment serves to hold the expected increase back, and if y_t is expected to decrease given y_{t-1}, the adjustment decreases the expected reduction. This (monotonicity) condition ensures that $a(y_t)$ is acting to correct the direction of the random walk rather than increase the walk in the same direction. There must also be a condition on the extent of the adjustment, otherwise a particular adjustment might induce a further larger adjustment (in absolute value) and so on; this is considered in more detail below.

The BRW adjustment function suggested by Nicolau (2002) is:

$$a(y_t) = e^{\alpha_0}(e^{-\alpha_1(y_{t-1}-\kappa)} - e^{\alpha_2(y_{t-1}-\kappa)}) \tag{5.41a}$$

$$= a(y_t)_{-ve} + a(y_t)_{+ve} \tag{5.41b}$$

where $a(y_t)_{-ve} = e^{[\alpha_0 - \alpha_1(y_{t-1}-\kappa)]}$ and $a(y_t)_{+ve} = -e^{[\alpha_0+\alpha_2(y_{t-1}-\kappa)]}$. As in the STAR family it is useful to define the 'deviation', in this case as $\xi_t = (y_{t-1} - \kappa)$; the parameter κ can also be interpreted as a threshold since deviations from it trigger adjustment. The adjustment function has two parts $a(y_t)_{-ve}$ and $a(y_t)_{+ve}$, corresponding (primarily), respectively, to adjustments arising from ensuring there is a lower bound, induced by large negative values of $(y_{t-1} - \kappa)$, and ensuring that there is an upper bound, induced by large positive values of $(y_{t-1} - \kappa)$.

Putting the adjustment function into (5.40), the BRW is obtained:

$$y_t = y_{t-1} + e^{\alpha_0}(e^{-\alpha_1(y_{t-1} - \kappa)} - e^{\alpha_2(y_{t-1} - \kappa)}) + \varepsilon_t \qquad (5.42a)$$

$$= y_{t-1} + a(y_t)_{-ve} + a(y_t)_{+ve} + \varepsilon_t \qquad (5.42b)$$

where $\alpha_0 < 0$, $\alpha_1 \geq 0$ and $\alpha_2 \geq 0$. It is evident from (5.42a) that:

$$E(\Delta y_t | y_{t-1}) = e^{\alpha_0}(e^{-\alpha_1(y_{t-1} - \kappa)} - e^{\alpha_2(y_{t-1} - \kappa)}) \qquad (5.43)$$

The pure random walk results in two cases. First, note that if $\alpha_1 = -\alpha_2$, then the two components of the BRW adjustment function cancel and $y_t = y_{t-1} + \varepsilon_t$. Also, for the more general case, that is, $\alpha_1 \neq -\alpha_2$, if $y_{t-1} = \kappa$ then there is no adjustment and $y_t = y_{t-1} + \varepsilon_t$. The idea of the term $a(y_{t-1} - \kappa)$ is to stop the process 'escaping' without limit, forcing it back to an interval around κ if it 'threatens' to get too far away.

For $\alpha_1 > 0$, the term $e^{-\alpha_1(y_{t-1} - \kappa)}$ works to return the process to an interval around κ if $y_{t-1} < \kappa$. In this case, for large (negative) deviations, $e^{-\alpha_1(y_{t-1} - \kappa)}$ becomes large positive, whereas the other term $e^{\alpha_2(y_{t-1} - \kappa)} \to 0$; this implies that $E(\Delta y_t | y_{t-1}) > 0$. Obversely, for $\alpha_2 > 0$ and $y_{t-1} > \kappa$, it is the term $e^{\alpha_2(y_{t-1} - \kappa)}$ that becomes large positive whilst $e^{-\alpha_1(y_{t-1} - \kappa)} \to 0$; taking the negative of $e^{\alpha_2(y_{t-1} - \kappa)}$ then ensures that $E(\Delta y_t | y_{t-1}) < 0$.

The responsiveness of the process is governed by the parameters α_0, α_1 and α_2. Symmetry of response, that is, equal effects from equal-sized but opposite-signed deviations of y_{t-1} about κ, follows from $\alpha_1 = \alpha_2$, Otherwise the process could be more responsive on one side than the other. For example, $\alpha_2 > \alpha_1$ implies greater responsiveness from positive deviations compared to the same size negative deviations.

5.5.2   The bounded random walk interval

The idea of a bounded interval can be formalised by finding the region for which $|a(y_{t-1} - \kappa)|$ is 'small', say $\zeta = 0.05$, and hence no force is present to drive the process outside the interval: this is the random walk interval, RWI(ε), defined as the interval on y_t such that:

$$\text{RWI}(\varepsilon) = y_t \in [|a(y_{t-1} - \kappa)| \leq \zeta] \qquad \text{random walk interval} \qquad (5.44)$$

First consider symmetric responses then, bearing in mind that if $\xi_t = (y_{t-1} - \kappa) < 0$, then the second term in $a(y_{t-1} - \kappa)$ will be virtually zero, it is only necessary to solve:

$$e^{\alpha_0}e^{-\alpha_1\xi_t} = \zeta \qquad (5.45a)$$

$$\alpha_0 - \alpha_1\xi_t = \ln\zeta \qquad \text{taking (natural) logarithms of both sides} \qquad (5.45b)$$

$$\xi_t = (\alpha_0 - \ln\zeta)/\alpha_1 \qquad \text{rearrange to solve for the deviation } \xi_t \qquad (5.45c)$$

To illustrate, consider $\alpha_1 = 3$, $\alpha_2 = 3$, $\alpha_0 = -15$, $\kappa = 100$, and $\zeta = 0.05$, then: $\xi_t = (-15 - \ln(0.05))/3 = -4.0$, which implies $y_{t-1} = 100 - 4.0$. By symmetry of argument, if $(y_{t-1} - \kappa) > 0$, then we obtain $y_{t-1} = 100 + 4.0$. Hence, RWI(0.05) $= 100\pm4.0 = [96.0, 104.0]$. For fixed α_0, increasing α_1 and α_2 reduces the random walk interval for given ζ; for example, if $\alpha_1 = 5$ and $\alpha_2 = 5$, then RWI(0.05) $= 100\pm2.40 = [97.6, 102.4]$. The RWIs for the symmetric cases with $\alpha_1 = \alpha_2 = 3$ and 5 are illustrated in Figures 5.13a and 5.14a respectively.

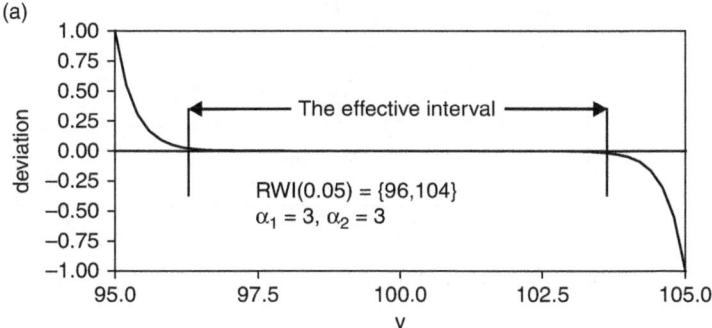

Figure 5.13a   Bounded random walk interval, $\alpha_1 = 3$, $\alpha_2 = 3$

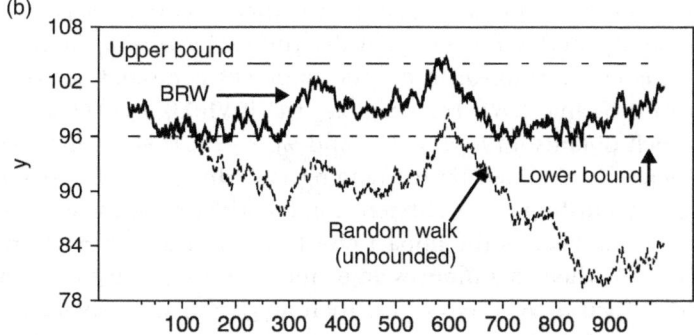

Figure 5.13b   Bounded and unbounded random walks, $\alpha_1 = 3$, $\alpha_2 = 3$

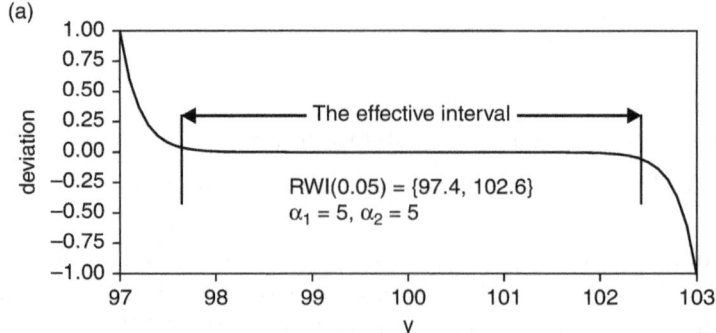

Figure 5.14a   Bounded random walk interval, $\alpha_1 = 5$, $\alpha_2 = 5$

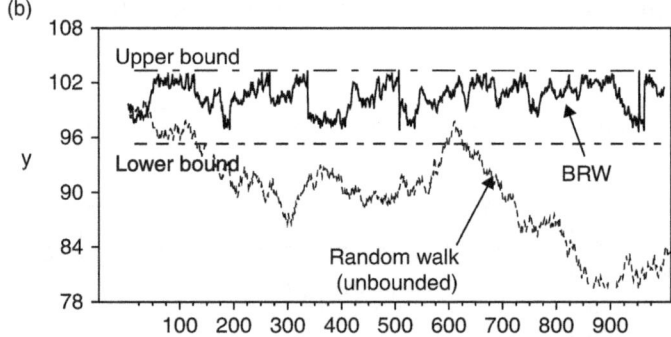

Figure 5.14b   Bounded and unbounded random walks, $\alpha_1 = 5$, $\alpha_2 = 5$

The asymmetric case is easily obtained by taking the union of the appropriate one-sided intervals. For example, if $\alpha_1 = 3$ and $\alpha_2 = 5$, indicating that positive deviations from κ result in a faster response, then RWI(0.05) = [96.0, 102.4].

Notice that the RWI is only uniquely determined by α_1 ($= \alpha_2$ in the symmetric case), if α_0 is taken as fixed. In general, different combinations of α_0 and α_1 can lead to the same RWI. For example, the RWI(0.05) $= 100 \pm 4.0 = $ [96.0, 104.0], which obtains for $\alpha_1 = \alpha_2 = 3$ and $\alpha_0 = -15$, is also generated by $\alpha_1 = \alpha_2 = 2$, $\alpha_0 = -5$ (obtained by rearranging $(\alpha_0 - \ln \zeta)/\alpha_1 = -4$ to solve for α_1 given α_0). What distinguishes different combinations of α_0 and α_1 ($= \alpha_2$), which lead to the same RWI, is the impact effect of a unit deviation. Consider $\xi_t = (y_{t-1} - \kappa) = -1$, then this induces adjustment of $a(y_t) = a(y_t)_{-ve} + a(y_t)_{+ve} = e^{\alpha_0 + \alpha_1} - e^{\alpha_0 - \alpha_1}$; if $\alpha_1 = 3$, $\alpha_0 = -15$, then $a(y_t) \approx 0$, whereas if $\alpha_1 = 2$, $\alpha_0 = -5$, then $a(y_t) = 0.05$. For $\xi_t = -4$, these become adjustments of 0.05 and 20, respectively.

5.5.3 Simulation of the BRW

To illustrate the effect of the bounds on the time series pattern for y_t, the BRW of (5.42a) was simulated for $t = 1, \ldots, 1000$, with $\varepsilon_t \sim \text{niid}(0, 0.16)$. The simulations are for the bounded random walk intervals of Figures 5.13a and 5.14a, and are placed below them in Figures 5.13b and 5.14b, respectively. The random walk, with the same sequence of innovations, is also shown for comparison. In each case the bound is evident as is the narrower interval of fluctuation for the case with larger values of α_1 and α_2. Figure 5.14b also indicates that adjustment at the boundary can be quite pronounced, with a sharp movement back into the interval.

To ensure that 'correcting' movements are not explosive, for example that a positive deviation does not induce a sufficiently large negative deviation, in absolute terms, that the positive deviation it then induces is larger than the initiating deviation, a further condition is required. That is, in addition to monotonicity, it is required that $|a(y)| < 2| \, y \, |$. For $a(y) = a(y_{t-1} - \kappa)$, the condition is $|a(y_{t-1} - \kappa)| < 2|(y_{t-1} - \kappa)|$, which is enforced in the simulations underlying Figures 5.13 and 5.14. This condition is more likely to be breached as σ_ε^2 increases.

5.6 Testing for nonstationarity when the alternative is a nonlinear stationary process

This section considers the problem of testing for a unit root, and hence nonstationarity, when the alternative is that the data are generated by a nonlinear, stationary process. Two cases, the ESTAR and BRW models, are considered for the alternative, although the principles easily generalise to other nonlinear specifications.

5.6.1 The alternative is an ESTAR model

5.6.1.i The KSS test for a unit root

A test for nonstationarity in the context of a stationary nonlinear alternative hypothesis has been suggested by Kapetanios et al. (2003), hereafter KSS. The framework is as follows. Start from the ESTAR(1) model with $\phi_1 = 1$, and assume that $c = 0$ (more general cases are considered below), thus:

$$\Delta y_t = \gamma_1 y_{t-1}(1 - e^{-\theta(y_{t-1})^2}) + z_t \tag{5.46}$$

The (initial) null and alternative hypotheses suggested by KSS, corresponding to linearity and ESTAR nonlinearity respectively, are $H_0 : \theta = 0$ and $H_A : \theta > 0$. Initially it is assumed that $z_t = \varepsilon_t \sim \text{iid}(0, \sigma_\varepsilon^2)$.

Note that there is an essential ambiguity in this set-up as the null hypothesis could equally be stated as $H_0' : \gamma_1 = 0$ and $H_A' : \gamma_1 < 0$. This ambiguity occurs because neither θ nor γ_1 are identified under the null hypothesis. That is, they could take any values without affecting the generation of the data under the null. Davies (1977, 1987) suggests a general procedure for dealing with such cases, which is considered later.

The procedure suggested by KSS is to take a first-order Taylor series approximation (see Chapter 4, Appendix) to the exponential part of the transition function, that is $\lambda_t = e^{-\theta(y_{t-1})^2}$, and for convenience of notation denote this as $f(\theta)$, so that $\Theta_t = (1 - f(\theta))$. Then evaluate the approximation at $\theta = 0$ and substitute the approximation into $\gamma_1 y_{t-1}(1 - f(\theta))$. The steps are as follows:

$$f(\theta) = f(\theta|\theta = 0) + \frac{\partial[f(\theta)|\theta = 0]}{\partial\theta}(\theta - 0) + R \tag{5.47a}$$

$$= 1 - \theta(y_{t-1})^2 + R \quad \text{using } e^0 \equiv 1 \tag{5.47b}$$

$$1 - f(\theta) = \theta(y_{t-1})^2 - R \tag{5.47c}$$

Therefore, on substitution of (5.47c) into (5.46):

$$\Delta y_t = \gamma_1 y_{t-1}(1 - e^{-\theta(y_{t-1})^2}) \tag{5.48a}$$

$$= \gamma_1\theta y_{t-1}(y_{t-1})^2 - (\gamma_1 y_{t-1})R + \varepsilon_t \tag{5.48b}$$

$$= \delta(y_{t-1})^3 + \varepsilon_t^* \tag{5.48c}$$

where $\delta = \gamma_1\theta$ and $\varepsilon_t^* = -(\gamma_1 y_{t-1})R + \varepsilon_t$.

The null and alternative hypotheses are now $H_0 : \delta = 0$ and $H_A : \delta < 0$. Notice that there is a change in sign of the alternative hypothesis compared to the original specification, which follows from $\delta = \gamma_1\theta$ with $\theta > 0$ and $\gamma_1 < 0$. Notice also that under the null $\varepsilon_t^* = \varepsilon_t$. The null hypothesis can now be tested by estimating (5.48c) by least squares and using the t-test statistic on $\hat\delta$. That is by the pseudo-t:

$$t_{NL} = \frac{\hat\delta}{\hat\sigma(\hat\delta)} \tag{5.49}$$

where $\hat\sigma(\hat\delta)$ is the estimated standard error of $\hat\delta$.

The distribution of t_{NL} under the null is non-standard and depends upon the deterministic terms in the model. In the simplest case, where y_t has a zero mean, the asymptotic null distribution of t_{NL} is:

$$t_{NL} \Rightarrow_D \frac{\left(\frac{1}{4}B(1)^4 - \frac{3}{2}\int_0^1 B(r)^2\,dr\right)}{\left(\int_0^1 B(r)^6\,dr\right)^{1/2}} \tag{5.50}$$

where $B(r)$ is standard Brownian motion. The test statistic is consistent under the alternative (see KSS, 2003, theorem 2.1). Rejection of the null hypothesis is rejection of linearity within the maintained model of a unit root.

KSS provide some percentiles for t_{NL} when $T = 1,000$. To assess the sensitivity of these to sample size, the critical values were also simulated for $T = 200$. The

Table 5.1   Critical values of the KSS test for nonlinearity t_{NL}

T = 200	1%	5%	10%	T = 1000	1%	5%	10%
Model 1	−2.77	−2.17	−1.87	Model 1	−2.82	−2.22	−1.92
Model 2	−3.47	−2.91	−2.63	Model 2	−3.48	−2.93	−2.66
Model 3	−3.93	−3.38	−3.11	Model 3	−3.93	−3.40	−3.13

Notes: The data is generated under the null, that is $y_t = y_{t-1} + \varepsilon_t$, with $\varepsilon_t \sim$ niid(0, 1), and the estimated model is: $\Delta y_t = \delta(y_{t-1})^3 + \varepsilon_t^*$. In Models 1, 2 and 3, y_t refers to raw data, de-meaned data and de-trended data, respectively. There were 50,000 simulations; critical values for T = 1,000 are from KSS (2003, Table 1).

critical values for nominal size of 1%, 5% and 10% are given in Table 5.1; it is evident from the table that there is relatively little variation in the critical values in the two cases.

KSS suggest accommodating possible serial correlation in the errors z_t in (5.46) by introducing p lags of the dependent variable, so that the estimated model is based on:

$$\Delta y_t = \gamma_1 y_{t-1}(1 - e^{-\theta(y_{t-1})^2}) + \sum_{j=1}^{p-1} c_j \Delta y_{t-j} + \varepsilon_t \qquad (5.51)$$

The maintained regression for the nonlinearity test is now:

$$\Delta y_t = \delta(y_{t-1})^3 + \sum_{j=1}^{p-1} c_j \Delta y_{t-j} + \varepsilon_t \qquad (5.52)$$

The asymptotic distribution of t_{NL} is unchanged by this amendment. As to choosing the lag length p, KSS suggest using standard lag selection criteria (for example information criteria or the marginal t procedure, see *UR, Vol. 1*, chapter 9).

5.6.1.ii   A joint test

The unit root and nonlinearity hypotheses could also be tested jointly by way of an F-type test. Thus, the maintained regression for the test is:

$$\Delta y_t = \gamma y_{t-1} + \delta(y_{t-1})^3 + \varepsilon_t^* \qquad (5.53)$$

The joint hypothesis of a unit root and *no* nonlinearity is then $H_0 : \gamma = \delta = 0$. In an F test context, the alternative is two-sided and consequently there would be a power loss against a one-sided test, as a result the F test was not recommended by KSS.

5.6.1.iii   An Inf-t test

An alternative to the test statistic t_{NL} can be obtained using the suggestion in Davies (1977, 1987) and see Chapter 1. The general idea is to calculate a test statistic for a grid of values for the non-identified parameter(s) and to form a single test statistic from the resulting test statistics. One possibility is to take the supremum, referred to as Sup (or infinum, Inf, if negative values lead to rejection), of the test statistics over the grid; sometimes these are just referred to as the 'Max' or 'Min' respectively. Although there are some cases in which the distribution of the Sup-test statistic under the null is standard, in the case of the unit root null this is not so and critical values are obtained by simulation. There are also alternatives to the Sup test, for example, taking the simple or exponential average of the test statistics over the grid. Whilst we only consider Sup/Inf-type tests here, the interested reader can refer to Andrews (1993) and Andrews and Ploberger (1994) for details of forms of the test statistic.

In the context of an ESTAR(1) model, suppose that, with known γ_1, there exists a test statistic for H_0, say $S(\theta|\gamma_1)$, conditional on γ_1, with distribution such that large positive values lead to rejection. As γ_1 is unknown but bounded, such that $\gamma_1 \in [L, U]_\gamma$, where L and U are practical lower and upper bounds for γ_1, compute the quantity $S(\theta|\gamma_1)$ over this range and denote the maximum of the computed test statistics as Sup-$(\theta|\gamma_1)$. Then reject H_0 for large *positive* values of Sup-(θ, γ).

Equally, suppose that with known θ, there exists a test statistic for $H_0' : \gamma_1 = 0$, conditional on $\theta \in [L, U]_\theta$, say $S(\gamma_1|\theta)$, with distribution such that large *negative* values lead to rejection in favour of H_A'. As θ is unknown but (finitely) bounded, such that $\theta \in [L, U]_\theta$, compute $S(\gamma_1|\theta)$ over this range and denote the minimum (rather than maximum) in this case as Inf-$S(\gamma_1|\theta)$, then reject $H_0' : \gamma_1 = 0$ for large negative values of Inf-$S(\gamma_1|\theta)$.

The second of these procedures may be more attractive as the estimation problem conditional on fixed θ is just that of a linear regression model, and an obvious test statistic is the pseudo-t, $t(\hat{\gamma}_1)$, based on the least squares estimate $\hat{\gamma}_1$ of γ_1. If the transition function is that of the ESTAR model, $\Theta_t = (1 - e^{-\theta(y_{t-1}-c)^2})$, then the grid is two-dimensional over θ and c, written as $[\theta \times C]$. The test statistic is:

$$\text{Inf-t} = \min[t(\hat{\gamma}_1)] \quad \text{for } (\theta, c) \in [\theta \times C], t(\hat{\gamma}_1) = \hat{\gamma}_1/\hat{\sigma}(\hat{\gamma}_1) \tag{5.54}$$

where $\hat{\sigma}(\hat{\gamma}_1)$ is the estimated standard error of $\hat{\gamma}_1$ conditional on $(\theta, c) \in [\theta \times C]$. The test statistic, Inf-t, is the minimum over all values in the two-dimensional grid. The problem is to determine the critical values and power of such a procedure. As the test statistic may be sensitive to certain parts of the parameter space, the choice of the grid is important; also power may be improved if the grid focuses on likely areas of rejection.

Initially two cases were considered, depending on whether both θ and c were estimated or just θ was estimated. However, as there was relatively little difference in the resulting distributions, the results are reported for the more likely case where both parameters are estimated. Preliminary simulations suggested

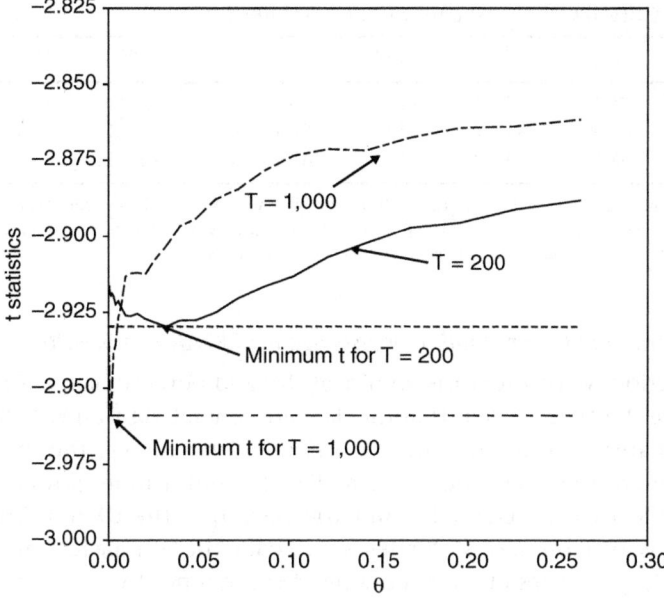

Figure 5.15   Inf-t statistics for a unit root

that, at least for finite samples, the minimum of the t statistics for $\hat{\gamma}_1$ occurs for grid points, θ^G, very close to 0. This is illustrated in Figure 5.15, which plots $t(\hat{\gamma}_1)$ against $\theta \in \theta^G$ using 50,000 simulations in which data are generated by $\Delta y_t = \varepsilon_t$, with $\varepsilon_t \sim$ niid(0, 1) for T = 200 and T = 1,000; the maintained regression is ESTAR(1) with demeaned data; and the data on $t(\hat{\gamma}_1)$ are from estimation in the grid with c = 0 (taken as an illustrative point). For T = 200, the fifth percentile for $t(\hat{\gamma}_1)$ has a minimum at –2.93 when $\theta = 0.032$; however, the variation is not great and at $\theta = 2.0$, the fifth percentile is –2.88. With T = 1,000, the fifth percentile for $t(\hat{\gamma}_1)$ has a minimum at –2.96 when $\theta = 0.0012$ and again the variation is not great.

The distributions were also found to be robust to changing the bounds of the grid and reasonable changes in the number of grid points within the grid. For example, because the minimum occurs early in the θ grid, a grid of 81 evenly spaced points, such that $\theta \in [0, 1]$, combined with a grid of 51 points, such that $c \in [-0.5, 0.5]$, resulted in virtually identical test fractiles (1%, 5% and 10%) compared to grids of $\theta \in [0, 2]$ and $\theta \in [0, 3]$. Also, the sensitivity of the test statistic to the c-grid was relatively minor.

Some percentiles for Inf-t are given in Table 5.2 for T = 200 and T = 1,000. For practical purposes there is almost no difference between the percentiles for these two sample sizes.

Table 5.2 Critical values of the Inf-t test for nonlinearity

T = 200	1%	5%	10%	T = 1,000	1%	5%	10%
Model 1	−2.88	−2.28	−1.97	Model 1	−2.87	−2.26	−1.99
Model 2	−3.69	−3.11	−2.81	Model 2	−3.73	−3.16	−2.92
Model 3	−4.17	−3.60	−3.32	Model 3	−4.12	−3.63	−3.37

Notes: The data are generated under the null, that is $y_t = y_{t-1} + \varepsilon_t$, with $\varepsilon_t \sim$ niid(0, 1). The estimated model is: $\Delta y_t = \gamma_1 y_{t-1}(1 - e^{-\theta(y_{t-1})^2}) + \varepsilon_t$. In models 1, 2 and 3, y_t refers to raw data, demeaned data and detrended data, respectively. There were 50,000 simulations.

5.6.1.iv *Size and power of unit root tests against ESTAR nonlinearity*

In this section we consider the empirical size and power of some tests of a unit root where the alternative is that the data are generated by an ESTAR(1) model. The test statistics of interest are the KSS test denoted t_{NL}, the Inf-t test and, to benchmark the tests, the standard DF test and a more powerful DF type test, specifically the weighted symmetric version of the DF test. The reported simulations focus on the tests for demeaned data, the DF type tests are, therefore, $\hat{\tau}_\mu$ and $\hat{\tau}_\mu^{ws}$; similar results were obtained for the other two cases, that is, cases using 'raw' data and detrended data, and are not reported.

5.6.1.v *Size*

KSS (2003, tables 2 and 3) report simulations for an assessment of size and power and these were supplemented by some additional simulations to enable consideration of the Inf-t test and the benchmarking of the results. The data generating process was $\Delta y_t = \gamma_1 y_{t-1}(1 - e^{-\theta(y_{t-1})^2}) + \varepsilon_t$, where $\varepsilon_t \sim$ niid(0, σ_ε^2). When $\gamma_1 = \theta = 0$, the data are generated under the null and the empirical size of the tests can be assessed. Otherwise, the data are generated under the alternative. Unlike the case where the alternative is linear, with a nonlinear alternative the results are sensitive to the innovation variance, which was set to $\sigma_\varepsilon^2 = 1$, as in KSS, and also, to assess sensitivity, $\sigma_\varepsilon^2 = 0.4^2 = 0.16$.

The critical values for the t_{NL} test and the Inf-t test were taken from Tables 5.1 and 5.2 respectively. The simulations suggest that although there is some undersizing or oversizing, this is slight and actual size remains reasonably faithful to nominal size; see Table 5.3. Thus, simulations for power will enable a meaningful comparison amongst test statistics. A sample size of T = 200 is used to illustrate the results from the simulations conducted here, which hold more generally. (The simulations in KSS (op. cit., table 2) involve a wider range of sample sizes, with raw data, demeaned data and detrended data.)

5.6.1.vi *Power*

The alternative to the linear unit root model is an ESTAR(1) model with a unit root in the inner regime, but which is globally ergodic; thus, under the

Table 5.3 Empirical size of unit root tests for ESTAR nonlinearity: 5% nominal size

	KSS:t_{NL}			Inf-t		
Nominal size	1%	5%	10%	1%	5%	10%
Simulated size	0.83	4.96	10.25	1.16	5.26	10.29

Notes: DGP is: $\Delta y_t = \gamma_1 y_{t-1}(1 - e^{-\theta(y_{t-1})^2}) + \varepsilon_t$, where $\varepsilon_t \sim$ niid$(0, \sigma_\varepsilon^2)$ and T = 200; data is demeaned. Table entries are the average of four simulations, with T = 200.

Table 5.4 Power of some unit root tests against an ESTAR alternative

σ_ε^2		t_{NL}	Inf-t	$\hat{\tau}_\mu$	$\hat{\tau}_\mu^{ws}$	σ_ε^2		t_{NL}	Inf-t	$\hat{\tau}_\mu$	$\hat{\tau}_\mu^{ws}$
$\gamma_1 =$											
-1.0	1.0					$\gamma_1 = -0.1$	1.0				
$\theta = 0.02$		100	100	100	100	$\theta = 0.02$		23.60	22.70	21.35	35.45
$\theta = 0.10$		100	100	100	100	$\theta = 0.10$		55.15	63.20	58.65	83.00
$\theta = 1.00$		100	100	100	100	$\theta = 1.00$		61.75	80.70	86.10	96.85
$\gamma_1 =$											
-1.0	0.4					$\gamma_1 = -0.1$	0.4				
$\theta = 0.02$		56.15	65.85	40.90	67.50	$\theta = 0.02$		9.15	11.25	9.75	13.85
$\theta = 0.10$		99.30	99.90	99.90	100	$\theta = 0.10$		20.40	22.30	17.65	28.25
$\theta = 1.00$		100	100	100	100	$\theta = 1.00$		60.20	70.80	69.50	90.15

Notes: DGP is: $\Delta y_t = \gamma_1 y_{t-1}(1 - e^{-\theta(y_{t-1})^2}) + \varepsilon_t$, where $\varepsilon_t \sim$ niid$(0, \sigma_\varepsilon^2)$ and T = 200; data is demeaned.

alternative $\gamma_1 \neq 0$ and $\theta \neq 0$. KSS (2003, table 3) report simulations that indicate that there are some gains in using t_{NL} relative to the DF t-type test; however, these gains are not generally very marked, nor uniform in the parameter space.

To consider these issues further, some simulations were undertaken, with the results summarised in Table 5.4 and accompanying Figures 5.16–5.19. The first set of figures, that is Figures 5.16 and 5.17, consider the case where γ_1 is fixed and θ varies. The second set of figures, Figures 5.18 and 5.19, consider the case where γ_1 varies and θ is fixed. The standard DF test $\hat{\tau}_\mu$ is included in this comparison, as is the weighted symmetric version of this test $\hat{\tau}_\mu^{ws}$, so that the benchmark is against a more realistically powerful alternative test. As the notation indicates, the data were demeaned but very similar results in terms of rankings were obtained with 'raw' data and detrended data. The DGP was:

$$\Delta y_t = \gamma_1 y_{t-1}(1 - e^{-\theta(y_{t-1})^2}) + \varepsilon_t$$

where $\varepsilon_t \sim$ niid$(0, \sigma_\varepsilon^2)$ and T = 200. Since the alternative is nonlinear, the results *are* sensitive to σ_ε^2, and to explore this sensitivity $\sigma_\varepsilon^2 = 1$ and $\sigma_\varepsilon^2 = 0.4^2$ were considered. Power is assessed with a 5% nominal size for each test; critical values for t_{NL} and Inf-t are from Tables 5.1 and 5.2.

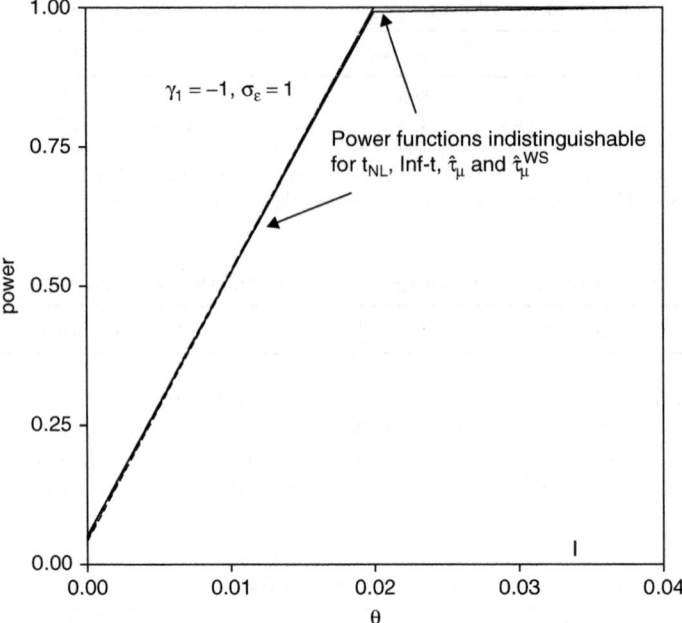

Figure 5.16a   ESTAR power functions, $\gamma_1 = -1$, $\sigma_\varepsilon = 1$

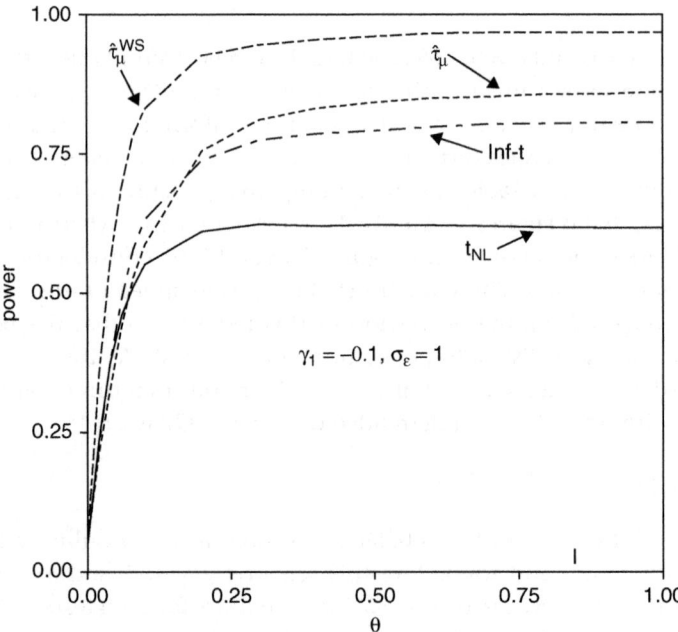

Figure 5.16b   ESTAR power functions, $\gamma_1 = -0.1$, $\sigma_\varepsilon = 1$

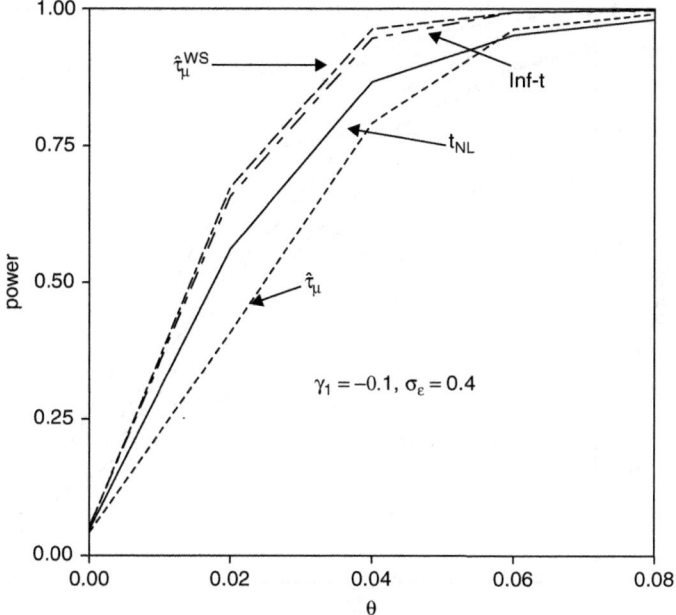

Figure 5.17a   ESTAR power functions, $\gamma_1 = -1$, $\sigma_\varepsilon = 0.4$

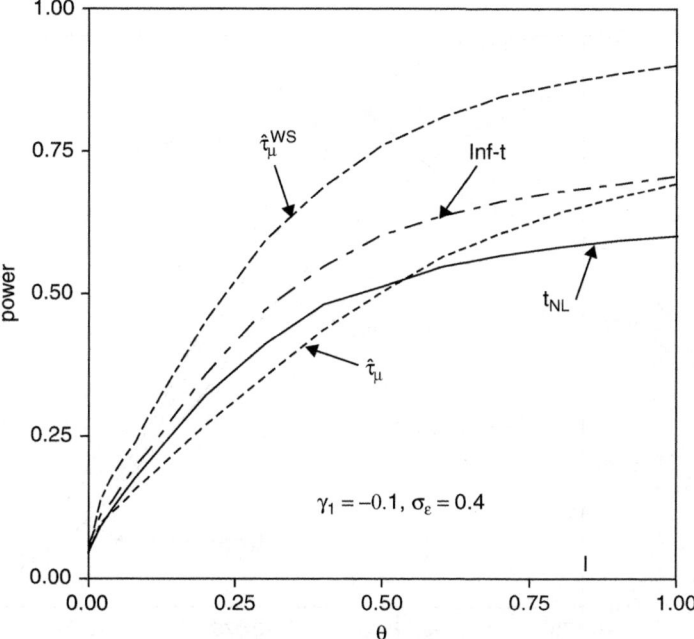

Figure 5.17b   ESTAR power functions, $\gamma_1 = -0.1$, $\sigma_\varepsilon = 0.4$

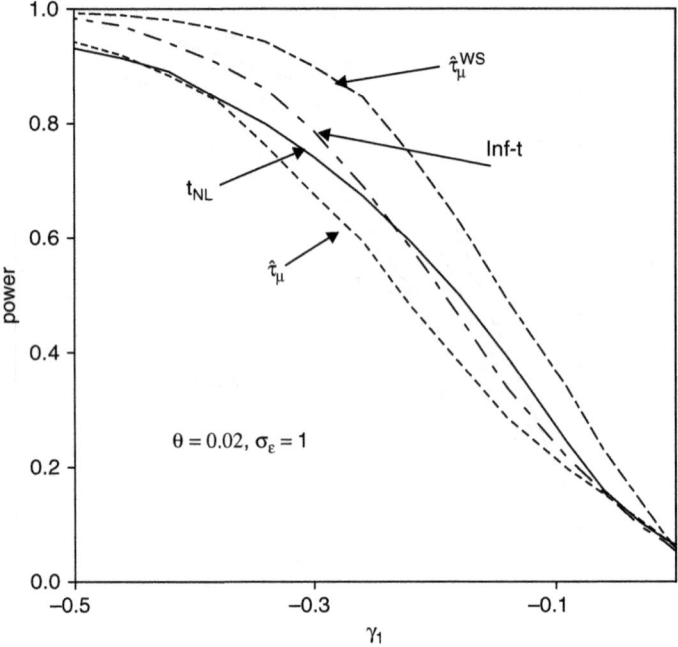

Figure 5.18a   ESTAR power functions, $\theta = 0.02$, $\sigma_\varepsilon = 1$

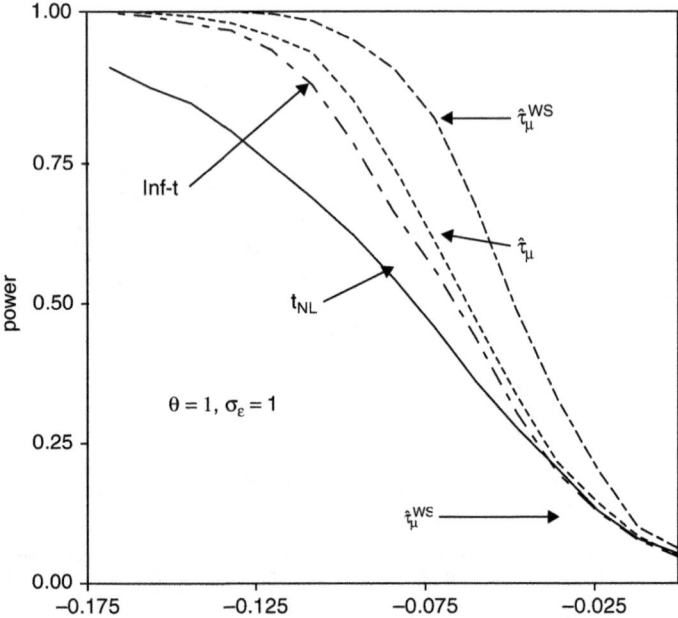

Figure 5.18b   ESTAR power functions, $\theta = 1$, $\sigma_\varepsilon = 1$

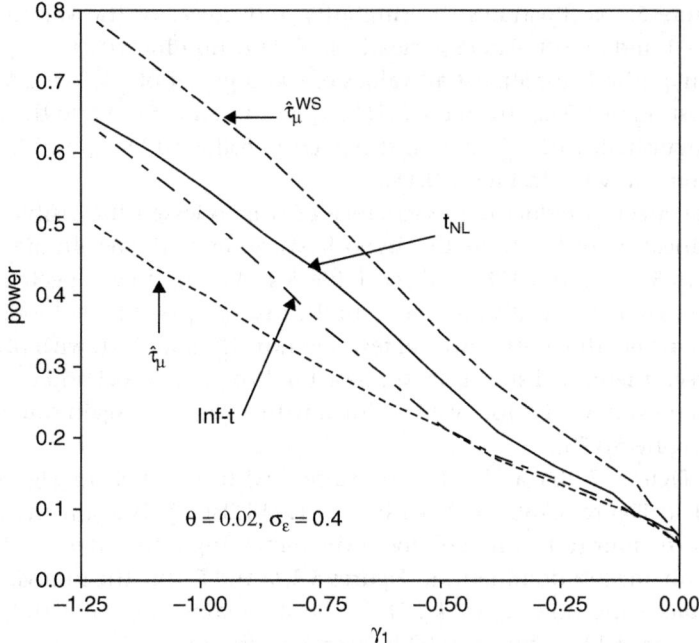

Figure 5.19a   ESTAR power functions, $\theta = 0.02$, $\sigma_\varepsilon = 0.4$

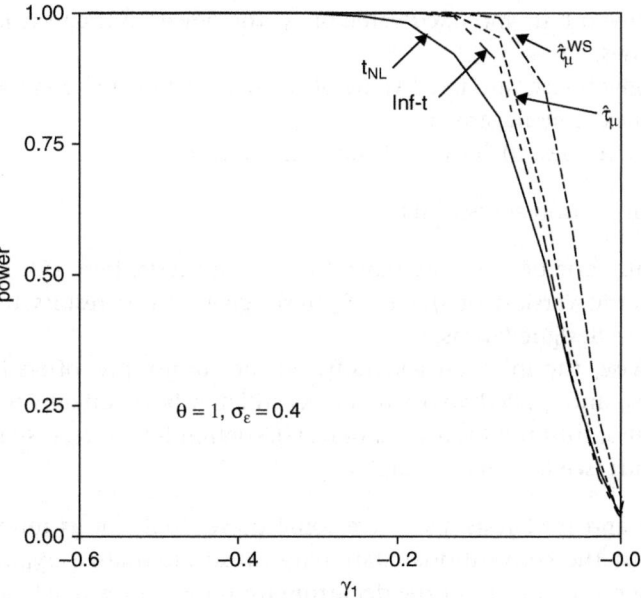

Figure 5.19b   ESTAR power functions, $\theta = 1$, $\sigma_\varepsilon = 0.4$

Figure 5.16a illustrates the similarity and power of the various tests when $\gamma_1 = -1$ and $\sigma_\varepsilon^2 = 1$. There is clearly sufficient nonlinearity for rejection of the null hypothesis for almost all values of θ and power of 100% is achieved for all the tests for $\theta = 0.02$. In Figure 5.16b, $\gamma_1 = -0.1$ and $\sigma_\varepsilon^2 = 1$ and the tests are now distinguished, with $\hat{\tau}_\mu^{ws}$ the most powerful, followed by $\hat{\tau}_\mu$ and Inf-t, but with t_{NL} some way behind if $\theta > 0.15$.

The effect of reducing the variance of ε_t is to lessen the visible effect of the nonlinearity in the alternative hypothesis. Some of the results are illustrated in Figures 5.17a and 5.17b, and see Table 5.4. As might be expected, in all cases power is reduced relative to $\sigma_\varepsilon^2 = 1$. In the case of $\gamma_1 = -1$ and $\sigma_\varepsilon^2 = 0.4^2$ see Figure 5.17a, where the most powerful tests are now $\hat{\tau}_\mu^{ws}$ and Inf-t, with little to choose between them, whilst $\hat{\tau}_\mu$ is least powerful. When $\gamma_1 = -0.1$ and $\sigma_\varepsilon^2 = 0.4^2$, $\hat{\tau}_\mu^{ws}$ is again most powerful followed by Inf-t, with $\hat{\tau}_\mu$ and t_{NL} again some way behind (see Figure 5.17b).

In Figures 5.18a and 5.18b γ_1 varies; with $\theta = 0.02$ in Figure 5.18a and $\theta = 1$ in Figure 5.18b, both with $\sigma_\varepsilon^2 = 1$. Whilst $\hat{\tau}_\mu^{ws}$ is again the most powerful of the four tests, the ranking of the remaining tests differs in the two cases, with no one test dominant. In Figures 5.19a and 5.19b, the innovation variance is reduced and now $\sigma_\varepsilon^2 = 0.4^2$. The $\hat{\tau}_\mu^{ws}$ test remains most powerful, whereas t_{NL} whilst second-best for $\theta = 0.02$ is least powerful for $\theta = 1$.

Overall, it is possible to distinguish conclusions that apply to each test from those that relate to a comparison amongst tests. First, for each test:

(i) for a given test, and fixed value of γ_1, the power of the tests increases as θ increases;

(ii) for a given test, and fixed value of θ, the power of the tests increases as γ_1 becomes more negative;

(iii) as σ_ε^2 increases, the power of the tests increases.

On a comparison across tests:

(i) choosing one of the more powerful standard tests, here $\hat{\tau}_\mu^{ws}$, the weighted symmetric version of the DF $\hat{\tau}_\mu$ test, gives better results than the two 'purpose-designed' tests;

(ii) otherwise, the Inf-t test generally, but not uniformly, offers better power than t_{NL} and $\hat{\tau}_\mu$. Perhaps better power still may be on offer by modifying the t_{NL} test or Inf-t test by, for example, GLS detrending or recursive detrending methods (see *UR, Vol. 1*, chapter 7).

The t_{NL} and Inf-t tests also have some power, but not as much as the DF test, against the conventional stationary linear alternative hypothesis; thus, depending on the extent of the departure from the null, a rejection of the null using t_{NL} or Inf-t can also follow if the alternative is of the standard linear form.

This suggests that the use of the 'purpose-designed' tests against nonlinearity should be interpreted as a diagnostic tool. It does not follow from rejection using t_{NL} or Inf-t that a nonlinear model has generated the data but, rather, that this is a possibility that should be explored.

Finally, KSS (2003, especially table 4) report simulation results for t_{NL} and the standard DF tests when the alternative is ESTAR(1) with a stationary inner regime. Neither form of test statistic is dominant over the parameter space, although the general effect of the stationary inner regime is to increase the power of t_{NL} relative to its DF counterpart. It seems likely, however, that, as in the case of a unit root in the inner regime, a more powerful version of the DF test will dominate t_{NL}.

5.6.2 Testing for nonstationarity when the alternative is a nonlinear BRW

In this section the stationary, nonlinear alternative to the unit root null hypothesis is the bounded random walk (BRW). The question of interest here is: what is the power of some standard unit root tests to detect this alternative? For convenience of reference, the BRW is restated:

$$y_t = y_{t-1} + e^{\alpha_0}(e^{-\alpha_1(y_{t-1}-\kappa)} - e^{\alpha_2(y_{t-1}-\kappa)}) + \varepsilon_t \tag{5.55}$$

To illustrate the power assessment, the adjustment function in the BRW is taken to be symmetric, so that $\alpha_1 = \alpha_2$. What do we expect a priori? It is evident that the random walk interval (RWI) decreases as the α_i increase; this occurs because the impact of even a fairly small deviation is larger, increasing the pressure for the realisations to move back into the RWI. On the other hand, for values of α_i close to zero, even relatively large deviations are tolerated without initiating any corrective action. Power is, thus, expected to increase as the α_i increase.

Another important characteristic of the BRW, as with STAR models, in terms of confining realisations to a limited interval, is the variance, σ_ε^2, of the innovations, ε_t. Consider two processes alike apart from their variance: process 1 with variance $\sigma_{\varepsilon,1}^2$ and process 2 with variance $\sigma_{\varepsilon,2}^2$, and $\sigma_{\varepsilon,1}^2 > \sigma_{\varepsilon,2}^2$. Compared to process 1, process 2, with the smaller variance, is not as likely to lead to realisations that 'excite' the adjustment function; hence, for more of the time it acts like a random walk without a bound. A unit root test is, therefore, less likely to discern a BRW with 'small' σ_ε^2 as a stationary process. On this basis power is anticipated to be an increasing function of σ_ε^2.

The dependence of power on α_i and σ_ε^2 is illustrated in Figure 5.20, which plots the power of the standard DF t-test, $\hat{\tau}_\mu$, and the weighted symmetric version of that test, $\hat{\tau}_\mu^{ws}$, applied to a maintained DF regression with a constant, using a 5% significance level. The data are generated by a BRW model, as in (5.55), with $T = 200$ and symmetry of response $\alpha_1 = \alpha_2 = \alpha(i)$, say, with $\alpha(i) \in [0, 5]$ in increments

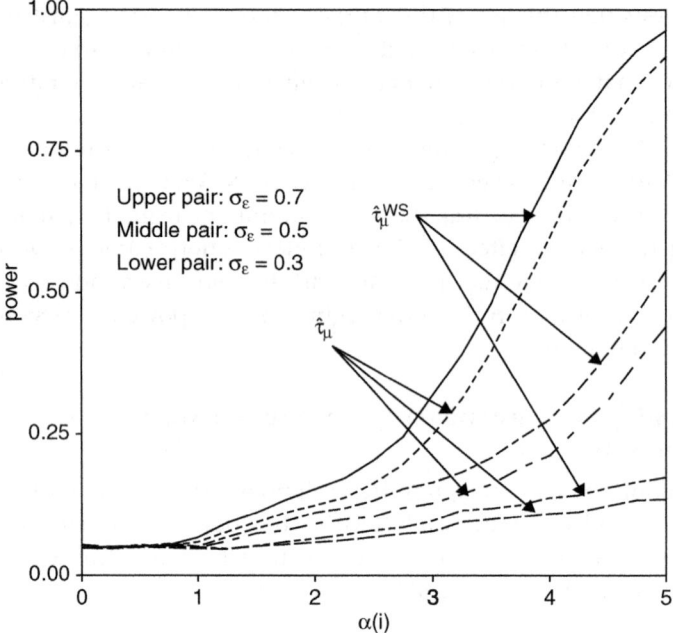

Figure 5.20   Power of unit root tests when DGP is BRW

Table 5.5   Power of $\hat{\tau}_\mu$ and $\hat{\tau}_\mu^{WS}$ against the BRW alternative for different values of $\alpha(i)$ and σ_ε^2.

	$\sigma_\varepsilon^2 = 0.3^2$		$\sigma_\varepsilon^2 = 0.5^2$		$\sigma_\varepsilon^2 = 0.7^2$	
	$\hat{\tau}_\mu$	$\hat{\tau}_\mu^{WS}$	$\hat{\tau}_\mu$	$\hat{\tau}_\mu^{WS}$	$\hat{\tau}_\mu$	$\hat{\tau}_\mu^{WS}$
$\alpha(i) = 1$	4.78	5.05	5.21	5.12	5.92	6.71
$\alpha(i) = 3$	7.82	9.69	12.78	16.46	24.83	31.92
$\alpha(i) = 5$	13.51	17.43	43.98	53.88	91.68	96.31

of 0. 25, and $\alpha_0 = -5$; σ_ε^2 is set to 0.3^2, 0.5^2 and 0.7^2, and there were 10,000 simulations. Some of the results are reported in Table 5.5.

Although six series are plotted in Figure 5.20, they are fairly easy to distinguish, being in pairs, with each pair corresponding to a different value of σ_ε^2, and power increasing as σ_ε^2 increases. The upper pair corresponds to $\sigma_\varepsilon^2 = 0.7^2$, the middle pair to $\sigma_\varepsilon^2 = 0.5^2$ and the lower pair to $\sigma_\varepsilon^2 = 0.3^2$.

It is nearly always the case that the WS version of the DF test is the more powerful of the two tests. Keeping σ_ε^2 constant at $\sigma_\varepsilon^2 = 0.5^2$, the power of $\hat{\tau}_\mu^{WS}$ for $\alpha(i) = 1$, 3, 5 was 5.1%, 16.5% and 53.9% respectively, whereas for the standard DF test, $\hat{\tau}_\mu$, it was 5.2%, 12.8% and 44.0% respectively. Note that when

$\alpha(i) = 1$, the power of the tests is close to the nominal size for $\sigma_\varepsilon^2 = 0.3^2$ and 0.5^2, suggesting no effective modification due to the BRW adjustment function.

To indicate the importance of σ_ε^2, the power of $\hat{\tau}_\mu^{ws}$ for $\alpha(i) = 3$ is approximately 32% for $\sigma_\varepsilon^2 = 0.7^2$, but nearly halved at 16.5% for $\sigma_\varepsilon^2 = 0.5^2$, and then 9.7% for $\sigma_\varepsilon^2 = 0.3^2$. Hence, in the case of 'moderate' values of σ_ε^2 and $\alpha(i)$, power is not high. Power for $\alpha(i) = 5$ and $\sigma_\varepsilon^2 = 0.7^2$ is high at just over 96% but, whilst practical experience of fitting the BRW is presently limited, values of this order seem unlikely judging from some of the simulations where, on occasion, the corresponding adjustments breached the non-explosivity condition.

5.7 Misspecification tests to detect nonlinearity

The results summarised in Section 5.6 suggest that at least in some circumstances nonlinearity may go undetected. Another approach is to test directly for the presence of nonlinearity, which is the concern of this section. One of the most popular approaches is due to Teräsvirta who suggested a means of discriminating between linear models and nonlinear STAR type models.

5.7.1 Teräsvirta's tests

Teräsvirta (1994) has suggested some tests, which are easy to apply, that address the issues both of whether a linear AR model is misspecified in the direction of an ESTAR or LSTAR model, and which of these alternatives is more likely. The tests may also be used to assist in the determination of the delay parameter d in the transition function.

The issue to which the misspecification tests are directed is that of testing whether the following linear AR(p) model is an adequate representation of the data:

$$y_t = \phi_0 + \sum_{i=1}^p \phi_i y_{t-i} + \varepsilon_t \tag{5.56}$$

The alternative is that the data are generated by a nonlinear STAR(p), given by:

$$y_t = \phi_0 + \sum_{i=1}^p \phi_i y_{t-i} + (\gamma_0 + \sum_{i=1}^p \gamma_i y_{t-i}) E(y_{t-d}) + \varepsilon_t \tag{5.57}$$

The transition function can be exponential $E(y_{t-d}) = \Theta(y_{t-d})$ or logistic, $E(y_{t-d}) = \Psi(y_{t-d})$, giving rise to an ESTAR or LSTAR nonlinear model respectively.

The testing problem is not straightforward, as has been demonstrated in the context of testing for a unit root. The problem is that there are parameters, γ_i, $i = 0, \ldots, p$, and those in $E(y_{t-d})$, that are not identified under the null hypothesis of linearity. A framework for testing, based on a Taylor series expansion of the transition function, has been suggested by Teräsvirta (1994); see also Luukkonen et al. (1988) and Eitrheim and Teräsvirta (1996). The tests are based on estimating an artificial regression of the following form:

$$y_t = \beta_0 + \sum_{i=1}^{p} \beta_{1i} y_{t-i}$$

$$+ \sum_{j=1}^{p} \beta_{2j} y_{t-j} y_{t-d} + \sum_{j=1}^{p} \beta_{3j} y_{t-j} y_{t-d}^2 + \sum_{j=1}^{p} \beta_{4j} y_{t-j} y_{t-d}^3 + \varepsilon_t \qquad (5.58)$$

This regression follows from a third-order approximation to a logistic transition function and also encompasses a first-order approximation to an exponential transition function. A set, or sequence, of coefficients β_{ij}, $j = 1, \ldots, p$, for given i is denoted $\{\beta_{ij}\}$. For later use, the following amended auxiliary regression with terms in y_{t-d}^4 is also defined here:

$$y_t = \beta_0 + \sum_{i=1}^{p} \beta_{1i} y_{t-i} + \sum_{j=1}^{p} \beta_{2j} y_{t-j} y_{t-d}$$

$$+ \sum_{j=1}^{p} \beta_{3j} y_{t-j} y_{t-d}^2 + \sum_{j=1}^{p} \beta_{4j} y_{t-j} y_{t-d}^3 + \sum_{j=1}^{p} \beta_{5j} y_{t-j} y_{t-d}^4 + \varepsilon_t \qquad (5.59)$$

The intuition behind these artificial regressions is that the first part is just a standard linear AR(p) model, which is augmented by terms that arise from approximating the nonlinear part of a STAR model. If the latter are not significant, then the AR(p) model is not mispecified in the direction of a STAR model.

The auxiliary regressions, (5.58) and (5.59), assume that $d \leq p$, and some modifications are required if this is not the case; initially we assume $d \leq p$. First consider a sequence of hypotheses (conditional on d) as follows. Note that in practice the index $t = 1$ would usually be assigned to the first observation so, notationally, the first effective observation in the sample is $p + 1$ and as the last is T, the effective sample size is $\tilde{T} = T - p$. The sequence is denoted S1 to S4.

S1. An overall test of the null of linearity against the alternative of nonlinearity of STAR form in the maintained regression model (5.58) is:

$H_0 : \beta_{2j} = \beta_{3j} = \beta_{4j} = 0$ for all j;

$H_A : \beta_{2j} \neq 0, \beta_{3j} \neq 0, \beta_{4j} \neq 0$ at least one of these holds for some j.

The suggested test statistic, based on the Lagrange Multiplier principle, is

$$LM_1 = \left(\frac{RSS_R - RSS_{UR}}{\tilde{\sigma}^2} \right) \text{ where } \tilde{\sigma}_\varepsilon^2 = \frac{RSS_R}{\tilde{T}} \qquad (5.60)$$

The restricted residual sum of squares, RSS_R, refers to the estimated version of the linear AR(p) model; see (5.56). The unrestricted residual sum of squares RSS_{UR} is that obtained from estimation of (5.58). The test statistic LM_1 is asymptotically distributed as $\chi^2(3p)$; however, it is preferable to correct for degrees of freedom and apply a standard F test, in which case the test statistic is:

$$F_1 = \left(\frac{RSS_R - RSS_{UR}}{RSS_{UR}} \right) \left(\frac{\tilde{T} - (4p + 1)}{3p} \right) \qquad (5.61)$$

This will be approximately distributed as an $F(3p, \tilde{T} - (4p + 1))$ variate, where $\tilde{T}$ is the effective sample size. The same χ^2 and F principles apply throughout the testing procedure. The notation RSS refers generically to the residual sum of squares, a subscript UR refers to an unrestricted regression, and a subscript R to a restricted regression, although the regressions referred to will, of course, change during the testing procedure.

This first test can be viewed as *misspecification* test, the idea being to consider the possibility that the linear regression is incorrectly specified in the general direction of a STAR model. The other tests in this sequence are intended to provide more guidance on *specification* issues.

S2. If H_0 of S1 is rejected, carry out a test that the set of terms with β_{4j} coefficients is zero. That is:

$H_{01} : \beta_{4j} = 0$ for all j against $H_{A1} : \beta_{4j} \neq 0$ for some j.

In line with the test in S1 an F test is suggested. This F test, say F_2, takes RSS_{UR} as in F_1, but RSS_R now refers to the model just restricted by $\beta_{4j} = 0$ for all j. F_2 will, therefore, be approximately distributed as $F(p, \tilde{T} - (4p + 1))$.

S3. Conditional on setting $\beta_{4j} = 0$ for all j, carry out a test that the set of terms with β_{3j} coefficients is zero. Thus, the maintained regression now omits all terms with β_{4j} coefficients, and this conditioning is indicated as $|\beta_{4j} = 0$.

$H_{02} : \beta_{3j} = 0|\beta_{4j} = 0$ for all j

$H_{A2} : \beta_{3j} \neq 0|\beta_{4j} = 0$ for some j.

This F test, say F_3, takes the RSS_{UR} as the restricted RSS from F_2, with the RSS_R for F_3 now referring to the model restricted by $\beta_{3j} = \beta_{4j} = 0$ for all j. F_3 will, therefore, be approximately distributed as $F(p, \tilde{T} - (3p + 1))$.

S4. Conditional on setting $\beta_{3j} = 0$ *and* $\beta_{4j} = 0$ for all j, carry out a test that the set of terms with β_{2j} coefficients is zero. Thus, the maintained regression now omits all terms with either β_{3j} or β_{4j} coefficients, and this conditioning is indicated by the notation $|\beta_{3j} = \beta_{4j} = 0$.

$H_{03} : \beta_{2j} = 0|\beta_{3j} = 0, \beta_{4j} = 0$ for all j against

$H_{A3} : \beta_{2j} \neq 0|\beta_{3j} = 0, \beta_{4j} = 0$ for some j.

This F test, say F_4, takes the RSS_{UR} as the restricted RSS from F_3, with the RSS_R for F_4 now referring to the model restricted by $\beta_{2j} = \beta_{3j} = \beta_{4j} = 0$ for all j. F_4 will, therefore, be approximately distributed as $F(p, \tilde{T} - (2p + 1))$.

Assuming that the general test of linearity, H_0, in S1 is rejected by use of F_1, the choice between ESTAR and LSTAR is based on relative strengths of rejection.

Table 5.6   Nonlinearity testing: summary of null and alternative hypotheses

	Null	Alternative	LSTAR pattern	ESTAR pattern
H_0	$\beta_2 = \beta_3 = \beta_4 = 0$	$\beta_2 \neq 0, \beta_3 \neq 0, \beta_4 \neq 0$, for at at least one element of these sets		
H_{01}	$\beta_4 = 0$	$\beta_4 \neq 0$	$\beta_4 \neq 0$	$\beta_4 = 0$
H_{02}	$\beta_3 = 0\mid \beta_4 = 0$	$\beta_3 \neq 0\mid \beta_4 = 0$	$\beta_3 = 0$ or $\beta_3 \neq 0$	$\beta_3 \neq 0$
H_{03}	$\beta_2 = 0\mid \beta_3 = 0, \beta_4 = 0$	$\beta_2 \neq 0\mid \beta_3 = 0, \beta_4 = 0$	$\beta_2 \neq 0$	$\beta_2 = 0$ or $\beta_2 \neq 0$

Notes: $\beta_2 = \{\beta_{2j}\}$, $\beta_3 = \{\beta_{3j}\}$, $\beta_4 = \{\beta_{4j}\}$; $\beta_i \neq 0$ indicates that at least one element of β_i is non-zero.

Recall that the respective adjustment functions are $\Theta(y_{t-d}) = (1 - e^{-\theta(y_{t-d}-c)^2})$ and $\Psi(y_{t-d}) = [1 + e^{-\psi(y_{t-d}-c)}]^{-1}$. The Teräsvirta rule is:

> If H_{01} and H_{03} are more strongly rejected (p-values are smaller) than H_{02} then choose LSTAR.

> If H_{02} more strongly rejected (p-value smallest) compared to H_{01} and H_{03} then choose ESTAR.

The basis of the rule is the likely magnitude of the sets of coefficients $\beta_2 = \{\beta_{2j}\}$, $\beta_3 = \{\beta_{3j}\}$, $\beta_4 = \{\beta_{4j}\}$. Based on the underlying coefficients, it can be observed that:

LSTAR pattern: $\beta_4 \neq 0$, $\beta_3 = 0$ occurs when $c = 0$ and $\gamma_0 = 0$ otherwise $\beta_3 \neq 0$, $\beta_2 \neq 0$.

ESTAR pattern: $\beta_4 = 0$, $\beta_3 \neq 0$, $\beta_2 = 0$ occurs when $c = 0$ and $\gamma_0 = 0$ otherwise $\beta_2 \neq 0$.

The various hypotheses are summarised in Table 5.6, along with the likely pattern of coefficients for LSTAR and ESTAR models. In practice it is important to bear in mind that the rule is intended as guidance, not for hard-and-fast application. As Teräsvirta (1994) notes, following rejection of H_0 it may be helpful to estimate both an LSTAR and an ESTAR model if the p-values for testing either H_{01} and H_{02} or H_{02} and H_{03} are close to each other.

There is a further use of the overall test of H_0 suggested by Teräsvirta (1994) in relation to choosing the delay parameter d. Although in many studies $d = 1$, in principle this need not be the case, although it might be unusual to find $d > p$, in which case one would need to look carefully at whether the AR lag order was sufficiently long. The suggestion is to carry out the test of H_0, usually with F_1, for the range of candidate values of d and then choose d such that the p-value of H_0 is at a minimum.

5.7.2   The Escribano-Jordá variation

There is a variation on the Teräsvirta framework due to Escribano and Jordá (2001) EJ, hereafter, in which the artificial regression is (5.59) rather than (5.58), from which it differs by including terms in y_{t-d}^4. EJ note that the first-order Taylor series expansion in the case where the transition function is ESTAR may be inadequate to capture the two inflexion points, and they suggest a second-order Taylor series expansion. This implies that terms involving the fourth power of the transition variable should be included in the auxiliary regression. The implied changes to the misspecification rule are straightforward.

The general linearity hypothesis, which is now framed in terms of the artificial regression given by (5.59), is:

$$H_0^{EJ} : \beta_{2j} = \beta_{3j} = \beta_{4j} = \beta_{5j} = 0 \qquad \text{for all j against}$$

$$H_A^{EJ} : \beta_{2j} \neq 0, \beta_{3j} \neq 0, \beta_{4j} \neq 0, \beta_{5j} \neq 0 \qquad \text{for some j.}$$

An F test on the same principle as for F_1 will be approximately distributed as $F(4p, \hat{T} - (5p + 1))$; the practical difference is that the residuals for RSS_{UR} are from estimation of (5.59) not (5.58).

As to specification, the following rule is suggested where it is helpful to define sub-hypotheses of H_0^{EJ} corresponding to odd and even first subscripts. Assuming that $\gamma_0 = c = 0$, then even powers of y_{t-d} should be absent if the STAR model is LSTAR, but odd powers of y_{t-d} should be absent if the STAR model is ESTAR. Hence, a finding of rejection of $H_{0,E}^{EJ}$: $\beta_{3j} = \beta_{5j} = 0$ in the maintained regression (5.59) against the two-sided alternative suggests ESTAR (nonzero even powers); and a rejection of $H_{0,L}^{EJ}$: $\beta_{2j} = \beta_{4j} = 0$ in (5.59) against the two-sided alternative suggests LSTAR (nonzero odd powers). The EJ rule is:

Choose ESTAR if the minimum p-value is for $H_{0,E}^{EJ}$; choose LSTAR if the minimum p-value is for $H_{0,L}^{EJ}$.

Since an argument against the EJ rule is that it involves p further, possibly redundant, regressors and may not, therefore, uniformly improve upon power relative to the Teräsvirta rule, there is also another possibility. In the ESTAR case, exclude the terms in β_{2j} and β_{4j} from the maintained regression and then test $H_{0,E}^{EJ,*}$: $\beta_{3j} = \beta_{5j} = 0$; and in the LSTAR case exclude the terms in β_{3j} and β_{5j} in the maintained regression, and then test $H_{0,L}^{EJ,*}$: $\beta_{2j} = \beta_{4j} = 0$. Whether this rule provides an improvement in power seems likely, but is not pursued here. As with the Teräsvirta rule, the EJ rule should be used for guidance since it may not be the case that $\gamma_0 = c = 0$.

Some examples of application of the Teräsvirta rule are taken from Skalin and Teräsvirta (1999) for Swedish data, 1861–1988, and summarised in Table 5.7, together with the conclusions drawn by the authors. (See also Teräsvirta and Anderson, 1992, for the application of STAR models to modelling business cycles.) In the case of GDP, linearity, H_0, is rejected, as is H_{01}; by itself this suggests an LSTAR model, which tends to be confirmed by nonrejection of H_{02}. The p-value for H_{01} is the smallest in the sequence, so an LSTAR is suggested. On

Table 5.7   Summary of illustrative tests for nonlinearity

	H_0	H_{01}	H_{02}	H_{03}	Conclude
GDP	0.019	0.0017	0.64	0.54	LSTAR
Industrial production	0.031	0.070	0.028	0.63	ESTAR
Exports	2.1×10^{-8}	0.00025	0.0039	0.00013	LSTAR
Productivity	0.025	0.045	0.036	0.47	ESTAR

Source: extracted from Skalin and Teräsvirta (1999); table entries are p-values for Teräsvirta's tests.

estimation, Skalin and Teräsvirta (op. cit.) found that the nonlinear model did not improve much over a linear model. For industrial production the smallest p-value is for H_{02}, and H_{03} is not rejected following rejection of H_{02}; this pattern suggests an ESTAR model. For exports, the smallest p-value is for H_{01}, and Skalin and Teräsvirta (op. cit.) opt for an LSTAR model. Finally, for productivity, the smallest p-value is for H_{02}, and H_{03} is not rejected, suggesting an ESTAR model.

5.8   An alternative test for nonlinearity: a conditional mean encompassing test

Chen (2003) suggests an alternative test for nonlinearity, based on Chen and Kuan (2002), related to Wooldridge (1990), which is designed to discriminate between two competing STAR models, for example LSTAR and ESTAR, but it also has wider applicability. Note that as neither the LSTAR nor ESTAR models are a special case of each other, discrimination between them is essentially a non-nested problem. In principle, a model encompassing approach could be used (see, for example, Mizon and Richard, 1986). Chen (op. cit.) suggests a pseudo-score encompassing (PSE) test, with the intention of avoiding the ambiguity that is present in the Teräsvirta rules. We follow the principle suggested in Chen (op. cit.), but in the interpretation here the test motivation is of the form suggested by Wooldridge (1990) as a conditional mean encompassing (CME) test.

5.8.1   Linear, nonnested models

Some reminders are in order first to establish the general principle of the suggested test. A simple linear regression illustrates the principles. Consider two specifications of the standard linear regression model with nonstochastic regressors and differing only in their specification of these regressors. Thus, these standard models are:

$$M_W : y = (X, W)(\beta_1, \beta_2)' + u_W \qquad\qquad E(u_W u_W' | M_W) = \sigma_W^2 I \qquad (5.62)$$

$$= X_w \beta + u_W$$

$$M_Z : y = (X, Z)(\lambda_1, \lambda_2)' + u_Z \qquad\qquad E(u_Z u_Z' | M_Z) = \sigma_Z^2 I \qquad (5.63)$$

$$= X_z \lambda + u_Z$$

where $W \not\subset Z$ and $Z \not\subset W$, so that the competing explanations are 'non-nested'. The regression models are allowed to have some regressors, X, in common. The dimensions of the matrices are: X is $T \times k_1$, Z is $T \times k_2$ and W is $T \times k_3$.

An F-encompassing test is obtained by forming the union of the regressors in the two models, here $V = X \cup Z \cup W$, and carrying out F tests for: (i) the significance of the set Z in V; (ii) the significance of the set W in V; and (iii) the significance of the set $Z \cup W$ in V. The CME test principle proceeds by asking whether the residuals from the null model M_W are orthogonal to the scores (partial derivatives of the log density function) according to the alternative model M_Z; the roles are then reversed so that M_Z becomes the null model and M_W the alternative model.

Intuitive justification for the CME test comes from considering the standard orthogonality conditions that arise from ML and LS estimation. In ML (and LS) estimation the derivatives of the LL function with respect to each of the regression coefficients from ML estimation are orthogonal to the residuals. In the case of a linear model $y = X\beta + u$, the derivatives are just the regressors and, therefore, $X'\hat{u} = \vec{0}$ where $\vec{0}$ is a $k_1 \times 1$ vector of zeros and $\hat{u}$ is the $T \times 1$ vector of ML (LS) residuals; the orthogonality condition for the i-th regressor is: $\sum_{t=1}^{T} X_{it} \hat{u}_t = 0$.

The idea of the CME (and PSE) test is to check whether the residuals of the null model $\hat{u}_W$ are close to being orthogonal to the regressors from the alternative model just as they are (identically) orthogonal to the variables from the null model. If $(X, Z)'\hat{u}_W$ is close to zero, in a well-defined sense, then M_Z adds nothing to M_W and M_W is said to (pseudo-score) encompass M_Z.

The roles of the null and alternative model are then reversed to see if M_Z encompasses M_W. In the former case, the test statistic is based on the distance $(X, Z)'\hat{u}_W$, but note that $X'\hat{u}_W = 0$ as X, of dimension $T \times k_1$, is common to both sets of regressors, and the residuals from the null model are orthogonal to X (and W); therefore, the first k_1 components are zero and the test will rest on the remaining k_2 components defined by $\hat{\delta}_W = Z'\hat{u}_W$, with $\hat{\delta}_W = 0$ under the null model. This quantity is a $k_2 \times 1$ vector with $k_2 \times k_2$ variance-covariance matrix $\text{Var}(\hat{\delta}_W) = E(Z'\hat{u}_W \hat{u}_W' Z) = Z'E(\hat{u}_W \hat{u}_W')Z$; to evaluate this quantity note that $\hat{u}_W = P_W u$ where $P_W = I - X_w(X_w' X_w)^{-1} X_w'$, $P_W' = P_W$ and $P_W P_W = P_W$; hence, $\text{Var}(\hat{\delta}_W) = Z'P_W E(u_W u_W')P_W' Z = \sigma_W^2 Z'P_W Z$.

We are now in a position to form a standard Wald test statistic based on $\hat{\delta}_W$:

$$CSE_{Z|W} = \hat{\delta}'_W[Var(\hat{\delta}_W)]^{-1}\hat{\delta}_W \qquad (5.64)$$

where $Var(\hat{\delta}_W) = \hat{\sigma}^2_W Z'P_W Z$ and $\hat{\sigma}^2_W = \hat{u}'_W \hat{u}_W/T$. $CSE_{Z|W}$ is asymptotically distributed as $\chi^2(k_2)$ under the null model M_W; the roles of M_W and M_Z are reversed to obtain $CSE_{W|Z}$. (The test statistic and its distribution are an application of a general principle; see, for example, Amemiya, 1985, section 1.5.2.)

Some illustrative checks on the size of this test were carried out on the grounds that the development of the test to the nonlinear case would receive encouragement from satisfactory properties in the linear case. This was found to be the case in a number of simulation designs, with two simple linear cases reported here for illustration. In both cases the set-up is as in (5.62). In the first case X, W and Z each comprise a single nonstochastic variable; these are generated by taking random draws from the three-dimensional multivariate normal distribution with unit variances and correlation coefficients ρ_{12}, ρ_{13}, ρ_{23}, over the sample path $t = 1, \ldots, T$; this generates the sequences $\{x_t\}_{t=1}^T$, $\{w_t\}_{t=1}^T$ and $\{z_t\}_{t=1}^T$, which are then fixed. The DGP is generated with M_W as the null model, where u_W is generated by draws from niid(0, 1).

The simulations then enable the generation of the distribution of the test statistic $CSE_{Z|W}$, with M_Z as the alternative model. In the first case, size is evaluated using the critical value from $\chi^2(1)$. In a variation to this case, M_Z comprises two regressors in Z and, hence, size is evaluated using the critical value from the $\chi^2(2)$ distribution. The sample sizes are alternately $T = 100$ and $T = 2,500$, the latter to approximate the asymptotic distribution; 10,000 simulations were generated in each case. A number of values of ρ_{ij} and β were used, with no qualitative variation in the results, which are illustrated for $\rho_{ij} = 0.5$ and $\beta = (\beta_0, \beta_1, \beta_2) = (1, 1, 4)$. For $T = 2,500$, the simulated sizes were 4.8% and 5.1%; and for $T = 100$, the simulated sizes were 4.4% and 5.0%. Thus, at least in the linear case, nominal and actual sizes are close enough to support the use of the test.

The purpose of this illustration with the linear model is to motivate the simplicity and intuition behind the CME-type test; its real value, at least potentially, is in offering a simple means to discriminate between alternative nonlinear, nonnested models, a leading example of which is the choice between an ESTAR model and an LSTAR model. (See also Question 5.7.)

5.8.2   Nonlinear, nonnested models: ESTAR and LSTAR

5.8.2.i   Linearisation

Linearisation of a nonlinear model and subsequent estimation by the method of Gauss-Newton is a frequently employed practical method of obtaining estimates that is used in commercial econometric software; the user usually specifies the required nonlinear routine rather than directly carrying out the linearisation.

Here the linearisation is of use in itself, so we spend some time on an outline for the present purposes; a more extensive treatment is given by Amemiya (1985) and an excellent textbook exposition is in Greene (2011).

Consider the nonlinear regression model given by:

$$y_t = G_t(x_t, \beta) + \varepsilon_t \tag{5.65}$$

where x_t is a $1 \times K$ vector of observations at time t on the explanatory variables and β is a $K \times 1$ parameter vector. This model can be linearised by taking a Taylor series expansion (TSE) about β_k^0 for $k = 1, \ldots, K$, then retaining only the first-order terms and evaluating the derivatives at $\beta^0 = (\beta_1^0, \ldots, \beta_K^0)'$. The first-order TSE is:

$$G_t(x_t, \beta) = G_t(x_t, \beta^0) + \sum_{k=1}^{K} \frac{\partial G_t(x_t, \beta)}{\partial \beta_k |\beta^0} \left(\beta_k - \beta_k^0\right) + R \tag{5.66}$$

where $|\beta^0$ indicates that the derivative is evaluated at β^0, and R is the remainder.

Collecting terms, rearranging, and ignoring the remainder, gives:

$$G_t(x_t, \beta) \approx G_t(x_t, \beta^0) - \sum_{k=1}^{K} \frac{\partial G_t(x_t, \beta)}{\partial \beta_k |\beta^0} \beta_k^0 + \sum_{k=1}^{K} \frac{\partial G_t(x_t, \beta)}{\partial \beta_k |\beta^0} \beta_k \tag{5.67}$$

Using the notation $\nabla G(x_t, \beta^0)_{kt} = \frac{\partial G_t(x_t, \beta)}{\partial \beta_k |\beta^0}$, the nonlinear model is linearised using just the first-order TS approximation as:

$$y_t \approx G_t(x_t, \beta) + \varepsilon_t$$

$$\approx G_t(x_t, \beta^0) - \sum_{k=1}^{K} \nabla G(x_t, \beta^0)_{tk} \beta_k^0 + \sum_{k=1}^{K} \nabla G(x_t, \beta^0)_{tk} \beta_k + \varepsilon_t \tag{5.68}$$

Collecting known terms to the left-hand side gives:

$$y_t^0 \approx \sum_{k=1}^{K} \nabla G(x_t, \beta^0)_{kt} \beta_k + \varepsilon_t \tag{5.69}$$

where:

$$y_t^0 = y_t - G_t(x_t, \beta^0) + \sum_{k=1}^{K} \nabla G(x_t, \beta^0)_{tk} \beta_k^0 \tag{5.70}$$

Given an estimate β^0, y_t^0 can be treated as the regressand in a regression with the regressors $\nabla G(x_t, \beta^0)_{tk}$, which are the derivatives of $G(.)$ with respect to β_k evaluated at β_k^0 and are often referred to as pseudo-regressors; denote the resulting estimates as β_k^1, then the method can be iterated replacing β_k^0 by β_k^1 to obtain β_k^2, say, and so on.

To illustrate this method, consider the simplest ESTAR(1) model and derivatives given by:

$$y_t = \phi_1 y_{t-1} + \gamma_1 y_{t-1} (1 - e^{-\theta(y_{t-1})^2}) + \varepsilon_{E,t}$$

$$= G_{E,t}(.) + \varepsilon_{E,t} \tag{5.71}$$

$$\partial G_{E,t}/\partial \phi_1 = y_{t-1} \tag{5.72a}$$

$$(\partial G_{E,t}/\partial \gamma_1 | \theta^0) = y_{t-1}(1 - e^{-\theta^0(y_{t-1})^2}) \tag{5.72b}$$

$$(\partial G_{E,t}/\partial \theta | \theta^0, \gamma_1^0) = (y_{t-1})^2 e^{-\theta^0(y_{t-1})^2}(\gamma_1^0 y_{t-1}) \tag{5.72c}$$

$$= \gamma_1^0 (y_{t-1})^3 e^{-\theta^0(y_{t-1})^2}$$

Taking a first-order TS approximation of $G_{E,t}(.)$, the linearisation results in:

$$y_t - [\phi_1^0 y_{t-1} + \gamma_1^0 y_{t-1}(1 - e^{-\theta^0(y_{t-1})^2})] + y_{t-1}\phi_1^0 +$$

$$y_{t-1}(1 - e^{-\theta^0(y_{t-1})^2})\theta^0 + \gamma_1^0 (y_{t-1})^3 e^{-\theta^0(y_{t-1})^2}\gamma_1^0$$

$$\approx y_{t-1}\phi_1 + [y_{t-1}(1 - e^{-\theta^0(y_{t-1})^2})]\theta + [\gamma_1^0 (y_{t-1})^3 e^{-\theta^0(y_{t-1})^2}]\gamma_1 + \varepsilon_{E,t} \tag{5.73}$$

Hence, given starting values ϕ_1^0 for ϕ_1, γ_1^0 for γ_1 and θ^0 for θ, the left-hand side variables become known, the regressors are empirically defined and estimates can be obtained, for example, by LS. The vector of pseudo-regressors at time t is apparent as:

$$[(y_{t-1}, y_{t-1}(1 - e^{-\theta^0(y_{t-1})^2}), \gamma_1^0 (y_{t-1})^3 e^{-\theta^0(y_{t-1})^2}]$$

The elements of this vector are known given the starting values. Letting $\kappa^i = (\phi_1^i, \gamma_1^i, \theta^i)'$ then the left-hand side of (5.73) starts with $i = 0$, and then estimation results in estimates corresponding to $i = 1$, which can be used to define a new left-hand side, with a second stage of estimation resulting in estimates corresponding $i = 2$. This iterative process can be judged to have converged by some criteria, for example, an 'inconsequential' change in the residual sum of squares.

5.8.2.ii　*Choosing between models in the STAR class*

The problem is now extended to a choice between two models in the STAR class, specifically the ESTAR and LSTAR models. The idea is to first illustrate the principle with the ESTAR(1) and LSTAR(1) versions of these models; the generalisation is straightforward.

The STAR(1) class is given by (an intercept is now included for greater realism):

$$y_t = \phi_0 + \phi_1 y_{t-1} + (\gamma_0 + \gamma_1 y_{t-1})E(y_{t-d}) + \varepsilon_t \tag{5.74}$$

$$y_t = G(1, y_{t-1}; \phi_0, \phi_1; \gamma_0, \gamma_1; \theta; c) + \varepsilon_{E,t} \qquad \text{ESTAR} \tag{5.75a}$$

$$y_t = G(1, y_{t-1}; \phi_0, \phi_1; \gamma_0, \gamma_1; \psi; c) + \varepsilon_{L,t} \qquad \text{LSTAR} \tag{5.75b}$$

The task is to discriminate between alternative specifications of the transition function, which is either exponential, $E(y_{t-d}) = \Theta(y_{t-d})$, or logistic, $E(y_{t-d}) = \Psi(y_{t-d})$, where $\Theta(y_{t-d}) = (1 - e^{-\theta(y_{t-d}-c)^2})$ and $\Psi(y_{t-d}) = (1 + $

$e^{-\psi(y_{t-d}-c)})-1$, respectively. In turn these specifications, with $d = 1$ for simplicity, are:

$$y_t = \phi_0 + \phi_1 y_{t-1} + (\gamma_0 + \gamma_1 y_{t-1})(1 - e^{-\theta(y_{t-1}-c)^2}) + \varepsilon_{E,t} \quad (5.76a)$$

$$= G_{E,t}(.) + \varepsilon_{E,t}$$

$$y_t = \phi_0 + \phi_1 y_{t-1} + (\gamma_0 + \gamma_1 y_{t-1})(1 + e^{-\psi(y_{t-1}-c)})^{-1} + \varepsilon_{L,t} \quad (5.76b)$$

$$= G_{L,t}(.) + \varepsilon_{L,t}$$

Let the fitted parts of these models be denoted $\hat{G}_E$ and $\hat{G}_L$, respectively, so that:

$$y_t = \hat{G}_{E,t}(1, y_{t-1}; \hat{\phi}_0, \hat{\phi}_1, \hat{\gamma}_0, \hat{\gamma}_1, \hat{\theta}, \hat{c}) + e_{E,t} \quad (5.77a)$$

$$y_t = \hat{G}_{L,t}(1, y_{t-1}; \hat{\phi}_0, \hat{\phi}_1, \hat{\gamma}_0, \hat{\gamma}_1, \hat{\psi}, \hat{c}) + e_{L,t} \quad (5.77b)$$

where $e_{E,t}$ and $e_{L,t}$ are residuals. The parameters are conveniently collected for each model as:

$$\kappa_E = (\phi_0, \phi_1, \gamma_0, \gamma_1, \theta, c)' \qquad \text{ESTAR} \qquad (5.78a)$$

$$\kappa_L = (\phi_0, \phi_1, \gamma_0, \gamma_1, \psi, c)' \qquad \text{LSTAR} \qquad (5.78b)$$

The CME/PSE test will be illustrated for the case where the null model is the LSTAR model and the alternative is the ESTAR model; reversing the roles of these models is a matter of simple substitution. The derivatives of $G_{E,t}(.)$ and $G_{L,t}(.)$ with respect to their respective parameters are as follows:

ESTAR(1) derivatives:

$$\partial G_{E,t}/\partial \phi_0 = 1 \quad (5.79a)$$

$$\partial G_{E,t}/\partial \phi_1 = y_{t-1} \quad (5.79b)$$

$$\partial G_{E,t}/\partial \gamma_0 = (1 - e^{-\theta(y_{t-d}-c)^2}) \quad (5.79c)$$

$$\partial G_{E,t}/\partial \gamma_1 = y_{t-1}(1 - e^{-\theta(y_{t-d}-c)^2}) \quad (5.79d)$$

$$\partial G_{E,t}/\partial \theta = (y_{t-1} - c)^2 e^{-\theta(y_{t-1}-c)^2}(\gamma_0 + \gamma_1 y_{t-1}) \quad (5.79e)$$

$$\partial G_{E,t}/\partial c = -2\theta(y_{t-1} - c) e^{-\theta(y_{t-1}-c)^2}(\gamma_0 + \gamma_1 y_{t-1}) \quad (5.79f)$$

LSTAR(1) derivatives:

$$\partial G_{L,t}/\partial \phi_0 = 1 \quad (5.80a)$$

$$\partial G_{L,t}/\partial \phi_1 = y_{t-1} \quad (5.80b)$$

$$\partial G_{L,t}/\partial \gamma_0 = (1 + e^{-\psi(y_{t-d}-c)})^{-1} \tag{5.80c}$$

$$\partial G_{L,t}/\partial \gamma_1 = y_{t-1}(1 + e^{-\psi(y_{t-d}-c)})^{-1} \tag{5.80d}$$

$$\partial G_{L,t}/\partial \psi = (y_{t-1}-c)(1 + e^{-\psi(y_{t-1}-c)})^{-2}e^{-\psi(y_{t-1}-c)}(\gamma_0 + \gamma_1 y_{t-1}) \tag{5.80e}$$

$$\partial G_{L,t}/\partial c = -\psi(1 + e^{-\psi(y_{t-1}-c)})^{-2}e^{-\psi(y_{t-1}-c)}(\gamma_0 + \gamma_1 y_{t-1}) \tag{5.80f}$$

A symbol ^ above the derivative will indicate substitution of an estimator for the coefficient.

The vectors of derivatives with respect to the estimated coefficients of the ESTAR and LSTAR models respectively, are:

$$\nabla\hat{G}_{E,t} = (\partial\hat{G}_{E,t}/\partial\hat{\phi}_0, \partial\hat{G}_{E,t}/\partial\hat{\phi}_1, \partial\hat{G}_{E,t}/\partial\hat{\gamma}_0, \partial\hat{G}_{E,t}/\partial\hat{\gamma}_1, \partial\hat{G}_{E,t}/\partial\hat{\theta}, \partial\hat{G}_{E,t}/\partial\hat{c})'$$

$$= (\nabla\hat{G}_{E,1t}, \nabla\hat{G}_{E,2t}, \nabla\hat{G}_{E,3t}, \nabla\hat{G}_{E,4t}, \nabla\hat{G}_{E,5t}, \nabla\hat{G}_{E,6t})' \tag{5.81a}$$

$$\nabla\hat{G}_{L,t} = (\nabla\hat{G}_{L,1t}, \nabla\hat{G}_{L,2t}, \nabla\hat{G}_{L,3t}, \nabla\hat{G}_{L,4t}, \nabla\hat{G}_{L,5t}, \nabla\hat{G}_{L,6t})' \tag{5.81b}$$

The notation explicitly recognises that the derivative is evaluated for each t in the sample range, hence the time subscript t. Note also that these vectors are $K \times 1$, where $K = 6$. Each vector partitions into three separate parts corresponding to the derivatives with respect to: (i) the basic AR coefficients, ϕ_0 and ϕ_1; (ii) the variational impacts on the AR coefficients γ_0 and γ_1; (iii) and the nonlinear coefficients, θ and c (or ψ and c in the case of the LSTAR model).

Define the $6 \times \tilde{T}$ matrix of all derivatives for the ESTAR case by $\nabla\hat{G}_E$ (where $\tilde{T}$ is the effective sample size):

$$\nabla\hat{G}_E = \begin{bmatrix} \nabla\hat{G}_{E,11} & \cdots & \nabla\hat{G}_{E,1\tilde{T}} \\ \vdots & \vdots & \vdots \\ \nabla\hat{G}_{E,61} & \cdots & \nabla\hat{G}_{E,6\tilde{T}} \end{bmatrix} \tag{5.82a}$$

$$= \begin{pmatrix} \nabla\hat{G}_{E,1} \\ \vdots \\ \nabla\hat{G}_{E,6} \end{pmatrix} \tag{5.82b}$$

Analogous expressions serve to define $\nabla\hat{G}_L$.

Define the following fitted values and associated residuals:

$$\hat{G}_{E,t} = \hat{\phi}_0 + \hat{\phi}_1 y_{t-1} + (\hat{\gamma}_0 + \hat{\gamma}_1 y_{t-1})(1 - e^{-\hat{\theta}(y_{t-1}-\hat{c})^2}) \tag{5.83}$$

$$e_{E,t} = y_t - \hat{G}_{E,t} \tag{5.84}$$

$$e_E = (e_{E,1}, e_{E,2}, \ldots, e_{E,\tilde{T}}) \tag{5.85}$$

$$\hat{G}_{Lt} = \hat{\phi}_0 + \hat{\phi}_1 y_{t-1} + (\hat{\gamma}_0 + \hat{\gamma}_1 y_{t-1})(1 + e^{-\hat{\psi}(y_{t-1} - \hat{c})})^{-1} \tag{5.86}$$

$$e_{L,t} = y_t - \hat{G}_{L,t} \tag{5.87}$$

$$e_L = (e_{L,1}, e_{L,2}, \dots, e_{L,\hat{T}}) \tag{5.88}$$

In this example, the null model is the LSTAR model so that the alternative is the ESTAR model (the roles can then be reversed). The test statistic is based on:

$$
\begin{aligned}
\hat{\Psi} &= \sum_{t=q}^{T} (\nabla\hat{G}_{E,t})(\hat{G}_{L,t} - \hat{G}_{E,t}) \\
&= -\sum_{t=q}^{T} (\nabla\hat{G}_{E,t})(e_{L,t} - e_{E,t}) \\
&= -\sum_{t=q}^{T} (\nabla\hat{G}_{E,t}) e_{L,t} \\
&= -(\nabla\hat{G}_E)(e_L)' \tag{5.89}
\end{aligned}
$$

Note that the right-hand side of the second line follows from the equality obtained by subtracting and adding y_t:

$$
\begin{aligned}
\hat{G}_{L,t} - \hat{G}_{E,t} &= -[(y_t - \hat{G}_{L,t}) - (y_t - \hat{G}_{E,t})] \\
&= -(e_{L,t} - e_{E,t}) \tag{5.90}
\end{aligned}
$$

The third line uses the result that, by construction, the residuals of a fitted model are orthogonal to the derivatives of the fitted values from that model; therefore, $(\nabla\hat{G}_E)e'_E = 0$. The 6×1 vector $\hat{\Psi}$, in effect, comprises an 'orthogonality' assessment for each of the derivatives of $\hat{G}_E$. The test is, therefore, based on assessing whether the residuals from the LSTAR model, which is the null model, are orthogonal to the derivatives from the ESTAR model, which is the alternative model.

Substituting the appropriate derivatives in $\nabla\hat{G}_{E,t}$ gives:

$$\nabla\hat{G}_{E,t} =$$
$$[1, y_{t-1}, (1 - e^{-\theta(y_{t-1} - c)^2}), y_{t-1}(1 - e^{-\theta(y_{t-1} - c)^2}),$$
$$(y_{t-1} - c)^2 e^{-\theta(y_{t-1} - c)^2}(\gamma_0 + \gamma_1 y_{t-1}), -2\theta(y_{t-1} - c) e^{-\theta(y_{t-1} - c)^2}(\gamma_0 + \gamma_1 y_{t-1})]' \tag{5.91}$$

It is convenient to split the 6×1 vector $\hat{\Psi}$ into its 3 sets of components corresponding to the partition in the vector of derivatives; thus:

$$\nabla\hat{G}_{E,t} : \hat{\Psi} = (\hat{\Psi}_1; \hat{\Psi}_2; \hat{\Psi}_3)'$$

where: $\hat{\Psi}_1 = (\hat{\Psi}_{1,1}, \hat{\Psi}_{1,2})$, $\hat{\Psi}_2 = (\hat{\Psi}_{1,3}, \hat{\Psi}_{1,4})$ and $\hat{\Psi}_3 = (\hat{\Psi}_{1,5}, \hat{\Psi}_{1,6})$.

The first two terms (p + 1, in general) in the sum are:

$$(\hat{\Psi}_{1,1}, \hat{\Psi}_{1,2}) = -\sum_{t=q}^{T} (1, y_{t-1}) e_{L,t}$$

$$= -\left(\sum_{t=q}^{T} e_{L,t}, \sum_{t=q}^{T} y_{t-1} e_{L,t}\right) \tag{5.92}$$

These terms are zero because it follows that the residuals from the alternative model (whichever that is) are orthogonal to the regressors common to both models, that is 1 and y_{t-1}.

This leaves two sets of components in the test vector $\hat{\Psi}_2$ and $\hat{\Psi}_3$, where:

$$\hat{\Psi}_2 = (\hat{\Psi}_{1,3} \ \hat{\Psi}_{1,4}) \quad \text{where}$$

$$\hat{\Psi}_{1,3} = -\sum_{t=q}^{T} (1 - e^{-\hat{\theta}(y_{t-1}-\hat{c})^2}) e_{Lt}$$

$$= -\nabla \hat{G}_{E,3} e'_L \tag{5.93a}$$

$$\hat{\Psi}_{1,4} = -\sum_{t=q}^{T} (1 - e^{-\hat{\theta}(y_{t-1}-\hat{c})^2}) y_{t-1} e_{Lt}$$

$$= -\nabla \hat{G}_{E,4} e'_L \tag{5.93b}$$

$$\hat{\Psi}_3 = (\hat{\Psi}_{1,5} \ \hat{\Psi}_{1,6}) \quad \text{where}$$

$$\hat{\Psi}_{1,5} = -\sum_{t=q}^{T} (y_{t-1} - \hat{c})^2 e^{-\hat{\theta}(y_{t-1}-\hat{c})^2} (\hat{\gamma}_0 + \hat{\gamma}_1 y_{t-1}) e_{L,t}$$

$$= -\nabla \hat{G}_{E,5} e'_L \tag{5.94a}$$

$$\hat{\Psi}_{1,6} = +2 \sum_{t=q}^{T} \hat{\theta}(y_{t-1} - \hat{c}) e^{-\hat{\theta}(y_{t-1}-\hat{c})^2} (\hat{\gamma}_0 + \hat{\gamma}_1 y_{t-1}) e_{L,t}$$

$$= -\nabla \hat{G}_{E,6} e'_L \tag{5.94b}$$

The PSE test suggested by Chen (2003), following Chen and Kuan (2002), is based on $\hat{\Psi}_2$, which checks the orthogonality condition for the derivatives with respect to γ_0 and γ_1.

The test statistic in the Wald form, which is being suggested here (see also 5.64) for the ESTAR(1) model against the null of the LSTAR(1) model, is:

$$\text{PSE}(\hat{\Psi}_2)_{E|L} = \hat{\Psi}_2 [\text{Var}(\hat{\Psi}_2)]^{-1} \hat{\Psi}'_2 \tag{5.95}$$

$\text{PSE}(\hat{\Psi}_2)_{E|L}$ is asymptotically distributed as $\chi^2(r)$, where r is the rank of $\text{Var}(\hat{\Psi}_2)$. In this example, $r = p + 1 = 2$ as $p = 1$; that is, in this case the $\text{PSE}(\hat{\Psi}_2)$ test checks two conditions for orthogonality. Equally, the test could be based on $\hat{\Psi}_2$ and $\hat{\Psi}_3$, in which case the test statistic would be asymptotically distributed as $\chi^2(4)$. Chen (op. cit.) allows for $\text{Var}(\hat{\Psi}_2)$ to be of less than full rank, in which case $[\text{Var}(\hat{\Psi}_2)]^{-1}$ is replaced by a generalised inverse. Changing the roles of the LSTAR and ESTAR models leads to the test statistic denoted $\text{PSE}(\hat{\Psi}_2)_{L|E}$.

The test statistic of (5.95) is, therefore, based on the derivatives $\partial G_{E,t}/\partial \gamma_0$ and $\partial G_{E,t}/\partial \gamma_1$, where estimated values replace unknown values of θ and c (a similar strategy is adopted in the procedure for testing for serial correlation suggested by Eitrheim and Teräsvirta (1996) and described in the following section). To make the construction of the test statistic analogous to the linear case, note that the pseudo-regressors replace the regressors. It is helpful to define the following notation: construct Z as the $\tilde{T} \times 2$ matrix of 'observations' on the pseudo-regressors, where the first column comprises $\partial \hat{G}_{E,t}/\partial \hat{\gamma}_0$, for example, the t-th element is $(1 - e^{-\hat{\theta}(y_{t-1}-\hat{c})^2})$, and the second column comprises $\partial \hat{G}_{E,t}/\partial \hat{\gamma}_1$ where the t-th element is $y_{t-1}(1 - e^{-\hat{\theta}(y_{t-1}-\hat{c})^2})$; both columns of Z are composed of entirely known elements given that estimates replace unknowns.

Next construct the equivalent of P_W (as in the linear case), but where $X_W = \nabla \hat{G}'_L$, which is a $\tilde{T} \times 6$ matrix comprising columns in which are evaluated the 6 derivatives of the LSTAR model. With these substitutions, the test statistic is in principle as in the linear case where, to complete the analogy, $\hat{\delta}_W$ is chosen to be $\hat{\psi}'_2$, which is a 2×1 vector, or choose $(\hat{\psi}_2, \hat{\psi}_3)'$, which tests all four conditions and is therefore a 4×1 vector.

In the more general case in which $p > 1$, the number of orthogonality conditions will increase one-for-one with the number of pseudo-regressors, as the residuals from the null model are checked for orthogonality against the pseudo-regressors of the alternative model.

Chen (2003) provides simulation results under some conditions that simplify estimation of the ESTAR and LSTAR models (specifically ESTAR(1) and LSTAR(1) models are used with $d = 1$, $c = 0$ and σ_ε^2 is assumed known), and see Chen and Kuan (2007) for a corrigendum; in this case, whilst the PSE test is in several cases more powerful than the Teräsvirta tests, there are some problems with the size not just of the PSE test but also the Teräsvirta tests, there being examples of undersizing and oversizing. More research would be useful in this case, especially for the situation likely to face the practitioner with more complex ESTAR and LSTAR models and, at least, where c and σ_ε^2 are unknown. In the next chapter, in the context of an empirical example, we take up a suggestion due to Amemiya (1985, section 4.3.4), that because the distributional results for the nonlinear ML (LS) estimator are asymptotic, a bootstrap approach may be useful (Amemiya, 1985, also suggests that the jackknife method could be considered).

5.8.2.iii *Illustration: ESTAR(1) or LSTAR(1) model for the US:UK real exchange rate*

In this illustration, the ESTAR(1) and LSTAR(1) models are used in deviation from threshold form, with monthly data on the bilateral real exchange rate between the US and the UK for the sample period 1973 m3 to 2000 m7. In fact, as the further empirical results reported in the next chapter suggest, a higher-order

STAR model may be warranted, but the essence here is to illustrate the principles rather than establish substantive results. The models to be considered are:

$$(y_t - c) = \phi_1(y_{t-1} - c) + \gamma_1(y_{t-1} - c)(1 - e^{-\theta(y_{t-1}-c)^2}) + \varepsilon_{E,t} \tag{5.96}$$

$$(y_t - c) = \phi_1(y_{t-1} - c) + \gamma_1(y_{t-1} - c)(1 + e^{-\psi(y_{t-1}-c)})^{-1} + \varepsilon_{L,t} \tag{5.97}$$

And, as the more detailed empirical work reported in the next chapter indicates, the outer regime restriction is imposed, that is $\phi_1 = -\gamma_1$. This requires some slight modifications to the derivatives presented earlier (5.79a–5.79f), which are now as follows.

ESTAR(1) derivatives:

$$\partial G_{E,t}/\partial \gamma_1 = -(y_{t-1} - c)e^{-\theta(y_{t-d}-c)^2} \tag{5.98a}$$

$$\partial G_{E,t}/\partial \theta = \gamma_1(y_{t-1} - c)^3 e^{-\theta(y_{t-1}-c)^2} \tag{5.98b}$$

$$\partial G_{E,t}/\partial c = 1 - [2\theta\gamma_1(y_{t-1} - c)^2 e^{-\theta(y_{t-1}-c)^2} - \gamma_1 e^{-\theta(y_{t-1}-c)^2}] \tag{5.98c}$$

LSTAR(1) derivatives:

$$\partial G_{L,t}/\partial \gamma_1 = (y_{t-1} - c)[(1 + e^{-\psi(y_{t-d}-c)})^{-1} - 1] \tag{5.99a}$$

$$\partial G_{L,t}/\partial \psi = \gamma_1(y_{t-1} - c)^2(1 + e^{-\psi(y_{t-1}-c)})^{-2}e^{-\psi(y_{t-1}-c)} \tag{5.99b}$$

$$\partial G_{L,t}/\partial c = 1 + \gamma_1$$
$$\qquad - \psi\gamma_1(y_{t-1} - c)(1 + e^{-\psi(y_{t-1}-c)})^{-2}e^{-\psi(y_{t-1}-c)}$$
$$\qquad - \gamma_1(1 + e^{-\psi(y_{t-1}-c)})^{-1} \tag{5.99c}$$

In these restricted versions of the models (the restrictions are often essential to obtain a model that is empirically identified) it seems sensible to base the CME/PSE test on all three derivatives, corresponding to three pseudo-regressors in each case. In the first case, the null model is taken to be the ESTAR model, and the test assesses the deviations from zero of the ESTAR residuals from the LSTAR pseudo-regressors; the roles are then reversed. The estimation results are reported in Table 6.6 in the following chapter; our interest here is in the test outcomes.

In each case, could be the test statistic is asymptotically distributed as $\chi^2(3)$. A bootstrap procedure could be used to obtain the p-value for each statistic; (in a related context, Amemiya (1985) suggested a bootstrap approach to approximating the distribution of the nonlinear estimator.) In the illustration used here, the sample values of the test statistics are:

$$PSE(\hat{\Psi}_2)_{E|L} = 2.21, \quad PSE(\hat{\Psi}_2)_{L|E} = 4.23$$

Neither test value is significant using $\chi^2_{0.05}(3) = 7.82$; but using a Teräsvirta-type rule, the ESTAR model would be preferred as the LSTAR model is less consistent with the sample data, that is, $PSE(\hat{\Psi}_2)_{L|E}$ results in a smaller p-value.

5.9 Testing for serial correlation in STAR models

A number of further tests of the adequacy of STAR models are given in Eitrheim and Teräsvirta (1996); see also van Dijk et al. (2002). One of the most important is the extension of testing for the absence of residual serial correlation, which is an aspect of model specification that has become standard in linear models and finds a natural extension to nonlinear models.

5.9.1 An LM test for serial correlation in the STAR model

Consider the STAR(p) model given by:

$$y_t = \phi_0 + \sum_{i=1}^{p} \phi_i y_{t-i} + (\gamma_0 + \sum_{i=1}^{p} \gamma_i y_{t-i}) E(y_{t-d}) + \varepsilon_t \tag{5.100a}$$

$$= G(1, y_{t-1}, \ldots, y_{t-p}; y_{t-d}; \phi_0, \phi_1, \ldots, \phi_p; \gamma_0, \gamma_1, \ldots, \gamma_p; \theta; c) + \varepsilon_t \tag{5.100b}$$

The $G(.)$ function in equation (5.100b) separates the variables and coefficients in the AR and ESTAR parts of the model. In the case of ESTAR model, then $E(y_{t-d}) = \Theta(y_{t-d}) = (1 - e^{-\theta(y_{t-d}-c)^2})$; if the model is LSTAR, then $E(y_{t-d}) = \Psi(y_{t-d}) = (1 + e^{-\theta(y_{t-d}-c)})^{-1}$; note that there are $2(p+1) + 2$ parameters of interest in $G(.)$ in (5.100b).

Before stating the LM test for serial correlation in the STAR model, it is useful to recall the nature of the test in the linear AR model where the function $G(.)$ is particularly simple, that is, $G(.) = G(1, y_t, \ldots, y_{t-p}; \phi_0, \phi_1, \ldots, \phi_p)$; see Godfrey (1978). An LM test of serial independence against either an MA(q) or AR(q) error process requires an auxiliary regression where the dependent variable is the residual $\hat{\varepsilon}_t$ from the linear AR(p) model regressed on q lags of $\hat{\varepsilon}_t$ and the regressors $(1, y_{t-1}, \ldots, y_{t-p})$. Although, by construction, the sequence $\{\hat{\varepsilon}_t\}_{t=p+1}^{T}$ has a zero mean, and so a constant would not be needed in the auxiliary regression if estimated over the same sample period, q further observations are 'lost' in forming the auxiliary regression and a constant is usually included to preserve the zero sum property for the residuals of that regression. A point to note is that the general principle is to include in the auxiliary regression q lags of $\hat{\varepsilon}_t$ and the $p + 1$ partial derivatives of $G(.)$ with respect to the parameters ϕ_i, $i = 0, \ldots, p$; in the linear case, the derivatives are just the regressors in the original model.

The LM test follows the same principle in the nonlinear model but, by its very nature, some of the partial derivatives are functions of the parameters. These unknowns are replaced by their estimates from the nonlinear regression, just as in the construction of a CME-type test. Consider the case where an ESTAR(p) model has been estimated. The derivatives of $\{\phi_i\}_{i=0}^{p}$ are just $(1, y_{t-1}, \ldots, y_{t-p})$ and the derivatives of $\{\gamma_i\}_{i=0}^{p}$ are $(1, y_{t-1}, \ldots, y_{t-p})\hat{\Theta}_t$, where $\hat{\Theta}_t$ is the transition function evaluated at the estimates of θ and c; also, two further derivatives required are $\partial G/\partial \theta$ and $\partial G/\partial c$. The latter is illustrated for an ESTAR(2) model, with the derivatives for the LSTAR model left to an exercise, see Q5.2.iii. The

ESTAR(2) model is given by:

$$y_t = \phi_0 + \phi_1 y_{t-1} + \phi_2 y_{t-2} + (\gamma_0 + \gamma_1 y_{t-1} + \gamma_2 y_{t-2})(1 - e^{-\theta(y_{t-d}-c)^2}) + \varepsilon_t \tag{5.101}$$

The required partial derivatives (see also 5.79 and 5.80 for the simpler case) are:

$$\partial G_{E,t}/\partial \phi_0 = 1; \partial G/\partial \phi_1 = y_{t-1}; \partial G_{E,t}/\partial \phi_2 = y_{t-2}; \tag{5.102a}$$

$$\partial G_{E,t}/\partial \gamma_0 = (1 - e^{-\theta(y_{t-d}-c)^2}); \partial G_{E,t}/\partial \gamma_1 = y_{t-1}(1 - e^{-\theta(y_{t-d}-c)^2}); \tag{5.102b}$$

$$\partial G_{E,t}/\partial \gamma_2 = y_{t-2}(1 - e^{-\theta(y_{t-d}-c)^2}); \tag{5.102c}$$

$$\partial G_{E,t}/\partial \theta = (y_{t-d} - c)^2 e^{-\theta(y_{t-d}-c)^2}(\gamma_0 + \gamma_1 y_{t-1} + \gamma_2 y_{t-2}) \tag{5.102d}$$

$$\partial G_{E,t}/\partial c = -2\theta(y_{t-d} - c)e^{-\theta(y_{t-d}-c)^2}(\gamma_0 + \gamma_1 y_{t-1} + \gamma_2 y_{t-2}) \tag{5.102e}$$

In general, the auxiliary regression for the test of serial correlation of order q in an ESTAR(p) model, is:

$$\hat{\varepsilon}_t = \delta_0 + \sum_{i=1}^{q} \delta_{1i}\hat{\varepsilon}_{t-i} + \sum_{i=1}^{P} \delta_{2i} y_{t-i} + \delta_{30}\hat{\Theta}(y_{t-d}) + \sum_{i=1}^{P} \delta_{3i}\hat{\Theta}(y_{t-d})y_{t-i}$$

$$+ \delta_{41}[(y_{t-d} - \hat{c})^2 \hat{\lambda}_t(\hat{\gamma}_0 + \sum_{i=1}^{P} \gamma_i y_{t-i})] +$$

$$\delta_{42}[2\hat{\theta}(y_{t-d} - \hat{c})\hat{\lambda}_t(\hat{\gamma}_0 + \sum_{i=1}^{P} \hat{\gamma}_i y_{t-i})] + \omega_t \tag{5.103}$$

where:

$$\hat{\Theta}(y_{t-d}) = (1 - e^{-\hat{\theta}(y_{t-d}-\hat{c})^2})$$

$$\hat{\lambda}_t = e^{-\hat{\theta}(y_{t-d}-\hat{c})^2}$$

and, as usual, ˆ above a coefficient indicates an estimator. The suggested test statistic, in the form of an F statistic, is:

$$F_{LM} = \left(\frac{RSS_R - RSS_{UR}}{RSS_{UR}}\right)\left(\frac{\check{T} - (n+q)}{q}\right) \tag{5.104}$$

where RSS_R is the residual sum of squares from (5.101) and RSS_{UR} is the residual sum of squares from the residuals defined from (5.103).

Note that q lagged residuals are needed in the specification of (5.103), hence the total number of observations available for estimation at this stage is $\check{T} = \tilde{T} - q$. The total number of regressors in (5.103) is n + q, where n = 2(p + 1) + 2; this would be reduced by 1 if c = 0 as there would be no term for $\partial G_{E,t}/\partial c$; and if r restrictions are imposed on the ESTAR (or LSTAR) model, for example relating to restrictions on the inner and outer regimes, then n = 2(p + 1) + 2 - r. Imposing restrictions on the STAR model has implications for the appropriate form of the auxiliary regression (5.103). (An example illustrates this point in a question at the end of the chapter.) For now, note that if there are K coefficients in the unrestricted regression, then the imposition of r restrictions reduces the number of freely estimated parameters to K − r, hence only K − r, not K, partial derivatives are required in (5.103).

In the context of the linear regression model, the zero sum property for $\hat{\varepsilon} = (\hat{\varepsilon}_{p+1}, \ldots, \hat{\varepsilon}_T)'$ is part of the orthogonality conditions (p + 1 in that case) that follow from the normal equations in LS minimisation. That is, construct the data matrix Z such that the t-th row is $(1, y_{t-1}, \ldots, y_{t-p})$; thus, the j-th column of Z comprises the T − p observations on the j-th regressor, and then $Z'\hat{\varepsilon} = 0$. The interpretation is that the LS residuals are orthogonal to each vector of observations of the partial derivatives in the regression where, in the linear model, the partial derivatives are just the regression variables (see also Section **5.8**).

However, in a nonlinear model, as Eitrheim and Teräsvirta (1996) note, the partial derivatives, for example (5.102a)–(5.102e) in the ESTAR(2) model, may not be exactly orthogonal to the residual vector; hence, they suggest an 'orthogonalisation' step as follows. Instead of using RSS_R in F_{LM}, first use the residual sum of squares, say RSS_{OR}, from the regression of $\hat{\varepsilon}_t$ on (just) the partial derivatives; and denote the residuals from this regression as $\tilde{\varepsilon}_t$. For example, in the ESTAR(p) model, the regression to obtain $\tilde{\varepsilon}_t$ is as (5.103), but excludes the lagged residuals $\hat{\varepsilon}_{t-i}$, i = 1, …, q; second, use these residuals $\tilde{\varepsilon}_t$, not $\hat{\varepsilon}_t$ in the auxiliary regression.

5.10   Estimation of ESTAR models

Although ESTAR models have been a popular choice for nonlinear modelling in a nonstationary context, some authors have reported difficulties in empirically identifying all of the parameters. For example, in an application of the ESTAR model to real exchange rates and ex-post real interest rates for 11 industrialised countries, KSS (2003) report that γ_1 was 'very poorly (empirically) identified' (my clarification in parentheses); as a result γ_1 was set equal to minus unity and other estimates were conditional on that value. Taylor et al. (2001) also reported similar problems in their application to real exchange rates.

The ESTAR model can be directly estimated using a standard nonlinear optimisation program (in the illustrations below the nonlinear routines in TSP and RATS were used). Possibly as a complement or an alternative to direct nonlinear estimation, the coefficients can be estimated by minimising the residual sum of squares in a grid search procedure; rather than grid search over the entire parameter space, the search can be focused on $(\theta, c) \in [\theta \times C]$.

5.10.1   A grid search

The grid search over $[\theta \times C]$ can avoid some of the difficulties encountered in direct estimation of all of the parameters jointly. Note that conditional on θ

and c, say $\tilde{\theta}$ and $\tilde{c}$, the transition function is just a function of y_{t-1}, say $\tilde{\Theta}_t = (1 - e^{-\tilde{\theta}(y_{t-1}-\tilde{c})^2})$, and the ESTAR model is then simply a modified linear AR model. For example, the ESTAR(1) is:

$$y_t = \phi_1 y_{t-1} + \gamma_1 y^*_{t-1} + \varepsilon_t \tag{5.105}$$

where $y^*_{t-1} = y_{t-1}\tilde{\Theta}_t$. Estimates of ϕ_1 and γ_1 are first obtained conditional on $\tilde{\theta}$ and $\tilde{c}$. The resulting (conditional) residual sum of squares is denoted $\tilde{\varepsilon}(\tilde{\theta}, \tilde{c})'\tilde{\varepsilon}(\tilde{\theta}, \tilde{c})$; and the procedure is repeated for all $(\tilde{\theta}, \tilde{c}) \in [\boldsymbol{\theta} \times \mathbf{C}]$. The final estimates of ϕ_1 and γ_1 are obtained as those relating to the overall minimum of $\tilde{\varepsilon}(\tilde{\theta}, \tilde{c})'\tilde{\varepsilon}(\tilde{\theta}, \tilde{c})$ for $(\tilde{\theta}, \tilde{c}) \in [\boldsymbol{\theta} \times \mathbf{C}]$. This principle extends quite simply to more complex ESTAR models. The focus here is on ESTAR models but the comments and procedures apply equally to LSTAR models.

5.10.2   Direct nonlinear estimation

As for direct nonlinear estimation, some guidance on starting values and the range of the grid $[\boldsymbol{\theta} \times \mathbf{C}]$ for the nonlinear parameters is as follows. The threshold parameter c should be in the range of y_t, and is unlikely to be too close to either end of the range. (In practice, there may be problems with short sample periods, where the realisations do not indicate the true range of y_t.) Thus, order the sequence $\{y_t\}$ such that $y_{(i)}$ is the i-th percentile of the sequence; that is, $y_{(1)} < y_{(2)} \cdots < y_{(i)} \cdots < y_{(100)} = \max(y_t)$. One might then take the 10-th and 90-th percentiles, or the inter-quartile range, as defining the likely outside limits of the grid. A fairly coarse grid could be used in the first instance, with a view to a finer interval in the second stage of the search. In direct nonlinear estimation, the median or mean of $\{y_t\}$ could be used as a starting value for c.

As to θ, this is the essential nonlinear term in the exponential in the transition function $\lambda_t = e^{-\theta(y_{t-1}-c)^2}$. Initially consider the case where c is known, for example, when it is assumed that $c = 0$, so that $\lambda_t = e^{-\theta y^2_{t-1}}$, or c is the sample mean of $\{y_t\}$ and the use of demeaned data is appropriate. The following argument is based on Ozaki (1992). Note that $e^{-\theta y^2_{t-1}} \to 0$ as $\theta y^2_{t-1} \to \infty$; then in the extreme $e^{-\theta y^2_{t-1}} \approx 0$ for $y_{t-1} = y_{(100)}$, where $\approx$ means that $e^{-\theta y^2_{t-1}} < \zeta$, for some ζ, where ζ is 'negligible'. Then taking (natural) logs, $\theta y^2_{(100)} < -\ln(\zeta)$, that is $\theta < -\ln(\zeta)/y^2_{(100)}$.

To take an example from the real exchange illustration used in the next chapter, suppose that $y^2_{(100)} = 0.5$ and, for practical purposes, $\zeta = 0.01$ is regarded as negligible, then $0 < \theta < 9.21$. This is admittedly still quite a large range, but a coarse grid search in the first instance should quickly indicate the likely range of θ; (and rescaling may reduce the range.) If $\zeta = 0.05$ then the range contracts to $0 < \theta < 5.99$, and if $\zeta = 0.20$ then the range contracts further to $0 < \theta < 3.21$.

In the case of a nonzero threshold c, the transition function is a function of $(y_{t-1} - c) = x_t$, and Ozaki's condition is $\theta < -\log(\zeta)/x^2_{(100)}$. A frequent empirical practice is to scale the variable in the transition function so that y_{t-1} is 'standardised' by subtracting c and dividing by the sample standard error, s_y, in the case of the LSTAR model, or the sample variance, s_y^2, for an ESTAR model; see Sections **5.2.9** and **5.3.1**. In the latter case, the transition function is, with $d = 1$, $\Theta(y_{t-1}) = (1 - e^{-\theta^*\{z_{t-1}\}^2})$, where $z_{t-1} = (y_{t-1} - c)/s_y$, so that z_{t-1} is scale-free, (changing the units of measurement of y_{t-1} will not change z_{t-1}). Now Ozaki's condition is: $\theta < -\log(\zeta)/z_{(100)}^2$.

5.10.3   Simulation set-up

Some of the estimation issues are assessed here by simulation. A full-scale evaluation would be quite complex as even in a simple ESTAR(1) model, the estimation results are potentially a function of the parameters θ, c, γ_1, ϕ_1 and σ_ε^2. In specifying the DGP, we concentrate on the case that has been quite widely used in empirical practice, that is, $\gamma_1 = -1$, $\phi_1 = 1$. These coefficient values characterise a unit root in the inner regime and white noise in the outer regime.

In the direct nonlinear estimation part of the simulation, starting values were required for the coefficients to be estimated and since, even with good starting values, the estimated values can occasionally go 'astray', it is necessary to adopt a rule for admissible estimates and also count the proportion of times that the estimates were inadmissible on the adopted rule.

Some preliminaries on starting values, grid selection and admissibility are considered in the first three subsections that follow and the simulation results are summarised in the fourth subsection.

5.10.3.i   Setting starting values for direct estimation

The results are illustrated for the procedure that initially estimates a linear AR(1) model and sets the starting value for ϕ_1 equal to the estimated coefficient on y_{t-1}, say $\hat{\phi}_1$ and the starting value for γ_1 was then set equal to $-\hat{\phi}_1$. The starting value of θ was set equal to zero, which is the default value in the non-linear routine; alternatives included starting θ at a non-zero value, for example the mid-point of the Ozaki range, but the nonlinear routines used here were found to be robust to these alternatives. The starting value for c was the mean of the realised sequence of y_t. The Gauss-Newton method was used in the nonlinear estimation routine with a relatively tight tolerance (0.0001). The results showed quite a significant difference between the cases with and without ϕ_1 estimated.

5.10.3.ii   Grid search

The grid search method used the Ozaki range for θ, (assuming $c = 0$) provided the range was at least $0 < \theta < 3$; the grid limits for c were taken to be the inter-quartile range of $\{y_t\}_{t=1}^T$; the respective ranges were divided into 50 and 30 equi-spaced points.

5.10.3.iii　The error variance and admissibility rules

As nonlinear processes are not invariant to the error variance, σ_ε^2, this was also varied in the simulation set-up, with $\sigma_\varepsilon^2 = 0.4^2$, 0.6^2, 0.8^2. Admissibility rules were applied to the estimates of γ_1 and θ; specifically an estimate $\hat{\gamma}_1$ of γ_1 was admissible if $(-2 \le \hat{\gamma}_1 \le 0)$, and an estimate $\hat{\theta}$ of θ was admissible if $(0 \le \hat{\theta} \le 4)$. These rules were applied jointly. Although these rules and starting values are to some extent arbitrary, they mimic what is likely to happen in practice, in the sense that some judgement is likely to be applied to the outcomes of nonlinear estimation.

In the simulation if an estimate was deemed inadmissible, another draw of the process was allowed to achieve the desired number of admissible samples. The initial number of simulations was set at $1,000 = NS$; but, say, IA samples were inadmissible, so that in all there were NS + IA simulations, then the ratio PAS $= NS/(NS + IA)$ is the proportion of admissible samples in the simulation. Such a procedure quite deliberately involves sample selection, intended to indicate the performance of direct estimation relative to a grid search. Direct nonlinear estimation did on occasion lead to quite unacceptable estimates; for example, in one instance $\hat{\theta}$ was well over 20,000!

5.10.3.iv　The set-up and simulation results

The DGP is an ESTAR(1) model as follows:

$$y_t = \phi_1 y_{t-1} + \gamma_1 y_{t-1}(1 - e^{-\theta(y_{t-1})^2}) + \varepsilon_t \tag{5.106}$$

with $\theta = 0.2$, 0.5, 1.0, 2.0; $c = 0$; $\phi_1 = 1$; $\gamma_1 = -1$; $\varepsilon_t \sim \text{niid}(0, \sigma_\varepsilon^2)$ and $\sigma_\varepsilon^2 = 0.4^2$, 0.6^2, 0.8^2. $T = 200$ and to avoid start-up problems, 100 pre-sample observations were generated and then discarded for the simulation samples.

There are a number of dimensions on which to summarise the simulation results. The first point to note is that, with the censoring of the estimates from direct nonlinear estimation, there is very little to choose between direct estimation and grid estimation; however, by itself this is misleading as to the efficacy of direct estimation. What becomes relevant is the proportion of samples that are judged admissible by the rules outlined above. This information is summarised in Figure 5.21, where θ is on the horizontal axis and PAS (proportion) is on the vertical axis. Note that the situation is worse for direct estimation when σ_ε^2 is small at 0.4^2, with less than 50% of samples then admissible if θ is also small. There are some improvements as θ increases and as σ_ε^2 increases; but not uniformly as far as the increase in θ is concerned. The grid search method dominates the direct estimation method. As θ increases, the degree of nonlinearity becomes more apparent in the data and this generally also leads to an increase in the proportion of admissible samples for the grid search method.

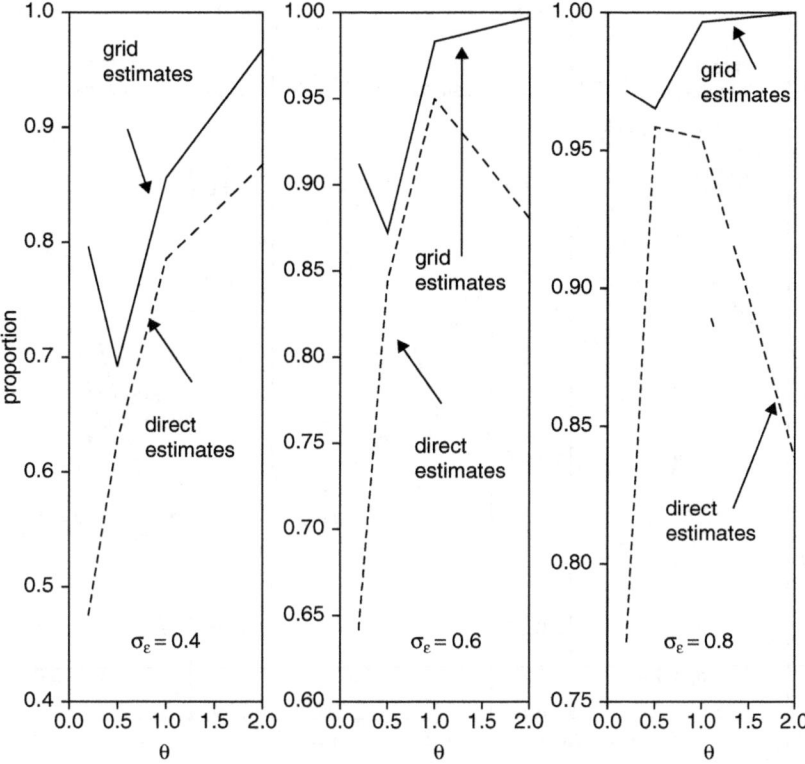

Figure 5.21   Admissible proportion of estimates

The next consideration is to assess the sensitivity of the mean bias and rmse of the two methods, and over the admissible samples, to changes in σ_ε^2 and θ. This is illustrated in Figure 5.22 for the mean bias and in Figure 5.23 for rmse. First, consider the simulation bias in Figure 5.22, where it is evident that there is a noticeable bias, which diminishes as both σ_ε^2 and θ increase. (Throughout grid estimates are shown by the solid line and direct estimates by the dashed line.) The 45° line in the figure shows where estimated equals actual and the difference between the plot of estimates and the 45° line is the (mean) bias. For θ less than 2, and particularly for smaller values of θ, the mean bias and rmse are substantial; and there is only a relatively slight reduction as θ increases. The mean bias in estimating θ is uniformly positive; that is, there is a systematic tendency to produce estimates of θ that are larger than the true values. Generally, the grid method is better in terms of accuracy. (Scaling the exponential did not change the bias and rmse results). Also, whilst not explicitly reported, using the estimates from the grid method as starting values for direct estimation did not lead to an improvement in bias or rmse compared to using the relatively

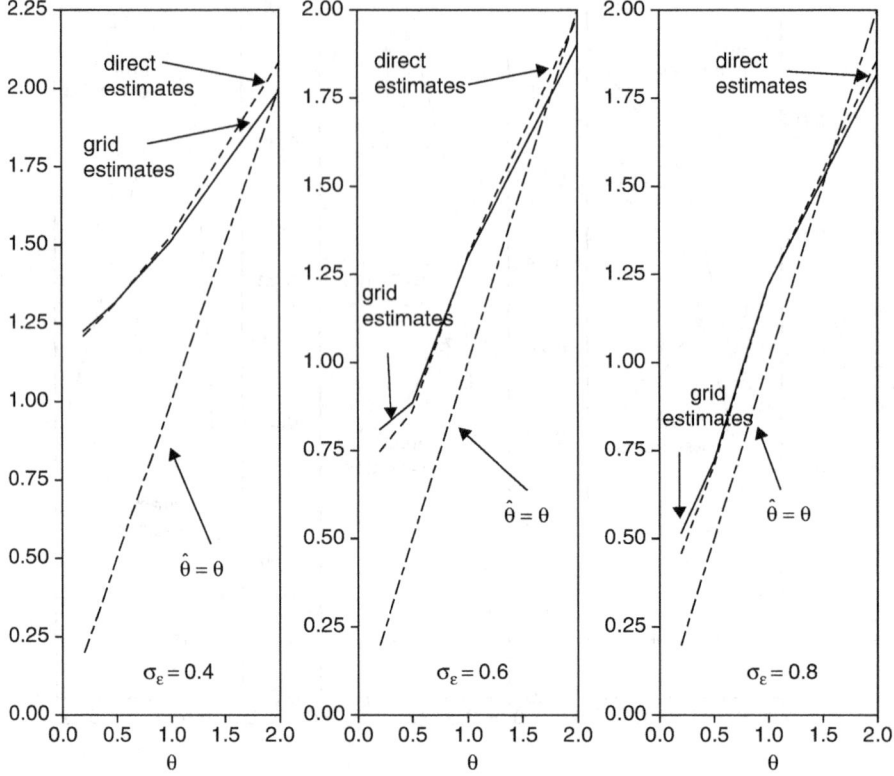

Figure 5.22   Mean of estimates of θ

uninformed starting values. The rmse corresponding to each value of σ_ε^2 is shown in Figure 5.23. Grid estimation tends to produce a smaller rmse than direct estimation. Generally rmse is smaller as σ_ε^2 increases, but there is not a uniform decline as θ increases.

The other estimated parameters are ϕ_1 and γ_1, with DGP values of 1 and −1 respectively. The mean estimates of ϕ_1 are shown in Figures 5.24, and the corresponding rmse is shown in Figure 5.25. The mean of the estimates of ϕ_1 are uniformly above 1 for both methods as θ varies, declining slightly as θ increases. There is a minor variation in the mean estimates as σ_ε^2 varies, but the rmse tends to increase with σ_ε^2 as shown clearly on Figure 5.25.

Figures 5.26 and 5.27 consider estimation of γ_1 with the mean of the estimates of γ_1 shown in the former and rmse in the latter. The estimates of γ_1 are sensitive to σ_ε^2, with accuracy depending on σ_ε^2 and θ. There is a tendency for an initial underestimation in absolute value (see Figure 5.26) but, as θ increases, the mean estimate of γ_1 gets closer to the true value as σ_ε^2 increases. The rmse of the

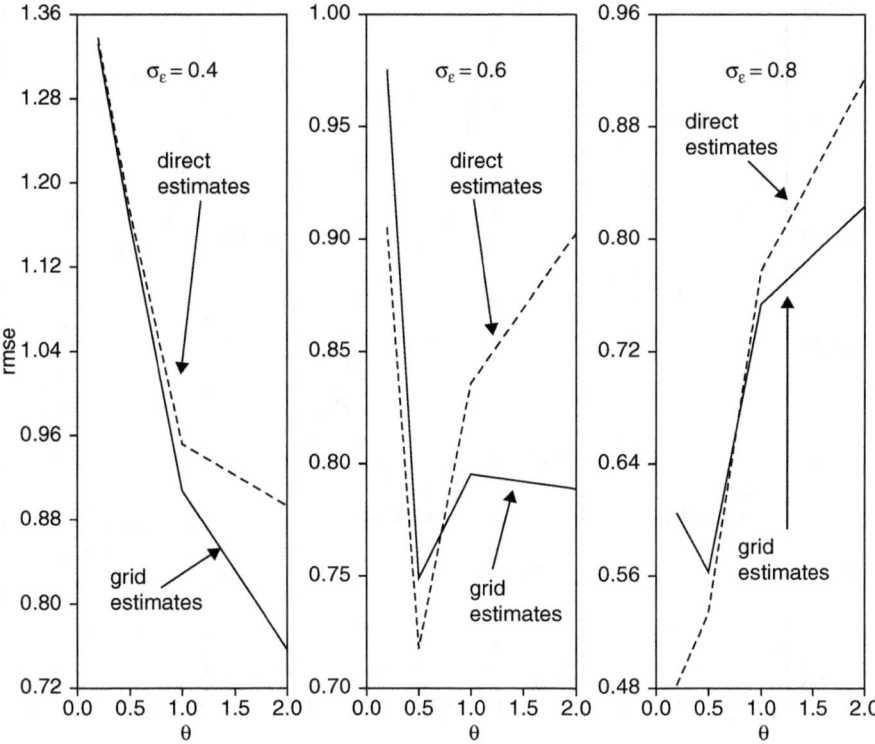

Figure 5.23   rmse of estimates of θ

estimates of γ_1 tends to increase with σ_ε^2 (see Figure 5.27); although note that for small values of θ, the rmse is largest for $\sigma_\varepsilon^2 = 0.4^2$.

In simulations not reported here in detail, the estimation results are better in the less realistic case where ϕ_1 is not estimated. (Luukkonen, 1990, also reports results from grid search estimation with ϕ_1 fixed that are reasonably favourable, but even so he still mentions problems in estimating γ_1 and θ even in quite large samples.) In this case the grid search is over θ, and γ_1 is estimated. Proportionately more of the estimates using the direct method are admissible; and for both methods, whilst θ is still overestimated, the mean bias is considerably less. For example, in the case of the grid search method, the mean estimates for $\theta = 0.2$, 0.5, 1.0 and 2.0 are, respectively, 0.33, 0.55, 1.01, 2.01 for $\sigma_\varepsilon^2 = 0.4^2$ and, respectively, 0.22, 0.52, 1.02, 2.00 for $\sigma^2 = 0.8^2$. The bias declines markedly as σ_ε^2 increases. The corresponding mean estimates of $\gamma_1 = -1$, are -0.86, -1.04, -1.05, -1.01 for $\sigma_\varepsilon^2 = 0.4^2$, and -1.03, -1.01, -1.01, -1.01 for $\sigma_\varepsilon^2 = 0.8^2$.

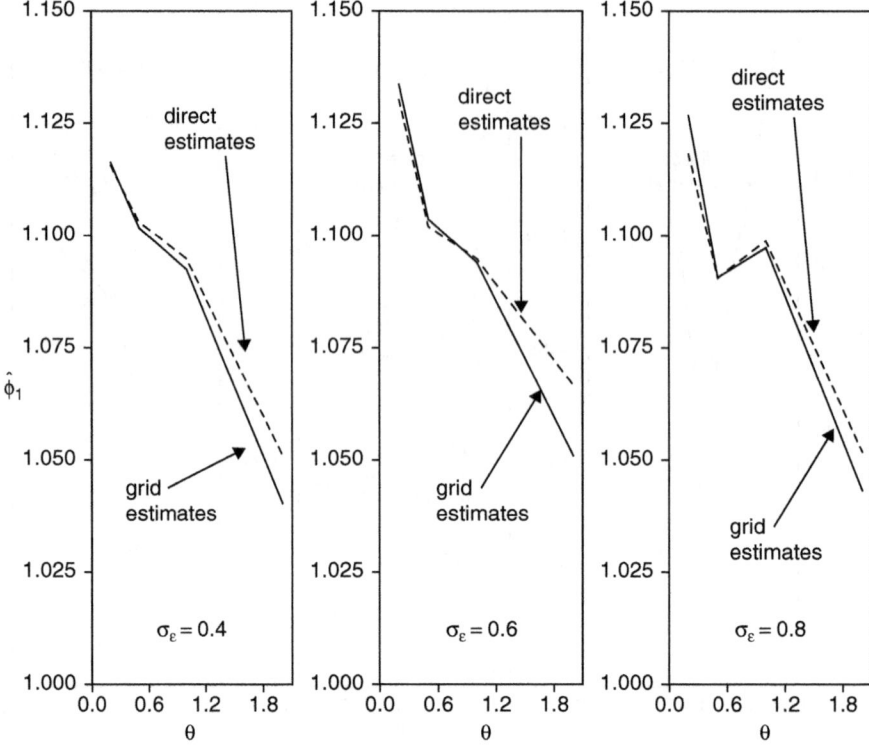

Figure 5.24   Mean of estimates of ϕ_1 as θ varies

Overall, the estimation results are better the more the nonlinearity is manifested in the data through an increase in θ or, generally, an increase in σ_ε^2. The bias in the more realistic case when ϕ_1 is estimated is of concern, and suggests some caution in assessing empirical results. Although not considered here, it would be of interest to see if the bias could be reduced by the methods described in *UR, Vol. 1*, chapter 4, the most attractive being the simulation method of MacKinnon and Smith (1998) or a bootstrap based method.

5.11   Estimation of the BRW

The BRW model was outlined in Section 5.5 and is relatively new; therefore, a small-scale simulation study was used to obtain an indication as to whether it could be directly estimated in a satisfactory way by nonlinear least squares. The

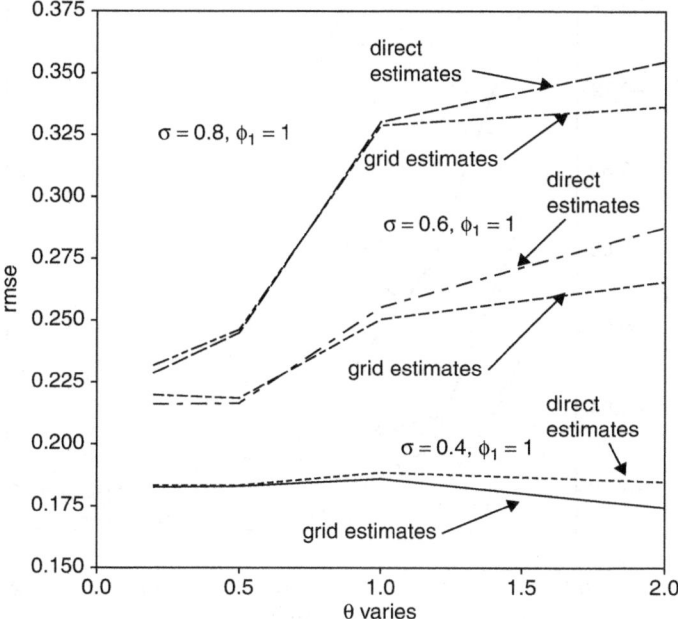

Figure 5.25   rmse of estimates of ϕ_1 as θ varies

basic BRW model is:

$$y_t = y_{t-1} + e^{\alpha_0}(e^{-\alpha_1(y_{t-1}-\kappa)} - e^{\alpha_2(y_{t-1}-\kappa)}) + \varepsilon_t \tag{5.107}$$

The focus was on the symmetric case, with $\alpha_1 = \alpha_2$. As in the case of estimation of the ESTAR model, the BRW model was estimated by nonlinear least squares (NLS) and also by a grid search method. The grid search method was informed by the NLS results, which indicated that there were few problems with the estimates of κ but that occasionally α_0, α_1 and α_2 were far from their true values. The grid search, therefore, concentrated on grids for α_0 and α_1 (= α_2 in the symmetric case), and κ was estimated. The NLS estimates are indicated by $\hat{\alpha}_i$ and the grid estimates by $\tilde{\alpha}_i$.

5.11.1   Simulation set-up

In the first set-up, the parameter values were chosen to match those in Nicolau (2002) who illustrated the path of a symmetric BRW with $\alpha_0 = -5$, $\alpha_1 = \alpha_2 = 3$, $\kappa = 0$ and $\sigma_\varepsilon^2 = 0.4^2$; there were 5,000 simulations, T = 200 and the default starting values of zero were used for NLS estimation. The grid limits were set at $\tilde{\alpha}_0 \in [-10, -0]$ and $\tilde{\alpha}_1 \in [0, 5]$, with 50 equally spaced grid points in each case. The grid search implicitly involves a censoring of acceptable values; therefore, for comparability, the mean estimates and rmse for direct NLS estimation are

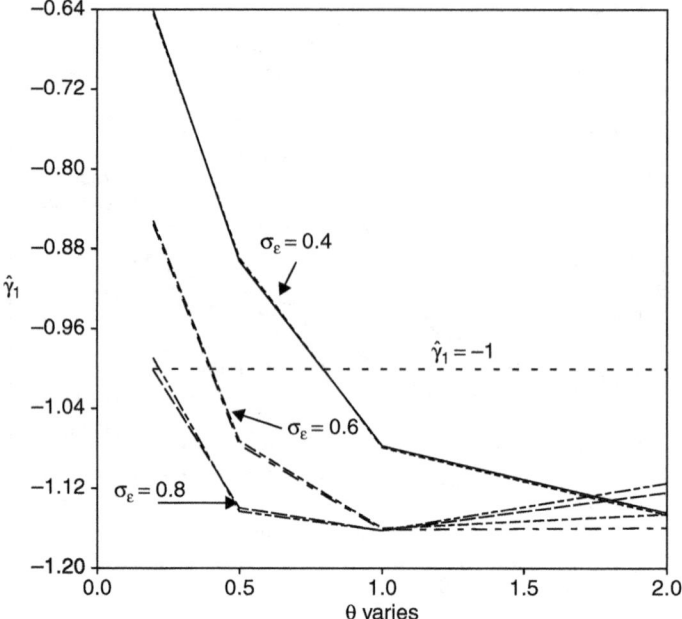

Figure 5.26   Mean of estimates of γ_1 as σ_ε and θ vary

also reported for the case where estimates outside this range are 'trimmed' out of the sample.

5.11.2   Simulation results

Without censoring of the sample, the grid search is better than NLS estimation in terms of a lower rmse for the estimated coefficients (see Table 5.8 for a summary of the results). For both methods, the mean bias is positive for estimation of α_1 and negative for estimation of α_0; the former is about 10%, whereas the latter varies with the estimation method used. If the 'raw' NLS estimates are censored, reflecting the use of prior views as in the grid search, there is little to choose between the two sets of results – see the column headed 'trimmed' in the table (hence, now the proportion of admissible samples can be below unity); if anything NLS estimation is now better in terms of rmse. For comparable accuracy, the grid search probably just needs refining by defining a finer grid close to the initial minimum. The advantage of imposing symmetry, if appropriate, is clear in terms of reducing the rmse and in controlling the number of samples for which the estimates are deemed inadmissible (on the basis of the same rule as used in the grid search).

   The distribution of the bias is illustrated in Figure 5.28 for the NLS estimates of $\hat{\alpha}_0$ (left panel) and $\hat{\alpha}_1$ (right panel), where symmetry is imposed, and Figure 5.29

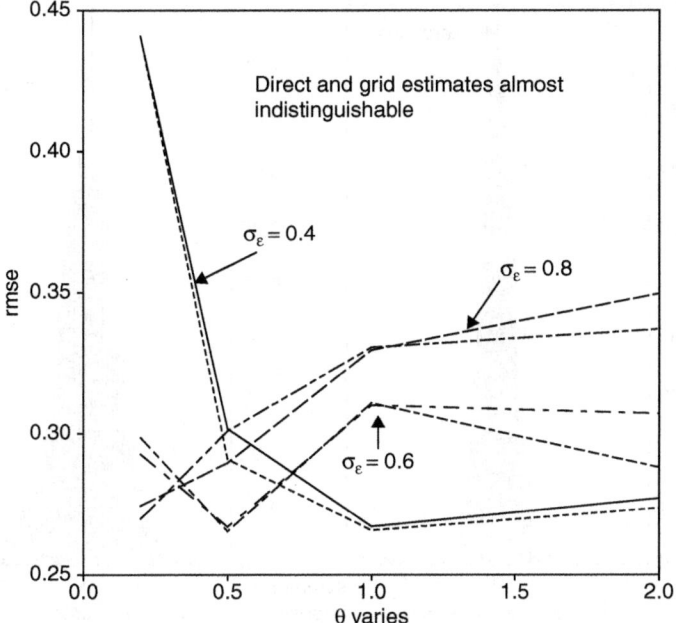

Figure 5.27 rmse of estimates of γ_1 as σ_ε and θ vary

Table 5.8 Estimation of the BRW model

Symmetry imposed	Mean estimates			rmse		
	NLS	'Trimmed'	Grid	NLS	'Trimmed'	Grid
$\alpha_0 = -5$	−5.193	−5.164	−5.208	1.072	0.864	0.913
$\alpha_1 = \alpha_2 = 3$	3.323	3.275	3.308	0.996	0.870	0.945
$\kappa = 0$	0.001	0.001	0.002	0.071	0.070	0.072
Admissible %		96%			96%	
Symmetry not imposed						
$\alpha_0 = -5$	−5.040	−4.882	**	1.036	0.769	**
$\alpha_1 = 3$	3.285	3.056	**	1.513	0.934	**
$\alpha_2 = 3$	3.255	3.043	**	1.413	0.914	**
$\kappa = 0$	0.001	0.000	**	0.185	0.164	**
Admissible %		87%			87%	

Notes: 'Trimmed' refers to the case where, for comparability with the grid estimates, direct NLS estimates have been subject to the same limits as in the grid search. **Without imposing symmetry a three-dimensional search would be required, which is not attempted here.

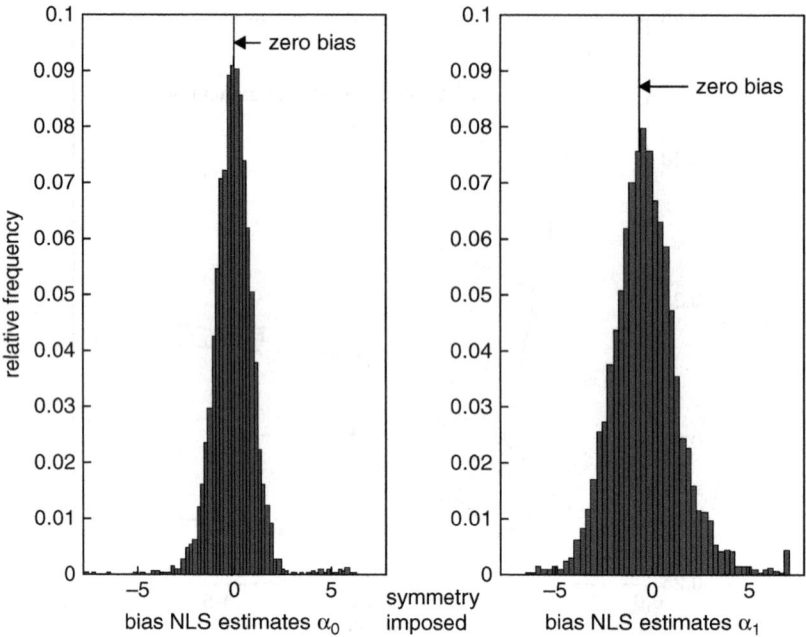

Figure 5.28   Distribution of bias of estimates of α_0 and α_1, symmetry imposed

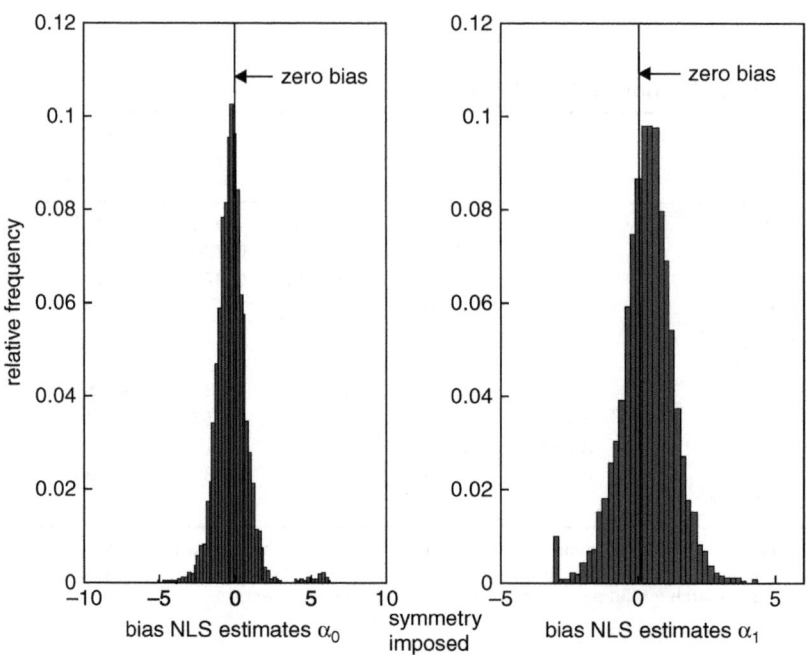

Figure 5.29   Distribution of bias of estimates of α_0 and α_1, symmetry not imposed

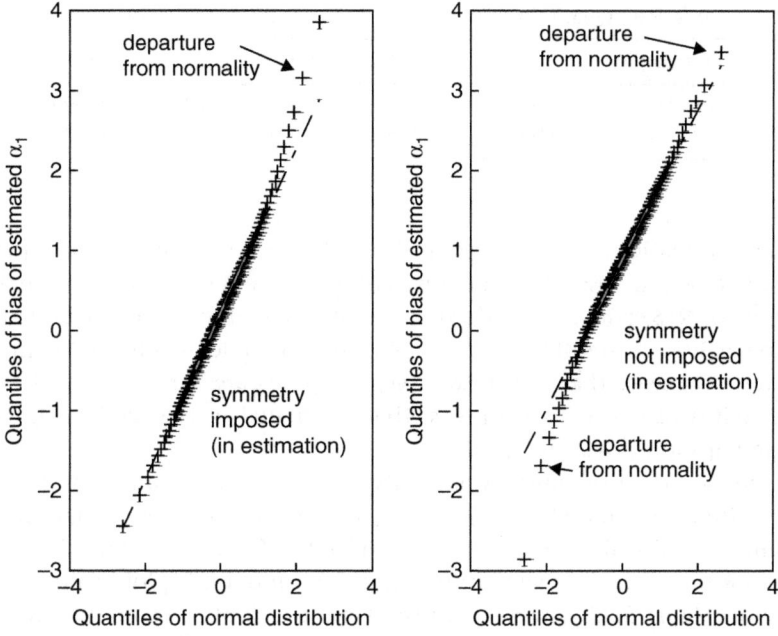

Figure 5.30   QQ plots of NLS estimates of bias of α_1

for the case where symmetry is not imposed (the results for $\hat{\alpha}_2$ were virtually identical to those for $\hat{\alpha}_1$ and are not shown). Not imposing symmetry increases the variance of both sets of estimates. There is a clear correlation between the under-(over-) estimation of α_1 and the over-(under-) estimation of the absolute value of α_0; thus, the coupling of a small value for $\hat{\alpha}_1$ and a large (negative) value for $\hat{\alpha}_0$ should serve to indicate that there are estimation problems. Note also that there are some outliers in both cases shown as small bars in the distributions outside the central part of the distribution.

Although broadly symmetric, the simulated distributions of the estimates are not entirely consistent with a normal distribution (as suggested by the outliers). This is illustrated for the bias in the (uncensored) NLS estimates in Figure 5.30, which shows a QQ plot of the quantiles of the estimates of α_1 against the quantiles of the normal distribution; the left panel gives the QQ plot for symmetry imposed in estimation and the right panel where symmetry is not imposed. Both simulated distributions show departures from normality in the right tail although this is less substantial for the symmetry not imposed case, and there is also a departure in the left tail when symmetry is not imposed. The grid estimates, not shown, partially correct this feature, as does the 'trimming' out of extreme values.

Table 5.9 Quantiles of the LR test statistic and quantiles of $\chi^2(1)$

%-quantile	5%	10%	90%	95%	99%
Simulated	0.0027	0.0123	2.332	3.257	6.099
$\chi^2(1)$	0.0039	0.0158	2.701	3.842	6.634

Other simulations, not reported, indicate that as the coefficients α_0, α_1 and α_2 were increased in absolute value it was necessary to decrease σ_ε^2 to ensure convergence of NLS estimation and to enable the grid method to define numerically sensible values of the BRW function. As a rough guide, it is likely in empirical practice to observe that if, for example, α_1 and α_2 are 'large', say greater than 10, then it is also likely to be the case that (estimated) σ_ε^2 is 'small', say less than, and perhaps much less than, 0.1.

Finally, some consideration was briefly given to the distribution of the LR test statistic for symmetry where $H_0: \alpha_1 = \alpha_2$ and $H_A: \alpha_1 \neq \alpha_2$. Some quantiles from the simulated distribution are reported in Table 5.9. The simulated distribution of the LR test statistic is shown in Figure 5.31 and a QQ plot against $\chi^2(1)$ is shown in Figure 5.32. The latter illustrates the lack of agreement between the simulated distribution of the LR test statistic and the $\chi^2(1)$ distribution, particularly in the 'right-hand' quantiles, which are those relevant for hypothesis testing. What the plot indicates is that a sample test statistic that is significant using, for example, the 95% quantile from $\chi^2(1)$, will be significant using the quantiles from the simulated distribution; but there will be under-rejection of the null if the right-tailed quantiles from $\chi^2(1)$ are used.

5.12 Concluding remarks

This chapter has considered a number of issues that arise in the practical implementation of some nonlinear models. The position is complicated by the need to distinguish between nonlinearity and nonstationarity. A typical procedure is to first test for a unit root but, in standard practice, the usual alternative is that the data are generated by a stationary AR(p) model, not a nonlinear (globally) stationary alternative. Hence, the scope of the alternative model has to be extended. Nonlinear models are attractive because they address potential internal inconsistencies in the application of unit root tests to some data series. Series that have a natural upper or lower bound, or both, are not good candidates for standard unit root tests, since under the null the series is unbounded. For example, a model fitted to the nominal interest rate should, in principle, recognise that negative nominal rates do not occur, or that the unemployment rate is bounded. A more likely alternative to a linear AR model with a unit root is a stationary, nonlinear model incorporating bounds. It is possible that such

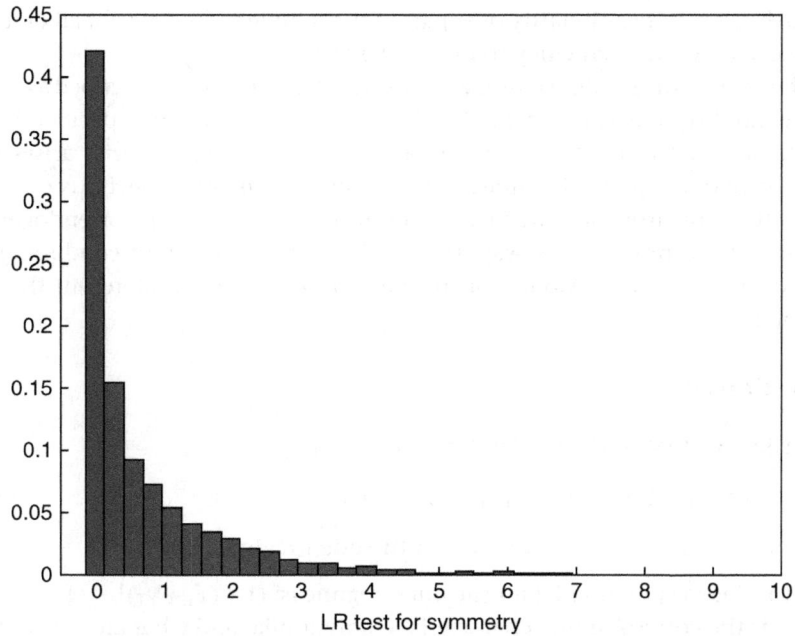

Figure 5.31 Distribution of the LR test statistic for symmetry

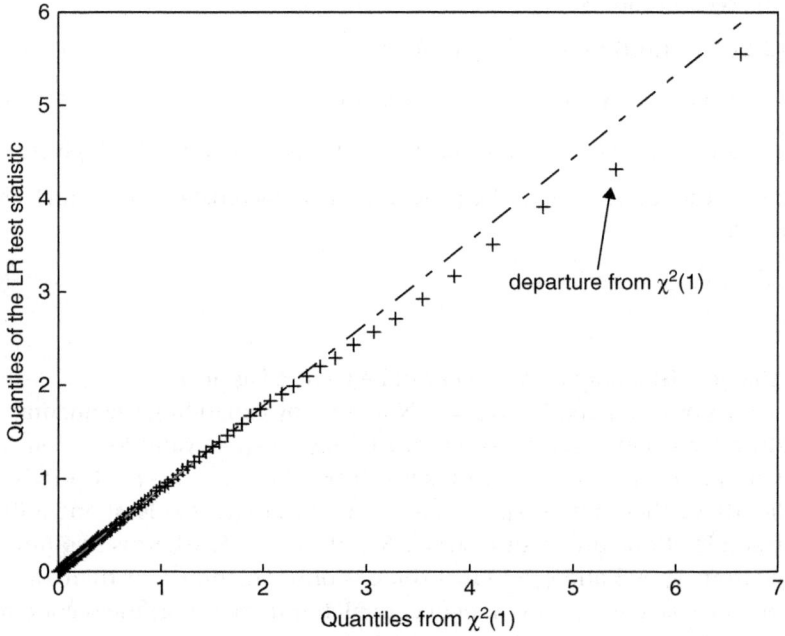

Figure 5.32 QQ plot of LR test statistic for symmetry against $\chi^2(1)$

models allow nonstationarity over part of the sample space, but not all of it and, hence, contain the tendency to nonstationarity.

The form of the nonlinear models considered here could be described as 'essentially' nonlinear as is illustrated from the transition functions specified by the ESTAR and LSTAR models; however, another from of nonlinearity arises from models that are 'piecewise' linear. That is they are linear for sections or conditions that are either specified externally to the linear sections or endogenised within the sections; either way, the nonlinearity comes from combining different linear sections. Models of this general form are considered in the next chapter.

Questions

Q5.1.i Show that in the ESTAR(2) model given by:

$$y_t = \phi_1 y_{t-1} + \phi_2 y_{t-2} + (\gamma_1 y_{t-1} + \gamma_2 y_{t-2})\Theta_t + \varepsilon_t \tag{A5.1}$$

if $\phi_i = -\gamma_i$, for $i = 1, 2$, then $y_t = \varepsilon_t$ in the outer regime.

Q5.1.ii The lag polynomial for the outer regime is: $(1 - (\phi_1 + \gamma_1)L - (\phi_2 + \gamma_2)L^2)$. What is the corresponding characteristic polynomial and what are the roots on the assumption that $\phi_i = -\gamma_i$, for $i = 1, 2$?

Q5.1.iii Show that if $\phi_1 + \phi_2 = 1$ then there is a unit root in the inner regime and the second root is $-\phi_2$.

A5.1.i By substitution of $\phi_i = \gamma_i$ in (A5.1)

$$y_t = -\gamma_1 y_{t-1} - \gamma_2 y_{t-2} + (\gamma_1 y_{t-1} + \gamma_2 y_{t-2})\Theta_t + \varepsilon_t \tag{A5.2}$$

In the outer regime $\Theta_t = 1$, and substituting this value into (A5.2) gives $y_t = \varepsilon_t$.

A5.1.ii In the inner regime, the process and characteristic polynomial, respectively, are:

$$y_t = \phi_1 y_{t-1} + \phi_2 y_{t-2} + \varepsilon_t \tag{A5.3}$$

$$\vartheta(\lambda) = \lambda^2 - \phi_1 \lambda - \phi_2 \tag{A5.4}$$

The characteristic polynomial $\vartheta(\lambda)$ can be factored as $\vartheta(\lambda) = (\lambda - \lambda_0)(\lambda - \bar{\lambda}_0)$, so that $\lambda = \lambda_0$ or $\lambda = \bar{\lambda}_0$ imply $\vartheta(\lambda) = 0$. Note that by expanding the quadratic and equating coefficients on like powers of λ then $\phi_1 = \lambda_0 + \bar{\lambda}_0$ and $\phi_2 = -\lambda_0 \bar{\lambda}_0$. Now by assumption $\phi_1 + \phi_2 = 1$ and on substitution $\lambda_0 + \bar{\lambda}_0 - \lambda_0 \bar{\lambda}_0 - 1 = 0$; but this implies that either or both $\lambda_0 = 1$, $\bar{\lambda}_0 = 1$; that is, there is at least one unit root. An example of one unit root is $\phi_1 = 1.5$ and $\phi_2 = -0.5$, whereas two unit roots follow from $\phi_1 = 2$ and $\phi_2 = -1$. If there is only one unit root then $\lambda_0 = 1$, so that $\phi_1 = 1 + \bar{\lambda}_0$ and $\phi_2 = -\bar{\lambda}_0$; hence, the latter implies that the second root is $\bar{\lambda}_0 = -\phi_2$.

A5.1.iii The characteristic polynomial is: $\vartheta(\lambda) = \lambda^2 - (\phi_1 + \gamma_1)\lambda - (\phi_2 + \gamma_2)$; the roots are obtained as the solution to $\vartheta(\lambda) = 0$. If $\phi_i = -\gamma_i$, for $i = 1, 2$, then note that the characteristic polynomial is just λ^2, and $\lambda^2 = 0$ is only satisfied by $\lambda = 0$. This confirms that $y_t = \varepsilon_t$ in the outer regime.

Q5.2.i Let $f(\theta) = e^{-\theta(y_{t-1})^2}$, obtain the first-to the fourth-order derivatives with respect to θ and evaluate them at $\theta = 0$. Hence, obtain the Taylor series approximation to fourth order.

Q5.2.ii What difference does it make if there is a nonzero threshold parameter, that is $f(\theta) = e^{-\theta(y_{t-1}-c)^2}$?

Q5.2.iii Compare this to the case where the transition function is logistic.

A5.2.i The first derivative has already been obtained – see Equation (5.47).

$$f^{(1)}(\theta) \equiv \partial f(\theta)/\partial\theta = -y_{t-1}^2 e^{-\theta(y_{t-1})^2}; \partial f(\theta|\theta = 0)/\partial\theta = -y_{t-1}^2;$$

$$f^{(2)}(\theta) = y_{t-1}^4 e^{-\theta(y_{t-1})^2}; f^{(2)}(0) = y_{t-1}^4;$$

$$f^{(3)}(\theta) = -y_{t-1}^6 e^{-\theta(y_{t-1})^2}; f^{(3)}(0) = -y_{t-1}^6.$$

The pattern is clear: $f^{(j)}(\theta) = (-1)^j y_{t-1}^{(2j)}$, where $f^{(j)}(\theta)$ is the j-th derivative of $f(\theta)$. Notice that there are *no odd-order powers* of y_{t-1} in the Taylor series.

For economy of notation let $f^{(j)}(0)$ denote the j-th derivative evaluated at $\theta = 0$. Then the fourth-order Taylor series approximation is:

$$f(\theta) = f(\theta|\theta = 0) + f^{(1)}(0)\theta + \frac{f^{(2)}(0)\theta^2}{2} + \frac{f^{(3)}(0)\theta^3}{6} + \frac{f^{(4)}(0)\theta^4}{24} + R_{(4)}$$

$$= 1 - y_{t-1}^2\theta + \frac{y_{t-1}^4}{2}\theta^2 - \frac{y_{t-1}^6}{6}\theta^3 + \frac{y_{t-1}^8}{24}\theta^4 + R_{(4)} \tag{A5.5}$$

The subscript on the remainder R indicates the number of terms in the Taylor series.

On substitution of $f(\theta)$ into $\gamma y_{t-1}(1 - f(\theta))$, we obtain:

$$\Delta y_t = \gamma y_{t-1}(1 - e^{-\theta(y_{t-1})^2})$$

$$= \gamma y_{t-1}\left[y_{t-1}^2\theta - \frac{y_{t-1}^4}{2}\theta^2 + \frac{y_{t-1}^6}{6}\theta^3 - \frac{y_{t-1}^8}{24}\theta^4\right] + \varepsilon_t^*$$

where $\varepsilon_t^* = -(\gamma y_{t-1})R_{(4)} + \varepsilon_t$. Rearranging we have:

$$\Delta y_t = \beta_2 y_{t-1}y_{t-1}^2 + \beta_4 y_{t-1}y_{t-1}^4 + \beta_6 y_{t-1}y_{t-1}^6 + \beta_8 y_{t-1}y_{t-1}^8 + \varepsilon_t^* \tag{A5.6}$$

where: $\beta_2 = \gamma\theta$, $\beta_4 = -\gamma\theta^2/2$, $\beta_6 = \gamma\theta^3/6$ and $\beta_8 = -\gamma\theta^4/24$.

A second-order approximation, therefore, yields:

$$\Delta y_t = \beta_2 y_{t-1}y_{t-1}^2 + \beta_4 y_{t-1}y_{t-1}^4 + \varepsilon_t^{**} \tag{A5.7}$$

where $\varepsilon_t^{**} = -(\gamma y_{t-1})R_{(2)} + \varepsilon_t$.

A5.2.ii With these derivations it is easy to go to the more general case where a unit root is not imposed, and there is a nonzero threshold parameter and a constant that might also be subject to change. For illustration take an ESTAR(1) model:

$$y_t = \phi_0 + \phi_1 y_{t-1} + (\gamma_0 + \gamma_1 y_{t-1})(1 - e^{-\theta(y_{t-1}-c)^2}) + \varepsilon_t \tag{A5.8}$$

Now $f(\theta) = e^{-\theta(y_{t-1}-c)^2}$, $f^{(1)}(\theta) = -(y_{t-1}-c)^2 e^{-\theta(y_{t-1}-c)^2}$, $f^{(1)}(0) = -(y_{t-1}-c)^2$; $f^{(2)}(\theta) = (y_{t-1}-c)^4 e^{-\theta(y_{t-1}-c)^2}$ and $f^{(2)}(0) = (y_{t-1}-c)^4$.

$$y_t = \phi_0 + \phi_1 y_{t-1} + (\gamma_0 + \gamma_1 y_{t-1})[(y_{t-1}-c)^2\theta - (y_{t-1}-c)^4\theta^2/2)]$$
$$- (\gamma_0 + \gamma_1 y_{t-1})R_{(4)} + \varepsilon_t \tag{A5.9}$$

A5.2.iii The LSTAR case is as follows. Start with $f(\theta) = \Theta(y_{t-1}) = (1 + e^{-\theta y_{t-1}})^{-1}$.

First note that $\partial(e^{-\theta y_{t-1}})/\partial\theta = -y_{t-1}e^{-\theta y_{t-1}}$, which is used throughout.

$$f(0) = (1 + e^0)^{-1} = \tfrac{1}{2}$$

$$f^{(1)}(\theta) = y_{t-1}e^{-\theta y_{t-1}}(1 + e^{-\theta y_{t-1}})^{-2}$$
$$f^{(1)}(0) = y_{t-1}/4$$

$$f^{(2)}(\theta) = 2y_{t-1}(y_{t-1}e^{-\theta y_{t-1}})(1 + e^{-\theta y_{t-1}})^{-3} - (1 + e^{-\theta y_{t-1}})^{-2}y_{t-1}^2 e^{-\theta y_{t-1}}$$
$$f^{(2)}(0) = 2y_{t-1}(y_{t-1}e^0)(1 + e^0)^{-3} - (1 + e^0)^{-2}y_{t-1}^2 e^0$$
$$= 2y_{t-1}(y_{t-1})(2)^{-3} - (2)^{-2}y_{t-1}^2 \qquad \text{now cancel the 2 in the first term}$$
$$= y_{t-1}(y_{t-1})(2)^{-2} - (2)^{-2}y_{t-1}^2$$
$$= 0$$

$$f^{(3)}(\theta) = 3y_{t-1}(2y_{t-1}(y_{t-1}e^{-\theta y_{t-1}}))(1 + e^{-\theta y_{t-1}})^{-4}$$
$$- (1 + e^{-\theta y_{t-1}})^{-3}2y_{t-1}^2(y_{t-1}e^{-\theta y_{t-1}})$$
$$- 2y_{t-1}(y_{t-1}^2 e^{-\theta y_{t-1}})(1 + e^{-\theta y_{t-1}})^{-3}$$
$$+ (1 + e^{-\theta y_{t-1}})^{-2}) \times y_{t-1}^3 e^{-\theta y_{t-1}}$$
$$= 6y_{t-1}^3 e^{-\theta y_{t-1}}(1 + e^{-\theta y_{t-1}})^{-4} - 4y_{t-1}^3 e^{-\theta y_{t-1}}(1 + e^{-\theta y_{t-1}})^3$$
$$+ y_{t-1}^3 e^{-\theta y_{t-1}}(1 + e^{-\theta y_{t-1}})^{-2}$$

$$f^{(3)}(0) = 6y_{t-1}^3 e^0 (1+e^0)^{-4} - 4y_{t-1}^3 e^0 (1+e^0)^{-3} + y_{t-1}^3 e^0 (1+e^0)^{-2}$$
$$= 3y_{t-1}^3 (2)^{-3} - y_{t-1}^3 (2)^{-1} + y_{t-1}^3 (2)^{-2}$$
$$= (2)^{-3} y_{t-1}^3$$
$$= y_{t-1}^3 / 8$$

The third-order Taylor series expansion is, therefore:

$$f(\theta) = f(\theta | \theta = 0) + f^{(1)}(0)\theta + \frac{f^{(2)}(0)\theta^2}{2} + \frac{f^{(3)}(0)\theta^3}{6} + R$$
$$= \frac{1}{2} + \frac{1}{4}\theta y_{t-1} + 0 + \frac{1}{48}\theta^3 y_{t-1}^3 + R \qquad (A5.10)$$

A little more algebra shows that all even-order derivatives evaluated at $\theta = 0$ are zero, which serves to distinguish the LSTAR and ESTAR models as far as the tests for nonlinearity are concerned.

On substitution of (A5.10) into $\Delta y_t = \gamma y_{t-1}(1+e^{-\theta y_{t-1}})^{-1} + \varepsilon_t$, the following is obtained:

$$\Delta y_t = \gamma y_{t-1}\left(\frac{1}{4}\theta y_{t-1} + \frac{1}{48}\theta^3 y_{t-1}^3\right) + \gamma y_{t-1}\left(R + \frac{1}{2}\right) + \varepsilon_t$$
$$= \beta_1 y_{t-1}^2 + \beta_3 y_{t-1}^3 + \varepsilon_t^* \qquad (A5.11)$$

where $\beta_1 = \gamma\theta/4$, $\beta_3 = \gamma\theta^3/48$ and $\varepsilon_t^* = \gamma y_{t-1}\left(R + \frac{1}{2}\right) + \varepsilon_t$. (The case $f(\theta) = (1+e^{-\theta(y_{t-1}-c)})^{-1}$ is left as an exercise.)

Q5.3 Write the ESTAR(p) in deviation from threshold form and obtain the restrictions, relative to the general ESTAR(p), that enable transitional effects on the constant.

A5.3 The ESTAR(p) with constant is:

$$y_t = \phi_0 + \sum_{i=1}^{p}\phi_i y_{t-i} + \left(\gamma_0 + \sum_{i=1}^{p}\gamma_i y_{t-i}\right)(1 - e^{-\theta(y_{t-1}-c)^2}) + \varepsilon_t \qquad (A5.12)$$

The ESTAR(p) in deviation from threshold form is:

$$y_t - c = \sum_{i=1}^{p}\phi_i(y_{t-i} - c) + \sum_{i=1}^{p}\gamma_i(y_{t-i} - c)(1 - e^{-\theta(y_{t-1}-c)^2}) + \varepsilon_t$$
$$y_t = \left(1 - \sum_{i=1}^{p}\phi_i\right)c - c\sum_{i=1}^{p}\gamma_i(1 - e^{-\theta(y_{t-1}-c)^2})$$
$$+ \sum_{i=1}^{p}\phi_i y_{t-i} + \sum_{i=1}^{p}\gamma_i y_{t-i}(1 - e^{-\theta(y_{t-1}-c)^2}) + \varepsilon_t \qquad (A5.13)$$

The deviation from threshold form implies the following two (testable) restrictions: $\phi_0 = (1 - \sum_{i=1}^{p}\phi_i)c$ and $\gamma_0 = -c\sum_{i=1}^{p}\gamma_i$. As the unit root is approached in the inner regime then $\sum_{i=1}^{p}\phi_i \to 1$ and $\phi_0 \to 0$.

Q5.4 Specify the LSTAR(1) model using the centred logistic transition function and compare it with equations (5.36) to (5.38), the three regimes of the LSTAR model.

A5.4 Using the centred logistic transition function, the LSTAR(1) model is:

$$y_t = \phi_1^* y_{t-1} + \gamma_1 y_{t-1}[(1 + e^{-\psi(y_{t-d}-c)})^{-1} - 0.5] + \varepsilon_t \tag{A5.14}$$

$$= \phi_1^* y_{t-1} + \gamma_1 y_{t-1} \Psi_C(y_{t-d}) + \varepsilon_t \tag{A5.15}$$

Observe that $\phi_1^* = \phi_1 + 0.5\gamma_1$.

The regimes generated by this LSTAR(1) model are obtained by setting $\Psi_C(-\infty) = -0.5$, $\Psi_C(0) = 0$ and $\Psi_C(\infty) = 0.5$, respectively:

Lower : $y_t = (\phi_1^* - 0.5\gamma_1)y_{t-1} + \varepsilon_t$ $y_{t-d} - c \to -\infty$; $\Psi_S(-\infty) = -0.5$

Middle : $y_t = \phi_1^* y_{t-1} + \varepsilon_t$ $y_{t-d} = c$; $\Psi_S(0) = 0$

Upper : $y_t = (\phi_1^* + 0.5\gamma_1)y_{t-1} + \varepsilon_t$ $y_{t-d} - c \to +\infty$; $\Psi_S(\infty) = 0.5$

On noting that $\Psi(-\infty) = 0$, $\Psi(0) = 0.5$, $\Psi(\infty) = 1$, the regimes correspond exactly to those obtained in the text (see Equations (5.36)–(5.38)):

Lower : $y_t = \phi_1 y_{t-1} + \varepsilon_t$ $y_{t-d} - c \to -\infty; \Psi(-\infty) = 0$

Middle : $y_t = (\phi_1 + 0.5\gamma_1)y_{t-1} + \varepsilon_t$ $y_{t-d} = c$; $\Psi(0) = 0.5$

Upper : $y_t = (\phi_1 + \gamma_1)y_{t-1} + \varepsilon_t$ $y_{t-d} - c \to +\infty; \Psi(\infty) = 1$

Q5.5.i Consider an ESTAR(2) model in deviations form, estimated subject to the restrictions that $\gamma_1 = -\phi_1$ and $\gamma_2 = -\phi_2$: specify the auxiliary regression and the test statistic for the Eitrheim and Teräsvirta extension of the LM test for serial correlation of the residuals.

Q5.5.ii Now also impose the unit root restriction in the inner regime and obtain the LM test.

A5.5.i The question can be approached in two ways, either to redefine the variables in the auxiliary regression or to impose (linear) restrictions on the coefficients in the auxiliary regression. The general principles should become clear from the example.

The 'general' ESTAR(2) model is given by:

$$y_t = \phi_0 + \phi_1 y_{t-1} + \phi_2 y_{t-2} + (\gamma_0 + \gamma_1 y_{t-1} + \gamma_2 y_{t-2})(1 - e^{-\theta(y_{t-d}-c)^2}) + \varepsilon_t \tag{A5.16}$$

There are 8 coefficients in this model, that is $2(p + 1) + 2 = 2(2 + 1) + 2 = 8$; and the unrestricted auxiliary regression for the serial correlation tests has q + 8 coefficients. However, we require the model in 'deviations' form, as follows:

$$y_t - c = \phi_1(y_{t-1} - c) + \phi_2(y_{t-2} - c)$$

$$+ [\gamma_1(y_{t-1} - c) + \gamma_2(y_{t-2} - c)](1 - e^{-\theta(y_{t-d} - c)^2}) + \varepsilon_t \tag{A5.17}$$

For the auxiliary regression, we require the partial derivatives with respect to: $\phi_1, \phi_2, \gamma_1, \gamma_2, \theta$ and c, which are as follows:

$$\partial G / \partial \phi_1 = (y_{t-1} - c) \qquad\qquad\qquad\qquad\qquad\qquad = G_{t,\phi1}$$

$$\partial G / \partial \phi_2 = (y_{t-2} - c) \qquad\qquad\qquad\qquad\qquad\qquad = G_{t,\phi2}$$

$$\partial G / \partial \gamma_1 = (y_{t-1} - c)(1 - e^{-\theta(y_{t-d} - c)^2}) \qquad\qquad\quad = G_{t,\gamma1}$$

$$\partial G / \partial \gamma_2 = (y_{t-2} - c)(1 - e^{-\theta(y_{t-d} - c)^2}) \qquad\qquad\quad = G_{t,\gamma2}$$

$$\partial G / \partial \theta = (y_{t-d} - c)^2 e^{-\theta(y_{t-d} - c)^2}[\gamma_1(y_{t-1} - c) + \gamma_2(y_{t-2} - c)] \quad = G_{t,\theta}$$

$$\partial G / \partial c = -2\theta(y_{t-d} - c)e^{-\theta(y_{t-d} - c)^2}[\gamma_1(y_{t-1} - c) + \gamma_2(y_{t-2} - c)]$$

$$- (\gamma_1 + \gamma_2)(1 - e^{-\theta(y_{t-d} - c)^2}) + (1 - \phi_1 - \phi_2) \qquad = G_{t,c}$$

The auxiliary regression for the test of q-th order serial correlation is:

$$\hat{\varepsilon}_t = \sum_{i=1}^{6} \delta_{1i}\hat{\varepsilon}_{t-i} + \delta_{21}G_{t,\phi1} + \delta_{22}G_{t,\phi2} + \delta_{31}G_{t,\gamma1}$$

$$+ \delta_{32}G_{t,\gamma2} + \delta_{41}G_{t,\theta} + \delta_{42}G_{t,c} + \omega_t \tag{A5.18}$$

We also now impose the 'outer regime' restrictions of the question: $\gamma_1 = -\phi_1$ and $\gamma_2 = -\phi_2$, so that the restricted ESTAR(2) is:

$$y_t - c = -\gamma_1(y_{t-1} - c) - \gamma_2(y_{t-2} - c) + [\gamma_1(y_{t-1} - c)$$

$$+ \gamma_2(y_{t-2} - c)](1 - e^{-\theta(y_{t-d} - c)^2}) + \varepsilon_t \tag{A5.19}$$

The required derivatives are now just those with respect to $\gamma_1, \gamma_2, \theta$ and c, as follows:

$$\partial G / \partial \gamma_1 = (y_{t-1} - c)(1 - e^{-\theta(y_{t-d} - c)^2}) - (y_{t-1} - c)$$

$$= -(y_{t-1} - c)(e^{-\theta(y_{t-d} - c)})$$

$$= G_{t,\gamma1}^*$$

$$\partial G/\partial\gamma_2 = (y_{t-2} - c)(1 - e^{-\theta(y_{t-d}-c)^2}) - (y_{t-2} - c)$$

$$= -(y_{t-2} - c)(e^{-\theta(y_{t-d}-c)})$$

$$= G^*_{t,\gamma2}$$

$$\partial G/\partial\theta = (y_{t-d} - c)^2 e^{-\theta(y_{t-d}-c)^2}[\gamma_1(y_{t-1} - c) + \gamma_2(y_{t-2} - c)]$$

$$= G^*_{t,\theta}$$

$$\partial G/\partial c = -2\theta(y_{t-d} - c) e^{-\theta(y_{t-d}-c)^2}[\gamma_1(y_{t-1} - c) + \gamma_2(y_{t-2} - c)]$$

$$- (\gamma_1 + \gamma_2)(1 - e^{-\theta(y_{t-d}-c)^2}) + (1 + \gamma_1 + \gamma_2)$$

$$= G^*_{t,c}$$

The auxiliary regression for the test of serial correlation of order q = 6 in the restricted ESTAR(2) model is:

$$\hat{\varepsilon}_t = \sum_{i=1}^{6} \delta_{1i}\hat{\varepsilon}_{t-i} + \delta_{31}G^*_{t,\gamma1} + \delta_{32}G^*_{t,\gamma2} + \delta_{41}G^*_{t,\theta} + \delta_{42}G^*_{t,c} + \omega_t \qquad (A5.20)$$

This auxiliary regression has q + 4 coefficients, 4 less than for the unrestricted model; otherwise the test statistic is as in (5.104), with RSS$_{UR}$ from the residuals defined in (A5.20) and RSS$_R$ from (A5.19) . Equivalently, note that (A5.20) and (A5.18) are the same if $\delta_{21} = -\delta_{31}$ and $\delta_{22} = -\delta_{32}$, so that an alternative to redefining the variables in the auxiliary regression is to impose (and test) these two restrictions on (A5.18).

A5.5.ii Imposing a unit root in the inner regime reduces the number of freely estimated coefficients by one, so the auxiliary regression must reflect this. The restriction implies $\gamma_1 + \gamma_2 = -1$ (equivalently $\phi_1 + \phi_2 = 1$), so that $\gamma_2 = -(1+\gamma_1)$. The relevant derivatives are $\partial G/\partial\gamma_1$, $\partial G/\partial\theta$ and $\partial G/\partial c$ as follows:

$$\partial G/\partial\gamma_1 = (y_{t-1} - c)(1 - e^{-\theta(y_{t-d}-c)^2}) - (y_{t-1} - c)$$

$$= -(y_{t-1} - c)(e^{-\theta(y_{t-d}-c)})$$

$$= G^*_{t,\gamma1}$$

$$\partial G/\partial\theta = (y_{t-d} - c)^2 e^{-\theta(y_{t-d}-c)^2}[\gamma_1(y_{t-1} - c) - (1+\gamma_1)(y_{t-2} - c)]$$

$$= G^+_{t,\theta}$$

$$\partial G/\partial c = -2\theta(y_{t-d} - c) e^{-\theta(y_{t-d}-c)^2}[\gamma_1(y_{t-1} - c) - (1+\gamma_1)(y_{t-2} - c)]$$

$$+ (1 - e^{-\theta(y_{t-d}-c)^2})$$

$$= G^+_{t,c}$$

The auxiliary regression uses these derivatives and will thus have q + 3 coefficients, 5 less than the unrestricted model. Note that the derivative $\partial G/\partial\gamma_2$ is not included because $\gamma_2 = -(1+\gamma_1)$ implies $\partial G/\partial\gamma_2 = -\partial G/\partial\gamma_1$.

Q5.6 Obtain the equilibrium solution to the ESTAR(1) model and derive the existence condition.

A5.6 The ESTAR(1) model written in 'Ozaki' form is:

$$y_t = (\varphi_1 + \pi_1 e^{-\theta y_{t-1}^2})y_{t-1} + \varepsilon_t \tag{A5.21}$$

Equilibrium implies the self-replicating property given by:

$$y^* = (\varphi_1 + \pi_1 e^{-\theta(y^*)^2})y^* \tag{A5.22}$$

In order to derive this, disregard ε_t in (A5.21) and substitute y^* for y_{t-1}. Now rearrange (A5.22) by first cancelling the common term y^*. Then:

$$e^{-\theta(y^*)^2} = \frac{1 - \varphi_1}{\pi_1} \tag{A5.23}$$

$$-\theta(y^*)^2 = \ln\left(\frac{1 - \varphi_1}{\pi_1}\right) \tag{A5.24}$$

$$y^* = \pm\theta^{-1/2}\left[\ln\left(\frac{1 - \varphi_1}{\pi_1}\right)\right]^{1/2} \tag{A5.25}$$

The existence condition is:

$$0 < \left(\frac{1 - \varphi_1}{\pi_1}\right) < 1$$

This condition arises because for y^* to be defined, the $\ln(.)$ term in (A5.25) must be positive; hence, the lower limit is 0 as it is not possible to take the log of a negative number. The upper limit arises because θ is positive and $(y^*)^2$ is positive, hence $-\theta(y^*)^2$ is negative; but, see (A5.24), for the $\ln(.)$ term to be negative, the argument $(1 - \varphi_1)/\pi_1$ must be less than 1.

Q5.7 Figure A5.1 shows a quantile-quantile plot of the bootstrap distributions of $PSE(\hat{\Psi}_3)_{E|L}$ and $PSE(\hat{\Psi}_3)_{L|E}$ relative to the $\chi^2(3)$ distribution (see Sections **5.8.1.** and **5.8.2.ii** for definitions of the terms). Interpret this figure.

A5.7 The figure shows a 45° line where the quantiles of $\chi^2(3)$ are matched with itself (the solid line); the quantiles for the two bootstrap distributions are then plotted against the quantiles of $\chi^2(3)$. The closer they are to the 45° line, the closer are the two bootstrap distributions to the $\chi^2(3)$ distribution. In the case of the bootstrap distribution of $PSE(\hat{\Psi}_3)_{E|L}$, the percentiles are reasonably close to $\chi^2(3)$, particularly around the 90% and 95% quantiles. For example, $\chi^2_{0.05}(3) = 7.81$ which corresponds to a bootstrap p-value of 95.4% for $PSE(\hat{\Psi}_3)_{E|L}$;

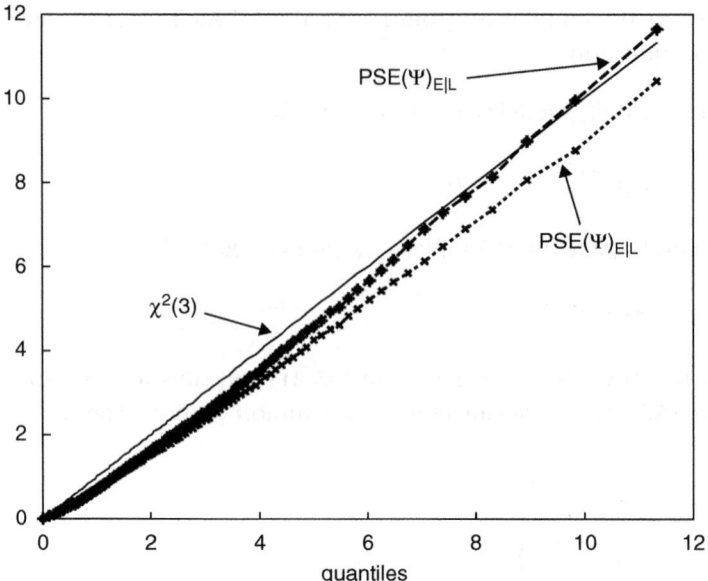

Figure A5.1   Plot of quantiles for STAR PSE test statistics

this is very close to 95%, which is evident in the figure. However, the tail is longer than in the case of $\chi^2(3)$, hence the distribution of $PSE(\hat{\Psi}_3)_{E|L}$ crosses the 45° line. In the case of the bootstrap distribution of $PSE(\hat{\Psi}_3)_{L|E}$, the corresponding quantiles lie everywhere below $\chi^2(3)$.

6
Threshold Autoregressions

Introduction

The previous chapter introduced a number of nonlinear (univariate) models, which had in common the idea of smooth transition between regimes. The most popular class of such models is the smooth transition autoregressive – or STAR – class, of which the exponential and logistic members are the most frequent in application, giving rise to the acronyms ESTAR and LSTAR. Another newer class of models, also considered in the previous chapter, motivated by similar considerations, was the bounded random walk, BRW.

In this chapter, the theme of nonlinearity is continued but along different lines. In this case, two or more linear AR models are discretely 'pieced' together, where each piece is referred to as a regime, and the indicator as to which regime is operating is based on the idea of a threshold parameter or threshold variable. This class of models is known as threshold AR, or TAR, models. As an indicator variable passes through a threshold value, the regimes switch, with certainty, from one to another. There may be more than two regimes, so that switching can occur at different points. The usual situation is that the variable indicating which regime is in operation is related to the 'dependent' variable, say y_t, most frequently a lag of the level or a lag of the first difference of y_t; the TAR model is then known as self-exciting, or SETAR. Other regime indicators are possible but then a case can be made for additionally including them directly into the model of the DGP. For example, if y_t is an interest rate variable and x_t is a measure of price inflation, a TAR for y_t could depend upon a threshold based on x_t; however, equally, since it seems likely that x_t generally contains information about y_t, a more satisfactory framework would be to model y_t and x_t jointly, with a threshold depending on either or both a function of y_t and x_t.

Considered in calendar time, the regimes are not constrained to be composed of adjacent observations; that is, regimes are not necessarily contiguous in the calendar time observations. In the latter case, the model falls into the class of structural breaks, which are considered in Chapters 7 and 8.

There are a number of issues that need to be determined if TAR models are to be a practical addition to the 'toolkit'. The generic issues relate to estimation of the threshold parameters, inference concerning the number of regimes and inference on the nonstationarity of the process. A problem that occurs in this context is that the distribution of standard tests, for example, a Wald or F test, for the number of regimes depends upon the stationarity or otherwise of the process, while a test for nonstationarity will depend upon the number of regimes. There are some testing and estimation procedures common to TAR models and STAR models, which relate to the threshold parameter(s). In both cases they are not identified under the null hypothesis of linearity; the common element of the solution involves a grid search, which is a variation on what has become known has the Davies problem after Davies (1977, 1987), see also Chapter 1.

This chapter starts in Section **6.1** with an introduction to multiple regime threshold autoregressions, including self-exciting TARs, momentum TARs, (MTARs), and two and three regime models. Section **6.2** deals with some general issues of estimation and testing in these models, with Section **6.3** containing more detail on estimation of TAR and MTAR models. How to test for a threshold effect in TAR and MTAR models is considered in Sections **6.5** and **6.6**, respectively. Testing for for a unit root when it is not known whether there is a threshold effect is considered in Section **6.7**. Section **6.8** brings together several of the techniques and issues of this and the previous chapter in the context of nonlinear modelling of the US dollar:pound sterling real exchange rate, and serves to highlight some of the tests and modelling issues that arise in a practical context.

6.1   Multiple regime TARs

Threshold autoregressive, or TAR, models have already been referred to as arising from LSTAR models with instantaneous transition once the threshold is reached. They are considered in greater detail in this section.

6.1.1   Self-exciting TARs

In the TAR set-up, two or more AR(1) autoregressions are distinguished by different parameters in the regime in which they operate. In the simplest case there are two regimes and the regime criterion is effectively endogenous, or 'self-exciting', being determined by a function of a lag of y_t, say y_{t-d} where d is, as in the STAR case, the delay parameter, relative to a threshold parameter (if constant) denoted κ or long-run attractor, if variable, denoted κ_t; see, for example, Enders and Granger (1998) hereafter EG. For default in illustrations we take $d = 1$. (The notational convention of referring to the delay parameter as d is long-standing and its use here should be distinguished from the long memory parameter in Chapters 3 and 4.) The self-exciting TAR is sometimes referred to by the acronym SETAR. Another prominent case is the momentum TAR, or MTAR (or MSETAR), in which the regime selector is based, in its self-exciting form, on $\Delta_r y_{t-d}$; usually $r = 1$ and $d = 1$. If y_t is I(1) then the regime selector is also I(1) in the SETAR case but I(0) in the MTAR case.

The idea in both types of TAR is to allow an asymmetry of response to differently signed deviations from a threshold parameter or threshold variable. In the case of the TAR, the threshold is an attractor, which can be interpreted as the equilibrium for a stationary process. The TAR model then allows different speeds of adjustment back to equilibrium depending on whether the previous level was above or below equlibrium. In the MTAR, with regime separation depending on the lagged change in the level, Δy_{t-1}, an asymmetric response allows, for example, a faster return to equilibrium following a negative change compared to a positive change.

6.1.1.i Two-regime TAR(1)

The simplest case to start with is a two-regime TAR(1) process, that is an AR(1) process with two regimes. The model, initially assuming that y_t is a zero mean variable, with a threshold parameter equal to 0, is as follows:

$$y_t = \varphi_1 y_{t-1} + z_t \quad \text{if } f(y_{t-1}) \leq 0, \text{ regime 1} \tag{6.1}$$

$$y_t = \phi_1 y_{t-1} + z_t \quad \text{if } f(y_{t-1}) > 0, \text{ regime 2} \tag{6.2}$$

If $\varphi_1 < 1$, $\phi_1 < 1$, then the generating process is stationary and y_t converges to 0. Subtracting y_{t-1} from the left-hand side, the usual way of writing the model is:

$$\Delta y_t = \alpha_1 y_{t-1} + z_t \quad \alpha_1 = \varphi_1 - 1 \quad \text{if } f(y_{t-1}) \leq 0, \text{ regime 1} \tag{6.3}$$

$$\Delta y_t = \alpha_2 y_{t-1} + z_t \quad \alpha_2 = \phi_1 - 1 \quad \text{if } f(y_{t-1}) > 0, \text{ regime 2} \tag{6.4}$$

For future reference, define $\phi = (\varphi_1 \ \phi_1)'$ and $\alpha = (\alpha_1 \ \alpha_2)'$. Also assume that $z_t = \varepsilon_t$ where ε_t is, as usual, distributed as iid$(0, \sigma_\varepsilon^2)$, so that the error is white noise; other cases are considered below. Note that the difference between regimes is in the 'persistence' parameters φ_1 and ϕ_1, rather than in the error variance in each regime, which is assumed to be constant. However, a difference in error variances between regimes is both plausible and easy to account for, and this issue is considered in Section **6.3.3**.

As there are two regimes in this process, it is referred to as the two-regime TAR, or 2RTAR. In the context of (6.3) and (6.4), the unit root null hypothesis is H_0, so that adjustment is symmetric, implying one regime not two, and $\varphi_1 = \phi_1 = 1$ implying H_0: $\alpha_1 = \alpha_2 = 0$. Other variations are of interest. For example, if $\varphi_1 \in (-1, 1)$ but $\phi_1 = 1$, then there is a unit root in the second regime, but the process is stationary in the first regime (referred to later as a 'partial' unit root case).

A leading case for the regime switching function is $f(y_{t-1}) = y_{t-1}$. The 2RTAR can then be written as a single equation using the indicator function defined as $I_{1t} = 1$ if $y_{t-1} \leq 0$ and $I_{1t} = 0$ if $y_{t-1} > 0$, and also as a standard regression model combining all observations:

$$\Delta y_t = \alpha_1 I_{1t} y_{t-1} + \alpha_2 (1 - I_{1t}) y_{t-1} + z_t \tag{6.5}$$

$$\Delta y = X_1 \alpha_1 + X_2 \alpha_2 + z \tag{6.6a}$$

$$= X\alpha + z \tag{6.6b}$$

where: $y = (y_1, \ldots, y_T)'$, $y_{-1} = (y_0, \ldots, y_{T-1})'$; X_1 and X_2 are $T \times 1$ vectors with typical elements $I_{1t} y_{t-1}$ and $(1 - I_{1t}) y_{t-1}$, respectively; $z = (z_1, \ldots, z_T)'$; and assume that $z_t = \varepsilon_t$. Define I_1 and I_2 as $T \times T$ selection matrices that select the observations appropriate to each regime: for regime 1, I_1 is such that the t-th diagonal is equal to 1 if $I_{1t} = 1$ and 0 otherwise, and for regime 2, $I_2 = I - I_1$, where I is the T-dimensional identity matrix. Using these matrices, the regressors can be defined as: $X_1 = I_1 y_{-1}$ and $X_2 = I_2 y_{-1}$, as in (6.6b); notice that $X_1' X_2 = 0$. Also $T_1 = \sum_{t=1}^{T} I_{1t}$ and $T_2 = T - T_1$ are the number of times that observations are generated by regimes 1 and 2 respectively.

An informative rearrangement of (6.5) emphasises the interpretation of the model as a base regime subject to change in the alternative regime:

$$\Delta y_t = \alpha_2 y_{t-1} + (\alpha_1 - \alpha_2) I_{1t} y_{t-1} + z_t$$

$$= \alpha_2 y_{t-1} + \gamma_1 I_{1t} y_{t-1} + z_t \tag{6.7}$$

where $\gamma_1 = (\alpha_1 - \alpha_2)$, and the hypothesis of symmetry, or equivalently one regime and, therefore, linearity, is $\gamma_1 = 0$.

The model so far is a rather special case because the equilibrium, or 'attractor', is a constant equal to zero. However, by prior adjustment of the data, the model with a zero attractor has wider relevance. There are two other cases of particular interest, although the principle easily generalises. First, suppose that the original variable of interest is y_t with attractor $\kappa \neq 0$, then define $w_t = y_t - \kappa$, so that equilibrium corresponds to $w_t = 0$; the TAR model, as in equations (6.1) and (6.2), then applies to w_t but can be translated back to the original variable y_t as follows:

$$y_t = \kappa + \varphi_1 (y_{t-1} - \kappa) + z_t \qquad \text{if } f(y_{t-1}) \leq \kappa, \text{ regime 1} \tag{6.8a}$$

$$y_t = \kappa + \phi_2 (y_{t-1} - \kappa) + z_t \qquad \text{if } f(y_{t-1}) > \kappa, \text{ regime 2} \tag{6.8b}$$

The other case of particular interest is where the attractor is a linear trend, $\kappa_t = \kappa_0 + \kappa_1 t$. In this case define $w_t = y_t - (\kappa_0 + \kappa_1 t)$; the 2RTAR model again applies to w_t with $w_t = 0$ representing equilibrium and implying $y_t = (\kappa_0 + \kappa_1 t)$. Translating back to the original variable, the 2RTAR model is:

Regime 1 if : $f(y_{t-1}) \leq \kappa_{t-1} = \kappa_0 + \kappa_1(t-1)$

$$y_t = (\kappa_0 + \kappa_1 t) + \phi_1(y_{t-1} - [\kappa_0 + \kappa_1(t-1)]) + z_t \tag{6.9a}$$

Regime 2 if : $f(y_{t-1}) > \kappa_{t-1} = \kappa_0 + \kappa_1(t-1)$

$$y_t = (\kappa_0 + \kappa_1 t) + \phi_1(y_{t-1} - [\kappa_0 + \kappa_1(t-1)]) + z_t \tag{6.9b}$$

Generally, κ or κ_t, as relevant, are not known and have to be estimated. EG have suggested prior demeaning or detrending of the data, so that in the former case κ is estimated by the (global) sample mean, $\bar{y}$, and in the latter case the coefficients κ_0 and κ_1 are estimated from an LS regression of y_t on a constant and linear time trend, resulting in LS estimators $\hat{\kappa}_0$ and $\hat{\kappa}_1$. Let $\tilde{y}_t$ denote the residuals from this prior process, that is $\tilde{y}_t = y_t - \hat{\kappa}$, where $\hat{\kappa} = \bar{y}$ or $\tilde{y}_t = y_t - \hat{\kappa}_t$, where $\hat{\kappa}_t = \hat{\kappa}_0 + \kappa_1 t$. Then $\tilde{y}_t$ replaces y_t in the 2RTAR specification, so that (6.5) becomes

$$\Delta\tilde{y}_t = \alpha_1 I_{1t}\tilde{y}_{t-1} + \alpha_2(1 - I_{1t})\tilde{y}_{t-1} + \tilde{z}_t \tag{6.10}$$

Notice that the error $\tilde{z}_t$ differs from z_t as it now includes the estimation error from the procedure of prior demeaning or detrending, otherwise H_0 is as before. As the sample mean is a biased estimate of κ, Chan (1993) considered the case of $\kappa_t = \kappa$ and suggested an alternative (consistent) estimator of κ (see Sections 6.2.1.ii and 6.3.2.ii below). We will retain the (generic) notation that $\tilde{y}_t$ denotes the demeaned or detrended variable.

The momentum version of the 2RTAR, denoted 2RMTAR, uses changes rather than the levels of y_t lagged once to determine the regime, for example $f(y_{t-1}) = \Delta y_{t-1}$. Then the indicator function for the 2RMTAR is defined as: $I_{1t} = 1$ if $\Delta y_{t-1} \leq \kappa$ and $I_{1t} = 0$ if $\Delta y_{t-1} > \kappa$. We have assumed here the likely case that the change is not trended, otherwise the general set-up is as in the levels 2RTAR. EG estimate a 2RMTAR for the spread between a short rate, R_t^S, and a long rate, R_t^L, for US government securities, so that $y_t = R_t^S - R_t^L$, and find adjustment to equlibrium, $\kappa = 0$, following a negative change, $\Delta y_{t-1} < 0$, to be faster than for a positive change.

At this point, it is worth highlighting what the 2RTAR specification does and does not do. The two regimes are just distinguished by their different speeds of adjustment, through the persistence coefficients φ_1 and ϕ_1. Given stationarity, there is a unique equilibrium to which y_t will return given a shock. This makes it possible, and unambiguous, to define the regime selector relative to 'the' attractor. However, it is possible that different equilibria will result from different regimes, in which case it is necessary to allow the implicit long run, which enters through the deterministic components in each of the two AR models, to be subject to change. This development is considered below.

Table 6.1   Integers for the ergodicity condition in a TAR model

d	1	2	3	4	12
f(d)	1	1	3	7	3
g(d)	0	2	4	8	4

Source: Extracted from Chen and Tsay (1991, p. 615).

6.1.1.ii   *Stationarity of the 2RTAR*

Previously Petruccelli and Woolford (1984) have shown that necessary and sufficient conditions for ergodicity of the 2RTAR with $f(y_{t-1}) = y_{t-1}$ and delay parameter $d = 1$ are: $\varphi_1 < 1$, $\phi_1 < 1$ and $\phi_1\varphi_1 < 1$. (Ergodicity implies stationarity whilst non-ergodicity implies nonstationarity; for related references see, for example, Cline and Pu, 2001, and Ling et al., 2007.) Notice that the 'standard' condition, $(\varphi_1, \phi_1) \in (-1, 1)$, is sufficient but not necessary (because φ_1 and ϕ_1 are allowed to be < -1). Chen and Tsay (1991) consider the conditions for the ergodicity of the 2RTAR(1) with $d \geq 1$ and show that two conditions must be added to the Petruccelli and Woolford conditions. Specifically, the complete conditions (the ergodic set) are:

$$\varphi_1 < 1, \phi_1 < 1 \text{ and } \phi_1\varphi_1 < 1, \phi_1^{f(d)}\varphi_1^{g(d)} < 1 \text{ and } \phi_1^{g(d)}\varphi_1^{f(d)} < 1 \tag{6.11}$$

where $f(d)$ and $g(d)$ are nonnegative integers depending on d. Some values are given in Table 6.1. For example, if $d = 2$, then the additional conditions are: $\phi_1\varphi_1^2 < 1$ and $\phi_1^2\varphi_1 < 1$, and these restrict the ergodicity region relative to $d = 1$, but only outside $(\varphi_1, \phi_1) \in (-1, 1)$. Note also that a point on the boundary for one parameter, for example $\phi_1 = 1$, the other parameter being inside the boundary, implies that y_t is nonstationary.

The conditions for stationarity are slightly different for an MTAR model. Lee and Shin (2000) note that if just one of φ_1 or ϕ_1 is on the boundary, that is equal to unity, and the other has an absolute value less than unity, then y_t *is* stationary – this contrasts with the TAR case. The following are the two 'partial' unit root cases, which imply nonstationarity for TAR models but stationarity for MTAR models: either

(i) $\varphi_1 = 1$ and $|\phi_1| < 1$; or   (6.12a)

(ii) $|\varphi_1| < 1$ and $\phi_1 = 1$ (6.12b)

6.1.2   **Generalisation of the TAR(1) and MTAR(1) models**

While the basic TAR(1) model, suggested by EG, is a good place to start, in practice a more general model may be needed to capture key features of the data. In this section some ways of extending the TAR model are discussed.

6.1.2.i EG augmentation

In the case that $z_t \neq \varepsilon_t$, to account for possible serial correlation in either the levels or momentum 2RTAR, EG suggested modifying (6.5) by adding $p - 1$ lagged differences of y_t, or $\tilde{y}_t$ when the data is demeaned or detrended. On the assumption that this augmentation removes the serial correlation, the resulting augmented model is:

$$\Delta \tilde{y}_t = \alpha_1 I_{1t} \tilde{y}_{t-1} + \alpha_2 (1 - I_{1t}) \tilde{y}_{t-1} + \sum_{j=1}^{p-1} c_j \Delta \tilde{y}_{t-j} + \tilde{\varepsilon}_t \tag{6.13}$$

In the case that there is no prior demeaning or detrending then $\tilde{y}_t = y_t$ and $\tilde{\varepsilon}_t = \varepsilon_t$. This specification is not a 2RTAR(p) because the threshold transition effect is not applied to all coefficients in an underlying AR(p) model.

6.1.2.ii 2RTAR(p) and 2RMTAR(p)

There is an alternative specification that is not equivalent to the EG augmentation, which serves to draw out the similarity of the TAR-type models with the STAR-type models (see Chapter 5). A generalisation of the 2RTAR(1) or 2RMTAR(1) models is to apply the threshold transition to the complete AR(p) model. Thus, start with an AR(p) model, here allowing for a nonzero mean through a constant, that is:

$$y_t = \phi_0 + \sum_{i=1}^{p} \phi_i y_{t-i} + \varepsilon_t \tag{6.14}$$

Now allow all of the AR(p) coefficients to potentially undergo a transition between regimes 1 and 2. The resulting 2RTAR(p) model is:

$$\begin{aligned} y_t &= (\varphi_0 + \sum_{i=1}^{p} \varphi_i y_{t-i}) I_{1t} + (\phi_0 + \sum_{i=1}^{p} \phi_i y_{t-i})(1 - I_{1t}) + \varepsilon_t \\ &= \phi_0 + \sum_{i=1}^{p} \phi_i y_{t-i} + [(\varphi_0 - \phi_0) + \sum_{i=1}^{p} (\varphi_i - \phi_i) y_{t-i}] I_{1t} + \varepsilon_t \\ &= \phi_0 + \sum_{i=1}^{p} \phi_i y_{t-i} + (\gamma_0 + \sum_{i=1}^{p} \gamma_i y_{t-i}) I_{1t} + \varepsilon_t \end{aligned} \tag{6.15}$$

where $\gamma_i = \varphi_i - \phi_i$ and the indicator function is defined as before, ($I_{1t} = 1$ for regime 1 and 0 otherwise). The model is a 2RTAR(p) model that is in regime 1 if $y_{t-1} \leq \kappa$, and in regime 2 if $y_{t-1} > \kappa$ then it is a 2RMTAR(p) model if the regimes are based on $\Delta y_{t-1} \leq \kappa$ and $\Delta y_{t-1} > \kappa$, respectively. For later reference, define $\gamma = (\gamma_0, \gamma_1, \ldots, \gamma_p)'$.

The interpretation of (6.15) is that the base regime is regime 2, with i-th coefficient ϕ_i, whereas in the alternative regime (1 in this case) the i-th coefficient undergoes a change of γ_i to $\phi_i + \gamma_i = \varphi_i$. The inclusion of the constant, which changes between regimes, means that the implied long run may also change.

The long-run constants in each regime, μ_1^{LR} and μ_2^{LR}, are:

$$\mu_1^{LR} = \frac{\phi_0 + \gamma_0}{1 - \sum_{i=1}^{P}(\phi_i + \gamma_i)}$$

$$= \frac{\varphi_0}{1 - \sum_{i=1}^{P} \varphi_i}$$

$$\mu_2^{LR} = \frac{\phi_0}{1 - \sum_{i=1}^{P} \phi_i}$$

It is possible that the long-run constants are equal despite changes in the underlying AR coefficients; the required condition is $\gamma_0(\sum_{i=1}^{P} \gamma_i)^{-1} = -\mu_2^{LR}$. For example, if $\phi_0 = 1$, $\phi_1 = 0.8$, then $\mu_2^{LR} = 5$, and $\gamma_0 \gamma_1^{-1} = -5$ ensures $\mu_1^{LR} = \mu_2^{LR}$; if $\gamma_0 = 0.2$, then $\gamma_1 = (-0.2/5) = -0.04$.

Notice that the generalisation in (6.15) does not lead to the suggestion that serial correlation be allowed for by adding p lagged differences to a base TAR(1) model, as in going from (6.10) to (6.13). The nature of difference is evident if (6.15) is written in ADF form as follows:

$$y_t = (\phi_0 + \sum_{i=1}^{P} \phi_i y_{t-i}) + (\gamma_0 + \sum_{i=1}^{P} \gamma_i y_{t-i})I_{1t} + \varepsilon_t \tag{6.16a}$$

$$= \phi_0 + \gamma_0 I_{1t} + \sum_{i=1}^{P}(\phi_j + \gamma_j I_{1t})y_{t-1} + \sum_{i=1}^{P-1}(c_j + d_j I_{1t})\Delta y_{t-j} + \varepsilon_t \tag{6.16b}$$

$$\text{where } c_j = -\sum_{i=j+1}^{P} \phi_i \quad \text{and } d_j = -\sum_{i=j+1}^{P} \gamma_i$$

$$= \phi_0 + \gamma_0 I_{1t} + \beta_t y_{t-1} + \sum_{j=1}^{P-1} \delta_{tj} \Delta y_{t-j} + \varepsilon_t \tag{6.16c}$$

$$\text{where } \beta_t = \sum_{i=1}^{P}(\phi_i + \gamma_i I_{1t}) \quad \text{and } \delta_{tj} = (c_j + d_j I_{1t})$$

$\Rightarrow$

$$\Delta y_t = \mu_t + \alpha_t y_{t-1} + \sum_{j=1}^{P-1} \delta_{tj} \Delta y_{t-j} + \varepsilon_t \tag{6.17}$$

where $\alpha_t = \beta_t - 1, \mu_t = \phi_0 + \gamma_0 I_{1t}$.

For reference back to the set-up and notation of (6.5), define: $\alpha_1 = \sum_{i=1}^{P} \varphi_i - 1 = \sum_{i=1}^{P}(\phi_i + \gamma_i) - 1$ and $\alpha_2 = \sum_{i=1}^{P} \phi_i - 1$. If there is a single regime, then (6.17) reduces to:

$$\Delta y_t = \phi_0 + \alpha_2 y_{t-1} + \sum_{j=1}^{P-1} c_j \Delta y_{t-j} + \varepsilon_t \tag{6.18}$$

Consider an AR(2) example, with zero mean or demeaned data so that $\phi_0 = 0$, and $\phi_1 = 0.5$, $\phi_2 = 0.1$; $\varphi_1 = 0.6$, $\varphi_2 = 0.3$; and, therefore, $\gamma_1 = 0.1$ and $\gamma_2 = 0.2$.

Then we have the following model:

$$y_t = (0.6y_{t-1} + 0.3y_{t-2})I_{1t} + (0.5y_{t-1} + 0.1y_{t-2})(1 - I_{1t}) + \varepsilon_t$$
$$= (0.5y_{t-1} + 0.1y_{t-2}) + [(0.6 - 0.5)y_{t-1} + (0.3 - 0.1)y_{t-2}]I_{1t} + \varepsilon_t$$
$$= (0.5 + 0.1I_{1t})y_{t-1} + (0.1 + 0.2I_{1t})y_{t-2} + \varepsilon_t$$
$$= [(0.5 + 0.1) + (0.1 + 0.2)I_{1t}]y_{t-1} + (-0.1 - 0.2I_{1t})\Delta y_{t-1} + \varepsilon_t$$

$\Rightarrow$

$$\Delta y_t = [-0.4 + (0.1 + 0.2)I_{1t}]y_{t-1} + (-0.1 - 0.2I_{1t})\Delta y_{t-1} + \varepsilon_t$$

$\Rightarrow$

Regime 1:

$$\Delta y_t = -0.1y_{t-1} - 0.3\Delta y_{t-1} + \varepsilon_t$$

Regime 2:

$$\Delta y_t = -0.4y_{t-1} - 0.1\Delta y_{t-1} + \varepsilon_t$$

Notice that the coefficients on both y_{t-1} and Δy_{t-1} change between regimes. As a variation consider the case where the change is solely on the first AR coefficient, that is:

$$y_t = (0.6y_{t-1} + 0.3y_{t-2})I_{1t} + (0.5y_{t-1} + 0.3y_{t-2})(1 - I_{1t}) + \varepsilon_t$$
$$= (0.8 + 0.1I_{1t})y_{t-1} - -0.3\Delta y_{t-1} + \varepsilon_t$$
$$\Delta y_t = (-0.2 + 0.1I_{1t})y_{t-1} - 0.3\Delta y_{t-1} + \varepsilon_t$$

Regime 1:

$$\Delta y_t = -0.1y_{t-1} - 0.3\Delta y_{t-1} + \varepsilon_t$$

Regime 2:

$$\Delta y_t = -0.2y_{t-1} - 0.3\Delta y_{t-1} + \varepsilon_t$$

In this case, the change is confined to the coefficient on y_{t-1}.

In this version of the 2RTAR(p) model, the regime change affects not just the coefficient on y_{t-1} but also the coefficients on the lagged dependent variables. The conditions that ensure that the coefficients on the Δy_{t-j} terms are unchanged between regimes, as imposed in the EG specification, are $d_j = 0$, $j = 1,..., p - 1$, where $d_j = -\sum_{i=j+1}^{p} \gamma_i$, which, in turn, imply $\gamma_i = 0$, that is, $\varphi_i = \phi_i$ for $i = 2, ..., p$. These conditions imply that the difference between the regimes is solely in the first AR coefficient so that $\varphi_1 \neq \phi_1$.

A further point to emphasise is that in the more general specification of the 2RTAR(p) it is possible to have the same root in both regimes, for example a unit root, so that the process is nonstationary, and have changes elsewhere so that the regime differences are not limited to asymmetric roots; Caner and Hansen (2001) give some examples of such changes.

It may also be reasonable to include a time trend in the model under the alternative hypothesis in the TAR(p) model; otherwise an alternative that allows a stationary but nonlinear process about a deterministic trend will be missed. If the time trend is included under H_A, the model is, first as an AR and then as an ADF:

$$y_t = (\phi_0 + \sum_{i=1}^{p} \phi_i y_{t-i}) + (\gamma_0 + \sum_{i=1}^{p} \gamma_i y_{t-i})I_{1t} + \phi_{p+1}t + (\gamma_{p+1}t)I_{1t} + \varepsilon_t$$

$$(6.19)$$

$$\Delta y_t = \mu_t + \alpha_t y_{t-1} + \sum_{j=1}^{p-1} \delta_{tj} \Delta y_{t-j} + \varepsilon_t \tag{6.20}$$

where, in addition to previously defined coefficients, $\mu_t = \phi_0 + \gamma_0 I_{1t} + \phi_{p+1}t + \gamma_{p+1}I_{1t}t$; rather than force the different trend through the same origin in the two regimes, it is likely that $\gamma_0 \neq 0$ if $\gamma_{p+1} \neq 0$. If this version of the model is used, then γ is redefined to include γ_{p+1}, so that $\gamma = (\gamma_0, \gamma_1, \ldots, \gamma_p, \gamma_{p+1})'$.

6.1.2.iii   Multiple regimes

Another direction in which the TAR model can be extended is to allow for more than two regimes. For example a TAR(p) model with m regimes, $m \geq 2$, is given by:

$$y_t = \sum_{j=1}^{m} \varphi_{0j}I_{jt} + \sum_{j=1}^{m} \left(\sum_{i=1}^{p} \varphi_{ij}y_{t-i}\right)I_{jt} + \upsilon_j I_{jt}\xi_t \tag{6.21}$$

where $I_{jt} = 1$ if $\kappa_{j-1} < f(y_{t-1}) \leq \kappa_j$ and 0 otherwise; κ_{j-1} is the 'floor' and κ_j is the 'ceiling' to $f(y_{t-1})$ in regime j; κ_0 is the overall floor and κ_m is the over-all ceiling, for example, $\kappa_0 = -\infty$ and $\kappa_m = \infty$ are often the defaults, but in some circumstances it makes more sense to bound the set of admissible threshold parameters. To enable further generality, the error variance is allowed to change across regimes. In this case ξ_t is iid(0, 1) and the role of $\upsilon_j I_{jt}$ is to capture this aspect of regime dependence as the error variance in the j-th regime is υ_j^2.

Multi-regime TAR models are nested since an AR(p) model is a restricted version of a 2RTAR(p) model; a 2RTAR(p) is a restricted version of a 3RTAR(p) model and so on. Hansen (1999) suggests a sequence of tests to determine the number of regimes. Also, some techniques have been suggested to ease the estimation problem, which involves a multi-dimensional grid search; for the case of $m > 2$, see Hansen (1999) and Gonzalo and Pitarakis (2002).

Kapetanios (2001) has considered the problem of selecting the lag order, p, in a multi-regime TAR; he considers a number of model selection criteria including the 'standard' information criteria of AIC, BIC and the Hannan-Quinn (HQ) version of the information criterion. Explicit derivations are given in this reference for the two-regime case, with the reasonable conjecture that the results extend to the multi-regime case. With some standard assumptions for the case where y_t is stationary and m and d are known, and the true order of the lag is in the set over which the search is made, Kapetanios (2001) is able to show that model selection by BIC and HQ results in weakly and strongly consistent estimation of the unknown true order. (Recall that weak consistency refers to convergence to the true value in probability and strong consistency to almost sure convergence.) These results are as in the linear AR model; also, as in the linear AR model, the AIC is neither weakly nor strongly consistent, summarised in the result that selection by AIC results in a positive probability of overestimating the true order (see, for example, Lütkepohl, 1993, section 4.3.2, and 2006).

These results relate to consistency, and hence to large samples. In the case of the linear AR model, it has been shown that although AIC leads to an inconsistent estimator of the lag order it nevertheless performs quite well in finite samples. Monte-Carlo results reported by Kapetanios (2001) do not find a clear overall 'winner', as the results depend on the set-up of the DGP. To illustrate, all information criteria perform better as the coefficients on the higher-order lags are more numerically significant; in such cases, generally, but not universally, BIC is better than AIC, notwithstanding its known tendency to select lag lengths too short in the linear AR; but AIC is better if the sample is 'small' (< 200) or the lag coefficients on the higher-order terms are small (for example 0.05).

Where the lag order is not common to all regimes, a two-stage process is suggested (Kapetanios, 2001, remark 5): in the first stage, assume a common order and use a consistent information criterion to determine the empirical order, $\hat{p}$; then, in the second stage, with threshold parameters set as estimated in the first stage, search within each regime over $p = 1, ..., \hat{p}$. The resulting estimators for each regime are consistent.

6.1.2.iv   Three-regime TAR: 3RTAR

One case that has received some attention is the TAR with three regimes; this is a special case of (6.21), which is now considered in greater detail.

A natural extension of the 2RTAR is to add a third regime, which then allows a middle, or inner, regime and two outer regimes; the model is denoted 3RTAR. Consider the case where the thresholds are indicated relative to two constants, κ_1 and κ_2 and, for simplicity, the error variances in the three regimes are assumed to be the same:

$$y_t = \varphi_1 y_{t-1} + z_t \qquad \text{if } f(y_{t-1}) \leq \kappa_1 \qquad\qquad \text{regime 1} \qquad (6.22)$$

$$y_t = \psi_1 y_{t-1} + z_t \qquad \text{if } \kappa_1 < f(y_{t-1}) \leq \kappa_2 \qquad \text{regime 2} \qquad (6.23)$$

$$y_t = \phi_1 y_{t-1} + z_t \qquad \text{if } f(y_{t-1}) > \kappa_2 \qquad\qquad \text{regime 3} \qquad (6.24)$$

where $\kappa_1 < \kappa_2$ and assume that $z_t = \varepsilon_t$. Notice that the AR coefficient in the inner regime is denoted ψ_1. The inner regime is sometimes referred to as the corridor regime, since it operates for y_t between κ_1 and κ_2. The case of symmetric outer regimes corresponds to $\varphi_1 = \phi_1$, otherwise the outer regimes are asymmetric $\varphi_1 \neq \phi_1$. The presence of three regimes, with two thresholds, means that the division between regimes under stationarity is not based on the long-run mean (for example κ or κ_t in the case of the 2RTAR model).

As in the 2RTAR, the regimes can be written to produce one model:

$$y_t = \varphi_1 I_{1t} y_{t-1} + \psi_1 (1 - I_{1t} - I_{2t}) y_{t-1} + \phi_1 I_{2t} y_{t-1} + \varepsilon_t \qquad (6.25)$$

The two indicator functions are defined as follows: set $I_{1t} = 1$ if $f(y_{t-1}) \leq \kappa_1$, and 0 otherwise, and set $I_{2t} = 1$ if $f(y_{t-1}) > \kappa_2$, and 0 otherwise. The base regime, here taken to be the inner regime, occurs when $I_{1t} = I_{2t} = 0 \Rightarrow (1 - I_{1t} - I_{2t}) = 1$, thus $y_t = \psi_1 y_{t-1} + z_t$. The other regimes can be expressed as deviations from the base, so that:

$$y_t = (\varphi_1 - \psi_1) I_{1t} y_{t-1} + \psi_1 y_{t-1} + (\phi_1 - \psi_1) I_{2t} y_{t-1} + \varepsilon_t$$

For example, setting $I_{1t} = 1$ results in the first regime:

$$y_t = (\varphi_1 - \psi_1) y_{t-1} + \psi_1 y_{t-1} + \varepsilon_t$$
$$= \varphi_1 y_{t-1} + \varepsilon_t$$

There are now three roots each with two possibilities if we confine the analysis to a root either being stationary or a unit root, thus there are $2^3 = 8$ distinct possibilities. To illustrate, we consider some of the resulting specifications. The case of no unit roots corresponds to (6.25) with $|\varphi_1| < 1$, $|\psi_1| < 1$, $|\phi_1| < 1$; this is still asymmetry and so regime change if the coefficients are not equal. The other limiting case is three unit roots and so symmetry and a single regime $\Delta y_t = \varepsilon_t$. Other cases are where there are either one or two unit roots. For example, with a unit root in each of the outer regimes:

$$y_t = (I_{1t} + I_{2t}) y_{t-1} + \psi_1 (1 - I_{1t} - I_{2t}) y_{t-1} + \varepsilon_t \quad |\psi_1| < 1$$

$$\Rightarrow$$

$$\Delta y_t = (\psi_1 - 1)(1 - I_{1t} - I_{2t}) y_{t-1} + \varepsilon_t$$
$$= \alpha_0 (1 - I_{1t} - I_{2t}) y_{t-1} + \varepsilon_t \quad \alpha_0 = \psi_1 - 1$$

Thus, $\Delta y_t = \varepsilon_t$ apart from the inner regime where $\Delta y_t = \alpha_0 y_{t-1} + \varepsilon_t$.

Another interesting and often studied case, motivated by consideration of the behaviour of real exchange rates, is a unit root in the inner regime:

$$y_t = \varphi_1 I_{1t} y_{t-1} + (1 - I_{1t} - I_{2t}) y_{t-1} + \phi_1 I_{2t} y_{t-1} + \varepsilon_t \qquad |\varphi_1| < 1, |\phi_1| < 1$$

$$\Rightarrow$$

$$\Delta y_t = (\varphi_1 - 1) I_{1t} y_{t-1} + (\phi_1 - 1) I_{2t} y_{t-1} + \varepsilon_t$$

$$= \alpha_1 I_{1t} y_{t-1} + \alpha_2 I_{2t} y_{t-1} + \varepsilon_t$$

Continuing with this case, define $\alpha = (\alpha_1, \alpha_2)'$. As in the case of the 2RTAR of (6.6b), the 3RTAR can then be written as:

$$\Delta y = X\alpha + \varepsilon \tag{6.26}$$

where the t-th row of X is $(I_{1t} y_{t-1}, I_{2t} y_{t-1})$, the t-th row of Δy is Δy_t and the t-th row of ε is ε_t, where $\varepsilon_t \sim \text{iid}(0, \sigma_\varepsilon)$.

A three regime TAR allows nonstationarity in one of the regimes while maintaining the global ergodicity of the process. For example, the 3RTAR(1) above with nonstationary in the inner regime is nonstationarity, but the process is globally stationary. For a more extensive discussion of the conditions for global ergodicity see Tweedie (1975), Chan et al. (1985), Chan and Tong (1985), Chan (1990), Cline and Pu (1999, 2001) and Bec, Ben Salem and Carrasco (2004).

If $\varphi_1 = \phi_1$ then the outer regimes have symmetric responses and this specification has some similarities with an ESTAR(1) model. This set-up has been proposed as a candidate model for the real exchange rate, which shows random walk-type behaviour for a large part of the time, but the theory of purchasing power parity suggests that there should be an equilibrium real exchange rate, so that the outer regimes serve to contain the extent of the random walk.

As in EG (1998), if $z_t \neq \varepsilon_t$, then KS (2006) suggest augmenting the basic model with p − 1 lagged differences of Δy_t to allow for serially correlated errors:

$$\Delta y_t = \alpha_1 I_{1t} y_{t-1} + \alpha_2 I_{2t} y_{t-1} + \sum_{j=1}^{p-1} c_j \Delta y_{t-j} + \varepsilon_t \tag{6.27}$$

This modification does not affect the asymptotic distributions of the test statistics given below. However, note that this is not the generalisation suggested by analogy with (6.15), where the latter allowed all of the AR coefficients to potentially undergo a change between regimes.

There are a number of specialisations of the 3RTAR and one that has been studied has been called the band TAR, motivated by an analysis of PPP. In this case there is a 'band' of inaction where deviations from PPP may exist without being arbitraged away. The band is $[-\kappa, \kappa]$, $\kappa > 0$, and the regime indicators are then: regime 1 if $y_{t-1} \leq -\kappa$, regime 2 if $|y_{t-1}| < \kappa$ and regime 3 if $y_{t-1} \geq \kappa$ (see Bec, Ben Salem and Carrasco, 2004).

6.1.2.v   TAR and LSTAR

The notation in (6.15) has been chosen to emphasise the correspondence between a 2RTAR(p) model and an LSTAR(p) model, the difference being that I_{1t} is a (discrete) indicator function, sometimes referred to as the Heaviside indicator function, taking the value 0 or 1 rather than a smooth transition function. In the LSTAR model, $\Psi(y_{t-1}) = [1 + e^{-\psi(y_{t-1}-c)}]^{-1}$ and as $\psi \to \infty$ the transition function tends to the discrete indicator function, with, for example, $\psi = 6$ sufficient for the transition function to look very like a switching function. The form in (6.15) also emphasises the 'traditional' 0/1 dummy variable interpretation, where the AR coefficients undergo a change at a number of points in the sample; for example, ϕ_i changes to $\phi_i + \gamma_i = \varphi_i$ if $y_{t-1} \leq \kappa$. The model might thus also be viewed as embodying a particular kind of structural break.

6.2   Hypothesis testing and estimation: general issues

The discussion so far has outlined a number of approaches to modelling non-linearity through regime change in an AR model. This section deals with some common issues, which arise in hypothesis testing and estimation, whether the model is the levels or momentum version of the TAR. There are two issues of particular interest in the context of these models. One is whether there are unit roots in one or more of the regimes. The other is whether there is a threshold effect; that is, is the model linear with just one regime or is there more than one regime? In the context of the EG version of the TAR(1) model, the threshold effect is just the asymmetry in the roots arising from $\varphi_1 \neq \phi_1$. However, more generally, the threshold effect can arise from changes in any of the AR(p) coefficients. As solutions to these problems involve common elements, it is helpful to outline the general approach before considering the specific details for a number of cases of interest.

6.2.1   Threshold parameter is not identified under the null

The problem in devising a test statistic for either testing for a threshold effect or a unit root, is that the threshold parameter κ is not identified under the null; in either case under the null there is only one regime so κ, or κ_t more generally, simply does not exist. The same problem obviously also occurs in estimation, as the coefficients of interest under the alternative depend on κ, which is generally unknown. The two problems have a common solution, which was outlined in Chapter 1.

6.2.1.i   Hypothesis testing

As far as the hypothesis testing problem is concerned, the solution for κ unknown is to 'integrate it out'; practically, this means generating a set of test statistics, say N, for $\kappa_{(i)}$ in a range likely to (preferably certain to) contain the true κ; say $\kappa_{(i)} \in \mathbf{K}$, where κ must be in the *interior* of $\mathbf{K}$. The N test statistics are each conditioned on a distinct value of $\kappa_{(i)}$; then a function of these N values is specified in order to generate a single test statistic for the purposes of testing the null against the alternative. An often-favoured choice is the Supremum value, or Sup, of the test statistic over the N realised values. If a Wald test statistic is used, the resulting test is the Sup-Wald test; equally one may apply the LM and LR principles to produce Sup-type tests. Some trimming of the range of the regime selector relative to the sample range is generally necessary to control the behaviour of the test statistic as fewer and fewer observations are included in one of the regimes. Typically, the suggested procedures leave not less than 10% or 15% of observations in each regime; that is they 'trim' 20% or 30%, respectively, overall. (For consideration of the theoretical issues in choosing an appropriate grid see KS, 2006.)

6.2.1.ii Estimation with κ unknown

The solution to the estimation problem is essentially the same as in the hypothesis testing case, and the same whether a TAR or MTAR is considered, whether a unit root is present and whether there are two or more regimes. Initially assume that the set of grid search values of $\kappa_{(i)}$ is as in the hypothesis testing case, that is $\kappa_{(i)} \in \mathbf{K}$ where $\mathbf{K}$ may have been trimmed. Each of these values results in a particular partition of the data into the two regimes, and conditional on this separation, the model is estimated and the conditional residual sum of squares noted. The preferred model is the one that results in the overall miminum, or Infimum or, simply, Inf, of the N values of the conditional residual sums of squares (RSS). No new problems of principle arise in the case of a multiple-regime TAR with $m \geq 2$; the search must of course be extended to $m - 1$ values of κ_i, such that $\kappa_1 < \kappa_2 < \cdots < \kappa_{m-1}$, and some care has to be taken to ensure that there are sufficient observations in each regime to empirically identify the regression coefficients. The next subsection considers some of the issues.

For simplicity the search for the minimum in the conditional RSS is assumed to be over the set $\mathbf{K}$; while this is the likely case, arguments for trimming in determining $\mathbf{K}$, which are motivated by considerations of test consistency and power, do not apply with such force just to estimation. Estimation requires the regression coefficients to be empirically identified; this is an absolute not a proportionate requirement. For example, in a sample of $T = 100$, then 10 or 15 observations may be sufficient (what matters is the variation in the data); ceteris paribus, doubling the sample size does not change this number. However, for simplicity the assumption that the set $\mathbf{K}$ serves both purposes is maintained.

6.2.1.iii Asymptotic distribution of the test statistic

A second problem common to variants of the hypothesis testing problem is that the asymptotic distribution of the chosen test statistic is generally either: (i) not known; or (ii) known but non-standard and so finite sample critical values have to be obtained by simulation; or (iii) is known but is a function of nuisance parameters, which make it impossible to tabulate critical values in generality.

All of these particular cases can, in principle, be solved by some form of bootstrapping, although not all have been in practice. A semiparametric bootstrap procedure will work when the starting estimates in the preliminary step of the bootstrap are obtained from consistent estimators and the asymptotic distribution of the test statistic is a continuous function of the model parameters. It will have some desirable properties if the test statistic is at least asymptotically pivotal, by which is meant that the asymptotic distribution of the test statistic is invariant for all possible DGPs that satisfy the null (for example, the same parameter restrictions but different data matrices); see Davidson and MacKinnon (2006) for an excellent summary. The parametric part of the bootstrap description refers to the need to estimate the unknown parameters of the true DGP. The bootstrap distribution of the test statistics described here are first-order asymptotically correct; see Hansen (1996) and Caner and Hansen (2001). The bootstrap procedure is particularly valuable in the third case where nuisance parameters prevent critical values from being obtained in any generality; and it is valuable in the first case provided, at least, that the conditions for a valid bootstrap can be established; again see Davidson and MacKinnon (2006).

6.3   Estimation of TAR and MTAR models

6.3.1   Known regime separation

If κ_t is known and φ_1 and ϕ_1 satisfy the stationarity conditions given in (6.12) for the TAR(1), (or for TAR(p) more generally), then the respective LS estimators $\hat{\varphi}_1$ and $\hat{\phi}_1$ are consistent, and are asymptotically distributed as multivariate normal (Chan, 1993, and Schoier, 1999). With regime separation known, this result extends to the multi-regime TAR, with m > 2. In this case not only is the estimation problem solved, so is hypothesis testing for a threshold effect since standard test statistics, for example an F test of the hypothesis $\varphi_1 = \phi_1$ (< 1), have standard distributions. However, generally κ_t is unknown and needs to be consistently estimated; also, the TAR may either be nonstationary or it is not known that it is stationary.

6.3.2   Unknown regime separation

6.3.2.i   The sample mean is not a consistent estimator of the threshold parameter

Consider the case where $\kappa_t = \kappa$, κ is unknown and the 2RTAR(1) is stationary. Then κ will not be consistently estimated by the (global) sample mean, $\bar{y}$, if adjustment is asymmetric. If adjustment is asymmetric, with $\varphi_1 > 0$, $\phi_1 > 0$ and $\varphi_1 > \phi_1$, then there is more persistence, that is, the process is slower to adjust back to equilibrium, κ, when $y_{t-1} \leq \kappa$ compared to $y_{t-1} > \kappa$. As a result the sequence will, ceteris paribus, spend more time in the first regime where $y_{t-1} \leq \kappa$, which will then be 'over-represented' if the sample mean is used to estimate κ. The same problem is still present in 2RTAR(p) as there is more persistence in one regime than the other. In a TAR with more than two regimes, clearly a more general method is in any case required, since at least two threshold parameters need to be specified. In MTAR models, while κ may be set to 0, this is often a default, rather than the result of knowledge or consistent estimation.

6.3.2.ii *Threshold parameters and delay parameter consistently estimated by conditional least squares (stationary two-regime TAR and MTAR)*

In the case of a stationary two-regime TAR(p), Chan (1993) – see also Klimko and Nelson (1978) – has shown that κ and d can be consistently estimated by conditional least squares, CLS. This result has been extended to the stationary multi-regime MTAR(p) (see Caner and Hansen, 2001), and to the stationary TAR(p) (see Gonzalo and Pitarakis, 2002).

The estimation procedure is as follows. Presently assume that d is known, say equal to 1, and the problem is to estimate κ and α. As outlined in Section **6.2.1.ii**, a consistent estimator of κ can be obtained in two steps: first, estimate α conditional on $\kappa = \kappa_{(i)}$, for $\kappa_{(i)} \in K$, i = 1, ..., N. Next, define the resulting conditional estimator and conditional residual sum of squares, RSS, as $(\hat{\alpha}|\kappa = \kappa_{(i)})$ and $(\hat{\varepsilon}'\hat{\varepsilon}|\kappa = \kappa_{(i)})$, respectively. Then, choose the estimator of κ as the value of $\kappa_{(i)}$ that corresponds to an overall minimum of the resulting N (conditional) residual sums of squares for estimating α. Denote this overall minimum as $(\hat{\varepsilon}'\hat{\varepsilon}|\kappa = \kappa_{(i)})_{inf}$, with the associated value of $\kappa_{(i)}$ providing the estimator $\hat{\kappa}$ of κ, and denote the corresponding value of $\hat{\alpha}$ as $\hat{\alpha}_{inf}$. Also, a consistent estimator of σ_ε^2 is given by $(\hat{\varepsilon}'\hat{\varepsilon}|\kappa = \kappa_{(i)})_{inf}/T$. For α in the ergodic set, the distribution of $\hat{\alpha}_{inf}$ is asymptotically normal; moreover, the asymptotic distribution of $\hat{\alpha}_{inf}$ is the same whether κ is known or consistently estimated by CLS. A similar grid search procedure to minimise the RSS can be used to determine the delay parameter, d.

The grid search over $\kappa_{(i)} \in K$ is based on the ordered (sorted) sequence of $\{y_t\}$, such that $y_{(1)} < y_{(2)} < ... < y_{(T)}$; however, the range is usually trimmed, whilst still providing sufficient observations to enable estimation in each regime. Asymptotic theory requires that $(T_j/T) \to \Lambda > 0$ as $T \to \infty$, where T_j is the number of observations in regime j. This implies defining the outer limits of the grid as the Λ-th and $(1 - \Lambda)$-th percentiles of y_{t-1}; for $\Lambda = 0.1$, these are the 10th and 90th percentiles. These basic results again extend to the multi-regime TAR, with $m > 2$, provided $T_j/T \to \Lambda > 0$ as $T \to \infty$ for each regime (this is actually an intricate matter and the statement here assumes that there are not any regimes of inaction for which $T_j/T \to 0$ is acceptable).

6.3.2.iii   *Estimation of a TAR or MTAR with three regimes*

An extension of the grid search procedure to a 3RTAR(1) or 3RTAR(p) model is straightforward in principle. In the case that d is known or imposed, for example $d = 1$, the grid search is extended to the two-dimensional grid with $N = n_1 \times n_2$ distinct points. For simplicity, we assume that the grid sets for hypothesis testing and estimation are the same but the arguments above suggest that this need not be the case.

In this three-regime case there are n_1 points for κ_1 and n_2 points for κ_2, say $(\kappa_{(1i)}, \kappa_{(2i)}) \in \mathbf{K}_1 \times \mathbf{K}_2$. Define the conditional estimate of α and the conditional residual sum of squares as $(\hat{\alpha}|\kappa_1, \kappa_2 = \kappa_{(1i)}, \kappa_{(2i)})$ and $(\hat{\varepsilon}'\hat{\varepsilon}|\kappa_1, \kappa_2 = \kappa_{(1i)}, \kappa_{(2i)})$, respectively. The unconditional LS estimator $\hat{\alpha}_{inf}$ is then that which minimises $(\hat{\varepsilon}'\hat{\varepsilon}|\kappa_1, \kappa_2 = \kappa_{(1i)}, \kappa_{(2i)})$ over $(\kappa_{(1i)}, \kappa_{(2i)}) \in \mathbf{K}_1 \times \mathbf{K}_2$.

Note that the nature of the 3RTAR model implies $\kappa_1 < \kappa_2$, hence not all points in the grid $\mathbf{K}_1 \times \mathbf{K}_2$ need to be evaluated. To illustrate, suppose that a two-dimensional grid (matrix) is constructed of the 80 percentiles between the 11-th and 90-th percentile of y_t, with $\kappa_{(1i)}$ on the rows and $\kappa_{(2i)}$ on the columns, so that $n_1 = n_2 = N$, then there are only $N(N-1)/2$ combinations for which $\kappa_1 < \kappa_2$, not N^2. In the example, this is 3,160 rather than 6,400. Even so, if the delay parameter is unknown, and has to be included in the search, and a bootstrap procedure is needed to obtain the p-value of a relevant test statistic, the overall number of evaluations starts to become impractical or, at least, in need of a technique that reduces the computational burden. Hansen (1999) shows shows that a sequential procedure due to Bai (1997) and Bai and Perron (1998, 2003a, b) can be adapted for this purpose, and Gonzalo and Pitarakis (2002) provide further details and proofs for some relevant cases.

6.3.2.iv   *Unknown delay parameter, d*

The indicator function can be generalised by allowing d, the delay parameter, to be an integer usually expected to be in the set $\mathbf{D} = [1, p]$; for example, in the case of a constant attractor, $I_{1t} = 1$ if $y_{t-d} \leq \kappa$ and $I_{1t} = 0$ if $y_{t-d} > \kappa$. Neither the extension to TAR(p) nor the generalisation of the indicator function alters the general principle of solving the estimation problem by a grid search; in the case that d is regarded as unknown in a 2RTAR model, the grid search becomes two-dimensional, now over $d \in \mathbf{D}$ and $\kappa_{(i)} \in \mathbf{K}_1$. The search thus seeks the minimum over $p \times n_1$ evaluations of the conditional residual sum of squares. In a 3RTAR the number of evaluations increases to $p \times (N(N-1)/2)$. If computational time is an issue (and it tends not to be with current processor speeds for one-off estimation), for example as may arise in bootstrapping, Hansen (1999) describes a procedure that substantially reduces the number of function evaluations.

6.3.2.v   Partial unit root cases

Some interesting cases are not explicitly covered by the discusion so far. Essentially these are the partial unit root cases. In a two-regime TAR or MTAR, a unit root in one regime only is a partial unit root. Lee and Shin (2000) note that the overall process is still stationary in a self-exciting MTAR even if there is partial unit root; and, similarly, Gonzalez and Gonzalo (1998, corollary 3) show that in an MTAR, without the restriction to self-exciting, and in the form specified by EG, the presence of one unit root and one stationary root still results in an overall process that is stationary.

Caner and Hansen (2001) have considered the 2RMTAR case where one of the alternative hypotheses is that of a partial unit root, and Gonzalo and Pitarakis (2002) have considered the multi-regime stationary case, which covers the MTAR. These results suggest that the LS estimator of the (remaining) unknown AR coefficients maintains consistency provided that the threshold parameters are identified, with the grid search including the true threshold parameter(s) and sufficient observations in each regime to enable estimation.

A three regime TAR, 3RTAR, allows an inner regime that may have a unit root, with outer regimes that are stationary. These outer regimes can then be viewed as bounding the extent of the nonstationarity and define an overall process that is globally ergodic. A grid search, satisfying the usual conditions, as in the MTAR case above, produces a consistent estimator of the threshold parameters. Although the conditions for global ergodicity are more complex in the case of multi-regime TARs, with $m > 3$, a reasonable surmise is that if the outer regimes are generated by stationary processes, but the inner regimes are characterised by unit roots, then the overall process 'contains' the nonstationarity sufficiently to remain stationary.

6.3.3   Regime dependent heteroscedasticity

The error variance in the basic 2RTAR(1) model of (6.1) and (6.2) was assumed homoscedastic across regimes. However, regime change might also be associated with different variances, say $\sigma_{1\varepsilon}^2$ for regime 1 and $\sigma_{2\varepsilon}^2$ for regime 2 and let $\sigma_{2\varepsilon}^2 = v^2 \sigma_{1\varepsilon}^2$. Because of regime separation, the overall conditional RSS, $(\hat{\varepsilon}'\hat{\varepsilon} | \kappa = \kappa_{(i)})$, comprises two additive residual sums of squares arising from the two regimes. Then:

$$(\hat{\varepsilon}'\hat{\varepsilon} | \kappa = \kappa_{(i)})_{\text{inf}} = (\hat{\varepsilon}' I_1 \hat{\varepsilon} | \kappa = \kappa_{(i)})_{\text{inf}} + (\hat{\varepsilon}' I_2 \hat{\varepsilon} | \kappa = \kappa_{(i)})_{\text{inf}}$$

Where I_1 and I_2 are the separation matrices defined earlier. A consistent estimator of the error variance for each regime is $\hat{\sigma}_{j\varepsilon}^2 = (\hat{\varepsilon}' I_j \hat{\varepsilon} | \kappa = \kappa_{(i)})_{\text{inf}}/T_j$ (see Chan, 1993, theorem 1). If homoscedasticty between regimes is incorrectly assumed, then $(\hat{\varepsilon}'\hat{\varepsilon} | \kappa = \kappa_{(i)})_{\text{inf}}/T$ is not a consistent estimator of either $\sigma_{1\varepsilon}^2$ or $\sigma_{2\varepsilon}^2$.

If the error variances are not constant across regimes, a simple transformation ensures a constant variance (necessary for most of the unit root tests). The GLS transformation involves premultiplying the model in (6.6a) by the transformation matrix P, where $P = I_1 + \upsilon^{-1}I_2$, the effect of which is just to divide the observations in the second regime by υ, that is:

$$P\Delta y = PX\gamma + Pz \qquad\qquad (6.28)$$

$$= PI_1\gamma_1 y_{-1} + PI_2\gamma_2 y_{-1} + Pz$$

$$= \gamma_1 I_1 y_{-1} + \gamma_2 I_2 \upsilon^{-1} y_{-1} + Pz \qquad \text{using } I_1 I_2 = 0$$

The *weighted* residual sum of squares, arising from the transformed model, which uses all of the residuals, can then be used to obtain a consistent estimator of the variance:

$$W(\hat{\varepsilon}'\hat{\varepsilon}|\kappa = \kappa_{(i)})_{\text{inf}} = (\hat{\varepsilon}'I_1\hat{\varepsilon}|\kappa = \kappa_{(i)})_{\text{inf}} + \upsilon^{-2}(\hat{\varepsilon}'I_2\hat{\varepsilon}|\kappa = \kappa_{(i)})_{\text{inf}}$$

$$\text{plim}(\hat{\varepsilon}'\hat{\varepsilon}|\kappa = \kappa_{(i)})_{\text{inf}} = T_1\sigma_{1\varepsilon}^2 + \upsilon^{-2}T_2\sigma_{2\varepsilon}^2$$

$$= T_1\sigma_{1\varepsilon}^2 + T_2\sigma_{1\varepsilon}^2$$

$$= T\sigma_{1\varepsilon}^2$$

Because of the orthogonality of the regressors in each regime, removing the heteroscedacity by way of the GLS transformation leaves the LS estimators of α unchanged. In practice υ is unknown but a consistent estimator is given by $\hat{\upsilon} = \hat{\sigma}_{2\varepsilon}/\hat{\sigma}_{1\varepsilon}$.

6.4   Testing for nonstationarity in TAR and MTAR models

6.4.1   Enders-Granger model, 2RTAR and 2RMTAR

We start with the simplest case, which is the EG model of equations (6.5) or (6.6), where the unit root hypothesis corresponds to H_0: $\alpha_1 = \alpha_2 = 0$. One possible alternative hypothesis is H_A: $\alpha_1 < 0, \alpha_2 < 0$; however, other alternatives may be of interest, for example, a unit root in one regime, say the first, but stationarity in the other, that is H_{A1}: $\alpha_1 = 0, \alpha_2 < 0$.

EG (1998) suggested an F-type test for the random walk null hypothesis, that is H_0: $\alpha_1 = \alpha_2 = 0$ against the two-sided alternative, H_A: $\alpha_1 \neq 0$ and/or $\alpha_2 \neq 0$. Note that the alternative for the F test is two-sided even though interest centres on $\alpha_1 < 0$ and/or $\alpha_2 < 0$. (A one-sided test is considered below.) The F test will be denoted $F(0,0)_\kappa$, the entries in (.) being the values corresponding to the null hypothesis and the subscript indicating the sample separation.

An interesting way of viewing the test statistic is to note that for a given regime separation, the regressors are orthogonal. This implies that the F-type test statistic is the average of the squared t statistics for the null hypothesis for each regime; of course in this context, the t statistics are the DF test statistics. So in regime 1 the null hypothesis is $\alpha_1 = 0$, resulting in a t-type test statistic, say $t(\hat{\alpha}_1)$, or in F form the square of this quantity, $t(\hat{\alpha}_1)^2$; and, analogously, $t(\hat{\alpha}_2)^2$ for the second regime. The F test can be constructed from these as the average of the two squared t statistics: $F(0,0)_\kappa = \frac{1}{2}[t(\hat{\alpha}_1)^2 + t(\hat{\alpha}_2)^2]$; the principle of averaging the squared t statistics extends to multiple regimes.

The large sample version of this test, which does not make a degrees of freedom correction, is the Wald-type test statistic, related to the F test as $W_2(0,0) = 2F(0,0)[\tilde{T}/(\tilde{T} - 2)]$, where $\tilde{T}$ is the effective sample size; thus, $W(0,0)_\kappa \approx 2F(0,0)_\kappa = t(\hat{\alpha}_1)^2 + t(\hat{\alpha}_2)^2$, that is (approximately), the sum of the two squared t statistics (see Chapter 1). Note that this is not the same as running the regressions for the two regimes separately and using the corresponding t-type statistics, as such a procedure does not impose a common error variance.

6.4.1.i Known threshold

Some critical values for the F-type test are given in Table 6.2. The design of the underlying process for generating these quantiles is worth noting at this stage. The generating process was $\Delta y_t = \varepsilon_t$, and in the 2RTAR, κ was assumed known and equal to zero. There were then three cases: use the unadjusted data, adjust the data by demeaning or detrending. Then determine the sample selection according to $\kappa = 0$ with the unadjusted, demeaned or detrended data. This resulted in three F-type test statistics denoted $F(0,0)_\kappa$, $F(0,0)_\kappa^\mu$ and $F(0,0)_\kappa^\beta$. The important point is that in these experiments κ is assumed known. This is obviously a rather specialised case as there is no need to 'integrate' out κ. Kapetanios and Shin (2006), hereafter KS, in a slightly different context, but one that is applicable here, obtained the asymptotic null distribution of $W(0,0)_\kappa$ from a 2RTAR model, which is as follows (for $\kappa = 0$):

$$W(0,0)_\kappa \Rightarrow_D \frac{\left(\int_0^1 I_{[B(s)\leq 0]}B(s)dB(s)\right)^2}{\int_0^1 I_{[B(s)\leq 0]}B(s)^2\,ds} + \frac{\left(\int_0^1 I_{[B(s)>0]}B(s)dB(s)\right)^2}{\int_0^1 I_{[B(s)>0]}B(s)^2\,ds} \qquad (6.29)$$

$$\equiv W(0,0)$$

Note that $\kappa = 0$ is not a restriction, since provided that κ is known the threshold can be adjusted so that the condition is met. $B(s)$ is standard Brownian motion defined on $s \in [0, 1]$ and the indicator variables for $B(s) \leq 0$ and $B(s) > 0$ are, respectively, $I_{[B(s)\leq 0]}$ and $I_{[B(s)>0]}$. This result also provides the asymptotic null distribution of the two-regime EG F-type test statistic, which is $W(0,0)/2$. In the demeaned or detrended cases, the Brownian motion in (6.29) is replaced by demeaned or detrended Brownian motion, respectively.

Enders (2001) gave some critical values for a revised test statistic denoted $F(0,0,c)_\kappa^\mu$, where $c = p - 1$, the order of the ADF, which took two additional factors into account. The first was to use Chan's (1993) suggestion to estimate the (constant) threshold κ by CLS. Second, the original EG test was based on

Table 6.2   Quantiles for testing for a unit root in a 2RTAR model and in a 2RMTAR model

2RTAR

	$F(0,0)_K$		$F(0,0)_K^{\mu}$		$F(0,0)_K^{\beta}$		$F(0,0,c)_K^{\mu}:90\%$			$F(0,0,c)_K^{\mu}:95\%$		
	90%	95%	90%	95%	90%	95%	(0)	(1)	(4)	(0)	(1)	(4)
100	3.18	3.95	3.79	4.64	5.27	6.30	5.08	5.39	5.38	6.06	6.34	6.32
250	3.10	3.82	3.74	4.56	5.18	6.12	5.11	5.26	5.36	6.03	6.12	6.29
1,000	3.04	3.75	3.74	4.56	5.15	6.08	n.a	n.a	n.a	n.a	n.a	n.a

2RMTAR

	$F(0,0)_K$		$F(0,0)_K^{\mu}$		$F(0,0)_K^{\beta}$		$F(0,0,c)_K^{\mu}:90\%$			$F(0,0,c)_K^{\mu}:95\%$		
	90%	95%	90%	95%	90%	95%	(0)	(1)	(4)	(0)	(1)	(4)
100	2.83	3.60	5.74	6.83	5.74	6.83	4.81	4.77	4.74	5.77	5.71	5.70
250	2.68	3.41	5.64	6.65	5.64	6.65	4.70	4.64	4.64	5.64	5.54	5.54
1,000	2.51	3.21	5.60	6.57	5.60	6.57	n.a	n.a	n.a	n.a	n.a	n.a

Source: Extracted from Enders and Granger (1998, table 1) and Enders (2001, tables 1 and 2); 2RTAR is the two-regime TAR, and 2RMTAR is the momentum 2RTAR; number in parentheses is the number of lagged changes of the dependent variable included in the maintained regression for Ender's variation of the test statistics; 'n.a' indicates not available; however, note that the variations in the test statsistics are fairly slight as T increases.

the maintained regression without lagged $\Delta \tilde{y}_t$, that is (6.6), rather than the augmented version given by (6.13), whereas Enders (2001) gives critical values for 0, 1 and 4 lagged $\Delta \tilde{y}_t$ (indicated by the columns with (0), (1) and (4), respectively, in Table 6.2). In addition, there are two cases depending on whether the regime indicator is $f(y_{t-1}) = y_{t-1}$ or $f(y_{t-1}) = \Delta y_{t-1}$, that is the TAR and MTAR, respectively.

EG note that the power of their suggested F-type tests is not generally greater than a standard DF test in a 2RTAR. For example, comparing $F(0,0)_K^{\mu}$ and $\hat{\tau}_{\mu}$, they found that the latter dominated except when the asymmetry was very pronounced (with α_1 of an order seven or eight times greater, in absolute value, than α_2). The situation was rather better for the F-type test in a 2RMTAR, where the asymmetry did not need to be so marked before $F(0,0)_K^{\mu}$ was more powerful than $\hat{\tau}_{\mu}$.

Note that the quantiles for Enders modified test statistic, $F(0,0,c)_K^{\mu}$, are uniformly larger than those for the orginal test statistics. In a power comparison Enders (op. cit.) finds that in a 2RTAR, the DF test, $\hat{\tau}_{\mu}$, is again generally more powerful, this time more so than both $F(0,0)_K^{\mu}$ and $F(0,0,c)_K^{\mu}$; the exception is where the asymmetry is marked, then $F(0,0,c)_K^{\mu}$ is more powerful than both alternatives. However, the revised test is not generally more powerful than the original test, the gain in the marked asymmetry case being the exception rather than the rule. The situation is more favourable to the F-type test in the 2RMTAR model, where the original test statistic $F(0,0)_\kappa^\mu$, not the revised test statistic $F(0,0,c)_\kappa^\mu$, is generally the most powerful of those tests considered.

In the case of the 2RTAR it seems likely that a unit root test with more power than the standard DF test will dominate the EG F-type tests; and this may be so even in the 2RMTAR case (for example, see Cook, 2004, for MTAR models). Part of the problem is that the F-type test is two-sided compared to the one-sided nature of DF t-type tests; therefore, as part of the size of the test is not needed as far as the direction of the alternative is concerned, so power in the required direction is lost. (This problem is addressed by the directional version of the F test.)

EG suggest that following rejection of the null hypothesis, it may be of interest to test for symmetric or asymmetric adjustment; that is, now the null hypothesis is $H_0: \alpha_1 = \alpha_2 (< 0)$ against $H_A: \alpha_1 \neq \alpha_2$, with a standard test statistic, for example an F or χ^2 statistic, which will, asymptotically have an F distribution or χ^2 distribution. There are two points of relevence here.

First, for TAR models, the F or Wald group of tests will have power against the two 'partial' unit root alternatives $H_{A1}: \alpha_1 = 0$, $\alpha_2 < 0$ and $H_{A2}: \alpha_1 < 0$, $\alpha_2 = 0$, which are of interest in their own right as ceiling or floor random walk models. Consider H_{A1} with a constant attractor, then

$$y_t = y_{t-1} + \varepsilon_t \qquad \text{if } y_{t-1} \leq 0, \quad \text{regime 1} \tag{6.30}$$

$$y_t = \phi_1 y_{t-1} + \varepsilon_t \quad \phi_1 < 1 \quad \text{if } y_{t-1} > 0, \quad \text{regime 2} \tag{6.31}$$

In regime 1, y_t can random walk without limit below and up to the 'ceiling' 0, thereafter it is 'contained' by a stationary regime. (If the threshold is κ, this translates to a random walk for y_t up to κ.) However, an MTAR model for the two 'partial' unit root cases is stationary (see Lee and Shin, 2000).

Second, even in the stationary case, as Hansen (1999) notes in a related context, in practice κ and d are not known and are part of the estimation procedure; as a result the distribution of, for example, the χ^2 version of the test in a TAR is not asymptotically χ^2.

Returning to the test or unit roots, it is possible that a 'one-sided' test would yield better power and the form of such a test has been suggested by Caner and Hansen (2001); in this case only negative values of the test statistic are taken as evidence against the null hypothesis of a unit root, so only square the i-th t statistic and count it as part of the test statistic if $\hat{\alpha}_i < 0$. This is the 'directional' test given by $F^D(0,0)_{\kappa_1,\kappa_2} = \frac{1}{2} [t(\hat{\alpha}_1)^2 I(\hat{\alpha}_1 < 0) + t(\hat{\alpha}_2)^2 I(\hat{\alpha}_2) < 0)]$, where I(.) is the usual indicator variable equal to one if the condition in (.) is true and zero otherwise. This is a principle that can be extended to multi-regime TARs, with the empirical quantiles obtained by boostrapping.

6.4.1.ii   Unknown threshold

The more realistic case in practice is when the threshold κ is not known, in which case κ can be estimated by CLS from a grid of possible values for κ, as outlined in Section **6.3.2.ii**. The test statistic can then be calculated for each point in the grid. Three versions of the test statistic have been suggested for such a case: the supremum of $W_2(0,0)_\kappa$ in the grid, say Sup-$W_2(0,0)_\kappa$, the simple average and the exponential average, defined respectively as:

$$W_2(0,0)_\kappa^{\text{ave}} = N^{-1} \sum_{i=1}^{N} W_2(0,0)_{\kappa,i}^{\text{ave}} \tag{6.32a}$$

$$W_2(0,0)_\kappa^{\text{exp}} = N^{-1} \sum_{i=1}^{N} \exp[W_2(0,0)_{\kappa,i}/2] \tag{6.32b}$$

KS (op. cit.) show that Sup-$W_2(0,0)_\kappa$ and $W_2(0,0)_\kappa^{\text{ave}}$ have the same limiting distribution as if κ is assumed to be known, that is (6.29); and $W_2(0,0)_\kappa^{\text{exp}}$ $\Rightarrow_D \exp[W(0,0)/2]$. (Note the definition of the exp version of this test follows KS, whereas the exp version of this test was defined in Chapter 1 as the log of what is here (6.32b), following Andrews, 1993.)

6.4.2   Testing for nonstationarity in a 3RTAR

6.4.2.i   Testing for two unit roots conditional on a unit root in the inner regime

As noted in Section **6.1.2.iv**, there are several models of interest in a three-regime TAR that contain unit roots in different combinations. One of particular interest is where there is a unit root in the inner regime and a test is designed to assess whether there are unit roots in the other regimes. The testing framework considered here is due to KS (2006), with the regime selector based on $f(y_{t-1}) = y_{t-1}$. The 3RTAR(1) is first restated for the t-th observation and for all T observations:

$$y_t = \alpha_1 I_{1t} y_{t-1} + \alpha_2 I_{2t} y_{t-1} + \varepsilon_t \tag{6.33a}$$

$$\Delta y = X\alpha + \varepsilon \tag{6.33b}$$

and $\varepsilon_t \sim \text{iid}(0, \sigma_\varepsilon^2)$ with finite $4+\xi$ moments, $\xi > 0$.

Conditional on a unit root in the inner regime, the null hypothesis corresponding to (overall) nonstationarity is H_0: $\alpha_1 = \alpha_2 = 0$; while the natural alternative to H_0 is H_A: $\alpha_1 < 0$ and/or $\alpha_2 < 0$. In practice, the suggested Wald test statistic, which is quadratic in α, effectively has (as in the EG F-type test) the two-sided alternative H_A: $\alpha_1 \neq 0$ and/or $\alpha_2 \neq 0$. Note that the testing problem here is conditional on a unit root in the inner regime and, intuitively, one can expect it to reduce to the problem of testing for two unit roots in a 2RTAR; for a relaxation of the assumption of a unit root in the inner regime see Bec, Ben Salem and Carrasco (2004) and Bec, Guay and Guerre (2008).

In constructing the test initially κ_1 and κ_2 are assumed known; whilst generally unrealistic this is a useful place to start. The LS estimator of α is $\hat{\alpha}$ and $\text{Var}(\hat{\alpha}) = \hat{\sigma}_\varepsilon^2 (X'X)^{-1}$, where $\hat{\sigma}_\varepsilon^2 = \hat{\varepsilon}'\hat{\varepsilon}/(\tilde{T}-2)$ and $\hat{\varepsilon}$ is the residual vector, and $\tilde{T}$ is the effective sample size. The Wald test statistic for the null hypothesis H_0: $\alpha = 0$ against the two-sided alternative H_A: $\alpha_1 \neq 0$ and/or $\alpha_2 \neq 0$, is:

$$W(0,0)_{\kappa_1,\kappa_2} = \hat{\alpha}'[\text{Var}(\hat{\alpha})]^{-1}\hat{\alpha}$$

$$= \frac{\hat{\alpha}'[X'X]^{-1}\hat{\alpha}}{\hat{\sigma}_\varepsilon^2} \tag{6.34}$$

The asymptotic distribution of $W(0,0)_{\kappa_1,\kappa_2}$ under the null is non-standard. It turns out that the distribution does not depend on the values of κ_1 and κ_2, so it is convenient initially to set these to zero. The Wald test statistic for this case, denoted $W(0,0)_{0,0}$, has the limiting distribution $W(0,0)$ given by (6.29).

KS suggest handling the deterministics as in EG, that is, as for linear models with prior demeaning or detrending, so that y_t can be interpreted as one of three possible cases: 'raw' data, demeaned data where an estimate of the mean, usually the global sample mean $\bar{y}$, has been removed, and detrended data where an estimate of the trend component, usually the LS estimate from a regression of the raw data on a constant and a deterministic trend, has been removed. As usual, in the latter two cases the Brownian motion in (6.29) is replaced by demeaned or detrended Brownian motion, respectively. The case where $z_t \neq \varepsilon_t$ and is weakly dependent is handled by adding lagged dependent variables to obtain an ADF of the form given by (6.27), but note that this is not as general as the model given by (6.17) extended to three regimes and preferred, for example, by Bec, Ben Salem and Carrasco (2004).

The distributional result in (6.29) then generalises to a 3RTAR(1) with κ_1 and κ_2 known and not equal to zero. Specifically, conditional on the null H_0: $\alpha_1 = \alpha_2 = 0$, with κ_1 and κ_2 given, and under a condition on the threshold discussed below, the Wald statistic $W(0,0)_{\kappa_1,\kappa_2}$ weakly converges to $W(0,0)$ as in (6.29). $W(0,0)_{\kappa_1,\kappa_2}$ is consistent in the sense that it diverges to infinity as $\alpha_1 < 0$ and $\alpha_2 < 0$. This result is unaffected by use of the ADF form if the ADF order is correct.

The problem that κ_1 and κ_2 are not generally known has already been considered (see Section **6.3.2.iii**). The solution is to 'integrate' them out by constructing a two-dimensional grid with N distinct points, in this case over κ_1 and κ_2, say $(\kappa_{(1i)}, \kappa_{(2i)}) \in K_1 \times K_2$ and calculate the Wald test for every combination in the grid. As in the one-dimensional search, the grid is based on the ordered sequence of $\{y_t\}$, but there is a distinction as to how the grid is arranged depending on whether its purpose is to provide settings for simulations with varying sample sizes or an empirical application with a single, fixed sample size. As to the simulation setting, the weak convergence result $W(0,0)_{\kappa_1,\kappa_2} \Rightarrow_D W(0,0)$ requires that the inner regime is of finite width under both under the null and under the alternative (KS, 2006, Assumption 3). Intuitively, as the inner regime parameter is set to zero, that is $\alpha_1 = 0$, then $P(\kappa_1 < y_{t-1} \leq \kappa_2) \to 0$ as $T \to \infty$.

One such scheme to set the grid limits is as follows:

$$(\Lambda_1, \Lambda_2) = \left(\bar{\Lambda} - \frac{c}{T^\varsigma}, \bar{\Lambda} + \frac{c}{T^\varsigma} \right)$$

where $\Lambda_1 = P(y_{t-1} < \kappa_{min}) > 0$ and $\Lambda_2 = P(y_{t-1} > \kappa_{max}) > 0$, κ_{min} is the minimum value of $\kappa_{(1i)}$ and κ_{max} is the maximum value of $\kappa_{(2i)}$. $\bar{\Lambda}$ is the sample quantile corresponding to zero, $\varsigma \geq \frac{1}{2}$ and c is a constant chosen to ensure sufficient observations to estimate the parameters; for example if $T = 100$, then $\varsigma = \frac{1}{2}$ and $c = 3$ gives a grid of the the the central 60% of observations; and if $T = 900$, then the grid comprises 20% of the observations. If the grid scheme always kept, say, the observations between the 10th and 90th percentiles, irrespective of increases in T, then the inner regime does not disappear as $T \to \infty$. However, it is not inconsistent in a single sample to adopt the latter scheme and that is, in fact, what KS (2006) use in their empirical examples.

The number of distinct points in the grid is $N = n_1 \times n_2$. The Wald test statistic $W(0,0)_{\kappa_1,\kappa_2}$ is necessarily non-negative and rejects for 'large' values relative to the critical values. Possible forms of the test statistics are as before, that is the Supremum of $W(0,0)_{\kappa_1,\kappa_2}$ in the grid, say Sup-$W(0,0)_{\kappa_1,\kappa_2}$, the simple average and the exponential average, defined respectively as.

$$W(0,0)^{ave}_{\kappa_1,\kappa_2} = N^{-1} \sum_{i=1}^{N} W(0,0)^{ave}_{\kappa_1,\kappa_2,i} \qquad N = n_1 \times n_2 \qquad (6.35a)$$

$$W(0,0)^{exp}_{\kappa_1,\kappa_2} = N^{-1} \sum_{i=1}^{N} \exp[W(0,0)^{ave}_{\kappa_1,\kappa_2,i}/2] \qquad N = n_1 \times n_2 \qquad (6.35b)$$

where $W(0,0)_{\kappa_1,\kappa_2,i}$ is the test statistic for the i-th point in the grid. The Sup-$W(0,0)_{\kappa_1,\kappa_2}$ test statistic is obtained as the Supremum over i of the N individual test statistics (although not all $N = n_1 \times n_2$ points in the grid need to be evaluated – see Section **6.3.2.iii**).

KS (2006) show that the test statistics Sup-$W(0,0)_{\kappa_1,\kappa_2}$ and $W(0,0)^{ave}_{\kappa_1,\kappa_2}$ converge weakly to $W(0,0)$, and $W(0,0)^{exp}_{\kappa_1,\kappa_2}$ converges weakly to $\exp[W(0,0)/2]$; they also provide some relevant percentiles of $W(0,0)$ and $\exp[W(0,0)/2]$ using 50,000 replications with $T = 5,000$. The critical values are obtained by generating data under the null $\Delta y_t = \varepsilon_t$, and estimating the maintained regression with observations specified as being in regime 1 if $y_{t-1} \leq 0$ and in regime 2 if $y_{t-1} > 0$; $\alpha = (\alpha_1, \alpha_2)'$ is then estimated by least squares from (6.33) and the Wald statistics are calculated. Note that, as expected because of the equivalence in the limiting distribution, the critical values for $W(0,0)$ are close to those extracted from Table 6.2 using the relationship $2\Phi = W(0,0)$ for $T = 1,000$, and these values are given in (.) parentheses after the critical values from KS (2002, Table 1).

In their Monte-Carlo study, with data generated by $\Delta y_t = \varsigma z_{t-1} + \varepsilon_t$, $\varsigma = 0$, 0.3, $T = 100$ and $T = 200$, KS (2006) find that $W(0,0)^{exp}_{\kappa_1,\kappa_2}$ has substantially better size properties than Sup-$W(0,0)_{\kappa_1,\kappa_2}$ and (generally) better size properties and better power properties than $W(0,0)^{ave}_{\kappa_1,\kappa_2}$. This conclusion also holds in the

Table 6.3   Simulated quantiles of $W(0,0)$ and $\exp[W(0,0)/2]$

	$W(0,0)$		
	no adjustment	demeaned	detrended
90%	6.01(6.08)	7.29(7.48)	10.35(10.30)
95%	7.49(7.50)	9.04(9.12)	12.16(12.16)
		$\exp[W(0,0)/2]$	
90%	20.18	38.28	176.80
95%	42.30	91.83	437.03

Source: Extracted from Kapetanios and Shin (2006, table 1) and numbers in (.) parentheses from Enders and Granger (1998).

presence of an AR(1) error. The poor size property of Sup-$W(0,0)_{\kappa_1,\kappa_2}$ relative to $W(0,0)^{\text{ave}}_{\kappa_1,\kappa_2}$ in finite samples is perhaps not surprising when the critical values of $W(0,0)$ are used, since Sup-$W(0,0)_{\kappa_1,\kappa_2} > W(0,0)^{\text{ave}}_{\kappa_1,\kappa_2}$, then the former will tend to be over-sized at least relative to the latter if the same critical value is used for both.

As to power, KS (2006) find that $W(0,0)^{\text{exp}}_{\kappa_1,\kappa_2}$ is more powerful than $W(0,0)^{\text{ave}}_{\kappa_1,\kappa_2}$, Sup-$W(0,0)_{\kappa_1,\kappa_2}$ being excluded on the basis of its poor empirical size. In the case of demeaned or detrended data, $W(0,0)^{\text{exp}}_{\kappa_1,\kappa_2}$ is more powerful than $\hat{\tau}_\mu$; and $\hat{\tau}_\mu$ tends to be more powerful than $W(0,0)^{\text{ave}}_{\kappa_1,\kappa_2}$, the more so as the threshold band, that is, $\kappa_1 \leftrightarrow \kappa_2$, is large. Ceteris paribus, the power of all the tests declines monotonically as the threshold band increases. For a more extensive power comparison using size-adjusted power see Maki (2009), who also included a one-sided t-type test due to Park and Shintani (2005), which performed well but did not dominate the other tests.

6.4.2.ii   Testing for three unit roots

The KS test focuses on the globally stationary process with a unit root inner regime bounded by two stationary regimes; however other cases may be of interest, for example, testing the null hypothesis of three unit roots. One approach is to generalise the F test constructed as the average of the squared t statistics. That is, first write the (simple) 3RTAR as:

$$\Delta y_t = \alpha_1 I_{1t} y_{t-1} + \alpha_0 I_{0t} y_{t-1} + \alpha_2 I_{2t} y_{t-1} + \varepsilon_t$$

where $I_{0t} \equiv (1 - I_{1t} - I_{2t})$, $\alpha_0 = \psi_1 - 1$, $\alpha_1 = \varphi_1 - 1$ and $\alpha_2 = \phi_1 - 1$. The case considered by Bec, Ben Salem and Carrasco (2004) corresponds to the possibility of three unit roots in the context of a band 3TAR with regimes distinguished by $y_{t-1} \leq -\kappa$, $|y_{t-1}| < \kappa$ and $y_{t-1} \geq \kappa$.

Allowing all regimes to generate observations even in the limit distinguishes this general set up from that considered above, which corresponds to setting

$\alpha_0 = 0$. The form of the test statistic is similar throughout these tests and is as in the 2RTAR model. Whilst the focus here is on the Wald form, test statistics can equally be derived from the LM or LR principles. The Wald test statistic is as $W(0,0)_{\kappa_1,\kappa_2}$ in (6.34), but α is defined such that $\alpha = (\alpha_0, \alpha_1, \alpha_2)'$, so that all three coefficients can be tested. In the case that the more general 3RTAR(p) is used, a selection matrix is defined that picks out the three coefficients in α.

One can again define three forms of the Wald test statistics being equivalents of the Sup, ave and exp forms of the underlying $W(\kappa_1, \kappa_2)$ statistic. To distinguish these from the KS set-up, they are denoted Sup-$W(0,0,0)_{\kappa_1,\kappa_2}$, $W(0,0,0)^{\exp}_{\kappa_1,\kappa_2}$ and $W(0,0,0)^{\text{ave}}_{\kappa_1,\kappa_2}$. Also, given the connection between the Wald test and the F test, the underlying test can also be obtained as the average of the squared t statistics on $\hat{\alpha} = (\hat{\alpha}_0, \hat{\alpha}_1, \hat{\alpha}_2)'$. As in the 2RTAR case, the test statistic can be calculated in a nondirectional or directional form, that is:

$$F(0,0,0)_{\kappa_1,\kappa_2} = \frac{1}{3}[t(\hat{\alpha}_0)^2 + t(\hat{\alpha}_1)^2 + t(\hat{\alpha}_2)^2]$$

$$F^D(0,0,0)_{\kappa_1,\kappa_2} = \frac{1}{3}[t(\hat{\alpha}_1)^2 I(\hat{\alpha}_0 < 0) + t(\hat{\alpha}_1)^2 I(\hat{\alpha}_1 < 0) + t(\hat{\alpha}_2)^2 I(\hat{\alpha}_2 < 0)]$$

As noted, it is important to ensure that there are observations in the inner regime as well as the outer regimes, even as $T \to \infty$. For example, Maki (2009) undertakes a simulation study of a number of unit root tests in a 3RTAR and searches from the 5% to the 45% quantiles for the lower threshold, κ_1, and from the 55% to the 95% quantile for the upper threshold, κ_2, so that there are never less than 10% of the observations in the inner regime. To the author's knowledge, the limiting null distribution of these Wald-type tests has not yet been obtained, so that quantiles for testing need to be obtained by simulation or bootstrapping, which ties in with Maki's (2009) recommendation to use size-adjusted critical values for the various tests. (See Bec, Ben Salem and Carrasco, 2004, theorem 2, for the general form of the limiting distribution.)

6.5   Test for a threshold effect in a stationary TAR model

Enders and Granger (1998) have suggested that following the rejection of the null hypothesis of a unit root in their version of a TAR(1), further tests could be carried out to examine the hypothesis of the existence of a threshold effect. Hansen (1997a) has also considered this case in the context of the generalised version of the 2RTAR, which allows all AR coefficients to potentially undergo a change between regimes; and Gonzalo and Gonzalez (1998) have considered the extension to case where the TAR is not necessarily self-exciting and with a regime indicator that is stationary.

6.5.1 A stationary 2RTAR

It is convenient to restate the relevant (general) 2RTAR(p) model to enable further discussion:

$$y_t = (\varphi_0 + \sum_{i=1}^{p} \varphi_i y_{t-i}) I_{1t} + (\phi_0 + \sum_{i=1}^{p} \phi_i y_{t-i})(1 - I_{1t}) + \varepsilon_t \tag{6.36a}$$

$$= \phi_0 + \sum_{i=1}^{p} \phi_i y_{t-i} + (\gamma_0 + \sum_{i=1}^{p} \gamma_i y_{t-i}) I_{1t} + \varepsilon_t \tag{6.36b}$$

For later developments define the following:

$$x_t = (1, y_{t-1}, y_{t-2}, \ldots, y_{t-p})'; \ \varphi = (\varphi_0, \varphi_1, \varphi_2, \ldots, \varphi_p)'; \ \phi = (\phi_0, \phi_1, \phi_2, \ldots, \phi_p)';$$

$$\omega = (\varphi_0, \varphi_1, \varphi_2, \ldots, \varphi_p, \phi_0, \phi_1, \phi_2, \ldots, \phi_p)'$$

Then (6.36a) can be expressed as:

$$y_t = (I_{1t} x_t, (1 - I_{1t}) x_t)' \begin{pmatrix} \varphi \\ \phi \end{pmatrix} + \varepsilon_t \tag{6.37a}$$

$$= X_t' \omega + \varepsilon_t \tag{6.37b}$$

The null hypothesis of no threshold effect is H_0: $\gamma_i = 0$, for $i = 0, 1, \ldots, p$, and this would usually be set against a two-sided alternative.

A variation on this theme of interest is to use the ADF version of the model and focus on the sum of the AR coefficients. That is:

$$\Delta y_t = \mu_t + \alpha_t y_{t-1} + \sum_{j=1}^{p-1} \delta_{tj} \Delta y_{t-j} + \varepsilon_t \tag{6.38d}$$

where $\alpha_t = \beta_t - 1$, $\mu_t = \phi_0 + \gamma_0 I_{1t}$, $\beta_t = \sum_{i=1}^{p} (\phi_i + \gamma_i I_{1t})$ and $\delta_{tj} = (c_j + d_j I_{1t})$. The null hypothesis is now H_0: $\sum_{i=1}^{p} \gamma_i = 0$; this may be joined with $\gamma_0 = 0$ so that the implied long run is the same under both regimes; again, generally, the alternative is two-sided.

The general principle of testing is the same whichever H_0 is being tested. Consider the case where H_0: $\gamma_i = 0$, $\forall i$, against the two-sided alternative, so that the number of restrictions being tested is $g = p + 1$. The suggested testing procedure follows Hansen (1997a). The residual sum of squares under H_0 is simply $(\hat{\varepsilon}'\hat{\varepsilon})_1$ where the residuals are obtained under the null model. Under H_A, the notation for the residual sum of squares needs to distinguish three cases depending on whether both κ and d are known. In the case that they are known, the notation is $(\hat{\varepsilon}'\hat{\varepsilon}|\kappa, d)$, whereas if neither are known it is $(\hat{\varepsilon}'\hat{\varepsilon}|\hat{\kappa}, \hat{d})_{inf}$, which is the minimum over the grid of values for κ and d. If d is known, but κ is not known, the latter is the residual sum of squares referred to earlier as $(\hat{\varepsilon}'\hat{\varepsilon}|\hat{\kappa}, d)_{inf}$.

The corresponding residual variances are:

$$\hat{\sigma}_1^2 = (\hat{\varepsilon}'\varepsilon)_1/\tilde{T} \qquad \text{under the null model}$$

$$\hat{\sigma}_2^2(\kappa, d)_{\text{inf}} = (\hat{\varepsilon}'\hat{\varepsilon}|\kappa, d)_2/\tilde{T} \qquad \text{under the alternative: } \kappa, d \text{ known}$$

$$\hat{\sigma}_2^2(\hat{\kappa})_{\text{inf}} = (\hat{\varepsilon}'\hat{\varepsilon}|\hat{\kappa}, d)_{2,\text{inf}}/\tilde{T} \qquad \text{under the alternative: } d \text{ known}$$

$$\hat{\sigma}_2^2(\hat{\kappa}, \hat{d})_{\text{inf}} = (\hat{\varepsilon}'\hat{\varepsilon}|\hat{\kappa}, \hat{d})_{2,\text{inf}}/\tilde{T} \qquad \text{under the alternative: } \kappa, d \text{ unknown}$$

where $\tilde{T}$ is the effective sample size. The estimators $\hat{\sigma}_2^2(\hat{\kappa}, \hat{d})_{\text{inf}}$ and $\hat{\sigma}_2^2(\hat{\kappa})_{\text{inf}}$ are sensible if the error variances are homoscedastic between regimes or have been transformed so as to be homoscedastic.

The first case to be considered is that of known regime seperation. Assume that regime selection is relative to the constant κ, so that regime 1 corresponds to $y_{t-d} \leq \kappa$ and regime 2 to $y_{t-d} > \kappa$; and initially assume that κ and the delay parameter d are known. A standard test statistic of Wald form is then:

$$W_{12}(\kappa, d) = \tilde{T}\left(\frac{\hat{\sigma}_1^2 - \hat{\sigma}_2^2(\kappa, d)}{\hat{\sigma}_2^2(\kappa, d)}\right)$$

$$= \tilde{T}\left(\frac{(\hat{\varepsilon}'\hat{\varepsilon})_1 - (\hat{\varepsilon}'\hat{\varepsilon}|k, d)_2}{(\hat{\varepsilon}'\hat{\varepsilon}|k, d)_2}\right) \tag{6.39a}$$

For given κ and d, $W_{12}(\kappa, d)$ is asymptotically distributed as $\chi^2(g)$, with large values rejecting linearity. In the special case that the threshold effect just concerns a scalar parameter, say H_0: $\gamma_i = 0$, then with κ known, the result above also implies that the corresponding t-type test does have a t distribution, at least asymptotically.

A corresponding F version of this test is obtained on noting that the m-regime model has $m(p + 1)$ coefficients, and there are $g = p + 1$ additional coefficients for each regime, with an effective sample size of $\tilde{T}$; in this case $m = 2$ is tested against $m = 1$, hence the $F_{12}(\kappa, d)$ test statistic is:

$$F_{12}(\kappa, d) = \left(\frac{(\hat{\varepsilon}'\hat{\varepsilon})_1 - (\hat{\varepsilon}'\hat{\varepsilon}|k, d)_2}{(\hat{\varepsilon}'\hat{\varepsilon}|k, d)_2}\right)\left(\frac{\tilde{T} - 2g}{g}\right) \tag{6.39b}$$

Note that $gF_{12}(\kappa, d) \approx W_{12}(\kappa, d)$, the approximation due to the degrees of freedom correction in the numerator, which is negligible as $T \to \infty$. Similar adjustments may be made to the other Wald statistics presented in this section to obtain their corresponding F versions.

The second case to be considered is that of unknown regime separation In practice κ and d are not known and need to be included in the estimation procedure. The resulting test statistic, for the estimated values $\hat{\kappa}$ of κ and $\hat{d}$ of d, is denoted $W_{12}(\hat{\kappa}, \hat{d})$. However, now, under these circumstances, the distribution of $W_{12}(\hat{\kappa}, \hat{d})$ is no longer $\chi^2(g)$. As Hansen

(1997a) notes, the problem is not with estimation of d, since for inferential purposes one may act as if d was known with certainty; in other words if κ was known but d was estimated, the test statistic would still be asymptotically $\chi^2(g)$. The problem arises because κ is 'not-identified' under the null hypothesis. Given the invariance with respect to d, we may first condition on d and then remove the conditioning by considering a range of values for d, that is, $d_{(i)} \in \mathbf{D}$. Thus, the test statistic with d known is:

$$W_{12}(\hat{\kappa}|d_{(i)}) = \tilde{T}\left(\frac{\hat{\sigma}_1^2 - \hat{\sigma}_2^2(\hat{\kappa}, d)}{\hat{\sigma}_2^2(\hat{\kappa}, d)}\right)$$

$$= \tilde{T}\left(\frac{(\hat{\varepsilon}'\hat{\varepsilon})_1 - (\hat{\varepsilon}'\hat{\varepsilon}|\hat{\kappa}, d)_{2,\inf}}{(\hat{\varepsilon}'\hat{\varepsilon}|\hat{\kappa}, d)_{2,\inf}}\right) \tag{6.40}$$

If d is also estimated, the test statistic is the maximum of $W_{12}(\hat{\kappa}|d_{(i)})$ over $d_{(i)} \in \mathbf{D}$, say $W_{12}(\hat{\kappa}, \hat{d})$. Note that as $(\hat{\varepsilon}'\hat{\varepsilon})_1$ is fixed for a particular sample, and $(\hat{\varepsilon}'\hat{\varepsilon}|\hat{\kappa}, \hat{d})_{2,\inf}$ is by definition the minimum over the set of $\kappa_{(i)}$ and $d_{(i)}$ values, then $W_{12}(\hat{\kappa}, \hat{d})$ is the Supremum of such test statistics over the $\kappa_{(i)}$ and $d_{(i)}$ values.

Whether d is known or estimated, the problem remains of determining the distribution of $W_{12}(\hat{\kappa}|d_{(i)})$, or equivalent asymptotically $W_{12}(\hat{\kappa}, \hat{d})$, in order to obtain critical values. The distribution of $W_{12}(\hat{\kappa}|d_{(i)})$ is shifted to the right relative to the distribution of $\chi^2(g)$, so that using the critical values from $\chi^2(g)$ will result in too many rejections for the nominal size of the test.

Hansen (1999) suggests a bootstrap procedure to obtain the p-value for a particular realisation $\hat{W}_{12}(\hat{\kappa}|d_{(i)})$ of $W_{12}(\hat{\kappa}|d_{(i)})$. In the bootstrap, the null model is the (one-regime) AR(p) model, which is estimated by LS and provides the initial AR coefficient estimates and residuals from which bootstrap draws can be taken. Bootstrap replicates are based on these quantities, with estimation under the null using the bootstrap data providing one realisation of $\hat{\sigma}_1^2$; the model under the alternative of a 2RTAR is also estimated with a grid search over the $k_{(i)}$ values (assuming for simplicity that d is known). This step provides $\hat{\sigma}_2^2(\kappa_{(i)})_{\inf}$, and a bootstrap realisation of the test statistic, say $W_{12}^{bs}(\hat{\kappa}|d_{(i)})$. Repeat this process B times and sort the bootstrap values of $W_{12}^{bs}(\hat{\kappa}|d_{(i)})$ to obtain the bootstrap distribution. The (asymptotic) p-value of $\hat{W}_{12}(\hat{\kappa}|d_{(i)})$ is the proportion of the B bootstrap samples in which $W_{12}^{bs}(\hat{\kappa}|d_{(i)}) > \hat{W}_{12}(\hat{\kappa}|d_{(i)})$. If d is unknown, the procedure is modified to search over the set of $d_{(i)}$ values.

6.5.2 Test for a threshold effect in a stationary multi-regime TAR, with unknown regime separation

An extension to a multiple regime TAR, with $m > 2$, whilst maintaining stationarity is straightforward (see Hansen, 1999). For example, a test statistic for the null of two regimes against the alternative of three regimes, assuming that d is known but the threshold parameters are unknown, is:

$$W_{2,3}(\kappa_{(1i)}, \kappa_{(2i)} | d_{(i)}) = \tilde{T} \left(\frac{\hat{\sigma}_2^2(\hat{k}_1) - \hat{\sigma}_3^2(\hat{k}_1, \hat{k}_2)}{\hat{\sigma}_3^2(\hat{k}_1, \hat{k}_2)} \right) \tag{6.41}$$

where, by extension, the residual variance estimator for the three-regime model is $\hat{\sigma}_3^2(\hat{k}_1, \hat{k}_2)$. (The form of the Wald test in (6.41) is often the easiest to compute; for the equivalence between the two forms of the test see Chapter 1, Q1.2.) To obtain the p-value of a realised value $\hat{W}_{2,3}(\kappa_{(1i)}, \kappa_{(2i)} | d_{(i)})$ of the test statistic, a bootstrap can be based on the null hypothesis of two regimes with the bootstrap repetitions based on the LS estimates under the null and sampling from the empirical distribution function of the residuals.

6.5.3 A test statistic for a null hypothesis on κ

It is also possible to test a hypothesis on κ, say $H_0: \kappa = \kappa_0$ against the two-sided alternative. The test statistic (for simplicity assuming d is known to be $d_{(i)}$) is:

$$W_\kappa = \tilde{T} \left(\frac{(\hat{\varepsilon}'\hat{\varepsilon} | \kappa = \kappa_0, d_{(i)})_2 - (\hat{\varepsilon}'\hat{\varepsilon} | \kappa = \kappa_{(i)}, d_{(i)})_{2,\inf}}{(\hat{\varepsilon}'\hat{\varepsilon} | \kappa = \kappa_{(i)}, d_{(i)})_{2,\inf}} \right) \tag{6.42}$$

In this case the null model is the TAR estimated with $\kappa = \kappa_0$, resulting in an RSS of $(\hat{\varepsilon}'\hat{\varepsilon} | \kappa = \kappa_0, d_{(i)})_2$. On some standard assumptions, primarily stationarity in both regimes, homoscedasticity between regimes and $\varepsilon_t \sim \text{iid}(0, \sigma_\varepsilon^2)$ with σ_ε^2 finite. In the self-exciting case considered here, Hansen obtains the asymptotic distribution of W_κ, specifically, $W_\kappa \Rightarrow_D \xi$, where ξ is a function of two independent Brownian motions. Of particular interest is the closed form of the density function of ξ, where:

$$P(\xi \leq x) = (1 - e^{-x/2})^2$$
$$\Rightarrow x = -2\ln(1 - \sqrt{P}) \tag{6.43}$$

Thus, it is easy to tabulate the asymptotic critical values by inverting this expression to obtain x for given P. For example, $P = 0.95$ gives $x = 7.352$, which is the 95% quantile (that is the upper 5% critical value); large sample values of W_κ relative to the critical value lead to rejection of the null hypothesis. For more on this topic, including the extension to the heteroscedastic case and constructing confidence intervals for κ based on test inversion, see Hansen (1997a, 2000).

6.6 Tests for threshold effects when stationarity is not assumed (MTAR)

The tests for a threshold effect described in the previous section assume that the TAR is stationary. Another relevant case allows for the possibility that either the TAR is nonstationary or it is not known to be stationary. Such a situation is possible in the context of the TAR(p) specification, although not in the case of the EG specification of the TAR, which confines change to the coefficient on y_{t-1}. In the former case, change can occur to any of the AR coefficients; this permits a threshold effect even if there is a unit root in each regime. For example, in an MTAR, Caner and Hansen (2001) find evidence that the threshold effect for a two regime model fitted to US unemployment data is in the change in the coefficient on Δy_{t-1} in the ADF form of the model, rather than in the dominant root. Thus, the first issue of interest is to test for a threshold effect when nonstationarity is permitted. The test procedures described here are due to Caner and Hansen (2001) and relate to the MTAR rather than TAR; however, the principles involved seem likely to extend to TAR models although as yet, as far as the author is aware, this extension has not been formally covered.

The model considered is the MTAR, with regime change variable z_{t-1}. In the self-exciting case $z_t = y_t - y_{t-m}$ so that z_t is I(0) given that y_t is I(1); this choice is not essential but it is necessary that z_t is a strictly stationary and ergodic variable. Regime 1 is in operation if $z_{t-d} \leq \kappa$, whereas regime 2 is in operation if $z_{t-d} > \kappa$. The data is not adjusted by demeaning or detrending beforehand, and it is not, therefore, assumed that $\kappa = 0$, rather, as above, κ is estimated along with the other parameters.

6.6.1 Known regime separation

If κ is known, Gonzalez and Gonzalo (1998) show that in the special case that the threshold effect just concerns a scalar parameter, as in the EG version of the TAR(1), say H_0: $\gamma_i = 0$, whether y_t is I(0) or y_t is I(1), the t statistic for the test of H_0 does have a t distribution, at least asymptotically. If the TAR does not include a constant, Gonzalez and Gonzalo (op. cit.) recommend including a constant to protect against the sensitivity of the finite sample distribution of the test statistic to the initial value y_0. It is a reasonable anticipation that this result will extend to the more general case considered here where the threshold effect concerns a vector of parameters, so that the corresponding test statistic is asymptotically distributed as $\chi^2(g)$ if κ is known.

6.6.2 Unknown regime separation

In the more realistic case κ is unknown and it is now material to the distribution of the test statistic whether y_t is I(0) or I(1). The null hypothesis of no threshold effect is as above, that is H_0: $\gamma_i = 0$, for $i = 0, 1, \ldots, p$; and the test statistics are also, in structure, as before. For example, $W_{12}(\hat{\kappa}|d_{(i)})$ is a test statistic for testing H_0. It is now necessary to distinguish whether the test statistic is conditioned on a unit root.

If a unit root is maintained, then the asymptotic distribution of $W_{12}(\hat{\kappa}|d_{(i)})$ is non-standard and Caner and Hansen (2001, theorem 4) show that it is the sum of two independent stochastic processes, one of which is a χ^2 process, and the other reflects the nonstationary regressor y_{t-1}. The former is the distribution that arises from the equivalent threshold test for stationary data and is derived in Hansen (1996, theorem 1); it is a χ^2 process as the test statistic has an asymptotic χ^2 null distribution for each $\kappa \in K$, but critical values cannot be obtained in general because of the dependence on the underlying data matrices. Either way, whether there is a unit root or not, critical values are not generally available.

As in the case of a stationary TAR, a bootstrap procedure is suggested to solve the problem of obtaining critical values for testing H_0. However, there are now two possible bootstrap set-ups depending on whether nonstationarity is permitted. These are referred to as the unrestricted (unit root not imposed) and restricted (unit root imposed) bootstraps. In each case, the bootstrap follows standard principles and care has to be taken in generating the bootstrap replicates where a unit root imposed is imposed under the null.

The preliminary step is to generate the parameter estimates and the residuals that will form the empirical distribution function from which bootstrap draws will be taken. In the unrestricted bootstrap, these estimates and residuals are obtained from estimation of the null model, that is a one-regime AR(p) model most conveniently estimated in ADF(p − 1) form. This imposes the null of no threshold effect but does not impose a unit root; the bootstrap set-up is just as described above (see Section **6.5.1**). In the initial step, LS estimation yields the coefficient estimates, residuals and empirical distribution function, that is $(\hat{\phi}_0, \hat{\alpha}, \hat{c}_1, \dots, \hat{c}_{p-1})$, e_t and $F(e)$, respectively, from the null model. After estimation of the alternative model, with grid searches over $k_{(i)}$ (and $d_{(i)}$ if d is not known) if necessary, a sample realisation of the test statistic $\hat{W}_{1,2}(\hat{\kappa}|d_{(i)})$ of $W_{1,2}(\hat{\kappa}|d_{(i)})$ is obtained. The bootstrap replicates are generated from:

$$\Delta y_t^b = \hat{\alpha} y_{t-1}^b + \sum_{j=1}^{p-1} \hat{c}_j \Delta y_{t-j}^b + e_t^b \tag{6.44}$$

where e_t^b is a random draw, with replacement, from $F(e)$, and initial values for y_{t-j}^b are set to sample values of the demeaned series. (The distribution of the test statistic is invariant to the value of the constant, which can, therefore, be omitted.) As before, the bootstrap distribution of the test statistic is obtained by sorting the B values of the bootstrap realisations of the test statistic, which then enables a bootstrap p-value to be determined for $\hat{W}_{1,2}(\hat{\kappa}|d_{(i)})$.

The restricted bootstrap differs from the above description in that the unit root is imposed by way of omitting the regressor y_{t-1} in the estimation of the ADF(p − 1) model, and then the unit root is imposed in generating the bootstrap data y_t^b, that is, impose $\alpha = 0$ in the initial LS estimation of the ADF model. Denote the relevant estimated coefficients, residuals and empirical distribution function as $(\tilde{\phi}_0, 0, \tilde{c}_1, \dots, \tilde{c}_{p-1})$, e_t^{ur} and $F(e^{ur})$, respectively, then these form the basis of generating the bootstrap replicates. The bootstrap recursion is now:

$$\Delta y_t^b = \sum_{j=1}^{p-1} \tilde{c}_j \Delta y_{t-j}^b + e_t^b \tag{6.45}$$

where e_t^b is now a random draw from $F(e^{ur})$. An alternative to re-estimating the model in the preliminary step is to set $\hat{\alpha} = 0$ from the unrestricted set of estimated coefficients, which becomes $(\hat{\phi}_0, 0, \hat{c}_1, \ldots, \hat{c}_{p-1})$; this is equivalent asymptotically, since if the unit root is part of the null DGP, the estimates will reflect this (as they are obtained from a consistent estimator). Imposing the unit root directly seems more in keeping with the spirit of the restricted bootstrap, which may be relevant in small samples. Other elements of the uncertainty in the process can be built into variants of these bootstrap procedures. These include searching for the order of the ADF model and d if not known.

The difference in the application of the two bootstrap procedures is that the restricted bootstrap is relevant for an I(1) process and the unrestricted bootstrap for an I(0) process; indeed, the unrestricted bootstrap will be invalid if the DGP is I(1). However, it is not usually known which of these assumptions is correct, and Caner and Hansen (2001) suggest calculating the p-value of the test statistic under both procedures and making a decision on the basis of the one with the larger p-value.

Caner and Hansen (op. cit.) find that even with B = 500, the empirical size of the suggested test procedure is close to the nominal size (with 5% nominal size and T = 100 being used in their simulations), whether the unrestricted or restricted bootstrap procedure is used and for values of α implying a unit root, $\alpha = 0$, or a stationary process, $\alpha = -0.25, -0.15$ and -0.05. The power of the test increases with the size of the threshold effect. Of particular interest is the case in which $\Delta\alpha := \alpha_2 - \alpha_1 \neq 0$; there is a noticeable increase in power if there is a unit root in one of the regimes, even for relatively small $\Delta\alpha$. For example in the context of a 2RMTAR(2) model and 5% nominally sized test with $\alpha_1 = -0.05$, power is 16.5% for $\Delta\alpha = -0.05$ but 82.1% with $\alpha_1 = 0$; as the threshold effect increases to $\Delta\alpha = -0.10$, the power increases to 43% for $\alpha_1 = -0.05$ and 93.1% for $\alpha_1 = 0$; see Caner and Hansen (op. cit., table II).

Gonzalez and Gonzalo (1998) also consider the case where the regime separation parameter κ has to be estimated. Their base model is close in spirit to the EG specification, but with regime separation not limited to the self-exciting case; the threshold effect, if present, is dependent on y_{t-1}. They show that the distribution of a supremum t-type test statistic for a threshold effect differs depending on whether a unit root is part of the maintained hypothesis, and they also provide asymptotic critical values for each case (see their test statistics S_1 and S_2, Gonzalez and Gonzalo, 1998, propositions 1 and 2). Interestingly there is very little difference between the critical values in the two cases, so that not knowing whether a unit root is present is not critical in this case (at least asymptotically).

6.7   Testing for a unit root when it is not known whether there is a threshold effect

The other case of interest in this more generalised framework is testing for non-stationarity by way of a unit root when there is a threshold effect, or it is not known whether there is a threshold effect. This case arises because, for example, in a two-regime model it is possible that $\alpha_1 = \alpha_2 = 0$, implying a unit root in each regime, but there is still a threshold effect which can be seen most easily as arising from changes to the coefficients on the Δy_{t-j} terms in the ADF spec-ification of the model. In the two-regime model, the test statistic is, as before, most simply viewed as the average of the squared t statistics on α_1 and α_2; that is as an F-type statistic, $\frac{1}{2} [t(\alpha_1)^2 + t(\alpha_2)^2]$; or as a one-sided (directional) F-type test statistic, $\frac{1}{2} [t(\alpha_1)^2(I(\hat{\alpha}_1) < 0) + t(\alpha_2)^2(I(\hat{\alpha}_2) < 0)]$; Wald versions of the tests, $W(0,0)_{\kappa_1,\kappa_2}$ and $W^D(0,0)_{\kappa_1,\kappa_2}$, are easily obtained (see Section **6.4.2**). The t-type test on individual coefficients may also be useful in testing for partial unit roots.

Caner and Hansen (2001) have obtained the asymptotic distributions for the two cases of interest in a self-exciting MTAR model: (i) no threshold effect and (ii) a threshold effect. In both cases the threshold parameter is estimated. The asymptotic distributions differ between the two cases, with asymptotic criti-cal values of $W(0,0)_{\kappa_1,\kappa_2}$, $W^D(0,0)_{\kappa_1,\kappa_2}$ and the individual t-type tests provided for the no threshold case in Caner and Hansen (2001, table III). For exam-ple, with 10% trimming at each end of the sample, the 10%, 5% and 1% critical values of the asymptotic distribution of $W(0,0)_{\kappa_1,\kappa_2}$ are 11.66, 6.59 and 17.85; and 11.09, 6.00 and 17.23 for $W^D(0,0)_{\kappa_1,\kappa_2}$. Dividing these crit-ical values by 2 obtains the critical values for the F versions of the tests. In the case of an identified threshold effect, the test statistic $W(0,0)_{\kappa_1,\kappa_2}$ has an asymptotic distribution that is bounded by the sum of a $\chi^2(1)$ distribution and the distribution of a squared DF t-type variate; Caner and Hansen (2001) calculate 10%, 5% and 1% critical values of 11.17, 6.12 and 17.29, respec-tively, for these bounds. An alternative to using these critical values is to bootstrap the critical values in the manner described for the tests of the threshold effect.

Of course, there is a problem in the sense that it is not known whether there is a threshold effect. However, based on a simulation experiment, Caner and Hansen (op. cit.) find size distortions using the asymptotic critical values in both cases, but these are worse using the upper bounds in the identified case and, on balance, suggest using the critical values from the unidentified case. Even better size fidelity is obtained using the bootstrap version of this test. As to power, generally the one-sided version of $W^D(0,0)_{\kappa_1,\kappa_2}$ has better power than the two-sided version and the t-type tests are useful in distinguishing particular cases.

6.8   An empirical study using different nonlinear models

This and the two previous chapters have introduced a number of alternatives to the simple linear AR model; these models have the common aim of offering a way of allowing some nonstationary behaviour in the time series under the alternative hypothesis. The design of this section is to deliberately fit different models to the same time series, in this case the logarithm of the dollar–sterling real exchange rate. As it transpires there is very little to choose between these different models based on traditional measures of the goodness of fit, but it might be possible to discriminate amongst the models using an encompassing type approach, although that is not pursued here.

6.8.1   An ESTAR model for the dollar–sterling exchange rate

Some of the specification and estimation issues for an ESTAR model are illustrated in this section for the (log) dollar–sterling real exchange rate using monthly data for the estimation period 1973m3 to 2000m7. The data is graphed in Figure 6.1, with some standard unit root tests given below in Table 6.4. The sample mean of 0.417 is also shown in Figure 6.1; note that whilst there is some random walk-like behaviour, the sample mean is crossed about 20 times over the period of 330 observations, which may suggest that the series is not clearly generated by a nonstationary process. (The expected number of sign changes (mean reversions) in the case of Gaussian inputs is $0.6363\sqrt{T} = 11$ in this case; see *A Primer*, chapter 5.)

6.8.1.i   Initial tests

Some unit root test statistics are given in Table 6.4. The decision as to whether to reject the unit root null hypothesis is borderline and depends on p (the lag augmentation parameter). The model selection criteria give different choices for the lag length: SIC favours $p = 2$, MAIC favours $p = 3$ and AIC favours $p = 4$. Hence, the test statistics in Tables 6.4 and 6.5 are for $p = 1, ..., 4$. At the 5% level, using τ_{μ}^{ws} and Inf-t, the two tests suggested to be more powerful by the simulation results in Section **5.6.1**, there is rejection for $p = 2$ and $p = 4$, but not for $p = 1$ or $p = 3$. It seems worth estimating a nonlinear model of STAR form, and to inform the specification a number of test statistics, due to Teräsvirta, and Escribano and Jordá, intended to guide the initial specification, are presented in Tables 6.5a and 6.5b.

The nonlinearity tests for $d = 1$ and $d = 2$ favour $d = 1$ (see tables 6.5a and 6.5b); the p-values for the Teräsvirta tests of H_0 being uniformly smaller for $p = 2, 3$ and 4. For $d = 1$, the p-values for the test of H_0 were 1.4% for $p = 2$, 5.7%

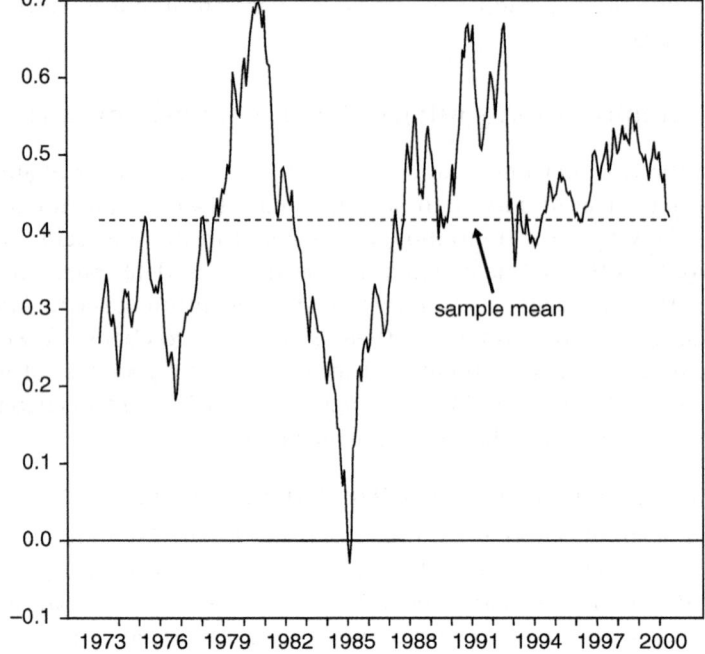

Figure 6.1   Dollar–sterling real exchange rate (logs)

Table 6.4   Unit root test statistics for the US:UK real exchange rate

p	t_{NL}	$\hat{\tau}_\mu$ (5% cv)	τ_μ^{WS} (5% cv)	Inf-t
1	−2.12	−1.86 (−2.870)	−2.00 (−2.527)	−2.37
2	−2.94	−2.58 (−2.867)	−2.69 (−2.527)	−3.31
3	−2.61	−2.26 (−2.864)	−2.38 (−2.526)	−2.94
4	−2.86	−2.49 (−2.862)	−2.61 (−2.524)	−3.15
5% cv	−2.91			−3.11

Notes: Demeaned data is used; 5% critical values (cv) for t_{NL} and Inf-t are from Table 6.1 for T = 200; the critical values for $\hat{\tau}_\mu$ and τ_μ^{WS} are calculated from the response surface coefficients in chapter 6 of *UR, Vol. 1*, using the exact sample size.

for p = 3 and 3.6% for p = 4, so all are strongly suggestive of nonlinearity of some form. (The EJ tests of H_0^{EJ} are not as uniform but, on balance, also favour d = 1.)

Consideration is now given to whether there is a pattern that might help to indicate whether an LSTAR or ESTAR model is more appropriate. The Teräsvirta test results do not make this choice transparent. Whether p = 2, 3 or 4 is chosen, the smallest p-value is in testing H_{01}, which would suggest an LSTAR model but

Table 6.5a   Tests for nonlinearity (delay parameter = 1)

p	H_0	H_{01}	H_{02}	H_{03}	H_0^{EJ}	$H_{0,E}^{EJ}$	$H_{0,L}^{EJ}$
		Teräsvirta tests			Escribano and Jordá tests		
1	1.27[0.284]	1.11[0.293]	2.50[0.116]	0.21[0.650]	1.12[0.344]	1.11[0.331]	1.47[0.231]
2	2.70[0.014]	4.58[0.011]	3.29[0.038]	0.15[0.859]	2.31[0.020]	3.29[0.011]	3.83[0.005]
3	1.86[0.057]	3.63[0.013]	1.72[0.163]	0.19[0.902]	1.55[0.106]	2.34[0.031]	2.54[0.020]
4	1.88[0.036]	3.75[0.054]	1.65[0.161]	0.17[0.952]	1.66[0.052]	2.52[0.011]	2.55[0.010]

Table 6.5b   Tests for nonlinearity (delay parameter = 2)

p	H_0	H_{01}	H_{02}	H_{03}	H_0^{EJ}	$H_{0,E}^{EJ}$	$H_{0,L}^{EJ}$
		Teräsvirta tests			Escribano and Jordá tests		
2	2.12[0.051]	2.87[0.058]	3.21[0.042]	0.23[0.797]	1.97[0.050]	2.43[0.048]	2.78[0.027]
3	1.47[0.159]	2.39[0.068]	1.78[0.151]	0.20[0.900]	1.72[0.062]	2.48[0.024]	1.98[0.068]
4	1.63[0.083]	3.01[0.018]	1.65[0.161]	0.17[0.954]	1.65[0.056]	2.38[0.017]	2.33[0.019]

Note: Asymptotic p-values in [.].

we would then not expect H_{03} to be not-rejected; on the other hand, reject-ing H_{02} but not rejecting H_{03}, which occurs when p = 2, suggests an ESTAR model.

The EJ tests can throw some light on this issue. With d = 1, there is a slightly stronger rejection of $H_{0,L}^{EJ}$ compared to $H_{0,E}^{EJ}$ (see for example p = 2 and d = 1) but, practically, there is little to choose between the rejections of $H_{0,L}^{EJ}$ and $H_{0,E}^{EJ}$. If we consider the alternative versions of these tests $H_{0,L}^{EJ,*}$ and $H_{0,E}^{EJ,*}$, which improve power by excluding variables redundant under the alternative from the maintained regression, the conclusion is clearer. For example with p = 2 and d = 1, the test statistic and p-value for $H_{0,L}^{EJ,*}$ are 1.28(0.276) whereas for $H_{0,E}^{EJ,*}$ they are 3.88(0.004); hence ESTAR is favoured on this rule.

The conclusion on the basis of these tests is rejection in favour of some form of nonlinearity, and specification issues may, in this case, be better addressed by estimating and assessing both types of model. Several empirical studies have favoured the symmetric adjustment implicit in the ESTAR model rather than the asymmetric adjustment of the LSTAR model; hence we start with an ESTAR specification.

6.8.1.ii   Estimation

Initial estimation suggested that an ESTAR(1) model was inadequate, as there was evidence of serial correlation in the residuals of that model. Thus, ESTAR

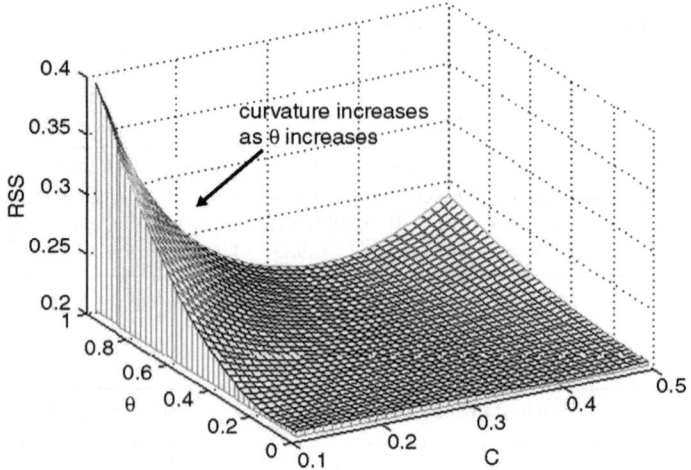

Figure 6.2   ESTAR(2): Residual sum of squares

models for lags 2 to 4 were estimated, and the residuals were assessed for serial correlation. The estimated model is in deviation from threshold form so that equilibrium implies that $y_t = c$. For example, an ESTAR(2) model is given by:

$$y_t - c$$

$$= \phi_1(y_{t-1} - c) + \phi_2(y_{t-2} - c) + [\gamma_1(y_{t-1} - c) + \gamma_2(y_{t-2} - c)](1 - e^{-\theta(y_{t-1}-c)^2}) + \varepsilon_t$$
$$(6.46)$$

Unrestricted estimation was hampered by the 'flatness' of the objective function, a feature noted by several researchers using ESTAR models, and two restrictions of interest were considered to see if the problem could be overcome. The restrictions were:

(1) there is a unit root in the inner regime, which implies $\phi_1 + \phi_2 = 1$;
(2) in the outer regime $y_t - c = \varepsilon_t$, which implies $\gamma_1 = -\phi_1$ and $\gamma_2 = -\phi_2$.

With these restrictions, which can be imposed without loss of fit, there is some curvature in the residual sum of squares function across both the dimensions θ and c. This is illustrated in Figure 6.2, where the function is quite flat when θ is close to zero but there is some curvature as θ increases with the minimum occurring at $\hat\theta = 0.466$ and $\hat c = 0.369$. Estimation details are summarised in Table 6.6.

The F version, F_{LM}, of the Eitrheim and Teräsvirta test for serial correlation is reported in the final column of Table 6.6 (this is the test as in Equation (5.104), there being only minor differences in the orthogonalised verion of the test).

Table 6.6 Estimation of ESTAR models for the dollar–sterling real exchange rate

ESTAR(1): Outer regime and unit root restrictions imposed: $\gamma_1 = -\phi_1$

	$\hat{\phi}_1$	$\hat{\phi}_2$	$\hat{\gamma}_1$	$\hat{\gamma}_2$	$\hat{\theta}$	$\hat{c}$	$\hat{\sigma}_\varepsilon \times 10^2$	F_{LM}
	1.098	–	–1.098	–	0.461	0.359	2.658	8.61
't'	(54.19)	–	(–)	–	(1.73)	(9.146)		[0.0]

ESTAR(1): Outer regime and unit root restrictions imposed: $\gamma_1 = -\phi_1$, $\phi_1 = 1$

	1	–	–1	–	0.359	0.370	2.655	8.66
't'	(–)	–	(–)	–	(2.23)	(8.62)		[0.0]

ESTAR(2): Outer regime restrictions imposed: $\gamma_1 = -\phi_1$, $\gamma_2 = -\phi_2$

	$\hat{\phi}_1$	$\hat{\phi}_2$	$\hat{\gamma}_1$	$\hat{\gamma}_2$	$\hat{\theta}$	$\hat{c}$	$\hat{\sigma}(\%)$	F_{LM}
	1.361	–0.351	–1.361	0.351	0.572	0.360	2.498	1.84
't'	(23.92)	(–6.59)	(–)	(–)	(2.29)	(12.32)		[0.091]

ESTAR(2): Outer regime and unit root restrictions imposed: $\gamma_1 = -\phi_1$, $\gamma_2 = -\phi_2$, $\phi_1 + \phi_2 = 1$

	1.349	–0.349	–1.349	0.349	0.466	0.369	2.495	2.10
't'	(25.49)	(–)	(–)	(–)	(3.05)	(11.86)		[0.053]

Notes: 't' values are shown in (.) parentheses; p-values are shown in [.] parentheses; F_{LM} is the F-test version of Eitrheim and Teräsvirta LM test for q-th order serial correlation where q = 6, see Chapter 5, Section **5.9.1**.

Note from (5.104) that this test has q degrees of freedom for the numerator and $\check{T} - (n + q)$ for the denominator, where $\check{T} = \tilde{T} - q$, $n = 2(p + 1) + 2 - r$, and r is the number of restrictions. The test statistic is reported for q = 6 and $\tilde{T} = 314$. To illustrate the denominator degrees of freedom, consider the ESTAR(2) with $\gamma_1 = -\phi_1$ and $\gamma_2 = -\phi_2$ imposed; then $n = 2(2 + 1) + 2 - r$ and r = 4, of which 2 restrictions follow from estimating the model in deviation form (see Chapter 5) and the 2 outer regime restrictions are directly imposed. Hence, the critical value is obtained from the F(6, 304) distribution; additionally, imposing a unit root in the inner regime reduces by one the number of freely estimated coefficients and the F distribution is for F(6, 305).

The values of F_{LM} clearly indicate that the ESTAR(1) model is inadequate with p-values of zero; these are improved in the ESTAR(2) model, with p-values above 5%. There is a further improvement in the ESTAR(3) model, but for illustrative purposes the estimated ESTAR(2) model is analysed here and the ESTAR(3) model is considered further below (there are no substantive differences between these two models).

Taking the reported ESTAR(2) model as illustrative, the real exchange rate is modelled with a unit root in the inner regime and, apart from a random error, as a constant in the outer regime, thus the estimated form is:

Inner regime : $y_t = y_{t-1} + 0.349 \Delta y_{t-1} + \hat{\varepsilon}_t$

Roots : $\lambda_0 = 1.0, \bar{\lambda}_0 = 1/0.349 = 2.865$

Outer regime : $y_t = 0.369 + \hat{\varepsilon}_t$

Roots : $\lambda_\infty = 0, \bar{\lambda}_\infty = 0$

In original units, the constant in the outer regime is estimated to be $e^{\hat{c}} = e^{0.369} = 1.446$ ($:£).

The contemporaneous roots are obtained from the characteristic polynomial

$$\vartheta(\lambda) = \lambda^2 - (\phi_1 + \gamma_1 \Theta_t)\lambda - (\phi_2 + \gamma_2 \Theta_t) \tag{6.47}$$

In this case, on imposing the restrictions, this results in:

$$\hat{\vartheta}(\lambda) = \lambda^2 - \hat{\phi}_1(1 - \hat{\Theta}_t)\lambda - \hat{\phi}_2(1 - \hat{\Theta}_t)$$
$$= \lambda^2 - -1.349(1 - \hat{\Theta}_t)\lambda + 0.349(1 - \hat{\Theta}_t)$$

where $\hat{\Theta}_t = (1 - e^{-0.466(y_{t-1} - 0.369)^2})$. A simple way of showing the time dependence of nonlinear responses is through the coefficient β_t on y_{t-1} in the ADF representation:

$$y_t = \hat{\beta}_t y_{t-1} + \hat{\delta}_t \Delta y_{t-1} + \hat{\varepsilon}_t \tag{6.48}$$

where $\hat{\beta}_t = \hat{\phi}_1 + \hat{\phi}_2 + (\hat{\gamma}_1 + \hat{\gamma}_2)\hat{\Theta}_t$. Note that in this case, because of the restrictions, $\hat{\beta}_t$ considerably simplifies, specifically:

$$\hat{\beta}_t = 1.349 - 0.349 - (1.349 - 0.349)(1 - e^{-0.466(y_{t-1} - 0.369)^2})$$
$$= 1 - (1 - e^{-0.466(y_{t-1} - 0.369)^2}) \tag{6.49}$$

Of course, this confirms that $\hat{\beta}_t = 1$ in the inner regime and $\hat{\beta}_t = 0$ in the outer regime, but what is also of interest is the pattern of $\hat{\beta}_t$ over the sample. The graph of $\hat{\beta}_t$ is shown in Figure 6.3, where the left-hand scale refers to $\hat{\beta}_t$ and the right-hand scale to the (log) real exchange rate as deviations from $\hat{c} = 0.369$, that is $y_t - \hat{c}$. The idea of this two-scale graph is to show the constraining behaviour implied by $\hat{\beta}_t$ as $y_t \gg \hat{c}$ or $y_t \ll \hat{c}$. The peaks and troughs in $y_t - \hat{c}$ are associated with $\hat{\beta}_t$ noticeably below 1; see, for example, the beginning of 1984 when $\hat{\beta}_t \approx 0.92 \ll 1$ and $y_t - \hat{c}$ is at its maximum absolute deviation of –0.4.

Ideally some adjustment should be made for the likely outcome that θ tends to be overestimated in finite samples. It would, therefore, be of interest to continue the example using a bias-corrected estimator. This is left for a future study.

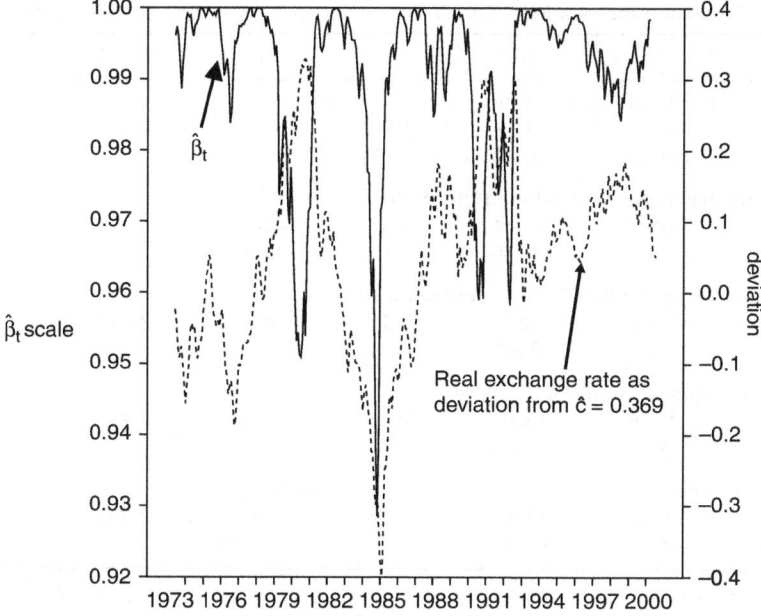

Figure 6.3   Two-scale figure of real \$:£ rate and $\hat{\beta}_t$

6.8.2   A bounded random walk model for the dollar–sterling exchange rate

To continue the illustration of fitting nonlinear models, the results in this section are reported for a BRW model for the dollar–sterling exchange rate. The basic BRW model is amended to allow lags of Δy_t and for estimation of the coefficient on y_{t-1}. The amended BRW model is:

$$y_t = \phi_1 y_{t-1} + e^{\alpha_0}(e^{-\alpha_1(y_{t-1}-\kappa)} - e^{\alpha_2(y_{t-1}-\kappa)}) + \sum_{j=1}^{J} \delta_i \Delta y_{t-j} + \varepsilon_t \qquad (6.50)$$

The estimation details are reported in Table 6.7. Although there is to date relatively little experience of estimating BRW models, especially directly by nonlinear LS, the results in Table 6.7 are quite promising; although, as in the case of the ESTAR model, it would be worth exploring the possibility of reducing the finite sample biases of the NLS estimators. Symmetry of response to positive and negative deviations was not initially imposed, nor was the coefficient on y_{t-1}, that is ϕ_1, imposed to be unity; however, in both cases, these restrictions could be imposed with virtually no loss of fit.

6.8.2.i   Estimation results

Initially the threshold κ was imposed to be the sample mean; however, relaxing this assumption resulted in $\hat{\kappa} = 0.355$ in the preferred model with $\hat{\phi}_1 = 1$ and

Table 6.7   Estimation of a BRW model for the dollar–sterling real exchange rate

$\hat{\phi}_1$	$\hat{\alpha}_0$	$\hat{\alpha}_1$	$\hat{\alpha}_2$	$\hat{\kappa}$	$\hat{\sigma}_\varepsilon \times 10^2$	LL
No symmetry imposed; threshold = sample mean						
1	−9.485	14.12	19.95	0.416	2.467	747.37
(-)	(−3.40)	(2.12)	(1.85)	(-)		
No symmetry imposed; estimated threshold						
1	−10.31	21.93	16.15	0.302	2.467	747.61
(-)	(−2.99)	(1.19)	(1.45)	(1.62)		
Symmetry imposed; threshold = sample mean						
1	−7.43	9.27	9.27	0.416	2.472	746.16
(-)	(−5.82)	(2.67)	(-)	(-)		
Symmetry imposed; estimated threshold						
1	−10.09	18.16	18.16	0.355	2.466	747.55
(-)	(−5.82)	(1.89)	(-)	(16.92)		
Symmetry imposed; estimated threshold; ϕ_1 estimated						
1.002	−9.10	15.48	15.48	0.353	2.468	747.56
{ 0.65}	(−5.82)	(1.89)	(-)	(16.92)		

Notes: Each regression also included Δy_{t-j}, j = 1, 2, 3; the basic regression $y_t = y_{t-1} + \sum_{j=1}^{3} \delta_i \Delta y_{t-j} + \varepsilon_t$, that is, excluding the nonlinear terms, had $\hat{\sigma}_\varepsilon\% = 2.498$ and LL = 741.74. 't' statistics are in (.) for the null that the coefficient is zero, but for $\hat{\phi}_1$, the null is that $\phi_1 = 1$. The italicised row corresponds to the reported model in (6.51).

symmetry imposed. Notice that this is very close to the estimated threshold of 0.369 in the ESTAR(2) model. The nonlinear adjustment function is important to the fit of the model; excluding it gave a loglikelihood (LL) of 741.74, compared to LL = 747.55 in the preferred model, resulting in an LR statistic of 11.62 compared to $\chi^2_{0.05}(3) = 7.81$. Of course, this makes a presumption about the distribution of the test statistic but, at least informally, it is clear that the nonlinear component is important.

The unit root can be imposed without loss of fit and freely estimating ϕ_1 results in $\hat{\phi}_1 = 1.002$. The estimated model, corresponding to a unit root, symmetry and an estimated threshold, is summarised as follows:

$$y_t = y_{t-1} + e^{-10.09}(e^{-18.16(y_{t-1}-0.355)} - e^{18.16(y_{t-1}-0.355)}) + \sum_{j=1}^{3} \hat{\delta}_i \Delta y_{t-j} + \hat{\varepsilon}_t$$

$$(6.51)$$

6.8.2.ii   An estimated random walk interval (RWI)

The estimated values of the nonlinear coefficients are considerably larger than those used for illustration in Chapter 5, but note that the estimated (conditional) standard error at approximately $\hat{\sigma}_\varepsilon = 0.025$ (that is, $\hat{\sigma}_\varepsilon^2 = 0.000625$) is very much smaller; hence, this combination works in a practical sense.

The implied random walk interval, RWI(ζ), is solved from (6.51) using the estimated coefficients; that is, for some choice of ζ solve:

$$RWI(\zeta) = y_t \in [|e^{-10.09}(e^{-18.16(y_{t-1}-0.355)} - e^{18.16(y_{t-1}-0.355)})| \leq \zeta] \qquad (6.52)$$

To illustrate, consider $\zeta = 0.01$, then this solves to RWI(0.01) $= y_t \in [0.355 \pm 0.302 = 0.052 \leftrightarrow 0.657]$. Recall that the interpretation of this interval is that for y_t within (0.052, 0.657), the adjustment impact of the nonlinear term in the BRW is less than 0.01. In context, the range of the (log) dollar–sterling rate, y_t, over the sample period is (–0.03, 0.70) and the RWI(0.01) is (approximately) the 1st through to the 96th percentile, that is RWI(0.01) occupies 95% of the range. Thus, the estimated nonlinear adjustment, which is significant statistically, is largely driven by relatively few of the outlying observations, with more on the positive than the negative side of the threshold.

As in the case of estimation of the ESTAR model for the exchange rate, the indications from the small simulation study reported in Chapter 5 are that in finite samples, α_1 and α_2 tend to be overestimated, perhaps by an order of 10%, whereas α_0 tends to be overestimated in absolute value (that is, estimates are more negative than the true value) by a slightly lesser order of magnitude. Thus, estimation problems are not likely to be as serious as in ESTAR models but, nevertheless, a further study could be addressed to obtaining a bias corrected estimator.

6.8.3 Two- and three-regime TARs for the exchange rate

To complete the estimation examples, 2RTAR and 3RTAR models are illustrated. One possibility this scheme allows is a corridor regime, which may exhibit unit root behaviour, bounded by two stationary regimes. A simple 3RTAR is restated below, with $f(y_{t-1}) = y_{t-1}$:

$$y_t = \phi_1 y_{t-1} + z_t \qquad \text{if } y_{t-1} \leq \kappa_1 \qquad \qquad \text{regime 1} \qquad (6.53)$$

$$y_t = \psi_1 y_{t-1} + z_t \qquad \text{if } \kappa_1 < y_{t-1} \leq \kappa_2 \qquad \text{regime 2} \qquad (6.54)$$

$$y_t = \varphi_1 y_{t-1} + z_t \qquad \text{if } y_{t-1} > \kappa_2 \qquad \qquad \text{regime 3} \qquad (6.55)$$

In practice it is usually necessary to allow for an intercept and the possibility of higher-order terms, which may themselves be subject to change. For example, in the case of the three-regime AR(3) model, the regimes can be written separately

Table 6.8   Estimation of the (linear) ADF(2) model for the dollar–sterling real exchange rate

$$\Delta y_t = 0.099 - 0.0234 y_{t-1} + 0.377 \Delta y_{t-1} - 0.125 \Delta y_{t-2} + \hat{\varepsilon}_t$$

$$(2.17)\quad(-2.24)\phantom{y_{t-1}}\quad(6.87)\phantom{\Delta y_{t-1}}\quad(-2.26)$$

$$\hat{\sigma}_\varepsilon \times 10^2 = 2.497; \text{RSS} = 0.200189$$

Notes: throughout the tables: t statistics are in (.) parentheses.

or in equivalent ADF(2) form. In the latter case:

$$\Delta y_t = \mu_t + \alpha_t y_{t-1} + \delta_{1t} \Delta y_{t-1} + \delta_{2t} \Delta y_{t-2} + \varepsilon_t \tag{6.56}$$

$$\mu_t = \psi_0 + (\phi_0 - \psi_0) I_{1t} + (\varphi_0 - \psi_0) I_{2t}$$

$$\alpha_t = [\psi_1 + (\phi_1 - \psi_1) I_{1t} + (\varphi_1 - \psi_1) I_{2t} - 1]$$

$$\delta_{1t} = (c_1 + (d_1 - c_1) I_{1t} + (\omega_1 - c_1) I_{2t})$$

$$\delta_{2t} = (c_2 + (d_2 - c_2) I_{1t} + (\omega_2 - c_2) I_{2t})$$

The data and sample period are as before and the results are illustrated with an ADF(2) model, which is robust to reasonable variations in this specification. Note that as the regime indicator is based on y_{t-1}, and not the change in the level, it is not necessarily I(0), and the results in Caner and Hansen (2001) do not strictly apply.

The linear model is first reported in Table 6.8 for reference and comparison with the multi-regime models. On the basis of the DF test statistic $\hat{\tau}_\mu = -2.24$, the null hypothesis of a unit root in this linear model would not be rejected at the usual significance levels. Estimation of the 2RTAR is reported in Table 6.9; the estimated threshold from this model is $\hat{\kappa}_1 = 0.345$, with 30.5% of sample observations in regime 1 and 69.5% in regime 2. (The grid was obtained as 290 equally spaced points between the 10th and 90th percentiles of y_t.) The implied long run is 0.269 for regime 1 and 0.486 for regime 2, and both are correctly placed as far as the regime indicator is concerned, although whether these are in fact achievable equilibria depends on the overall stability of the system. Although the implied equilibria differ, the difference between the two regimes is slight as far as proximity to the unit root is concerned, as both show considerable persistence. An assessment of the unit root null is considered below in Section 6.8.5.

The estimation of the 3RTAR model was conditioned on the estimated threshold from the 2RTAR. Initial estimation allowed κ_1 and κ_2 to be determined from an unrestricted search over both dimensions; however, this resulted in only nine observations in the middle regime and, consequently, some difficulty in 'pinning down' that regime. An alternative estimation strategy is based on arguments from Bai (1997), Bai and Perron (1998), and see also Hansen (1999), albeit

Table 6.9 Estimation of the two regime TAR(3) model for the dollar–sterling real exchange rate

---

Regime 1: $y_{t-1} \leq \hat{\kappa}_1 = 0.345$; $\hat{\sigma}_\varepsilon \times 10^2 = 2.513$

$$\Delta y_t = 0.016 - 0.0592 y_{t-1} + 0.409 \Delta y_{t-1} - 0.096 \Delta y_{t-2} + \hat{\varepsilon}_t$$
$$\quad (1.83) \quad (-1.82) \qquad (4.11) \qquad\quad (-0.96)$$

Regime 2: $y_{t-1} > \hat{\kappa}_1 = 0.345$; $\hat{\sigma}_\varepsilon \times 10^2 = 2.475$

$$\Delta y_t = 0.028 - 0.0585 y_{t-1} + 0.373 \Delta y_{t-1} - 0.116 \Delta y_{t-2} + \hat{\varepsilon}_t$$
$$\quad (2.82) \quad (-2.86) \qquad (5.69) \qquad\quad (-1.74)$$

Overall (pooled): $\hat{\sigma}_\varepsilon \times 10^2 = 2.487$, RSS $= 0.196043$
't' statistics are from the pooled regression.

Sup-W$(0,0)_\kappa = 11.38$, W$(0,0)_\kappa^{\text{ave}} = 4.29$, W$(0,0)_\kappa^{\text{exp}} = 21.94$
Regime details
Percentage of observations in each regime:
Regime 1: 30.5%; Regime 2: 69.5%.
Implied long run in each regime:
Regime 1: $y_1^* = 0.269$; Regime 2: $y_2^* = 0.486$

Note: * indicates an equilibrium value.

Table 6.10 Estimation of the three-regime TAR(3) model for the dollar–sterling real exchange rate

---

Regime 1: $y_{t-1} \leq \hat{\kappa}_1 = 0.345$; $\hat{\sigma}_\varepsilon \times 10^2 = 2.513$

$$\Delta y_t = 0.016 - 0.059 y_{t-1} + 0.409 \Delta y_{t-1} - 0.096 \Delta y_{t-2} + \hat{\varepsilon}_t$$
$$\quad (1.84) \quad (-1.83) \qquad (4.15) \qquad\quad (-0.96)$$

Regime 2: $0.346 = \hat{\kappa}_1 < y_{t-1} \leq \hat{\kappa}_2 = 0.482$; $\hat{\sigma}_\varepsilon \times 10^2 = 2.217$

$$\Delta y_t = 0.052 - 0.114 y_{t-1} + 0.172 \Delta y_{t-1} - 0.039 \Delta y_{t-2} + \hat{\varepsilon}_t$$
$$\quad (1.71) \quad (-1.63) \qquad (1.68) \qquad\quad (-0.42)$$

Regime 3: $y_{t-1} > \hat{\kappa}_2 = 0.481$; $\hat{\sigma}_\varepsilon \times 10^2 = 2.706$

$$\Delta y_t = 0.027 - 0.057 y_{t-1} + 0.505 \Delta y_{t-1} - 0.174 \Delta y_{t-2} + \hat{\varepsilon}_t$$
$$\quad (1.29) \quad (-1.48) \qquad (5.98) \qquad\quad (-1.86)$$

Overall (pooled): $\hat{\sigma}_\varepsilon \times 10^2 = 2.4751$; RSS $= 0.191741$

Sup-W$(0,0,0)_{\kappa_1,\kappa_2} = 12.73$, W$(0,0,0)_{\kappa_1,\kappa_2}^{\text{ave}} = 6.28$, W$(0,0,0)_{\kappa_1,\kappa_2}^{\text{exp}} = 69.0$
Regime details
Percentage of observations in each regime:
Regime 1: 30.5%; Regime 2: 36.9%; Regime 3: 32.6%.
Implied long run in each regime:
Regime 1: $y_1^* = 0.269$; Regime 2: $y_2^* = 0.451$; Regime 3: $y_3^* = 0.488$

Note: * indicates an equilibrium value.

Table 6.11 Test statistics for threshold effects and boostrapped p-values in the three-regime and two-regime TAR models

	3 to 1	3 to 1 (unit root null)	3 to 2	2 to 1	2 to 1 (unit root null)
$F(df_1, df_2)$	$F(8, 313)$	$F(9, 313)$	$F(4, 313)$	$F(4, 317)$	$F(5, 317)$
Test value	1.72	2.10	1.76	1.68	2.35
p-value					
F	[0.09]	[0.03]	[0.14]	[0.16]	[0.04]
Bootstrap	[0.44]	[0.19]	[0.53]	[0.50]	[0.21]

Note: To obtain the corresponding (approximate) Wald test statistics, multiply the F value by the first degrees of freedom parameter, for example the '3 to 1' Wald statistic has a value of 8 times $1.72 = 13.76$ (approximately).

in a stationary context, that is, if the correct model has three regimes but only two are allowed then $\hat{\kappa}_1$ from the 2RTAR is consistent for one threshold in the pair (κ_1, κ_2), say κ_1 for convenience; thus, one only has to search for the other threshold parameter to obtain, say $(\hat{\kappa}_2|\hat{\kappa}_1)$. The method can be iterated to obtain a second round estimate of κ_1, say $[\hat{\kappa}_1|(\hat{\kappa}_2|\hat{\kappa}_1)]$; however, in this case one iteration did not improve the fit of the equation. The resulting estimated threshold parameters are $\hat{\kappa}_1 = 0.345$ and $\hat{\kappa}_2 = 0.481$, and note that each implied long run is within the respective regimes' selection regions, with an estimated 30.5%, 36.9% and 32.6% of observations in the respective regimes.

6.8.4 A threshold effect?

The next issue to be considered is, having estimated the two- and three-regime models, what the balance of evidence is for the presence of threshold effects. This question is complicated by the presence or absence of a unit root, with the correct order of integration unknown in general. In this case, the test statistics are bootstrapped both with and without the null restricted by the imposition of a unit root; that is, practically, the lagged level y_{t-1} is omitted from the ADF version of the model for the restricted case. The test statistics and their bootstrapped p-values are given in Table 6.11.

The interpretation of the test statistics is as follows. The first test assesses whether the three regimes can be collapsed to one, first in the case that the unit root is not imposed and second in the case that it is imposed in the null hypothesis (see Section **6.6.2** for discussion of the restricted bootstrap procedure). In the first case, there are 8 restrictions as the 3RTAR in ADF(2) form has 12 coefficients, whereas the single regime ADF(2) has 4 coefficients; imposing a unit root in the latter adds a further restriction to test. Assuming standard distributions apply then the null hypothesis is not rejected in the first case with p-value = 9.2%, but is rejected in the second with p-value 2.9% However, the 95% bootstrap quantiles are sufficiently to the right of their standard counterparts that both bootstrap p-values are in excess of 5% at 44% and 19%, and taking the more conservative (larger) p-value does not, therefore, lead to a change in inference (do not reject the null of one regime).

The third column refers to the question of whether three regimes (the alternative) can be collapsed to two (the null). The test statistic is 1.76 with a 'standard' p-value of 14% and a bootstrapped p-value of 53%, suggesting that the null should not be rejected.

The last two columns take the alternative as the two-regime model and consider whether they can be collapsed into a single regime. As before, assuming standard distributions apply, the null hypothesis is not rejected (test statistic 1.68, p-value 16%) if the unit root is not imposed in the null, but is rejected otherwise (test statistic 2.35, p-value 4.1%); however, as in the 'three to one' case, the bootstrapped p-values are much larger at 50% and 21%, respectively, and the evidence is not against a single regime with a unit root.

The bootstrap samples have been generated on the basis of no heteroscedasticity amongst regimes. This assumption seems broadly justified on the basis of the estimated $\hat{\sigma}_{\varepsilon i}$ for each regime; however, if an adjustment is needed there are two possibilities. First, as described in Section **6.3.3**, an adjustment could be made by a simple weighted least squares procedure. Alternatively, the random draws from the empirical distribution function (EDF) of the residuals within the bootstrap procedure can be regime-dependent; in effect for a three-regime TAR, three EDFs are defined, one for each regime and residual draws are then made from the appropriate regime.

6.8.5 The unit root null hypothesis

Given that there is little evidence in favour of a TAR, at least at conventional significance levels, this section is largely illustrative. The null hypothesis of a unit root is tested using the Wald-type tests or their F equivalents. In the case of the 2RTAR, the test values, reported in Table 6.9, are Sup-$W(0,0)_\kappa = 11.38$, $W(0,0)_\kappa^{ave} = 4.29$ and $W(0,0)_\kappa^{exp} = 21.94$, each of which is less than the corresponding asymptotic 95% quantile, being respectively 9.04 for the first two tests and 91.83 for the third test (see tables 6.2 and 6.3). Hence the null of a unit root in each regime is not rejected in the 2RTAR model.

In the 3RTAR, the Wald-type test values, reported in Table 6.10, are Sup-$W(0,0,0)_\kappa = 12.73$, $W(0,0,0)_\kappa^{ave} = 6.28$ and $W(0,0,0)_\kappa^{exp} = 69.0$, each of which is less than the corresponding (simulated) 95th percentile (see Table 6.12). As a reminder, the Wald test is a multiple of the F version of the test statistic which, for the null hypothesis of a unit root, is the average of the sum of the squared t statistics on the coefficient of y_{t-1} in each of the regimes; the directional version of the F test is the same in this case as the t statistics are negative at the supremum (and indeed at all possible sample separations). However, the p-values may

Table 6.12   Simulated quantiles for unit root tests in 3RTAR

	Sup-W$(0,0,0)_{\kappa_1,\kappa_2}$	W$(0,0,0)^{ave}_{\kappa_1,\kappa_2}$	W$(0,0,0)^{exp}_{\kappa_1,\kappa_2}$
95%	20.86	7.58	200.5
99%	26.25	9.41	1327

differ slightly due to the occurrence of some positive $\hat{\alpha}_i$ in the simulation or bootstrap samples. To illustrate, the distributions of the test statistics F$(0,0,0)_{\kappa_1,\kappa_2}$ and F$^D(0,0,0)_{\kappa_1,\kappa_2}$ were bootstrapped and resulted in p-values of 30% and 22%, respectively; thus both suggest non-rejection of the unit root null hypothesis. If no threshold effect is allowed, which is the case referred to as an 'unidentified threshold' in Caner and Hansen (2001), then the bootstrapped p-values were 14% and 13%, respectively. Examining the 2RTAR case is left to Q6.3.

6.9   Concluding remarks

The first part of this chapter continued the theme of the previous chapter, considering developments of the simple linear AR model to allow for the possibility of local nonstationarity. In particular, consideration was given to a popular class of models that allow structural breaks in the standard autoregressive model. These breaks take the form of piecewise linearity, where each 'piece' is defined as a regime but the regimes are not necessarily contiguous in time, and movement across or between regimes is triggered by a regime selector passing through a threshold. For obvious reasons these models are known as threshold AR models (TARs).

The particular nature of the breaks differs from the kind of models introduced by Perron (1989), which are considered in Chapters 7 and 8, but are similar in some ways. The Perron break-type models generate regimes which comprise observations necessarily adjacent in time; for example, a break induced by the events of the Great Crash (GC) induces two regimes, one for the pre-GC phase and the other for the post-GC regime; in these cases, the regime selector is simply the index of time. The TAR framework differs in that the regimes may switch back and forth over time, so even if there are only two regimes this will not necessarily induce two regimes of (time)-adjacent observations.

In the case of TAR models there are two related issues that dominate the testing literature. First, is there really more than one regime? Second, if there is more than one regime, what are the complications for testing for nonstationarity? The answer to the first issue is vital to the second that presupposes the first, but the problem is that the resolution of the second issue affects the testing procedure for the first: 'Catch 22' is alive and well! The problem, which should be obvious enough at this stage, is that the presence of unit roots alters the asymptotic distribution of standard test statistics. Thus a standard test, for example a Wald or F test for multiple regimes, will not have the same distribution when there is nonstationarity in the regimes compared to the case when the regimes are stationary. One solution is to compute the p-values both ways and base inference on the larger(est) p-value; this is referred to as a conservative procedure.

When testing for a unit root, the distribution of the test statistic will depend on whether there is a threshold effect, so that there will be one distribution in the unidentified threshold case and another in the identified threshold case. One approach is to condition the unit root testing procedure on the outcome of the (more conservative) threshold test (with due regard to the cumulative type 1 error since at least two tests combine into the overall procedure). However, in a simulation study, where the regime selector was I(0), Caner and Hansen (2001) found that the critical values for (non-directional) unit root test conditional on no threshold effect were better at maintaining size fidelity, even if there was a threshold effect. In both cases they also reported that the bootstrap versions of the finite sample distributions, whilst still showing size distortions, improve upon using the asymptotic critical values with an advantage, now more marginal, to the unidentified case.

TAR models offer an alternative to the STAR models of the previous chapter, but in some cases capture similar characteristics; for example, a STAR model will look like a TAR model, with a discontinuous and abrupt jump, if the response to a deviation from the threshold is very sensitive. On the other hand, the smooth transition aspect of STAR models has been used as an alternative to the discrete jump of the TAR model; in a modified TAR, the transition between regimes is modelled as a smooth adjustment process; see, for example, Harvey and Mills (2002) for a double-regime model with smooth transition.

The chapter concluded by considering different approaches to modelling the \$:£ real exchange rate, illustrating the modelling methods of this and the previous chapter. The models considered captured nonlinear effects in different ways, although they had in common the general notion of regime change. Although the emphasis here is on fitting different models, two crucial issues deserve further attention. The first relates to a generic problem in econometrics: that of deciding which amongst competing models is preferred The encompassing principle offers a potential general solution to this problem, and the CME and PSE type tests of Wooldridge (1990) and Chen and Kuan (2002), respectively, are particularly simple to apply.

The second is fundamental from an economic perspective; that is, whilst nonlinear models in different forms are attractive because they address the obvious drawback of the implication of the potential unboundedness of random walk type behaviour, what is the economic rationale behind different regimes? This question is particularly relevant to TAR models: what features of the economic system lead to regime-dependent behaviour; and can economic sense be made of the particular estimates of the threshold parameters?

This area is one that continues to attract research in applications and methods. For example, see Maki (2009) for a review and assessment of unit root tests in a three regime model, and Battaglia and Protopapas (2010), who considered both nonlinearity and nonstationarity and allow the regime switching to be self-exciting, smooth transition or piecewise linear. Threshold models can be extended to the variance as in TAR versions of ARCH (TARCH) models (see, for example, Baragona and Cucina, 2008). Kejriwal and Perron (2010) extend the work of Bai (1997) and Bai and Perron (1998), considering a sequential procedure to determine the number and timing of structural breaks. Maki (2010) considers the problem of deciding whether nonlinearity arises from a TAR or from a series of structural breaks. A particularly important framework for analysing the latter is considerd in the next chapter.

Questions

Q6.1 Consider the 2RTAR with different variances between regimes. Obtain a permutation matrix that reorders the observations in the model so that the regressor matrix is block diagonal.

A6.1 The substance of the answer is straightforward and intuitive; however, the mechanics of the set-up is something of a 'five-fingered' exercise in the detail. The non-singular $T \times T$ permutation matrix, say P_1, is based on the regime separation matrices I_1 and I_2. Let $P_1 = [P_{11} : P_{21}]'$ comprise two parts, one for each regime, with P_{11} of dimension $T_1 \times T$ and P_{21} of dimension $T_2 \times T$. Consider I_1, this has the t-th diagonal element, $I_{1t} = 1$ if $y_{t-1} \leq \kappa$; then starting at $i = 1$, let the i-th row of P_{11} have a 1 in the t-th column if $I_{1t} = 1$, and 0 otherwise; follow a similiar procedure for P_{21} based on I_2. The 2RTAR model is invariant to multiplication by the non-singular matrix P_1, which reorders the observations in y and z $(= \varepsilon)$ and (block) diagonalises the regressors:

$$y = I_1 \phi_1 y_{-1} + I_2 \varphi_1 y_{-1} + z$$
$$P_1 y = P_1 I_1 \phi_1 y_{-1} + P_1 I_2 \varphi_1 y_{-1} + P_1 z$$
$$= P_{11} \phi_1 y_{-1} + P_{21} \varphi_1 y_{-1} + P_1 z$$

Example:

$$
\begin{pmatrix} y_1 \\ y_2 \\ y_3 \\ y_4 \end{pmatrix} =
\begin{bmatrix} 1 & 0 & 0 & 0 \\ 0 & 0 & 0 & 0 \\ 0 & 0 & 1 & 0 \\ 0 & 0 & 0 & 0 \end{bmatrix}
\begin{pmatrix} y_0 \\ y_1 \\ y_2 \\ y_3 \end{pmatrix} \phi_1 +
\begin{bmatrix} 0 & 0 & 0 & 0 \\ 0 & 1 & 0 & 0 \\ 0 & 0 & 0 & 0 \\ 0 & 0 & 0 & 1 \end{bmatrix}
\begin{pmatrix} y_0 \\ y_1 \\ y_2 \\ y_3 \end{pmatrix} \varphi_1 +
\begin{pmatrix} \varepsilon_1 \\ \varepsilon_2 \\ \varepsilon_3 \\ \varepsilon_4 \end{pmatrix}
$$

$$
\mathbf{P}_1 = \begin{bmatrix} 1 & 0 & 0 & 0 \\ 0 & 0 & 1 & 0 \\ 0 & 1 & 0 & 0 \\ 0 & 0 & 0 & 1 \end{bmatrix}
$$

$$
\begin{bmatrix} 1 & 0 & 0 & 0 \\ 0 & 0 & 1 & 0 \\ 0 & 1 & 0 & 0 \\ 0 & 0 & 0 & 1 \end{bmatrix} \begin{pmatrix} y_1 \\ y_2 \\ y_3 \\ y_4 \end{pmatrix} = \begin{bmatrix} 1 & 0 & 0 & 0 \\ 0 & 0 & 1 & 0 \\ 0 & 1 & 0 & 0 \\ 0 & 0 & 0 & 1 \end{bmatrix} \begin{bmatrix} 1 & 0 & 0 & 0 \\ 0 & 0 & 0 & 0 \\ 0 & 0 & 1 & 0 \\ 0 & 0 & 0 & 0 \end{bmatrix} \begin{pmatrix} y_0 \\ y_1 \\ y_2 \\ y_3 \end{pmatrix} \phi_1
$$

$$
+ \begin{bmatrix} 1 & 0 & 0 & 0 \\ 0 & 0 & 1 & 0 \\ 0 & 1 & 0 & 0 \\ 0 & 0 & 0 & 1 \end{bmatrix} \begin{bmatrix} 0 & 0 & 0 & 0 \\ 0 & 1 & 0 & 0 \\ 0 & 0 & 0 & 0 \\ 0 & 0 & 0 & 1 \end{bmatrix} \begin{pmatrix} y_0 \\ y_1 \\ y_2 \\ y_3 \end{pmatrix} \varphi_1
$$

$$
+ \begin{bmatrix} 1 & 0 & 0 & 0 \\ 0 & 0 & 1 & 0 \\ 0 & 1 & 0 & 0 \\ 0 & 0 & 0 & 1 \end{bmatrix} \begin{pmatrix} \varepsilon_1 \\ \varepsilon_3 \\ \varepsilon_2 \\ \varepsilon_4 \end{pmatrix}
$$

$$
\begin{pmatrix} y_1 \\ y_3 \\ y_2 \\ y_4 \end{pmatrix} = \begin{pmatrix} y_0 \\ y_2 \\ 0 \\ 0 \end{pmatrix} \phi_1 + \begin{pmatrix} 0 \\ 0 \\ y_1 \\ y_3 \end{pmatrix} \varphi_1 + \begin{pmatrix} \varepsilon_1 \\ \varepsilon_3 \\ \varepsilon_2 \\ \varepsilon_4 \end{pmatrix}
$$

$$
= \begin{bmatrix} y_0 & 0 \\ y_2 & 0 \\ 0 & y_1 \\ 0 & y_3 \end{bmatrix} \begin{pmatrix} \phi_1 \\ \varphi_1 \end{pmatrix} + \begin{pmatrix} \varepsilon_1 \\ \varepsilon_3 \\ \varepsilon_2 \\ \varepsilon_4 \end{pmatrix}
$$

$$
\begin{pmatrix} Y_1 \\ Y_2 \end{pmatrix} = \begin{bmatrix} X_1 & 0_1 \\ 0_1 & X_2 \end{bmatrix} \begin{pmatrix} \phi_1 \\ \varphi_1 \end{pmatrix} + \begin{pmatrix} E_1 \\ E_2 \end{pmatrix}
$$

where $Y_1 = (y_1, y_3)'$, $Y_2 = (y_2, y_4)'$, $X_1 = Y_{1,-1}$, $X_2 = Y_{2,-1}$, $E_1 = (\varepsilon_1, \varepsilon_3)'$ and $E_2 = (\varepsilon_2, \varepsilon_4)'$. The block diagonality in the regressor matrix is now apparent. (Extension: show that the LS and GLS estimators are identical.)

Q6.2 An ESTAR(3) model was also estimated for the exchange rate example of section **6.8.1**. The estimation details are given below.

Q6.2.i Use a likelihood ratio test to test the unit root restriction.

Q6.2.ii Does the LR test of the previous question have an asymptotic χ^2 distribution under the null?

Table A6.1   ESTAR(3) models for the dollar–sterling real exchange rate

ESTAR(3): Outer regime restriction imposed: $\gamma_1 = -\phi_1$, $\gamma_2 = -\phi_2$, $\gamma_3 = -\phi_3$

	$\hat{\phi}_1$	$\hat{\phi}_2$	$\hat{\phi}_3$	$\hat{\gamma}_1$	$\hat{\gamma}_2$	$\hat{\gamma}_3$	$\hat{\theta}$	$\hat{c}$	$10^2 \times \hat{\sigma}_\varepsilon$	F_{LM}	LL
	1.396	–0.504	0.117	–1.396	0.504	–0.117	0.524	0.359	2.485	1.71	744.51
't'	23.63	–5.58	–2.10	(–)	(–)	(–)	2.10	11.23		[0.119]	

ESTAR(3): Outer regime and unit root restrictions imposed:
$\gamma_1 = -\phi_1$, $\gamma_2 = -\phi_2$, $\gamma_3 = -\phi_3$, $\phi_1 + \phi_2 + \phi_3 = 1$

	$\hat{\phi}_1$	$\hat{\phi}_2$	$\hat{\phi}_3$	$\hat{\gamma}_1$	$\hat{\gamma}_2$	$\hat{\gamma}_3$	$\hat{\theta}$	$\hat{c}$	$10^2 \times \hat{\sigma}_\varepsilon$	F_{LM}	LL
	(**)	–0.502	0.117	(**)	0.502	–0.117	0.418	0.369	2.480	1.52	744.37
't'	(–)	–5.59	–2.10	(–)	(–)	(–)	2.72	10.70		[0.171]	

Note: See Table 6.6; LL is the log-likelihood.

Q6.2.iii What are the missing coefficients indicated by (**) in the second part of the table?

Q6.2.iv In the general case, the time-varying AR coefficient is $\beta_t = \sum_{i=1}^{p} (\phi_i + \gamma_i \Theta_t)$; show how this is obtained for the estimated ESTAR(3) model.

A6.2.i The LR test statistic is $2(744.51 - 744.37) = 0.28$.

A6.2.ii The LR test for a unit root does not have an asymptotic χ^2 distribution under the null. However, 0.28 is clearly not significant.

A6.2.iii The missing coefficients indicated by (**) are obtained as follows: the restriction $\phi_1 + \phi_2 + \phi_3 = 1$ implies $\phi_1 = 1 - (\phi_2 + \phi_3)$, hence in terms of the estimates $\hat{\phi}_1 = 1 - (\hat{\phi}_2 + \hat{\phi}_3) = 1 - ((-0.502) + 0.117) = 1.385$; note that this compares with $\hat{\phi}_1 = 1.396$ when the unit root is not imposed. The other missing coefficient is $\hat{\gamma}_1$, which is obtained from the restriction $\gamma_1 = -\phi_1$, thus $\hat{\gamma}_1 = -1.385$.

A6.2.iv The time-varying coefficient appropriate to this model is $\beta_t = \sum_{i=1}^{3} (\phi_i + \gamma_i \Theta_t)$; refer back to sections **5.2.5** and **5.2.6** and note that $\Theta_t = (1 - e^{-\theta(y_{t-1}-c)^2})$ is the transition function. In the ESTAR(3) case, the time-varying coefficient β_t is constructed as:

$$\beta_t = \phi_1 + \gamma_1 \Theta_t + \phi_2 + \gamma_2 \Theta_t + \phi_2 + \gamma_3 \Theta_t$$
$$= \phi_1 + \phi_2 + \phi_2 + \Theta_t(\gamma_1 + \gamma_2 + \gamma_3)$$

If the outer regime restriction is imposed, then $\gamma_1 + \gamma_2 + \gamma_3 = -(\phi_1 + \phi_2 + \phi_3)$ and:

$$\beta_t = (\phi_1 + \phi_2 + \phi_2)(1 - \Theta_t)$$

and if, in addition, the unit root restriction is imposed, then β_t simplifies further to:

$$\beta_t = (1 - \Theta_t)$$

$$= e^{-\theta(y_{t-1}-c)^2}$$

Since $\theta > 0$ and $(y_{t-1} - c)^2 \geq 0$, then $0 < \beta_t \leq 1$, and is equal to 1 when $(y_{t-1} - c) = 0$, that is, the process is in the inner regime and $\beta_t \to 0$ as $(y_{t-1} - c) \to \pm\infty$. (Extension: obtain the inner and outer regimes and roots of the ESTAR(3) model and compare them with the ESTAR(2) model reported in the text.)

Q6.3.i Consider the 2RTAR of Table 6.9: write the two equations as one equation in ADF form by defining an appropriate indicator variable.

Q6.3.ii Calculate and interpret the test for one regime rather than two.

A6.3.i The ADF form of a 2RTAR is given by Equation (6.20). One of the regimes is chosen to be the base regime, in this case the second. In ADF form the 2RTAR is:

$$\Delta y_t = \mu_t + \alpha_t y_{t-1} + \sum_{j=1}^{2} \delta_{tj} \Delta y_{t-j} + \varepsilon_t \tag{A6.1}$$

The empirical form of this equation, using the estimates from Table 6.9, is:

$$\Delta y_t = (0.028 - 0.012I_{1t}) - (0.0585 - 0.0007I_{1t})y_{t-1} + (0.373 + 0.036I_{1t})\Delta y_{t-1}$$

$$- (0.116 - 0.020I_{1t})\Delta y_{t-2} + \hat{\varepsilon}_t$$

The indicator variable is $I_{1t} = 1$ if $y_{t-1} \leq 0.345$ and 0 if $y_{t-1} > 0.345$. For example, in regime 1, $\alpha_t = -(0.0585 - 0.0007) = -0.0592$.

Q6.3.ii The test statistic is given in Wald form by:

$$W_{12}(\hat{\kappa}|d_{(i)}) = \tilde{T}\left(\frac{(\hat{\varepsilon}'\hat{\varepsilon})_1 - (\hat{\varepsilon}'\hat{\varepsilon}|\hat{\kappa}, d)_{2,\inf}}{(\hat{\varepsilon}'\hat{\varepsilon}|\hat{\kappa}, d)_{2,\inf}}\right)$$

See Equation (6.39a) and d is here assumed known, $d = 1$. The F version of this test, $F_{12}(\hat{\kappa}, d)$, makes a degrees of freedom adjustment. The relevant residual sums of squares from Tables 6.9 and 6.10 are: $(\hat{\varepsilon}'\hat{\varepsilon})_1 = 0.200189$ and $(\hat{\varepsilon}'\hat{\varepsilon}|\hat{\kappa}, \hat{d})_{2,\inf} = 0.196043$. Hence with $\tilde{T} = 325$, $g = 4$ and $\hat{T} = 325 - 8 = 317$, the Wald and F test statistics are:

$$W_{12}(\hat{\kappa}, d) = 325\left(\frac{0.200189 - 0.196043}{0.196043}\right) = 6.872$$

$$F_{12}(\hat{\kappa}, d) = \left(\frac{0.200189 - 0.196043}{0.196043}\right)\left(\frac{325 - 8}{4}\right) = 1.676$$

See also Table 6.11. Note that approximately $W_{12}(\hat{\kappa}, \hat{d}) = 4F_{12}(\hat{\kappa}, \hat{d})$ and exactly $W_{12}(\hat{\kappa}, d) = 4(325/317)F_{12}(\hat{\kappa}, d)$. The bootstrapped 95th percentile was 3.68, with a p-value of 50% indicating non-rejection of the null hypothesis of a single regime. A variation on this statistic is obtained by imposing the unit root in the model corresponding to the null hypothesis. In this case, with the RSS = 0.203315 from the single regime ADF(2) with unit root imposed, the F version of the test is:

$$F_{12}(\hat{\kappa}, d) = \left(\frac{0.203315 - 0.196043}{0.196043} \right) \left(\frac{325 - 8}{5} \right) = 2.352$$

Again see Table 6.11. The 95% bootstrap quantile was 3.39 giving a bootstrap p-value of 21%, which is closer to the rejection region than not including the unit root restriction but still sufficiently far away to indicate non-rejection of the null of a single regime.

7
Structural Breaks in AR Models

Introduction

Threshold autoregressions are a way of characterising structural changes associated with the degree of persistence of shocks in particular regimes. If the regime indicator is taken to be some function of y_t itself the process is said to be 'self-exciting' leading to the acronym SETAR for self-exciting threshold autoregression. Another form of structural change is when there are distinct break points, leading to different regimes, at particular points of time. For example in the simplest case, regime 1 operates up to time T_1 and regime 2 operates thereafter, where the regimes are distinguished in some way, for example by the degree of persistence in the autoregression (but other characteristics of the process such as the variance could change). The particular nature of the effect of the structural change could also vary from transient to permanent. There may be multiple change (or break) points, which may occur with certainty (probability of break = 1) or uncertainty (probability of break greater than 0 but less than 1) and the actual or potential breakpoints may or may not be known to the researcher. Transition may be abrupt, suddenly from one regime to another, or smooth, taking place gradually over several periods.

Two seminal articles in this area by Perron (1989) and Hamilton (1989), who dealt with different models for the nature of the structural change, have led to important developments with, even now, ongoing research into issues with practical implications that have yet to be resolved (see Perron, 2006, for a critical assessment).

Perron (1989) suggested that time series that were found to be nonstationary on the basis of DF tests, for example, the macroeconomic time series in the classic study by Nelson and Plosser (1982), might well be stationary around a (deterministic) trend if the trend was allowed to vary over time. In the simplest case, a one-time break in the trend may be enough to 'confuse' the unit root tests into a finding of nonstationarity. Thus, to ignore the structural break would lead to spurious nonrejections of the unit root null hypothesis.

Examples of events momentous enough to be considered as structural breaks include the Great Depression, the Second World War, the 1973 OPEC oil crisis and, more recently, 9/11 in 2001 and the financial crisis (the 'credit crunch') of 2008. Under the alternative hypothesis of stationarity, trended time series affected by such events could be modelled as stationary around a broken or split trend. Equally time series that display no obvious trend could be subject to a level shift under the alternative so that the long-run mean is affected by the special event. An example of the latter is provided in Perron and Vogelsang (1992), who consider the bilateral real exchange rate between the US and the UK using annual data for 1892 to 1988, where there appears to be a level shift during the later part of the Second World War period.

In the Perron approach, although the date of the break or breaks may be unknown, there is no uncertainty as to which regime the economy is in during each of the periods defined by the breaks. Hamilton (op. cit.) introduced a different development of the basic model. An alternative to the 'known' regime framework is to attach a probability to each of the regimes; naturally, the probabilities sum to 1 but as probabilities of 0 and 1 are generally excluded for each regime, which regime is in operation at a particular point in time is not known with certainty. Developments of this approach could allow the transition probabilities to vary over time.

Because of the scope of the structural break literature and variety of models that have been suggested, this chapter has necessarily to limit what is covered here; however, it is undoubtedly the case that the framework due to Perron (1989) has been particularly influential and so that approach forms the basis of this chapter. The next chapter takes and develops various issues raised here.

First note that this is a necessarily detailed, although not technically complex, chapter; this arises because the typology of structural breaks includes at least three different cases, which are themselves combined with differences that may arise under the null and alternative hypotheses. This chapter deals with the case of a single known breakpoint, the key unknown factor being the impact of the break, whether, for example, it is distributed over time and if such a break is present under the null of a unit root as well as in the alternative view of trend stationarity, albeit with a 'broken' trend (or mean in the case of non-trended data). The following chapter, which requires this one as background, considers the case, more likely to occur in practice, where the break point or points are not known with certainty.

Section **7.1** presents the basic underlying and familiar model from which one can introduce the idea of a structural break, which is followed up in Section **7.2**. That section outlines some key issues, such as: is the impact of a break immediate and complete or distributed over time? It also develops the basic distinctions of the additive outlier (AO) and the innovation outlier (IO) in the modelling framework for structural breaks. Higher-order dynamics are introduced in Section **7.3**, which also provides the 'encompassing' principle of the distributed outlier (DO) model. Matters of selection between the various models are considered in Section **7.4**, with an example in Section **7.5**. The unit root null hypothesis is introduced in Section **7.6**, with the testing framework outlined in Section **7.7** and critical values in Section **7.8**. The implications of structural breaks for unit root tests, for example the spurious non-rejection of the unit root hypothesis, are considered in Section **7.9**.

7.1 Basic structural break models

It is easiest to start with the case of a single known breakpoint and then to generalise that starting case. We briefly set out a basic model in order to understand the terminology and key issues in this area.

In this development we assume that the breakpoint is known, as is the nature of the impact of the break (whether it is one-off or lasting). The basic model is then easily amended to model further developments. By way of reminder the basic model, with no structural change, is restated.

7.1.1 Reminder

We first revisit the basic model in the simplest set-up, as follows:

$$y_t = \mu_t + z_t \tag{7.1}$$

$$(1 - \rho L)z_t = \varepsilon_t \qquad \varepsilon_t \sim \text{iid}(0, \sigma_\varepsilon^2) \tag{7.2}$$

Typical specifications for μ_t are $\mu_t = \mu_1$ and $\mu_t = \mu_1 + \beta_1 t$, that is, the constant mean and deterministic trend specifications. A word on notation at this stage compared to previous chapters. Previously, y_0 has been used rather than μ_1 to denote μ_t at $t = 0$, in part because y_0 is the presample value of y_t. To accommodate a break in the impact of the deterministic components, the established way chosen in this area of the literature is to use the notation $\mu_t = \mu_1$ or $\mu_t = \mu_1 + \beta_1 t$, and then change μ_1 to μ_2 and/or β_1 to β_2 in the event of a break.

Returning to the substantive matter – see (7.1) and (7.2) – two cases arise depending on the value of ρ. In the case of a unit root $\rho = 1$, so that (7.2) is $(1 - L)z_t = \varepsilon_t$, then (7.1) and (7.2) imply:

$$(1 - L)y_t = (1 - L)\mu_t + \varepsilon_t \tag{7.3}$$

Note that if $\mu_t = \mu_1$ then $(1 - L)\mu_t = 0$ and if $\mu_t = \mu_1 + \beta_1 t$ then $(1 - L)\mu_t = \beta_1$. As a result, the models under the null of a unit root are, respectively:

$$y_t = y_{t-1} + \varepsilon_t, \qquad \text{if } \mu_t = \mu_1 \tag{7.4a}$$

$$\Rightarrow$$

$$\Delta y_t = \varepsilon_t \tag{7.4b}$$

$$y_t = \beta_1 + y_{t-1} + \varepsilon_t, \qquad \text{if } \mu_t = \mu_1 + \beta_1 t \tag{7.5a}$$

$$\Rightarrow$$

$$\Delta y_t = \beta_1 + \varepsilon_t \tag{7.5b}$$

Under the alternative hypothesis $|\rho| < 1$, and the resulting models under the two specifications of μ_t are, respectively:

$$y_t = (1 - \rho)\mu_1 + \rho y_{t-1} + \varepsilon_t, \qquad \mu_t = \mu_1 \tag{7.6}$$

$$\Rightarrow$$

$$y_t - \mu_1 = \rho(y_{t-1} - \mu_1) + \varepsilon_t \tag{7.7}$$

$$y_t = (1 - \rho)\mu_1 + \rho\beta_1 + (1 - \rho)\beta_1 t + \rho y_{t-1} + \varepsilon_t, \ \mu_t = \mu_1 + \beta_1 t \tag{7.8}$$

$$\Rightarrow$$

$$y_t - \mu_1 - \beta_1 t = \rho(y_{t-1} - \mu_1 - \beta_1(t - 1)) + \varepsilon_t \tag{7.9}$$

Note that equations (7.7) and (7.9) write the models in deviation from mean/trend form. In the event that that the limiting case $\rho = 0$ holds, the models under the alternative hypothesis reduce to deterministic constants plus noise, that is, respectively:

$$y_t = \mu_1 + \varepsilon_t \qquad \text{if } \mu_t = \mu_1 \tag{7.10}$$

$$y_t = \mu_1 + \beta_1 t + \varepsilon_t \quad \text{if } \mu_t = \mu_1 + \beta_1 t \tag{7.11}$$

There are two ways of estimating the parameters of interest. For example, in the case of the constant mean, a two-stage procedure consists of a first stage in which an estimate $\hat{\mu}_1$ of μ_1, usually obtained from the regression of y_t on a constant, is used to define the demeaned data $y_t - \hat{\mu}_1$ and $y_{t-1} - \hat{\mu}_1$; then, in the second stage, the demeaned data is used in a regression of the form (7.7) to obtain an estimate of ρ.

Alternatively, the regression (7.7) can be directly estimated from an LS regression of y_t on a constant and y_{t-1}; the constant is $\mu_1^* = (1 - \rho)\mu_1$, interpreted as the 'short-run' constant, and an estimate of μ_1 is obtained on dividing $\hat{\mu}_1^*$ by 1 minus the estimated coefficient, $\hat{\rho}$, on y_{t-1}, where $\hat{\ }$ indicates the estimate from the direct LS regression. In the case of the with-trend specification, the two-stage procedure first involves prior estimation of μ_1 and β_1 by the regression of

y_t on a constant and a time trend, and then the second stage is estimation of a regression of the form of (7.9) but using the de-trended data; alternatively, the regression (7.8) can be directly estimated and estimates of the parameters μ_1, β_1 and ρ uniquely recovered.

7.2 Single structural break

We consider how to model a structural break that reflects temporal instability, with a deliberately simple approach taken in this section to emphasise the key issues. The discussion will serve to highlight that even within a simple dynamic structure there are a number of issues to be considered apart from the central issue of whether there is a unit root. The plan for this section is to introduce the basic distinctions with minimal dynamics and complications, and then in the following section elaborate on these more detailed points.

Immediate or distributed impact of a break?

The first point to note is how the effect of the break is incorporated into the autoregressive process; the full impact of the break may be immediate or distributed over time. This distinction has given rise to two forms of the structural break model due to Perron (1989), known as the Additive Outlier (AO) and the Innovation Outlier (IO) models, and each of these is considered in turn in this section. In the next section a slight generalisation of the AO/IO framework is introduced, which is also extended to cover the more realistic cases of higher order dynamics.

Is there a break under the null hypothesis of a unit root?

The second point to note is that the break may be present under both the alternative and the null or it may be present under the alternative but not the null. This distinction will have an effect on the distribution of standard test statistics such as the normalised-bias and pseudo-t test statistics.

What is the nature of the break?

The third point is the nature of the break and here there are three standard distinctions, each with differing interpretations depending on whether the data is generated under the null hypothesis of a unit root or the alternative hypothesis of stationarity. There could be a change in the intercept, a change in the slope or a change in both the intercept and the slope.

Is the true break date known?

In the first instance, the assumption of a known break date is maintained. This was the situation considered by Perron (1989) but subsequently relaxed by a number of authors. Initially we assume that the break date is known, generalising that assumption later in this and the following chapter.

7.2.1   Additive outlier (AO)

This section describes a class of models known as additive outliers to be distinguished from innovation outliers, the distinction being how the impact of the break is distributed over time. In the AO model, the impact of the structural break is complete within the period: there is no distinction between the short-run and the long-run impact of the break.

We start with the simplest case. Suppose the trend function allows a change in the intercept parameter, the effective date of the change being at observation $T_b^c + 1$ in a sample of 1 to T observations; the date of the break is T_b^c in the sense that the break splits the sample into two periods comprising observations $1, \ldots, T_b^c$ and $T_b^c + 1, \ldots, T$, respectively. Here it is assumed that T_b^c is known and the superscript c serves to indicate that the break date is correct; the corresponding (correct) break fraction is $\lambda_b^c = T_b^c / T$. (When the break date is assumed unknown, the superscript c is omitted so that the notation is λ_b.) There are three models of interest corresponding to a change in the intercept, a change in the slope (that is the trend), and a change in the intercept and a change in the slope; these are referred to as Models 1, 3 and 2, the terminology following Vogelsang and Perron (1998).

7.2.1.i   AO model specification

The change in the intercept is captured by the following specification, referred to as Model 1:

Model 1

$$\mu_t = \mu_1 + \beta_1 t + \mu_2 DU_t \quad DU_t = 1 \text{ if } t > T_b^c \text{ and 0 otherwise} \tag{7.12a}$$

$$\mu_t = \mu_1 + \beta_1 t \qquad\qquad \text{if } t \leq T_b^c \tag{7.12b}$$

$$\mu_t = \mu_1 + \mu_2 + \beta_1 t \qquad \text{if } t \geq T_b^c + 1 \tag{7.12c}$$

This model is sometimes referred to as the 'crash' model following Perron's use of the model to capture the negative effects of the Great Crash of 1929 on a number of macroeconomic variables (thus, in this case, μ_2 would be negative). The model is referred to in Perron (1989) as Model A and in Vogelsang and Perron (1998) as Model 1. Use has also been made of this model by Perron and Vogelsang (1992) without the term $\beta_1 t$ in μ_t and that specification is here referred to as Model 1a, which we state here for future reference.

Model 1a

$$\mu_t = \mu_1 + \mu_2 DU_t \quad DU_t = 1 \text{ if } t > T_b^c \text{ and 0 otherwise} \tag{7.13a}$$

$$\mu_t = \mu_1 \qquad\qquad \text{if } t \leq T_b^c \tag{7.13b}$$

$$\mu_t = \mu_1 + \mu_2 \qquad \text{if } t \geq T_b^c + 1 \tag{7.13c}$$

A change in the slope (trend) is captured by the following specification, referred to as Model 3:

Model 3

$$\mu_t = \mu_1 + \beta_1 t + \beta_2 DT_t \qquad\qquad DT_t = t - T_b^c \text{ and } 0 \text{ otherwise} \qquad (7.14a)$$

$$\mu_t = \mu_1 + \beta_1 t \qquad\qquad\qquad \text{if } t \le T_b^c \qquad\qquad\qquad\qquad (7.14b)$$

$$\mu_t = \mu_1 + \beta_1 t + \beta_2(t - T_b^c) \qquad \text{if } t \ge T_b^c + 1 \qquad\qquad\qquad (7.14c)$$

$$= (\mu_1 - \beta_2 T_b^c) + (\beta_1 + \beta_2)t$$

This is referred to as the 'changing growth' model, which is Model B in Perron (1989) and Model 3 in Vogelsang and Perron (1998). (Notice that if $\beta_2 = 0$, then μ_t just reverts to $\mu_t = \mu_1 + \beta_1 t$, that is, no change; some authors write the coefficient on DT_t explicitly as a difference from β_1; see, for example, Montañés and Reyes, 1998; the same comment applies to μ_2.)

Combining the intercept and slope changes leads to Model 2:

Model 2

$$\mu_t = \mu_1 + \beta_1 t + \mu_2 DU_t + \beta_2 DT_t \qquad\qquad\qquad\qquad\qquad (7.15a)$$

$$\mu_t = \mu_1 + \beta_1 t \qquad\qquad\qquad \text{if } t \le T_b^c \qquad\qquad\qquad (7.15b)$$

$$\mu_t = \mu_1 + \mu_2 + \beta_1 t + \beta_2(t - T_b^c) \qquad \text{if } t \ge T_b^c + 1 \qquad\qquad (7.15c)$$

$$= \mu_1 + (\mu_2 - \beta_2 T_b^c) + (\beta_1 + \beta_2)t$$

This is referred to as the 'mixed model' model, which is Model C in Perron (1989) and Model 2 in Vogelsang and Perron (1998). (See also Q7.1 which examines the distinction between the notation used in Perron, 1989, and the notation used here.)

In addition to the dummy variables DU_t and DT_t, it is useful to define the one-time dummy variable $DT_t^b = 1$ if $t = T_b + 1$, 0 otherwise. Note that $\Delta DU_t = DT_t^b$ and $\Delta DT_t = DU_t$; also $DU_t = \Delta^{-1} DT_t^b$, where $\Delta^{-1} = (1 - L)^{-1}$ is the summation operator and is without ambiguity since DT_t^b has a starting value of 0.

Now consider the model as in (7.1) and (7.2) but with μ_t as in Model 1. Under the alternative hypothesis $|\rho| < 1$, the model resulting from allowing an intercept break is:

$$y_t = \mu_t + z_t \qquad\qquad\qquad\qquad\qquad\qquad\qquad \text{again} \qquad (7.1)$$

$$= \mu_1 + \beta_1 t + \mu_2 DU_t + z_t \qquad \text{using } \mu_t = \mu_1 + \beta_1 t + \mu_2 DU_t$$

$$(1 - \rho L)z_t = \varepsilon_t \qquad\qquad\qquad\qquad\qquad\qquad\qquad \text{again} \qquad (7.2)$$

$$(1 - \rho L)y_t = (1 - \rho L)(\mu_1 + \beta_1 t + \mu_2 DU_t) + \varepsilon_t \qquad\qquad\qquad (7.16)$$

$$y_t - \mu_1 - \beta_1 t - \mu_2 DU_t = \rho(y_{t-1} - \mu_1 - \beta_1(t-1) - \mu_2 DU_{t-1}) + \varepsilon_t \qquad (7.17)$$

whereas with a unit root, the specification is:

$$y_t = \mu_t + z_t \qquad\qquad\qquad\qquad \text{again} \qquad (7.1)$$

$$(1 - L)z_t = \varepsilon_t \qquad\qquad\qquad\qquad \text{again} \qquad (7.3)$$

$$(1 - L)y_t = (1 - L)\mu_t + \varepsilon_t \qquad\qquad \text{substitute for } z_t$$

$$= (1 - L)(\mu_1 + \beta_1 t + \mu_2 DU_t) + \varepsilon_t \qquad\qquad (7.18)$$

$$\text{substitute for } \mu_t = \mu_1 + \beta_1 t + \mu_2 DU_t$$

$$= \beta_1 + \mu_2 \Delta DU_t + \varepsilon_t$$

$$y_t = \beta_1 + y_{t-1} + \mu_2 DT_t^b + \varepsilon_t \qquad \text{using } \Delta DU_t = DT_t^b \qquad (7.19)$$

Thus, if the intercept break is present under the null, then the one-off dummy variable DT_t^b is included.

Comparison of the no-change case, that is Equation (7.9), and the intercept change case, that is Equation (7.17), suggests that estimation could proceed in two stages. First a prior regression of y_t on a constant, time trend, t, and DU_t serves to define data that is detrended *and* adjusted for a change in the intercept, that is:

$$\hat{\tilde{y}}_t = y_t - \hat{\mu}_1 - \hat{\beta}_1 t - \hat{\mu}_2 DU_t \qquad\qquad (7.20)$$

In the second stage regression $\hat{\tilde{y}}_t$ is the dependent variable, which is regressed on $\hat{\tilde{y}}_{t-1}$ to provide an estimate of ρ. That is, the regression is:

$$\hat{\tilde{y}}_t = \rho \hat{\tilde{y}}_{t-1} + \tilde{\varepsilon}_t \qquad\qquad (7.21)$$

where $\tilde{\varepsilon}_t$ is a linear combination of ε_t and the error in estimating μ_1, β_1 and μ_2 by $\hat{\mu}_1$, $\hat{\beta}_1$ and $\hat{\mu}_2$. In fact a slight modification of this regression is required and the one-off dummy variable DT_t^b is added to ensure that the pseudo-t statistic is not dependent on nuisance parameters (Perron and Vogelsang, 1992a, 1992b). Thus (7.21) becomes:

$$\hat{\tilde{y}}_t = \rho \hat{\tilde{y}}_{t-1} + \delta_0 DT_t^b + \tilde{\varepsilon}_t \qquad\qquad (7.22)$$

This regression can be estimated by LS and tests for a unit root can be based on the LS estimate of ρ. It is usually convenient, as in the standard case, to set up the null hypothesis so that the restricted coefficient is zero, hence $\hat{\tilde{y}}_{t-1}$ is subtracted from both sides of (7.22) to obtain:

$$\Delta \hat{\tilde{y}}_t = \gamma \hat{\tilde{y}}_{t-1} + \delta_0 DT_t^b + \tilde{\varepsilon}_t \qquad\qquad (7.23)$$

where $\gamma = \rho - 1$. The unit root null then corresponds to $\gamma = 0$. Notice from the specification in (7.23) that there is a break under the null as well as the alternative, which will affect the critical values. The critical values for the n-bias and pseudo-t tests have been provided in a number of sources, which are detailed below (see also chapter 6 of *UR, Vol. 1*). Applied research has tended to almost exclusively use the pseudo-t test, which is denoted $\tilde{\tau}_\gamma^{(i)}$(AO) for the AO Model; if emphasis is needed that this statistic is calculated for the correct break date/fraction, the notation is extended to $\tilde{\tau}_\gamma^{(i)}(\lambda_b^c, AO)$.

7.2.1.ii Estimation

Although the nature of the break differs between the three models, the general approach is the same. In the first stage, regress y_t on the components of μ_t and form the residuals $\hat{\tilde{y}}_t = y_t - \hat{\mu}_t$, interpreted as the deviations from trend; then in the second stage, regress $\Delta\hat{\tilde{y}}_t$ on $\hat{\tilde{y}}_{t-1}$ and DT_t^b. This approach assumes that the error dynamics are AR(1) – see equation (7.2); more complex dynamics are introduced in the following section.

7.2.2 Innovation outlier (IO)

The innovation outlier approach involves a slightly different development of how to model the impact of the break compared to the AO specification. Rather than the impact of the break being complete within the period $T_b^c + 1$, it has an effect that is distributed over time $T_b^c + h, h = 1, 2, \ldots$. Unlike the AO case, there is a distinction between the short-run impact of the break, defined as the impact at time $T_b^c + 1$, and the long-run impact, defined as the sum of all the impacts. The term innovation outlier is motivated by the analogy with the effect of a change in an innovation, ε_t, on y_t.

7.2.2.i IO Model 1

The alternative hypothesis

In this case the model is developed as follows, first under the alternative hypothesis of stationarity:

$$y_t = \mu_1 + \beta_1 t + z_t \tag{7.24}$$

$$(1 - \rho L)z_t = \mu_2 DU_t + \varepsilon_t \tag{7.25}$$

Notice that in this specification the impact of the break at $T_b^c + 1$, that is $\mu_2 DU_t$, is in effect not distinguished from the innovation ε_t. Thus, multiplying (7.25) by $(1 - \rho L)^{-1}$ gives z_t as follows:

$$z_t = (1 - \rho L)^{-1}(\mu_2 DU_t + \varepsilon_t) \tag{7.26}$$

and substituting (7.26) for z_t in (7.24), the equation for y_t can be written as:

$$y_t = \mu_1 + \beta_1 t + (1 - \rho L)^{-1}(\mu_2 DU_t + \varepsilon_t) \tag{7.27}$$

As $(1 - \rho L)^{-1}$ is an infinite polynomial in L, the specification in (7.27) makes it apparent that the effect of the structural break is distributed over time and, in this respect, it is modelled in the same manner as if it was a shock to the innovation, ε_t. Compare this specification to the AO specification that led to equation (7.17).

Practically, although (7.27) is useful for interpretation it is not used directly for estimation. Rather multiply (7.27) through by $(1 - \rho L)$ to obtain:

$$(1 - \rho L)y_t = (1 - \rho L)(\mu_1 + \beta_1 t) + (\mu_2 DU_t + \varepsilon_t) \tag{7.28}$$

Now rearrange, as in the no-break case of (7.8), to obtain:

$$y_t = (1 - \rho)\mu_1 + \rho\beta_1 + (1 - \rho)\beta_1 t + \rho y_{t-1} + \mu_2 DU_t + \varepsilon_t \tag{7.29}$$

In contrast to the AO model, there is a distinction between the short-run and long-run effect, associated with the structural break. Consider (7.27) which, apart from the term in ε_t, is the trend function, written here for convenience and denoted y_t^{tr}, that is:

$$y_t^{tr} = \mu_1 + \beta_1 t + (1 - \rho L)^{-1}\mu_2 DU_t \tag{7.30}$$

As $|\rho| < 1$ then:

$$(1 - \rho L)^{-1}\mu_2 = (1 + \rho L + \rho^2 L^2 + \ldots)\mu_2 \tag{7.31}$$

from which we can read off the immediate impact of the break, that is at $T_b^c + 1$, as μ_2, the cumulative impact at $T_b^c + 2$ is $(1 + \rho)\mu_2$, and in the long run ($T_b^c + h$ as $h \to \infty$) the cumulative impact is $(1 - \rho)^{-1}\mu_2$. Thus, μ_t can now be interpreted as the long-run trend function, which in this case is:

$$\mu_t = \mu_1 + \beta_1 t + (1 - \rho)^{-1}\mu_2 DU_t \tag{7.32}$$

The null hypothesis

Under the null hypothesis $\rho = 1$, and the impact of the break is modelled as a permanent change to the *level* of the series for $t \geq T_b^c + 1$.

$$(1 - L)z_t = \mu_2 DT_t^b + \varepsilon_t \tag{7.33}$$

$$\Rightarrow$$

$$z_t = (1 - L)^{-1}\mu_2 DT_t^b + (1 - L)^{-1}\varepsilon_t \tag{7.34a}$$

$$= \mu_2 DU_t + (1 - L)^{-1}\varepsilon_t \tag{7.34b}$$

$$= \sum_{i=0}^{t} \varepsilon_i \qquad \text{for } t \leq T_b^c \tag{7.34c}$$

$$= \mu_2 + \sum_{i=0}^{t} \varepsilon_i \qquad \text{for } t \geq T_b^c + 1 \tag{7.34d}$$

For simplicity the summation in (7.34c) assumes that the process starts with ε_0 (which might, for example, be set to zero).

With the break modelled by equation (7.33) and following the same steps through as in the development above gives:

$$y_t = \mu_1 + \beta_1 t + z_t \tag{7.35}$$

$$(1 - L)z_t = \mu_2 DT_t^b + \varepsilon_t \tag{7.36}$$

$$(1 - L)y_t = (1 - L)(\mu_1 + \beta_1 t) + \mu_2 DT_t^b + \varepsilon_t \tag{7.37}$$

To obtain equation (7.37), multiply (7.35) by $(1 - L)$ and use (7.36); notice that the first term on the right-hand side of (7.37) simplifies because $(1 - L)(\mu_1 + \beta_1 t) = 0 + \beta_1$. As a result (7.37) can be written as:

$$\Delta y_t = \beta_1 + \mu_2 DT_t^b + \varepsilon_t \tag{7.38}$$

This specification makes it clear that a break is being assumed under the null as well as under the alternative hypothesis. If no break is allowed under the null, then the term in DT_t^b is redundant and is deleted from equation (7.38).

Maintained regression

The method suggested by Perron (1989) is, as in the standard DF case, to form a maintained regression that allows the null and alternative hypotheses to be determined as special cases. This involves the union of the regressors under the null and alternative hypotheses:

$$\Delta y_t = \kappa_1 + \kappa_2 t + \gamma y_{t-1} + \kappa_3 DT_t^b + \kappa_4 DU_t + \varepsilon_t \tag{7.39}$$

Then the specification under the null implies: $\gamma = 0$, $\kappa_1 = \beta_1$, $\kappa_2 = 0$, $\kappa_4 = 0$, and $\kappa_3 \neq 0$ if there is a break under the null. Under the alternative hypothesis: $\gamma < 0$, $\kappa_1 = (1 - \rho)\mu_1 + \rho\beta_1$, $\kappa_2 = (1 - \rho)\beta_1$, $\kappa_3 = 0$ and $\kappa_4 = \mu_2$. The idea is to estimate (7.39) by LS and to form the usual test statistics based on the LS estimate of γ. The pseudo-t test is denoted $\tilde{\tau}_\gamma^{(1)}(\text{IO})$; critical values are tabulated in detail in a number of sources and are discussed below.

7.2.2.ii   IO Model 2

The alternative hypothesis

In this case there are simultaneous breaks in the intercept and the slope which, as in the case of IO Model 1, are treated in the same manner as the innovation. Thus (7.25) is modified to:

$$(1 - \rho L)z_t = \mu_2 DU_t + \beta_2 DT_t + \varepsilon_t \tag{7.40}$$

Following the steps as before, the resulting model is:

$$y_t = (1 - \rho)\mu_1 + \rho\beta_1 + (1 - \rho)\beta_1 t + \rho y_{t-1} + \mu_2 DU_t + \beta_2 DT_t + \varepsilon_t \tag{7.41}$$

This is as (7.29) with the addition of the term $\beta_2 DT_t$ capturing the slope change.

The null hypothesis

The interpretation of the break under the null is of a simultaneous change in the level of the series and in the drift. This is modelled by:

$$(1 - L)z_t = \mu_2 DT_t^b + \beta_2 DU_t + \varepsilon_t \tag{7.42}$$

$$\Rightarrow$$

$$z_t = (1 - L)^{-1}\mu_2 DT_t^b + (1 - L)^{-1}\beta_2 DU_t + (1 - L)^{-1}\varepsilon_t \tag{7.43a}$$

$$= \mu_2 DU_t + \beta_2 DT_t + (1 - L)^{-1}\varepsilon_t \tag{7.43b}$$

$$= \sum\nolimits_{i=0}^{t} \varepsilon_i \qquad\qquad \text{for } t \leq T_b^c \tag{7.43c}$$

$$= \mu_2 + \beta_2(t - T_b^c) + \sum\nolimits_{i=0}^{t} \varepsilon_i \qquad \text{for } t \geq T_b^c + 1 \tag{7.43d}$$

Note that (7.43b) uses $(1 - L)^{-1}DU_t = DT_t$.

Using (7.42) in (7.35), and simplifying, results in IO Model 2 under the null:

$$\Delta y_t = \beta_1 + \mu_2 DT_t^b + \beta_2 DU_t + \varepsilon_t \tag{7.44}$$

Maintained regression

The maintained regression is formed from the union of the regressors in the null and alternative models as follows:

$$\Delta y_t = \kappa_1 + \kappa_2 t + \gamma y_{t-1} + \kappa_3 DT_t^b + \kappa_4 DU_t + \kappa_5 DT_t + \varepsilon_t \tag{7.45}$$

which is as for IO Model 1 with, in addition, $\kappa_4 = \beta_2$ and $\kappa_5 = 0$ under the null, whereas under the alternative $\kappa_4 = \mu_2$ and $\kappa_5 = \beta_2$. Otherwise, the principle is as before of estimating (7.45) by LS and basing test statistics on the LS estimate of γ.

7.2.2.iii IO Model 3

This is a special case of Model 2 so the outline can be brief going straight to the specifications under the alternative and null hypotheses, and then the maintained regression.

Alternative hypothesis

$$y_t = (1 - \rho)\mu_1 + \rho\beta_1 + (1 - \rho)\beta_1 t + \rho y_{t-1} + \beta_2 DT_t + \varepsilon_t \tag{7.46}$$

Null hypothesis

$$\Delta y_t = \beta_1 + \beta_2 DU_t + \varepsilon_t \tag{7.47}$$

Maintained regression

$$\Delta y_t = \kappa_1 + \kappa_2 t + \gamma y_{t-1} + \kappa_4 DU_t + \kappa_5 DT_t + \varepsilon_t \tag{7.48}$$

In the maintained regression, under the null $\kappa_4 = \beta_2$ and $\kappa_5 = 0$, whereas under the alternative $\kappa_4 = 0$ and $\kappa_5 = \beta_2$.

A point to note for later occurs from the comparison of (7.45) and (7.48): the latter differs from the former only by the absence of the term in DT_t^b, a one-off dummy variable which, consequently, has no impact on the asymptotic distribution of the standard test statistics. Perron (1989) suggests modifying the maintained regression to match the case where there is no break under the null hypothesis, thus DU_t is deleted from (7.47) and also from (7.48), resulting in the following modified maintained regression:

$$\Delta y_t = \kappa_1 + \kappa_2 t + \gamma y_{t-1} + \kappa_5 DT_t + \varepsilon_t \tag{7.49}$$

7.3   Development: higher-order dynamics

An important distinction in the theory and application of models with structural breaks concerns the distribution of the impact of the break. A key distinction between the additive outlier and innovation outlier approaches was outlined in the previous section. In this section we consider this distinction further, starting with the problem of how to model the impact of a change in the intercept in the stationary model (that is, under the alternative hypothesis). An important part of the development is the extension of the error dynamics to processes of higher order than AR(1).

7.3.1   Stationary processes (the alternative hypothesis)

In the general form there is a distributed lag on the dummy variable and a distributed lag on the innovation, with associated lag polynomials $\Omega(L)$ and $\Lambda(L)$ respectively. The Model 1 set-up, that is, a shift in the intercept, is considered first.

7.3.1.i   Model 1 set-up

The trend stationary model, that is the model under the alternative hypothesis, is:

$$y_t = y_t^{tr} + z_t \tag{7.50}$$

$$y_t^{tr} = \mu_1 + \beta_1 t + \Omega(L)\mu_2 DU_t \tag{7.51}$$

$$\psi(L)z_t = \theta(L)\varepsilon_t \tag{7.52}$$

The components of the model are as follows. Equation (7.50) states that y_t comprises a trend function, y_t^{tr}, and a stochastic component, z_t. The trend function y_t^{tr} comprises components that are entirely deterministic, these are a constant and a time trend as in the standard case, plus a term that captures the effect of the structural break in the intercept; the latter includes the possibility of a distributed lag to allow a distinction between the short-run and the long-run effects of the break. One possibility is that the lag is redundant, in which case $\Omega(L) = 1$. In any event setting $L = 1$ in $\Omega(L)$ obtains the long-run trend function μ_t, that is:

$$\mu_t = \mu_1 + \beta_1 t + \Omega(1)\mu_2 DU_t \tag{7.53}$$

The model for the stochastic component z_t is a standard ARMA($p + 1$, q) process, with AR polynomial $\psi(L)$ and MA polynomial $\theta(L)$. As we are considering the model under the alternative, it is assumed that $\psi(L)$ can be factored so that $\psi(L) = (1 - \rho L)\varphi(L)$, with $|\rho| < 1$ and the roots of $\varphi(L)$ outside the unit circle. Thus (7.52) can be written as:

$$(1 - \rho L)\varphi(L)z_t = \theta(L)\varepsilon_t \tag{7.54}$$

Below we also require that $\theta(L)$ is invertible.

Invertibility of $\varphi(L)$ implies that z_t can be expressed as an infinite order autoregression, that is:

$$z_t = (1 - \rho L)^{-1}\varphi(L)^{-1}\theta(L)\varepsilon_t \tag{7.54a}$$

$$= \Lambda(L)\varepsilon_t \tag{7.54b}$$

where:

$$\Lambda(L) = (1 - \rho L)^{-1}\varphi(L)^{-1}\theta(L) \text{ and } \Lambda(0) = 1 \tag{7.55}$$

The next step is to substitute y_t^{tr} from (7.51) and z_t from (7.54b) into the equation for y_t, that is (7.50), this results in:

$$y_t = \mu_1 + \beta_1 t + \Omega(L)\mu_2 DU_t + \Lambda(L)\varepsilon_t \tag{7.56}$$

This model is the general one from which three particular cases of interest can be generated. So far $\Omega(L)$ is undefined; in principle, it could be an infinite order polynomial with $\Omega(0) = 1$. Specifically consider the following three possibilities for the specification of $\Omega(L)$:

Additive outlier (AO) : $\Omega(L) = 1$ (7.57a)

Innovation outlier (IO) : $\Omega(L) = \Lambda(L) \neq 1$ (7.57b)

Distributed outlier (DO) : $\Omega(L) \neq \Lambda(L), \Omega(L) \neq 1$ (7.57c)

where, for example, $\Omega(L) = 1$ just indicates a redundant polynomial.

7.3.1.ii   Model 2 set-up

The structural break model is now generalised to the 'mixed model' (Model 2). The trend function (7.51) is replaced by:

$$y_t^{tr} = \mu_1 + \beta_1 t + \Omega(L)\mu_2 DU_t + \Omega(L)\beta_2 DT_t \tag{7.58}$$

As before, three cases can be distinguished as follows.

Additive outlier (AO) : $\quad \Omega(L) = 1$ $\hspace{4cm}$ (7.59a)

Innovation outlier (IO) : $\quad \Omega(L) = \Lambda(L) \neq 1$ $\hspace{3cm}$ (7.59b)

Distributed outlier (DO) : $\quad \Omega(L) \neq \Lambda(L), \Omega(L) \neq 1$ $\hspace{1.5cm}$ (7.59c)

7.3.1.iii   Model 3 set-up

This structural break model is a special case of the last model. The trend function (7.51) is replaced by:

$$y_t^{tr} = \mu_1 + \beta_1 t + \Omega(L)\beta_2 DT_t \hspace{3cm} (7.60)$$

Similarly three cases are distinguished as follows:

Additive outlier (AO) : $\quad \Omega(L) = 1$ $\hspace{4cm}$ (7.61a)

Innovation outlier (IO) : $\quad \Omega(L) = \Lambda(L) \neq 1$ $\hspace{3cm}$ (7.61b)

Distributed outlier (DO) : $\quad \Omega(L) \neq \Lambda(L), \Omega(L) \neq 1$ $\hspace{1.5cm}$ (7.61c)

The following discussion concentrates in detail on the case of Model 1, the intercept shift; Model 2, the intercept and slope shift, follows by extension; and Model 3, the slope shift, is a special case of Model 2.

7.3.2   Higher-order dynamics: AO models

In the AO case the impact of a break is immediate and complete, hence there is no distributed lag response and, therefore, $\Omega(1) = 1$. Consider Model 1, used as an illustration:

$y_t = y_t^{tr} + z_t$ $\quad$ again $\hspace{5.5cm}$ (7.50)

$y_t^{tr} = \mu_1 + \beta_1 t + \mu_2 DU_t$ $\quad$ same short-run and long-run trends $\hspace{1cm}$ (7.62)

$z_t = \Lambda(L)\varepsilon_t$ $\quad$ again $\hspace{5.5cm}$ (7.54b)

Hence, on substituting (7.54b) and (7.62) into (7.50) results in:

$$y_t = \mu_1 + \beta_1 t + \mu_2 DU_t + \Lambda(L)\varepsilon_t \hspace{3cm} (7.63a)$$

$$A(L)(y_t - \mu_1 - \beta_1 t - \mu_2 DU_t) = \varepsilon_t \hspace{3cm} (7.63b)$$

Equation (7.63b) follows by rearranging (7.63a) and noting the definition $A(L) \equiv \Lambda(L)^{-1}$.

For convenience of notation, define $\tilde{y}_t \equiv y_t - \mu_1 - \beta_1 t - \mu_2 DU_t$, which is familiar as the deviation from trend. Also note the Dickey-Fuller decomposition of the lag polynomial $A(L) = (1 - \alpha L) - C(L)(1 - L)$, where $\alpha = 1 - A(1)$. (For details of the DF decomposition, see *UR, Vol. 1*, chapter 3.) Then starting with (7.63b) and using the definitions of $A(L)$ and $\tilde{y}_t$, we have:

$$A(L)\tilde{y}_t = \varepsilon_t \qquad\qquad A(L) \equiv \Lambda(L)^{-1} \tag{7.64}$$

$$[(1-\alpha L) - C(L)(1-L)]\tilde{y}_t = \varepsilon_t \quad \text{using the DF decomposition} \tag{7.65}$$

$$\tilde{y}_t = \alpha\tilde{y}_{t-1} + C(L)\Delta\tilde{y}_t + \varepsilon_t \quad \text{rearranging terms to the r.h.s} \tag{7.66}$$

$$\Delta\tilde{y}_t = \gamma\tilde{y}_{t-1} + C(L)\Delta\tilde{y}_t + \varepsilon_t \qquad \gamma = (\alpha - 1)$$

$$= \gamma\tilde{y}_{t-1} + \sum_{j=1}^{\infty} c_j \Delta\tilde{y}_{t-j} + \varepsilon_t \tag{7.67}$$

In practice, the infinite summation in the $C(L)$ polynomial is truncated usually by a data dependent rule, (see, for example, Ng and Perron, 1996 and *UR, Vol. 1*, chapter 9). Even if the upper limit on the summation is finite, which occurs when the MA polynomial is redundant, in practice the order of the AR polynomial is unknown and the data-dependent rule is still relevant. Also, $\tilde{y}_t$ is replaced by the residuals, $\hat{\tilde{y}}_t$, from a regression of y_t on a constant, a time trend and DU_t, resulting in parameter estimates $\hat{\mu}_1$, $\hat{\beta}_1$ and $\hat{\mu}_2$. Thus (7.67) is approximated by:

$$\Delta\hat{\tilde{y}}_t = \gamma\hat{\tilde{y}}_{t-1} + \sum_{j=1}^{k-1} c_j \Delta\hat{\tilde{y}}_{t-j} + \tilde{\varepsilon}_t \tag{7.68}$$

where $\tilde{\varepsilon}_t$ is a linear combination of ε_t, the error in using the residuals, $\hat{\tilde{y}}_t$, rather than the actual deviations from trend, $\tilde{y}_t$, and the truncation error, if present, in approximating $C(L)$ by a finite order polynomial; note the use of the notation k for the order of the (AR) lag polynomial where a truncation rule has been applied.

Details on estimation and inference for this model are dealt with after specification of the null hypothesis. For the moment the emphasis is on the different ways of modelling the structural break. Finally note, for emphasis, that $y_t^{tr} = \mu_t$ and this is estimated by:

$$\tilde{y}_t^{tr} = \hat{\mu}_1 + \hat{\beta}_1 t + \hat{\mu}_2 DU_t \tag{7.69}$$

Thus, the estimated short-run and long-run effects of the structural break, in this case just an intercept shift, are the same and estimated to be $\hat{\mu}_2$.

The development so far has dealt with Model 1. The specifications of Models 1a, 2 and 3 are straightforward given the development so far: each lead to the same form of the estimating equation, that is (7.68), but differ in the specification of the trend function $y_t^{tr} = \mu_t$. For convenience of reference, they are stated below.

7.3.2.i Summary of models: AO case

AO Model specifications under the alternative of stationarity

Generic AO estimating equation

$$\text{General}: \Delta\hat{\tilde{y}}_t = \gamma\hat{\tilde{y}}_{t-1} + \sum_{j=1}^{k-1} c_j \Delta\hat{\tilde{y}}_{t-j} + \tilde{\varepsilon}_t \tag{7.70}$$

Possible trend specifications

$$\text{Model 1}: \quad y_t^{tr} = \mu_t = \mu_1 + \beta_1 t + \mu_2 DU_t \tag{7.71}$$

$$\text{Model 1a}: \quad y_t^{tr} = \mu_t = \mu_1 + \mu_2 DU_t \tag{7.72}$$

$$\text{Model 2}: \quad y_t^{tr} = \mu_t = \mu_1 + \beta_1 t + \mu_2 DU_t + \beta_2 DT_t \tag{7.73}$$

$$\text{Model 3}: \quad y_t^{tr} = \mu_t = \mu_1 + \beta_1 t + \beta_2 DT_t \tag{7.74}$$

where $\tilde{y}_t = y_t - y_t^{tr}$ is the deviation from trend, and $\hat{\tilde{y}}_t$ is the estimated version of $\tilde{y}_t$ usually obtained as the residuals from a LS regression of y_t on the (deterministic) regressors in y_t^{tr}.

7.3.3 Higher order dynamics: IO models

In the IO case it is assumed that the structural break shock is not distinguished from a normal random shock, thus $\Omega(L) = \Lambda(L)$. Again using Model 1 as an illustration, (7.50) and (7.51) become:

$$y_t = \mu_1 + \beta_1 t + \Lambda(L)(\mu_2 DU_t + \varepsilon_t) \tag{7.75}$$

The resulting model development is then as follows:

$$\Lambda(L)^{-1}(y_t - \mu_1 - \beta_1 t) = \mu_2 DU_t + \varepsilon_t \quad \text{multiply through by } \Lambda(L)^{-1} \tag{7.76}$$

$$A(L)y_t = A(L)(\mu_1 + \beta_1 t) + \mu_2 DU_t + \varepsilon_t \quad A(L) \equiv \Lambda(L)^{-1}, \text{rearrange terms}$$

$$= [A(1)\mu_1 + \beta_1 \sum_{i=1}^{\infty} ia_i] + A(1)\beta_1 t + \mu_2 DU_t + \varepsilon_t \tag{7.77}$$

$$\Rightarrow$$

$$\Delta y_t = [A(1)\mu_1 + \beta_1 \sum_{i=1}^{\infty} ia_i] + A(1)\beta_1 t + \gamma y_{t-1} + \sum_{j=1}^{\infty} c_j \Delta y_{t-j} + \mu_2 DU_t + \varepsilon_t \tag{7.78}$$

The last line uses the DF decomposition of $A(L)$ and $\gamma = (\alpha - 1)$.

In practice the infinite summations in (7.78) are truncated, so that the resulting model is:

$$\Delta y_t = \breve{\mu}_1 + \breve{\beta}_1 t + \gamma y_{t-1} + \sum_{j=1}^{k-1} c_j \Delta y_{t-j} + \mu_2 DU_t + \tilde{\varepsilon}_t \tag{7.79}$$

$$\breve{\mu}_1 = [A(1)\mu_1 + \beta_1 \sum_{i=1}^{k} ia_i] \tag{7.80}$$

$$\breve{\beta}_1 = A(1)\beta_1 \tag{7.81}$$

The error $\tilde{\varepsilon}_t$ in (7.79) is a linear combination of the original innovation ε_t, the trend estimation error and the approximation error induced by truncating the infinite lag polynomials in (7.78).

The next question of interest is how to recover the trend functions for the IO model. In the AO case there is no problem because there is no distinction between the short-run and long-run trend functions, see (7.71)–(7.74). However, in the IO case $\Omega(L) = \Lambda(L)$ and, hence, the trend function embodies a dynamic adjustment to the shock of the structural break. The trend function incorporating a dynamic adjustment to the shock is:

$$y_t^{tr} = \mu_1 + \beta_1 t + \Lambda(L)\mu_2 DU_t \tag{7.82}$$

Note that $\Lambda(L) = A(L)^{-1}$, and that $A(L)$ can be recovered from $C(L)$ and $\gamma = \alpha - 1$; a numerical example is given below. The long-run trend function is:

$$\mu_t = \mu_1 + \beta_1 t + \Lambda(1)\mu_2 DU_t \tag{7.83}$$

where $\Lambda(1) = A(1)^{-1}$. Note that $a_i = c_i - c_{i-1}$, for $i = 2,\ldots,k-1$, and $a_k = -c_{k-1}$ (see the DF decomposition in *UR, Vol. 1*, chapter 3), which means that a_i, for $i = 2,\ldots,k$, can be identified. Further, a_1 can then be identified using $\gamma = (\alpha - 1) = [1 - A(1)] - 1$, which implies that $A(1) = -\gamma$ and $\alpha = 1 + \gamma$, therefore $a_1 = \alpha - \sum_{i=2}^{k} a_i$. From (7.81), $\beta_1 = \tilde{\beta}_1/A(1)$ and (7.80) can be used to obtain μ_1. Lastly, whilst the impact effect of the break is μ_2, which can be obtained directly from (7.79), the long-run effect is $\mu_2\Lambda(1)$, where use is made of $\Lambda(1) = A(1)^{-1}$.

Example

This example is based on a simplified version of the IO(1) model estimated for the log of annual real US GNP (1932–2001), reported in the following chapter. The estimated model is:

$$\Delta y_t = 1.91 + 0.01t - 0.29y_{t-1} + 0.49\Delta y_{t-1} + 0.08DU_t + \hat{\tilde{\varepsilon}}_t \tag{7.84}$$

where $T_b^c = 1937$, which is assumed to be known.

In terms of the previous notation:

$$\hat{\tilde{\mu}}_1 = 1.91; \hat{\tilde{\beta}}_1 = 0.01; \hat{\gamma} = -0.29; \hat{c}_1 = 0.49; \hat{\mu}_2 = 0.08;$$

$$\Rightarrow$$

$$\hat{\alpha} = 1 + \hat{\gamma} = 1 - 0.29 = 0.71;$$

$$\hat{A}(1) = 0.29, \hat{\Lambda}(1) = \hat{A}(1)^{-1} = 1/0.29 = 3.45;$$

$$\hat{\beta}_1 = \tilde{\hat{\beta}}_1/\hat{A}(1) = (0.01/0.29) = 0.0344 \text{ (that is approximately 3.4\% p.a)};$$

$$\hat{a}_2 = -0.49; \hat{a}_1 = 0.71 - (-0.49) = 1.20;$$

$$\hat{\mu}_1 = [1.91 - 0.0344(1.20 + 2(-0.49)]/0.29 = 6.56;$$

$$\hat{\mu}_2\hat{\Lambda}(1) = (0.08/0.29) = 0.276.$$

The estimated A(L) polynomial is:

$$\hat{A}(L) = 1 - 1.2L + 0.49L^2 \tag{7.85}$$

with roots $1.224 \pm 0.737i$, each having an absolute value of 1.429 and implying that $\hat{A}(L)$ can be inverted to obtain $\hat{\Lambda}(L) = \hat{A}(L)^{-1}$; $\hat{\Lambda}(L)$ is an infinite polynomial in L, the first four terms of which are given explicitly in the following empirical version of (7.82) (and see Q7.2 for details of the calculation):

$$y_t^{tr} = 6.56 + 0.0344t + 0.08(1.0DU_t + 1.2DU_{t-1} + 0.95DU_{t-2} + 0.552DU_{t-3}\ldots) \tag{7.86}$$

The estimated long-run trend function is:

$$\hat{\mu}_t = 6.56 + 0.0344t + 0.276DU_t \tag{7.87}$$

$$= 6.56 + 0.0344t \qquad \text{for } t \le t_b$$

$$= 6.84 + 0.0344t \qquad \text{for } t \ge t_b + 1$$

The sum to 10 terms of the expansion of $\hat{\mu}_2\hat{\Lambda}(L)$ is $0.08(3.391) = 0.271$, which is very close to the eventual long-run response of 0.276.

7.3.3.i Summary of models: IO case

For convenience of reference, as in the AO case, the models for four different specifications of the trend function are stated below.

IO model specifications under the alternative of stationarity

Model 1 : $\quad y_t^{tr} = \mu_1 + \beta_1 t + \mu_2 \Lambda(L)DU_t$

$$\mu_t = \mu_1 + \beta_1 t + \mu_2 \Lambda(1)DU_t$$

$$\Delta y_t = \breve{\mu}_1 + \breve{\beta}_1 t + \gamma y_{t-1} + \sum_{j=1}^{k-1} c_j \Delta y_{t-j} + \mu_2 DU_t + \tilde{\varepsilon}_t \tag{7.88}$$

Model 1a : $\quad y_t^{tr} = \mu_1 + \mu_2 \Lambda(L)DU_t$

$$\mu_t = \mu_1 + \mu_2 \Lambda(1)DU_t$$

$$\Delta y_t = \breve{\mu}_1 + \gamma y_{t-1} + \sum_{j=1}^{k-1} c_j \Delta y_{t-j} + \mu_2 DU_t + \tilde{\varepsilon}_t \tag{7.89}$$

Model 2 : $\quad y_t^{tr} = \mu_1 + \beta_1 t + \mu_2 \Lambda(L)DU_t + \beta_2 \Lambda(L)DT_t$

$$\mu_t = \mu_1 + \beta_1 t + \mu_2 \Lambda(1)DU_t + \beta_2 \Lambda(1)DT_t$$

$$\Delta y_t = \breve{\mu}_1 + \breve{\beta}_1 t + \gamma y_{t-1} + \sum_{j=1}^{k-1} c_j \Delta y_{t-j} + \mu_2 DU_t + \beta_2 DT_t + \tilde{\varepsilon}_t \tag{7.90}$$

Model 3 : $y_t^{tr} = \mu_1 + \beta_1 t + \beta_2 \Lambda(L)DT_t$

$\qquad \mu_t = \mu_1 + \beta_1 t + \beta_2 \Lambda(1)DT_t$

$$\Delta y_t = \breve{\mu}_1 + \breve{\beta}_1 t + \gamma y_{t-1} + \sum_{j=1}^{k-1} c_j \Delta y_{t-j} + \beta_2 DT_t + \tilde{\varepsilon}_t \qquad (7.91)$$

Models $1, 2, 3 : \breve{\mu}_1 = [A(1)\mu_1 + \beta_1 \sum_{i=1}^{k} i a_i]; \breve{\beta}_1 = A(1)\beta_1$

Model 1a : $\breve{\mu}_1 = A(1)\mu_1$

7.3.4   Higher-order dynamics: distributed outlier (DO) models

In this case the effect of the shock is distributed over time but not in the same way as an 'ordinary' shock, thus in Model 1 $\Omega(L) \neq \Lambda(L)$; this may be plausible if there are particular characteristics of the shock that distinguish it from 'run-of-the-mill' shocks *and* the effect of the shock is distributed over time.

7.3.4.i   Model 1

The starting point is now:

$y_t = y_t^{tr} + z_t$ $\qquad\qquad\qquad\qquad\qquad$ again $\qquad\qquad$ (7.50)

$y_t^{tr} = \mu_1 + \beta_1 t + \Omega(L)\mu_2 DU_t$ $\qquad\qquad \Omega(L) = A(L)^{-1}B(L)$ $\qquad$ (7.92)

$A(L)z_t = \varepsilon_t$ $\qquad\qquad\qquad\qquad\qquad A(L) = \theta(L)^{-1}\varphi(L)$ $\qquad$ (7.93)

$z_t = A(L)^{-1}\varepsilon_t$ $\qquad\qquad\qquad\qquad\qquad\qquad\qquad\qquad\qquad$ (7.94)

$\Omega(L)\mu_2 DU_t + z_t = A(L)^{-1}B(L)\mu_2 DU_t + A(L)^{-1}\varepsilon_t$ $\qquad$ (7.95)

$B(L) = 1 + b(L)$ is a polynomial that serves to distinguish between the AO and IO cases; an illustration is given below. Using (7.95) and (7.92) in (7.50) results in:

$y_t = \mu_1 + \beta_1 t + \Omega(L)\mu_2 DU_t + A(L)\varepsilon_t$ $\qquad\qquad\qquad\qquad$ (7.96)

$A(L)(y_t - \mu_1 - \beta_1 t) = B(L)\mu_2 DU_t + \varepsilon_t$ $\qquad\qquad\qquad$ (7.97)

Note that if $B(L) = A(L)$ then $\Omega(L) = 1$ and the AO model results; and if $B(L) = 1$ then $\Omega(L) = A(L)^{-1}$ and the IO model results. The DO model thus nests or encompasses both AO and IO models. By following steps as in the IO case, the resulting DO model is:

$$\Delta y_t = \breve{\mu}_1 + \breve{\beta}_1 t + \gamma y_{t-1} + \sum_{j=1}^{k-1} c_j \Delta y_{t-j} + B(L)\mu_2 DU_t + \tilde{\varepsilon}_t \qquad (7.98a)$$

$$= \breve{\mu}_1 + \breve{\beta}_1 t + \gamma y_{t-1} + \sum_{j=1}^{k-1} c_j \Delta y_{t-j} + \mu_2 DU_t + b(L)\mu_2 DU_t + \tilde{\varepsilon}_t \qquad (7.98b)$$

where $B(L) = A(L)\Omega(L) = 1 + b(L)$. The specification (7.98b) differs from (7.78) in involving lags of DU_t, represented in the polynomial $b(L)$, which are redundant when $\Omega(L) = \Lambda(L)$, in which case the IO model results. Model 1a is a special case of (7.98) with the time trend omitted.

In principle, the adequacy of the IO assumption that the lag response to shocks originating from structural breaks and 'normal' random shocks is the same, can be tested by comparing (7.90) and (7.98b); the IO model results if the lag coefficients in $b(L)$ are jointly zero. Possible tests include a misspecification-type test of LM form based on a regression of the residuals from the IO model on the regressors in (7.90) and lags of DU_t; alternatives include Wald type tests based on estimating (7.98b) (see also Vogelsang, 1997), and the use of information criteria. We return to this issue below.

Illustration

An example will serve to illustrate the general principle. In the example $A(L)$ is AR(1) and $B(L)$ is MA(1); thus, $B(L) = 1 - b_1 L$ and $A(L) = (1 - \rho L)$. Note that $b(L) = -b_1 L$ and $a_1 = \rho$. Then the DO model is as follows:

$$(1 - \rho L)(y_t - \mu_1 - \beta_1 t) = (1 - b_1 L)\mu_2 DU_t + \varepsilon_t \tag{7.99}$$

and on multiplying through by $(1 - \rho L)^{-1} \Rightarrow$

$$(y_t - \mu_1 - \beta_1 t) = (1 - \rho L)^{-1}(1 - b_1 L)\mu_2 DU_t + (1 - \rho L)^{-1}\varepsilon_t \tag{7.100}$$

now rearrange in terms of $y_t \Rightarrow$

$$y_t = \mu_1 + \beta_1 t + (1 - \rho L)^{-1}(1 - b_1 L)\mu_2 DU_t + (1 - \rho L)^{-1}\varepsilon_t \tag{7.101}$$

$$(1 - \rho L)y_t = (1 - \rho L)(\mu_1 + \beta_1 t) + (1 - b_1 L)\mu_2 DU_t + \varepsilon_t$$

$$y_t = \rho y_{t-1} + (1 - \rho L)(\mu_1 + \beta_1 t) + (1 - b_1 L)\mu_2 DU_t + \varepsilon_t$$

$$= \rho y_{t-1} + \breve{\mu}_1 + \breve{\beta}_1 t + \mu_2 DU_t - b_1\mu_2 DU_{t-1} + \varepsilon_t \tag{7.102}$$

The last equation is the DO Model 1, which differs from the AO and IO versions of the model.

The trend and long-run trend functions are, respectively:

$$y_t^{tr} = \mu_1 + \beta_1 t + (1 - \rho L)^{-1}(1 - b_1 L)\mu_2 DU_t \tag{7.103}$$

$$\mu_t = \mu_1 + \beta_1 t + [(1 - b_1)/(1 - \rho)]\mu_2 DU_t \tag{7.104}$$

Note that if $b_1 = \rho$ then $B(L) = A(L)$ and the model is AO, not DO, and if $b_1 = 0$ then the model is IO, not DO.

7.3.4.ii   Model 2

In this case y_t^{tr} of (7.51) is replaced by (7.58) so that the complete model is now:

$$y_t = y_t^{tr} + z_t \qquad\qquad \text{again} \qquad\qquad (7.50)$$

$$y_t^{tr} = \mu_1 + \beta_1 t + \Omega(L)\mu_2 DU_t + \Omega(L)\beta_2 DT_t \qquad (7.105)$$

$$z_t = A(L)^{-1}\varepsilon_t \qquad (7.106)$$

Then substituting (7.105) and (7.106) into (7.50) results in:

$$y_t = \mu_1 + \beta_1 t + \Omega(L)\mu_2 DU_t + \Omega(L)\beta_2 DT_t + \Lambda(L)\varepsilon_t \qquad (7.107)$$

Multiplying through by A(L) gives:

$$A(L)y_t = A(L)(\mu_1 + \beta_1 t) + B(L)\mu_2 DU_t + B(L)\beta_2 DT_t + \varepsilon_t \qquad (7.108)$$

Hence, the form of the model to be estimated, with detailed steps as before, is:

$$
\begin{aligned}
\Delta y_t &= \breve{\mu}_1 + \breve{\beta}_1 t + \gamma y_{t-1} + \sum_{j=1}^{k-1} c_j \Delta y_{t-j} + B(L)\mu_2 DU_t + B(L)\beta_2 DT_t + \tilde{\varepsilon}_t \\
&= \breve{\mu}_1 + \breve{\beta}_1 t + \gamma y_{t-1} + \sum_{j=1}^{k-1} c_j \Delta y_{t-j} + \mu_2 DU_t + \beta_2 DT_t \\
&\quad + b(L)\mu_2 DU_t + b(L)\beta_2 DT_t + \tilde{\varepsilon}_t \qquad (7.109)
\end{aligned}
$$

In this case, lags of DU_t and DT_t are included in the specification of the model under the alternative of stationarity; however, the links between the dummy variables imply that (7.109) is not estimated as it is written. Specifically, for example, $DT_t = DU_t + DT_{t-1}$ and $DT_{t-1} = DU_{t-1} + DT_{t-2}$ imply $DT_t = DU_t + DU_{t-1} + DT_{t-2}$. Thus, there are redundancies in the dummy variables and what is required is the (minimal) representation of the space spanned by DU_t, DT_t, $b(L)DU_t$ and $b(L)DT_t$.

Consider the case where A(L) and B(L) are second-order polynomials, so that (7.109) is:

$$
\begin{aligned}
\Delta y_t &= \gamma y_{t-1} + c_1 \Delta y_{t-1} + \breve{\mu}_1 + \breve{\beta}_1 t + \mu_2 DU_t - b_1\mu_2 DU_{t-1} - b_2\mu_2 DU_{t-2} \\
&\quad + \beta_2 DT_t - b_1\beta_2 DT_{t-1} - b_2\beta_2 DT_{t-2} + \tilde{\varepsilon}_t \\
&= \gamma y_{t-1} + c_1 \Delta y_{t-1} + \breve{\mu}_1 + \breve{\beta}_1 t + (\mu_2 + \beta_2)DU_t + [(1-b_1)\beta_2 - b_1\mu_2]DU_{t-1} \\
&\quad - b_2\mu_2 DU_{t-2} + (1 - b_1 - b_2)\beta_2 DT_{t-2} + \varepsilon_t \qquad (7.110)
\end{aligned}
$$

This example also indicates why in this set-up, as in the AO and IO set-ups, the lag polynomial on the structural break dummies is constrained to be the same for intercept and slope breaks. Specifically, there are four, not six, dummy variables in (7.110), which serve to identify the four coefficients: μ_2, β_2, b_1 and b_2.

7.3.4.iii　*Model 3*

DO Model 3 is a special case of (7.109), with the terms in DU_t omitted, that is:

$$\Delta y_t = \breve{\mu}_1 + \breve{\beta}_1 t + \gamma y_{t-1} + \sum_{j=1}^{k-1} c_j \Delta y_{t-j} + \beta_2 DT_t + b(L)\beta_2 DT_t + \tilde{\varepsilon}_t \qquad (7.111)$$

As DU_t is not included in (7.111) there is no 'overlap' in the dummy variables, although it is possible to express the set of DT_t variables in terms of one of its values, and DU_t and its lags (see the discussion following (7.109)).

7.3.4.iv　*Summary of models: DO case*

For convenience of reference the DO models for four different specifications of the trend function are stated below.

DO model specifications under the alternative of stationarity

Model 1 :　$y_t^{tr} = \mu_1 + \beta_1 t + \mu_2 \Omega(L)DU_t$

$\qquad\qquad \mu_t = \mu_1 + \beta_1 t + \mu_2 \Omega(1)DU_t$

$$\Delta y_t = \breve{\mu}_1 + \breve{\beta}_1 t + \gamma y_{t-1} + \sum_{j=1}^{k-1} c_j \Delta y_{t-j} + \mu_2 DU_t + b(L)\mu_2 DU_t + \tilde{\varepsilon}_t \qquad (7.112)$$

Model 1a :　$y_t^{tr} = \mu_1 + \mu_2 \Omega(L)DU_t$

$\qquad\qquad \mu_t = \mu_1 + \mu_2 \Omega(1)DU_t$

$$\Delta y_t = \breve{\mu}_1 + \gamma y_{t-1} + \sum_{j=1}^{k-1} c_j \Delta y_{t-j} + \mu_2 DU_t + b(L)\mu_2 DU_t + \tilde{\varepsilon}_t \qquad (7.113)$$

Model 2 :　$y_t^{tr} = \mu_1 + \beta_1 t + \mu_2 \Omega(L)DU_t + \beta_2 \Omega(L)DT_t$

$\qquad\qquad \mu_t = \mu_1 + \beta_1 t + \mu_2 \Omega(1)DU_t + \beta_2 \Omega(1)DT_t$

$$\Delta y_t = \breve{\mu}_1 + \breve{\beta}_1 t + \gamma y_{t-1} + \sum_{j=1}^{k-1} c_j \Delta y_{t-j} + \mu_2 DU_t + \beta_2 DT_t$$
$$+ b(L)\mu_2 DU_t + b(L)\beta_2 DT_t + \tilde{\varepsilon}_t \qquad (7.114)$$

Model 3 :　$y_t^{tr} = \mu_1 + \beta_1 t + \beta_2 \Omega(L)DT_t$

$\qquad\qquad \mu_t = \mu_1 + \beta_1 t + \beta_2 \Omega(1)DT_t$

$$\Delta y_t = \breve{\mu}_1 + \breve{\beta}_1 t + \gamma y_{t-1} + \sum_{j=1}^{k-1} c_j \Delta y_{t-j} + \beta_2 DT_t + b(L)\beta_2 DT_t + \tilde{\varepsilon}_t \qquad (7.115)$$

7.4 AO or IO?

The question of which type of structural break model, AO or IO, should be estimated has received relatively minor attention in an otherwise voluminous literature. Vogelsang and Perron (1998) note from their simulation results that size and power for the AO model when the data are generated by an IO model are similar to the case where data are generated by an AO model, and vice versa. However, due to problems with the IO model, but not the AO model, when there is a break under the null (as well as the alternative), they recommend using the AO model if large slope shifts are suspected under the null. The framework outlined above offers a way of distinguishing between these models. The following discussion takes Model 1 as the illustrative case; the method for other models is a simple extension of the argument.

7.4.1 DO to AO

A view on the AO/IO distinction follows from (7.63), which is now re-arranged as if it was in IO form, by taking the terms in DU_t to the right-hand side:

$$A(L)(y_t - \mu_1 - \beta_1 t) = A(L)\mu_2 DU_t + \varepsilon_t \tag{7.116}$$

After some manipulation, as before, the resulting model is:

$$\Delta y_t = \breve{\mu}_1 + \breve{\beta}_1 t + \gamma y_{t-1} + \sum_{j=1}^{k-1} c_j \Delta y_{t-j} + A(L)\mu_2 DU_t + \tilde{\varepsilon}_t \tag{7.117}$$

Now compare this with the DO specification of Model 1, for convenience restated here:

$$\Delta y_t = \breve{\mu}_1 + \breve{\beta}_1 t + \gamma y_{t-1} + \sum_{j=1}^{k-1} c_j \Delta y_{t-j} + B(L)\mu_2 DU_t + \tilde{\varepsilon}_t \quad \text{again} \tag{7.98a}$$

Comparing (7.98a) and (7.117) it is apparent that if $B(L) = A(L)$, which follows from $\Omega(L) = 1$, then the AO model results; in effect it is these restrictions that allow the dummy variable DU_t to be taken into the trend function, y_t^{tr}, and thus the deviation from trend can be estimated in a prior regression. On the other hand if $B(L) = 1$, which follows from $\Omega(L) = A(L)$, then the IO model results.

The restrictions could be tested using an F or LR test, but as we shall see, there are some complications. For the moment we outline the principle. As noted, the test that the AO model is a valid reduction of a DO model corresponds to restricting $B(L) = A(L)$, where $A(L)$ refers to an autoregressive lag polynomial, which is either finite by specification or a finite approximation. Thus, the order of $B(L)$ is implicitly constrained to be the same as the order of $A(L)$; we assume that this restriction is valid (if the order of $B(L)$ exceeds that of $A(L)$, an AO reduction is invalid, without the need for further testing).

It is easier to work with (7.98a) in its AR, not ADF, form, thus:

$$y_t = \breve{\mu}_1 + \breve{\beta}_1 t + \sum_{j=1}^{k} a_j y_{t-j} + \mu_2 DU_t - \mu_2 \sum_{j=1}^{k} b_j DU_{t-j} + \tilde{\varepsilon}_t \qquad (7.118)$$

The AO restrictions are that the coefficient on the j-th lag of DU_t is (the negative of) the product of the coefficient on the j-th lag of y_t and the coefficient on DU_t, for $j = 1, \ldots, k$. The resulting F statistic is:

$$F_2 = \frac{[\hat{\tilde{\varepsilon}}_t(AO)'\hat{\tilde{\varepsilon}}_t(AO) - \hat{\tilde{\varepsilon}}_t(DO)'\hat{\tilde{\varepsilon}}_t(DO)]}{\hat{\tilde{\varepsilon}}_t(DO)'\hat{\tilde{\varepsilon}}_t(DO)} \frac{(T^+ - k^+)}{k} \qquad (7.119)$$

where $T^+ = T - k$, $k^+ = 2k + 3$, and $\hat{\tilde{\varepsilon}}_t(.)'\hat{\tilde{\varepsilon}}_t(.)$ is the residual sum of squares for the model indicated within parentheses. The notation F_1 is used for the corresponding F type test where the time trend is excluded, that is, Model 1a.

7.4.2 DO to IO

The IO model is obtained from (7.118) by imposing the restrictions $b_j = 0$ for $j = 1, \ldots, k$. For $\mu_2 \neq 0$, (and if $\mu_2 = 0$ there is no structural break) these restrictions can be tested by testing the joint hypothesis that the coefficients on DU_{t-j}, $j = 1, \ldots, k$, are zero. The corresponding F test is denoted F_4 and defined as follows:

$$F_4 = \frac{[\hat{\tilde{\varepsilon}}_t(IO)'\hat{\tilde{\varepsilon}}_t(IO) - \hat{\tilde{\varepsilon}}_t(DO)'\hat{\tilde{\varepsilon}}_t(DO)]}{\hat{\tilde{\varepsilon}}_t(DO)'\hat{\tilde{\varepsilon}}_t(DO)} \frac{(T^+ - k^+)}{k} \qquad (7.120)$$

where $\hat{\tilde{\varepsilon}}_t(IO)'\hat{\tilde{\varepsilon}}_t(IO)$ is the residual sum of squares from the IO model. The notation F_3 is used for the corresponding F-type test where the time trend is excluded, that is, Model 1a.

7.4.3 Distributions of test statistics

The question now arises as to the distribution of the F-type statistics in this context. It is known that in the case of nonlinear hypotheses and the presence of lagged dependent variables amongst the regressors, F_2, for example, will only be approximately distributed as $F(k, T^+ - k^+)$. What is of importance for accurate inference is how close one can get to the unit root whilst maintaining a reasonable degree of fidelity of the actual size of the test to its nominal size. In practice, a further problem, not considered yet, is that the break date is generally unknown; some solutions to this problem are considered in the section below and in the next chapter; see also Vogelsang (1997) and Andrews and Ploberger (1994).

To illustrate the problem, the simple first order case was simulated. In unrestricted and restricted forms, respectively, these are:

DO form

$$y_t = a_1 y_{t-1} + \breve{\mu}_1 + \breve{\beta}_1 t + \mu_2 DU_t - b_1 \mu_2 DU_{t-1} + \varepsilon_t \tag{7.121}$$

AO form

$$y_t = a_1 y_{t-1} + \breve{\mu}_1 + \breve{\beta}_1 t + \mu_2 DU_t - a_1 \mu_2 DU_{t-1} + \varepsilon_t \tag{7.122}$$

IO form

$$y_t = a_1 y_{t-1} + \breve{\mu}_1 + \breve{\beta}_1 t + \mu_2 DU_t + \varepsilon_t \tag{7.123}$$

where, in (7.121–7.123), $a_1 = \rho$. (There is no approximation error so the notation uses ε_t rather than $\tilde{\varepsilon}_t$.)

First, two cases are considered corresponding to the reduction of the DO model to AO using the test statistics F_1 and F_2 respectively, in the no-trend specification (Model 1a omits the time trend from (7.117) and (7.118)) and the with trend specification (Model 1). The second case is the reduction of the DO model to IO using the test statistics F_3 and F_4. In the simulations $T = 100$, $\mu_1 = 7$, $\mu_2 = 3$, $\beta_1 = 0.05$; $\varepsilon_t \sim niid(0, 1)$ and, therefore, the intercept shift is 3 in units of the standard error of the innovation; the results are based on 5,000 replications. The break date is assumed known and taken to be $\lambda_b^c = 0.5$.

A key parameter is ρ, with the unit root approached as $\rho \to 1$. The impact of ρ is shown in Figure 7.1, which graphs the 90% and 95% quantiles of the simulated distributions of the test statistics F_1 and F_2 for $\rho \in [0, 1)$; also graphed for comparison are the 90% and 95% quantiles of the F distribution. In the case of F_1, the simulated and correct F quantiles are practically close for $\rho < 0.7$ but start to diverge thereafter, becoming quite noticeably divergent for $\rho > 0.8$; a similar comparison for F_2 suggests that the divergence occurs earlier being noticeable for $\rho > 0.7$. As the unit root is approached the simulated quantiles increase; this implies that the AO restriction would be rejected too often using the F quantiles. This finding is not surprising since (as shown in UR, Vol. 1, chapter 5) the quantiles of the empirical distribution function for a standard t-type test in an AR(1) model, with $T = 100$, do not match those of the appropriate t distribution well before ρ approaches the unit root.

Figure 7.2 shows the 90% and 95% quantiles of the simulated distributions of the test statistics F_3 and F_4 for $\rho \in [0, 1)$, together with the 90% and 95% quantiles of the F distribution. The departure of the simulated and F quantiles, and hence the implied actual size from the nominal size, is not as marked as in the cases of F_1 and F_2, in particular the quantiles do not increase (almost) monotonically as the unit root is approached; however, the simulated quantiles are generally above the F quantiles.

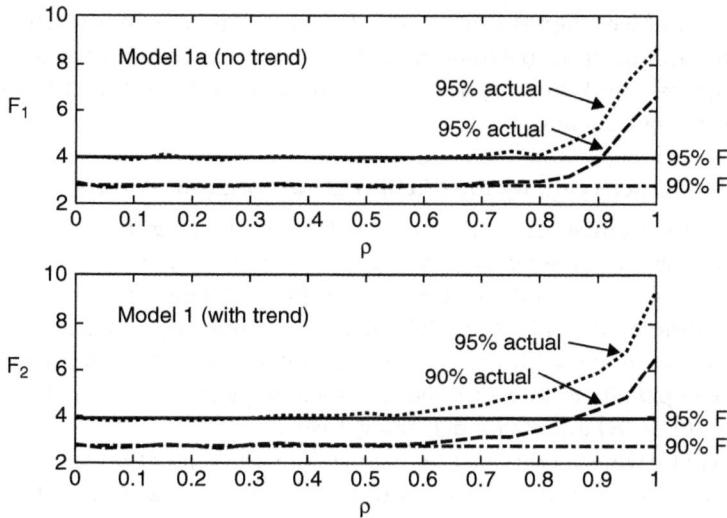

Figure 7.1   F test of AO restriction

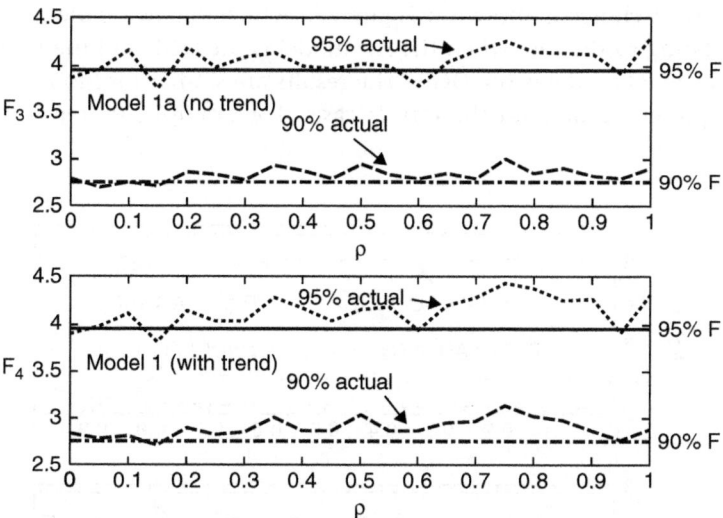

Figure 7.2   F test of IO restriction

Given the dependence of the quantiles on ρ, there are two (related) possibilities that should achieve a somewhat better fidelity of actual and nominal size, given that many macroeconomic times series exhibit persistence, so that the relatively interesting region of the parameter space is $\rho \in [0.6, 1)$. One possibility is to simulate the quantiles based on an estimate, or, preferably, a bias – adjusted estimate, of ρ (see chapter 4 of *UR, Vol. 1*). The second possibility, which builds upon this approach, is to extend this procedure and bootstrap the test statistic (see chapters 5 and 8 of *UR, Vol. 1*); this procedure is illustrated below in an example using US real GNP.

7.4.4   Model selection by log-likelihood/information criteria

Note that the number of restrictions involved in reducing the DO model to IO is the same as the number in reducing the DO model to an AO model. Hence, choosing between AO and IO models on the basis of standard information criteria, for example AIC and BIC, which impose a 'complexity' penalty, is, in effect, to choose the model with the greater maximised log-likelihood (max-LL). The complexity penalty is, however, relevant for a pairwise comparison between the DO and AO or DO and IO models respectively.

To illustrate whether this is a useful procedure, the simulation set-up described in the previous section was also used, first to examine whether AIC and BIC could discriminate between DO and AO, and between DO and IO, and second to compare AO and IO directly using the maximum of the LL. As to the first of these assessments, starting from the DO model, the probability of correct and incorrect choice is shown in Figure 7.3, which has two panels, the upper panel relating to the case when the AO model is the DGP and the lower panel to when the IO model is the DGP. (The results are similar for Models 1 and 1a and are presented here for the former case.) For example, P(AO| AO, AIC), is

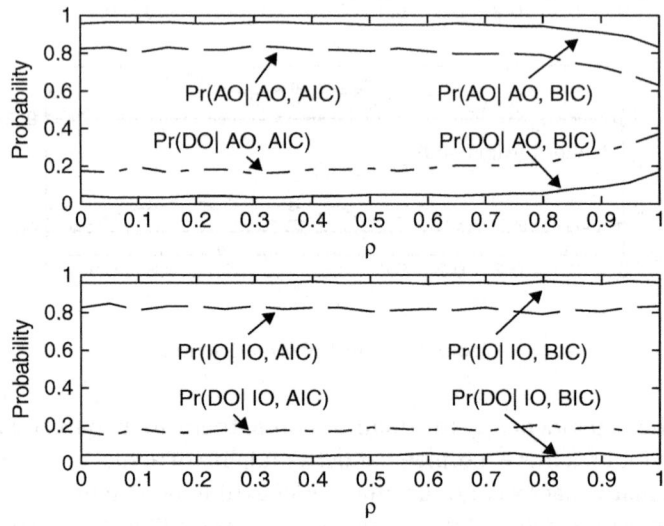

Figure 7.3   Selection by AIC/BIC (relative to DO)

the probability (relative frequency in the simulations) of choosing AO by AIC when AO is the DGP; and P(DO|AO, AIC) is the complementary probability of choosing DO by AIC when AO is the DGP. For a given criterion these two probabilities sum to unity.

It is clear from both panels that BIC is better than AIC at making the correct choice. For example, when the DGP is the AO model, (see the upper panel) the probability of the correct choice for BIC is close to 1 except as $\rho \to 1$, and even then the probability does not fall below 0.8. In contrast the probability of correct choice by AIC is always below that for BIC. If the DGP is the IO model, the probability of correct choice by BIC is even higher at around 0.96 throughout the range $\rho \in [0, 1]$; and again the probability of correct choice by AIC is always below that for BIC.

Figure 7.4 shows the situation when the choice is by max-LL between AO and IO. Again there are two panels, with the upper and lower panels, respectively referring to the AO and IO models being the DGP. In both cases the probability of correct and incorrect choice is ½ for ρ close to zero. (The additional term in the AO model is $-\rho \mu_2 DU_{t-1}$ which $\to 0$ as $\rho \to 0$ and, thus, the models become indistinguishable.) If the AO model is the DGP, it is more likely to be selected as $\rho \to 1$, but may be incorrectly selected with a probability not below 0.2 for $\rho < 0.7$ and nowhere in $\rho \in [0, 1]$ is the probability of correct choice equal to unity. The situation is markedly better if the DGP is the IO model; then the probability of incorrect selection falls to zero for $\rho \in [0.6, 1]$.

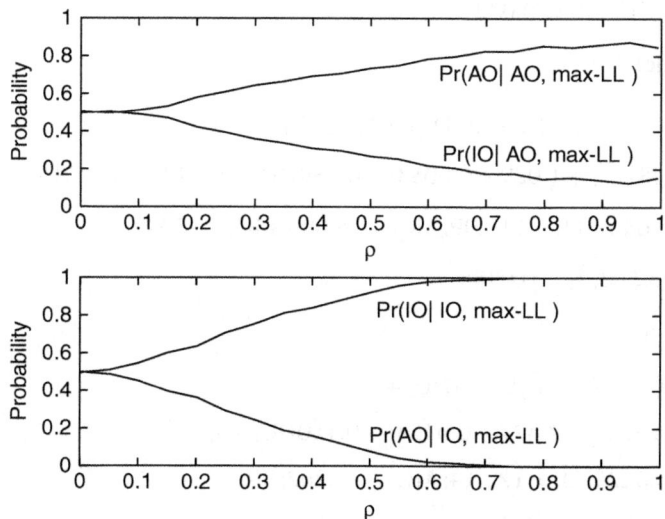

Figure 7.4   Selection by max-LL, AO and IO models

7.5   Example AO(1), IO(1) and DO(1)

To illustrate, an example of the various models is based on the estimation results reported in the next chapter relating to annual data on the log of real US GNP; the effective sample period comprises $69(= T^+ = T - k)$ observations, 1934–2001, which is small enough to suggest that some care should be taken with the critical values. The aim of the illustration is to assess the corresponding AO and IO models relative to their immediate generalisation as a DO model. Model 1 is used, although a more complete study would also have to assess whether this choice is justified (this issue is dealt with in the next chapter). To simplify we assume that the break date is 1940 (this date is in fact suggested for the AO model by selecting the break date that maximises the evidence of a break based on the t statistic on μ_2).

7.5.1   Estimation

The models and their empirical counterparts are as follows:

DO model

$$y_t = a_1 y_{t-1} + \breve{\mu}_1 + \breve{\beta}_1 t + \mu_2 DU_t - b_1 \mu_2 DU_{t-1} + \varepsilon_t \tag{7.124}$$

$$= 0.637 y_{t-1} + 2.358 + 0.0119t + 0.163 DU_t - 0.132(0.163) DU_{t-1} + \hat{\varepsilon}_t$$

$$\hat{\sigma}_\varepsilon = 0.0427; LL = 122.24; AIC = -6.236; BIC = -6.074$$

$$\Rightarrow \hat{\mu}_1 = 6.438, \hat{\beta}_1 = 0.0327$$

AO model

$$y_t = a_1 y_{t-1} + \breve{\mu}_1 + \breve{\beta}_1 t + \mu_2 DU_t - a_1 \mu_2 DU_{t-1} + \varepsilon_t \tag{7.125}$$

$$= 0.848 y_{t-1} + 1.029 + 0.0053t + 0.159 DU_t - 0.848(0.159) DU_{t-1} + \hat{\varepsilon}_t$$

$$\hat{\sigma}_\varepsilon = 0.0457; LL = 117.09; AIC = -6.086; BIC = -5.986$$

$$\Rightarrow \hat{\mu}_1 = 6.568, \hat{\beta}_1 = 0.0348$$

IO model

$$y_t = a_1 y_{t-1} + \breve{\mu}_1 + \breve{\beta}_1 t + \mu_2 DU_t + \varepsilon_t \tag{7.126}$$

$$= 0.619 y_{t-1} + 2.477 + 0.0124t + 0.150 DU_t + \hat{\varepsilon}_t$$

$$\hat{\sigma}_\varepsilon = 0.0425; LL = 122.14; AIC = -6.262; BIC = -6.132$$

$$\Rightarrow \hat{\mu}_1 = 6.444, \hat{\beta}_1 = 0.0327$$

First some comments on the estimated models. The AO model has been written to emphasise that the coefficient on DU_{t-1} is $-a_1\mu_2$ (note here that $a_1 = \rho$); it is then written in the more familiar (estimated) deviation from trend form. The estimate of b_1 from the DO model is $\hat{b}_1 = 0.132$ compared to $\hat{a}_1 = 0.848$ from the AO model; this difference and the reduction in the log-likelihood (LL) suggests, at an 'eyeballing' level, that the AO model is an invalid reduction. The test statistic for this reduction is $F_2 = 10.30$, which at face value level strongly suggests rejection of the reduction from DO to AO.

The other comparison of interest is between the DO and IO models. At an informal level, note that the empirical differences between the two models are slight, with minor differences in the comparable coefficients. The difference in max-LL is slight and the test statistic is $F_4 = 0.18$, indicating that the reduction from DO to IO is supported. Also of interest are the values of BIC, which lead to selection of IO relative to DO but DO relative to AO; this result implies that, in terms of max-LL, AO is inferior to IO, confirming the choice of IO indicated by the F tests.

7.5.2 Bootstrapping the critical values

Although the conclusion here is clearly in favour of the IO model, it is nevertheless of interest to bootstrap the test statistics, partly for the principle of the procedure. In the bootstrap procedure for F_2, the estimated coefficients and residuals, $\hat{\varepsilon}_t$, from the AO model (7.125) form the basis of the bootstrap data generation process. Random draws are taken from the residuals resulting in the sequence $\{e_t\}_1^T$, with equal probability of occurrence. Within each bootstrap run, with data generated under the AO null, the unrestricted model – the DO model – is estimated and the test statistic F_2 is calculated. The bootstrap runs are replicated B times (here B = 4,999), each value of F_2 is saved and then the B values are sorted to obtain the empirical (bootstrap) distribution of F_2. It is this distribution that provides the quantiles of interest, in this case the 90% and 95% quantiles. To obtain the bootstrap distribution of F_4, the data are generated from estimates of the IO model as the null hypothesis; otherwise the practice is as for F_2. Bootstrap p-values given a particular realisation of the test statistic can also be calculated within the routine. For example, suppose the realised value of F_2 is F_2, then count the number of times that F_2 is exceeded in the bootstrap distribution, say $\# F_2$, and then normalise $\# F_2$ by the number of bootstrap replications to obtain the bootstrap p-value for F_2.

On this basis the bootstrapped 90% and 95% quantiles for F_2 are 4.20 and 5.78 respectively, and for F_4 they are 3.08 and 5.33 respectively. By comparison the 90% and 95% quantiles of the F(1, 64) distribution are 2.78 and 3.99. This comparison answers two questions. First, by comparison with the bootstrap quantiles, the realised values of the test statistics $F_2 = 10.3$ and $F_4 = 0.18$ firmly suggest rejection of the AO specification in favour of the IO specification.

Second, the bootstrap quantiles are sufficiently different from the quantiles of the F distribution to suggest that the latter should not be used as a matter of routine. The bootstrap p-values are, as could be anticipated in this case, zero for $F_2 = 10.3$ and (virtually) one for $F_4 = 0.18$. In practice, such clear-cut decisions are unlikely and the bootstrap p-values offer a way of distinguishing between models.

7.6   The null hypothesis of a unit root

The discussion so far has assumed that a stationary process generates the data. In this section the null hypothesis of a unit root and, hence, nonstationarity is introduced. There are two situations of interest that have been considered in the literature. In Perron's (1989) seminal article a break was allowed under the unit root null; other authors, notably Zivot and Andrews (1992), did not allow a break under the null in the context of IO models; Vogelsang and Perron (1998) take up issues related to whether there is a break under the null.

7.6.1   A break under the null hypothesis

To each of the three specifications under the assumption of stationarity there is a corresponding specification of $\Omega(L)$ under the null of a unit root. The three cases are particular cases of the following specifications:

Model 1

$$y_t = \mu_1 + \beta_1 t + \Omega(L)\mu_2 D1_t + \Lambda(L)\varepsilon_t \tag{7.127}$$

Model 2

$$y_t = \mu_1 + \beta_1 t + \Omega(L)\mu_2 D1_t + \Omega(L)\beta_2 D2_t + \Lambda(L)\varepsilon_t \tag{7.128}$$

Model 3

$$y_t = \mu_1 + \beta_1 t + \Omega(L)\beta_2 D2_t + \Lambda(L)\varepsilon_t \tag{7.129}$$

where $D1_t$ and $D2_t$ indicate dummy variables particular to the model. General to all models is $\Lambda(L) = (1 - L)^{-1}\varphi(L)^{-1}\theta(L)$, where $(1 - L)^{-1}$ is the summation operator. Note that the unit root is imposed in $\Lambda(L)$.

7.6.2   AO with unit root

The additive outlier models with a unit root and break under the null have the same general form $\Delta\tilde{y}_t = a^*(L)\Delta\tilde{y}_t + \varepsilon_t$, where $\tilde{y}_t$ is the deviation from trend and $a^*(L)$ is a lag polynomial.

7.6.2.i *AO Model 1*

In the case of the additive outlier model $\Omega(L) = 1$ and $D1_t = DU_t$. These imply:

$$y_t = \mu_1 + \beta_1 t + \mu_2 DU_t + z_t \tag{7.130}$$

$$(1 - L)z_t = \Lambda^*(L)\varepsilon_t \qquad \Lambda^*(L) = \varphi(L)^{-1}\theta(L) \tag{7.131}$$

$$(1 - L)\tilde{y}_t = \Lambda^*(L)\varepsilon_t \qquad \text{substituting for } z_t \text{ from (7.130)}$$

$$\tilde{y}_t = y_t - \mu_1 - \beta_1 t - \mu_2 DU_t$$

$$\Rightarrow A^*(L)\Delta\tilde{y}_t = \varepsilon_t \qquad A^*(L) \equiv \Lambda^*(L)^{-1}$$

$$\Rightarrow \Delta\tilde{y}_t = a^*(L)\Delta\tilde{y}_t + \varepsilon_t \qquad a^*(L) \equiv A^*(L) - 1 \tag{7.132}$$

The unit root has been imposed in (7.131) and, consequently, only terms in $\Delta\tilde{y}_t$ appear in (7.132), where $\tilde{y}_t$ is the deviation from trend.

7.6.2.ii *AO Model 2*

In this case $\Omega(L) = 1$, $D1_t = DU_t$ and $D2_t = DT_t$, and following the steps as for Model 1 this gives:

$$y_t = \mu_1 + \beta_1 t + \mu_2 DU_t + \beta_2 DT_t + z_t \tag{7.133}$$

$$\Rightarrow \Delta\tilde{y}_t = a^*(L)\Delta\tilde{y}_t + \varepsilon_t \tag{7.134}$$

where $\tilde{y}_t = y_t - \mu_1 - \beta_1 t - \mu_2 DU_t - \beta_2 DT_t$.

7.6.2.iii *AO Model 3*

This is a special case of Model 2 with $\mu_2 = 0$, $\Omega(L) = 1$ and $D2_t = DT_t$. The general form is as in (7.132) and (7.134) leading to:

$$\Delta\tilde{y}_t = a^*(L)\Delta\tilde{y}_t + \varepsilon_t \tag{7.135}$$

where $\tilde{y}_t = y_t - \mu_1 - \beta_1 t - \beta_2 DT_t$.

7.6.3 IO with unit root

7.6.3.i *IO Model 1*

In the IO model, $\Omega(L) = \Lambda(L)$ and $D1_t = DT_t^b$. The specification under the null with a break is:

$$y_t = \mu_1 + \beta_1 t + \Lambda(L)(\mu_2 DT_t^b + \varepsilon_t) \tag{7.136}$$

Multiplying (7.136) through by $(1 - L)$ to cancel $(1 - L)^{-1}$ gives:

$$\Delta y_t = \beta_1 + \Lambda^*(L)(\mu_2 DT_t^b + \varepsilon_t) \tag{7.137}$$

$$A^*(L)\Delta y_t = A^*(1)\beta_1 + \mu_2 DT_t^b + \varepsilon_t \qquad \text{multiply through by } A^*(L) \tag{7.138}$$

$$\Rightarrow \Delta y_t = a^*(L)\Delta y_t + A^*(1)\beta_1 + \mu_2 DT_t^b + \varepsilon_t \qquad \text{rearrange (7.138)} \tag{7.139}$$

In the absence of a break under the null, the term in DT_t^b is omitted from (7.139).

7.6.3.ii   IO Model 2

In IO Model 2, $\Omega(L) = \Lambda(L)$, $D1_t = DT_t^b$ and $D2_t = DU_t$. The specification under the null with a break is:

$$y_t = \mu_1 + \beta_1 t + \Lambda(L)(\mu_2 DT_t^b + \beta_2 DU_t + \varepsilon_t) \tag{7.140}$$

$$\Delta y_t = \beta_1 + \Lambda^*(L)(\mu_2 DT_t^b + \beta_2 DU_t + \varepsilon_t) \tag{7.141}$$

$$A^*(L)\Delta y_t = A^*(1)\beta_1 + \mu_2 DT_t^b + \beta_2 DU_t + \varepsilon_t \tag{7.142}$$

$$\Rightarrow \Delta y_t = a^*(L)\Delta y_t + A^*(1)\beta_1 + \mu_2 DT_t^b + \beta_2 DU_t + \varepsilon_t \tag{7.143}$$

The end result is simple enough: it is clear from (7.143) that under the null there is a step change captured by $\mu_2 DT_t^b$ and a change in the drift captured by $\beta_2 DU_t$; the IO specification does not include lags of these variables. If there is no break under the null, then the terms in DT_t^b and DU_t are not present in (7.143).

7.6.3.iii   IO Model 3

This is the special case of IO Model 2 with $\mu_2 = 0$. The resulting specification is:

$$y_t = \mu_1 + \beta_1 t + \Lambda(L)(\beta_2 DU_t + \varepsilon_t) \tag{7.144}$$

This leads to:

$$\Delta y_t = a^*(L)\Delta y_t + A^*(1)\beta_1 + \beta_2 DU_t + \varepsilon_t \tag{7.145}$$

Under the null there is just a change in the drift captured by $\beta_2 DU_t$, which is absent if there is no break under the null.

7.6.4   DO with unit root

The idea of the distributed outlier models is that, whilst they are of interest in their own right, they include the AO and IO models as special cases.

7.6.4.i   DO Model 1

In this model $\Omega(L) \neq \Lambda(L)$, $\Omega(L) \neq 1$ and $D1_t = DT_t^b$; it is the first of these conditions that serves to distinguish the DO model from the AO and IO models. The specification under the null with a break is:

$$y_t = \mu_1 + \beta_1 t + \Omega(L)\mu_2 DT_t^b + \Lambda(L)\varepsilon_t \tag{7.146}$$

Define g(L) implicitly as follows:

$$\Omega(L) = (1 - L)^{-1}[\varphi(L)^{-1}\theta(L) + g(L)]$$

$$= \Lambda(L) + (1 - L)^{-1}g(L)]$$

and then substitute for $\Omega(L)$ in (7.146):

$$y_t = \mu_1 + \beta_1 t + \mu_2(1-L)^{-1}g(L)DT_t^b + \Lambda(L)(\mu_2 DT_t^b + \varepsilon_t) \tag{7.147}$$

$$\Delta y_t = \beta_1 + \mu_2 g(L)DT_t^b + \Lambda^*(L)(\mu_2 DT_t^b + \varepsilon_t) \tag{7.148}$$

on multiplying (7.147) through by $(1-L)$

$$\Delta y_t = a^*(L)\Delta y_t + A^*(1)\beta_1 + \mu_2 DT_t^b + \mu_2 h(L)DT_t^b + \varepsilon_t \tag{7.149}$$

where $h(L) = A^*(L)g(L)$. The final specification (7.149) differs from (7.139) in including lags of DT_t^b through a nondegenerate $h(L)$, and differs from the corresponding AO model in that $h(L)$ differs from $A^*(L)$. All terms in DT_t^b are omitted if there is no break under the null.

7.6.4.ii DO Model 2
In DO Model 2 $\Omega(L) \neq \Lambda(L)$, $\Omega(L) \neq 1$; also, as in the IO case, $D1_t = DT_t^b$ and $D2_t = DU_t$. The specification under the null with a break is:

$$y_t = \mu_1 + \beta_1 t + \Omega(L)\mu_2 DT_t^b + \Omega(L)\mu_2 DU_t + \Lambda(L)\varepsilon_t \tag{7.150}$$

Define $g(L)$ implicitly as following (7.146):

$$\Omega(L) = (1-L)^{-1}[\varphi(L)^{-1}\theta(L) + g(L)]$$
$$= \Lambda(L) + (1-L)^{-1}g(L)] \tag{7.151}$$

Substituting for $\Omega(L)$ in (7.150) gives:

$$y_t = \mu_1 + \beta_1 t + \Lambda(L)(\mu_2 DT_t^b + \beta_2 DU_t + \varepsilon_t) + \mu_2(1-L)^{-1}g(L)DT_t^b$$
$$+ \beta_2(1-L)^{-1}g(L)DU_t \tag{7.152}$$

and rearranging in the manner of previous models:

$$\Delta y_t = a^*(L)\Delta y_t + A^*(1)\beta_1 + \mu_2 DT_t^b + \mu_2 h(L)DT_t^b + \beta_2 DU_t$$
$$+ \beta_2 h(L)DU_t + \varepsilon_t \tag{7.153}$$

where $h(L) = A^*(L)g(L)$. Comparing the resulting model with the IO specification, it is apparent that, as in the specification under stationarity, lags of the dummy variables, not just the current values, are included. However, note that, as indicated in the discussion following (7.109), there are redundancies among the dummy variables, reflecting the number of coefficients in $h(L)$ plus μ_2 and

β_2. In particular note that $DU_{t-1} = DU_t - DT_t^b$, $DU_{t-2} = DU_{t-1} - DT_{t-1}^b = DU_t - DT_t^b - DT_{t-1}^b$ and so on, implying $DU_{t-i} = DU_t - (DT_t^b + DT_{t-1}^b + \cdots + DT_{t-(i-1)}^b)$. Hence, it is only necessary to include DU_t, and DT_t^b and its lags through to DT_{t-H}^b, where H is the longest lag in $h(L)$, a total of $H + 2$ variables. If there is no break under the null then all terms in DT_t^b and DU_t are omitted.

7.6.4.iii DO Model 3

As in the IO case, this model is the special case obtained from Model 2 by omitting the terms in DT_t^b. The resulting model is:

$$\Delta y_t = a^*(L)\Delta y_t + A^*(1)\beta_1 + \beta_2 DU_t + \beta_2 h(L)DU_t + \varepsilon_t \tag{7.154}$$

This differs from the corresponding IO model by including lags of DU_t. All terms in DU_t are omitted if there is no break under the null.

7.7 Testing framework for a unit root

If it is known which regime, the null or the alternative, generated the data the next step is to estimate the appropriate model. However, usually the critical problem is to first assess whether the null hypothesis of a unit root is consistent with the data relative to the stationary alternative. The approach adopted by Perron (1989) is to extend the DF method to estimate a maintained model that allows either of the specifications, the null or the alternative, to be favoured within a hypothesis-testing framework. In this approach, the maintained model includes the specifications under the null and alternative as special cases.

7.7.1 The AO models

The standard procedure is to estimate a maintained regression for the AO models in deviation from trend form, without the restriction that reduces the model to the unit root case, and then test whether the null hypothesis of a unit root can be rejected. Thus consider again the 'generic' AO estimating equation, initially introduced in the context of the stationary model, and reproduced here for ease of reference:

$$\Delta \hat{\tilde{y}}_t = \gamma \hat{\tilde{y}}_{t-1} + \sum_{j=1}^{k-1} c_j \Delta \hat{\tilde{y}}_{t-j} + \tilde{\varepsilon}_t \tag{7.155}$$

where $\hat{\tilde{y}}_t$ is the LS estimate of the deviation from trend, and the trend specifications depend on the particular model specification; see (7.71)–(7.74). The unit root null hypothesis is $H_0: \gamma = 0$, whereas, as usual, the alternative hypothesis is $H_A: \gamma < 0$.

Perron (1989) initially suggested (7.155) as the basis for inference; however as he pointed out later (see Perron and Vogelsang, 1992a, 1992b), the regression needs a minor modification for AO Models 1 and 2 to ensure that the pseudo-t statistic is not dependent on nuisance parameters. Specifically, the required maintained regression is:

AO Models 1, 2

$$\Delta \hat{\tilde{y}}_t = \gamma \hat{\tilde{y}}_{t-1} + \sum_{j=1}^{k-1} c_j \Delta \hat{\tilde{y}}_{t-j} + \sum_{j=0}^{k-1} \delta_j DT_{t-j}^b + \tilde{\varepsilon}_t \tag{7.156}$$

The maintained regression for AO Models 1 and 2 thus includes current and lagged values of DT_t^b to the same order as the lag on $\Delta \hat{\tilde{y}}_t$. In the case of AO Model 3, equation (7.155) is used.

The pseudo-t test statistics for the AO models are denoted $\tilde{\tau}_\gamma^{(i)}(AO)$ for $i = 1, 2, 3$; implicitly each is a function of λ_b^c or λ_b. This dependence is suppressed unless it has particular relevance. Sources for the critical values are detailed below. The one-time dummies in Models 1 and 2 ensure that the limiting distributions of the test statistics are the same as in the IO case. The deviation from trend approach to estimating the AO models implicitly assumes that a break is allowed under the null.

7.7.2   IO Models

7.7.2.i   IO Model 1

For ease of reference the specifications under the null and alternative hypotheses are restated:

Null with break

$$\Delta y_t = a^*(L)\Delta y_t + A^*(1)\beta_1 + \mu_2 DT_t^b + \varepsilon_t \tag{7.157}$$

Null without break

$$\Delta y_t = a^*(L)\Delta y_t + A^*(1)\beta_1 + \varepsilon_t \tag{7.158}$$

Alternative

$$\Delta y_t = \gamma y_{t-1} + C(L)\Delta y_t + \breve{\mu}_1 + \breve{\beta}_1 t + \mu_2 DU_t + \varepsilon_t \tag{7.159}$$

In the DF case without structural breaks, all that is needed in the maintained regression model are: a constant, which is present, but with different interpretations, under both the null and alternative hypotheses; a time trend, present only under the alternative; lags of Δy_t, present under the null and alternative hypotheses; and, critically in terms of distinguishing the unit root, the lag of y_t, that is y_{t-1}. Comparing (7.157) and (7.159), when a potential structural break is present under both the null and alternative hypotheses, the maintained model should also include DT_t^b, present under the null, and DU_t present under the alternative. The resulting maintained model is:

Maintained regression with break under null

$$\Delta y_t = \gamma y_{t-1} + \sum_{j=1}^{k-1} c_j \Delta y_{t-j} + \kappa_1 + \kappa_2 t + \kappa_3 DT_t^b + \kappa_4 DU_t + \tilde{\epsilon}_t \tag{7.160}$$

whereas if there is no break under the null, then the term in DT_t^b should be omitted:

Maintained regression no break under null

$$\Delta y_t = \gamma y_{t-1} + \sum_{j=1}^{k-1} c_j \Delta y_{t-j} + \kappa_1 + \kappa_2 t + \kappa_4 DU_t + \tilde{\epsilon}_t \tag{7.161}$$

The maintained regressions are written assuming that the order of C(L) is uncertain, which induces an additional approximation error, so that $\tilde{\epsilon}_t$ replaces ϵ_t.

The interpretation of the κ_i coefficients depends on whether the data are generated under the null or the alternative hypothesis. If the null is correct, then $\kappa_1 = A^*(1)\beta_1$, whereas if the alternative is correct, then $\kappa_1 = \breve{\mu}_1$; similarly under the null $\kappa_3 = \mu_2$ as in (7.157). If the alternative is correct, $\kappa_2 = \breve{\beta}_1$ and $\kappa_4 = \mu_2$ as in (7.159). The usual test statistic is the pseudo-t on γ, denoted $\tilde{\tau}_\gamma^{(1)}(IO)$.

7.7.2.ii   IO Model 2

Following the principle established with IO Model 1, the relevant specifications are:

Null with break

$$\Delta y_t = a^*(L)\Delta y_t + A^*(1)\beta_1 + \mu_2 DT_t^b + \beta_2 DU_t + \epsilon_t \tag{7.162}$$

Null without break

$$\Delta y_t = a^*(L)\Delta y_t + A^*(1)\beta_1 + \epsilon_t \tag{7.163}$$

Alternative

$$\Delta y_t = \gamma y_{t-1} + C(L)\Delta y_t + \breve{\mu}_1 + \breve{\beta}_1 t + \mu_2 DU_t + \beta_2 DT_t + \epsilon_t \tag{7.164}$$

Maintained regression with break under null

$$\Delta y_t = \gamma y_{t-1} + \sum_{j=1}^{k-1} c_j \Delta y_{t-j} + \kappa_1 + \kappa_2 t + \kappa_3 DT_t^b + \kappa_4 DU_t + \kappa_5 DT_t + \tilde{\epsilon}_t \tag{7.165}$$

Maintained regression no break under null

$$\Delta y_t = \gamma y_{t-1} + \sum_{j=1}^{k-1} c_j \Delta y_{t-j} + \kappa_1 + \kappa_2 t + \kappa_4 DU_t + \kappa_5 DT_t + \tilde{\epsilon}_t \tag{7.166}$$

In the maintained model when there is a break under the null, the regression includes DT_t^b and DU_t from the null and, additionally, DT_t from the alternative. Note that the coefficient on DU_t has an interpretation that differs depending on whether its justification originates from the null or the alternative; in the former case $\kappa_4 = \beta_2$ as in (7.162), whereas in the latter case, $\kappa_4 = \mu_2$ as in (7.164). The pseudo-t test statistic is denoted $\tilde{\tau}_\gamma^{(2)}(\text{IO})$.

7.7.2.iii IO Model 3

If the same approach is taken as for IO Model 2, a problem becomes apparent. First, the set-ups for the null, alternative and maintained regressions are as follows:

Null with break

$$\Delta y_t = a^*(L)\Delta y_t + A^*(1)\beta_1 + \beta_2 DU_t + \varepsilon_t \tag{7.167}$$

Null without break

$$\Delta y_t = a^*(L)\Delta y_t + A^*(1)\beta_1 + \varepsilon_t \tag{7.168}$$

Alternative

$$\Delta y_t = \gamma y_{t-1} + C(L)\Delta y_t + \breve{\mu}_1 + \breve{\beta}_1 t + \beta_2 DT_t + \varepsilon_t \tag{7.169}$$

Maintained regression with break under null

$$\Delta y_t = \gamma y_{t-1} + \sum_{j=1}^{k-1} c_j \Delta y_{t-j} + \kappa_1 + \kappa_2 t + \kappa_3 DU_t + \kappa_4 DT_t + \tilde{\varepsilon}_t \tag{7.170}$$

Maintained regression without break under null

$$\Delta y_t = \gamma y_{t-1} + \sum_{j=1}^{k-1} c_j \Delta y_{t-j} + \kappa_1 + \kappa_2 t + \kappa_4 DT_t + \tilde{\varepsilon}_t \tag{7.171}$$

Comparing the maintained regression for IO Model 2 with a break (7.165) with IO Model 3, then (7.170) only differs by omitting DT_t^b; because this is a one-time dummy variable, it will not affect the asymptotic distribution of the test statistic which is, therefore, the same for IO Model 3 and IO Model 2. Within this approach it is not possible, therefore, to distinguish the maintained regression with a 'joined' slope change from the step change plus slope change set-up of the mixed model. However, if no break is allowed under the null, then the maintained regression (7.171), which excludes DU_t, can be used. The pseudo-t test statistic for this latter case is denoted $\tilde{\tau}_\gamma^{(3)}(\text{IO})$.

7.7.3   DO Models

7.7.3.i   DO Model 1

Following the principles illustrated in the previous cases, we have the following DO specifications:

Null with break

$$\Delta y_t = a^*(L)\Delta y_t + A^*(1)\beta_1 + \mu_2 DT_t^b + \mu_2 h(L)DT_t^b + \varepsilon_t \tag{7.172}$$

Null without break

$$\Delta y_t = a^*(L)\Delta y_t + A^*(1)\beta_1 + \varepsilon_t \tag{7.173}$$

Alternative

$$\Delta y_t = \gamma y_{t-1} + C(L)\Delta y_t + \tilde{\mu}_1 + \tilde{\beta}_1 t + \mu_2 DU_t + g(L)\mu_2 DU_t + \varepsilon_t \tag{7.174}$$

The specification under the alternative involves lags of DU_t, whereas the null in (7.172) involves lags of DT_t^b. However, given the discussion following (7.153), the dummy variable space is spanned by DU_t, and DT_t^b and its lags through to DT_{t-H}^b, where H is the longest lag in h(L); if the null is specified without a break, then (7.174) can be interpreted as the maintained regression.

Maintained regression with break under null

$$\Delta y_t = \gamma y_{t-1} + \sum_{j=1}^{k-1} c_j \Delta y_{t-j} + \kappa_1 + \kappa_2 t + \kappa_3 DT_t^b + \kappa_4 DU_t$$
$$+ f(L)DT_t^b + \tilde{\varepsilon}_t \tag{7.175}$$

Maintained regression without break under null

$$\Delta y_t = \gamma y_{t-1} + \sum_{j=1}^{k-1} c_j \Delta y_{t-j} + \kappa_1 + \kappa_2 t + \kappa_3 DU_t + g(L)DU_t + \tilde{\varepsilon}_t \tag{7.176}$$

7.7.3.ii   DO Model 2

Null with break

$$\Delta y_t = a^*(L)\Delta y_t + A^*(1)\beta_1 + \mu_2 DT_t^b + \mu_2 h(L)DT_t^b + \beta_2 DU_t$$
$$+ \beta_2 h(L)DU_t + \varepsilon_t \tag{7.177}$$

Null without break

$$\Delta y_t = a^*(L)\Delta y_t + A^*(1)\beta_1 + \varepsilon_t \tag{7.178}$$

Alternative

$$\Delta y_t = \gamma y_{t-1} + C(L)\Delta y_t + \tilde{\beta}_1 t + \mu_2 DU_t + g(L)\mu_2 DU_t$$
$$+ \beta_2 DT_t + g(L)\beta_2 DT_t + \varepsilon_t \tag{7.179}$$

The specification under the alternative involves lags of DT_t and DU_t, whereas the specification under the null involves lags of DU_t and DT_t^b. What is required in the maintained regression is a set of dummy variables that spans the dummy variable space. We have already seen in the discussion following (7.153) that the specification under the null only requires DU_t, DT_t^b and its lags through to DT_{t-H}^b, a total of $H + 2$ variables. On the same principle, under the alternative all that is required is DT_t, DU_t and its lags through to DU_{t-H}. The union of these two sets is spanned by DU_t, DT_t, DT_t^b and its lags through to DT_{t-H}^b.

In the case of a null without a break the maintained regression is effectively just that under the alternative hypothesis; this is written in (7.181) using one of the minimal sets of dummy variables.

Maintained regression with break under null

$$\Delta y_t = \gamma y_{t-1} + \sum_{j=1}^{k-1} c_j \Delta y_{t-j} + \kappa_1 + \kappa_2 t + \kappa_3 DT_t^b + \kappa_4 DU_t + \kappa_5 DT_t$$
$$+ f(L)DT_t^b + \tilde{\varepsilon}_t \tag{7.180}$$

Maintained regression without break under null

$$\Delta y_t = \gamma y_{t-1} + \sum_{j=1}^{k-1} c_j \Delta y_{t-j} + \kappa_1 + \kappa_2 t + \kappa_4 DU_t + \kappa_5 DT_t + k(L)DU_t + \tilde{\varepsilon}_t$$
$$\tag{7.181}$$

7.7.3.iii DO Model 3

Null with break

$$\Delta y_t = a^*(L)\Delta y_t + A^*(1)\beta_1 + \beta_2 DU_t + \beta_2 h(L)DU_t + \varepsilon_t \tag{7.182}$$

Null without break under null

$$\Delta y_t = a^*(L)\Delta y_t + A^*(1)\beta_1 + \varepsilon_t \tag{7.183}$$

Alternative

$$\Delta y_t = \gamma y_{t-1} + \sum_{j=1}^{k-1} c_j \Delta y_{t-j} + \tilde{\mu}_1 + \tilde{\beta}_1 t + \beta_2 DT_t + b(L)\beta_2 DT_t + \tilde{\varepsilon}_t \tag{7.184}$$

The maintained regression should include the union of variables in the null and alternative models excluding redundant variables. A minimal set is given by DT_t,

DU_t and its lags through to DU_{t-H} (where H is the longest lag). When there is no break under the null, there are no terms (directly) in DU_t and the model under the alternative can also be interpreted as the maintained regression.

Maintained regression with break under null

$$\Delta y_t = \gamma y_{t-1} + \sum_{j=1}^{k-1} c_j \Delta y_{t-j} + \kappa_1 + \kappa_2 t + \kappa_4 DU_t + \kappa_5 DT_t + k(L)DU_t + \tilde{\varepsilon}_t$$

(7.185)

Maintained regression without break under null

$$\Delta y_t = \gamma y_{t-1} + \sum_{j=1}^{k-1} c_j \Delta y_{t-j} + \kappa_1 + \kappa_2 t + \kappa_5 DT_t + k(L)DT_t + \tilde{\varepsilon}_t \qquad (7.186)$$

Note that a problem analogous to that for the IO versions of Models 2 and 3 occurs for the corresponding DO models. Specifically, the maintained regression (7.185) only differs from (7.180) by the omission of DT^b_{t-H} (note that $DU_t, \ldots, DU_{t-H}$, in (7.185) can be spanned by DU_{t-H} and $DT^b_t, \ldots, DT^b_{t-(H-1)}$); and DT^b_{t-H} is a one-time dummy that does not affect the asymptotic distribution of the test statistic. On the other hand, the maintained regressions for the case of no break under the null do differ (in finite samples and asymptotically) for DO Models 2 and 3. Thus, as in the IO case, the no-break maintained regression (7.186) for DO Model 3 could be used if the case at hand warrants the assumption of no break under the null.

7.8   Critical values

7.8.1   Exogenous versus endogenous break

The critical values for the unit root tests depend upon whether the date of the break is (assumed) known. Broadly, if the break date is unknown and has to be estimated this introduces greater uncertainty into the inference procedure, which is reflected in larger negative values being required for a rejection of the null hypothesis of a unit root. This section considers the case of a known break date; the case of an unknown break date is considered in the next chapter.

7.8.2   Critical values

Critical values for the AO and IO models are available in several sources, which are detailed below in Table 7.1. The critical values are available as asymptotic critical values in all cases and for some finite sample sizes. The table number references should facilitate recourse to the appropriate source for a finite sample critical value.

Table 7.1   Asymptotic critical values for AO and IO models 1, 2 and 3, with $\lambda_b = \lambda_b^c$

	λ_b^c								
	0.1	0.2	0.3	0.4	0.5	0.6	0.7	0.8	0.9
size↓	$\tilde{\tau}_\gamma^{(1)}(AO), \tilde{\tau}_\gamma^{(1)}(IO)$					P(89,IV.B), ZA(92,2)			
5%	−3.68	−3.77	−3.76	−3.72	−3.76	−3.76	−3.80	−3.75	−3.69
10%	−3.40	−3.47	−3.46	−3.44	−3.46	−3.47	−3.51	−3.46	−3.38
	$\tilde{\tau}_\gamma^{(2)}(AO), \tilde{\tau}_\gamma^{(2)}(IO)$					P(89,VI.B), ZA(92,4)			
5%	−3.75	−3.99	−4.17	−4.22	−4.24	−4.24	−4.18	−4.04	−3.80
10%	−3.45	−3.66	−3.87	−3.95	−3.96	−3.95	−3.86	−3.69	−3.46
	$\tilde{\tau}_\gamma^{(3)}(AO)$					PV(92,I), PV(93,1)			
5%	−3.52	−3.72	−3.85	−3.91	−3.93	−3.94	−3.89	−3.83	−3.72
10%	−3.23	−3.41	−3.54	−3.61	−3.65	−3.65	−3.60	−3.55	−3.42
IO w/o break	$\tilde{\tau}_\gamma^{(3)}(IO)$					P(89,V.B), ZA(92,3)			
5%	−3.65	−3.80	−3.87	−3.94	−3.96	−3.95	−3.85	−3.82	−3.68
10%	−3.36	−3.49	−3.58	−3.66	−3.68	−3.66	−3.57	−3.50	−3.35

Notes: Except for $\tilde{\tau}_\gamma^{(3)}(IO)$, the maintained regressions include the dummy variables required under the null to capture the break.

Sources: P(89) is Perron (1989); PV(92) is Perron and Vogelsang (1992); PV(93) is Perron and Vogelsang (1993); the table number in the source is indicated after the date reference. Some finite sample critical values are also included in the sources. Response surface calibrations are available in Carrion-i-Silvestre et al. (1999). Note that the relevant limiting distributions (and hence critical values) for Model 3 in the AO and IO cases (with known break date) are not the same.

7.9   Implications of a structural break for DF tests

7.9.1   Spurious non-rejection of the null hypothesis of a unit root

An important outcome of Perron's (1989) study was the finding that if the data are generated by a trend stationary model but with a break in the intercept, trend or both, then standard unit root test statistics, such as the DF pseudo-t and the normalised bias (n-bias) will tend not to find in favour of the alternative. That is, there will be spurious non-rejection of the null hypothesis of a unit root. The results we cite here are due to Perron (1989, 1990) and Montañés and Reyes (1998) hereafter MR, and see also MR (1999), in their development of Perron's (1989) arguments. The behaviour of Dickey-Fuller F-type tests under the crash model alternative (Model 3) are considered by Sen (2001, 2003).

7.9.2   The set-up

The set-up is as follows: under the alternative, either Model 1, or respectively Model 1a, 2 or 3, the DGP is trend stationary in its simplest AO form, with deterministic variables that are required under the different specifications.

DGP and alternative hypothesis

DGP : Model 1 $y_t = \mu_1 + \beta_1 t + \mu_2 DU_t + \varepsilon_t$ (7.187)

DGP : Model 1a $y_t = \mu_1 + \mu_2 DU_t + \varepsilon_t$ (7.188)

DGP : Model 2 $y_t = \mu_1 + \beta_1 t + \mu_2 DU_t + \beta_2 DT_t + \varepsilon_t$ (7.189)

DGP : Model 3 $y_t = \mu_1 + \beta_1 t + \mu_2 DT_t + \varepsilon_t$ (7.190)

Except for Model 1a, the maintained regression is the standard 'with trend' regression for a simple DF test.

Maintained regression: Models 1, 2, 3

$$\Delta y_t = \mu_1 + \beta_1 t + \gamma y_{t-1} + \tilde{\varepsilon}_t \qquad (7.191)$$

Maintained regression: Model 1a

$$\Delta y_t = \mu_1 + \gamma y_{t-1} + \tilde{\varepsilon}_t \qquad (7.192)$$

where $\gamma = \rho - 1$; the break fraction λ_b^c is regarded as a parameter in the asymptotic results; $\lambda_b^c T$ is assumed to be an integer and the pre-break and post-break samples increase at the same rate as T. In the case of Model 1a, the time trend is omitted.

7.9.3   Key results

The key results from Perron (1989, 1990) and MR (1998, 1999) are summarised below, where $\rightarrow_p$ denotes convergence in probability (see *A Primer*, chapter 4); for simplicity the results are given for $\varepsilon_t \sim iid(0, \sigma_\varepsilon^2)$; extensions for non-iid innovations are generally available in the cited sources.

The DGP is Model 1 (MR, 1999, proposition 1B; see Perron 1989, theorem 1, for extension):

$$1(i) \ \hat{\rho} \rightarrow_p \frac{\mu_2^2 [\lambda_b^c - 4(\lambda_b^c)^2 + 6(\lambda_b^c)^3 - 3(\lambda_b^c)^4]}{\mu_2^2 [\lambda_b^c - 4(\lambda_b^c)^2 + 6(\lambda_b^c)^3 - 3(\lambda_b^c)^4] + \sigma_\varepsilon^2} \qquad (7.193)$$

$$1(ii) \ T^{-1}[T(\hat{\rho} - 1)] \rightarrow_p - \frac{\sigma_\varepsilon^2}{\mu_2^2 [\lambda_b^c - 4(\lambda_b^c)^2 + 6(\lambda_b^c)^3 - 3(\lambda_b^c)^4] + \sigma_\varepsilon^2} \qquad (7.194)$$

$$1(iii) \ T^{-1/2}\hat{\tau}_\beta \rightarrow_p - \frac{\sigma_\varepsilon}{[2\mu_2^2 [\lambda_b^c - 4(\lambda_b^c)^2 + 6(\lambda_b^c)^3 - 3(\lambda_b^c)^4] + \sigma_\varepsilon^2]^{1/2}} \qquad (7.195)$$

DGP is Model 1a (Perron, 1990):

1a(i) $\hat{\rho} \rightarrow_p \dfrac{\mu_2^2(1 - \lambda_b^c)\lambda_b^c}{\mu_2^2(1 - \lambda_b^c)\lambda_b^c + \sigma_\varepsilon^2}$ (7.196)

1a(ii) $T^{-1}[T(\hat{\rho} - 1)] \rightarrow_p -\dfrac{\sigma_\varepsilon^2}{\mu_2^2(1 - \lambda_b^c)\lambda_b^c + \sigma_\varepsilon^2}$ (7.197)

1a(iii) $T^{-1/2}\hat{\tau}_\beta \rightarrow_p -\dfrac{\sigma_\varepsilon}{(2\mu_2^2(1 - \lambda_b^c)\lambda_b^c + \sigma_\varepsilon^2)^{1/2}}$ (7.198)

DGP is Model 2 (MR, 1998, theorem 2):

2(i) $\hat{\rho} \rightarrow_p 1$ (7.199)

2(ii) $T(\hat{\rho} - 1) \rightarrow_p -\dfrac{3[1 + \lambda_b^c + 4(\lambda_b^c)^2]}{2[1 - 2\lambda_b^c + 4(\lambda_b^c)^2]\lambda_b}$ (7.200)

2(iii) $\hat{\tau}_\beta \rightarrow_p -\dfrac{(1 + \lambda_b + 4(\lambda_b^c)^2)[3\{\lambda_b^c - 3(\lambda_b^c)^2 + 6(\lambda_b^c)^3 - 4(\lambda_b^c)^4\}]^{1/2}}{2(1 - 2\lambda_b^c + 4(\lambda_b^c)^2)\lambda_b^c}$ (7.201)

DGP is Model 3 (MR, 1998, theorem 1):

3(i) $\hat{\rho} \rightarrow_p 1$ (7.202)

3(ii) $T(\hat{\rho} - 1) \rightarrow_p -\dfrac{3(1 - 2\lambda_b^c)}{2(1 - \lambda_b^c)\lambda_b^c}$ (7.203)

3(iii) $T^{-1/2}\hat{\tau}_\beta \rightarrow_p -(1 - 2\lambda_b^c)\left(\dfrac{3(\beta_2)^2(1 - \lambda_b^c)\lambda_b^c}{8\sigma_\varepsilon^2 + (\beta_2)^2(1 - \lambda_b^c)\lambda_b^c}\right)^{1/2}$ (7.204)

As MR observe, results 3(i) and 3(ii) in the case of Model 3, and result 2(ii) in the case of Model 2, differ from those in Perron (1989), but the practical effects are very much the same. For further developments and discussion of these results see also Montañés, Olloqui and Calvo (2005).

7.9.3.i Data generated by Model 1

When the data are generated by Model 1, Perron (1989) showed that $\hat{\rho}$ tends to a limit that gets closer to 1 as the magnitude of the break, μ_2, increases, but $\hat{\rho}$ does not converge to 1. The n-bias test statistic $T(\hat{\rho} - 1)$ diverges to $-\infty$ at rate T so, at least asymptotically, the null hypothesis is rejected. The pseudo-t $\hat{\tau}_\beta$ also diverges to $-\infty$, but at the slower rate of $T^{1/2}$; see results 1(ii) and 1(iii), respectively.

To assess the finite sample behaviour of $\hat{\tau}_\beta$, which is the appropriate DF test statistic for the with trend model, the DGP of Model 1 (see Equation (7.187)), was simulated with $T = 200$ and $z_t = \varepsilon_t$, $\varepsilon_t \sim \text{niid}(0, 1)$, $\mu_1 = \beta_2 = 1$, and 10,000

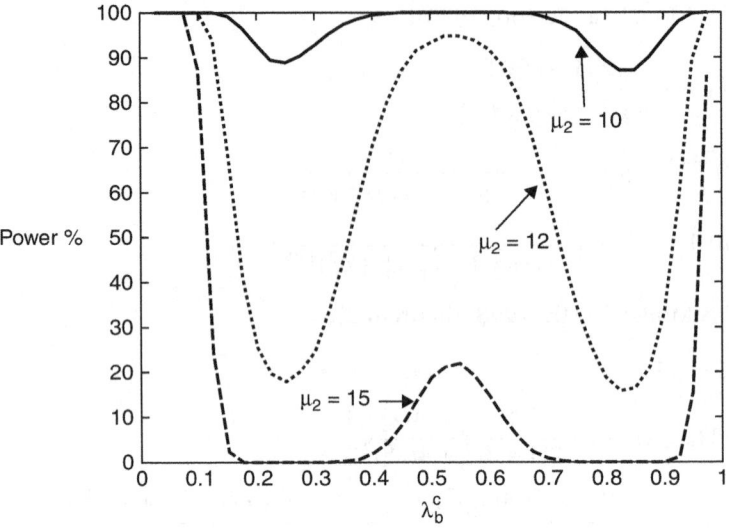

Figure 7.5   Power of DF test: Model 1 level break

replications for different values of $\mu_2 \in [0.5, 20]$. Thus the data are generated as stationary, in fact white noise about a trend with a break in the intercept, where the timing of the break varies; thus the presence of a unit root should be rejected. Some of the results are presented in Figure 7.5 for power as a function of λ_b^c and selected values of μ_2; the graphs show the % of rejections of the null hypothesis of a unit root at the 5% significance level using $\hat{\tau}_\beta$. Thus μ_2 is fixed for each curve shown, which is then a function of λ_b^c. Note that there are non-rejections, but only for 'large' values of μ_2. The percentage of rejections of the unit root null is 100% or close thereto for $\mu_2 \leq 9$; however, power becomes lower for a given break fraction as μ_2 increases; for example, with $\mu_2 = 15$ there is no detection of stationarity for λ_b^c between approximately 0.2 and 0.4 and (symmetrically) between 0.7 and 0.9. Power as a function of λ_b^c is highest at the beginning and ends of the sample and, relatively, when the break occurs approximately in the middle of the sample.

7.9.3.ii   Data generated by Model 1a

Model 1a (see Equation (7.188)) is a special case of Model 1 without the time trend in the stationary alternative. Some of the simulation results for T = 200 are illustrated in Figure 7.6, with the set-up as before but now the DGP is Model 1a and the maintained regression does not include a time trend. The figure shows power using $\hat{\tau}_\mu$. The general pattern is that the null is rejected if the break is early in the sample but not otherwise; the trigger point is noticeably sharp and increases with μ_2. Even when μ_2 is relatively small, for example $\mu_2 = 1$, the

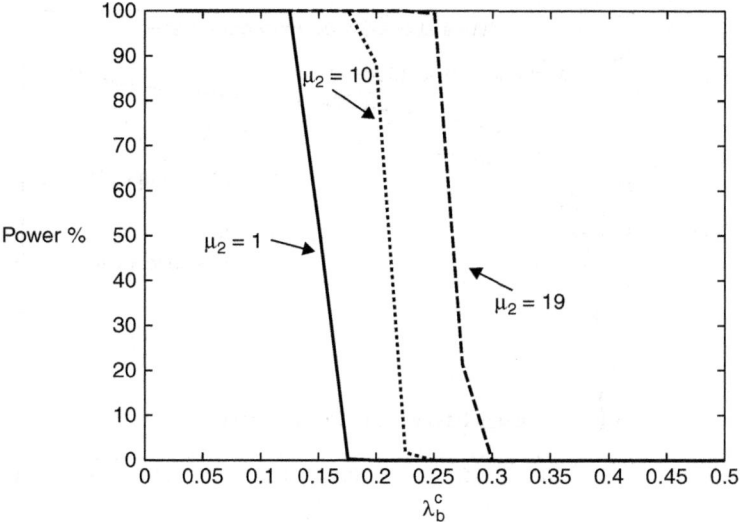

Figure 7.6   Power of DF test: Model 1a level break

null will not be rejected if the break occurs later than about the first 15% of the sample. Even with a substantial value for μ_2, for example $\mu_2 = 19$, the null is not rejected if $\lambda_b^c > 0.3$. This raises the question of whether including a trend in the maintained regression, even when it is not in the DGP, would protect against spurious non-rejection.

7.9.3.iii   Data generated by Model 2

When the DGP is Model 2, it is again the case that $\hat{\rho}$ tends to 1 in probability. The limiting n-bias test statistic $T(\hat{\rho} - 1)$ (see (7.200)), just depends upon λ_b^c and only allows rejection of the null when λ_b^c is at the beginning of the sample. This is illustrated in Figure 7.7 as the dotted line, which shows the graph of $T(\hat{\rho} - 1)$, as in 2(ii), against λ_b^c and plots the 5% critical value, cv, to indicate when the null hypothesis would be rejected. The null hypothesis of a unit root is not rejected when $T(\hat{\rho} - 1)$ is above the 5% cv line (that is $T(\hat{\rho} - 1) > -21.39$) but is rejected when it is below the 5% cv line. Note that there is a cross-over point, $\lambda_b^c = 0.093$, so that rejection of the null on the basis of the n-bias test statistic occurs (asymptotically) when the break occurs in the first 9% of the sample.

The pseudo t-statistic, $\hat{\tau}_\beta$, converges to negative values and, again, does not depend upon the magnitude of the break, just the break fraction λ_b^c. At the 5% significance level, rejection of the null occurs for $\lambda_b^c < 0.085$; that is, for a break in the first 8% of the sample. To assess this asymptotic result are finite samples, the DGP for Model 2 was again simulated with $T = 200$, and the results are presented in Figure 7.8. The basic simulation details are as before but with the DGP

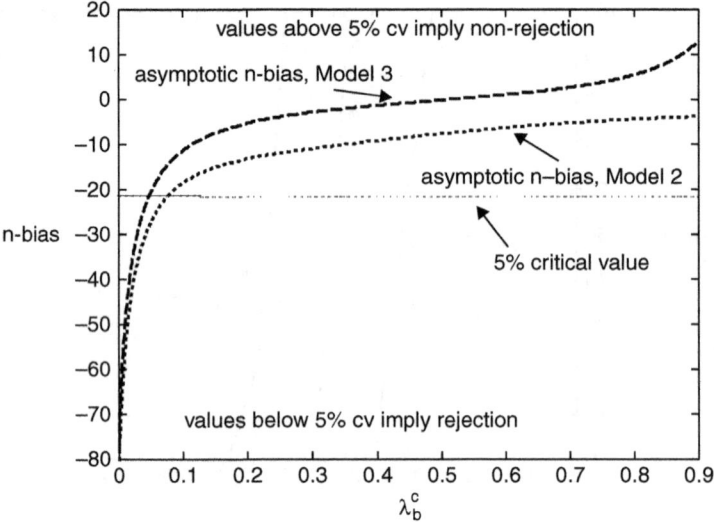

Figure 7.7   Asymptotic n-bias: Models 2 and 3

for Model 2, and now there is a slope break so that $\beta_2 \neq 0$. Whilst the asymptotic result does not depend upon the magnitude of the break, the finite sample results clearly do. The power curves move *in* as β_2 increases, becoming increasingly close to vertical and, therefore, 100% rejection at close to the asymptotic breakpoint of $\lambda_b^c = 0.085$.

7.9.3.iv   *Data generated by Model 3*

In the case of Model 3, $\hat{\rho}$ tends to 1 in probability, but as result 3(ii) shows, this does not mean that the n-bias of $\hat{\rho}$ tends to 0; rather it has a nonzero limit that depends on λ_b^c. This is also illustrated in Figure 7.7, which plots the limit is n-bias test statistic of (7.203) against λ_b^c and the 5% asymptotic critical value (cv) of -21.39. The cross-over point is at $\lambda_b^c = 0.065$; that is, when the break occurs in the first 6% of the sample (asymptotically) the null hypothesis is rejected at the 5% significance level.

Result 3(iii) shows that the pseudo t-statistic, $\hat{\tau}_\beta$, diverges, implying (asymptotic) rejection of the null for $\lambda_b^c < 0.5$. To assess the finite sample behaviour of $\hat{\tau}_\beta$, the DGP of Model 3 was simulated with essential details as before but with the DGP for Model 3, and with a slope break so that $\beta_2 \neq 0$. The results are presented in Figure 7.9. The common pattern for different values of β_2 is that at some point for $\lambda_b^c \in (0, 0.5)$ there is a sharp decline from 100% to 0% rejections; for example, when $\beta_2 = 1$ this is triggered at $\lambda_b^c = 0.15$, substantially less than 0.5. For $\beta_2 = 5$, the trigger point is $\lambda_b^c = 0.375$. As β_2 increases the decline is steeper and the trigger point approaches $\lambda_b = 0.5$.

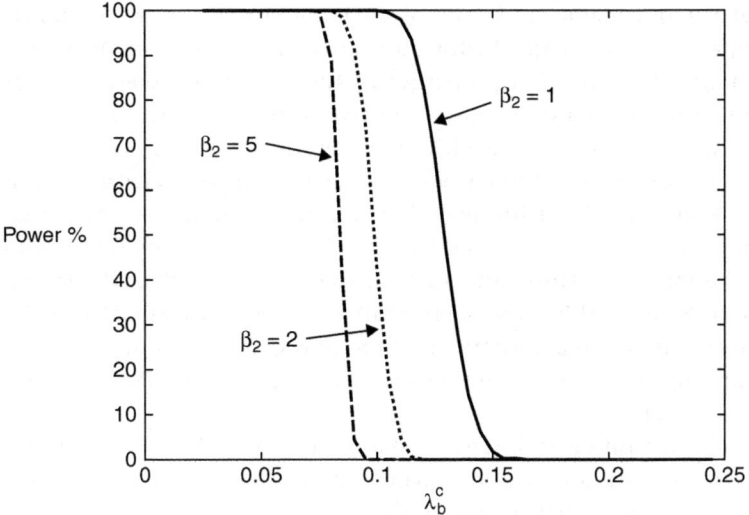

Figure 7.8 Power of DF test: Model 2 slope break

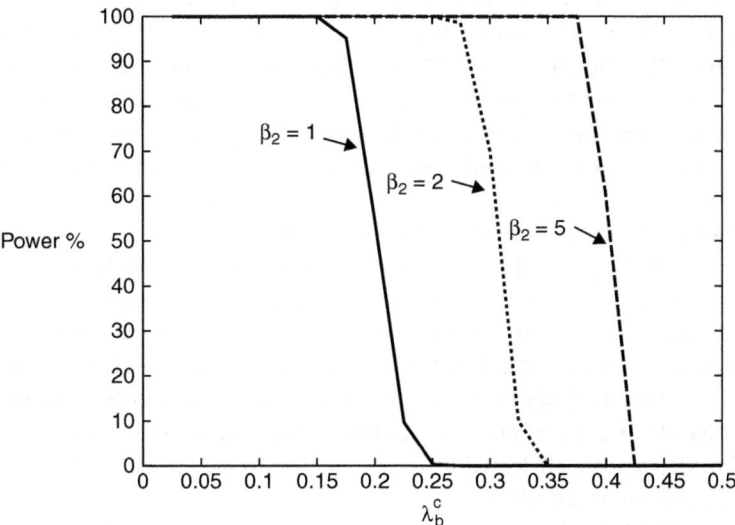

Figure 7.9 Power of DF test: Model 3 slope break

7.10 Concluding remarks

The debate about whether economic time series are better characterised as being generated by stationary or nonstationary processes would almost certainly find the balance of evidence in favour of the latter view if it were not for the possible impact of some major historical events. In the absence of such changes, determining the long-run properties of the process generating a time series becomes a more secure inferential procedure the larger the sample of realisations which one has to work with; however, the longer the sample in calendar time, the more likely is it that the sample includes a period with a 'major' event. Almost every decade in the twentieth century includes an event that would fall into this categorisation; and the opening of the twenty-first century, with 9/11, the Iraq War, Hurricane Katrina, the credit crunch and the Euro crisis, serves to reinforce this view. Such events are certainly associated with notable movements in key macroeconomic time series, and confound the problem of determining whether these time series are generated by DGPs that are stationary or nonstationary.

Theory is definitive: if the DGP is stationary, but includes 'trend breaks', then when DF-type tests are used there will tend to be an unjustified non-rejection of the null hypothesis that the generating process is nonstationary. The tests will too often *not* reject the null of a unit root. This is the 'spurious' non-rejection problem.

Of course, the view of trend breaks taken in this chapter (and the next) is at one level overly simplistic. Banerjee et al. (1992, p. 271) note that 'Interpreted literally, the single-shift/deterministic-trend model has little economic appeal, but interpreted more broadly it can be thought of as a metaphor for there being a few large events that determine the growth path of output over a decade or two – in the United States, the Depression or, later, the productivity slowdown'.

The simplicity can be criticised on several levels. An event such as the Great Crash of 1929 is quite likely to have had differential impacts in time on the various components of GDP, with some components, for example consumers' expenditure, reacting faster than others, for example exports. To some extent, the innovation outlier model captures the implied distributed lag response of differential and gradual adjustment, but within the context of linear adjustment. Other models of adjustment that capture nonlinear response may be relevant, for example, a smooth transition type of nonlinear adjustment in the move from one regime to another.

There are still a number of issues to be considered before the techniques of this chapter will be of use in a practical study. There will almost inevitably be some doubt about the precise dating of a structural break; it might be argued that in some cases this will not be the case. For example, 9/11 happened on 11th September 2001 and that is unarguable. However, the processes that led to the various economic impacts of that 'shock' were not all concentrated at that time; the realisation of what had happened was not such a clearly categorical event and, at least, whether such a shock was, for example, dated in the third quarter or the fourth quarter of 2001 is more likely to be an empirical matter rather than a datum.

Not only is there likely, in practice, to be some doubt over the precise timing of shocks, but there is also likely to be doubt about the best form to model the impact of the shocks. The issue of model selection was considered in Section **7.4**. Also, the assumption that there is a single break is often a convenient working assumption, particularly over a long historical period. These qualifications suggest that there are, in practice, a large number of related and quite critical issues to be considered and these are the subject of the next chapter.

Questions

Q7.1 The notation used in this chapter for the mixed model follows Vogelsang and Perron (1998), which differs from Perron (1989, p. 1364) (where Model 2 is referred to as Model C). Specifically, the difference is that the dummy variable on the slope change is defined as $DT_t = t$ if $t > T_b$ and 0 otherwise, whereas we have used $DT_t = t - T_b^c$ and 0 otherwise. Do the two ways of writing the model lead to the same implications?

A7.1 The answer is, of course, yes but there is a difference in the interpretation of the coefficients. Specifically, the 'Model C' version of the mixed model, Model 2, is:

$$y_t = \mu_1 + \beta_1 t + \mu_2^+ DU_t + \beta_2 DT_t + \varepsilon_t \quad \text{'Model C'} \tag{A7.1a}$$

$$y_t = \mu_1 + \beta_1 t + \mu_2 DU_t + \beta_2 DT_t + \varepsilon_t \quad \text{Model 2} \tag{A7.1b}$$

Note that if $t > T_b$ then $t = T_b + t - T_b$ and $\beta_2 DT_t = \beta_2(T_b + t - T_b) = \beta_2 T_b + \beta_2 DT_t$, where $\beta_2 T_b$ is a constant; thus, for the equivalence of (A7.1a) and (A7.1b), $\mu_2^+ = \mu_2 - \beta_2 T_b$. This means that care has to be taken in the interpretation of the intercept change parameter in the Model C version of the mixed model; on the other hand for $t > T_b$ there is the convenience that the model can be written as $y_t = (\mu_1 + \mu_2^+) + (\beta_1 + \beta_2)t + \varepsilon_t$, so that the post-break trend coefficient can just be read off as the coefficient on the time trend.

Q7.2 Following on from equation (7.84), show how to obtain, in principle, the inverse of $\hat{A}(L) = 1 - 1.2L + 0.49L^2$, that is $\hat{\Lambda}(L) = \hat{A}(L)^{-1}$; and then obtain the numerical values of the coefficients up to order L^5.

A7.2 Note that $\hat{\Lambda}(L) = \hat{A}(L)^{-1}$ is an infinite series in the lag operator L. Hence we may write for a second-order polynomial A(L):

$$(1 - a_1 L - a_2 L^2)^{-1} = (\omega_0 + \omega_1 L + \omega_2 L^2 + \omega_2 L^3 + \cdots) \tag{A7.2}$$

The problem is to obtain the lag coefficients ω_i. The solution method was shown in *UR, Vol. 1*, chapter 3. Multiply both sides of (A7.2) through by A(L):

$$1 = (1 - a_1 L - a_2 L^2)(1 + \omega_1 L + \omega_2 L^2 + \omega_3 L^3 + \omega_4 L^4 + \cdots +) \qquad \text{(A7.3)}$$

$$= (1 + \omega_1 L + \omega_2 L^2 + \omega_3 L^3 + \omega_4 L^4 + \cdots +)$$

$$- (a_1 L + a_1 \omega_1 L^2 + a_1 \omega_2 L^3 + a_1 \omega_3 L^4 + \cdots)$$

$$- (a_2 L^2 + a_2 \omega_1 L^3 + a_2 \omega_2 L^4 + a_2 \omega_3 L^5 + \cdots +)$$

$$= 1 +$$

$$(\omega_1 - a_1)L + (\omega_2 - a_1 \omega_1 - a_2)L^2 + (\omega_3 - a_1 \omega_2 - a_1 \omega_2)L^3$$

$$+ (\omega_4 - a_1 \omega_3 - a_2 \omega_3)L^4 + \cdots)$$

The unknown coefficients ω_i can be solved for recursively by noting that the left-hand side of (A7.3) involves no powers of L, hence all the coefficients on powers of L on the right-hand side are equal to zero. Thus:

$$\omega_0 = 1; \omega_1 = a_1; \omega_2 = a_1 \omega_1 + a_2; \omega_3 = a_1 \omega_2 + a_2 \omega_1; \omega_4 = a_1 \omega_3 + a_2 \omega_2; \ldots;$$

$$\omega_i = a_1 \omega_{i-1} + a_2 \omega_{i-2}$$

Using the estimated values of $a_1 = 1.2$ and $a_2 = -0.49$, the first six coefficients in the sequence are:

$$\omega_0 = 1, \omega_1 = 1.2, \omega_2 = 1.2^2 - 0.49 = 0.95, \omega_3 = 1.2(0.95) + (-0.49)1.2 = 0.552,$$

$$\omega_4 = 1.2(0.552) + (-0.49)0.95 = 0.197,$$

$$\omega_5 = 1.2(0.197) + (-0.49)0.552 = -0.034.$$

The sum of the ω_i to this point is 3.865, compared to the long-run sum of 3.448.

Q7.3 Show that AO Model 1 under the null hypothesis of a unit root and a break can be written in the following way:

$$\Delta y_t = a^*(L)\Delta y_t + A^*(1)\beta_1 + \mu_2 A^*(L)DT_t^b + \varepsilon_t \qquad \text{(A7.4)}$$

A7.3 AO Model 1 with a unit root and break is:

$$y_t = \mu_1 + \beta_1 t + \mu_2 DU_t + \Lambda(L)\varepsilon_t \qquad \text{(A7.5)}$$

where $\Lambda(L) \equiv A(L)^{-1} = (1-L)^{-1}\varphi(L)^{-1}\theta(L)$. Operate on (A7.5) with $(1-L)$ to obtain:

$$(1-L)y_t = (1-L)(\mu_1 + \beta_1 t + \mu_2 DU_t) + \Lambda^*(L)\varepsilon_t \qquad (A7.6)$$

$$= \beta_1 + \mu_2 DT_t^b + \Lambda^*(L)\varepsilon_t$$

$$\Rightarrow A^*(L)\Delta y_t = A^*(1)\beta_1 + \mu_2 A^*(L)DT_t^b + \varepsilon_t \qquad A^*(L) \equiv \Lambda^*(L)^{-1}$$

$$\Rightarrow \Delta y_t = a^*(L)\Delta y_t + A^*(1)\beta_1 + \mu_2 A^*(L)DT_t^b + \varepsilon_t$$

$$(1-L)A^*(L)(y_t - \mu_1 - \beta_1 t - \mu_2 DU_t) = \varepsilon_t \qquad (A7.7)$$

$$A^*(L)(\Delta y_t - \beta_1 - \mu_2 DT_t^b) = \varepsilon_t \qquad (A7.8)$$

Hence taking terms other than Δy_t to the right hand side and defining $A^*(L) = 1 - a^*(L)$, results in (A7.4).

Q7.4 Consider the following example of an AR(2) model of structural change:

$$y_t = \sum_{i=1}^{2} a_i y_{t-i} + \breve{\mu}_1 + \breve{\beta}_1 t + (1 - a_1 L - a_2 L^2)(\mu_2 DU_t + \beta_2 DT_t) + \varepsilon_t \qquad (A7.9)$$

Confirm that this equation can be estimated as:

$$\Delta y_t = \gamma y_{t-1} + c_1 \Delta y_{t-1} + \bar{\mu}_1 + \bar{\beta}_1 t + \delta_1 DU_t + \delta_2 DT_t + \eta_1 DT_t^b + \eta_2 DT_{t-1}^b + \varepsilon_t \qquad (A7.10)$$

A7.4 The key point is to use the relationships amongst the dummy variables as follows:

$$DU_{t-1} = DU_t - DT_t^b$$

$$DU_{t-2} = DU_{t-1} - DT_{t-1}^b$$

$$= DU_t - DT_t^b - DT_{t-1}^b$$

$$DT_{t-1} = DT_t - DU_t$$

$$DT_{t-2} = DT_{t-1} - DU_{t-1}$$

$$= DT_t - DU_t - DU_{t-1}$$

$$= DT_t - DU_t - (DU_t - DT_t^b)$$

$$= DT_t - 2DU_t + DT_t^b$$

Substitution of the lagged values of DU_t and DT_t into the lag polynomials gives:

$$(1 - a_1 L - a_2 L^2)DU_t = ((1 - a_1 - a_2)DU_t + (a_1 + a_2)DT_t^b + a_2 DT_{t-1}^b)$$

$$(1 - a_1 L - a_2 L^2)DT_t = ((1 - a_1 - a_2)DT_t + (a_1 + 2a_2)DU_t - a_2 DT_t^b)$$

These are then substituted into A(7.9) and collecting coefficients on like terms results in A(7.10).

Q7.5 Consider the following model:

$$y_t = \mu_1 + \beta_1 t + z_t \tag{A7.11}$$

$$(1 - \rho L)(1 + \varphi_1 L) z_t = \varepsilon_t \tag{A7.12}$$

i. Write the model in terms of y_t rather than the residuals from the regression of y_t on the deterministic trend components.
ii. Write this model in the case that there is a unit root.

A7.5.i In this case $\varphi^*(L) = (1 - \rho L)(1 + \varphi_1 L) = (1 - (\rho - \varphi_1)L - \rho\varphi_1 L^2)$. Now obtain the ADF factorisation of $\varphi^*(L)$; that is $A(L) = (1 - \alpha L) - C(L)(1 - L)$, where $A(L) = (1 - a_1 L - a_2 L^2)$. In this case:

$$C(L) = c_1 L; c_1 = -a_2 = -\rho\varphi_1; \text{ and } \alpha = a_1 + a_2 = (\rho - \varphi_1) + \rho\varphi_1.$$

$$[(1 - \alpha L) - c_1 L(1 - L)]y_t = [(1 - \alpha L) - c_1 L(1 - L)](\mu_1 + \beta_1 t) + \varepsilon_t$$

$$\Rightarrow$$

$$\Delta y_t = (\alpha - 1)y_{t-1} + c_1 \Delta y_{t-1} + [(1 - \alpha L) - c_1 L(1 - L)](\mu_1 + \beta_1 t) + \varepsilon_t$$

$$= (\alpha - 1)y_{t-1} + c_1 \Delta y_{t-1} + (1 - \alpha)(\mu_1 + \beta_1 t) + (\alpha - c_1)\beta_1$$

$$= (\alpha - 1)y_{t-1} + c_1 \Delta y_{t-1} + [(1 - \alpha)\mu_1 + (\alpha - c_1)\beta_1] + (1 - \alpha)\beta_1 t \tag{A7.13}$$

$$= \gamma y_{t-1} + c_1 \Delta y_{t-1} + \mu + \beta t \tag{A7.14}$$

where $\mu = (1 - \alpha)\mu_1 + (\alpha - c_1)\beta_1$ and $\beta = (1 - \alpha)\beta_1$; α, c_1, μ_1 and β_1 are uniquely identified from (A7.13).

A7.5.ii In the case of a unit root $\rho = 1$, $\varphi^*(L) = (1 - (1 - \varphi_1)L - \varphi_1 L^2)$, $\alpha = (1 - \varphi_1) + \varphi_1 = 1$ and $c_1 = -\varphi_1$. Therefore, the ADF factorisation is $(1 - L) + \varphi_1(1 - L)$ and the resulting model is:

$$\Delta y_t = -\varphi_1 \Delta y_{t-1} + [(1 - L) + \varphi_1(1 - L)](\mu_1 + \beta_1 t) + \varepsilon_t$$

$$= -\varphi_1 \Delta y_{t-1} + [(1 + \varphi_1)(1 - L)](\mu_1 + \beta_1 t) + \varepsilon_t$$

$$= -\varphi_1 \Delta y_{t-1} + (1 + \varphi_1)\beta_1 + \varepsilon_t \tag{A7.15}$$

Note that $(1 - L)(\mu_1 + \beta_1 t) = 0 + \beta_1(t - (t - 1)) = \beta_1$. Alternatively set $\rho = 1$ in c_1, and note $\alpha = 1$, and then specialise (A7.13).

Q7.6 What does a DO(2) model look like?

A7.6 In this case, the model is:

$$(1 - \rho L)(y_t - \mu_1 - \beta_1 t) = (1 + \delta_1 L + \delta_2 L^2)\mu_2 DU_t + \varepsilon_t \tag{A7.16}$$

This results in:

$$\Delta y_t = \gamma y_{t-1} + \breve{\mu}_1 + \breve{\beta}_1 t + \mu_2 DU_t + \delta_1 \mu_2 DU_{t-1} + \delta_2 \mu_2 DU_{t-2} + \varepsilon_t \tag{A7.17}$$

8
Structural Breaks with Unknown Break Dates

Introduction

This chapter extends the last in several ways. Perron's (1989) article was seminal in the sense of transforming the view that it was sensible to consider economic time series, particularly macroeconomic time series, as either generated by a process with a unit root (or roots) or generated by a process that was stationary about a (simple) linear trend. The idea of regime change that could affect either or both of these processes led to a fundamental re-evaluation of the simplicity of these opposing mechanisms, and an increase in the complexity of choosing between them.

In practice, although there are likely to be some contemporaneous and, later, historical indications of regime changes, there is almost inevitably likely to be uncertainty not only about the dating of such changes but also the nature of the changes. This poses another set of problems for econometric applications. If a break is presumed, when did it occur? Which model captures the nature of the break? If multiple breaks occurred, when did they occur? Have one or more breaks occurred: (a) under the alternative; (b) under the null?

In his original study, Perron (1989) considered thirteen of the macroeconomic time series originally considered by Nelson and Plosser NP, (1982), and data for post-war quarterly real GNP, where the breakpoints were informed by preliminary investigations. In general, for example, one might graph the data and run some deterministic trend regressions, with appropriate dummy variables, and examine the fit around the dates of possible key events. For eleven of the annual series in the NP data set, by prior inspection Perron (1989) used a break date of 1929, and the crash model (Model 1) in IO form. For the remaining two series, real wages and common stock prices, the mixed model (Model 2) was specified in IO form also with a break date of 1929. For quarterly real GNP, the break date of 1973 was selected for a changing growth model (Model 3) in AO form. Subsequent research has queried the nature of the breaks, for example should a crash or mixed model be used? What is the nature of the response to the breaks: is it, for example, AO or IO? And what is the number of breaks?

In some cases it may be evident that one particular model should be chosen; for example, if the data is clearly not trended the question simplifies to whether there has been a break in the level and Model 1a is appropriate (no trend is included under the alternative). In practice, as we illustrate below with US GNP, such issues are rarely so clear-cut and a strategy is required. One possibility is to estimate the most general model, that is Model 2, but in addition estimate the other two models (three if Model 1a is included) and assess the robustness of the results on the unit root tests. Whilst useful, this procedure may not be definitive enough in practice. Montañés, Olloqui and Calvo (2005), hereafter MOC, consider this issue further and suggest that information criteria may provide a better way of selecting the model on which to base unit root inference. Sen (2003) makes the point that if the time of the break is unknown, it is also quite likely that the form of the break model under the alternative is also unknown, and Sen (2007) suggests basing inference on the mixed model, that is Model 2, for a consistent estimate of the break date.

Whilst much of the unit root plus structural break literature has emphasised inference on the unit root issue, it is as logical to start with the question of whether there has been a structural break, and what is the nature of the break. Of course, inference on this issue is likely to be complicated by whether there is a unit root, just as inference on the latter is complicated by whether there is a structural break. A possible resolution of this conundrum is to consider inference jointly on the unit root and the break.

This chapter is arranged as follows. Section **8.1** starts with consideration of a break in the trend function, testing for such a break at an unknown date and including the unit root in a joint null hypothesis. Section **8.2** deals with the realistic situation in which the break date is unknown and what criteria have been suggested to estimate the break date. This discussion is continued in Section **8.3** to consider some simulation evidence on whether the suggested criteria result in a good estimator of the break date. Section **8.4** considers another critical issue, rather taken for granted so far, that is, which of the various models of the break process is the most appropriate, for example, a break whether should be modelled as a one-off change in the level, a permanent change in the level, a one-off change in the growth rate and so on. Section **8.4** recognises that in a long historical period there may well be more than one break and considers how the models are extended to the multiple break case, a good place to start being the extension to two break points. Consideration of this problem is continued in Section **8.5**, which focuses on estimation and selecting two break dates. Finally, some of the issues are illustrated in Section **8.6** using a time series on US GNP.

8.1   A break in the deterministic trend function

If there is no break in the trend function, then unit root tests have their usual interpretation; however, as indicated in the previous chapter, the application of unit root tests when there is a break in the trend function is generally misleading, with size and power likely to be affected. Thus, a possible alternative strategy is to test directly for a break in the trend function, although this will itself be complicated by whether the DGP is stationary or nonstationary, just as the problem of testing for a unit root is complicated by whether there is a break in the trend function. Another possibility is to combine into one step the 'pre-test' for a trend break and a test of the unit root hypothesis itself. These approaches are discussed in this section, starting with a pre-test for a break in the trend function.

In this chapter we move to the more realistic case where the break date is unknown so that the correct break fraction and break date, λ_b^c and T_b^c, respectively, are distinguished from the assumed values, λ_b and T_b.

8.1.1   Testing for a break in the deterministic trend function: Sup-W(λ_b)

Vogelsang (1997) considered the problem of testing for a break in the deterministic trend function in the AO set-up. The illustration used here is a particular case of Vogelsang's general framework applied to AO Model 2. The set-up is:

$$y_t = (1,t)\begin{pmatrix}\mu_1\\\beta_1\end{pmatrix} + I(t > T_b)(1, t - T_b^c)\begin{pmatrix}\mu_2\\\beta_2\end{pmatrix} + z_t \tag{8.1}$$

$$A(L)z_t = \varepsilon_t \tag{8.2}$$

where $I(t > T_b)$ is the indicator function equal to 1 if $t > T_b$ and 0 otherwise. The indicator function is a convenient way of representing a number of dummy variables; for present purposes note that $I(t > T_b^c)1 = DU_t$, $I(t > T_b^c)(t - T_b^c) = DT_t$ and, also, $I(t = T_b^c + 1)1 = DT_t^b$.

The dynamics, and possible unit root, are present in z_t. For simplicity the dynamics are assumed to be pure AR but, as a conjecture, the general approach extends to ARMA errors, provided that the fitted order of the regression expands at an appropriate rate. As before, $A(L)$ can be factored as $A(L) = (1 - \rho L)\varphi(L)$ and if $\rho = 1$, then the process generating y_t has a unit root.

The trend break model can be written in familiar form as:

$$y_t = \mu_1 + \beta_1 t + \mu_2 DU_t + \beta_2 DT_t + z_t \tag{8.3}$$

Thus, (8.3), with AR dynamics as in (8.2), is a particular case of AO Model 2. The null hypothesis is now that there is no break in the deterministic function, H_0: $\mu_2 = \beta_2 = 0$, against the alternative that at least one of these coefficients is non-zero, H_A: $\mu_2 \neq 0$ and/or $\beta_2 \neq 0$. Whilst the development of this model in terms of analysing the unit root is familiar, here the focus is on the trend break coefficients.

Multiplying (8.3) through by $A(L)$, and rearranging into ADF form, gives:

$$A(L)y_t = A(L)(\mu_1 + \beta_1 t + \mu_2 DU_t + \beta_2 DT_t) + \varepsilon_t \tag{8.4}$$

$$\Rightarrow$$

$$\Delta y_t = \gamma y_{t-1} + C(L)\Delta y_t + A(L)(\mu_1 + \beta_1 t + \mu_2 DU_t + \beta_2 DT_t) + \varepsilon_t \tag{8.5}$$

$$= \gamma y_{t-1} + C(L)\Delta y_t + \tilde{\mu}_1 + \tilde{\beta}_1 t + A(L)(\mu_2 DU_t + \beta_2 DT_t) + \varepsilon_t \tag{8.6}$$

$$= \gamma y_{t-1} + C(L)\Delta y_t + \tilde{\mu}_1 + \tilde{\beta}_1 t + \delta_1 DU_t + \delta_2 DT_t + \eta(L)DT_t^b + \varepsilon_t \tag{8.7}$$

where $\eta(L) = \eta_1 + \eta_2 L + \cdots + \eta_{k-1}L^{k-1}$. The last line is a consequence of the connections between the different dummy variables (an example below illustrates this point). It follows that $H_0: \mu_2 = \beta_2 = 0$ implies that $\delta_1 = \delta_2 = 0$, so that a test can be based on (8.7) or, as Vogelsang (1997) notes, the one-off dummy variables can be omitted without loss asymptotically, so the asymptotic distribution of the test statistic can be determined from:

$$\Delta y_t = \gamma y_{t-1} + C(L)\Delta y_t + \tilde{\mu}_1 + \tilde{\beta}_1 t + \delta_1 DU_t + \delta_2 DT_t + \tilde{\varepsilon}_t \tag{8.8}$$

where $\tilde{\varepsilon}_t = \eta(L)DT_t^b + \varepsilon_t$.

8.1.1.i Example

An example where $A(L)$ is a second order AR polynomial illustrates the key points of (8.4)–(8.7). The AR and then ADF versions of the model are:

$$y_t = \sum_{i=1}^{2} a_i y_{t-i} + \tilde{\mu}_1 + \tilde{\beta}_1 t + (1 - a_1 L - a_2 L^2)(\mu_2 DU_t + \beta_2 DT_t) + \varepsilon_t$$

$$= \sum_{i=1}^{2} a_i y_{t-i} + \tilde{\mu}_1 + \tilde{\beta}_1 t + \delta_1 DU_t + \delta_2 DT_t + \eta_1 DT_t^b + \eta_2 DT_{t-1}^b + \varepsilon_t \tag{8.9}$$

$$\Rightarrow$$

$$\Delta y_t = \gamma y_{t-1} + c_1 \Delta y_{t-1} + \tilde{\mu}_1 + \tilde{\beta}_1 t + \delta_1 DU_t + \delta_2 DT_t + \eta_1 DT_t^b + \eta_2 DT_{t-1}^b + \varepsilon_t \tag{8.10}$$

where: $\gamma = (a_1 + a_2) - 1$; $c_1 = -a_2$;

$\tilde{\mu}_1 = A(1)\mu_1 + (a_1 + 2a_2)\beta_1$; $\tilde{\beta}_1 = A(1)\beta_1$; $A(1) = 1 - a_1 - a_2$; and

$\delta_1 = A(1)\mu_2 + (a_1 + 2a_2)\beta_2$; $\delta_2 = A(1)\beta_2$;

$\eta_1 = (a_1 + a_2)\mu_2$; $\eta_2 = a_2(\mu_2 - \beta_2)$.

Equation (8.10) is asymptotically equivalent to:

$$\Delta y_t = \gamma y_{t-1} + c_1 \Delta y_{t-1} + \tilde{\mu}_1 + \tilde{\beta}_1 t + \delta_1 DU_t + \delta_2 DT_t + \tilde{\varepsilon}_t \tag{8.11}$$

where $\tilde{\varepsilon}_t = \eta_1 DT_t^b + \eta_2 DT_{t-1}^b + \varepsilon_t$. However, it may be advisable to include the one-off dummy variables in finite sample applications.

8.1.1.ii   Unknown break date

A problem common to testing for a trend break or a unit root is that so far, and somewhat unrealistically, the break fraction has been assumed known. To illustrate how this problem can be solved, first rewrite the dummy variables to explicitly recognise their dependence on the break fraction, which is now *not* assumed to be known.

$$\Delta y_t = \gamma y_{t-1} + C(L)\Delta y_t + \tilde{\mu}_1 + \tilde{\beta}_1 t + \delta_1 DU_t(\lambda_b) + \delta_2 DT_t(\lambda_b) + \eta(L)DT_t^b(\lambda_b) + \varepsilon_t \tag{8.12}$$

There is a set of dummy variables for each permissible value of $\lambda_b \in \Theta = (0,1)$; note that in this framework λ_b is assumed present only under the alternative hypothesis. Some 'trimming' of the range of λ_b is required so that $\Theta = (\lambda_b^*, 1 - \lambda_b^*)$, $\lambda_b^* > 0$; for example, $\lambda_b^* = 0.01$ or 0.15. The idea is then to obtain the distribution of a test statistic for H_0, where the test statistic is evaluated for all $\lambda_b \in \Theta$; for the general principles of such a procedure see Davies (1977, 1987) and Andrews and Ploberger (1994); and note that the problem and its solution have already occurred in a different context in Chapters 1, 5 and 6.

The test statistic considered by Vogelsang (1997) is the Wald test for H_0 in the model given by:

$$\Delta y_t = \gamma y_{t-1} + \sum_{j=1}^{k-1} c_j \Delta y_{t-j} + \tilde{\mu}_1 + \tilde{\beta}_1 t + \delta_1 DU_t(\lambda_b) + \delta_2 DT_t(\lambda_b) + \tilde{\varepsilon}_t \tag{8.13}$$

say

$$w_t = x_t(\lambda_b)'\beta + \tilde{\varepsilon}_t \tag{8.14}$$

where:

$$x_t(\lambda_b)' = (y_{t-1}, \Delta y_{t-1}, \dots, \Delta y_{t-k}, 1, t, DU_t(\lambda_b), DT_t(\lambda_b)); \text{ and}$$

$$w_t = \Delta y_t; \beta = (\gamma, c_1, \dots, c_{k-1}, \tilde{\mu}_1, \tilde{\beta}_1, \delta_1, \delta_2)'.$$

Collecting all T observations, the model can be written as:

$$w = X(\lambda_b)\beta + \varepsilon \tag{8.15}$$

where $w = (w_1, \dots, w_T)'$ and $x_t(\lambda_b)'$ is the t-th row of $X(\lambda_b)$; the notation assumes that k pre-sample values of y_t are available, otherwise the first element in w is w_{k+1}). The Wald statistic is:

$$W(\lambda_b) = \frac{(R\hat{\beta} - s)'[R(X(\lambda_b)'X(\lambda_b))R']^{-1}(R\hat{\beta} - s)}{\hat{\sigma}_\varepsilon^2(\lambda_b)} \tag{8.16}$$

The restrictions matrix R is of dimension $g \times K$, where $K = (k+4)$ and $g = 2$; in this case R sets the $(k+3)$-rd and $(k+4)$-th elements of β to zero; s is a 2×1 zero vector; and the variance estimator $\hat{\sigma}_\varepsilon^2(\lambda_b)$ is based on the regression model

(8.13). See Chapter 1 for further details. Practically, it is usually somewhat easier to calculate an F test statistic for H_0 using a direct comparison of the residual sums of squares under the unrestricted and restricted regressions, in which case $F(\lambda_b) \approx W(\lambda_b)/g$, which can be used to either to obtain the Wald test statistic or to transform the Wald test critical values.

To obtain a test statistic, the Wald test $W(\lambda_b)$ is evaluated for G points in a grid over $\lambda_b \in \Theta$. The test statistic variants are the mean, exponential mean and supremum over the G resulting values of $W(\lambda_b)$, defined as follows (see also Chapter 6, Section **6.4.1**):

Mean: $$W(\lambda_b)^{ave} = G^{-1} \sum\nolimits_{\lambda_b \in \Theta} W(\lambda_b) \qquad (8.17a)$$

Exponential mean: $$W(\lambda_b)^{exp} = \log \left(G^{-1} \sum\nolimits_{\lambda_b \in \Theta} \exp[W(\lambda_b)/2] \right) \qquad (8.17b)$$

Supremum: $$\text{Sup-}W(\lambda_b) = \sup\nolimits_{\lambda_b \in \Theta} W(\lambda_b) \qquad (8.17c)$$

(Note that there is a slight difference in the exponential version of this test compared to Chapter 6, where here the logarithm is taken.)

Critical values depend on whether there is a unit root in the DGP, and whether the maintained regression is for Model 2 or for Model 1a. Some asymptotic critical values extracted from Vogelsang (1997, tables 1 and 2) are given below.

8.1.1.iii Conservative critical values

Vogelsang (op. cit.) notes that there is a problem in deciding which critical values to use when it is not known whether there is a unit root; and there is evidently a substantial difference between the critical values for the same statistics that are alike apart from the distinction between the stationary and nonstationary cases. One possibility is to use a conservative approach, which implies using the critical values for the nonstationary case. Alternatively, in the next section, we describe a development of the Sup-$W(\lambda_b)$ test due to Murray and Zivot (1998), which incorporates the unit root as part of the null hypothesis.

Using a conservative approach ensures the correct size asymptotically, but entails a loss of power if the unit root is not exact; that is, in local-to-unity asymptotics $\rho = 1 - \bar{\rho}/T$ with $\bar{\rho} > 0$, and the unit root is approached but is not exactly 1 for finite samples. From Vogelsang's (1997) asymptotic results we see that asymptotic size is approximately halved for $\bar{\rho} \approx 9$, implying $\rho = 0.9$ for $T = 100$ and $\rho = 0.95$ for $T = 200$; with this more conservative actual size, power is then affected.

The theoretical and finite sample ($T = 100$) simulation results in Vogelsang (1997) suggest the following in the case of a one-time break in the level or the slope, that is, μ_2 or $\beta_2 \neq 0$. (The statistics are invariant to μ_1 and β_1, which can therefore be set to 0 in the simulations; conservative asymptotic critical values are used, that is, assuming a unit root.)

(i) $W(\lambda_b)^{exp}$ and Sup-$W(\lambda_b)$ exhibit power that is similar and increases montonically with μ_2 and β_2.

(ii) Power increases as persistence, measured by γ, decreases.

(iii) $W(\lambda_b)^{ave}$ exhibits non-monotonic power in μ_2 and β_2.

On this basis if a one-time break is suspected the advice is to use either $W(\lambda_b)^{exp}$ or Sup-$W(\lambda_b)$.

For further developments of the testing framework see Bunzel and Vogelsang (2005), Sayginsoy and Vogelsang (2011) and Perron and Yabu (2009).

8.1.1.iv *A double break suggests a loss of power*

In the case of two breaks, for example a double slope change, there is a substantial loss of power, and power can be non-monotonic for all versions of the test statistic; the test statistics may also miss the breaks almost entirely in some parts of the parameter space. In Vogelsang's simulation results, $W(\lambda_b)^{exp}$ had the best chance of correctly indicating the existence of two breaks if either of the break magnitudes is large ($> 2\sigma_\varepsilon$) and γ is close to zero, or if the break magnitude is slight and there is little persistence, $\gamma \approx -1$. A possible amelioration of this problem, if two breaks are suspected, is to split the sample and examine each part of the sample separately; if a break is found in the first part of the sample, then the complete sample test can be conditioned on the presence of one break when testing for, in effect, a second break.

There are two developments of this approach. First, rejection of the null hypothesis $H_0: \mu_2 = \beta_2 = 0$ does not answer the question of whether it is the intercept or slope or both that have changed. We consider this problem next. Then, in the section following, consideration is given to an extension of the null hypothesis that includes the unit root as part of the null hypothesis.

8.1.2 Rejection of the null hypothesis – partial parameter stability: $LR(\hat{\lambda}_b)_i$

When the alternative hypothesis is a joint hypothesis on the intercept and slope parameters, rejection is ambiguous as to the cause. If rejection arises through a change in the intercept, Model 1 or 1a is suggested; a change in the slope suggests Model 3; and a change in both the intercept and slope suggests Model 2. It is tempting to consider two further tests (separately), one for the non-constancy of the intercept and the other for the non-constancy of the slope within the context of the maintained regression given by (8.13). The test statistics would be straightforward extensions of the procedure already outlined for the joint case; however, despite its intuitive attraction, Hsu and Kuan, hereafter HK (2001), show that, in the context of a deterministically trending regressor and stationary errors, this leads to a procedure with size distortions; in particular, the null hypothesis of constancy will be rejected too often.

8.1.2.i The Hsu and Kuan (2001) procedure

The essential problem is that the limiting distribution of the Wald test for, say, the intercept break parameter, depends upon the slope break parameter. The problem is not peculiar to the Wald test, applying equally to the LR and LM versions of the tests; however, it does not apply to tests with stationary (non-trending) regressors, in which case the tests maintain their correct size, at least asymptotically. Note though that, unlike Vogelsang (1997), the error assumptions in HK do not allow for a unit root; this may not be too restrictive, since it could be argued that, at least if no break is allowed under the null, it is just the alternative under stationarity that guides the selction of the break type and date.

It is convenient to set the model up as follows:

$$y_t = \mu_1 + \mu_2 DU_t(\lambda_b) + \beta_1 t + \beta_2 DT_t(\lambda_b) + z_t \tag{8.18}$$

$$= \mu_t + \beta_t + z_t$$

where in the second line $\mu_t = \mu_1 + \mu_2 DU_t(\lambda_b)$ and $\beta_t = \beta_1 t + \beta_2 DT_t(\lambda_b)$.

Fairly standard (stationarity) assumptions apply to z_t. In particular, HK (2001, assumption A), assume that z_t is a linear process in the white noise inputs ε_t, with finite $(4 + \varsigma)$-th moments for $\varsigma > 0$, $z_t = \sum_{j=1}^{\infty} \psi_j \varepsilon_{t-j}$ and $\sum_{j=1}^{\infty} j |\psi_j| < \infty$. The long-run variance of z_t is defined by $\sigma_z^2 = \lim_{T \to \infty} T^{-1} E\left(S_{z,T}^2\right)$, where $S_{z,T} = \sum_{t=1}^{T} z_t$.

Rather than proceed by way of an ADF transformation, the analysis in HK is based directly on (8.18), with a semi-parametric estimator of the serial correlation structure in z_t used in empirical applications. A reasonable conjecture, supported by a suggestion in HK, is that the results will be essentially the same if the ADF approach is followed.

The hypotheses relevant to Models 1, 2 and 3, respectively, can be expressed as:

$$H_{01}: \mu_2 = 0 \qquad H_{02}: \mu_2 = \beta_2 = 0 \qquad H_{03}: \beta_2 = 0$$

Now, consider the following variations of the regression model (8.18) with associated residual sum of squares (RSS). These are written as a function of an estimated break fraction $\hat{\lambda}_b$, which is discussed below.

$$y_t = \mu_1 + \beta_1 t + \beta_2 DT_t(\hat{\lambda}_b) + z_t \qquad \text{RSS}_1 \tag{8.19}$$

$$y_t = \mu_1 + \beta_1 t + z_t \qquad \text{RSS}_2 \tag{8.20}$$

$$y_t = \mu_1 + \mu_2 DU_t(\hat{\lambda}_b) + \beta_1 t + z_t \qquad \text{RSS}_3 \tag{8.21}$$

Notice that RSS_1 is the residual sum of squares that obtains by imposing the null hypothesis H_{01}; similarly RSS_2 follows by imposing H_{02}; and RSS_3 by imposing

H_{03}. Also, define by RSS_{UC} the residual sum of squares from the unconstrained regression (8.18) and, for reference below, the associated estimator of the variance $\tilde{\sigma}^2(\hat{\lambda}_b)_{UC} = RSS_{UC}/T$. Clearly $RSS_i \geq RSS_{UC}$ for $i = 1, 2, 3$. The basis of the suggested procedure for testing either H_{01} or H_{03} is the LR statistic given by:

$$LR(\hat{\lambda}_b)_i = T\left[\frac{RSS_i - RSS_{UC}}{RSS_{UC}}\right] \text{ for } i = 1, 3 \qquad (8.22)$$

The LR principle is preferred by HK (2001) because, as they point out, it is a pivotal function in the presence of unidentified nuisance parameters, in this case the break fraction, whereas the Wald test statistic is not. An LR test statistic could also be defined in an analogous way for $LR(\lambda_b)_2$, with $RSS_i = RSS_2$. The test statistic for Model 1a is with RSS_i as in (8.19) but with the term in $\beta_1 t$ omitted.

There are now two cases to consider that affect the test procedure. Consider testing H_{01} (the same reasoning applies to testing H_{03}). The null H_{01} is non-specific concerning β_2, hence under the null either (i) $\mu_2 = 0$ and $\beta_2 \neq 0$ or (ii) $\mu_2 = 0$ and $\beta_2 = 0$. Under (i), the limiting distribution of $LR(\lambda_b)_1$ is $\chi^2(1)$ and $LR(\lambda_b)_1$ is a consistent test statistic with power depending on μ_2. However, if (ii) is correct then $LR(\lambda_b)_1$ does not have a limiting $\chi^2(1)$ distribution; simulations in HK suggest that it has fatter tails leading to over-rejection if critical values from $\chi^2(1)$ are used. As a first step, therefore, it is necessary to establish whether there has been change of *some* kind; then if there has been change, then test H_{01} and/or H_{03}. To operationalise this procedure requires an estimator $\hat{\lambda}_b$ of λ_b, which is taken as the value of λ_b that leads to a minimum of RSS_{UC}, equivalent to maximising $LR(\lambda_b)_2$.

8.1.2.ii   *The two-step procedure*

The two steps in the HK procedure are, therefore:

(1) Perform a test of H_{02}, which could be done by using the Vogelsang test or the supremum version of $LR(\lambda_b)_2$, say Sup-$LR(\lambda_b)_2$. If non-rejection is the result, then there is no more to be done. Noting that the asymptotic distributions of Sup-$W(\lambda_b)$ and Sup-$LR(\lambda_b)_2$ are the same, the critical values in this step can be obtained from Vogelsang (1997, table 1, p = 1 case), extracted below in Table 8.1; refer also to the stationary case and Model 2. An estimator $\hat{\lambda}_b$ of λ_b is simultaneously obtained from the value of λ_b which results in the supremum of the test statistic

(2) If there is rejection of H_{02}, then test H_{01}, and/or H_{03}, using $LR(\hat{\lambda}_b)_1$, and/or $LR(\hat{\lambda}_b)_3$, respectively, with critical values from $\chi^2(1)$.

As is usual in multiple tests, the overall size of the procedure will generally exceed the size of any component test, hence the overall size can be controlled by 'allocating' the size amongst the tests; a typical procedure is to divide the overall size equally amongst the components (using the Bonferroni inequality),

Table 8.1 Asymptotic 95% quantile for Wald test statistics for a break in the trend

	$W(\lambda_b)^{ave}$		$W(\lambda_b)^{exp}$		Sup-$W(\lambda_b)$	
Stationary case	Model 1a	Model 2	Model 1a	Model 2	Model 1a	Model 2
$\lambda_b^* = 0.01$	2.66	4.42	2.20	3.52	10.85	15.44
$\lambda_b^* = 0.15$	2.20	3.50	1.89	3.13	9.00	13.29
Unit root case						
$\lambda_b^* = 0.01$	3.91	8.22	4.84	8.18	18.20	25.27
$\lambda_b^* = 0.15$	3.43	7.19	4.71	8.12	17.88	25.10

Source: Vogelsang (1997, tables 1 and 2).
Notes: p = 0, 1, respectively, corresponding to an intercept and an intercept and linear trend); the critical values are for the 95% quantile, that is the (upper) 5% critical value. The extent of trimming of the examined break points is given by λ_b^*.

for example, two tests of 2.5% for a maximum overall size of 5%. Alternatively use $\alpha_i = 1 - (1 - \alpha)^{1/n}$, where α is the overall size of the test procedure and n is the number of tests.

Finally, for application of the tests, it is necessary to consider modelling potential serial correlation in the z_t. HK adopt a Newey-West/Andrews-type semi-parametric estimator of σ_z^2 based on a window of length ϑ, for the unconstrained regression (8.18), that is:

$$\hat{\sigma}_z^2 = T^{-1}\Sigma_{t=1}^T \hat{z}_t^2 + 2T^{-1}\sum_{j=1}^{\vartheta} \omega_j(\vartheta)\sum_{t=j+1}^T \hat{z}_t\hat{z}_{t-j} \tag{8.23}$$

where $\omega_j(\vartheta) = 1 - j/(1+\vartheta))$, $\{\hat{z}_t\}$ are the LS residuals from (8.18) corresponding to RSS$_{UC}$ and ϑ is determined by Andrews (1991) data-dependent method. Other methods of estimating $\hat{\sigma}_z^2$ are possible and indeed may be preferable, for example using an AR approximation (see *UR Vol. I* chapters 6 and 9).

8.1.3   Including the unit root in the joint null: Sup-$W(\lambda_b, \gamma)$

In Vogelsang's approach, there are two sets of critical values depending on whether there is a unit root in the DGP. Murray (1998) and Murray and Zivot, 1998 (MZ) have suggested a similar approach, but include the unit root along with the structural break parameters in the null hypothesis. This results in a Sup-Wald test for a joint null similar in spirit to the DF joint F-type tests that test for a null of a unit root and the absence of deterministic terms in the standard case. The general specification of the test statistic is as before, that is, $W(\lambda_b)$ or its F version, but the restrictions matrix is now 3 x K, with γ in (8.13) also set equal to zero. To distinguish this test statistic from the Vogelsang test, it is denoted Sup-$W(\lambda_b, \gamma)$. In principle, other versions of the test statistic could also be calculated, for example the Mean and Exp versions as in Section **8.1.1.ii.**

Table 8.2   Model 2 asymptotic and finite sample 5% critical values for Sup-W(λ_b, γ) for the joint null hypothesis of a break in the trend function and unit root

$\lambda_b^* = 0.05$	$T = 50$	$T = 100$	$T = 200$	$T = \infty$
$k = 0$	31.923	30.428	29.328	28.628
k (t-sign)	36.858	32.626	31.227	28.628

Notes: The first row refers to simulations where the known lag length, $k = 0$, is imposed, and the second row to where the lag length is chosen by a 10% two-sided marginal t procedure. There is some slight variation in the critical values depending on the extent of trimming; $\lambda_b^* = 0.05$ indicates 5% trimming. To calculate the critical values for the F version of the test use $F = W(\lambda_b)/3$.
Source: Sen (2003, Table 1).

Sen (2003) includes the F version of the Sup-W(λ_b, γ) test in his analysis of the effects of misspecification of the type of structural break. The asymptotic and some finite sample 5% critical values (with 5% trimming) for IO Model 2 are given in Table 8.2.

In principle, the MZ approach could be applied to other variations of combining a unit root test with a test for a break in the trend function. For example, in Model 1 the unit root hypothesis could be jointly considered with the intercept break parameter, and in Model 3 it could be jointly considered with slope break parameter.

8.2   Break date selection for unit root tests

A key practical aspect of the application of unit root tests that allows for the possibility of a structural break is to isolate the break date. Some guidance may be available at an informal level by first estimating the candidate model(s) and examining the residuals by graphical or more structured means for the possibility of outliers that may indicate a break or narrow down the search to a particular period. Other more formal methods have been suggested and some are considered in this section.

8.2.1   Criteria to select the break date

Historically, one of the most influential criteria for choosing λ_b is due to Zivot and Andrews (1992), hereafter ZA, who observed that Perron's choice of breakpoint was data-dependent and, therefore, endogenous rather than exogenously determined. From this perspective, a unit root test from such a procedure is *conditional* on a particular choice of breakpoint.

8.2.1.i Minimising the psuedo-t statistic $\tilde{\tau}_\gamma^{(i)}(\lambda_b)$

To remove this conditionality ZA suggested choosing the minimum of the psuedo-t statistic(s), $\tilde{\tau}_\gamma^{(i)}(\lambda_b)$, over $\lambda_b \in \Theta$. Thus, the ZA selection of the break date is that date which gives most weight to the trend stationary alternative, that being the least favourable to the null hypothesis and the most favourable to the alternative hypothesis.

More formally the dependence of the t statistic for a unit root on the break fraction is indicated by writing the $\tilde{\tau}_\gamma^{(i)}$ statistic as a function of λ_b, that is $\tilde{\tau}_\gamma^{(i)}(\lambda_b)$; recall that the notation λ_b^c is reserved for the correct break fraction, that is $T_b^c = \lambda_b^c T$. The minimum over $\lambda_b \in \Theta$ is denoted $\inf_{\lambda_b \in \Theta}[\tilde{\tau}_\gamma^{(i)}(\lambda_b)]$. Where there is no ambiguity this is referred to more simply as $\inf\text{-}\tilde{\tau}_\gamma^{(i)}(\lambda_b)$; thus for λ_b in the set Θ, the test statistic is the minimum of all $\tilde{\tau}_\gamma^{(i)}$, and at this minimum the implied estimator of λ_b^c is denoted $\hat{\lambda}_b(\inf\text{-}\tilde{\tau}_\gamma^{(i)})$.

ZA adopt the null hypothesis that the series $\{y_t\}$ is I(1) with drift, specifically $y_t = \mu + y_{t-1} + \varepsilon_t$; this null differs from, for example, Perron (1989), which allows structural breaks under the null. On the ZA view, an event like the Great Crash occurs as a result of a realisation from the left tail (that is, large negative error in this case) of the DGP. The alternative hypothesis is that $\{y_t\}$ is generated by a trend stationary process with a one-time break at an unknown point in time.

ZA use the IO version of the model set-up, but no break is allowed under the null hypothesis; the maintained regressions are as specified in Chapter 7, Section 7.6. As there is no break under the null, the break coefficients have their interpretation under the alternative. The maintained regressions are (7.161, 7.166 and 7.171) from Chapter 7 reproduced below:

$$\Delta y_t = \gamma y_{t-1} + \sum_{j=1}^{k-1} c_j \Delta y_{t-j} + \kappa_1 + \kappa_2 t + \kappa_4 DU_t(\lambda_b) + \tilde{\varepsilon}_t \tag{8.24}$$

$$\Delta y_t = \gamma y_{t-1} + \sum_{j=1}^{k-1} c_j \Delta y_{t-j} + \kappa_1 + \kappa_2 t + \kappa_4 DU_t(\lambda_b) + \kappa_5 DT_t(\lambda_b) + \tilde{\varepsilon}_t \tag{8.25}$$

$$\Delta y_t = \gamma y_{t-1} + \sum_{j=1}^{k-1} c_j \Delta y_{t-j} + \kappa_1 + \kappa_2 t + \kappa_4 DT_t(\lambda_b) + \tilde{\varepsilon}_t \tag{8.26}$$

The dummy variables are defined conditional on the assumed time of the break as the notation now indicates, for example, $DU_t(\lambda_b)$.

Whilst the critical values reported by ZA are for the IO model set-up with no break permitted under the null, they have a slightly more general application. First, inclusion of the one-off dummy variable(s), $DT_t^b(\lambda_b)$, as in the maintained regressions (7.160) and (7.165) for IO Models 1 and 2 respectively, does not affect the asymptotic null distribution. Second, in the cases of Models 1 and 2, the critical values are the same for the IO and AO versions of the models (see Vogelsang and Perron, 1998).

8.2.1.ii   The decision rule

The ZA decision rule is to reject the null of a unit root at the α significance level if:

$$\inf -\tilde{\tau}_\gamma^{(i)}(\lambda_b) < \kappa_{\inf,\gamma}^{(i)}(\alpha) \tag{8.27}$$

where $\kappa_{\inf,\gamma}^{(i)}(\alpha)$ is the left-tail critical value of the distribution of $\inf -\tilde{\tau}_\gamma^{(i)}(\lambda_b)$ at the α significance level. The left-tail critical values for the distribution of $\inf -\tilde{\tau}_\gamma^{(i)}(\lambda_b)$, given α, cannot be greater (that is, they are at least as negative) than those for a fixed λ_b; hence, using critical values for exonegous λ_b, when λ_b is *not* exogenous, will bias the unit root test toward rejection of the unit root null; the extent of the bias will depend on the difference between critical values from the conditional and unconditional distributions for $\tilde{\tau}_\gamma^{(i)}(\lambda_b)$.

To illustrate, asymptotic critical values for the left tail of the null distribution at the 5% and 10% significance levels are given in Table 8.3 (second column; the table also includes critical values for a number of other tests). Compared to the exogenous break case critical values (see Table 7.1), the critical values are larger negative than when λ_b is treated as exogenously correct. For example, the 5% critical value for IO Model 1 is –4.80 compared to an average of –3.74 for λ_b^c fixed in the set $\Theta = [0.1, 0.9]$.

8.2.1.iii   Critical values

Note for practical purposes that the finite sample distributions for $\inf -\tilde{\tau}_\gamma^{(i)}(\lambda_b)$ are shifted further left than the asymptotic distributions; this suggests that there is enough variation due to finite sample effects to treat the asymptotic critical values only as a broad guide, with finite sample simulations necessary for inference with better size properties. ZA take this finite sample sensitivity into account in their re-evaluation of the Nelson and Plosser data set; for example, they report that in the case of (log) real US GNP, with 62 observations and fitting Model 1, the 5% finite sample critical value is –5.35 compared to the asymptotic critical value of –4.80.

8.2.2   Signficance of the coefficient(s) on the dummy variable(s)

A rather natural alternative to the ZA procedure is to choose the break date by considering the sample evidence for a break of level or slope in the form of the significance of the coefficient(s) on the corresponding dummy variable(s).

8.2.2.i   The 'VP criteria'

Perron (1997) considers the IO model and choosing the break date by using the t test on the intercept change dummy in Model 1 or the slope change dummy in Models 2 or 3. Information, if available, on the likely direction of the change can be used to improve the power of the test; for example, in the case of a 'crash'

Table 8.3 Asymptotic critical values for $\tilde{\tau}_\gamma^{(i)}(\lambda_b)$, AO and IO models 1, 2 and 3, with λ_b selected by different criteria

	Inf-$\tilde{\tau}_\gamma^{(i)}$	Min/Max [t(μ_2)]	Max [|t(μ_2)|]	Max [t(β_2)]	Max| [t|β_2|]	Max F(μ_2,β_2)
			$\tilde{\tau}_\gamma^{(1)}$ (AO)			
5%	−4.80 ZA(92,2), P(97,1a), VP(98,1a)	−4.01 VP(98,1b)	−4.17 VP(98,1c)	n.r	n.r	n.r
10%	−4.58	−3.74	−3.90	n.r	n.r	n.r
			$\tilde{\tau}_\gamma^{(1)}$ (IO)			
5%	−4.80 ZA(92,2), P(97,1a), VP(98,1a)	−4.64 P(97,1b)	−4.84 P(97,1c)	n.r	n.r	n.r
10%	−4.58	−4.37	−4.59	n.r	n.r	n.r
			$\tilde{\tau}_\gamma^{(2)}$ (AO)			
5%	−5.08 ZA(92,4), P(97,1d), VP(98,2a)	n.a	n.a	−4.28 VP(98,2b)	−4.50 VP(98,2c)	−4.61 VP(98,2d)
10%	−4.82	n.a	n.a	−3.95	−4.20	−4.31
			$\tilde{\tau}_\gamma^{(2)}$ (IO)			
5%	−5.08 ZA(92,4), P(97,1c), VP(98,2a)	n.a	n.a	−4.62 P(97,1e), VP(98,4)	−4.91 P(97,1f), VP(98,4)	−5.16 VP(98,3)
10%	−4.82	n.a	n.a	−4.28	−4.59	−4.86
			$\tilde{\tau}_\gamma^{(3)}$ (AO)			
5%	−4.36 PV(93,3), P(97,1g)	n.r	n.r	−4.08 P(97,1h) VP(98,4)	−4.34 P(97,1i) VP(98,4)	n.r
10%	−4.07	n.r	n.r	−3.77	−4.04	n.r
IO* w/o break dvs			$\tilde{\tau}_\gamma^{(3)}$ (IO)			
5%	−4.42 ZA(92,3)	n.r	n.r	−4.13	−4.39 BLS(92,2)	n.r
10%	−4.11	n.r	n.r	−3.85	−4.12	n.r

Notes: sources: ZA(92) is Zivot and Andrews (1992); BLS(92) is Banerjee, Lumsdaine and Stock (1992); PV(93) is Perron and Vogelsang (1993); P(97) is Perron (1997); and VP(98) is Vogelsang and Perron (1998); in each case table numbers follow the date; * indicates without one-off break dummy variables under the null; n.a is not available and n.r is not relevant.

effect in Model 1, select the break date by the *minimum* of the t statistics on the intercept change dummy; in the case of a 'boom', select the break date by the *maximum* of the t statistics; if there is no prior view, select the break date by the maximum of the absolute value of the t statistics on the break parameter. This approach was extended by VP (1998) to the AO model and, in the case of Model 2, to the F test for the joint null that both the level and slope break parameters are zero against the two-sided alternative.

The criteria, referred to as the VP criteria, are:

Model 1: use the t statistic on the coefficient of DU_t: $\text{Min}[t(\mu_2)]$, $\text{Max}[t(\mu_2)]$, $\text{Max}[|t(\mu_2)|]$.

Model 2: use the t statistic on the coefficient of DT_t: $\text{Min}[t(\beta_2)]$, $\text{Max}[t(\beta_2)]$, $\text{Max}[|t(\beta_2)|]$.

Model 2: use the F statistic for the null hypothesis that coefficients on DU_t and DT_t: $\beta_2 = \mu_2 = 0$: $\text{Max}[F(\mu_2, \beta_2)]$.

Model 3: use the t statistic on the coefficient of DT_t: $\text{Min}[t(\beta_2)]$, $\text{Max}[t(\beta_2)]$, $\text{Max}[|t(\beta_2)|]$.

It should be understood throughout that 'Min' or 'Max' (sometimes written 'Inf' or 'Sup') refer to an evaluation over all $\lambda_b \in \Theta$; the resulting unit root test statistic is the psuedo-t $\tilde{\tau}_\gamma^{(i)}(\lambda_b)$ which, for emphasis, may then be written to indicate which selection criterion has been used, for example, $\tilde{\tau}_\gamma^{(1)}(\lambda_b, \text{Min}[t(\mu_2)])$ if the t statistic on the coefficient of DU_t was used in Model 1. These various criteria, based on the significance of the appropriate dummy variables, imply different distributions under the null and hence lead to different critical values of $\tilde{\tau}_\gamma^{(i)}(\lambda_b)$.

8.2.2.ii   *Critical values*

As far as the asymptotic distributions are concerned, there are some points to note in comparing these to the corresponding ones when the break date is assumed known. In the latter case, the asymptotic distributions for the AO and IO cases are the same in Model 1; they are also the same in Model 2; but the asymptotic distributions for the AO and IO versions of Model 3 are different.

When the break date is endogenised, the AO/IO equivalence still holds for Model 1 for choice by $\text{inf-}\tilde{\tau}_\gamma^{(i)}(\lambda_b)$ using the asymptotic critical values, and near enough for the finite sample critical values. It holds for neither for choice by the relevant t statistic on the coefficient on the break dummy (compare, for example, VP (1998), table 1 (a), (b), (c) with Perron (1997), table 1 (a), (b), (c) for Model 1.)

As to Model 2, with choice by $\inf -\tilde{\tau}_\gamma^{(i)}(\lambda_b)$, the AO/IO equivalence holds for the asymptotic critical values but not for the finite sample critical values (compare VP, table 2(a) with Perron (1997, table 1(d))). It holds for neither for choice by the relevant t statistic on the coefficient on the break dummy. Neither does the equivalence hold for choice by $\text{Max}[F(\mu_2, \beta_2)]$ in Model 2 (compare VP, table 2 (d) and table 3). All of this means that not only does great care have to be exercised in selecting the type of break and break date, but that once these are chosen the correct set of critical values has to be located. See Table 8.3 for a summary of the 5% and 10% asymptotic critical values.

8.2.3 A break under the null hypothesis

So far it has been assumed that the null asymptotic distributions, and so tables of critical values, have *not* been generated with a break under the null. However, when there is a break under the null the situation is quite complex, and some of the key results are summarised below (see, especially, VP, 1998).

8.2.3.i Invariance of standard test statistics

First, note when the correct break date is used:

the psuedo-t, $\tilde{\tau}_\gamma^{(1)}(\lambda_b)$, in the AO and IO models, is exactly invariant to μ_2;

in AO and IO Models 2, $\tilde{\tau}_\gamma^{(2)}(\lambda_b)$ is exactly invariant to μ_2;

in AO Models 2 and 3, and IO Model 2, $\tilde{\tau}_\gamma^{(i)}(\lambda_b)$ is exactly invariant to β_2;

in IO Model 3, $\tilde{\tau}_\gamma^{(3)}(\lambda_b)$ is not exactly invariant to β_2.

In finite samples, with an incorrect break date, asymptotic invariance replaces exact invariance for μ_2; in IO Model 3, $\tilde{\tau}_\gamma^{(3)}(\lambda_b)$ is not even asymptotically invariant.

The results now summarised focus on AO and IO Models 2 and 3 for their invariance or lack of it to β_2. We return to the extent of the lack of finite sample invariance for Model 1 (and 1a) in Section **8.3.2** below.

With a break under the null the data are generated by:

$$y_t = \beta_2 DT_t + z_t \tag{8.28}$$

$$z_t = z_{t-1} + \varepsilon_t \Rightarrow \tag{8.29}$$

$$y_t = y_{t-1} + \beta_2 DU_t + \varepsilon_t \tag{8.30}$$

The key results, highlighting problems, include the following.

(1) Even with the correct break date, as well as for an incorrect break date, testing by the psuedo-t in IO Model 3 has the problem that $\tilde{\tau}_\gamma^{(3)}(\lambda_b^c)$ diverges to $-\infty$ with positive probability and size is distorted.

(2) Equally, but only with an incorrect break date, $\inf\text{-}\tilde{\tau}_\gamma^{(i)}(\lambda_b)$, for AO and IO Models 2 and 3, diverges to $-\infty$.

(3) For AO and IO Models 2 and 3, if the correct break date is not chosen, then the asymptotic distribution of $\inf\text{-}\tilde{\tau}_\gamma^{(i)}(\lambda_b)$ is not invariant to the slope break parameter, β_2. Size is distorted if the no-break critical values are used, (see, for example, VP, Tables 6A, 6B). As β_2 increases the asymptotic null distribution shifts to the left; hence, using critical values for the distribution for $\beta_2 = 0$, will lead to over-sized tests (more false rejections of the null than the nominal size); size approaches 1 for $\beta_2 \neq 0$ and $T \to \infty$.

(4) Even where the asymptotic distributions converge, that is for $\tilde{\tau}_\gamma^{(i)}(\lambda_b^c)$, $i = 2$, 3 for AO, in the case of the correct break date and where the correct break date is chosen asymptotically (as is the case with the VP criteria), these distributions differ from the no break under the null case. These limiting distributions are the same as if the break date was known; hence the sources for critical values are Perron (1989) for AO, $i = 2$, and Perron and Vogelsang (1993) for AO, $i = 3$, summarised in Table 7.1.

(5) For AO Models 2 and 3, the correct break date is chosen asymptotically by the VP criteria.

(6) For IO Models 2 and 3, the VP criteria do not choose the correct break date asymptotically (strictly this statement relies on simulation results for $\text{Max}[F(\mu_2, \beta_2)]$. The problem here is that practically the IO Model 2 is perhaps *the* leading model of choice. Simulations in VP suggest that the problem of size distortion becomes acute for $\mu_2 > 5$ and $\beta_2 > 5$, and that these are quite large magnitudes (measured in units of standard deviation of the innovation errors).

(7a) Use a method for choosing the break date that gets the right date, at least asymptotically. When this latter condition is satisfied then, for the AO models, the limiting distributions are the same as if the break date was known if either the t or F criteria on β_2 are used (VP, 1998, equations 30 and 31).

(7b) Nevertheless, there is still a practical problem in AO Models 2 and 3, because these distributions differ from the no break under the null case and have critical values that are less negative compared to when $\beta_2 = 0$. In the case of the VP criteria, the critical values for $\beta_2 = 0$ could be used knowing that this will lead to under-sizing (said to be conservative) if $\beta_2 \neq 0$. This will reduce power because a sample value of, say, $\tilde{\tau}_\gamma^{(3)}(\lambda_b, \text{AO})$, between the critical values appropriate for $\beta_2 = 0$ and $\beta_2 \neq 0$ would lead to incorrect non-rejection if it was actually the case that $\beta_2 \neq 0$. Note that this contrasts with the over-sizing for the $\inf\text{-}\tilde{\tau}_\gamma^{(i)}(\lambda_b)$ case. The likely impact of this problem also depends upon λ_b^c because the critical values from the asymptotic distribution of $\tilde{\tau}_\gamma^{(2)}(\lambda_b^c, \text{AO})$ are maximised (in absolute value) at $\lambda_b^c = 0.5$ and are least in absolute value at the beginning and end of the sample; for example the 5% cv varies between -3.75 for $\lambda_b^c = 0.1$ and -4.20 for $\lambda_b^c = 0.5$. The 5% cvs from the corresponding asymptotic distributions using $\text{Max}[t(\beta_2^A)]$, if the direction of the break is known, $\text{Max}[|t(\beta_2^A)|]$ or $\text{Max}[F(\mu_2, \beta_2)]$ are -4.28, -4.50 and -4.61, respectively; thus the problem is not so acute when λ_b^c gives a break date close to the middle of the sample.

(7c) VP (1998) suggest a pretest of a $\beta_2 = 0$ that is valid when the order of integration is unknown, but note that this will also affect the overall size of the unit root test. (See the Vogelsang procedure described in Section **8.1** and Harvey, Leybourne and Newbold, 2004). An alternative procedure based on bootstrapping the finite sample critical values was suggested by Nunes, Newbold and Kuan (1997). Note from *UR, Vol. 1*, chapter 8, that a key part of the bootstrap procedure is to generate bootstrap replications under the null, which in this case will necessarily involve imposition of a unit root; hence, this is the point at which to incorporate the possibility of a break. Of course this also involves estimation of the break date. In order to do this Nunes et al. (1997) estimate an ARMA model in first differences (hence the unit root is imposed) and select the break date by minimising BIC; this specification then forms the basis of the bootstrap replications and hence the bootstrap finite sample critical values.

In summary, when there is a break under the null, or it is not known whether such a possibility can be ruled out, the situation is more problematical than when there is no break under the null. In addition to problems of model choice and the more realistic case of an unknown break date, some of the models and selection criteria lead variously to divergent test statistics, under sizing, over sizing and incorrect selection of the break date!

In the admittedly unlikely event that the break date is known and/or there is no break under the null, the situation is straightforward as to convergence of the distributions of the relevant test statistics. However, there are still two pervasive questions: which type of break model to choose, and is there a break under the null? Both are questions that apply to the more likely case that the break date is unknown and these are considered below.

8.3 Further developments on break date selection

In addition to the criteria based on statistical significance, Lee and Strazicich, (2001) hereafter LSt, provide simulation evidence which suggests that the BIC has good properties in selecting the correct break date with high probability. Harvey, Leybourne and Newbold (2001) show that a simple modification to the break date selected by $\max|t(\mu_2)|$ in the cases of the IO versions of Models 1 and 1a leads to selection of the correct break date with high probability, and they provide some finite sample as well as asymptotic critical values for a modified psuedo-t test based on this choice.

8.3.1   Lee and Strazicich (2001): allowing a break under the null

LSt suggest choosing the break date where BIC is minimised; they also provide simulation results to enable an evaluation of some different criteria for break selection. They consider IO Models 1 and 2 and the following breakpoint selection criteria:

BIC:	$\ln[\tilde{\sigma}_i^2(\lambda_b)] + K(i)(\ln T)/T$	(8.31)
Zivot-Andrews:	$\inf\text{-}\tilde{\tau}_\gamma^{(i)}(\lambda_b)$	(8.32)
VP:	$\max|t(\beta_2)|$	(8.33)

where $K(i)$ is the number of estimated coefficients in the i-th maintained regression, $\tilde{\sigma}_i^2(\lambda_b) = RSS(\lambda_b)_i/T$ is the ML estimator of the corresponding error variance and $RSS(\lambda_b)_i$ is the residual sum of squares from the maintained regression; these quantities are functions of the chosen breakpoint through the fraction λ_b. (Note that Nunes, Newbold and Kuan, 1997, use $K(i) + 1$, the additional parameter referring to estimation of the break date.) With the breakpoint chosen by BIC it is then possible to use the psuedo-t statistic at that point to test the null hypothesis; this is referred to as $\tilde{\tau}_\gamma^{(i)}(\lambda_b, BIC)$.

The maintained regressions include the one-time dummy DT_t^b. The set-up is that there is a break under the alternative *and* under the null. Critical values for $\inf\text{-}\tilde{\tau}_\gamma^{(i)}(\lambda_b)$ are from ZA (1992) and Perron (1997), and from Perron (1997) for $\max|t(\beta_2)|$; alternately LSt (2001) also use the critical values assuming exogeneity of the break from Perron (1989) (this is known to be the limiting distribution for the AO version of the model when the break date is asymptotically correct; see VP (1998, equation 30)).

8.3.1.i   *Simulation results*

In the case of IO Model 1, the break under the null means a change in the 'intercept', not a change in the drift. At the correct break date, $\tilde{\tau}_\gamma^{(1)}(\lambda_b^c, IO)$ is exactly invariant to the break parameter; at the incorrect break date $\tilde{\tau}_\gamma^{(1)}(\lambda_b, IO)$ is only asymptotically invariant to the break parameter. The simulations will, therefore, indicate the extent of the finite sample invariance and the behaviour of the test statistics under the alternative.

In the case of IO Model 2, there are two problems. First, whilst the psuedo-t $\tilde{\tau}_\gamma^{(2)}(\lambda_b)$ is exactly invariant to the slope break parameter when the correct break date is known, it is not even asymptotically invariant when the correct break date is unknown; and, second, the break date is not correctly chosen by any of the VP criteria if there is a break under the null. Hence, the simulations indicate how serious these problems are in finite samples.

The null hypothesis (= DGP) for each case is:

Model 1 : $$y_t = y_{t-1} + \mu_1 + \mu_2 DT_t^b + \varepsilon_t \tag{8.34}$$

Model 2 : $$y_t = y_{t-1} + \mu_1 + \mu_2 DT_t^b + \beta_2 DU_t + \varepsilon_t \tag{8.35}$$

LSt (2001) consider $T = 100$, $T_b^c = 50$, $\mu_2 \in [0, 4, 6, 8 \ 10]$, $\beta_2 \in [0.04, 0.06, 0.08, 0.10]$.

The main findings are summarized below, firstly under the null hypothesis (NH), with $\inf$-$\tilde{\tau}_y^{(i)}(\lambda_b)$, $\text{Max}[|t(\mu_2)|]$ for Model 1 or $\text{Max}[|t(\beta_2)|]$ for Model 2 (5% critical values are taken from ZA (1992) for the former and Perron (1997) for the latter two), and then, secondly, under the alternative hypothesis (AH).

NH:1.i. Using $\inf$-$\tilde{\tau}_y^{(i)}(\lambda_b)$, $\text{Max}[|t(\mu_2)|]$ or $\text{Max}[|t(\beta_2)|]$ leads to more rejections than the nominal size of the test; these spurious rejections increase as the break magnitude increases. For example, 16.9% empirical size for 5% nominal size for $\inf$-$\tilde{\tau}_y^{(i)}(\lambda_b)$ at $\mu_2 = 6$ in Model 1.

NH:1.ii. Use of the exogenous break critical values (Perron, 1989) leads to considerably oversized tests. For example, 55.8% size for $\inf$-$\tilde{\tau}_y^{(i)}(\lambda_b)$ at $\mu_2 = 6$ in Model 1.

NH:2. Using $\inf$-$\tilde{\tau}_y^{(i)}(\lambda_b)$, $\max[|t(\mu_2)|]$ or $\max[|t(\beta_2)|]$, the breakpoint tends to be incorrectly estimated at $T_b^c - 1$. For example, at $\mu_2 = 6$ in Model 1 and using $\inf$-$\tilde{\tau}_y^{(i)}(\lambda_b)$, 36.7% of breaks are located at $T_b^c - 1$ and only 0.3% at T_b^c; the corresponding numbers for $\max[|t(\mu_2)|]$ are 37.7% and 0% .

NH:3.i. BIC is substantially better than either $\inf$-$\tilde{\tau}_y^{(i)}(\lambda_b)$, $\max[|t(\mu_2)|]$ or $\max[|t(\beta_2)|]$ at locating the correct breakpoint. For example at $\mu_2 = 6$ in Model 1, 98.7% of breaks are correctly located at T_b^c.

NH:3.ii. However, if $\tilde{\tau}_y^{(i)}(\lambda_b, \text{BIC})$ is used in combination with the critical values for $\inf$-$\tilde{\tau}_y^{(i)}(\lambda_b)$, then the test is undersized at the 5% nominal level, and the extent of the undersizing increases as the break increases. For example, in Model 1 there is an actual size of 3.8% when $\mu_2 = 0$ and 0.6% when $\mu_2 = 6$; this will obviously affect power.

NH:3.iii. If $\tilde{\tau}_y^{(i)}(\lambda_b, \text{BIC})$, $i = 1, 2$, is used, with the critical values from Perron's exogenous break tests, then the tests are initially substantially oversized but approach the correct nominal size as the magnitude of the break increases. For example, in Model 1, 27.1% size with $\mu_2 = 0$ but 5.9% with $\mu_2 = 6$.

The picture is much the same under AH, with $\rho = 0.8$:

AH:1. Using $\inf\text{-}\tilde{\tau}_\gamma^{(i)}(\lambda_b)$, $\text{Max}[|t(\mu_2)|]$ or $\text{Max}[|t(\beta_2)|]$ leads to incorrect location of the break date at $T_b^c - 1$ and almost never gets it right.

AH:2.i. BIC is very much better at selecting the correct break date; for example, in Model 1 around 80% correct for $\mu_2 = 4$ and 99% correct for $\mu_2 = 6$.

AH:2.ii. Power is difficult to assess because size is distorted when $\mu_2 \neq 0$ or $\beta_2 \neq 0$; however, power increases for $\inf\text{-}\tilde{\tau}_\gamma^{(i)}(\lambda_b)$, $\text{Max}[|t(\mu_2)|]$ or $\text{Max}[|t(\beta_2)|]$ as the break magnitude increases.

AH:2.iii. In the case of $\tilde{\tau}_\gamma^{(i)}(\lambda_b, \text{BIC})$, $i = 1, 2$, size decreases with the break magnitude, and the correct size is obtained using Perron's exogenous break critical values for 'large' breaks; thus power will only be size-adjusted for large breaks. For example, power of 53.7% for Model 1 with $\mu_2 = 10$ and 36.8%.

The two central messages from this simulation set-up (which includes a break under the null) are:

(i) choosing the break point using $\inf\text{-}\tilde{\tau}_\gamma^{(i)}(\lambda_b)$, $\text{Max}[|t(\mu_2)|]$ or $\text{Max}[|t(\beta_2)|]$ will tend to mislocate it at $T_b^c - 1$;
(ii) using BIC, on the other hand, has a very high probability of correct location.

8.3.1.ii   Use BIC to select the break date?

These conclusions point towards the use of BIC to locate the break date. However, there is a practical problem. The distribution of the psuedo-t associated with the BIC choice of break point, $\tilde{\tau}_\gamma^{(i)}(\lambda_b, \text{BIC})$, only approaches the distribution of $\tilde{\tau}_\gamma^{(i)}(\lambda_b)$ for $i = 1, 2$, assuming an exogenous choice of break point as the break magnitude increases. However, from a practical perspective this still leaves a problem of obtaining correctly sized critical values when the magnitude of the break is not large. Use of $T_b^c - 1$ rather than T_b^c has implications for the bias in estimating ρ, which is maximised at $T_b^c - 1$, and the oversizing of $\inf\text{-}\tilde{\tau}_\gamma^{(i)}(\lambda_b)$, which is also maximised at $T_b^c - 1$; see LSt (2001, table 3).

LSt also consider the case where there is a break under the null and the one-time dummy DT_t^b is (incorrectly) excluded from the maintained regression; recall that DT_t^b should be included, otherwise the maintained regression does not nest the null; further, since DT_t^b is a one-time dummy its presence will not affect critical values relevant to the case where there is no break under the null. The key result is that omitting DT_t^b and using $\inf\text{-}\tilde{\tau}_\gamma^{(i)}(\lambda_b)$ leads to a much better probability of correct location of the break point than when DT_t^b is included.

However, this is not a panacea as far as using other key parameters are concerned; because of the misspecification of the maintained regression, the end result for the bias in estimating ρ and oversizing of $\inf\text{-}\tilde{\tau}_\gamma^{(i)}(\lambda_b)$ is about the same as if the maintained regression is correctly specified but leads to a choice of

$T_b^c - 1$. Nevertheless, both the correctly and (deliberately) misspecified maintained regressions could be run; the probability is that they will differ by 1 period in their selections of T_b; if so, choose T_b from the misspecified regression but choose the corresponding value of $\tilde{\tau}_\gamma^{(i)}(\lambda_b)$ from the correctly specified regression. By its nature the chosen $\tilde{\tau}_\gamma^{(i)}(\lambda_b)$ will not now be the minimum of all $\tilde{\tau}_\gamma^{(i)}(\lambda_b)$. The problem here is that the critical values for $\inf\text{-}\tilde{\tau}_\gamma^{(i)}(\lambda_b)$ do not relate to this method of choice, so this is not yet the complete solution if there is a break under the null.

8.3.2 Harvey, Leybourne and Newbold (2001)

Harvey, Leybourne and Newbold (2001) (HLN), also address the problem of correct breakpoint selection and suggest a solution with practical effect, at least for Models 1 and 1a. (The asymptotics underlying this result are slightly different from VP, in that the break magnitude is proportional to the sample size rather than fixed as the sample size increases; this has the consequence that it is possible to derive a limiting null distribution that is independent of the break magnitude.)

Recall from **8.2.3** that with an incorrect break date $\tilde{\tau}_\gamma^{(1)}(\lambda_b)$, in AO and IO models, whilst asymptotically invariant is no longer exactly invariant to μ_2 (that is finite sample critical values are sensitive to the value of μ_2), and the same result holds for $\tilde{\tau}_\gamma^{(1a)}(\lambda_b)$ from Model 1a. However, HLN show by simulation that finite sample effects can be important for both AO and IO Models 1a when selection is by $\inf\text{-}\tilde{\tau}_\gamma^{(1a)}(\lambda_b)$ or $\text{Max}[|t(\mu_2)|]$; as μ_2 increases, oversizing become severe for AO with $\inf\text{-}\tilde{\tau}_\gamma^{(1a)}(\lambda_b)$ and for IO with either $\inf\text{-}\tilde{\tau}_\gamma^{(1a)}(\lambda_b)$ or $\text{Max}[|t(\mu_2)|]$.

8.3.2.i The break date is incorrectly chosen

The root of the problem, as HLN show by simulation and theoretical argument, is that the break date is incorrectly chosen (confirming the results of LSt, who adopted conventional asymptotics where the break magnitude is fixed). The t statistic on the break parameter, $t(\mu_2)$, is maximised at $T_b^c - 1$ not T_b^c. Let $\tilde{T}_b$ denote the estimate of T_b^c using the $\text{Max}|t(\mu_2)|$ rule and then add 1 to get the estimate $\breve{T}_b = \tilde{T}_b + 1$; this removes the bias in $\tilde{T}_b$. The unit root test is then the psuedo-t, $\tilde{\tau}_\gamma^{(1)}(\lambda_b)$, corresponding to $\breve{T}_b$, say $\breve{\tau}_\gamma^{(1)}(\breve{\lambda}_b)$. To solve the problem of appropriate critical values, Harvey et al (op. cit.) simulate a driftless random walk without breaks, with trimming, so that $\lambda_b \in (0.2, 0.8)$. The two cases considered have the maintained regressions for Model 1 and Model 1a respectively. 5% critical values for some finite sample sizes and the asymptotic case are given in Table 8.4.

8.3.3 Selection of break model under the alternative

How to approach the question of selecting the correct model depends in part on the focus of an enquiry. It is possible that choosing the wrong model under the

Table 8.4   5% critical values for the $\tilde{\tau}_\gamma^{(i)}(\breve{\lambda}_b)$ test, $i =$ Model 1, Model 1a; break date $= \breve{T}_b = \tilde{T}_b + 1$

	Model 1	Model 1a
T = 50	−4.42	−3.93
T = 100	−4.44	−4.01
T = 200	−4.50	−4.07
T = 500	−4.59	−4.16
T = ∞	−4.72	−4.33

Notes: extracted from HLN (2001, Table 2).
Source: The maintained regressions are:

$$\Delta y_t = \mu_1 + \beta_1 t + \gamma y_{t-1} + \mu_2 DT_t^b(\lambda_b) + \beta_2 DU_t(\lambda_b) + \varepsilon_t \quad \text{Model 1}$$
$$\Delta y_t = \mu_1 + \gamma y_{t-1} + \delta DT_t^b(\lambda_b) + \mu_2 DU_t(\lambda_b) + \varepsilon_t \qquad \text{Model 1a}$$
$$\text{DGP: } y_t = y_{t-1} + \varepsilon_t$$

alternative might still lead to the correct conclusion that a series is generated by a stationary DGP with a structural break; the form of the break is incorrectly chosen but stationarity is correctly concluded. In this case, the probability of interest is the (unconditional) probability of rejecting the null hypothesis (whichever maintained regression is used).

Alternatively, the form of the break may well be of interest in its own right; that is, not only do we want to discover whether a unit root is present, but if not we also want to know whether the break has affected the level or slope or both. In this case we are interested in the conditional probability of choosing model i given that model i is the DGP, and then, given model i, the probability of rejecting the unit root.

In this section the results of two studies, which relate to the implications of choosing the wrong model, are summarised. The first is due to Montañés, Olloqui and Calvo (2005) hereafter MOC, who consider the theoretical implications, with some simulation results, and the second is the simulation-based study by Sen (2003). The theoretical results are summarised first, followed by the simulation results.

8.3.3.i   *Misspecifying the break model: what are the consequences?*

MOC consider the implications of choosing the wrong type of break model for the behaviour of the psuedo-t statistic $\tilde{\tau}_\gamma^{(i)}(\lambda_b)$. Their analysis is based on the IO model but qualitatively similar results obtain for the AO model. With three types of break model, misspecifications can arise in six ways; for example Model 1 is correct, but Model 2 is used; Model 1 is used but Model 3 is correct; and so on. Some cases may, however, be more relevant than others; for example, it seems a reasonable view, a priori, that using Model 2 when Model 1 is correct may not distort inference too much, as nothing that is in Model 1 is omitted from Model 2 (this also follows the line of approach suggested by Sen (2003) to use the most general model).

A summary of the essence of their theoretical results, first for the case of no break under the null, follows. In each case, the data is generated by the corresponding model under the trend-break alternative, whereas the estimated model is the IO version of the model, specified as in Perron (1989) with the one-off dummy variable in Models 1 and 2 (although this is not critical to the asymptotic results). The errors are permitted to follow a stationary and invertible ARMA process and the aysmptotics assume that the break is of fixed size rather than proportional to the sample size.

The results are divided into two cases, first when the break is correctly (and exogenously) selected as correct, and the second when the break is incorrectly located. This latter case is relevant because criteria such as $\inf\text{-}\tilde{\tau}_\gamma^{(1)}(\lambda_b)$, $\text{Max}[|t(\mu_2)|]$ and $\text{Max}[|t(\beta_2)|]$ involve estimation at incorrect break dates.

8.3.3.ii No break under the null

Model 2 correct; Model 1 estimated
If $\lambda_b = \lambda_b^c$ then $\tilde{\tau}_\gamma^{(1)}(\lambda_b^c) \to_p 0 \Rightarrow$ the unit root is not rejected and the test has no power.
If $\lambda_b \neq \lambda_b^c$, then $\tilde{\tau}_\gamma^{(1)}(\lambda_b^c) \to_p < 0$, that is, the limit of $\tilde{\tau}_\gamma^{(1)}(\lambda_b^c)$ is always negative and depends only on λ_b and λ_b^c.

Model 2 correct; Model 3 estimated
If $\lambda_b = \lambda_b^c$ then $\tilde{\tau}_\gamma^{(3)}(\lambda_b^c) \to_p -(3/2)[(1 - \lambda_b^c)/\lambda_b^c]^{0.5} \equiv t(\lambda_b^c) < 0$, the test statistic $t(\lambda_b^c) \to_p -\infty$ as $\lambda_b^c \to 0$ and $t(\lambda_b^c) \to_p 0$ as $\lambda_b^c \to 1$; this implies that rejection of the unit root is only possible for a break at the beginning of the sample, for example $t(\lambda_b^c) \to_p -6.54$ for $\lambda_b^c = 0.05$ and $t(\lambda_b^c) \to_p -4.5$ for $\lambda_b^c = 0.1$, and only the first of these implies rejection. If $\lambda_b \neq \lambda_b^c$, the qualitative conclusions are essentially the same, with rejection of the unit root occurring if the break point is at the beginning of the sample.

Model 3 correct; Model 1 estimated
If $\lambda_b = \lambda_b^c$, then $\tilde{\tau}_\gamma^{(1)}(\lambda_b^c) \to_p 0 \Rightarrow$ the unit root is not rejected and the test has no power.
If $\lambda_b \neq \lambda_b^c$, then $\tilde{\tau}_\gamma^{(1)}(\lambda_b^c) \to_p \pm\infty \Rightarrow$ thus either sign is possible depending on the error in locating the break: the unit root may be rejected $\tilde{\tau}_\gamma^{(1)}(\lambda_b^c) \to_p -\infty$, but as $\tilde{\tau}_\gamma^{(1)}(\lambda_b^c) \to_p +\infty$ is also possible, the test may erroneously indicate explosive behaviour.

Model 3 correct; Model 2 estimated
If $\lambda_b = \lambda_b^c$, then $\tilde{\tau}_\gamma^{(2)}(\lambda_b^c) \to_p -\infty$, diverging at the rate $T^{1/2}$, and the unit root is rejected.

If $\lambda_b \neq \lambda_b^c$, then $\tilde{\tau}_\gamma^{(2)}(\lambda_b^c) \to_p \pm\infty$, diverging at the rate $T^{1/2}$, and either sign is possible depending on the error in locating the break; for example, when $\lambda_b \in [(\lambda_b < \lambda_b^c) \cap (0.5 < \lambda_b^c < 1.0) \cap (\lambda_b < 2\lambda_b^c - 1)]$, then $\tilde{\tau}_\gamma^{(2)}(\lambda_b^c) > 0$, so that rejection of the null is not possible; for example, if $\lambda_b^c = 0.8$, then $\lambda_b < 0.6$ satisfies these conditions. This is an illustration that estimating the more general model does not safeguard against an incorrect decision.

8.3.3.iii   Break under the null

The second set of cases considered by MOC is when the data is generated under the null with a break and, as before, the maintained regressions are the IO versions of the models. Under the null, the models are:

$$\text{Model 1:} \qquad y_t = y_{t-1} + \mu_1 + \mu_2 DT_t^b(\lambda_b) + \varepsilon_t \qquad\qquad (8.36)$$

$$\text{Model 2:} \qquad y_t = y_{t-1} + \mu_1 + \mu_2 DT_t^b(\lambda_b) + \beta_2 DU_t(\lambda_b) + \varepsilon_t \qquad (8.37)$$

$$\text{Model 3:} \qquad y_t = y_{t-1} + \mu_1 + \beta_2 DU_t(\lambda_b) + \varepsilon_t \qquad\qquad (8.38)$$

However, since the one-off dummy variable can be deleted without loss, asymptotically these models reduce to:

$$\text{Model 1:} \qquad y_t = y_{t-1} + \mu_1 + \varepsilon_t \qquad\qquad (8.39)$$

$$\text{Model 2:} \qquad y_t = y_{t-1} + \mu_1 + \beta_2 DU_t(\lambda_b) + \varepsilon_t \qquad (8.40)$$

$$\text{Model 3:} \qquad y_t = y_{t-1} + \mu_1 + \beta_2 DU_t(\lambda_b) + \varepsilon_t \qquad (8.41)$$

Given the asymptotic irrelevance of DT_t^b, the case of interest, which represents a misspecification, is when Model 2 (or 3) generates the data but the maintained regression for Model 1 is estimated. In this case the following result.

Model 2 correct; Model 1 estimated
If $\lambda_b = \lambda_b^c$ then $\tilde{\tau}_\gamma^{(1)}(\lambda_b^c) \to_p 0 \Rightarrow$ the unit root null is not rejected; whilst this is the correct decision given that the data are generated under the null, the nominal size of the test is not respected.
If $\lambda_b > \lambda_b^c$ then $\tilde{\tau}_\gamma^{(1)}(\lambda_b)$ diverges and for some parts of the parameter space the unit root null will be rejected, particularly where there is a large break at the beginning of the sample. The null is not likely to be rejected when $\lambda_b^c > 0.5$.

8.3.3.iv   Simulation results

MOC carry out a simulation analysis with $T = 100$, $\lambda_b^c = 0.3, 0.5, 0.7$ and errors generated by an ARMA(1, 1) model. First, consider the misspecification arising from Model 1 being correct but either Models 2 or 3 are used. Whilst some of the results are to be anticipated, some are not.

Using Model 2 does not lead to a serious or noticeable loss of power for a moderate intercept break ($\mu_2 = 1$) even if the break point is mislocated, $\lambda_b \neq \lambda_b^c$.

However, if the intercept break is large ($\mu_2 = 10$ in the simulations) power is sustained for a correct break date, but there can be a serious loss of power even for incorrect break dates close to the correct date; for example, with both ARMA parameters equal to zero, for a location error within 5% of $\lambda_b^c = 0.5$ power drops, quite remarkably, from 100% to 1%.

If Model 3 is estimated when Model 1 is correct, then power for $\mu_2 = 1$ is comparable to when Model 2 is estimated, broadly between 70% and 90% depending upon the ARMA parameters. However, generally power falls away to zero as μ_2 increases, the exceptions being perverse in the sense that power can be higher when there is an error in the location of the break.

The other 'nested' case of interest in the simulation analysis is when the correct model is Model 3 but Model 2 is estimated (when Model 1 is estimated $\tilde{\tau}_\gamma^{(1)}(\lambda_b^c)$ has virtually no power). Here higher power can occur for $\lambda_b < \lambda_b^c$; for example with λ_b 5% less than λ_b^c, so that whilst power is maintained the break date may be incorrectly indicated.

Sen (2003) suggests the strategy of using the most general model, which is the mixed model of Model 2, as the alternative. As in the case of MOC, the argument is that whilst there may be some loss of power if either Model 1 or Model 3 is correct, the specification of the maintained regression using Model 2 will, at least, allow for inference based on the correct model. Not all empirical studies have reflected this choice. For example, the 'preliminary investigation' view may involve some initial assessment of the likely break model through, for example, graphical inspection. Sen's study complements MOC in the sense that he considers a number of choice criteria for the unknown break fraction.

As to the loss of power, Sen (2003) generates some simulations for $T = 100$ and $T_b^c = 50$, that is, $\lambda_b^c = 0.5$, with the DGP an ARMA(2, 1) version of IO Model 2 given by:

$$(1 - (\rho + \varphi)L + \varphi L^2)y_t = (1 + \theta_1 L)(\mu_2 DU_t + \beta_2 DT_t + \varepsilon_t) \tag{8.42}$$

Variations on this DGP allow Models 1 and 3 to be generated as special cases. The test statistics considered are: inf-$\tilde{\tau}_\gamma^{(i)}(\lambda_b)$; $\tilde{\tau}_\gamma^{(i)}(\lambda_b)$ corresponding to Max[|t(μ_2)|] for Model 1, Max[|t(β_2)|] for Models 2 and 3, Max[F(μ_2, β_2)] for Model 2; and the Murray and Zivot (1998), Sup-W(λ_b, γ) test; see Sections **8.1.3** and **8.2.2** for details of these tests. The lag length is treated as unknown and estimated using the t-sign procedure with a 10% two-sided significance level; the null hypothesis is assumed not to include a break.

Noting that empirical and nominal size was close for the test statistics considered, apart from the case of a negative MA component and, hence, empirical power can be compared, Sen's main findings are as follows. (These results are for the IO model; however, Sen (op. cit.) reports that the qualitative results also hold for the AO form of the models.)

Model 1 correct; Model 1 estimated
If the correct model is Model 1, inf-$\tilde{\tau}_\gamma^{(1)}(\lambda_b)$ and $\tilde{\tau}_\gamma^{(1)}(\text{Max}[|\mu_2|])$ have higher power than for $i = 2, 3$; that is, using the correct maintained regression for Model 1 results in these test statistics having the best power.

Model 1 correct; Model 2 estimated
The power of the Sup-W(λ_b, γ) test, which assumes Model 2 by construction, is maintained and generally comparable with the power using inf-$\tilde{\tau}_\gamma^{(1)}(\lambda_b)$ and $\tilde{\tau}_\gamma^{(1)}(\text{Max}[|\mu_2|])$. Overall, the 'over-specification' of using Model 2 when Model 1 is the DGP is not too harmful to power.

Model 1 correct; Model 3 or 2 estimated
Using the test statistics assuming Model 3 as the maintained regression results in virtually zero power, whereas using the test statistics assuming Model 2 generally results in a moderate loss of power.

Model 2 correct; Model 1 estimated
The power of inf-$\tilde{\tau}_\gamma^{(1)}(\lambda_b)$ and $\tilde{\tau}_\gamma^{(1)}(\text{Max}[|\mu_2|])$ is generally close to zero except when the level coefficient is large relative to the slope coefficient (that is when Model 2 'looks' like Model 1).

Model 2 correct; Model 2 estimated
If the correct model is Model 2 (mixed model), then the power of inf-$\tilde{\tau}_\gamma^{(2)}(\lambda_b)$ and $\tilde{\tau}_\gamma^{(2)}(\text{Max}[F(\mu_2, \beta_2)])$ generally increase with increases in the level and slope coefficients. The Sup-W(λ_b, γ) test is generally more powerful than both inf-$\tilde{\tau}_\gamma^{(2)}(\lambda_b)$ and $\tilde{\tau}_\gamma^{(2)}(\text{Max}[F(\mu_2, \beta_2)])$, and its power increases with increases in the level and slope coefficients.

Model 2 correct; Model 3 estimated
The power of inf-$\tilde{\tau}_\gamma^{(3)}(\lambda_b)$ and $\tilde{\tau}_\gamma^{(3)}(\text{Max}[|\beta_2|])$ decrease with increases in μ_2 for given β_2; the power increases with increases in β_2 for given μ_2. The power of inf-$\tilde{\tau}_\gamma^{(3)}(\lambda_b)$ and $\tilde{\tau}_\gamma^{(3)}(\text{Max}[|\beta_2|])$ can be larger than the power of inf-$\tilde{\tau}_\gamma^{(2)}(\lambda_b)$ and $\tilde{\tau}_\gamma^{(2)}(\text{Max}[F(\mu_2, \beta_2)])$, but this advantage decreases as μ_2 increases for given β_2. Consider the case of positive serial correlation, that is, $\varphi > 0$ in (8.42) and no MA component, which is a possible practical situation. The power of $\tilde{\tau}_\gamma^{(3)}(\text{Max}[|\beta_2|])$, which is monotonic in β_2, initially exceeds that of $\tilde{\tau}_\gamma^{(2)}(\text{Max}[F(\mu_2, \beta_2)])$ for given μ_2 as β_2 increases, but then falls behind because the power of $\tilde{\tau}_\gamma^{(2)}(\text{Max}[F(\mu_2, \beta_2)])$ increases faster in β_2. Overall, if Model 2 is the DGP there are some parts of the parameter space where test statistics for the misspecified model, Model 3, do better than the test statistics for the correctly specified model. Without prior knowledge of the break coefficients, Sup-W(λ_b, γ) is the most consistent performer.

Model 3 correct: Model 1 estimated
Assuming Model 1 as the maintained regression results in virtually zero power as might be expected from the absence of a slope change dummy variable; this was the situation found by MOC.

Model 3 correct; Model 2 estimated   Assuming Model 2 results in a loss of power of the order of 20%–50% for the psuedo-t tests, depending upon the model parameters. There is also a loss of power of the Sup-W(λ_b, γ) test assuming Model 2, but the loss is generally not as great as for inf-$\tilde{\tau}_{\gamma}^{(2)}(\lambda_b)$ and $\tilde{\tau}_{\gamma}^{(2)}(\text{sup-}F(\lambda_b))$.

Model 3 correct; Model 3 estimated
If the correct model is Model 3 (changing growth), inf-$\tilde{\tau}_{\gamma}^{(3)}(\lambda_b)$ and $\tilde{\tau}_{\gamma}^{(3)}(\text{Max}\{F(\mu_2, \beta_2)\})$ have higher power than for i = 1, 2. Overall, there is a clear advantage in terms of power to correctly identifying Model 3, rather than using the more general model, Model 2, except when β_2 is quite large, practically $\beta_2 > 0.5$.

Sen's results indicate two key points. First, while using Model 2 when the DGP is Model 1 is not too costly in terms of power; using Model 2 when Model 3 is the DGP has potentially greater costs. Thus, the strategy of ensuring that the wrong model is not used as the basis for inference, for example do not use Model 1 when Model 3 is correct by using the maintained regression for Model 2, involves trade-off costs in terms of lost power that are generally lower when Model 1 is correct than when Model 3 is correct. Second, Sup-W(λ_b, γ) test has power properties that make it a very useful addition to the set of tests for a structural break.

8.3.3.v   Implications of choosing the wrong model
Overall, bringing together the theoretical and simulation results, choosing the wrong type of model can have serious consequences. At worst power is low, possibly zero, and the break date is incorrectly located. This suggests that it would be useful to consider some criteria for selecting the model as well as, as in the previous sections, criteria for selecting the break date.

8.4   What kind of break characterises the data?

The question of which kind of break characterises the data is important in itself. For example, under the alternative that real GNP is generated by a trend stationary process, did the Second World War result in a one-off change in level and/or a change in the growth rate? There are two possibilities here. Since the crash and changing growth models (Models 1 and 3 respectively) are special cases of the mixed model (Model 2), in principle the problem of selection could be approached by testing the break coefficients by Sup-Wald type tests along the lines suggested by Vogelsang (1997) – see Section **8.1.1** – and then assessing the p-values of the restrictions corresponding to the three models.

For example, calculate the test statistic for the null that the coefficients on the intercept and slope dummies are jointly zero (the maintained regression is Model 2) and obtain the Sup-Wald test value; also calculate the Sup-Wald (in this case Sup-t) test statistics for Models 1 and 3 that each involve one restriction on Model 2. Choose the model with the smallest p-value (least consistent with the null of a zero coefficient(s)). The practical problem with this approach is that the null distributions have to be obtained by means of simulation, so obtaining the required p-values, while straightforward, is time-consuming. Alternatively, a simple option is to apply some commonly used model selection criteria that are based on imposing a penalty for additional coefficients. These are considered next along with a variation on the inf-$\tilde{\tau}_\gamma^{(i)}(\lambda_b)$ selection rule.

Note the distinction between the question of which model best characterises the data, which is considered here, and the question of, given a particular model, which is the best form of that model. In the former case the choice, in the framework considered here, is between Models 1, 1a, 2 and 3, and in the latter case, given one of these models, the question is what is the nature of the dynamic adjustment: AO, IO or DO? Although these questions have been addressed separately (see Section **7.4** for the latter case) in principle both could be considered jointly.

8.4.1   Methods of selecting the type of break

In this section four criteria for selecting the model are considered: AIC, BIC, the log of the LS residual variance (or equivalently minimising the LS standard error se) and minimising inf-$\tilde{\tau}_\gamma^{(i)}(\lambda_b)$ over the set of models.

The AIC and BIC are frequently used model selection criteria, defined as follows:

$$\text{AIC} = \ln[\tilde{\sigma}_{\varepsilon,i}^2(\lambda_b)] + 2K(i)/T \tag{8.43}$$

$$\text{BIC} = \ln[\tilde{\sigma}_{\varepsilon,i}^2(\lambda_b)] + K(i)(\ln T)/T \tag{8.44}$$

where $\tilde{\sigma}_{\varepsilon,i}^2$ is the ML estimator of the variance and $K(i)$ is the number of estimated coefficients, both in model i.

The LS estimator of the variance, $\hat{\sigma}_{\varepsilon,i}^2 = \text{RSS}_i/(T-K) = \tilde{\sigma}_i^2[T/(T-K)]$, includes a degrees of freedom adjustment and can be put in a similar framework to the two information criteria as follows:

$$\text{se} = \ln[\tilde{\sigma}_{\varepsilon,i}^2(\lambda_b)] + \ln[T/(T-K)] \tag{8.45}$$

Evidently AIC, BIC and se differ only in the complexity penalty imposed. Each of these criteria is calculated for $\lambda_b \in \Theta$ and then for i = 1, 2, 3; the chosen model is then that for which the criterion is at an overall minimum.

Another criterion, this time specific to the unit root issue, is to extend the minimum psuedo-t criterion, as applied to possible break fractions, to the three possible models:

$$\text{min-t} = \inf^{(i)}(\inf_{\lambda_b}\{\tilde{\tau}_{\gamma}^{(i)}(\lambda_b)\}) \tag{8.46}$$

As with the information criteria, for this criterion the minimum psuedo-t is calculated for $\lambda_b \in \Theta$ and then for Model i, i = 1, 2, 3; the model chosen corresponds to the overall minimum.

The simulation experiment is kept simple to see if the model selection criteria work well enough to be considered further. Each of the three break models is alternately taken as the DGP; thus for, say, Model 1 as the DGP, the maintained regressions in IO form corresponding to the three models are run; the proportion of times that each model is chosen is then recorded. This is repeated for the alternate models.

$$\text{Model 1}: y_t = \rho y_{t-1} + \mu_1 + \beta_1 t + \mu_2 DU_t(\lambda_b) + \varepsilon_t \tag{8.47}$$

$$\text{Model 2}: y_t = \rho y_{t-1} + \mu_1 + \beta_1 t + \beta_2 DT_t(\lambda_b) + \varepsilon_t \tag{8.48}$$

$$\text{Model 3}: y_t = \rho y_{t-1} + \mu_1 + \beta_1 t + \mu_2 DU_t(\lambda_b) + \beta_2 DT_t(\lambda_b) + \varepsilon_t \tag{8.49}$$

Throughout T = 200 and $\mu_2 = 5$, $\beta_1 = 1$; β_2, $\mu_2 \in [0.05, 0.1, \ldots, 0.9, 0.95, 1, 2, \ldots, 10]$; the grid for the change parameters is finer in the lower part of the grid; $\varepsilon_t \sim \text{niid}(0, \sigma_\varepsilon)$ and $\sigma_\varepsilon = 1$, so that the change parameters are in units of the innovation standard error; $\rho \in [0, 0.9]$; $\lambda_b \in \Theta = [0.05, 0.1, \ldots, 0.9, 0.95]$. Note that the possibility of correlated but stationary deviations around the deterministic components is allowed; although $\rho = 0$ in the original Perron specification, in practice $\rho > 0$ is likely. The maintained regressions are as specified in section **7.6** and include the one-off dummy variable DT^b.

The results are presented graphically in Figures 8.1, 8.2 and 8.3 for the cases where Model i, i = 1, 2 and 3, respectively, is the DGP; each figure contains 4 x 3 subfigures. Each column of figures refers to the selected model and each row to the criterion used. The break fraction is on the left-hand of the two horizontal axes and either μ_2 or β_2 is on the right-hand axis. The probability of selection (simulated relative frequency) is shown on the vertical axis. In the case of Model 3, the results are presented for $\mu_2 = 5$ (as used by Vogelsang and Perron, 1997, in their simulations); for all models in these figures $\rho = 0.9$, with $\rho = 0$ considered separately below.

For Figure 8.1 interest concerns whether the criteria select Model 1, which is the DGP. All criteria improve as the size of the break increases. Overall, the min-t criterion does best, followed by BIC, AIC and then se. To interpret the figures, note that a good selection criterion is one where the graph in the first column looks most like a cube over the whole surface with the 'top' at 1; this implies a joint probability mass that lies close to the 'floor' for graphs in the other two

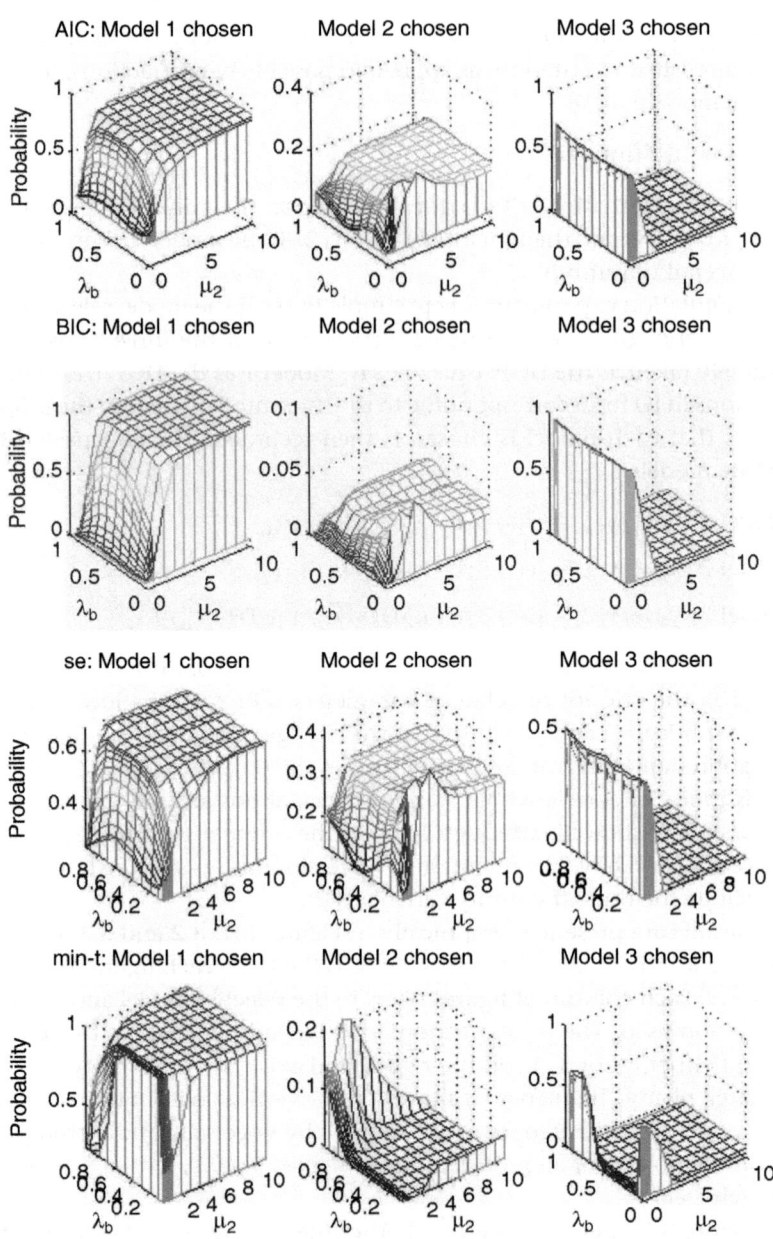

Figure 8.1   Model 1 correct

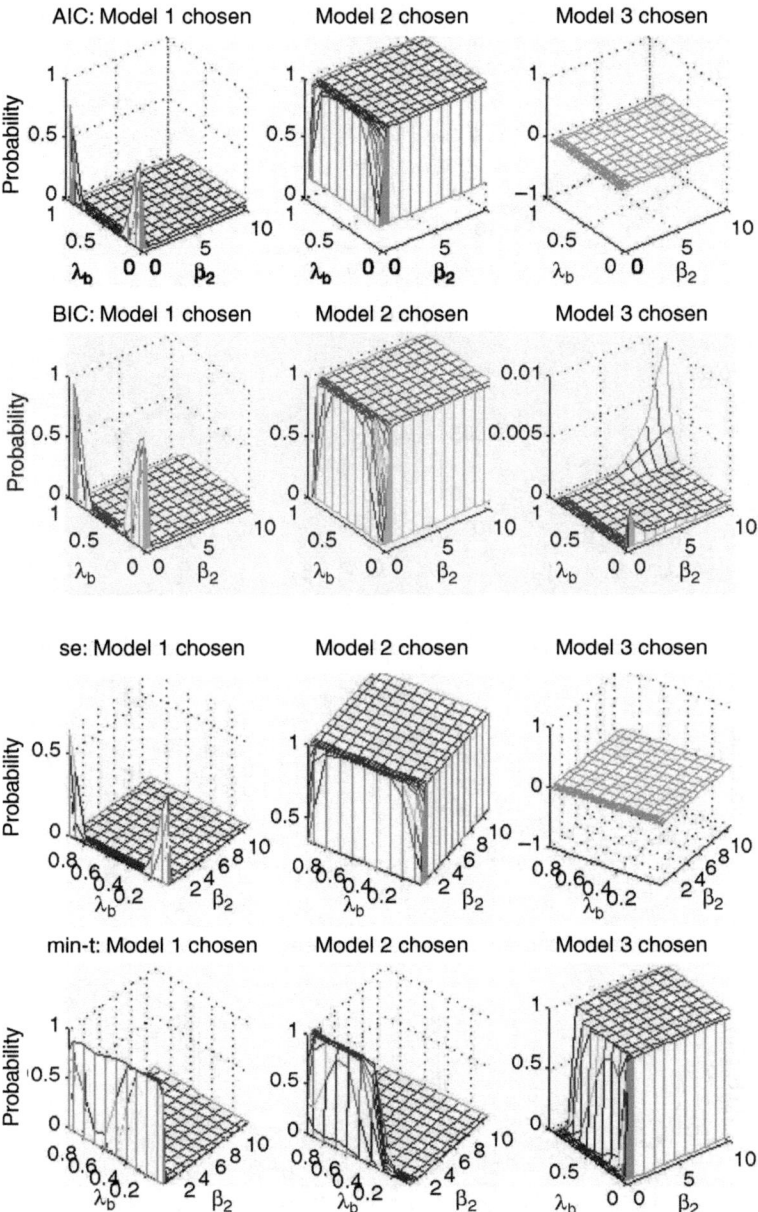

Figure 8.2   Model 2 correct

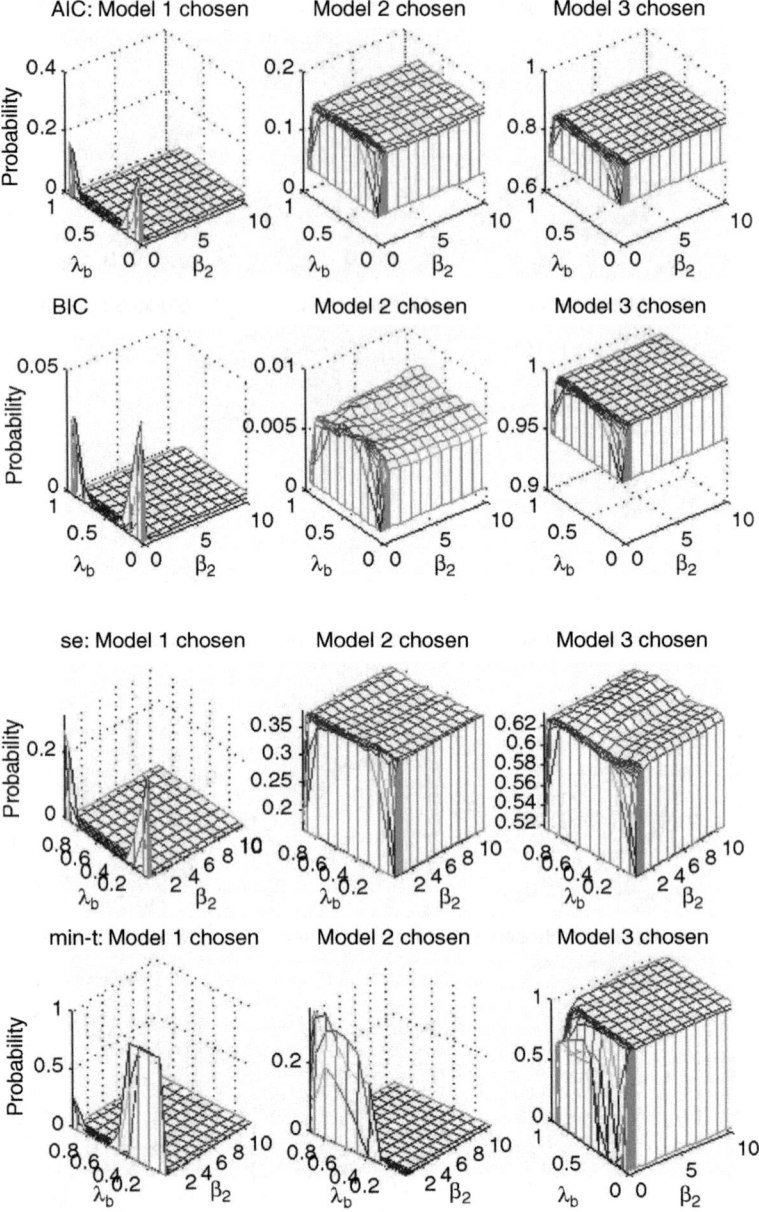

Figure 8.3   Model 3 correct

columns. Min-t does particularly well relative to the other criteria for $0.1 < \lambda_b < 0.6$ and $\mu_2 < 2$; none of the other criteria do well when $\mu_2 < 2$, with Model 3 tending to be chosen in such cases, see especially BIC. Both se and AIC have non-negligible probabilities of Model 2 being chosen throughout the parameter range.

For Figure 8.2 interest concerns whether the criteria select Model 2, which is the DGP. Now AIC, BIC and se all do much better with little to choose between them, and it is min-t that tends to indicate the wrong model, predominantly choosing Model 3 rather than Model 2. Even with AIC, BIC and se, as well as with min-t, there is a problem in selecting the right model if β_2 is not large and the breakpoints are at the beginning or end of the sample (see the sub-figures in the first column that show Model 1 being chosen in these circumstances).

Turning to Figure 8.3, where the DGP is Model 3, the best criterion is BIC followed by min-t. Using AIC or se leads to a non-negligible probability of incorrectly choosing Model 2 even with β_2 large. BIC does particularly well in this situation, with no probability below 0.95 even with breaks near the beginning or end of the sample and β_2 not large.

The qualitative results in terms of rankings for the different criteria are generally maintained for different values of ρ, but there is an important qualification for the min-t criterion. To illustrate consider $\rho = 0$ and $\rho = 0.9$ (as above) for Model 2 using BIC and alternately min-t, with the results shown in the left-hand panel of Figure 8.4 for $\rho = 0$ and the right-hand panel for $\rho = 0.9$. BIC is virtually unaffected by ρ, getting the choice right provided the break is neither too early nor too late in the sample nor too small; however, min-t does very much better when $\rho = 0$, deteriorating markedly for $\rho = 0.9$.

Overall, no single criterion performs best for all three models; however, BIC is the most consistent performer, but problems with λ_b close to 0 or 1 with β_2 or μ_2 less than 1 are pervasive. Of course, a model selection criterion cannot be expected to be right in finite samples all of the time; in a hypothesis testing framework this is controlled by the type I error, which does not have an explicit role in the criteria that have been considered.

8.5 Multiple breaks

In a particular period there may, of course, may be more than one event that could potentially give rise to a structural break in the process generating the data. A starting point to the more general case is to allow for two breakpoints in the sample, the dates of which are denoted $T_{b1}^c = \lambda_{b1}^c T$ and $T_{b2}^c = \lambda_{b2}^c T$ for the correct break dates, and the convention is adopted that $\lambda_{b1}^c < \lambda_{b2}^c$. When there is no presumption that the break dates are known, the notation is: $T_{b1} = \lambda_{b1} T$ and $T_{b2} = \lambda_{b2} T$, respectively. As in the single break case, the usual practical case is that the break dates are unknown. Next, the most general of the single

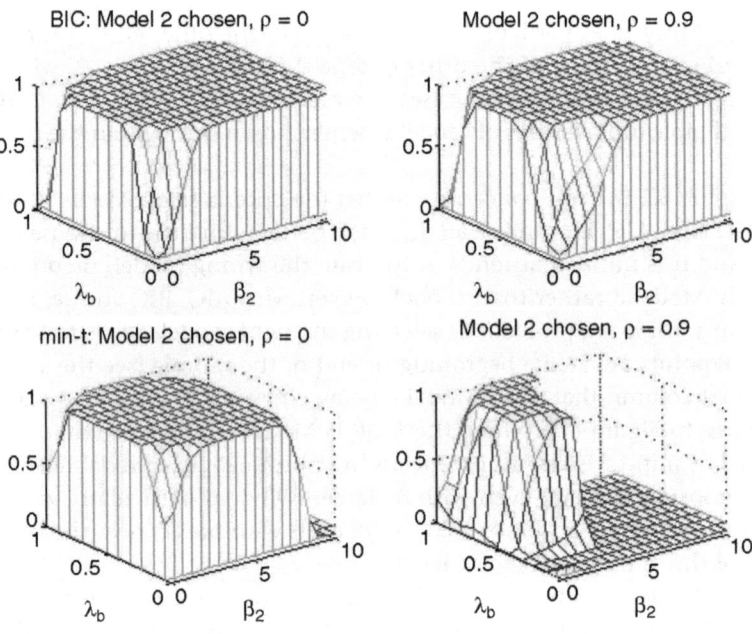

Figure 8.4   Model 2 correct: $\rho = 0$, $\rho = 0.9$

breakpoint models is considered, that is Model 2, and then generalised to allow two breakpoints, first in the AO framework and under the alternative hypothesis; the notation now reflects the possibility of two breaks. The simpler AO models can be obtained as special cases of Model 2.

8.5.1   AO Models

AO Model (2, 2)

$$y_t = \mu_1 + \beta_1 t + \mu_{21} DU1_t(\lambda_{b1}) + \mu_{22} DU2_t(\lambda_{b2})$$
$$+ \beta_{21} DT1_t(\lambda_{b1}) + \beta_{22} DT2_t(\lambda_{b2}) + \omega_t \tag{8.50}$$

$DU1_t = 1$ if $t > T_{b1}$ and 0 otherwise; $DT1_t = t - T_{b1}$ if $t > T_{b1}$ and 0 otherwise; $DU2_t = 1$ if $t > T_{b2}$ and 0 otherwise; $DT2_t = t - T_{b2}$ if $t > T_{b2}$ and 0 otherwise.

As in the case of the one-break version of Model 2, the level and slope change, if present, are constrained to occur at the same time, that is T_{b1} and T_{b2}, respectively. In the AO framework, the data is detrended under the alternative hypothesis; that is, by running the regression in (8.50) and using the residuals denoted $\tilde{\tilde{y}}_t^{(i)}$ in a subsequent test for a unit root.

Various special cases of (8.50) can be obtained by setting particular coefficients to zero; those most directly of interest are:

AO Model (1, 1): Double-level/intercept shift; $\beta_{21} = \beta_{22} = 0$

$$y_t = \mu_1 + \beta_1 t + \mu_{21} DU1_t(\lambda_{b1}) + \mu_{22} DU2_t(\lambda_{b2}) + \omega_t \qquad (8.51a)$$

AO Model (1, 2): level shift plus level plus slope change; $\beta_{21} = 0$

$$y_t = \mu_1 + \beta_1 t + \mu_{21} DU1_t(\lambda_{b1}) + \mu_{22} DU2_t(\lambda_{b2}) + \beta_{22} DT2_t(\lambda_{b2}) + \omega_t \qquad (8.51b)$$

AO Model (3, 3): Double-slope change; $\mu_{21} = \mu_{22} = 0$

$$y_t = \mu_1 + \beta_1 t + \beta_{21} DT1_t(\lambda_{b1}) + \beta_{22} DT2_t(\lambda_{b2}) + \omega_t \qquad (8.51c)$$

The notation indicates the coupling of models in general, Model (i, h); for example, Model (1, 2) couples the level change of Model 1 with the simultaneous level plus slope change of Model 2. Notice that Model (i, h) $\neq$ Model (h, i) except when i = h.

Relaxing the constraint that level and slope changes must occur simultaneously gives Model (1, 3) as follows:

AO Model (1, 3): level shift plus slope change; $\mu_{22} = 0 = \beta_{21} = 0$:

$$y_t = \mu_1 + \beta_1 t + \mu_{21} DU1_t(\lambda_{b1}) + \beta_{22} DT2_t(\lambda_{b2}) + \omega_t \qquad (8.51d)$$

The simplest specification under the null hypothesis is that there is no break; alternately, the Perron (1989) approach to specification of the null implies that at each of the two breakpoints there is potentially a change in the level and a change in the drift, so that under the null:

$$y_t = y_{t-1} + \beta_1 + \mu_{21} DT1_t^b(\lambda_{b1}) + \mu_{22} DT2_t^b(\lambda_{b2})$$
$$+ \beta_{21} DU1_t(\lambda_{b1}) + \beta_{22} DU2_t(\lambda_{b2}) + \varepsilon_t \qquad (8.52)$$

where $DU1_t(\lambda_{b1})$ and $DU2_t(\lambda_{b2})$ are as defined above; in addition, $DT1_t^b(\lambda_{b1}) = 1$ for $t = T_{b1} + 1$ and 0 otherwise; and $DT2_t^b(\lambda_{b2}) = 1$ for $t = T_{b2} + 1$ and 0 otherwise.

The estimation problem is to determine the two break dates and estimate the corresponding test statistic(s); the inference problem is to determine the critical values from the null distribution and whether the sample data are consistent with the null. In addition, as in the single break case, there is also the problem of determining the model specification.

The AO Model with two breaks just requires the prior estimation of the appropriate version of (8.50), which then serves to define the residuals for the second-stage regression; these are denoted $\tilde{\tilde{y}}_t^{(i,h)}$ for AO Model (i, h). In the case of two breaks and model combinations involving Models 1 and 2, current, and lagged, values of $DTi_t^b(\lambda_{bi})$ as necessary should be included in the second-stage regression. Recall that in the case of AO Model 3 and a single break, no one-off dummy variables were required; however, only in the AO Model (3, 3) combination would that now seem to be warranted, and in the combinations AO Models (1, 3), (3, 1), (2, 3) and (3, 2) dummy variables are included. (Strictly, these arguments are by analogy with the single break case; the author is not aware that the relevant asymptotic theory has been presented for the multiple break case.)

In summary, the second stage regressions for the AO models are:

$$\Delta\tilde{y}_t^{(i,h)} = \gamma\tilde{y}_{t-1}^{(i,h)} + \sum_{j=1}^{k-1} c_j \Delta\tilde{y}_{t-j}^{(i,h)} + \sum_{j=0}^{k} \delta_{1j} DT1_{t-j}^b(\lambda_{b1})$$

$$+ \sum_{j=0}^{k} \delta_{2j} DT2_{t-j}^b(\lambda_{b2}) + \tilde{\varepsilon}_t \qquad i,h = 1,2 \tag{8.53}$$

$$\Delta\tilde{y}_t^{(i,3)} = \gamma\tilde{y}_{t-1}^{(i,3)} + \sum_{j=1}^{k-1} c_j \Delta\tilde{y}_{t-j}^{(i,3)} + \sum_{j=0}^{k} \delta_{1j} DT1_{t-j}^b(\lambda_{b1})$$

$$+ \sum_{j=0}^{k} \delta_{2j} DT2_{t-j}^b(\lambda_{b2}) + \tilde{\varepsilon}_t \qquad\qquad\qquad i = 1,2 \tag{8.54}$$

$$\Delta\tilde{y}_t^{(3,i)} = \gamma\tilde{y}_{t-1}^{(3,i)} + \sum_{j=1}^{k-1} c_j \Delta\tilde{y}_{t-j}^{(3,i)} + \sum_{j=0}^{k} \delta_{1j} DTi_{t-j}^b(\lambda_{bi}) + \tilde{\varepsilon}_t \qquad i = 1,2 \tag{8.55}$$

$$\Delta\tilde{y}_t^{(3,3)} = \gamma\tilde{y}_{t-1}^{(3,3)} + \sum_{j=1}^{k-1} c_j \Delta\tilde{y}_{t-j}^{(3,3)} + \tilde{\varepsilon}_t \tag{8.56}$$

8.5.2 IO Models

Maintained regressions for the IO models also follow by extension of the single break case. The simplest cases are as follows, allowing for a break under the null hypothesis (other cases follow by setting the appropriate coefficients to zero in IO Model (2, 2)):

IO Model (1, 1)

$$\Delta y_t = \mu_1 + \beta_1 t + \gamma y_{t-1} + \sum_{j=1}^{k-1} c_j \Delta y_{t-j} + \delta_{21} DT1_t^b(\lambda_{b1}) + \delta_{22} DT2_t^b(\lambda_{b2})$$

$$+ \mu_{21} DU1_t(\lambda_{b1}) + \mu_{22} DU2_t(\lambda_{b2}) + \varepsilon_t \tag{8.57}$$

IO Model (2, 2)

$$\Delta y_t = \mu_1 + \beta_1 t + \gamma y_{t-1} + \sum_{j=1}^{k-1} c_j \Delta y_{t-j} + \delta_{21} DT1_t^b(\lambda_{b1}) + \delta_{22} DT2_t^b(\lambda_{b2})$$

$$+ \mu_{21} DU1_t(\lambda_{b1}) + \mu_{22} DU2_t(\lambda_{b2}) + \beta_{21} DT1_t(\lambda_{b1})$$

$$+ \beta_{22} DT2_t(\lambda_{b2}) + \varepsilon_t \tag{8.58}$$

IO Model (3, 3)

$$\Delta y_t = \mu_1 + \beta_1 t + \gamma y_{t-1} + \sum_{j=1}^{k-1} c_j \Delta y_{t-j} + \delta_{21} DT1_t^b(\lambda_{b1}) + \delta_{22} DT2_t^b(\lambda_{b2})$$

$$+ \beta_{21} DT1_t(\lambda_{b1}) + \beta_{22} DT2_t(\lambda_{b2}) + \varepsilon_t \tag{8.59}$$

If there is no break under the null hypothesis the terms in $DT1_t^b$ and $DT2_t^b$ are omitted from the regressions.

8.5.3 Grid search over possible break dates

As to estimation with unknown break dates, first define a two-dimensional grid over possible breakpoints T_{b1}, $T_{b2} \in [T_{iF}, T_{iL}]$, where $T_{iF} > 1$ is the first date on which a break could have taken place and $T_{iL} < T$ is the last date. The omitted points, as in the single break case, are 'trimmed' out of the sample, usually by an equal number of observations at either end of the sample; here a break is assumed not to have occurred either at the beginning or end of the sample and typically $T_{iF} = 2$ and $T_{iL} = T - 1$, in which case there are $T - 2$ admissible breakpoints.

Clearly $T_{b1} \neq T_{b2}$ and the model nomenclature implies without loss that $T_{b1} < T_{b2}$; it is also assumed that $T_{b1} < T_{b2} - 1$, so that the two breaks do not occur on consecutive dates (implying at least two observations in each regime; see Lumsdaine and Pappell, 1997). When this is not the case, it may be more appropriate to view the deterministics as arising from outliers, see, for example, Franses and Haldrup (1994).

These restrictions on the break dates can be used to inform the search. First consider Models (h, h). Suppose the trimming is $T_{iF} = 2$ and $T_{iL} = T - 1$, then let $\tilde{T} = T - 2$ and define a square matrix Ω of dimension $\tilde{T}$ by $\tilde{T}$. Then in combination $T_{b1} < T_{b2}$ and the $T_{b1} < T_{b2} - 1$ rule out all pairings below and including the diagonal in Ω. The search is, therefore, restricted to the upper triangular section of Ω, a search over $\tilde{T}(\tilde{T} - 1)/2 - (\tilde{T} - 1)$ permissible pairs. Now consider the hybrid models, for example (1, 2), which assumes a level shift followed in time by a simultaneous level and slope shift and, thus, the $T_{b1} < T_{b2} - 1$ rule applies. The hybrid model (2, 1) assumes that there is first a simultaneous level and slope shift then a simple level shift, so that $T_{b1} < T_{b2} - 1$ rule again applies.

The extension of the grid search procedure to more than two breaks is potentially costly and Kejriwal and Perron (2010) consider a sequential procedure for such cases; see also Bai and Perron (2003a, b).

8.5.3.i Selecting the break dates

To select the two break dates, Lumsdaine and Papell (1997) suggest using the ZA (1992) criterion of choosing the minimum psuedo-t statistic over the two-dimensional search (they also provide the asymptotic distribution theory for the two-break case with no break under the null). However, doubts over the use of this criterion in the single break case seem likely to carry over to the two-break case, especially if breaks are admitted under the null.

In principle, other selection criteria discussed in the case of one break are applicable here; for example, there is some simulation evidence that BIC works well in locating a one-time break (see Section **8.4.1**, and also Nunes et al. 1997 and

LSt, 2001). In the case that the model specification and lag lengths are fixed, this will give the same result as maximising the likelihood or minimising the residual sum of squares. Also, although it seems that an extension of the procedure based on test statistics for the significance of the breaks should work well, Vogelsang (1997) finds that the power of Wald-type tests is very poor, for example, even two large breaks in the same direction have little chance of detection.

8.5.3.ii   Test statistics

Test statistics also follow by extension of the single break case. The most obvious is simply, as before, the psuedo-t on the estimate of γ, which is denoted $\tilde{\tau}_\gamma^{(i,h)}(\lambda_{b1}, \lambda_{b2})$ in either AO or IO form, where the minimum now refers to the minimum over the two-dimensional search. Similarly, extensions on the Murray-Zivot (MZ) Wald test that includes the unit root and coefficients on the break parameters in the joint null hypothesis are straightforward, but it properties are presently not known. Given the large number of possibilities that arise in even the two-break case, critical values could be obtained on a case-by case-basis. One ubiquitous procedure is to extend the bootstrap, which could also take into account the difficulties arising from not knowing whether to allow a break under the null as well as under the alternative hypothesis. Bootstrapping to obtain the critical values has been used by ZA (1992) and Nunes et al. (1997) in the one-break case and Lumsdaine and Papell (1997) in the two-break case.

8.6   Illustration: US GNP

Some of the issues and methods of this and the preceding chapter are illustrated with (log) US GNP in constant prices for the period 1932 to 2001. The data are graphed in Figure 8.5. The series is clearly positively trended and, if characterised by a deterministic trend, shows a number of possible breakpoints. A preliminary visual check suggests a number of possibilities for structural breaks and, to aid this interpretation, the growth rate (% p.a.) is overlaid on the right-hand scale of Figure 8.5. For example, while recovery from the Great Crash occurred in the 1930s, there was a relative setback in 1938, which was then followed by strong growth until a larger setback in 1946.

Notice that after 1950 the annual growth rate is much more narrowly contained compared to the 1930s and 1940s. This suggests, at least visually, that even aside from the Great Crash, which much of the literature in this area has concentrated upon, there followed a turbulent period, which was initially one of strong recovery followed by a substantial fall in the growth rate, potentially characterised as a structural break or breaks. This might have a straightforward explanation in terms of increased expenditure on armaments due to the Second World War; if so the pre- and post-war periods might be better compared net

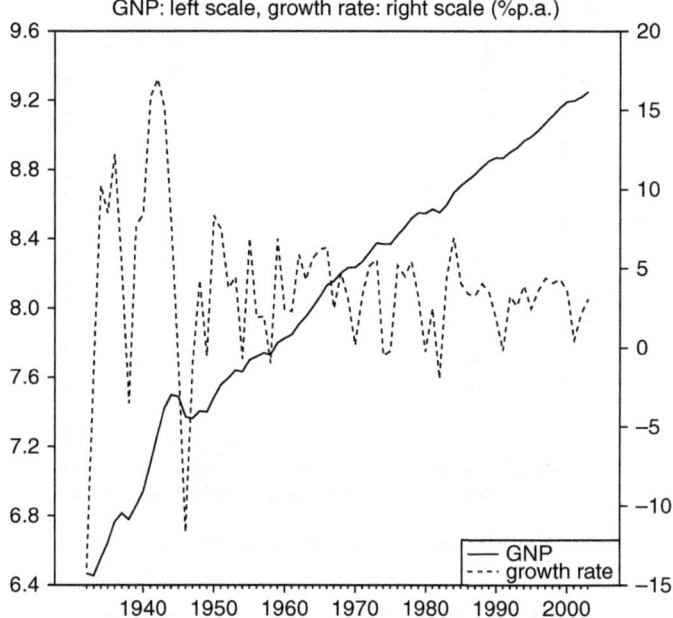

Figure 8.5   US GNP (log and growth rate)

of defence spending, but for present purposes the illustration uses the conventional measure of GNP. Initially, it is assumed that there is a *single* structural break but its form and date are unknown. A range of tests is reported here to encourage some familiarity with use of the relevant statistics and obtaining critical values from the correct sources; it would not generally be necessary to report all of the following test statistics in a particular study.

8.6.1   A structural break?

The first step is to apply a number of the Vogelsang (1997) diagnostic-type tests for a (single) structural break together with the Murray-Zivot version of the test that also includes the unit root as part of the null hypothesis – see Section **8.1.1**. Recall that Vogelsang provided critical values for the mean, $W(\lambda_b)^{ave}$, exponential, $W(\lambda_b)^{exp}$, and supremum, Sup-$W(\lambda_b)$, versions of his tests, but favoured the latter two in terms of power. Model 1 allows for a break in the intercept, Model 2 allows for breaks in the intercept and slope, and Model 3 allows a break in the slope. The critical values depend upon whether the US GNP series is generated by a stationary process (Vogelsang, op. cit., table 1) or a nonstationary process, (Vogelsang, op. cit., table 2); extracts from these tables were provided in Table 8.1.

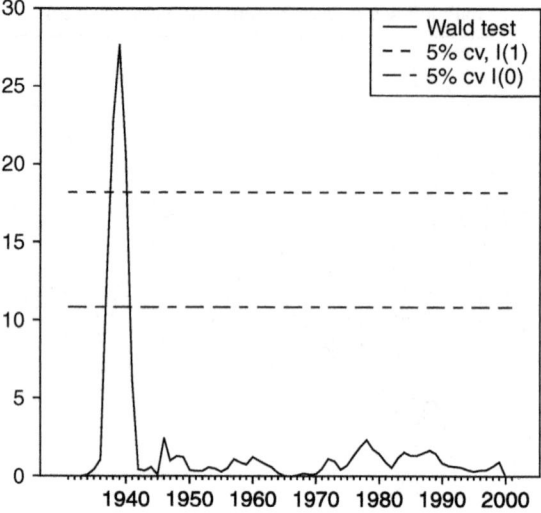

Figure 8.6a   Sup-W test, null: no break in intercept

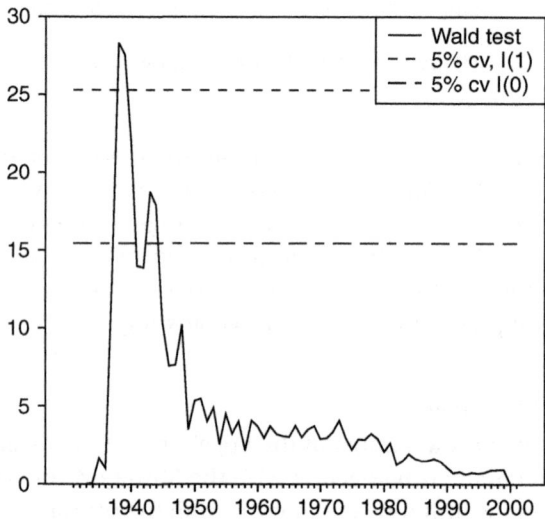

Figure 8.6b   Sup-W test, null: no break in intercept + slope

In this illustration, the MZ test, Sup-W(λ_b, γ), is used in its form for IO Model 2, so that it tests jointly for a unit root and no breaks in the intercept or slope. Some critical values are extracted in Table 8.2. An ADF(1) version of the three models is suggested by a marginal t criterion for lag length selection; and note that the data are strongly trended so that Model 1a is not considered.

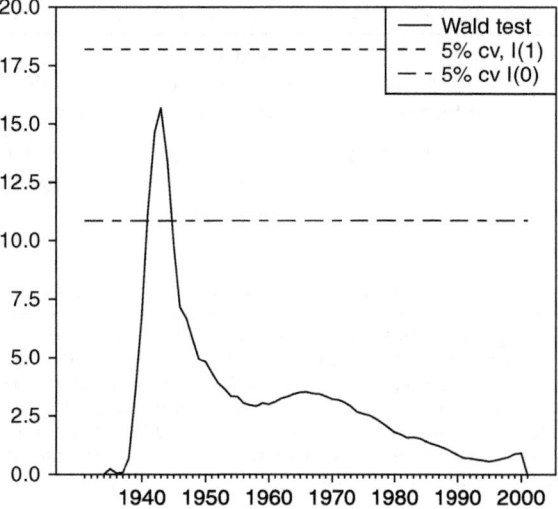

Figure 8.6c   Sup-W test, null: no break in slope

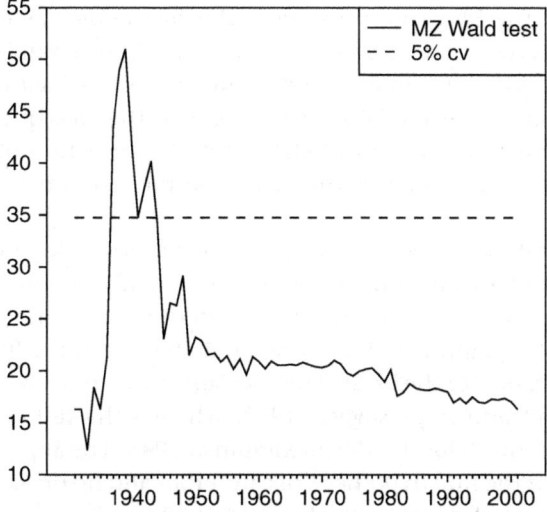

Figure 8.6d   ZM Sup-W test: unit root and no breaks

The test results are presented in Figures 8.6a–d where, in each case, the complete set of statistics for $\lambda_b \in \Theta$ is graphed against the 5% (asymptotic) critical values (that is, the 95% quantiles of the limiting null distribution) assuming stationarity and nonstationarity (referred to as 'conservative' critical values), respectively. The $W(\lambda_b)^{exp}$ version of the Vogelsang test leads to a

Table 8.5   Summary of Wald-type diagnostic break tests for one break

	Model 1 Intercept break	Model 2 Intercept and slope breaks	Model 3 Slope break
Sup-W(λ_b)	27.69 (18.20)	28.28 (25.27)	8.70 (18.20[+])
Break date	*1939*	*1939*	*1943*
Sup-W(λ_b, γ)		50.98 (34.74)	
Break date		*1939*	
W$(\lambda_b)^{exp}$	9.65 (4.84)	10.40 (8.18)	4.33 (4.84[+])

Notes: Sup-W(λ_b) and W$(\lambda_b)^{exp}$ are the Vogelsang (1997) tests (see Section **8.1.1**); (asymptotic) quantiles are in (.) parentheses and are the 'conservative' values, that is they assume a unit root, from Vogelsang (table 2) and see Table 8.1; the cvs assuming stationarity are less than half those shown; [+] indicates assumed value for one restriction test, alternatively take the 95% quantile in the middle column. Sup-W(λ_b, γ) is the MZ version of the Sup-Wald test; quantiles/critical values are interpolated from T = 50 and T = 100 in Sen (2003, Table 1).

single test statistic and this, along with the maximum values of the Wald tests, is summarised in Table 8.5.

Compared to the 5% conservative critical values (which assume nonstationarity), the Sup-W(λ_b) statistics that involve a break in the intercept, or breaks in the intercept and slope, indicate a structural break (see Figures 8.6a and 8.6b, where in both cases the Sup-W(λ_b) value exceeds the 95% quantile). Also, the W$(\lambda_b)^{exp}$ version of the test in Model 2 indicates rejection of the null, with W$(\lambda_b)^{exp}$ = 10.40 compared to the 95% quantile (conservative, asymptotic) value of 8.18.

The interpretation of the Sup-W(λ_b) test for a slope alone depends on the critical value (not included in Vogelsang, 1997). If the critical value for the one-restriction test applies (that is, as for an intercept break only), the supremum is below the 95% quantile in the I(1) case but above it in the I(0) case.

As to an indication of the break date, the Sup-W(λ_b) tests involving the intercept or intercept and slope suggest 1939, whereas the test involving a slope change only (Model 3) locates the maximum at 1943. The Sup-W(λ_b, γ) test suggests rejection of the null hypothesis of a unit root and no breaks; the maximum of these test values also locates the break in 1939 (see Figure 8.6d).

At this stage, the results of the Vogelsang tests support a break, probably in both the intercept and slope. The MZ Wald test also supports this view and, additionally, leads to rejection of the unit root null. Both versions of the test statistics suggest 1939 as the break date.

8.6.2   Which break model?

The BIC values were also calculated to assist selection of the appropriate model; that is, the minimum value of BIC across the three models of the minimum BIC

Table 8.6   BIC for the AO and IO versions of Models 1, 2, and 3

	Model 1	Model 2	Model 3
AO	99.5	100	99.2
IO	100	99.3	98.5

Notes: For IO models: BIC $= \ln[\bar{\sigma}_i^2(\lambda_b)] + K(i)(\ln T)/T$, where K(i) is the number of regressors in Model i and T is the effective number of observations; for AO models, K(i) is the number of regressors plus the number of trend coefficients (3 for Models 1 and 3, and 4 for Model 2); entries are relative to the overall minimum for AO and IO, respectively, where these are: AO Model 2, BIC $= -6.696$; IO Model 1, BIC $= -6.785$.

in each model for $\lambda_b \in \Theta$. This resulted in an overall minimum for Model 2 in the AO version and for Model 1 in the IO version. See Table 8.6, where the BIC values are expressed as a ratio to the minimum BIC for the AO and IO models, respectively.

The earlier simulation results (see Section **8.4**) applied in the context of IO models suggested that BIC would be a reliable criterion for model selection provided that the break did not occur too close to either end of the sample. If the IO results apply to the AO model, the probability of choosing Model 2 when it is not Model 2 that generated the data under the alternative are slight (see Figure 8.1 for Model 1 correct and Figure 8.3 for Model 3 correct), and probably less than 5%. However, the probability of selecting IO Model 1 when Model 2 is correct is quite large (above 0.5) when the break date is close to the beginning of the sample (as may be the case here) and the change in the slope parameter is relatively small. In view of this possibility, details of both IO Models 1 and 2 are reported along with the preferred AO model in Tables 8.6 and 8.7. Also, the simulation results in Sen (2003) suggest that the reduction in power using Model 2 is slight when the DGP is actually IO Model 1.

8.6.3   The break date?

The diagnostic tests of Section **8.6.1** lead to the suggestion that 1939 is the break date if either Model 1 or Model 2 is used. However, other criteria are not entirely in agreement, a situation that is not unusual in a number of empirical studies (see, for example, the extensive literature on the Nelson and Plosser data series). The results are summarised in Table 8.7.

8.6.3.i   Break dates suggested by different criteria

In AO Model 2, the break date estimated by minimising BIC is 1945, whereas it is 1937 for IO Model 1. Both contrast with the dates selected by the tests of section **8.6.1**; there is also a contrast with the date selected using inf-$\tilde{\tau}_\gamma^{(i)}(\lambda_b)$ for AO Model 2, which is 1938, but if inf-$\tilde{\tau}_\gamma^{(i)}(\lambda_b)$ is used for IO Model 1, the break is

Table 8.7   Break dates suggested by various criteria in Models 1, 2 and 3

	Model 1		Model 2		Model 3	
	AO	IO	AO	IO	AO	IO
BIC date	1938	1937	1945	1937	1944	1943
Break date $\inf\text{-}\tilde{\tau}_\gamma^{(i)}(\lambda_b)$	1938 −7.29	1939 −7.10	1938 −6.54	1939 −6.82	1947 −5.15	1943 −5.68
Break date $\text{Max}[|t(\mu_2^A)|]$	1940 14.61	1939 5.26	n.a	n.a	n.r	n.r
Break date $\text{Max}[|t(\beta_2^A)|]$	n.r	n.r	1945 15.83	1943 6.39	1943 16.18	1943 3.96
Break date Sup-F(μ_2) Sup-F(μ_2,β_2) Sup-F(β_2)	n.r	1939 27.69	1945 161.68	1939 28.28	n.r	1943 15.69
Break date Sup-W(λ_b,γ)	n.r	1939 51.44	n.r	1939 50.98	n.r	1943 36.35

Notes: $\text{Max}[|t(\mu_2^A)|]^2 = \text{Sup-F}(\mu_2)$; $\text{Max}[|t(\beta_2^A)|]^2 = \text{Sup-F}(\beta_2)$; $\text{Sup-W}(\lambda_b) = q$ times Sup-F(μ_2,β_2), where q is number of restrictions; nr in non relevant.

located at 1939. For $\lambda_b \in \Theta$, Figure 8.7a graphs $\tilde{\tau}_\gamma^{(2)}(\lambda_b, AO)$ and BIC, and Figure 8.7b is the graph of $\tilde{\tau}_\gamma^{(1)}(\lambda_b, IO)$ and BIC. The left scale is for the DF-type test (referred to as the Perron DF) and the right scale for BIC.

8.6.3.ii   *AO Model 2 and IO Models 1 and 2: estimation results with BIC date of break*

For illustration some of the detailed estimation results for the break date selected by BIC are presented in Tables 8.8 and 8.9.

8.6.3.iii   *The estimated long-run trend functions*

Also of interest are the implied trend and long-run trend functions associated with each model (see Section 7.3 for details on the general derivation). These are summarised below for the BIC selected break dates of 1945 for AO Model 2 and 1937 for IO Models 1 and 2. The long-run trends look different because of the different break dates; however, they have in common that the trend after the break indicates an average growth rate of about 3.33% p.a. IO Model 2 fits the data with a negative trend before 1938.

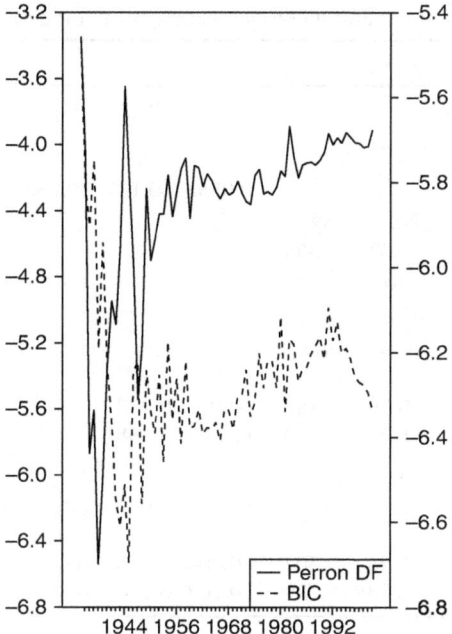

Figure 8.7a   AO Model 2: break date varies

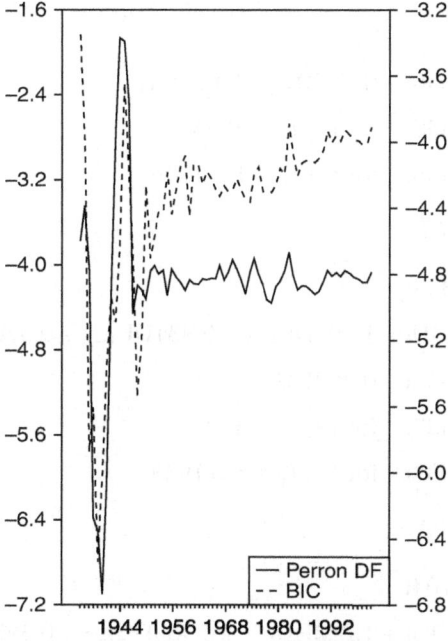

Figure 8.7b   IO Model 1: break date varies

Table 8.8   Estimation of AO and IO Models (single break model)

AO Model 2

Break = 1945	constant			t	DU_t	DT_t			
$\hat{\sigma}_\varepsilon$									
$\tilde{y}_t$	6.117			0.0858	–0.137	–0.052			
$\Delta\tilde{y}_t$		$\tilde{y}_{t-1}$	$\Delta\tilde{y}_{t-1}$				DT_t^b	DT_{t-1}^b	0.0295
		–0.339	0.355				0.020	–0.044	
't'		–4.164	3.078				0.633	–1.470	

IO Model 1

Break = 1937		y_{t-1}	Δy_{t-1}	t	DU_t		DT_t^b		$\hat{\sigma}_\varepsilon$
$\Delta\hat{y}_t$	1.908	–0.290	0.486	0.009	0.082		–0.158		0.0292
't'	6.596	–6.384	5.689	6.072	3.488		–4.963		

IO Model 2

Break = 1937		y_{t-1}	Δy_{t-1}	t	DU_t	DT_t	DT_t^b		$\hat{\sigma}_\varepsilon$
$\Delta\hat{y}_t$	1.936	–0.282	0.504	–0.003	0.098	0.012	–0.157		0.0293
't'	6.644	–6.088	5.739	–0.208	3.316	0.896	–4.884		

8.6.3.iii.a   AO Model 2

$$y_t^{tr} = \mu_t$$
$$= 6.117 + 0.0858t - 0.137DU_t - 0.052DT_t \tag{8.60}$$
$$= 6.117 + 0.0858t \quad \text{for } t \le T_b = 1945$$
$$= 5.980 + 0.0336t \quad \text{for } t \ge T_b + 1 = 1946$$

8.6.3.iii.b   IO Model 1

$$y_t^{tr} = 6.548 + 0.0333t$$
$$+ 0.082(1.0DU_t + 1.191DU_{t-1} + 0.931DU_{t-2} + 0.528DU_{t-3}\dots) \tag{8.61a}$$
$$\mu_t = 6.548 + 0.0333t + 0.282DU_t \tag{8.61b}$$
$$= 6.548 + 0.0333t \quad \text{for } t \le T_b = 1937$$
$$= 6.830 + 0.0333t \quad \text{for } t \ge T_b + 1 = 1938$$

8.6.3.iii.c   IO Model 2

$$y_t^{tr} = 6.867 - 0.0104t$$
$$+ 0.098(1.0DU_t + 1.222DU_{t-1} + 0.988DU_{t-2} + 0.592DU_{t-3}\dots)$$
$$+ 0.012(1.0DT_t + 1.222DT_{t-1} + 0.988DT_{t-2} + 0.592DT_{t-3}\dots) \tag{8.62a}$$

Table 8.9 Estimation of two breaks, IO (2, 3) Model

Break dates 1940, 1945 chosen by BIC

	Const	y_{t-1}	Δy_{t-1}	t	$DU1_t$	$DU2_t$	$DT2_t$	$DT1_t^b$	$DT2_t^b$	$\hat{\sigma}_\varepsilon$
$\Delta \hat{y}_t$	1.764	-0.265	0.205	0.0095	0.228	-0.033	0.032	-0.084	-0.110	0.0256
't'		-3.638	2.054	1.383	4.999	-3.319	3.232	-2.064	-3.843	

Break dates 1939, 1981 chosen by $\inf\text{-}\tilde{\tau}_\gamma^{(2,3)}(\lambda_{b1}, \lambda_{b2})$

	Const	y_{t-1}	Δy_{t-1}	t	$DU1_t$	$DU2_t$	$DT2_t$	$DT1_t^b$	$DT2_t^b$	$\hat{\sigma}_\varepsilon$
$\Delta \hat{y}_t$	3.015	-0.470	0.501	0.0212	0.130	-0.005	-0.003	-0.107	-0.059	0.0282
't'	7.824	-7.509	6.007	2.591	4.862	-0.686	-2.378	-3.136	-2.028	

$$\mu_t = 6.867 - 0.0104t + 0.347DU_t + 0.0437DT_t$$

$$= 6.867 - 0.0104t \quad \text{for } t \leq T_b = 1937 \tag{8.62b}$$

$$= 7.214 + 0.0333t \quad \text{for } t \geq T_b + 1 = 1938$$

The split trend of AO Model 2 is graphed in Figure 8.8a where the BIC selected date of 1945 is used and in Figure 8.8b where the inf-$\tilde{\tau}_\gamma^{(2)}(\lambda_b, AO)$ selected date of 1938 is used. At least by eye, the split trend with a break date of 1938 seems to occur too early (this is also the case if 1939 is used) and misses the continued increase in GNP and then the subsequent decline in 1946, which is well captured by the BIC date of 1945. The trends after 1946 are indistinguishable so the difference is essentially in whether the break occurred before or at the end of the Second World War. The split trends from IO Models 1 and 2, respectively, are shown in Figures 8.9a and b; again the timing of the break occurs too early at least for IO Model 2, which would capture the key characteristics of the data better if the break date was six or so years later.

8.6.4   Inference on the unit root null hypothesis

In this section the focus is on the unit root null hypothesis and some of the difficulties that arise in interpreting the estimated models.

8.6.4.i   AO Model 2

In the case of the AO model, VP (1998) derived that in the case of a break under the null, then in Model 2 the correct break date is selected asymptotically by Max[|t(β_2)|] and Sup-W(λ_b) = 2Max[F(μ_2, β_2)]; the resulting asymptotic distribution of $\tilde{\tau}_\gamma^{(2)}(\lambda_b, AO)$ is then the same as when the break date is (exogenously) known (and critical values can be obtained from Perron, 1989).

When it is not known that there is a break under the null, the suggested conservative procedure is to use the critical values assuming *no* break under the null. These are $\tilde{\tau}_\gamma^{(2)}(\lambda_b, AO)$ using Max[|t(β_2)|] and Max[F(μ_2, β_2)] – see also Table 8.3 for asymptotic critical values. The critical values for $\tilde{\tau}_\gamma^{(2)}(\lambda_b, AO)$ (with k chosen by t-sign) are approximately –4.88 and –4.95, respectively. The break dates chosen by these two criteria are the same at 1945, and thus the same as that chosen with BIC. At this break date, the test statistic is $\tilde{\tau}_\gamma^{(2)}(\lambda_b, AO, 1945) = -4.16$, which does not lead to rejection of the null at the 5% level. However, using the asymptotic distribution in Perron (1989, Table VI.B), assuming the break date is known, and λ_b is approximately 0.2, gives a 5% (10%) cv of –3.99 (–3.66); even with some finite sample allowance for an increase in the absolute value of the cv, it thus also seems likely that the unit root null would be rejected if a break is allowed under the null.

8.6.4.ii   IO Models

8.6.4.ii.a   IO Model 1   We now consider the estimated IO Model 1. At the break date of 1939, which is selected by all of inf-$\tilde{\tau}_\gamma^{(1)}(\lambda_b)$, Max[|t($\mu_2$)|] and MZ Sup-W($\lambda_b$, γ), the unit root test statistic is $\tilde{\tau}_\gamma^{(1)}(\lambda_b, IO) = -7.10$. The 5% cv for selection

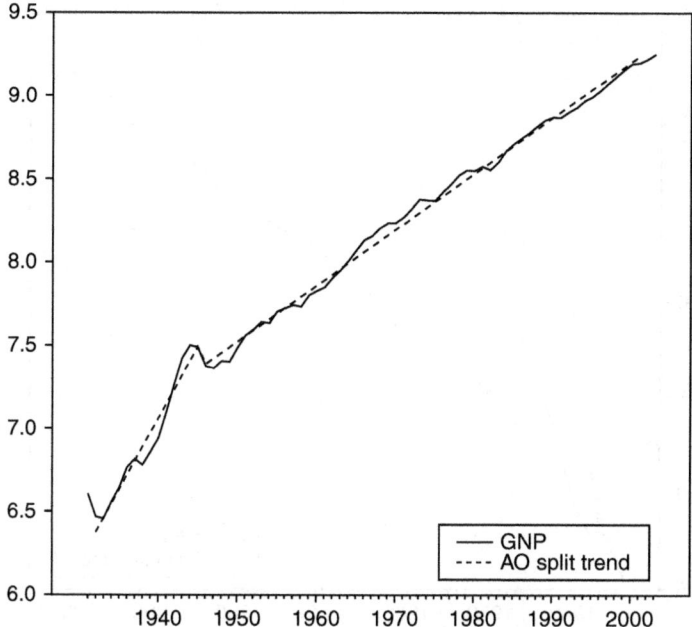

Figure 8.8a   US GNP, split trend, AO Model 2, break = 1945

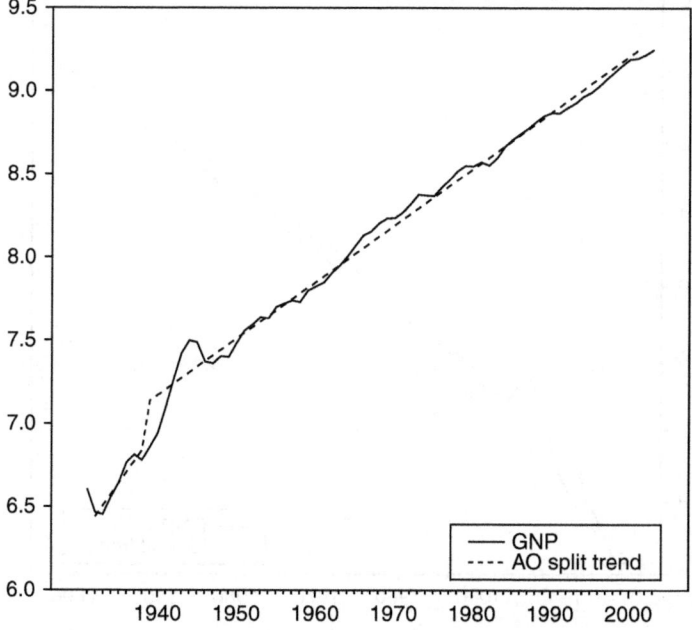

Figure 8.8b   US GNP, split trend, AO Model 2, break = 1938

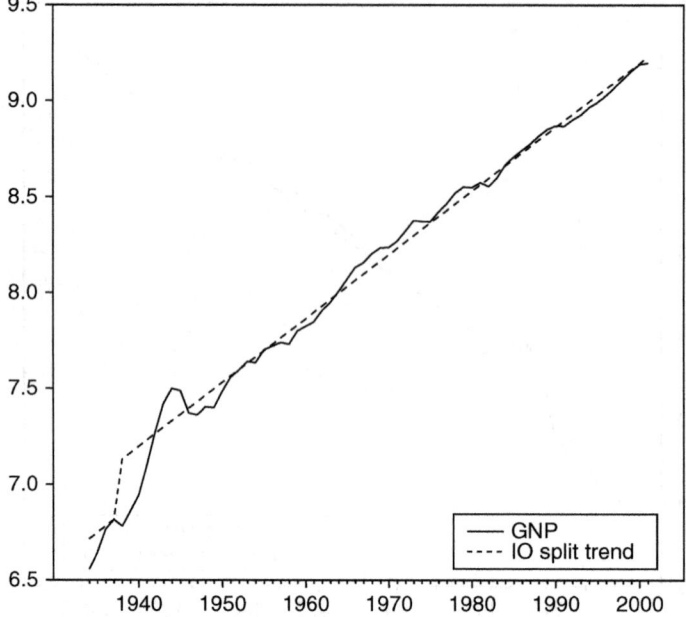

Figure 8.9a   US GNP, split trend, IO Model 1, break = 1937

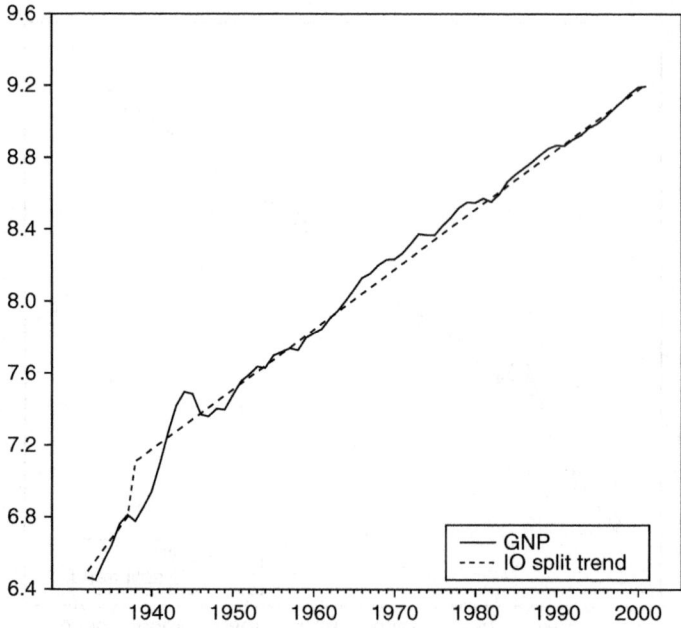

Figure 8.9b   US GNP, split trend, IO Model 2, break = 1937

by inf-$\tilde{\tau}_\gamma^{(i)}(\lambda_b) = -5.15$, and is obtained from VP, table 1, panel (a), alternatively Perron, 1997, table 1 (a). (Note that in the case of selection by inf-$\tilde{\tau}_\gamma^{(i)}(\lambda_b)$, the asymptotic distributions of $\tilde{\tau}_\gamma^{(i)}(\lambda_b, IO)$, $i = 1, 2$, are identical to $\tilde{\tau}_\gamma^{(i)}(\lambda_b, AO)$ and VP, table 1, panel (a), is headed AO.) In the case of selection by Max[|t(μ_2)|], the 5% cv obtained from Perron, 1997, Table 1 (c) is (approximately) -5.13. Both critical values assume no break under the null. On the basis of these results, the null hypothesis is rejected.

Also relevant are the results of Harvey, Leybourne and Newbold (2001), referred to in Section **8.3.2**, who noted that whilst the distribution of $\tilde{\tau}_\gamma^{(1)}(\lambda_b, IO)$ is asymptotically invariant to the magnitude of a break in the intercept under the null, it is not invariant in finite samples; as a result an incorrect choice of break date (by the VP criteria) can lead to size distortions. They suggested moving the break date forward by one period compared to that indicated by Max[|t(μ_2)|], supporting the simulation results of LSt (2001), and provided critical values of $\tilde{\tau}_\gamma^{(1)}(\lambda_b, IO)$ for such a procedure – see table 8.4 and HLN (table 2). Following this procedure moves the break date from 1939 (see Table 8.7) to 1940, with $\tilde{\tau}_\gamma^{(1)}(\lambda_b, IO, 1940) = -6.13$ compared to a 5% cv of (approximately) -4.43 from Table 8.4. In this case also, the null hypothesis is rejected.

8.6.4.ii.b   IO Model 2   To consider how sensitive these results are to the choice of Model 1, IO Model 2 is also considered; recall that arguments in Sen (2003) and MOC (2005) favour using the most general model.

First, consider the situation if it is assumed that there is no break under the null. The break date selected using inf-$\tilde{\tau}_\gamma^{(2)}(\lambda_b, IO)$, Max[F($\mu_2, \beta_2$)] and MZ Sup-W($\lambda_b, \gamma$) is 1939, as in Model 1, whereas with Max[|t(β_2)|] it is 1943 (see Table 8.7). The relevant test statistics are: $\tilde{\tau}_\gamma^{(2)}(\lambda_b, IO, 1939) = -6.82$; $\tilde{\tau}_\gamma^{(2)}(\lambda_b, IO, 1943) = -6.09$; Sup-W($\lambda_b, \gamma, 1939) = 50.98$. Critical values are obtained as follows: using inf-$\tilde{\tau}_\gamma^{(2)}(\lambda_b, IO)$, the 5% cv is -5.41, from VP, table 2 panel (a), (and note that IO = AO in this case); using Max[|t(β_2)|], the asymptotic 5% cv is -4.91, from VP, Table 4; using Max[F(μ_2, β_2)], the finite sample 5% cv is approximately -5.47 from VP, Table 3; and the 5% cv for Sup-W(λ_b, γ) is approximately 34.74 from Sen (2003, Table 1). *All* cases lead to rejection of the null hypothesis.

8.6.4.ii.c   A break under the null   Now consider if there are complications arising if the null allows a break in the intercept and/or drift parameters. Recall the asymptotic results noted in Section **8.2.3**. In particular, if inf-$\tilde{\tau}_\gamma^{(2)}(\lambda_b, IO)$ is used, then size tends to 1 as T increases but, practically, for T = 100, size distortions are not problematic for a slope break of less than one unit of the residual standard error, $\hat{\sigma}_\varepsilon$; however, there are quite serious distortions for a slope break of $2\hat{\sigma}_\varepsilon$ (VP, Table 7B), and for $|\mu_2| > 5\hat{\sigma}_\varepsilon$ in the case of IO Model 1. Further, the situation is *not* remedied if the break date is selected by Max[|t(β_2)|] or Sup-F(μ_2, β_2), since neither of these criteria results in the true break date being chosen asymptotically. The simulation results in VP show that size distortions are generally more severe using these criteria than using $\inf\text{-}\tilde{\tau}_\gamma^{(2)}(\lambda_b,\text{IO})$ and could be problematical even for a slope break of one $\hat{\sigma}_\varepsilon$.

Estimation imposing a unit root but allowing a break in the intercept, that is IO Model 1, resulted in an estimate of μ_2 of approximately $3.8\hat{\sigma}_\varepsilon$; using IO Model 2 resulted in a similar estimate of μ_2 of $3.5\hat{\sigma}_\varepsilon$; additionally the estimated break in the drift, β_2, was $1.6\hat{\sigma}$. The latter is large enough to suggest that there *are* likely to be size distortions using $\inf\text{-}\tilde{\tau}_\gamma^{(i)}(\lambda_b)$, $\text{Max}[|t(\beta_2)|]$ or $\text{Sup-F}(\mu_2,\beta_2)$.

In the case of IO Models 1 and 2, the simulations in LSt (2001), allowing a break under the null, suggested that choosing the break date using BIC gave better results than using $\inf\text{-}\tilde{\tau}_\gamma^{(i)}(\lambda_b)$, $\text{Max}[|t(\mu_2)|]$ or $\text{Max}[|t(\beta_2)|]$ (but problems that might occur with break dates close to either end of the sample were not considered). If the magnitude of the break parameter is large, then using critical values for Perron's exogenous tests for the BIC-chosen break date resulted in accurately sized tests.

The BIC-selected date in both IO Models 1 and 2 is 1937, resulting in $\tilde{\tau}_\gamma^{(1)}(\lambda_b,\text{IO},1937)) = -6.38$ and $\tilde{\tau}_\gamma^{(2)}(\lambda_b,\text{IO},1937)) = -6.09$. (However, this does suggest a break close to the beginning of the sample, which could be a problematic choice.) In the former case, the 5% cv for $\mu_2 = 4$ in LSt (2001, table 1) is -4.33 and hence the null is rejected. (Note that as μ_2 increases this tends to -3.75 virtually identical to the 'exogenous' 5% cv of -3.74 from Perron (1989, table IV.B.) Whilst this cv is based on $\lambda_b^c = 0.5$, note that the cvs in Perron (op. cit.) *decrease* in absolute value as λ_b^c moves away from 0.5.

A similar argument applies for IO Model 2. Choosing by BIC, the 5% cvs vary between -4.97 and -4.24 for an exogenous break date Perron, 1989, table VI.B); hence, the null is also rejected on this comparison. Whilst there is perhaps a slight doubt here in relying on BIC when the selected date is so close to the beginning of the sample, the results from the IO models consistently suggest rejection of the null hypothesis whether a break is allowed under the null or not.

8.6.5 Overall

Overall, on the basis of a single break, the various test statistics suggest rejection of the null hypothesis of a unit root. However, there is little agreement on the precise timing of the break. Some of the criteria place the break before the start of the Second World War, others at the end of the war; at least in terms of the fit of the split trend, a break in 1945 seems closer to the central characteristics of the data.

The lack of agreement on the timing of the break can, however, be critical. The visual evidence from the fitted long-run trends suggests that a (single) break date before the Second World War misses the timing of major changes in the series. In the next section, the analysis is extended to allow for two breaks, in part to assess whether a single break is distorting the empirical analysis.

8.6.6 Two breaks

Whilst much of the research on trend breaks allows for a single break, some of the issues that arise when extending this framework to allow for the possibility of two breaks are illustrated in this section. The best of the two-break IO models in terms of minimising BIC was the IO(2, 3) specification, which allowed an intercept and slope break (referred to as a double break) coupled with a slope break (a single break). As far as estimation was concerned there was no presumption in terms of the order in time of the double break and the single break and, apart from the obvious restriction that the break dates be distinct (and $T_{b1} \neq T_{b2} \pm 1$), the search was unrestricted.

8.6.6.i IO(2, 3) model

Within the IO(2, 3) model, the selected break dates were chosen by BIC and inf-$\tilde{\tau}_\gamma^{(2,3)}(\lambda_{b1}, \lambda_{b2})$, respectively, where the minimum is taken over a two-dimensional search. Selection by BIC gave a double break, that is, Model 2 in 1940 followed by a slope break, that is, Model 3 in 1945; whereas selection by inf-$\tilde{\tau}_\gamma^{(2,3)}(\lambda_{b1}, \lambda_{b2})$ gave a double-break model in 1939 and a slope break model in 1981.

The nature of the surface generated by the search is illustrated in Figures 8.10a and 8.10b for BIC and Figures 8.11a and 8.11b for inf-$\tilde{\tau}_\gamma^{(2,3)}(\lambda_{b1}, \lambda_{b2})$. In both cases the negative of the criteria is plotted as this gives a better picture of the 3-D surface. In the first instance, two 'slices' of the 3-D surface are shown in Figures 8.10a and Figure 8.11a, respectively; these correspond to the criterion for break date 1 conditional on break date 2 in the left panel and vice versa in the right panel. It is clear that the activity in terms of possible break dates for BIC selection is concentrated in the 1940s. In the 3-D plot for selection by BIC (see Figure 8.10b) there are two 'ridges' locating the two breaks, one at the beginning and the other at the end of the Second World War. The picture with selection by inf-$\tilde{\tau}_\gamma^{(2,3)}(\lambda_{b1}, \lambda_{b2})$ is not entirely the same. From Figures 8.11a and b note that the first break, the double break, is 1939 and one year behind the BIC selected date, but the second break, the slope break, is in 1981 and, therefore much later than the BIC selection.

The estimation details are summarised in Table 8.9. In order to assess these values of the test statistic, the critical values are obtained by bootstrapping the process. A similar procedure was adopted by ZA in the single-break case and Lumsdaine and Pappell (1997) in the two-break case.

The first stage in the bootstrap run is to estimate the empirical model, under the null hypothesis of a unit root, over all permitted break dates using a particular two-break date selection criterion, for example selection by BIC; there are two versions of the null depending on whether a break is permitted. In each case, the empirical model serves to define the empirical distribution function, EDF, for the residuals in the bootstrap run that follows. Within the bootstrap

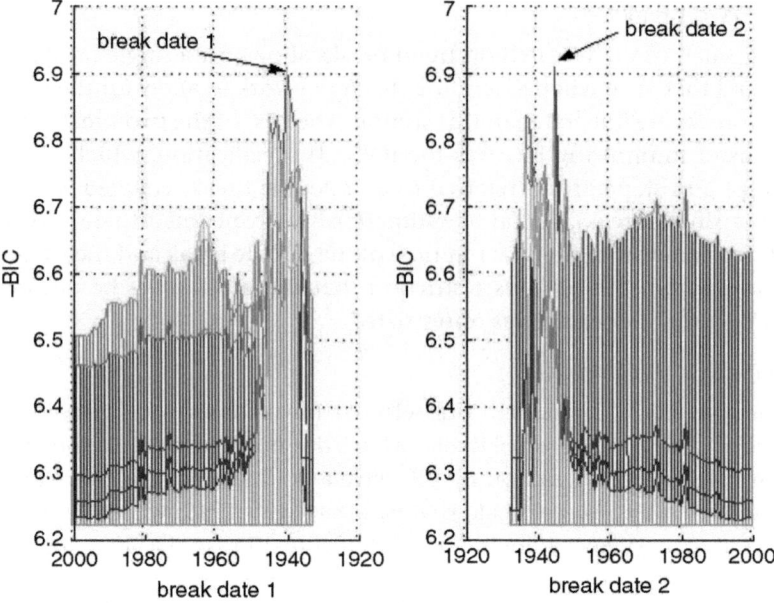

Figure 8.10a   IO Model(2, 3): BIC, first and second breaks

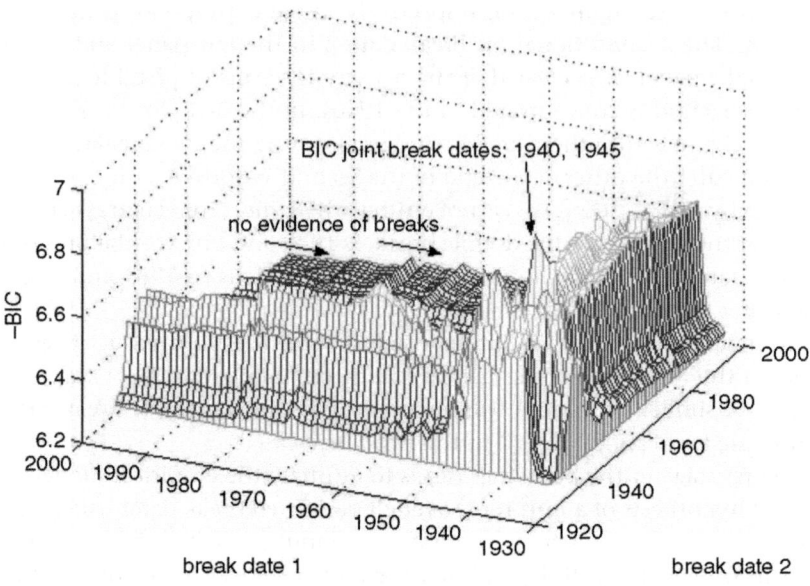

Figure 8.10b   IO Model(2, 3): BIC, 3-D view

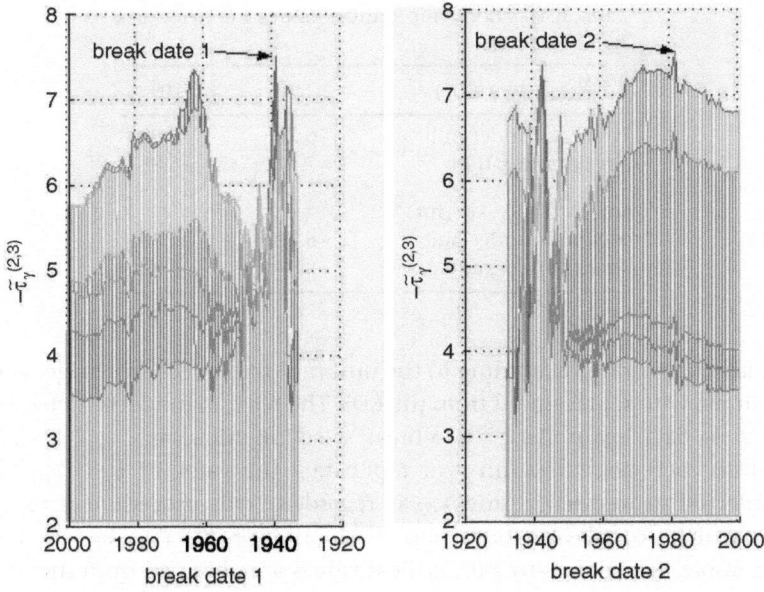

Figure 8.11a   IO Model(2, 3): $\tilde{\tau}_\gamma^{(2,3)}$, first and second breaks

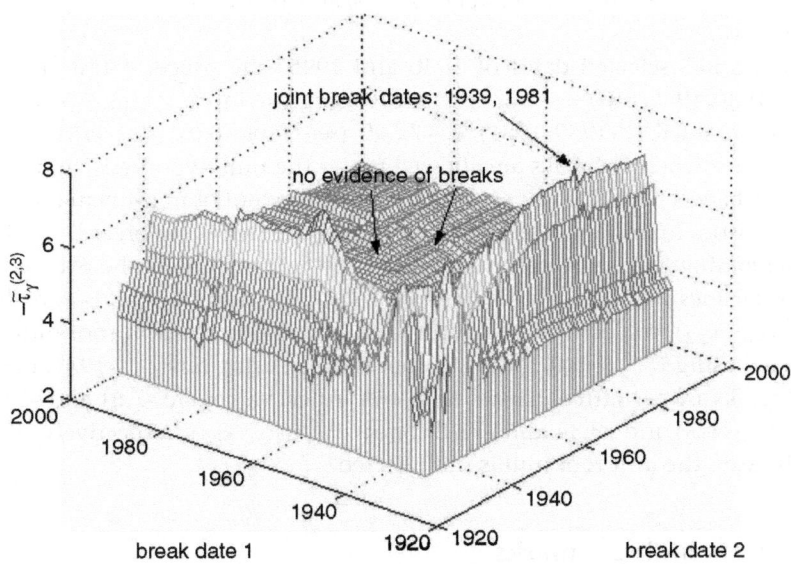

Figure 8.11b   IO Model(2, 3): $\tilde{\tau}_\gamma^{(2,3)}$, 3-D view

Table 8.10   Bootstrap critical values for two-break
IO (2, 3) model

$\tilde{\tau}_\gamma^{(2,3)}(\lambda_{b1}, \lambda_{b2}, \text{BIC})$	5%	10%
No break under null	−6.26	−5.80
Breaks under null	−7.61	−6.88
$\text{inf-}\tilde{\tau}_\gamma^{(2,3)}(\lambda_{b1}, \lambda_{b2}, \text{inf})$		
No break under null	−6.72	−6.39
Breaks under null	−8.34.	−7.78

run, data is generated according to the unit root null of the first stage, with random draws with replacement from the EDF. The IO(2, 3) model is then estimated with these data, again using a two-break selection criterion.

Within each bootstrap run – or replicate – the value of $\tilde{\tau}_\gamma^{(2,3)}(\lambda_{b1}, \lambda_{b2})$ for the selected values of λ_{b1} and λ_{b2} is recorded. This procedure is repeated B times, which provides the bootstrap distribution of $\tilde{\tau}_\gamma^{(2,3)}(\lambda_{b1}, \lambda_{b2}, \text{BIC})$, where, for example, selection is by BIC; critical values are obtained from the quantiles for this distribution. The bootstrap distributions are obtained for B = 4,999. In the first stage, where the null hypothesis is imposed, break dates are always selected by BIC; the bootstrap runs use either BIC or $\text{inf-}\tilde{\tau}_\gamma^{(2,3)}(\lambda_{b1}, \lambda_{b2})$ to select the break dates, with resultant test statistics denoted $\tilde{\tau}_\gamma^{(2,3)}(\lambda_{b1}, \lambda_{b2}, \text{BIC})$ and $\text{inf-}\tilde{\tau}_\gamma^{(2,3)}(\lambda_{b1}, \lambda_{b2}, \text{inf})$, respectively. The 5% and 10% critical values are given in Table 8.10.

At the BIC selected dates of 1940 and 1945, the psuedo-t test statistic is $\tilde{\tau}_\gamma^{(2,3)}(1940, 1945, \text{BIC}) = -3.638$, whereas using the $\text{inf-}\tilde{\tau}_\gamma^{(2,3)}(\lambda_{b1}, \lambda_{b2})$ criterion results in $\text{inf-}\tilde{\tau}_\gamma^{(2,3)}(1939, 1981) = -7.509$ (see Table 8.9). The critical values depend on whether breaks are allowed under the null hypothesis, and the difference between critical values will indicate the extent of non-invariance of the test statistics to drift change parameters under the null. If no breaks are allowed in generating the empirical model and bootstrap replicates, the 5% and 10% critical values are −6.26 and −5.80 for $\tilde{\tau}_\gamma^{(2,3)}(\lambda_{b1}, \lambda_{b2})$ and −6.79 and −6.39 for $\text{inf-}\tilde{\tau}_\gamma^{(2,3)}(\lambda_{b1}, \lambda_{b2})$, respectively. With these critical values, the unit root null is not rejected using $\tilde{\tau}_\gamma^{(2,3)}(1940, 1945, \text{BIC})$ but rejected using $\text{inf-}\tilde{\tau}_\gamma^{(2,3)}(1939, 1981)$.

If breaks are permitted, the 5% and 10% critical values are −7.61 and −6.88 for $\tilde{\tau}_\gamma^{(2,3)}(\lambda_{b1}, \lambda_{b2})$ and −8.34 and −7.78 for $\text{inf-}\tilde{\tau}_\gamma^{(2,3)}(\lambda_{b1}, \lambda_{b2})$, respectively. Now in both cases, the unit root null is not rejected.

8.7   Concluding remarks

This chapter has developed a number of areas related to the practical application of some relatively simple structural break models. Even though these have been kept simple, the complexities that arise indicate just how difficult it is to distinguish the apparently simple dichotomy of whether a unit root is or is not present in the process generating the data. Important isues arise as to the nature of the break: does it have an impact on the level of the series or the growth of the series or both? How is the impact distributed over time? Is there just one breakpoint or several? The practical resolution of these issues is, in turn, complicated by what some might see as the central point of enquiry, that is, whether the data are generated by a stationary or a nonstationary process.

There are several other aspects of the 'break' literature that have not been considered here. For example, Perron and Rodríguez (2003) consider unit root tests other than the standard DF pseudo-t applied to the structural break model; in particular, they extend the tests of Elliott, Rothenberg and Stock (1996), which are based on GLS detrending, to allow a one time change in the trend function. These tests generally have advantages over the DF tests in reducing size distortions when errors are serially correlated and improving upon power for 'nearly' integrated alternatives. Another possible dimension of a break is a change in the innovation variance; see, for example, Hamori and Tokihisa (1997), Kim, Leybourne and Newbold (2002) and Cavaliere and Taylor (2006, 2008). The area of structural breaks generally still attracts much research; see, for example, Cavaliere and Iliyan (2007), Harvey, Leybourne and Taylor (2008, 2009) and Harris et al. (2009).

Practical applications of structural break tests have often been limited to the situation in which there are just one or two potential breaks. Arguably this is for theoretical and practical simplicity since as data periods lengthen, and with more not less turbulence in world affairs, in the context of increasingly interconnected national economic systems, if the structural break paradigm has credence then methods and practice should reflect this increasingly more complex situation. Bai and Perron (1998, 2003a, b) have considered the more general case and routines are available that compute their procedures. See http://www.estima.com/procs_perl/mainproclistwrapper.shtml for RATS code, and for an interesting application focusing on 9/11 and other related events see Enders and Sandler (2005)

One of the messages from this and the previous chapter is that a central part of the complexity of this area arises from the combinations of the types of models used to characterise a 'break', or more generally a structural change, with the possibly unknown number of breakpoints. This gives rise to uncertainty on several dimensions, not just the single dimension of whether there is a unit root. The reader could now usefully consult Perron (2006) for a directed overview of key issues in the estimation, testing and computation of structural change models including univariate, multivariate and cointegrated time series.

Questions

Q8.1 Explain how to obtain the Wald test of Section **8.1.3** that includes the unit root in the joint null hypothesis.

A8.1 In effect the resulting test statistic is just a variant of Equation (8.16), the framework for which is restated here for convenience. The (unrestricted) regression model is:

$$w = X(\lambda_b)\beta + \varepsilon \tag{A8.1}$$

where $w = (w_1, \ldots, w_T)'$ and $x_t(\lambda_b)'$ is the t-th row of $X(\lambda_b)$, where:

$$x_t(\lambda_b)' = (y_{t-1}, \Delta y_{t-1}, \ldots, \Delta y_{t-k}, 1, t, DU_t(\lambda_b), DT_t(\lambda_b)) \tag{A8.2}$$

$$w_t = \Delta y_t$$

$$\beta = (\gamma, c_1, \ldots, c_{k-1}, \mu_1, \beta_1, \delta_2, \delta_2)'$$

The Wald statistic takes its usual form:

$$W(\lambda_b) = \frac{(R\hat{\beta} - r)'[R(X(\lambda_b)'X(\lambda_b))R']^{-1}(R\hat{\beta} - r)}{\hat{\sigma}_\varepsilon^2(\lambda_b)} \tag{8.16}$$

The restrictions matrix R is now of dimension $3 \times K$, where $K = (k + 4)$; in this case R sets the 1st, $(k + 3)$-rd and $(k + 4)$-th elements of β to zero; r is a 3×1 zero vector; and the variance estimator $\hat{\sigma}_\varepsilon^2(\lambda_b)$ is based on the regression model (A8.1). As before, it is usually somewhat easier to calculate an F test statistic for H_0 using a direct comparison of the residual sums of squares (RSS) under the unrestricted and restricted regressions, in which case $3F(\lambda_b) = W(\lambda_b)$. The unrestricted RSS is obtained from LS estimation of (A8.1), whereas the restricted RSS is obtained from estimation of

$$w = X_r(\lambda_b)\beta_r + \varepsilon_r \tag{A8.3}$$

where $w = (w_1, \ldots, w_T)'$ and $x_{r,t}(\lambda_b)'$ is the t-th row of $X_r(\lambda_b)$, where:

$$x_{r,t}(\lambda_b)' = (\Delta y_{t-1}, \ldots, \Delta y_{t-k}, 1, t,) \tag{A8.4}$$

and $\beta = (c_1, \ldots, c_{k-1}, \mu_1, \beta_1)'$.

Q8.2 Consider Figure 8.2, which shows the results of some simulation experiments to evaluate AIC, BIC, se (standard error) and min-t as selection criteria where Model 2 is correct. Remember that min-t is actually min-t $= \inf^{(i)}(\inf_{\lambda_b}\{\hat{\tau}_\gamma^{(i)}(\lambda_b)\})$ considered across the three models. Compare the criteria and highlight the 'danger' regions, that is, where there is a high probability of the wrong model being selected.

A8.2 Generally, the information criteria (including se in that group) do better than min-t. Notice that for the information criteria, the bulk of the probability is for Model 2 correctly chosen; however, the use of min-t results too often in Model 3 being selected, except when β_2 is small, practically $\beta_2 < 2$. The 'danger' regions are where there is some probability mass for the wrong model. Consider AIC and BIC and note from Figure 8.2 that there is some probability for small β_2 when λ_b is at the beginning or end of the sample period and note that the distribution is symmetric in this respect. Of the two, AIC and BIC, in this comparison AIC does slightly better than BIC, with less probability in the danger region; in this respect se is comparable and indeed, if anything, slightly better than both AIC and BIC.

Q8.3 The following example is based on Hansen (2001), who considers the problem of dating possible breaks in US productivity. The underlying model involves y_t as labour productivity, measured as the growth rate of the ratio of output, taken as the index of Industrial Production of manufacturing and durables, to average weekly hours, February 1947 to April 2001. The regression model is a simple AR(1):

$$\Delta y_t = \mu_1 + \gamma_1 y_{t-1} + \varepsilon_t \quad t =, 1 \ldots, 651$$

Hansen (2001) reports the Chow test computed for all possible periods allowing for 5% trimming at the beginning and end of the sample.

Q8.3.i What is the model and the null and alternative hypotheses underlying the Chow test?

Q8.3.ii Should the researcher use critical values from the standard F distribution with appropriate degrees of freedom?

Q8.3.iii How does Hansen's model relate to the models of this and previous chapters? (see also Chapter 1.)

Q8.3.iv Modify the testing procedure to include a unit root in the null hypothesis.

A8.3.i. The basic idea of the Chow test is to compute the regression parameters for two sub-periods of the total period, with T_1 and T_2 observations respectively, such that $T_1 + T_2 = T$. Conditional on a constant variance for the two periods, the test can be interpreted as leading to rejection if there is a change in one or both of the parameters, which in this case are μ_1 and γ_1. A convenient way of setting up the Chow test is to use dummy variables which are familiar from this chapter, that is:

$$\Delta y_t = \mu_1^* + \gamma_1 y_{t-1} + I(t > T_b)\mu_2 + I(t > T_b)\gamma_2 y_{t-1} + \varepsilon_t$$

Initially ssume that T_b is known and serves to distinguish the first T_1 observations from the remaining T_2 observations for which $I(t > T_b)$. The Chow test is then just the F test of the joint null hypothesis, H_0: $\mu_2 = \gamma_2 = 0$ against H_0:

$\mu_2 = 0$ and/or $\gamma_2 = 0$; as such provided that T_b is known, it is distributed as $F(2, T - 4)$ under H_0.

A8.3.ii. However, T_b is not, in general, known and Hansen (2001) colourfully calls his construction a 'devilish trickster' who computes all possible Chow tests (that is subject to trimming) with a view to finding the break date that is most favourable to the existence of a break having occurred. Having done this, the distribution of the test statistic for H_0 no longer has a standard F distribution, not even asymptotically. In effect the trickster is undertaking a Sup-F test, taking the maximum F over all possible sub-periods. Hansen (2001) uses the Chow test in its LM form, which if T_b is known, is distributed as $\chi^2(2)$ under H_0; since $gF(g, T - K) \Rightarrow_D \chi^2(g)$, we can use $2F_{0.05}(2, T - K) \approx \chi^2_{0.05}(2) = 5.99$. However, the correct quantile allowing for the computation over all possible sub-periods is 12.9 (from Andrews, 1993). As might be expected, there are many fewer, but still a noticeable number, of breakpoints using the correct quantile. Of course, if there is a unit root, that is $\gamma_1 = 0$, then even this procedure is not correct.

A8.3.iii. To see how this basic model relates to those considered here, note that if we leave $\rho_1 = \gamma_1 + 1$ to be a free parameter, then the null hypothesis is one-dimensional and relates just to a break in the intercept, that is:

$$y_t = \mu_1^* + \rho_1 y_{t-1} + I(t > T_b)\mu_2 + \varepsilon_t$$

Consider the IO version of Model 1a, that is:

$$y_t = \mu_1 + z_t$$

$$(1 - \rho_1 L)z_t = \mu_2 DU_t + \varepsilon_t$$

$$y_t = (1 - \rho_1)\mu_1 + \rho_1 y_{t-1} + \mu_2 DU_t + \varepsilon_t$$

Therefore, the models are equivalent interpreting μ_1^* as $\mu_1^* = (1 - \rho_1)\mu_1$, equivalently $\mu_1 = \mu_1^*/(1 - \rho_1)$, which is the long-run mean (if stationary).

The extension to allow $\gamma_2 \neq 0$ can be viewed as an application of a two-regime threshold autoregression, but with the regime separator taken as simple temporal breakpoint rather than as some function of lagged y_t.

Q8.3.iv Next consider the model imposing $\gamma_2 = 0$, so that:

$$\Delta y_t = \mu_1^* + \gamma_1 y_{t-1} + I(t > T_b)\mu_2 + \varepsilon_t$$

The joint null with a unit root is H_0: $\gamma_1 = \mu_2 = 0$ against H_A: $\gamma_1 = 0$ and/or $\mu_2 = 0$. The resulting test statistic is is an example of the Murray and Zivot (MZ) testing procedure (see Section **8.1.3**), in which the unit root null is included with the structural break parameter here in a breaking mean function. Relative to the original model H_0: $\gamma_1 = \gamma_2 = \mu_2 = 0$. In either case, the presence of a unit root under H_0 changes the distribution compared to assuming that it is not present.

9
Conditional Heteroscedasticity and Unit Root Tests

Introduction

The two types of model introduced in this chapter are linked by the theme of (conditional) heteroscedasticity. The first model relates to the development of AR models to allow for random rather than fixed coefficients, where the particular focus is on a stochastic unit root (StUR) in contrast to the deterministic unit root of the conventional approach. A number of economic models can be viewed as generating a StUR; for example, a model of efficient markets can generate a root that is stochastic and differs from a unit root by the expected return on a financial asset (see LeRoy, 1982, and Leybourne, McCabe and Mills, 1996); another example is based on the permanent income hypothesis, which leads to an equation for consumption with a stochastic root that differs from unity by the ratio of the difference between the rate of inter-temporal time preference and the real rate of interest to the rate of interest (see Granger and Swanson, 1997).

An interesting way of reparameterising the StUR model is as a conventional fixed unit root plus heteroscedasticity that depends on the scale of the lagged dependent variable. DF-type test statistics are not robust in finite samples or asymptotically to the presence of a StUR, so it is useful to first test for a StUR and conditional on the outcome of this test, then test for a conventional unit root.

Another form of conditional heteroscedasticity is of the familiar ARCH/GARCH form. Whilst it is known from the theoretical work of Phillips (1987) and Phillips and Perron (1988) that the asymptotic distribution of the DF family of test statistics is not affected by this form of conditional heteroscedasticity, there is finite sample evidence that points generally to over-sizing, the extent of which depends on the 'strength' of the ARCH or GARCH effect (see for example Kim and Schmidt, 1993). Thus, there is the possibility that a test statistic that uses some information about the presence of heteroscedasticity will improve upon the conventional test statistics. This information can be in the rather neutral form that accompanies the construction of robust standard errors as suggested by White (1980), and now in a number of forms (see Cribari-Neto, 2004), leading to 'robust' unit root tests; or it can involve more structure as in the ML estimation of, for example, an AR-GARCH model. Equally there may be a role for the method of recursive mean/trend adjustment (RMA), which proves to be close to the power envelope in the conventional case (see *UR, Vol. 1*, Chapter 7), the rationale of which is to remove the correlation between the lagged dependent variable and the error term.

This chapter proceeds as follows. Section **9.1** outlines some models with a stochastic unit root, illustrates the problem this causes for DF-type test statistics and suggests some tests for a StUR. Section **9.2** is concerned with the AR-GARCH model, illustrating the over-sizing that is generally induced by the GARCH effect and the loss of (size-adjusted) power. Some alternative test statistics are suggested that have better size retention and better power and, as in Section **9.1**, this section concludes with an example.

9.1   Stochastic unit roots

9.1.1   The basic framework

In a standard AR (or ARMA) model the roots of the AR polynomial are taken as fixed. An extension of this class of models is to allow the AR coefficients to be random, giving rise to a random coefficient autoregression, RCAR, and hence a stochastic root or roots. For example, consider the AR(1) model with a stochastic root, then the process generating the data is:

$$y_t = \rho_t y_{t-1} + \varepsilon_t \qquad t = 1, \ldots, T \tag{9.1}$$

where, for simplicity, $\varepsilon_t \sim \text{iid}(0, \sigma_\varepsilon^2)$ and $y_0 = 0$ (otherwise y_t is redefined to subtract y_0). The AR(1) coefficient, ρ_t, is written with a t subscript to indicate that it is not necessarily constant as time varies.

A possible scheme to model the variation in ρ_t is as follows:

$$\rho_t = \rho + \delta_t \tag{9.2}$$

where $\delta_0 = 0$ and $\delta_t \sim \text{iid}(0, \omega^2)$. Note that the variance of δ_t is conventionally denoted ω^2, rather than σ_δ^2, and ε_t and δ_t are assumed to be independent random variables (for a relaxation of this assumption see Su and Roca, 2012). In evaluating covariance stationarity it can be shown (see Leybourne, McCabe and

Tremayne 1996, hereafter LMT, and Distaso, 2008), that:

$$E(y_t) = 0 \tag{9.3a}$$

$$\sigma_{y,t}^2 = \sigma_\varepsilon^2 \left[\sum_{j=0}^{t-1} (\rho^2 + \omega^2) \right]^j \tag{9.3b}$$

$$\text{cov}(y_t y_{t+s}) = \rho^s \sigma_{y,t}^2 \tag{9.3c}$$

Hence, if $\rho^2 + \omega^2 \geq 1$ then the variance of y_t increases with t; for example, if $\rho^2 + \omega^2 = 1$, then $\sigma_{y,t}^2 = t\sigma_\varepsilon^2$, which is as in the standard unit root case but without the requirement that $\rho = 1$, and it implies that if $\omega^2 > 0$, then a fixed root less than unity is not sufficient to ensure stationarity. Whether y_t is generated by a stationary or nonstationary process depends on how close ρ is to unity and the scale of the variations in ρ_t. Whilst the familiar I(d) notation refers to the case with $\rho = 1$ as I(1), the stochastic root case, with $\rho_t = 1 + \delta_t$ and $\delta_t \sim$ iid(0, ω^2), $\omega^2 > 0$, will be referred to as $I_t(1)$.

The idea of the stochastic root model is that ρ is subject to generally small but frequent shocks; at a first approach these shocks are assumed to be iid, but subsequently some dependence is allowed in the dynamics of the shocks. Stochastic root models may also arise from the structure of economic fundamentals. Leybourne, McCabe and Mills (1996), hereafter LMM, give the following example. Let y_t denote the price of a (financial) stock at time t, then the one-period return on the stock is $r_t = \Delta y_t / y_{t-1}$ and the expected return at $t - 1$ is:

$$E_{t-1}(r_t) = [E_{t-1}(y_t) - y_{t-1}]/y_{t-1} \tag{9.4}$$

Note that a time subscript is here added to the expectation operator for clarity us to when the expectation is formed. Multiplying through by y_{t-1} and rearranging gives:

$$E_{t-1}(y_t) = [1 + E_{t-1}(r_t)]y_{t-1} \tag{9.5}$$

Letting $\varepsilon_t = y_t - E_{t-1}(y_t)$ be the expectational error and defining $\delta_t \equiv E_{t-1}(r_t)$, then:

$$y_t = (1 + \delta_t)y_{t-1} + \varepsilon_t \tag{9.6}$$

This is of the form (9.2) with $\rho_t = 1 + \delta_t$ and var(δ_t) = $\omega^2 > 0$. Since the underlying, or fixed, root is unity, this is an example of a stochastic unit root, which is the case considered extensively by McCabe and Tremayne (1995), LMT (1996), LMM (1996) and Granger and Swanson (1997). Note that after subtracting y_{t-1} from both sides of (9.6), the equation can then be rearranged in the usual DF form but with a random coefficient, that is:

$$\Delta y_t = \delta_t y_{t-1} + \varepsilon_t \tag{9.7}$$

An interesting interpretation of the stochastic root model arises by moving the stochastic component into the error term. Consider the case in which (9.2) holds with $\delta_t \sim$ iid$(0, \omega^2)$, so that:

$$y_t = \rho y_{t-1} + \xi_t \tag{9.8a}$$

$$\xi_t = \delta_t y_{t-1} + \varepsilon_t \tag{9.8b}$$

Then

$$E(\xi_t|y_{t-1}) = E(\delta_t y_{t-1}) + E(\varepsilon_t)$$

$$= y_{t-1}E(\delta_t) + E(\varepsilon_t) \quad \text{because } y_{t-1} \text{ is taken as given}$$

$$= 0$$

$$\text{var}(\xi_t|y_{t-1}) = \text{var}(\delta_t y_{t-1} + \varepsilon_t)$$

$$= y_{t-1}^2 \omega^2 + \sigma_\varepsilon^2 \quad \text{cov}(\delta_t y_{t-1}, \varepsilon_t) = 0$$

Note that the model can now be interpreted as a conventional AR(1) model where the error ξ_t is heteroscedastic depending on the lagged level of y_t. If $\rho = 1$ then the model is referred to as possessing heteroscedastic integration (HI), (McCabe and Smith, 1998).

An important implication of the presence of a stochastic unit root is that y_t cannot be differenced to stationarity, which is the usual procedure to achieve stationarity in the conventional unit root case. The reason is evident from consideration of (9.7): in the case of a fixed unit root, this would just be $\Delta y_t = \varepsilon_t$ and since by assumption ε_t is stationary, then so is Δy_t; however, in the case that δ_t is not a constant, because its variance ω^2 is non-zero, the process generating Δy_t is not stationary.

Some illustrations of the data generated by a conventional unit root and a stochastic unit root are provided in Figure 9.1, which has four panels. The 'size' of the stochastic part of the unit root is indexed by the variance, ω^2, of δ_t, here taken to be $\omega^2 = 0.0001$, 0.001, 0.010, 0.050, where $\delta_t \sim$ iid$(0, \omega^2)$. Each panel shows the conventional random walk and the corresponding stochastic random walk (both start at zero) with the same sequence of innovations ε_t. Even though the two series start together they begin to depart even for small values of ω^2, although it can be difficult to tell the difference in the early part of the sample. The larger is ω^2 the greater the difference between the two series. An interesting feature of the stochastic unit root series becomes evident as ω^2 increases and that is the clustering of 'spikes' of volatility in a way that is similar to the clustering that ARCH and GARCH models seek to capture, and emphasises the heteroscedastic integration interpretation of McCabe and Smith (1998).

McCabe and Smith (1998) show that the effect of a stochastic unit root on the asymptotic distribution of the DF t-type test statistics is to add a negative stochastic term; for small deviations (that is 'small' ω^2) under homoscedastic

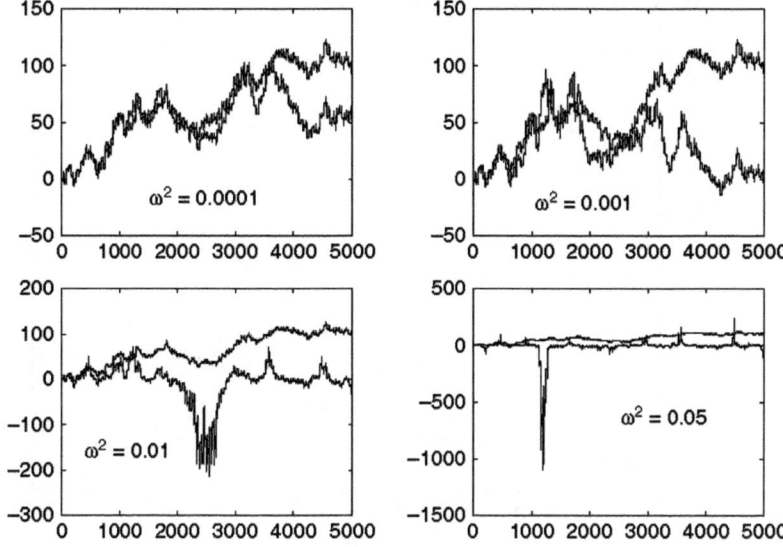

Figure 9.1 Illustrative stochastic random walks

integration, the effect on the asymptotic distribution is a scale shift rather than a left displacement. Together these results imply that the DF t-type tests will not have power (which requires the left displacement) for small heteroscedastic integration but will gain power as the deviations become substantial. To illustrate, data were generated by the process (9.7), where $\varepsilon_t \sim niid(0, 1)$ and $\omega^2 \in [0, 0.25)$, with $T = 100$ and 10,000 replications. The standard DF t-type statistics, $\hat{\tau}_\mu$ and $\hat{\tau}_\beta$, were calculated and the results are summarised in Figure 9.2. The known effect of superfluous deterministic terms on the relative power of $\hat{\tau}_\mu$ and $\hat{\tau}_\beta$ is also apparent in this context, with the power function for $\hat{\tau}_\mu$ above that for $\hat{\tau}_\beta$. The theoretical arguments of LMT are clearly supported with very little power for either $\hat{\tau}_\mu$ or $\hat{\tau}_\beta$ until about $\omega^2 = 0.15$, which in this context is a large variance generating a lot of nonstationarity that should be evident from a plot of the time series. Either way the implications from using $\hat{\tau}$-type tests are misleading if there is HI: the null hypothesis of the DF test is a fixed unit root, so nonrejection seems to provide support for a unit root, whereas rejection seems to provide support for a stationary root; however, the correct decision is in favour of a stochastic unit root.

9.1.2 Tests designed for a StUR

9.1.2.i Background

Following on from the observation that standard DF tests have very little power for HI processes, it would be useful to have a test that does have such power.

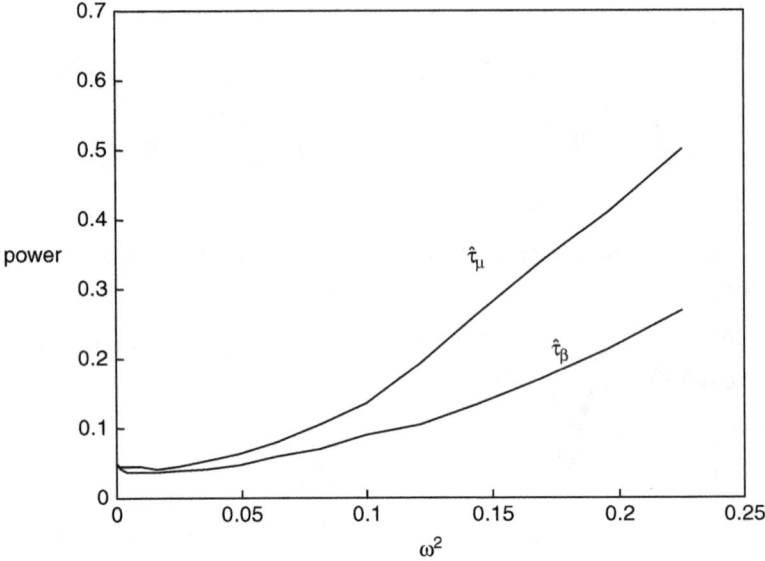

Figure 9.2 Power of $\hat{\tau}$-type tests in the stochastic unit root model

There are two situations to consider, depending on the dynamic structure of δ_t. In the simplest case δ_t is assumed to be iid, giving rise to test statistics denoted below as Z_i, where the i subscript indexes the included deterministic components. As the shocks are iid, it is possible that there are frequent changes of state from an explosive to a stationary state and vice versa. In some cases it may be more reasonable to model a gradual change between states, which can, for example, be captured by allowing δ_t to follow an AR(1) process as in the following specification:

$$(1 - \varphi_1 L)\delta_t = \eta_t \qquad\qquad (9.9a)$$

$$\Rightarrow$$

$$\delta_t = \varphi_1 \delta_{t-1} + \eta_t \qquad \varphi_1 < 1 \qquad\qquad (9.9b)$$

$$\delta_t = \delta_{t-1} + \eta_t \qquad \varphi_1 = 1 \qquad\qquad (9.9c)$$

where $\eta_t \sim$ iid$(0, \sigma_\eta^2)$ and is independent of ε_t; notice that $\omega^2 = (1 - \varphi_1^2)^{-1}\sigma_\eta^2 < \infty$ for $|\varphi_1| < 1$, whereas ω^2 is not defined for $\varphi_1 = 1$ when δ_t is a random walk. A key case is the joint hypothesis $\omega^2 = 0$ and $\rho = 1$, in which case the generating model reduces to the standard unit root (pure random walk) model:

$$y_t = y_{t-1} + \varepsilon_t$$

This is the case we will concentrate on here, but note that Distaso (2008) has considered the case where $E(\rho_t) \neq 1$ (assuming that ρ_t is iid).

The suggested test statistics (see McCabe and Tremayne, 1995 and LMM, 1996), depend upon whether it is maintained under the alternative that (1), $|\varphi_1| < 1$ or (2), $\varphi_1 = 1$, the difference being whether the departures from the unit root are generated by a stationary AR(1) process, as in (9.4), or a random walk as in (9.5); note that case (1) includes the iid case for which $\varphi_1 = 0$. Before stating the test statistics, two generalisations of the model comprising (9.1), (9.3) and either (9.4) or (9.5) are considered. These are to incorporate deterministic components and allow for higher-order dynamics, so that under the alternative hypothesis:

$$y_t^* = \rho_t y_{t-1}^* + \varepsilon_t \qquad t = 1, \ldots, T \tag{9.10}$$

$$y_t^* \equiv y_t - \mu_t - \sum_{j=1}^{p} \varphi_j y_{t-j} \tag{9.11}$$

$$\rho_t = 1 + \delta_t \qquad \sigma_\delta^2 \equiv \omega^2 > 0$$

The cases considered by LMT and LMM are a quadratic trend $\mu_t = \beta_0 + \beta_1 t + \beta_2 t(t+1)/2$ and a linear trend $\mu_t = \beta_0 + \beta_1 t$, designated $i = 1$ and $i = 2$, respectively (the case $\mu_t = \beta_0$ is not distinguishable in this approach from $i = 2$; the notation follows convention in this area, although a more intuitive notation would use $i = 1$ for the linear trend).

The principle of the test statistics is LM (Lagrange multiplier), which has the advantage in this context of only requiring estimation under the null hypothesis, which is that here there is a (fixed) unit root. Notice that under the null, (9.10) and (9.11) reduce to the following:

$$\Delta y_t^* = \varepsilon_t \tag{9.12}$$

$$\Delta y_t^* \equiv \Delta y_t - \Delta \mu_t - \sum_{j=1}^{p} \varphi_j \Delta y_{t-j} \tag{9.13}$$

Hence, the first stage is to regress Δy_t on the components of $\Delta \mu_t$ and p lags of Δy_t allowing definition of the following residuals:

$$\hat{\varepsilon}_t = \Delta y_t - \Delta \hat{\mu}_t - \sum_{j=1}^{p} \hat{\varphi}_j \Delta y_{t-j} \tag{9.14}$$

where $\wedge$ indicates a LS estimate and

$$\Delta \hat{\mu}_t = \hat{\beta}_1 + \hat{\beta}_2 t \quad \text{for } i = 1, \beta_0 + \beta_1 t + \beta_2 t(t+1)/2$$

$$\Delta \hat{\mu}_t = \hat{\beta}_1 \qquad \text{for } i = 2, \mu_t = \beta_0 + \beta_1 t \text{ (and for } \mu_t = \beta_0)$$

9.1.2.ii The test statistics

The null hypothesis is H_0: $\omega^2 = 0$, whereas the alternative hypothesis is H_A: $\omega^2 > 0$; and the tests are conditional on $\rho = 1$. The case where the test statistic is not conditioned on $\rho = 1$ is considered by Distaso (2008) assuming that δ_t is iid. The test statistics are right-sided, that is, they reject for large sample values relative to the selected quantiles.

Having defined $\hat{\varepsilon}_t$, the test statistics are as follows distinguishing: (i), $|\varphi_1| < 1$ and (ii), $\varphi_1 = 1$, where in the former case the test statistics are designated Z_i and in the latter case E_i, $i = 1, 2$.

$|\varphi_1| < 1$

$$Z_i = T^{-3/2} \hat{\sigma}_\varepsilon^{-2} \hat{\kappa}^{-1} \sum_{t=2}^{T} \left(\sum_{j=1}^{t-1} \hat{\varepsilon}_j \right)^2 (\hat{\varepsilon}_t^2 - \hat{\sigma}_\varepsilon^2) \tag{9.15}$$

$\varphi_1 = 1$

$$E_i = T^{-3} \hat{\sigma}_\varepsilon^{-4} \sum_{i=2}^{T} \left\{ \left[\sum_{t=i}^{T} \hat{\varepsilon}_t \left(\sum_{j=1}^{t-1} \hat{\varepsilon}_j \right) \right]^2 - \hat{\sigma}_\varepsilon^2 \sum_{t=i}^{T} \left(\sum_{j=1}^{t-1} \hat{\varepsilon}_j \right)^2 \right\} \tag{9.16}$$

where:

$\hat{\varepsilon}_t = \Delta y_t - \Delta \hat{\mu}_t$ in general

$\hat{\varepsilon}_t = \Delta y_t - \hat{\beta}_1 - \hat{\beta}_2 t$ for $i = 1$

$\hat{\varepsilon}_t = \Delta y_t - \hat{\beta}_1$ for $i = 2$

$\hat{\sigma}_\varepsilon^2 = T^{-1} \sum_{t=1}^{T} \hat{\varepsilon}_t^2$

$\hat{\kappa}^2 = T^{-1} \sum_{t=1}^{T} (\hat{\varepsilon}_t^2 - \hat{\sigma}_\varepsilon^2)^2$

For convenience of notation, pre-sample values of y_t are assumed to exist, so that the available sample is always labelled as starting at $t = 1$ and the sample size is T. This is just a matter of convention; for example if $p = 2$, then one observation is 'lost' in obtaining Δy_t and a further two are 'lost' in obtaining Δy_{t-1} and Δy_{t-2}, hence an overall sample of 103 results in a usable sample conveniently labelled from $t = 1$ to $100 = T$ (so the pre-sample observations are $t = -2, -1, 0$).

The asymptotic distribution of Z_1 is given in LMT (1996) and for E_i in LMM (1996). Some quantiles for testing are given in Table 9.1.

There are several practical issues in the application of these tests. One is the selection of the lag length p (see *UR, Vol. 1*, chapter 9); LMM found that nominal size was maintained with a general-to-specific strategy using the marginal t-statistic at a 5% two-sided significance level, with under-fitting causing more serious size distortions than over-fitting; further, power was also maintained using this strategy. A second issue is which test to use. LMM (1996) found that the power of Z_1 was poor when δ_t followed a random walk (for which E_1 was appropriate), whereas E_1 had low power when δ_t was iid. In the intermediate case δ_t followed a stationary AR(1) process and E_1 again had low power, with the exception as $\psi_1 > 0.8$, whereas Z_1 had relatively good power (and much better than that for E_1) for all values of $\psi_1 \in (0.20, 0.95)$, although power fell away (relatively) as $\varphi_1 \to 1$.

Table 9.1   Quantiles of test statistics: Z_1, Z_2, E_1 and E_2

	Z_1			Z_2		
	90%	95%	99%	90%	95%	99%
100	0.142	0.192	0.320	0.142	0.192	0.320
200	0.127	0.176	0.299	0.127	0.176	0.299
500	0.114	0.161	0.278	0.114	0.161	0.278
1,000	0.104	0.149	0.261	0.104	0.149	0.261
	E_1			E_2		
100	0.067	0.072	0.081	0.052	0.060	0.072
200	0.065	0.069	0.077	0.051	0.059	0.069
500	0.064	0.068	0.074	0.051	0.058	0.067
1,000	0.064	0.068	0.073	0.050	0.057	0.067

Sources: Z_1, LMT (1996); Z_2, author's calculations based on 50,000 replications; E_1 and E_2, LMM (1996).

9.1.2.iii   Some examples

LMM revisited the original Nelson and Plosser (1982) data set comprising 13 macroeconomic variables. Lag lengths were selected by the general-to-specific (G-t-S) strategy using the t statistic on the marginal lag, with the maximum lag set at 5. On the basis of the Z_1 test, LMM found that the I(1) null was rejected in favour of the $I_t(1)$ alternative at the 5% level for nominal GNP, the GNP deflator and nominal wages, and that these results were robust to the choice of lag length. As LMM note, the variables for which the I(1) null was more likely to be rejected were denominated in nominal terms, for example nominal GNP rather than real GNP, reflecting the greater volatility of the nominal series compared to their real counterparts.

LMT (1996) also gave a number of examples and the illustration chosen here is for (logarithm of) the London Stock Exchange market index FTSE 350 (daily from 1 January 1986 to 28 November 1994). The fitted model used $p = 4$ and the test statistics were $E_1 = 0.065$, $E_2 = 0.063$, $Z_1 = 0.199$ and $Z_2 = 0.008$. The first two are significant at the 5% level (although E_1 is only marginally so), Z_1 is significant at better than the 5% level, whilst Z_2 is not significant. LMM estimated the StUR model under the $I_t(1)$ alternative, using the scheme with $\delta_t = \delta_{t-1} + \eta_t$, and found a marked improvement in fit compared to the case where a fixed unit root was imposed (further estimation details are provided in LMM, 1996). In practice there are a number of other issues that it is useful to consider in an empirical specification. These include, as Distaso (2008) notes, extending the hypothesis tests to allow for $\rho_t = \rho + \delta_t$, where the fixed root ρ is not necessarily equal to unity; allowing for dependence between ε_t and δ_t as in Su and Roca (2012); and determining a criterion to choose an appropriate model for δ_t; in the context outlined here this implies estimating φ_1 in the model $\delta_t = \varphi_1 \delta_{t-1} + \eta_t$.

9.2   DF tests in the presence of GARCH errors

To focus ideas in an exposition of DF tests it is convenient to start with the very simple AR(1) model $y_t = \rho y_{t-1} + \varepsilon_t$, where $\varepsilon_t \sim$ iid$(0, \sigma_\varepsilon^2)$, with $\sigma_\varepsilon^2 < \infty$, in which case the null hypothesis to be tested is $H_0 : \rho = 1$ and the alternative is $H_A : |\rho| < 1$. The resulting test statistics $\delta = T(\hat{\rho} - 1)$ and $\hat{\tau} = (\hat{\rho} - 1)/\hat{\sigma}(\hat{\rho})$, where $\hat{\rho}$ is the LS estimator of ρ and $\hat{\sigma}(\hat{\rho})$ is its estimated standard error, have limiting distributions that are often referred to as DF distributions (see *UR, Vol. 1*, chapter 6). The error assumptions have been generalised in a number of ways resulting in variations to the estimated model or test statistic such that the DF test statistics maintain their respective DF distributions (see, for example, Phillips, 1987, and Phillips and Perron, 1998).

This section is concerned with error processes that are ARCH (autoregressive conditional heteroscedasticity) or GARCH (generalised ARCH). This specification is widely used in modelling financial time series as it gives rise to a clustering phenomenon recognised as periods of high volatility, followed by relatively quiescent periods. ARCH and GARCH specifications have also been used in modelling other types of series, and Engle's (1982) seminal article used a time series on the UK inflation rate to illustrate an ARCH model. Whilst the essential result is that the DF test statistics maintain their respective limiting DF distributions in the presence of ARCH/GARCH errors, provided the error process is not itself integrated and has a finite second moment, there remain several important issues that arise in the context of finite samples and in improving the power of the DF unit root tests.

9.2.1   Conditional and unconditional variance

As a reminder of the distinction between the conditional and the unconditional variance (of y_t) start with a simple AR(1) process:

$$y_t = \rho y_{t-1} + \varepsilon_t \qquad |\rho| < 1 \tag{9.17}$$

where $\varepsilon_t \sim (0, \sigma_\varepsilon^2)$. The unconditional variance of y_t, σ_y^2, takes the variation in y_{t-1} into account, so that:

$$\sigma_y^2 = \text{var}(\rho y_{t-1} + \varepsilon_t)$$
$$= \rho^2 \sigma_y^2 + \sigma_\varepsilon^2 + 2\rho \text{cov}(y_{t-1}\varepsilon_t) \qquad \text{using var}(y_{t-1}) = \sigma_y^2$$
$$= (1 - \rho^2)^{-1}\sigma_y^2 \qquad \text{using cov}(y_{t-1}\varepsilon_t) = 0$$

σ_ε^2 is the variance of y_t conditional on y_{t-1}, $\text{var}(y_t|y_{t-1})$; that is holding y_{t-1} fixed, the only source of variation is ε_t which has variance σ_ε^2 and this is referred to as the conditional variance of y_t. If σ_ε^2 is constant then $\text{var}(y_t|y_{t-1}) = \text{var}(y_{t+1}|y_t) = \text{var}(y_{t+2}|y_{t+1}) = \ldots = \sigma_\varepsilon^2$, that is, the same degree of uncertainty is attached to the one-step-ahead forecast whichever point in the sequence of y_t

is being considered; in particular the one-step-ahead forecast variance is invariant to past values of ε_t or ε_t^2. ARCH and GARCH processes relax this assumption and, therefore, enable the possibility of clustering, so that heteroscedasticity is allowed in the conditional variance.

The GARCH(r, s) specification, which nests the ARCH(r,) specification for conditional heteroscedasticity, is as follows:

$$\varepsilon_t = u_t \sigma_t \tag{9.18}$$

$$\sigma_t^2 = \alpha_0 + \alpha(L)\varepsilon_t^2 + \gamma(L)\sigma_t^2 \tag{9.19}$$

where $u_t \sim$ iid(0, 1), $\alpha(L) = \sum_{i=1}^{s} \alpha_i L^i$ and $\gamma(L) = \sum_{j=1}^{r} \gamma_j L^j$. Engle's original ARCH(r) specification results when $\gamma(L) = 0$. Sufficient conditions for $\sigma_t^2 > 0$ (an essential property of a variance) are $\alpha_0 > 0$, $\alpha_i \geq 0$ and $\gamma_j \geq 0$. A necessary and sufficient condition for stationarity is $\alpha(1) + \gamma(1) < 1$ (Bollerslev, 1986); this condition together with the assumption that all of the roots of the polynomials $1 - \alpha(L) - \gamma(L)$ and $1 - \gamma(L)$ lie outside the unit circle is sufficient for strict stationarity and ergodicity of $\{\varepsilon_t\}$, with finite variance (see Bougerol and Picard, 1992, Ling and McAleer, 2002, Wang, 2006 and Wang and Mao, 2008). The unconditional variance of ε_t is $\sigma^2 = \alpha_0 (1 - \alpha(1) - \gamma(1))^{-1}$, hence for σ^2 to be positive given $\alpha_0 > 0$, it must be the case that $1 - \alpha(1) - \gamma(1) > 0$.

9.2.2 Illustrative AR(1) + GARCH(1, 1) series

As noted, a key characteristic of AR models combined with ARCH or GARCH errors is that they generate volatility clustering and, therefore, are particularly relevant for modelling financial time series and, for example, commodity prices, which exhibit periods of intense activity followed by quiet periods. Some simulated series generated from an AR(1) model with unit root combined with GARCH(1, 1) errors are shown in Figure 9.3. The simulated series start in the top left panel with a pure random walk, followed by series with the same random inputs but with a GARCH(1, 1) error structure, with pairs of α_1 and γ_1 such that their sum is 0.95. The GARCH(1, 1) sequence then starts in the top right panel with $\alpha_1 = 0.9$, $\gamma_1 = 0.05$; subsequently α_1 decreases by 0.1 and so γ_1 increases by 0.1 (moving from left to right on the figure). It is evident that the key parameter is α_1, the volatility parameter, since as this declines the series becomes more like the initial random walk, with the prominence of the volatility clusters declining accordingly.

9.2.2.i *Asymptotic distributions of DF tests with GARCH errors*

Next consider the asymptotic distributions of the DF test statistics $T(\hat{\rho}_i - 1)$ and $\hat{\tau}_i$ for i = 0, μ, β, that is μ_t contains, respectively, no deterministic terms, a constant, and a constant and a linear trend (collectively referred to as the DF tests). Let the data be generated by $y_t = y_{t-1} + \varepsilon_t$, where ε_t follows the GARCH

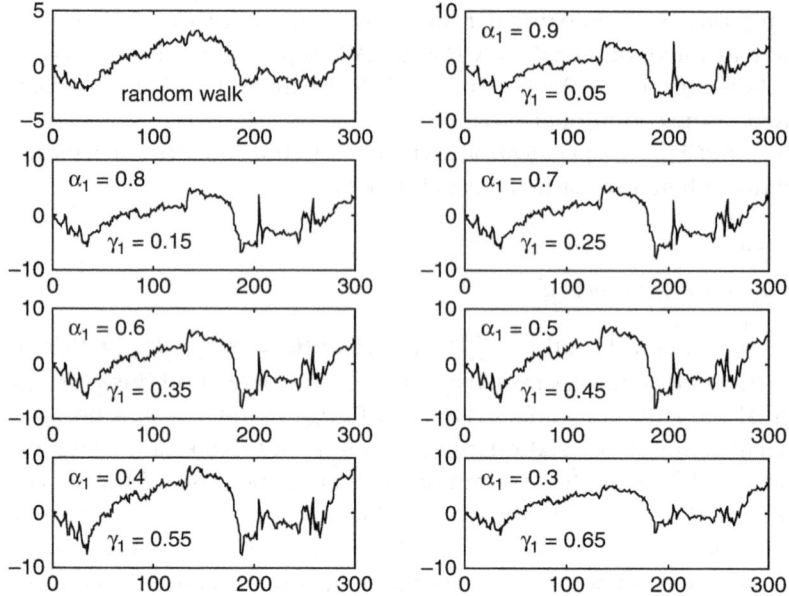

Figure 9.3   Simulated time series with GARCH(1, 1) errors

(r, s) process defined above; see (9.18) and (9.19). Assume that $E(u_t^2) = 1$ and $\alpha(1) + \gamma(1) < 1$ (which implies that $\sigma_\varepsilon^2 < \infty$), then Wang and Mao (2008, theorem 1) show that asymptotic distributions of the DF test statistics with GARCH errors are the same as in the case with $\varepsilon_t \sim \text{iid}(0, \sigma_\varepsilon^2)$. (Strictly the theorem referred to relates to the cases $i = 0$ and $i = \mu$; however $i = \beta$, follows by extension.)

Also note from Remark 7 of Wang and Mao (2008) that their theorem 1 remains valid even if the third and higher moments of u_t, that is $E(u_t^{2+\upsilon})$ for $\upsilon > 0$, are infinite, which removes the previously thought need for a finite fourth moment condition. Theorem 2 of Wang and Mao (2008) extends their theorem 1 to an $AR(p) = ADF(p - 1)$ generation process. The data is now assumed to be generated by:

$$\Delta y_t = \gamma y_{t-1} + \sum_{j=1}^{p-1} c_j \Delta y_{t-j} + \varepsilon_t \tag{9.20}$$

$$\varepsilon_t = u_t \sigma_t$$

$$\sigma_t^2 = \alpha_0 + \alpha(L)\varepsilon_t^2 + \gamma(L)\sigma_t^2$$

with the usual stationarity condition for the GARCH(r, s) errors, and a unit root corresponds to $\gamma = 0$ in (9.20). As elsewhere, deterministic components may be included by replacing y_t with $\tilde{y}_t \equiv y_t - \mu_t$ and estimating μ_t by means of a prior regression.

9.2.2.ii Illustration with GARCH(1, 1)

To illustrate some of the issues, the widely applied GARCH(1, 1) model was combined with an AR(1) model, the AR(1)-GARCH(1, 1) model, so that the generating process was:

$$\tilde{y}_t = \rho \tilde{y}_{t-1} + \varepsilon_t \qquad \text{AR(1) model} \qquad (9.21a)$$

$$\tilde{y}_t \equiv y_t - \mu_t \qquad \text{demean or detrend} \qquad (9.21b)$$

$$\varepsilon_t = u_t \sigma_t \qquad u_t \sim N(0,1), t(4), \text{Cauchy} \qquad (9.21c)$$

$$\sigma_t^2 = \alpha_0 + \alpha_1 \varepsilon_{t-1}^2 + \gamma_1 \sigma_{t-1}^2 \qquad \text{GARCH } (1,1) \qquad (9.21d)$$

$$\sigma^2 = \frac{\alpha_0}{(1 - \alpha_1 - \gamma_1)} \qquad \text{unconditional variance} \qquad (9.21e)$$

Initial values were set to zero. The fitted regressions allowed for deterministic terms, so that the specification of $y_t - \mu_t$ differed to distinguish the different $\hat{\tau}$-type test statistics $\hat{\tau}$, $\hat{\tau}_\mu$ and $\hat{\tau}_\beta$.

The experimental design was based on Kim and Schmidt (1993) and Wang and Mao (2008), with $\alpha_0 = 0.05$ and combinations of α_1 and γ_1 with the same sum $\alpha_1 + \gamma_1 = 0.95$ and, for sensitivity to the IGARCH specification, $\alpha_1 + \gamma_1 = 1$. Setting $\alpha_0 = 1 - \alpha_1 - \gamma_1$ implies that the unconditional variance is constant at unity. The scaled conditional errors, u_t, are variously taken to be draws from $N(0, 1)$, the 't' distribution with 4 degrees of freedom, t(4), (as in Wang and Mao, 2008) and the Cauchy distribution; note that the variance of a random variable distributed as t(n) is $n/(n-2)$, and the draws from t(4) were standardised by $2^{1/2}$. The Cauchy distribution does not have any moments and so represents a failure of the conditions required for the $\hat{\tau}$ tests to have a limiting DF distribution and indeed, for the case where $\alpha_1 = \gamma_1 = 0$, also a failure of the finite second moment condition required in the simple case where there is conditional homoscedasticity. The case $u_t \sim N(0, 1)$ obviously satisfies the second moment condition, as does $u_t \sim t(4)$, but the latter generates much fatter tails to the distribution. When $\sigma_u^2 < \infty$ but $E(u_t^4) = \infty$, the errors are referred to as being heavy-tailed. The t(4) distribution was chosen because of the key result in Wang and Mao (2008) that relaxes the finite fourth moment condition. (Note that a random variable distributed as t(k) has finite moments for $h < k$, hence t(4) does not have a finite fourth moment.)

The results reported in Table 9.2 are for $T = 200$ and a nominal size of 5%, with the critical values for the $\hat{\tau}_i$ tests obtained from the extension of the MacKinnon (1991) response surface coefficients reported in *UR, Vol. 1*, chapter 6 (appendix). The results in Table 9.2 for $u_t \sim N(0, 1)$ and $u_t \sim t(4)$ show that the size of the DF tests tends to increase as the volatility parameter, α_1, increases; taking $\hat{\tau}_\mu$ as an example, when $u_t \sim N(0, 1)$ the empirical size for $\alpha_1 = 0.05$ and $\gamma_1 = 0.9$ is 5.36%, but 8.38% for $\alpha_1 = 0.3$ and $\gamma_1 = 0.65$; the corresponding percentages

Table 9.2   Size of $\hat{\tau}$-type tests in the presence of GARCH(1, 1) errors from different distributions; $T = 200$, nominal 5% size

α_1	0.0	0.05	0.10	0.15	0.20	0.25	0.30	0.90	0.95
γ_1	0.0	0.90	0.85	0.80	0.75	0.70	0.65	0.05	0.05
$N(0, 1)$									
$\hat{\tau}$	5.02	5.12	4.92	5.40	5.58	5.50	5.54	5.60	6.20
$\hat{\tau}_\mu$	5.06	5.36	6.00	6.72	7.48	8.16	8.38	8.72	11.80
$\hat{\tau}_\beta$	5.40	5.50	5.98	6.60	7.48	8.48	8.84	9.44	14.20
$t(4)$									
$\hat{\tau}$	4.94	5.58	5.72	5.92	6.80	6.14	6.44	5.84	5.94
$\hat{\tau}_\mu$	5.58	6.50	6.48	7.88	8.52	8.84	8.96	9.62	11.36
$\hat{\tau}_\beta$	4.76	6.10	7.32	8.50	9.24	9.60	10.24	10.84	14.34
Cauchy									
$\hat{\tau}$	3.19	43.2	46.0	47.3	47.7	47.7	47.7	37.2	39.80
$\hat{\tau}_\mu$	7.65	34.6	38.7	40.4	41.2	41.5	41.8	30.6	34.42
$\hat{\tau}_\beta$	5.96	31.9	36.3	38.3	39.4	40.0	40.1	31.0	34.74

for $u_t \sim t(4)$ are 6.35% and 9.52%. The more extensive results in Wang and Mao (2008) show that the degree of the over-sizing reduces as the sample size increases; thus, although the limiting result that the asymptotic distributions of the DF tests without and without GARCH errors are the same is a powerful result, there is an opportunity in finite samples to improve the fidelity of the actual to the nominal size. The final column in Table 9.2 illustrates the IGARCH case, for which it is known that the DF tests no longer have the respective DF distributions. For $u_t \sim N(0, 1)$, the power of $\hat{\tau}_\mu$ is now about double its nominal size of 5%, whilst that for $\hat{\tau}_\beta$ is over 13%; broadly similar over-sizing characterises the case where $u_t \sim t(4)$. When the u_t are drawn from the Cauchy distribution, although the effect on the standard case (errors are iid) is moderate, the introduction of GARCH errors leads to gross over-sizing.

The simulations were also designed to consider the power of the DF $\hat{\tau}$ tests when the errors are generated by a GARCH(1, 1) process compared to the homoscedastic case. Given the differences in finite sample size, a comparison of 'raw' power would give a power advantage to over-sized tests; power is therefore reported on a size-adjusted basis. The results are summarised in Table 9.3 for $\rho = 0.95$ and $\rho = 0.9$ and $u_t \sim N(0, 1)$. The case with $\hat{\tau}$, which shows a slight deterioration in power moving from left to right across the table, is reported for reference only as it is unlikely to be used in practice as an allowance for a constant is the minimum specification of the deterministic terms. More realistic are the cases with $\hat{\tau}_\mu$ and $\hat{\tau}_\beta$ and in each case power deteriorates as the volatility parameter α_1 increases; for example, for $\alpha_1 = 0.3$, $\gamma_1 = 0.65$ and $\rho = 0.95$, the power of $\hat{\tau}_\mu$ declines from 30.7% in the iid case to 26.3% and for $\hat{\tau}_\beta$ from 18.3% to 14.8%.

Table 9.3 Power of $\hat{\tau}$-type tests in the presence of GARCH (1, 1) errors; $\rho = 0.95$, 0.9, $T = 200$ and $u_t \sim N(0, 1)$, nominal 5% size

α_1	0	0.05	0.10	0.15	0.20	0.25	0.30	0.35
γ_1	0	0.90	0.85	0.80	0.75	0.70	0.65	0.60
$\rho = 0.95$								
$\hat{\tau}$	80.06	78.98	78.04	76.28	75.34	73.92	75.34	73.50
$\hat{\tau}_\mu$	30.72	30.74	30.06	29.42	28.86	28.20	26.28	25.66
$\hat{\tau}_\beta$	18.26	17.92	16.82	16.42	15.90	15.12	14.84	14.58
$\rho = 0.90$								
$\hat{\tau}$	99.72	99.54	99.18	98.94	98.62	98.36	98.34	97.76
$\hat{\tau}_\mu$	85.98	85.12	82.16	80.52	78.08	76.00	72.44	70.70
$\hat{\tau}_\beta$	63.96	61.70	58.00	55.12	52.80	50.00	48.20	46.16

9.2.3 Tests that may improve size retention and power with GARCH errors

Whilst the standard DF test statistics are asymptotically robust to conditional heteroscedasticity, there is some size infidelity in finite samples, mainly in the form of over-sizing, and power deteriorates quite substantially in the presence of GARCH errors. There are several possible ways of improving this aspect of their performance. A simple and quite popular possibility is to continue to use the LS-based test, that is $\hat{\tau}_i$, but combine it with robust standard errors, resulting in a robustified version of the $\hat{\tau}$ family of DF test statistics (see Haldrup, 1994, and Demetrescu, 2010). This approach is considered next; subsequently tests that use recursive mean adjustment, RMA, are considered. RMA is a simple alternative to the standard DF tests, and is known to increase power close to the power envelope in the case of iid errors. Some other tests are also considered, but more briefly, in Section **9.2.3.iv**.

9.2.3.i *Using heteroscedasticity consistent standard errors*

Consider a simple AR(1) model:

$$\tilde{y}_t = \rho \tilde{y}_{t-1} + \varepsilon_t \tag{9.22a}$$

$$\Rightarrow$$

$$\Delta \tilde{y}_t = \gamma \tilde{y}_{t-1} + \varepsilon_t \tag{9.22b}$$

The heteroscedasticity robust $\hat{\tau}$-type test is:

$$\hat{\tau}_i^w = \frac{\hat{\rho} - 1}{\hat{\sigma}_w(\hat{\rho})} \tag{9.23}$$

$$= \frac{\hat{\gamma}}{\hat{\sigma}_w(\hat{\gamma})} \qquad i = 0, \mu, \beta$$

where $\hat{\sigma}_w(\hat{\rho})$ is White's heteroscedasticity robust standard error and the w superscript indicates the robust $\hat{\tau}_i$ test. In its simplest version this is derived as follows. It is convenient first to use the framework of the standard linear regression model $y = X\beta + \varepsilon$, where $E(\varepsilon\varepsilon'|X) = \Omega$ and Ω is diagonal but not a scalar diagonal matrix, the typical diagonal element being $\sigma^2_{\varepsilon,t}$. The (conditional) variance-covariance matrix of $\hat{\beta}$, the LS estimator of β, is:

$$\text{var}(\hat{\beta}|X) = E[(X'X)^{-1}X'\varepsilon\varepsilon'X(X'X)^{-1}|X] \tag{9.24}$$

$$= (X'X)^{-1}E(X'\varepsilon\varepsilon'X|X)(X'X)^{-1}$$

$$= (X'X)^{-1}E\left(\sum_{t=1}^{T}\varepsilon_t^2 X_t'X_t|X\right)(X'X)^{-1}$$

$$= (X'X)^{-1}\left(\sum_{t=1}^{T}\sigma^2_{\varepsilon,t}X_t'X_t|X\right)(X'X)^{-1}$$

where $X_t = (1, X_{t2}, X_{t3}, \ldots, X_{tk})$ and the equality in the third line follows as, by assumption, $E(\varepsilon_t\varepsilon_s|X) = 0$ for $t \neq s$, and, therefore, $E(\varepsilon\varepsilon'|X)$ is diagonal (but not scalar diagonal).

White (1980) shows that $T^{-1}\left(\sum_{t=1}^{T}\hat{\varepsilon}_t^2 X_t'X_t\right)$ is a consistent estimator of $T^{-1}E(X'\varepsilon\varepsilon'X)$, where $\hat{\varepsilon}_t$ are the LS residuals defined using $\hat{\beta}$, which is consistent under the usual conditions; for example, in the case of stochastic regressors that they are asymptotically uncorrelated with the ε_t, even with an heteroscedastic error variance-covariance matrix. The resulting estimator of the robust (conditional) variance-covariance matrix of $\hat{\beta}$ is, therefore:

$$\text{var}_w(\hat{\beta}) = (X'X)^{-1}X'\hat{\Omega}^{-1}X(X'X)^{-1} \tag{9.25}$$

where $\hat{\Omega} = \text{diag}(\hat{\varepsilon}_1^2, \ldots, \hat{\varepsilon}_T^2)$.

In the usual way, the standard error of $\hat{\beta}_j$ is the square root of the j-th diagonal of the variance matrix, but in this case $\text{var}_w(\hat{\beta})$ is used. The standard errors may be used directly to obtain robust t tests and to invert the t statistics to obtain robust confidence intervals.

Turning to the problem at hand in the AR(1) case $X_t = \tilde{y}_{t-1}$ and $X = (\tilde{y}_2, \ldots, \tilde{y}_{T-1})'$, that is, in this case there is a single regressor, hence:

$$\text{var}_w(\hat{\rho}) = (X'X)^{-1}\left(\sum_{t=2}^{T}\hat{\varepsilon}_t^2\tilde{y}_{t-1}^2\right)(X'X)^{-1} \tag{9.26}$$

$$= \left(\sum_{t=2}^{T}\tilde{y}_{t-1}^2\right)^{-1}\left(\sum_{t=2}^{T}\hat{\varepsilon}_t^2\tilde{y}_{t-1}^2\right)\left(\sum_{t=2}^{T}\tilde{y}_{t-1}^2\right)^{-1}$$

Since $\text{var}_w(\hat{\rho})$ is a scalar it may be denoted $\hat{\sigma}_w^2(\hat{\rho})$ and the required standard error for the $\hat{\tau}$ test is the square root of this quantity, that is $\hat{\sigma}_w(\hat{\rho})$ where, in practice, $\hat{\tilde{y}}_{t-1}$ replaces $\tilde{y}_{t-1}$.

The basic framework extends easily to the ADF($p-1$) maintained regression, where the underlying model is AR(p), resulting in:

$$\Delta \tilde{y}_t = \gamma \tilde{y}_{t-1} + \sum_{j=1}^{p-1} \alpha_j \Delta \tilde{y}_{t-j} + \varepsilon_t \tag{9.27}$$

In this case X is a matrix with typical row given by $X_t = (\tilde{y}_{t-1}, \Delta \tilde{y}_{t-1}, \ldots, \Delta \tilde{y}_{t-(p-1)})$ and $\beta = (\gamma, \alpha_1, \ldots, \alpha_{p-1})'$. With this interpretation of the rows of X, the required standard error is the square root of the first diagonal element of $\text{var}_w(\hat{\beta})$, where, in practice, $\hat{\tilde{y}}_{t-j}$ replaces $\tilde{y}_{t-j}$.

There are other related versions of this robust standard error; see, for example, MacKinnon and White (1985), Davidson and MacKinnon (1993) and Cribari-Neto (2004). To generalise the framework $\text{var}_w(\hat{\beta})$ is now rewritten with a weight ω_t on the squared residual ε_t^2, so that:

$$\text{var}_w(\hat{\beta}) = (X'X)^{-1} \left(\sum_{t=1}^{T} \omega_t \hat{\varepsilon}_t^2 X_t' X_t \right) (X'X)^{-1} \tag{9.29}$$

The weights and the different designations are as follows: HC0, $\omega_t = 1$; HC1, $\omega_t = T/(T-k)$; HC2, $\omega_t = 1/(1-p_t)$; HC3, $\omega_t = 1/(1-p_t)^2$; HC4, $\omega_t = 1/(1-p_t)^{\eta_t}$; where $\eta_t = \min(4, p_t/\bar{p})$ and p_t is the t-th diagonal element of $P = X(X'X)^{-1}X'$, which is related to the least squares projection matrix from $\hat{\varepsilon} = (I-P)\varepsilon$. The variations to the standard White case, HC0, are motivated by an attempt to improve the small sample performance relative to HC0 (see MacKinnon and White, 1985).

Long and Ervin (2000) found HC3 to be the best of adjusted HC estimators HC0–HC3. Andrews and Guggenberger (2011) use HC3 in their construction of confidence intervals with asymptotically correct size with a near or unit root when the errors are GARCH; and Demetrescu (2010) uses HC3 in evaluating some DF tests with GARCH errors. Subsequent to Long and Ervin (2000), Cribari-Neto (2004) suggested HC4 for a further improvement in finite sample performance. The differences for the simulations reported in this chapter were marginal and the results reported below use HC4.

9.2.3.ii A recursive mean/trend approach

In the case of standard DF tests there are several variants that lead to better power and an improvement in size retention when the errors are weakly dependent. A particularly simple, but effective, approach in the homoscedastic case is that of recursive mean or trend adjustment (for simplicity referred to as RMA); see So and Shin (1999), Shin and So (2002) and chapters 4, 7 and 13 of *UR, Vol. 1*. The question here is whether an improvement can also be achieved in the case of conditionally heteroscedastic errors.

The idea of RMA is simple enough. First consider the case where a mean is removed from the data for the case in which $\mu_t = \mu$. The 'global' mean is either

$\bar{y}^G = \sum_{t=1}^{T} y_t/T$ over all T observations, or $\bar{y} = \sum_{t=2}^{T} y_t/(T-1)$ over $T-1$ observations; the data is then demeaned by subtracting $\bar{y}^G$, or $\bar{y}$, as an estimator of μ, from y_t, so that $y_t - \bar{y}^G$ or $y_t - \bar{y}$, respectively, are used in any subsequent regressions. In contrast, the recursive mean only uses observations prior to period t in the calculation of the mean and by so doing reduces the correlation between the regressor y_{t-1} and the error term.

The RMA procedure gives a *sequence* of estimators of the mean, say $\{\bar{y}_t^r\}_{t=1}^{T}$:

$$\bar{y}_1^r = y_1; \quad \bar{y}_2^r = 2^{-1} \sum_{i=1}^{2} y_i; \quad \bar{y}_3^r = 3^{-1} \sum_{i=1}^{3} y_i; \quad \ldots; \quad \bar{y}_t^r = t^{-1} \sum_{i=1}^{t} y_i; \quad \ldots;$$

$$\bar{y}_T^r = T^{-1} \sum_{i=1}^{T} y_i.$$

Note that $\bar{y}_T^r = \bar{y}^G$, so that in only one case do the two means coincide. The RMA means are related by the recursion (see *UR, Vol. 1*, chapter 4):

$$\bar{y}_t^r = \bar{y}_{t-1}^r (t-1)/t + y_t/t \tag{9.30}$$

The recursive means can differ quite markedly from the global mean, especially in the early part of the sample and for an integrated process; of course, $\bar{y}_t^r \to \bar{y}^G$ as $t \to T$.

Now, in contrast to the usual procedure, rather than demean the data using $\bar{y}^G$, or $\bar{y}$, the recursive mean at time $t-1$, $\bar{y}_{t-1}^r$, is used. For example, in the simplest case, this results in:

$$y_t - \bar{y}_{t-1}^r = \rho(y_{t-1} - \bar{y}_{t-1}^r) + \zeta_t \tag{9.31}$$

where $\zeta_t = (1 - \rho)(\mu - \bar{y}_{t-1}^r) + \varepsilon_t$. The null hypothesis is, as usual, $H_0: \rho = 1$. Counterparts to the DF tests are straightforward, being:

$$\hat{\delta}_\mu^{rma} = T(\hat{\rho}_{rma} - 1) \tag{9.32}$$

$$\hat{\tau}_\mu^{rma} = (\hat{\rho}_{rma} - 1)/\hat{\sigma}(\hat{\rho}_{rma}) \tag{9.33}$$

where $\hat{\rho}_{rma}$ is the LS estimator from (9.31) and $\hat{\sigma}(\hat{\rho}_{rma})$ is its estimated standard error.

A similar procedure can be applied if the deterministic components comprise a constant and a linear trend. In the simplest case, the resulting DF regression is:

$$(y_t - \hat{\mu}_{t-1}^r) = \rho(y_{t-1} - \hat{\mu}_{t-1}^r) + \zeta_t \tag{9.34}$$

where $\hat{\mu}_t^r = \hat{\beta}_{0,t} + \hat{\beta}_{1,t}t$, and $\hat{\beta}_{0,t}$ and $\hat{\beta}_{1,t}$ are the LS estimators based on a regression of y_t on a constant and t, including observations to period t; $\hat{\mu}_t^r$ is then lagged once to avoid correlation with the error term. The term $y_{t-1} - \hat{\mu}_{t-1}^r$ is just the recursive residual from the preliminary de-trending regression at $t-1$. The RMA counterparts to the DF test are denoted $\hat{\delta}_\beta^{rma}$ and $\hat{\tau}_\beta^{rma}$ and are as in (9.32) and (9.33), respectively, but with $\hat{\rho}_{rma}$ obtained from LS estimation of (9.34). Robust versions of the RMA test statistics are also possible and these are indicated with an additional w superscript, $\hat{\tau}_\mu^{\text{rma},w}$ and $\hat{\tau}_\beta^{\text{rma},w}$.

The extension to ADF versions of the RMA tests will often be relevant in practical cases. First consider an AR(p) model for y_t with a constant mean $\mu_t = \mu$, then this can be rearranged into ADF form as follows:

$$(y_t - \mu) = \sum_{j=1}^p \phi_j(y_{t-j} - \mu) + \varepsilon_t \tag{9.35}$$

$$= \left(\sum_{j=1}^p \phi_j\right)(y_{t-1} - \mu) + \sum_{j=1}^{p-1} c_j[(y_{t-j} - \mu) - (y_{t-j-1} - \mu)] + \varepsilon_t$$

$$= \rho(y_{t-1} - \mu) + \sum_{j=1}^{p-1} c_j \Delta y_{t-j} + \varepsilon_t \qquad \text{as } \Delta\mu = 0$$

where $\rho = \sum_{j=1}^p \phi_j$ and $c_j = -\sum_{i=j+1}^p \phi_i$; and as in the simple case, a unit root corresponds to $\rho = 1$. Since this latter condition will also be satisfied by multiple unit roots, it is assumed that there is only one unit root in $\phi(L) = 1 - \sum_{j=1}^p \phi_j L^j$, other roots being outside the unit circle.

The next step is to replace μ by its recursive mean estimator, $\bar{y}_{t-1}^r$, resulting in:

$$(y_t - \bar{y}_{t-1}^r) = \rho(y_{t-1} - \bar{y}_{t-1}^r) + \sum_{j=1}^{p-1} c_j \Delta y_{t-j} + \zeta_t \tag{9.36}$$

where $\zeta_t = (1 - \rho)(\mu - \bar{y}_{t-1}^r) + \varepsilon_t$. Thus, in essence, all that is required is to augment the simple RMA model with lags of Δy_t, as in the standard case.

The next case to consider is that of a linear trend, so that the AR(p) model with trend is:

$$(y_t - \mu_t) = \sum_{j=1}^p \phi_j(y_{t-j} - \mu_{t-j}) + \varepsilon_t \tag{9.37}$$

$$= \rho(y_{t-1} - \mu_{t-1}) + \sum_{j=1}^{p-1} c_j[(y_{t-j} - \mu_{t-j}) - (y_{t-j-1} - \mu_{t-j-1})] + \varepsilon_t$$

The empirical analogue of this results from replacing $y_{t-j} - \mu_{t-j}$ by the recursive residual $\hat{u}_{t-j}^r = y_{t-j} - \hat{\mu}_{t-j}^r$, so that the ADF is formed as follows:

$$(y_t - \hat{\mu}_{t-1}^r) = \rho \hat{u}_{t-1}^r + \sum_{j=1}^{p-1} c_j \Delta \hat{u}_{t-j}^r + \zeta_t \tag{9.38}$$

where $\zeta_t = (\mu_t - \hat{\mu}_{t-1}^r) - \sum_{j=1}^p \phi_j(\mu_{t-j} - \hat{\mu}_{t-j}^r) + \varepsilon_t$. Some critical values for $\hat{\tau}_\mu^{\text{rma}}$ and $\hat{\tau}_\beta^{\text{rma}}$ are presented in Table 9.4.

9.2.4 Simulation results

This section considers whether the $\hat{\tau}$-type tests with variants using robust standard errors or RMA are able to improve upon the standard DF tests in terms of size retention and power when the errors are conditionally heteroscedastic. The simulation set-up is as for Tables 9.5 (size) and 9.6 (power).

Table 9.4   Critical values for $\hat{\tau}_{\mu}^{rma}$ and $\hat{\tau}_{\beta}^{rma}$

	$\hat{\tau}_{\mu}^{rma}$			$\hat{\tau}_{\beta}^{rma}$		
	1%	5%	10%	1%	5%	10%
100	−2.53	−1.88	−1.53	−2.47	−1.81	−1.46
200	−2.53	−1.89	−1.54	−2.50	−1.85	−1.49
500	−2.52	−1.88	−1.55	−2.47	−1.82	−1.47

Source: UR, *Vol. 1*, Chapter 7 and Shin and So (2002) for $\hat{\tau}_{\mu}^{rma}$.

9.2.4.i   Size

The simulation set up is as for the size calculations reported in section **9.2.2**. The results for the simplest version of $\hat{\tau}$ are reported just for the record, as this is an unlikely situation in practice. In this case all of the reported test statistics remain faithful to their nominal size. The more realistic cases are where the deterministic components comprise a mean or linear trend. In the former case, using a robust standard error induces a slight under-sizing as α_1 increases. The strategy suggested by Demetrescu (2010), of averaging $\hat{\tau}_i$ and $\hat{\tau}_i^w$ (with the test statistics here denoted $\hat{\tau}_i^{ave}$), works in the sense of resulting in a test statistic with empirical size close to the nominal size. The RMA versions of the test statistics, that is $\hat{\tau}_{\mu}^{rma}$ and $\hat{\tau}_{\beta}^{rma}$, have slightly better size retention than their standard counterparts, but tend to be over-sized; using the RMA tests in combination with a robust standard error returns the empirical size close to the nominal size and so there is no further advantage to considering averaging the robust RMA test with the standard test. Thus, overall, the better (systematic) size performance is delivered by $\hat{\tau}_i^{ave}$ and the robust versions of the RMA test statistics.

9.2.4.ii   Power

The simulation set-up is as for the size calculations but the data are now generated under the alternative hypothesis with $\rho = 0.95$, $\rho = 0.9$ and $u_t \sim N(0, 1)$. Given the differences in finite sample size a comparison of 'raw' power would give an advantage to over-sized tests; power is therefore reported on a size-adjusted basis. The illustrative results are reported in Table 9.6.

In all cases, $\hat{\tau}_i^w$, $i = 0, \mu, \beta$, there is no systematic advantage in (size-adjusted) power in using a robust standard error. This result implies that there is no power advantage to averaging $\hat{\tau}_i$ and $\hat{\tau}_i^w$, so that the advantage of using a robust standard error lies solely with better size retention. On the other hand the RMA-based test statistics have much better power not only when the errors are iid, as expected (see Chapter 7 of UR, *Vol. 1*), but also when the errors follow a GARCH (1, 1) process, although power does decline as the volatility parameter increases. Additionally, with one minor exception, using a robust standard

Table 9.5 Size of various $\hat{\tau}$-type tests in the presence of GARCH(1, 1) errors; $T = 200$, nominal 5% size

α_1	0	0.05	0.10	0.15	0.20	0.25	0.30	0.35
γ_1	0	0.90	0.85	0.80	0.75	0.70	0.65	0.60
$\hat{\tau}$	5.02	5.12	4.92	5.40	5.58	5.50	5.54	5.60
$\hat{\tau}^w$	5.20	5.18	4.98	5.04	4.94	4.58	4.30	4.34
$\hat{\tau}^{ave}$	5.22	5.06	4.76	5.14	5.12	5.06	4.88	4.74
$\hat{\tau}_\mu$	5.06	5.36	6.00	6.72	7.48	8.16	8.38	8.72
$\hat{\tau}_\mu^w$	5.00	4.96	5.06	4.74	4.48	4.08	3.92	3.80
$\hat{\tau}_\mu^{ave}$	4.84	5.16	5.50	5.44	5.56	5.56	5.34	5.34
$\hat{\tau}_\mu^{rma}$	5.30	5.50	6.14	6.46	7.14	7.74	7.84	8.22
$\hat{\tau}_\mu^{rma,w}$	5.18	4.74	5.12	4.98	4.88	4.70	4.76	4.70
$\hat{\tau}_\beta$	5.40	5.50	5.98	6.60	7.48	8.48	8.84	9.44
$\hat{\tau}_\beta^w$	5.00	4.46	4.06	3.64	3.40	3.12	2.84	2.74
$\hat{\tau}_\beta^{ave}$	4.50	4.36	4.24	4.18	4.36	4.42	4.56	4.56
$\hat{\tau}_\beta^{rma}$	5.00	5.36	5.62	6.20	7.10	7.78	8.58	9.32
$\hat{\tau}_\beta^{rma,w}$	5.38	5.40	5.14	5.10	5.12	5.40	5.18	5.12

error does not further improve power. To illustrate consider $\hat{\tau}_\mu$ and $\hat{\tau}_\mu^{rma}$ when $\rho = 0.95$, which have empirical power of about 30.7% and 50.6% respectively, when the errors are niid; when $\alpha_1 = 0.2$ and $\gamma_1 = 0.75$ this reduces to about 29% and 45% respectively, and using a robust standard error for $\hat{\tau}_\mu^{rma}$ reduces the latter to about 40%.

Whilst in terms of size the advantage lies with the RMA tests combined with a robust standard error, the advantage in terms of power lies with the RMA tests with a conventional standard error. Compared with the standard DF tests, it is, however, still better to use the RMA tests combined with a robust standard error.

9.2.4.iii Other methods (ML and GLS)

A natural extension of the maximum likelihood (ML) approach to testing for a unit root (see Shin and Fuller, 1998), is to jointly estimate the AR(p)-GARCH(r, s) or ARMA(p, q)-GARCH(r, s) model by ML, and use the ML equivalent of the standard unit root test statistic. Assuming normality ML is generally more efficient than (O)LS; see Engle (1982, section 7) for the fixed design regression model and Li et al. (2002) for an overview of the required moment conditions in the ARMA-GARCH case. This efficiency gain should benefit the unit root test.

9.2.4.iii.a Maximum likelihood Maximum likelihood estimation of the AR(p)-GARCH(r, s) model, with the focus on providing a unit root test, has been considered by, amongst others, Ling and Li (1998), Seo (1999), Ling et al. (2002),

Table 9.6 Power of $\hat{\tau}$-type tests in the presence of GARCH (1, 1) errors; $\rho = 0.95, 0.9$, $T = 200$ and $u_t \sim N(0,1)$, nominal 5% size

α_1	0	0.05	0.10	0.15	0.20	0.25	0.30	0.35
γ_1	0	0.90	0.85	0.80	0.75	0.70	0.65	0.60
$\rho = 0.95$								
$\hat{\tau}$	**80.06**	**78.98**	**78.04**	**76.28**	**75.34**	**73.92**	**75.34**	**73.50**
$\hat{\tau}^w$	79.22	74.50	68.64	65.66	62.52	57.82	56.00	54.96
$\hat{\tau}^{ave}$	80.00	77.36	74.30	72.16	68.02	67.76	66.16	66.52
$\hat{\tau}_\mu$	30.72	30.74	30.06	29.42	28.86	28.20	26.28	25.66
$\hat{\tau}_\mu^w$	30.62	28.70	26.42	25.58	23.42	22.08	21.24	20.40
$\hat{\tau}_\mu^{ave}$	31.78	30.04	28.84	28.38	26.58	25.68	24.70	23.84
$\hat{\tau}_\mu^{rma}$	**50.56**	**48.84**	**46.66**	**45.74**	**45.02**	**43.92**	**42.16**	**40.62**
$\hat{\tau}_\mu^{rma,w}$	48.36	**48.84**	45.22	42.16	39.98	37.42	35.26	33.20
$\hat{\tau}_\beta$	18.26	17.92	16.82	16.42	15.90	15.12	14.84	14.58
$\hat{\tau}_\beta^w$	17.44	15.08	14.68	14.90	13.70	12.92	13.28	13.72
$\hat{\tau}_\beta^{ave}$	17.82	16.60	15.68	15.70	15.34	15.30	14.96	15.24
$\hat{\tau}_\beta^{rma}$	**21.82**	**21.40**	**21.32**	**20.76**	**20.42**	**20.08**	**19.72**	**19.24**
$\hat{\tau}_\beta^{rma,w}$	20.96	20.10	19.20	18.12	16.56	16.38	16.20	16.24
$\rho = 0.90$								
$\hat{\tau}$	**99.72**	**99.54**	**99.18**	**98.94**	**98.62**	**98.36**	**98.34**	**97.76**
$\hat{\tau}^w$	99.62	98.54	95.92	92.68	89.22	84.38	81.68	79.14
$\hat{\tau}^{ave}$	99.70	99.26	98.62	97.72	96.24	95.56	94.30	93.80
$\hat{\tau}_\mu$	85.98	85.12	82.16	80.52	78.08	76.00	72.44	70.70
$\hat{\tau}_\mu^w$	84.34	77.84	69.76	63.72	56.76	51.84	48.30	45.82
$\hat{\tau}_\mu^{ave}$	86.46	82.70	78.02	74.08	69.54	66.28	63.68	61.02
$\hat{\tau}_\mu^{rma}$	**96.16**	**94.50**	**92.90**	**91.28**	**89.54**	**88.04**	**86.16**	**84.50**
$\hat{\tau}_\mu^{rma,w}$	94.82	93.06	88.68	84.14	79.72	74.76	70.58	66.30
$\hat{\tau}_\beta$	63.96	61.70	58.00	55.12	52.80	50.00	48.20	46.16
$\hat{\tau}_\beta^w$	60.08	52.34	46.16	43.28	38.48	34.86	33.28	33.26
$\hat{\tau}_\beta^{ave}$	63.28	57.54	52.68	49.50	46.50	44.68	42.92	42.46
$\hat{\tau}_\beta^{rma}$	**69.72**	**67.02**	**64.42**	**61.02**	**58.60**	**56.86**	**55.04**	**53.10**
$\hat{\tau}_\beta^{rma,w}$	67.72	64.32	59.92	55.64	50.72	47.84	45.94	44.16

Note: Bold entries indicate best performance.

Ling and Li (2003) and Ling et al. (2003). The GARCH process is assumed to be stationary and a unit root, if present, lies in the AR part of the model, the other roots of the AR polynomial being outside the unit circle. It is convenient to consider the AR(p) model in ADF(p − 1) form, so that the model specification, reproduced here for convenience, is:

$$\Delta y_t = \gamma y_{t-1} + \sum_{j=1}^{p-1} c_j \Delta y_{t-j} + \varepsilon_t \tag{9.39a}$$

$$\varepsilon_t = u_t \sigma_t \tag{9.39b}$$

$$\sigma_t^2 = \alpha_0 + \alpha(L)\varepsilon_t^2 + \gamma(L)\sigma_t^2 \tag{9.39c}$$

A unit root corresponds to $\gamma = 0$.

The (average) loglikelihood function (apart from a constant), assuming normality and conditional on initial values (including those for ε_t and σ_t^2) is:

$$CLL(\psi) = T^{-1} \sum_{t=1}^{T} LL_t \tag{9.40a}$$

$$\sum_{t=1}^{T} LL_t = -\frac{1}{2} \sum_{t=1}^{T} \ln \sigma_t^2 - \frac{1}{2} \sum_{t=1}^{T} \left(\frac{\varepsilon_t^2}{\sigma_t^2} \right) \tag{9.40b}$$

where $\psi = (\gamma, c_1, \ldots, c_{p-1}, \delta')'$ and $\delta = (\alpha_0, \alpha_1, \ldots, \alpha_r, \beta_1, \ldots, \beta_s)'$, δ comprises the parameters of the GARCH part of the model, and σ_t^2 is considered as a function of ε_t as specified in (9.39c). Estimation based on the likelihood function has been considered by Engle (1982) for ARCH models and Bollerslev (1986) for GARCH models.

The approach outlined here is due to Seo (1999), who assumed, inter-alia, that u_t is iid, $E(u_t^3) = 0$ and $E(\varepsilon_t^8) < \infty$; the latter, in particular, is a restriction that will bind on some GARCH processes (for example, it would rule out some fat-tailed distributions such as t(v), $v \leq 8$; it is, however, a sufficient assumption and may be stronger than necessary).

Seo (1999) distinguishes three test statistics depending on the method of estimating the covariance matrix: the information matrix, the inverse of the outer product of the gradient matrix or White's (1980) robust covariance matrix. The methods result in the same estimator of the covariance matrix if the fourth moment of the scaled error, $E(u_t^4) = \kappa = 3$, which is the case for the normal distribution (see White, 1982). The central case is conveniently taken as using White robust standard errors. Then the limiting distribution is as follows (Seo, 1999, Theorem 1 and Lemma 3):

$$\hat{\tau}_i^{ml,w} \Rightarrow_D \lambda \frac{\int_0^1 B_1(s)dB_1(s)}{\left(\int_0^1 B_1(s)^2 ds \right)^{1/2}} + (1 - \lambda^2)^{1/2} \frac{\int_0^1 B_1(s)dB_2(s)}{\left(\int_0^1 B_1(s)^2 ds \right)^{1/2}} \tag{9.41a}$$

$$= \lambda \frac{\int_0^1 B_1(s)dB_1(s)}{\left(\int_0^1 B_1(s)^2 ds \right)^{1/2}} + (1 - \lambda^2)^{1/2} N(0, 1) \tag{9.41b}$$

where B_1 and B_2 are independent standard Brownian motions. If the AR model uses demeaned or detrended data, then the standard Brownian motions are replaced by demeaned or detrended Brownian motions. The distribution of $\hat{\tau}_i^{ml,w}$ is a convex combination with weights λ and $(1 - \lambda^2)^{1/2}$ of the corresponding DF distribution, obtained when $\lambda = 1$, and a standard normal, obtained when $\lambda = 0$. In the case that the errors are iid, the standard DF distribution results. The test statistics using the alternative methods for estimating the covariance matrix have limiting distributions that are scalar multiples of the distribution for $\hat{\tau}_i^{ml,w}$; the scalars equal unity when the distribution of u_t is normal or $\kappa = 3$.

Whilst the quantiles from the DF distribution could be used for testing, the difference between the typically used critical values increases as more deterministic terms are included at the detrending stage; this conservative procedure results in a loss of power in cases known anyway to suffer low power. Seo (1999), therefore, gives a method for estimating the nuisance parameter λ, which depends in part upon $E(u_t^4)$, and provides a set of tables of critical values that depend upon $\hat{\lambda}$.

As the distribution of the test statistic $\hat{\tau}_i^{ml,w}$ under the near-unit root alternative depends upon the GARCH effect, which shifts the distribution to the left, power increases relative to the standard DF distribution which has no GARCH effect. This is confirmed in simulations reported in Seo (1999) for $\hat{\tau}^{ml,w}$ (that is, no deterministics), which suggest a notable increase in power using ML relative to the standard DF test.

An ML procedure was also suggested by Ling and Li (2003) and Ling et al. (2003), who considered the AR(1)-GARCH(1, 1) model. Assuming that u_t is symmetrically distributed and $E(u_t^4) < \infty$, they showed that for the one-step and iterated (quasi) ML estimator based on LS starting values, the moment restrictions, $E(\varepsilon_t^4) < \infty$ of Ling and Li (1998) and $E(\varepsilon_t^8) < \infty$ of Seo (1999), can be relaxed to $E(\varepsilon_t^2) < \infty$, for which the required condition is $\alpha_1 + \beta_1 < 1$. Normality of u_t is not required, so the resulting ML estimator is referred to as a quasi-ML estimator (if the distribution is known the efficiency of the ML estimator can be restored by using the appropriate likelihood function). In the AR(1)-GARCH(1, 1) model $\Psi = (\gamma, \alpha_0, \alpha_1, \beta_1)' = (\gamma, \delta')'$. The first step in the procedure is to estimate γ by LS, say $\hat{\gamma} = (\hat{\rho} - 1)$, obtain the residuals and then estimate δ by ML, say $\tilde{\delta}^{(1)}$. The likelihood function is then maximised conditional on $\tilde{\delta}^{(1)}$ to update $\hat{\gamma}$ in one iterative step to obtain $\tilde{\gamma}$. Further iterations would be possible, as suggested in Ling et al. (2003, equation 3.2), as follows (but do not alter the asymptotic distribution):

$$\tilde{\rho}^{(i+1)} = \tilde{\rho}^{(i)} - \left[\sum_{t=1}^{T} \frac{\partial^2 LL_t}{\partial \rho^2} \right]_{\psi^{(i)}}^{-1} \left[\sum_{t=1}^{T} \frac{\partial LL_t}{\partial \rho} \right]_{\psi^{(i)}}^{-1} \tag{9.42}$$

$$\tilde{\delta}^{(i+1)} = \tilde{\delta}^{(i)} - \left[\sum_{t=1}^{T} \frac{\partial^2 LL_t}{\partial \delta^2} \right]_{\psi^{(i)}}^{-1} \left[\sum_{t=1}^{T} \frac{\partial LL_t}{\partial \delta} \right]_{\psi^{(i)}}^{-1} \tag{9.43}$$

where the scheme is started with $\tilde{\rho}^{(1)} = \hat{\rho}$. A t-type test statistic can then be obtained, with nuisance parameters that can be estimated consistently, which has the same limiting distribution as for $\hat{\tau}_i^{ml,w}$ in Seo (1999) detailed above and hence use can be made of the critical values tabulated therein (Seo, 1999, table 3). The simulation results reported in Ling et al. (2003), show that the ML-based test statistics are at least as good as their LS versions in maintaining size and generally better, and are more powerful. A reasonable conjecture is that these results extend to the ARMA(p, q)-GARCH(r, s) case. As to the ARMA part of the model, there are two possibilities, which follow from the standard approach. One is to specify an ADF(k) model allowing k to increase with T by a data-dependent rule; alternatively, as in the exact ML approach of Shin and Fuller (1998), specify an ARMA model, which is estimated by ML. Both of these approaches allow the addition of GARCH(r, s) errors.

9.2.4.iii.b GLS Guo and Phillips (2001a, b) suggest a GLS approach to constructing a unit root test statistic for the general case where there is conditional heteroscedasticity but it is not of a specific form (see also Andrews and Guggenberger (2012)). The end result is a test statistic which, like the ML unit root tests, has an asymptotic distribution that is a weighted combination of the corresponding DF distribution and a standard normal, with the weight depending on the strength of the heteroscedasticity.

Ley y be the vector with typical element y_t and X be a matrix with typical row $X_t = (\tilde{y}_{t-1}, \Delta\tilde{y}_{t-1}, \ldots, \Delta\tilde{y}_{t-(p-1)})$ and $\beta = (\rho, c_1, \ldots, c_{p-1})'$. With this interpretation of y and X, GLS is applied to $y = X\beta + \varepsilon$, where $E(\varepsilon\varepsilon'|X) = \Omega = \text{diag}(\sigma_{\varepsilon,1}^2, \ldots, \sigma_{\varepsilon,T}^2)$. A number of relatively standard assumptions are made concerning the properties of Ω (see Guo and Phillips, 2001a, Assumption 1), notably only requiring $E(\varepsilon_t^2) < \infty$ as far as the variance is concerned.

The infeasible GLS estimator and its variance matrix, conditional on X, are:

$$\hat{\beta}_{gls} = (X'\Omega^{-1}X)^{-1}X'\Omega^{-1}y \tag{9.44}$$

$$\text{var}(\hat{\beta}_{gls}) = (X'\Omega^{-1}X)^{-1} \tag{9.45}$$

Note that this estimator shares with ML that part of the objective function which minimises the weighted residual sum of squares (that is, what would obtain if the likelihood function just maximised the second term in (9.40b) treating σ_t^2 as known).

In the case of stationary regressors (and Ω known), $\hat{\beta}_{gls}$ is never less efficient than (O)LS and is generally more efficient. Guo and Phillips (2001a, Remark 6.3) show that this result also extends to the coefficient on the nonstationary regressor (under the unit root H_0, $\tilde{y}_{t-1}$ is I(1)). For example, in some illustrative calculations the ratio of expected variances of GLS to OLS is as low as 0.4, increasing to 1 as conditional homoscedasticity is approached.

The practical question, as in most applications of GLS, is how to make the estimator feasible. For the present, assume that a consistent estimator of Ω is available, say $\hat{\Omega} = \text{diag}(\hat{\sigma}^2_{\varepsilon,1}, \ldots, \hat{\sigma}^2_{\varepsilon,T})$, then the feasible GLS (FGLS), counterparts to the GLS estimators are:

$$\hat{\beta}_{\text{fgls}} = (X'\hat{\Omega}^{-1}X)^{-1}X'\hat{\Omega}^{-1}y \tag{9.46}$$

$$\text{var}(\hat{\beta}_{\text{fgls}}) = (X'\hat{\Omega}^{-1}X)^{-1} \tag{9.47}$$

The corresponding GLS $\hat{\tau}$-type unit root test statistic is then:

$$\hat{\tau}_i^{\text{fgls}} = \frac{(\hat{\rho}_{\text{fgls}} - 1)}{\hat{\sigma}(\hat{\rho}_{\text{fgls}})} \tag{9.48}$$

$$\hat{\sigma}(\hat{\rho}_{\text{fgls}}) = \text{var}(\hat{\beta}_{\text{fgls}})^{1/2}_{(1)} \tag{9.49}$$

where $\hat{\rho}_{\text{fgls}}$ is the first element of $\hat{\beta}_{\text{fgls}}$ and the subscript on the variance matrix $\text{var}(\hat{\beta}_{\text{fgls}})$ indicates the first diagonal element. The asymptotic null distribution of $\hat{\tau}_i^{\text{fgls}}$ is a convex mixture of the corresponding DF distribution and the standard normal (Guo and Phillips, 2001a, Theorem 6.1):

$$\hat{\tau}_i^{\text{fgls}} \Rightarrow_D \kappa \frac{\int_0^1 B(s)dB(s)}{\left(\int_0^1 B(s)^2 ds\right)^{1/2}} + (1-\kappa^2)^{1/2}N(0,1) \tag{9.50}$$

where $B(s)$ is standard Brownian motion. $0 \leq \kappa \leq 1$ is a weight that depends upon the 'strength' of the conditional heteroscedasticity, with $\kappa = 1$ in the case of conditional homoscedasticity, in which case the distribution reduces to that in the standard DF case. As usual, in the case of demeaned or detrended data, $B(s)$ is replaced by demeaned or detrended Brownian motion. This clearly has the same form as for the distribution of $\hat{\tau}_i^{\text{ml,w}}$ (and the quantiles only differ slightly from those of Seo, 1999).

The next question is to obtain a practical solution to estimating Ω. There are two ways to proceed depending on how much structure is given to the model of conditional heteroscedasticity. One possibility is to parameterise it as, for example, GARCH(r, s), and then proceed, as in the first step of the iterative estimator of Ling et al. (2003), by first carrying out LS estimation of the ADF part of the model to obtain the residuals, which are then used in the estimation of the GARCH part of the model, providing estimates $\hat{\sigma}^2_t$ of σ^2_t. However, Guo and Phillips (2001a) consider a nonparametric approach, suggesting a local linear kernel estimator of σ^2_t with a Gaussian kernel based on q lagged values of the OLS residuals and give conditions for this estimator to be consistent. This method has the advantage that it does not depend on a particular parametric structure for the GARCH errors, although the bandwidth q has to be chosen. The simulation results in Guo and Phillips (2001a) show that the resulting FGLS-based unit root

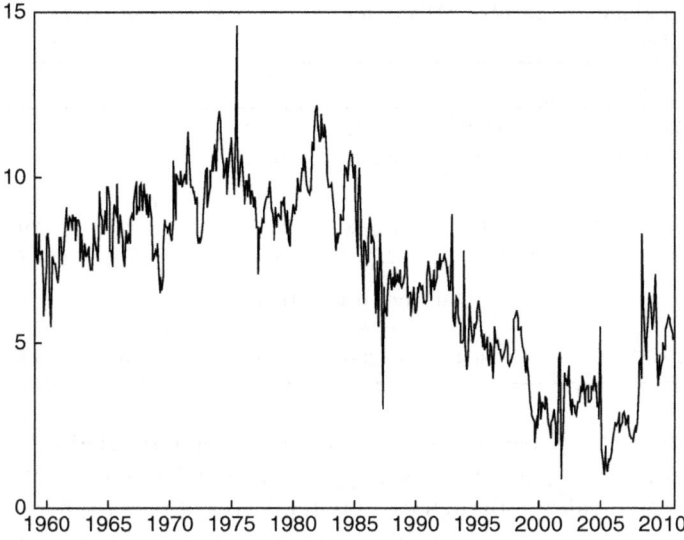

Figure 9.4   US personal savings ratio (%)

test is better than the standard DF test $\hat{\tau}$ (only the case with no deterministic terms and ARCH(1) errors is considered) at retaining size and is more powerful.

9.2.5   An ADF(p)-GARCH(1, 1) example

This example uses data on the US personal savings ratio, with monthly series from 1959m1 to 2011m7, a sample comprising 631 observations. The time series is graphed in Figure 9.4, which shows a variation between 1% in 2005m4 and 12.2% in 1981m11, and a strong suggestion of several ARCH/GARCH clusters of volatility. In this case the appropriate AR model allows for a mean, and lag selection by BIC resulted in an ADF(3).

The GARCH coefficients were significant (see Table 9.7) confirming the visual impression from Figure 9.4; note that $\hat{\alpha}_1 + \hat{\gamma}_1 \approx 0.9$, indicating a stationary GARCH process. The standard DF test $\hat{\tau}_\mu = -1.807$ suggests non-rejection of the unit root hypothesis; the adjustments due to the robust standard errors were slight, resulting in $\hat{\tau}_\mu^w = -1.804$ and $\hat{\tau}_\mu^{ave} = -1.806$. The RMA versions of the test statistics were $\hat{\tau}_\mu^{rma} = -1.337$ and $\hat{\tau}_\mu^{rma,w} = -1.314$, again leading to non-rejection of the unit root hypothesis. For reference the ML test statistic is also reported $\hat{\tau}_\mu^{ml} = -1.114$ (using the outer product of the gradient of the likelihood function) and, similarly, supports non-rejection of the null hypothesis.

It is possible to simulate the p-value associated with each test statistic and this can take one of two forms, in the first case taking the null distribution corresponding to errors that are iid, which is the standard case; alternatively, allowing

Table 9.7 Estimation details: ADF(3) for the US savings ratio, with and without GARCH(1, 1) errors

	$\hat{\rho}/\hat{\gamma}$	$\hat{c}_1$	$\hat{c}_2$	$\hat{c}_3$	$\hat{\alpha}_0$	$\hat{\alpha}_1$	$\hat{\gamma}_1$
LS	0.981/−0.019	−0.306	−0.222	−0.249	–	–	–
't'	(−1.807)	(−7.79)	(−5.56)	(−6.42)			
	$\hat{\rho}/\hat{\gamma}$				$\hat{\alpha}_0$	$\hat{\alpha}_1$	$\hat{\gamma}_1$
ML	0.987/−0.013	−0.262	−0.168	−0.186	0.109	0.299	0.502
't'	(−1.114)	(−3.39	(−2.79	(−3.56)	(3.05)	(3.62)	(5.33)

Unit root test statistics					
$\hat{\tau}_\mu$	$\hat{\tau}_\mu^w$	$\hat{\tau}_\mu^{ave}$	$\hat{\tau}_\mu^{rma}$	$\hat{\tau}_\mu^{rma,w}$	$\hat{\tau}_\mu^{ml}$
−1.807	−1.804	−1.806	−1.337	−1.314	−1.114

for a GARCH effect, basing the simulations on the estimated GARCH coefficients. The latter suggests the bootstrapping the null distribution and obtaining the bootstrap p-value, although this alternative is not pursued here. In the first case, the resulting p-values are 36.4% and 14.6% for $\hat{\tau}_\mu$, and for $\hat{\tau}_\mu^{rma}$ respectively, and in the second case 36.8% and 14.9% for $\hat{\tau}_\mu$ and for $\hat{\tau}_\mu^{rma}$ respectively. These are almost the same for each test as the sample size is here large enough for the GARCH errors not to have an impact on the empirical distribution. Note that the p-value for $\hat{\tau}_\mu^{rma}$ is substantially lower than that for $\hat{\tau}_\mu$, but not small enough to suggest rejection of the unit root null; this difference does, however, support the higher power of $\hat{\tau}_\mu^{rma}$ (and the ML unit root test).

9.3 Concluding remarks

This chapter has focused on two topics related to the impact of heteroscedasticity on standard DF-type statistics. In the first case, conditional heteroscedasticity is induced by a stochastic unit root, which has an impact on the limiting distribution of the test statistics, necessitating a test statistic different from the standard DF type. In the second case, the conditional heteroscedasticity arises from GARCH errors. Whilst, under certain conditions, this does not affect the limiting distributions of the DF-type tests, it leads to over-sizing (in general) in finite samples and a loss of power. Thus, different tests can be constructed to exploit the weaknesses of the LS-based DF tests, and these have the potential to render a more faithful size and improve power. A number of possibilities have been suggested including using robust (to heteroscedasticity) standard errors, recursive mean adjustment, ML estimation and (feasible) GLS rather than (O)LS. The simulation results reported here and in the relevant literature have shown that these aims can be achieved.

Continuing with the theme of the impact of the specification of the variance on unit root tests a related area, not covered in this chapter, is to consider the case of an infinite variance (see Chan and Tran, 1989, and Phillips, 1990). In that case and when ε_t is iid or is weakly dependent, with common distribution function that belongs to the domain of attraction of a stable law with index $\varsigma \in (0, 2)$, the limiting distributions of $T(\hat{\rho} - 1)$ and $\hat{\tau}$ are functions of a Lévy process for index ς, which takes the place of Brownian motion in the standard case, the latter corresponding to $\varsigma = 2$. Although the DF tests are reasonably robust to fat-tailed distributions, they are sensitive to extreme fat tails, as in the Cauchy distribution (which fails the finite second moment condition), and the simulation evidence in Patterson and Heravi (2003) for some finite samples indicated that the null hypothesis of equality in the distributions of the quantiles (taking niid as the base, both under the null and under the alternative) could be rejected for a number of fat-tailed distributions.

Rather than seek to identify a particular distribution, an alternative procedure is to use an estimator that is robust to fat-tailed distributions and may also be robust to outliers; see, for example, Lucas (1995a, b), Hoek et al. (1995), Herce (1996) and Rothenberg and Stock (1997). The class of M estimators minimizes a function of the residuals, as in LS, but that function will not necessarily be the residual sum of squares. Particular examples of M estimators include the least absolute deviations (LAD) estimator, the ML estimator based on a t(v) distribution with fairly small v, the Huber (1981) function $\psi(\varepsilon) = \min[c, \max(-c, \varepsilon_t)]$ where $c = 1.345$ or is chosen by the researcher. As in the ML and GLS methods outlined above, the limiting distributions of the resulting test statistics are functions of a Brownian motion, a standard normal distribution and some nuisance parameters, where the latter can be estimated consistently to operationalise the test statistic, (see Thompson, 2004).

Questions

Q9.1 Consider the test statistic Z_2 where $\Delta\mu_t = 0$ and there are no additional dynamics for y_t, so that the null model is $y_t = y_{t-1} + \varepsilon_t$, show that Z_2 simplifies to:

$$Z_2 = T^{-3/2}\hat{\sigma}_\varepsilon^{-2}\hat{\kappa}^{-1}\sum_{t=1}^{T} y_{t-1}^2((\Delta y_t)^2 - \sigma_{\Delta y}^2)$$

$$\hat{\sigma}_\varepsilon^2 = \hat{\sigma}_{\Delta y}^2 = T^{-1}\sum_{t=1}^{T}(\Delta y_t)^2$$

$$\hat{\kappa}_\varepsilon^2 = T^{-1}\sum_{t=1}^{T}((\Delta y_t)^2 - \sigma_\varepsilon^2)^2$$

A9.1 First note that, in general, Z_2 is given by (see 9.15):

$$Z_2 = T^{-3/2}\hat{\sigma}_\varepsilon^{-2}\hat{\kappa}^{-1}\sum_{t=2}^{T}\left(\sum_{j=1}^{t-1}\hat{\varepsilon}_j\right)^2(\hat{\varepsilon}_t^2 - \hat{\sigma}_\varepsilon^2)$$

Note that in this case, there are no parameters to be estimated prior to constructing the test statistic, so that $\hat{\varepsilon}_t = \Delta y_t$ and, given $y_0 = 0$, therefore, $y_{t-1} = \sum_{j=1}^{t-1}\varepsilon_j$

and $\hat{\varepsilon}_t^2 = (\Delta y_t)^2$ and $\sigma_{\Delta y}^2 = \hat{\sigma}_\varepsilon^2$, so that on making the substitutions, the test statistics are equivalent.

Q9.2 The unconditional variance of y_t was subscripted with t in (9.3b), $\sigma_{y,t}^2$, to indicate that it is not constant in the RCAR(1) model. Why?

A9.2 For convenience of reference $\sigma_{y,t}^2$ is given below.

$$\sigma_{y,t}^2 = \sigma_\varepsilon^2 \left[\sum_{j=0}^{t-1} (\rho^2 + \omega^2) \right]^j$$

Note that the convergence of the asymptotic variance $\lim_{t \to \infty} \sigma_{y,t}^2$ depends on $\rho^2 + \omega^2$, and the number of terms in the summation (the process is assumed to have started with an initial value of zero). Both components of the sum are positive and the limiting sum exists for $0 < \rho^2 + \omega^2 < 1$. As in the standard case, stationarity is usually taken to refer to the asymptotic variance because the number of terms on the right-hand side of $\sigma_{y,t}^2$ depends on t, so $\sigma_{y,t}^2$ is not constant. It is possible that ω^2 is sufficiently large to 'excite' a stationary fixed root into the nonstationary region; equally, the root may then drop back into the stationary region. For example, if $\rho = 0.98$, then there will be nonstationarity if $\omega^2 \geq 0.0396$ as $0.98^2 + 0.0396 = 1$; this does not mean that ρ_t always equals (or exceeds) unity, but there is a positive probability that the boundary will be reached. Indeed if we add that δ_t is niid($0, \omega^2$), then that probability can be calculated; for example, the probability of δ_t exceeding 0.02 with $\omega^2 = 0.0396$, and therefore ρ_t exceeding unity, is $P(z \geq 0.02/\sqrt{0.0396}) = 0.54$.

Q9.3 Write the likelihood function for the AR(1) model with a stochastic unit root and indicate in general terms how to obtain an LM test statistic.

A9.3 The model on which the likelihood function is based is (see 9.8a and 9.8b):

$$y_t = \rho y_{t-1} + \xi_t$$
$$\xi_t = \delta_t y_{t-1} + \varepsilon_t$$

so that $\sigma_{\xi,t}^2 = y_{t-1}^2 \omega^2 + \sigma_\varepsilon^2$. The loglikelihood can be obtained by reinterpreting the AR-GARCH formulation, so that referring back to (9.38a) and (9.38b), then apart from a constant and assuming normality (or referring to the resulting estimator and test statistics as quasi-ML), then:

$$CLL(\psi) = T^{-1} \sum_{t=1}^{T} LL_t$$

$$\sum_{t=1}^{T} LL_t = -\frac{1}{2} \sum_{t=1}^{T} \ln \sigma_{\xi,t}^2 - \frac{1}{2} \sum_{t=1}^{T} \left(\frac{\varepsilon_t^2}{\sigma_{\xi,t}^2} \right)$$

where in this case $\psi = (\rho, \omega^2, \sigma_\varepsilon^2)'$. For convenience of notation let $\ell(\psi) \equiv CLL(\psi)$, then the score vector is the vector of derivatives of $\ell(\psi)$ with respect to ψ, that is:

$$S(\psi)' \equiv \frac{\partial \ell(\psi)}{\partial \psi} = \left(\frac{\partial \ell(\psi)}{\partial \rho}, \frac{\partial \ell(\psi)}{\partial \omega^2}, \frac{\partial \ell(\psi)}{\partial \sigma_\varepsilon^2} \right)$$

The Hessian, in this case a 3×3 matrix, is $H(\psi) = \frac{\partial^2 \ell(\psi)}{\partial \psi \partial \psi'}$. From standard results, (see, for example, Hayashi, 2000, chapter 7), we know that $\sqrt{T}S(\psi) \Rightarrow_D N(0, \Sigma)$. Let $\widetilde{\psi}$ denote the parameters evaluated at some hypothetical values with corresponding score vector and Hessian, $S(\widetilde{\psi})$ and $H(\widetilde{\psi})$. An LM test statistic is obtained by setting $\widetilde{\psi}$ equal to its value under H_0, say $\widetilde{\psi}_0$, so that the LM test statistic takes the following form:

$$LM(\widetilde{\psi}_0) = TS(\widetilde{\psi}_0)' \widetilde{\Sigma}(\widetilde{\psi}_0)^{-1} S(\widetilde{\psi}_0)$$

where $\widetilde{\Sigma}(\widetilde{\psi}_0)$ is a consistent estimator of Σ evaluated at $\widetilde{\psi}_0$. In the stationary case $LM(\widetilde{\psi}_0) \Rightarrow_D \chi^2(q)$, where q is the number of restrictions being tested (assuming that none are redundant). However, in the nonstationary case, $LM(\widetilde{\psi}_0)$ has a non-standard distribution, which is a function of Brownian motion (see Distaso, 2008, for the case at hand.)

This approach enables Distaso (2008) to distinguish several specifications of the null and alternative hypotheses in the context of $\rho_t \sim iid(\rho, \omega^2)$, namely: (i) $\omega^2 = 0$ against $\omega^2 > 0$, that is a stochastic root but not necessarily a stochastic *unit* root (no hypothesis on ρ); (ii) $\rho = 1$ against $|\rho| < 1$, the null is a unit root process, which may be stochastic (no hypothesis on ω^2); and (iii) $\rho = 1$ and $\omega^2 = 0$ against either $|\rho| < 1$ or $\omega^2 > 0$ or both. The corresponding test statistic is then obtained by an appropriate specification of $\widetilde{\psi}_0$ according to the null hypothesis, where hypotheses (i) and (ii) have $q = 1$, whereas $q = 2$ for hypothesis (iii). See Distaso (2008) for the details, such as the role of the Hessian matrix and the distributions of the resulting test statistics.

References

Agiakloglou, C., and P. Newbold. (1994) Lagrange multiplier tests for fractional difference, *Journal of Time Series Analysis* 15, 253–262.

Amemiya, T. (1985) *Advanced Econometrics*, Harvard: Harvard University Press.

Anderson, T. W. A. (1971) *The Statistical Analysis of Time Series*, New York: John Wiley & Sons.

Andrews, D. W. K. (1991) Heteroskedasticity and autocorrelation consistent covariance matrix estimation, *Econometrica* 59, 817–858.

Andrews, D. W. K. (1993) Tests for parameter instability and structural change with unknown change point, *Econometrica* 61, 821–856.

Andrews, D. W. K., and P. Guggenberger. (2003) A bias-reduced log-periodogram regression estimator of the long-memory parameter, *Econometrica* 71, 675–712.

Andrews, D. W. K., and P. Guggenberger. (2011) A conditional-heteroskedasticity-robust confidence interval for the autoregressive parameter, *Cowles Foundation Discussion Paper* No. 1812.

Andrews, D. W. K., and P. Guggenberger. (2012) Asymptotics for LS, GLS, and feasible GLS statistics in an AR(1) model with conditional heteroskedaticity, forthcoming in the *Journal of Econometrics*.

Andrews, D. W. K., and W. Ploberger. (1994) Optimal tests when a nuisance parameter is present only under the alternative, *Econometrica* 62, 1383–1414.

Andrews, D. W. K., and Y. Sun. (2004) Adaptive local polynomial whittle estimation of long-range dependence, *Econometrica* 72, 569–614.

Aparicio, F., Escribano, A., and A. Sipols. (2006) Range unit-root (RUR) tests: robust against nonlinearities, error distributions, structural breaks and outliers, *Journal of Time Series Analysis* 27, 545–576.

Bai, J. (1997) Estimating multiple breaks one at a time, *Econometric Theory* 13, 315–352.

Bai, J., and P. Perron. (1998) Estimating and testing linear models with multiple structural changes, *Econometrica* 66, 47–78.

Bai, J., and P. Perron. (2003a) Critical values for multiple structural change tests, *Econometrics Journal* 6, 72–78.

Bai, J., and P. Perron. (2003b) Computation and analysis of multiple structural change models, *Journal of Applied Econometrics* 18, 1–22.

Banerjee, A., Lumsdaine, R. L., and J. H. Stock. (1992) Recursive and sequential tests of the unit-root and trend-break hypotheses: theory and international evidence, *Journal of Business & Economic Statistics* 10, 271–287.

Baragona, R., and D. Cucina. (2008) Double threshold autoregressive conditionally heteroscedastic model building by genetic algorithms, *Journal of Statistical Computation and Simulation* 78, 541–558.

Barkoulas, J., and C. F. Baum. (2006) Long-memory forecasting of US monetary indices, *Journal of Forecasting* 25, 291–302.

Battaglia, F., and M. K. Protopapas. (2010) Multi-regime models for nonlinear nonstationary time series, *COMISEF Working Paper Series* WPS-26.

Beaudry, P. and G. Koop. (1993) Do recessions permanently change output? *Journal of Monetary Economics* 31, 149–163.

Bec, F., Ben Salem, M., and M. Carrasco. (2004) Tests for unit-root versus threshold specification with an application to the purchasing power parity relationship, *Journal of Business & Economic Statistics* 22, 382–395.

Bec, F., Guay, A., and E. Guerre. (2008) Adaptive consistent unit-root tests based on autoregressive threshold model, *Journal of Econometrics* 142, 94–133.

Beran, J. (1994) *Statistics for Long Memory Processes*, New York: Chapman and Hall.

Bod, P., Blitz, D., Franses, P. H., and R. Kluitman. (2002) An unbiased variance estimator for overlapping returns, *Applied Financial Economics* 12, 155–158.

Bollerslev, T. (1986) Generalized autoregressive conditional heteroskedasticity, *Journal of Econometrics* 31, 307–327.

Bougerol, P., and N. Picard. (1992) Stationarity of GARCH processes and of some nonnegative time series, *Journal of Econometrics* 52, 115–127.

Box, G. E. P., and G. M. Jenkins. (1970) *Time Series Analysis: Forecasting and Control*, San Francisco: Holden-Day.

Breitung, J. (2002) Nonparametric tests for unit roots and cointegration, *Journal of Econometrics* 108, 343–363.

Breitung, J., and C. Gourieroux. (1997) Rank tests for unit roots, *Journal of Econometrics* 81, 7–27.

Breitung J., and U. Hassler. (2002) Inference on the cointegration rank in fractionally integrated processes, *Journal of Econometrics* 110, 167–185.

Breusch, Trevor S. (1978) Testing for autocorrelation in dynamic linear models, *Australian Economic Papers* 17, 334–335.

Brockwell, P. J., and R. A. Davis. (2006) *Time Series: Theory and Methods*, 2nd edition, New York: Springer.

Bunzel, H., and T. J. Vogelsang. (2005) Powerful trend function tests that are robust to strong serial correlation, with an application to the Prebisch-Singer hypothesis, *Journal of Business & Economic Statistics* 23, 381–394.

Calvet, L., and A. Fisher. (2002) Multifractality in asset returns: theory and evidence, *The Review of Economics and Statistics* 84, 381–406.

Caner, M., and B. E. Hansen. (2001) Threshold autoregression with a unit root, *Econometrica* 69, 1555–1596.

Carrion-i-Silvestre, J. L., Sanso-i-Rossello, A., and M. A. Ortuno. (1999) Response surfaces estimates for the Dickey-Fuller unit root test with structural breaks, *Economics Letters* 63, 279–283.

Cavaliere, G., and G. Iliyan. (2007) Testing for unit roots in autoregressions with multiple level shifts, *Econometric Theory* 23, 1162–1215.

Cavaliere, G., and A. M. R. Taylor. (2006) Testing for a change in persistence in the presence of a volatility shift, *Oxford Bulletin of Economics and Statistics* 68, 761–781.

Cavaliere, G., and A. M. R. Taylor. (2008) Testing for a change in persistence in the presence of non-stationary volatility, *Journal of Econometrics* 147, 84–98.

Chambers, M. (1998) Long memory and aggregation in macroeconomic time series, *International Economic Review* 39, 1053–1072.

Chan, K. S. (1990) Deterministic stability, stochastic stability, and ergodicity, appendix 1, in H. Tong (ed.) *Non-linear Time Series Analysis: A Dynamical System Approach*, Oxford: Oxford University Press.

Chan, K. S. (1993) Consistency and limiting distribution of the least squares estimator of a threshold autoregressive model, *The Annals of Statistics* 21, 520–533.

Chan, K. S., Petruccelli, J. D., Tong, H., and S. W. Woolford. (1985) A multiple threshold AR(1) model, *Journal of Applied Probability* 22, 267–279.

Chan, K. S. and H. Tong. (1985). On the use of the deterministic Lyapunov function for the ergodicity of stochastic difference equations, *Advances in Applied Probability* 17, 666–678.

Chan, N. H., and L. T. Tran. (1989) On the first order autoregressive process with infinite variance, *Econometric Theory* 5, 354–362.

Charles, A., and O. Darné. (1999) Variance ratio tests of random walk: an overview, *Journal of Economic Surveys* 23, 503–527.

Chen, Y-T. (2003) Discriminating between competing STAR models, *Economics Letters* 79, 161–167.

Chen, Y-T., and C-M, Kuan. (2002) The pseudo-true score encompassing test for non-nested hypotheses, *Journal of Econometrics* 106, 271–295.

Chen, Y-T., and C-M., Kuan. (2007) Corrigendum to 'The pseudo-true score encompassing test for non-nested hypotheses' (*Journal of Econometrics* 106, 271–295), *Journal of Econometrics* 141, 1412–1417.

Chen, R., and R. S. Tsay. (1991) On the ergodicity of TAR(1) processes, *Annals of Applied Probability* 1, 613–634.

Chow, G. C. (1960) Tests of equality between sets of coefficients in two linear regressions, *Econometrica* 28, 591–605.

Cioczek-Georges, R., and B. B. Mandelbrot. (1995) A class of micropulses and antipersistent fractional Brownian motion, *Stochastic Processes and their Applications* 60, 1–18.

Cioczek-Georges, R., and B. B. Mandelbrot. (1996) Alternative micropulses and fractional Brownian motion, *Stochastic Processes and their Applications* 64, 143–152.

Clements, M. P., and D. F. Hendry. (1998) *Forecasting Economic Time Series*, Cambridge, Cambridge University Press.

Clements, M. P. and D. F. Hendry. (1999) *Forecasting Non-Stationary Economic Time Series*, Cambridge, Cambridge University Press.

Cline, D. B. H., and H. H. Pu. (1999) Geometric ergodicity of nonlinear time series, *Statistica Sinica* 9, 1103–1118.

Cline D. B. H., and H. H. Pu. (2001) Geometric transience of nonlinear time series, *Statistica Sinica* 11, 273–287.

Cochrane, J. H. (1988) How big is the random walk in GNP?, *Journal of Political Economy* 96, 893–920.

Cook, S. (2004) A momentum-threshold autoregressive unit root test with increased power, *Statistics and Probability Letters* 67, 307–310.

Cribari-Neto, F. (2004) Asymptotic inference under heteroskedasticity of unknown form, *Computational Statistics and Data Analysis* 45, 215–233.

Davidson, J. (2000) *Econometric Theory*, Oxford: Blackwell Publishers.

Davidson, J., and N. Hashimzade. (2009) Type I and type II fractional Brownian motions: a reconsideration, *Computational Statistics and Data Analysis* 53, 2089–2106.

Davidson, R., and J. G. MacKinnon (1993) *Estimation and Inference in Econometrics*, Oxford: Oxford University Press.

Davidson, R., and J. G. MacKinnon. (2004) *Econometric Theory and Methods*, Oxford: Oxford University Press.

Davidson, R., and J. G. MacKinnon. (2006) Bootstrap methods in econometrics, chapter 25, in T. C. Mills and K. D. Patterson (eds), *The Palgrave Handbook of Econometrics: Volume 1: Econometric Theory*, Basingstoke: Palgrave-Macmillan.

Davies, R. B. (1977) Hypothesis testing when a nuisance parameter is present only under the alternative, *Biometrika* 64, 247–254.

Davies, R. B. (1987) Hypothesis testing when a nuisance parameter is present only under the alternatives, *Biometrika* 74, 33–43.

Davies, R. B., and D. S. Harte. (1987) Tests for Hurst effect, *Biometrika* 74, 95–101.

Delgado, M. A., and C. Velasco. (2005) Sign tests for long-memory time series, *Journal of Econometrics* 128, 215–251.

Demetrescu, M. (2010) On the Dickey Fuller test with White standard Errors, *Statistical Papers* 51, 11–25.

Deo, R. S., and M. Richardson. (2003) On the asymptotic power of the variance ratio test, *Econometric Theory* 19, 231–239.

Dickey, D. A. (1976) Estimation and hypothesis testing in nonstationary time series, PhD thesis, Iowa State University, Ames Iowa.

Dickey, D. A., and W. A. Fuller. (1979) Distribution of the estimators for autoregressive time series with a unit root, *Journal of the American Statistical Association* 74, 427–431.

Dickey, D. A., and W. A. Fuller. (1981) Likelihood ratio statistics for autoregressive time series with a unit root, *Econometrica* 49, 1057–1022.

Diebold, F. X., and C. Chen. (1996) Testing structural stability with endogenous breakpoint. A size comparison of an analytic and bootstrap procedures, *Journal of Econometrics* 70, 221–241.

Diebold, F. X., and G. D. Rudebusch. (1991) On the power of Dickey-Fuller tests against fractional alternatives, *Economics Letters* 35, 155–160.

Distaso, W. (2008) Testing for unit root processes in random coefficient autoregressive models, *Journal of Econometrics* 142, 581–609.

Dolado, J., Gonzalo, J., and Mayoral, L. (2002). A fractional Dickey–Fuller test for unit roots, *Econometrica* 70, 1963–2006.

Dolado, J., Gonzalo, J., and L. Mayoral. (2008) Wald tests of I(1) against I(d) alternatives: some new properties and an extension to processes with trending components, *Studies in Nonlinear Dynamics and Econometrics* 12, 1–32.

Dolado, J. J., Gonzalo, J., and L. Mayoral. (2009) Simple Wald tests of the fractional integration parameter: an overview of new results, in J. Castle and N. Shepard (eds), *The Methodology and Practice of Econometrics* (A Festschift in Honour of David Hendry), Oxford: Oxford University Press.

Dougherty, C. (2011) *Introduction to Econometrics*, 4th Edition, Oxford: Oxford University Press.

Eitrheim, Ø., and T. Teräsvirta. (1996) Testing the adequacy of smooth transition autoregressive models, *Journal of Econometrics* 74, 59–75.

Elliott, G., Rothenberg, T. J., and J. H. Stock. (1996) Efficient tests for an autoregressive root, *Econometrica* 64, 813–836.

Enders, W. (2001) Improved critical values for the Enders-Granger unit-root test, *Applied Economic Letters* 8, 257–261.

Enders, W., and C. W. J. Granger. (1998) Unit-root tests and asymmetric adjustment with an example using the term structure of interest rates, *Journal of Business & Economic Statistics* 16, 304–311.

Enders, W., and T. Sandler. (2005) After 9/11: Is it all different now?, *Journal of Conflict Resolution* 49, 259–277.

Engle, R. F. (1982). Autoregressive conditional heteroscedasticity with estimates of the variance of United Kingdom inflation, *Econometrica* 50, 987–1007.

Escribano, Á., and O. Jordá. (2001) Testing nonlinearity: decision rules for selecting between logistic and exponential STAR models, *Spanish Economic Review* 3, 193–209.

Fillol, J. (2007) Estimating long memory: scaling function vs Andrews and Guggenberger GPH, *Economics Letters* 95, 309–314.

Fillol, J., and F. Tripier. (2003) The scaling function-based estimator of the long memory parameter: a comparative study, *Economics Bulletin* 3, 1–7.

Fong, W. M., Koh, S. F., and S. Ouliaris. (1997) Joint variance ratio tests of the martingale hypothesis for exchange rates, *Journal of Business & Economic Statistics* 15, 51–59.

Fotopoulos, S. B., and S. K. Ahn. (2003) Rank based Dickey-Fuller tests, *Journal of time Series Analysis* 24, 647–662.

Franses, P. H. and N. Haldrup. (1994) The effects of additive outliers on tests for unit roots and cointegration, *Journal of Business and Economic Statistics* 12, 471–478.

Fuller, W. (1976) *An Introduction to Statistical Time Series*, 1st edition, New York: John Wiley.

Fuller, W. (1996) *An Introduction to Statistical Time Series*, 2nd edition, New York: John Wiley.

Geweke, J. F., and S. Porter-Hudak. (1983) The estimation and application of long memory time series models, *Journal of Time Series Analysis* 4, 221–238.

Godfrey, L. G. (1978) Testing against general autoregressive and moving average error models when the regressors include lagged dependent variables, *Econometrica* 46, 1293–1302.

Gonzalez, M., and J. Gonzalo. (1998) Threshold unit root models, *Working Paper*, Universidad Carlos III de Madrid.

Gonzalo, J., and J-Y. Pitarakis. (2002) Estimation and model selection based inference in single and multiple threshold models, *Journal of Econometrics*, 110, 319–352.

Granger, C. W. J. (1980) Long memory relationships and the aggregation of dynamic models, *Journal of Econometrics* 14, 227–238.

Granger, C. W. J. (1990) Aggregation in time series variables: a survey, in T. Barker and M. H. Pesaran (eds), *Disaggregation in Economic Modeling*, London: Routledge.

Granger, C. W. J., and J. Hallman. (1991) Nonlinear transformations of integrated time series, *Journal of Time Series Analysis* 12, 207–224.

Granger, C. W. J. and R. Joyeux. (1980) An introduction to long memory time series models and fractional differencing, *Journal of Time Series Analysis* 1, 15–29.

Granger, C. W. J., and P. Newbold. (1977) *Forecasting Economic Time Series*, New York: Academic Press.

Granger, C. W. J., and N. R. Swanson. (1997) An introduction to stochastic unit root processes, *Journal of Econometrics* 80, 35–62.

Greene, W. H. (2011) *Econometric Analysis*, 7th edition, New York: Prentice Hall.

Guggenberger, P., and Y, Sun. (2006) Bias-reduced log-periodogram and Whittle estimation of the long-memory parameter without variance inflation, *Econometric Theory* 22, 863–912.

Guo, B. B., and P. C. B. Phillips. (2001a) Testing for autocorrelation and unit roots in the presence of conditional heteroskedasticity of unknown form, unpublished working paper, Department of Economics, University of California, Santa Cruz.

Guo, B. B., and P. C. B. Phillips. (2001b) Efficient estimation of second moment parameters in ARCH models, unpublished manuscript.

Haldrup, N. (1994) Heteroscedasticity in non-stationary time series, some Monte Carlo evidence, *Statistical Papers* 35, 287–307.

Hamilton, J. D. (1989) A new approach to the economic analysis of nonstationary time series and the business cycle, *Econometrica* 57, 357–384.

Hamori, S., and A. Tokihisa. (1997) Testing for a unit root in the presence of a variance shift, *Economics Letters* 57, 245–253.

Hanselman, D.C., and B. Littlefield. (2005) *Mastering MATLAB 7*, New York: Prentice Hall.

Hansen, B. E. (1996) Inference when a nuisance parameter is not identified under the null hypothesis, *Econometrica* 64, 413–430.

Hansen, B. E. (1997a) Inference in TAR models, *Studies in Nonlinear Dynamics & Econometrics* 2, 119–131.

Hansen, B. E. (1997b) Approximate asymptotic p values for structural change tests, *Journal of Business & Economic Statistics* 15, 60–67.

Hansen, B. E. (1999) Testing for linearity, *Journal of Economic Surveys* 13, 551–576.

Hansen, B. E. (2000) Sample splitting and threshold estimation, *Econometrica* 68, 575–603.

Harris, D., Harvey, D. I., Leybourne, S. J., and A. M. R. Taylor. (2009) Testing for a unit root in the presence of a possible break in trend, *Econometric Theory* 25, 1545–1588.

Harvey, D. I., Leybourne, S. J., and P. Newbold. (2001) Innovational outlier unit root tests with an endogenously determined break in level, *Oxford Bulletin of Economics and Statistics* 63, 359–375.

Harvey, D., Leybourne, S. L., and P. Newbold. (2004) Tests for a break in level when the order of integration is unknown, *Oxford Bulletin of Economics and Statistics* 66, 133–146.

Harvey, D. I., Leybourne, S. J., and A. M. R. Taylor. (2008) Erratum to 'A simple, robust and powerful test of the trend hypothesis' [*Journal of Econometrics*, 141(2), (2007), 1302–1330], *Journal of Econometrics* 143, 396–397.

Harvey, D. I., Leybourne, S. J., and A. M. R. Taylor. (2009) Simple, robust, and powerful tests of the breaking trend hypothesis, *Econometric Theory* 25, 995–1029.

Harvey, D., and T. C. Mills. (2002) Unit roots and double smooth transitions, *Journal of Applied Statistics* 29, 675–683.

Hasan, M. N. and R. W. Koenker. (1997) Robust rank tests of unit root hypothesis, *Econometrica* 65, 131–161.

Hassler, U., and J. Wolters. (1994) On the power of unit roots tests against fractional alternatives, *Economics Letters* 45, 1–5.

Hatanaka, M. (1996) *Time-Series-Based Econometrics*, Oxford: Oxford University Press.

Hayashi, F. (2000) *Econometrics*, Princeton: Princeton University Press.

Henry, M. (2001) Robust automatic bandwidth for long memory, *Journal of Time Series Analysis* 22, 293–316.

Herce, M. A. (1996) Asymptotic theory of LAD estimation in a unit root process with finite variance errors, *Econometric Theory* 12, 129–153.

Hoek, H., Lucas, A., and H. K. van Dijk. (1995) Classical and Bayesian aspects of robust unit root inference, *Journal of Econometrics* 69, 27–59.

Holt, M. T., and L. A. Craig. (2006) Nonlinear dynamics and structural change in U.S. Hog-Corn cycle: a time-varying STAR approach, *American Journal of Agricultural Economics* 88, 215–233.

Hosking, J. R. M. (1981) Fractional differencing, *Biometrica* 68,165–176.

Hosking, J. R. M. (1984) Modeling persistence in hydrological time series using fractional differencing, *Water Resources Research* 20, 1898–1908.

Hsu, C.-C., and C.-M. Kuan. (2001) Distinguishing between trend break models: Method and empirical evidence, *Econometrics Journal* 4, 171–190.

Huber, P. J. (1981) *Robust Statistics*, New York: John Wiley and Sons.

Hurst, H. (1951) Long term storage capacity of reservoirs, *Transactions of the American Society of Civil Engineers* 116, 770–799.

Hurvich, C. M., and Beltrao, K. I. (1993) Asymptotics for the low-frequency ordinates of the periodogram of a long-memory time series, *Journal of Time Series Analysis* 14, 455–472.

Hurvich, C. M., and J. Brodsky. (2001) Broadband semiparametric estimation of the memory parameter of a long memory time series using fractional exponential models, *Journal of Time Series Analysis* 22, 221–249.

Hurvich, C. M., and W. W. Chen. (2000) An efficient taper for potentially overdifferenced long-memory time series, *Journal of Time Series Analysis* 21, 155–180.

Hurvich, C. M., and B. K. Ray. (1995) Estimation of the memory parameter for nonstationary or noninvertible fractionally integrated processes, *Journal of Time Series Analysis* 16, 17–42.

Hurvich, C. M., R. Deo., and J. Brodsky. (1998) The mean squared error of Geweke and Porter-Hudak's estimator of the memory parameter of a long memory time series, *Journal of Time Series Analysis* 19, 19–46.

Hurvich, C., E. Moulines., and P. Soulier. (2001) An adaptive broadband estimator of the fractional differencing coefficient, acoustics, speech and signal processing, Proceedings 2001 IEEE International Conference 6, 3417–3420.

Hurvich, C. M., E. Moulines, and P. Soulier. (2002) The FEXP estimator for potentially non-stationary linear time series, *Stochastic Processes and Their Applications* 97, 307–340.

Iouditsky, A., Moulines, E., and P. Soulier. (2001) Adaptive estimation of the fractional differencing coefficient, *Bernoulli* 7, 699–731.

Kalia, R. N. (ed.) (1993) *Recent Advances in Fractional Calculus*, New York: Global Publishing Company.

Kapetanios, G. (2001) Model selection in threshold models, *Journal of Time Series Analysis* 22, 733–754.

Kapetanios, G., and Y. Shin. (2006) Unit root tests in three-regime SETAR models, *Econometrics Journal* 9, 252–278.

Kapetanios, G., Shin, Y., and A. Snell. (2003) Testing for a unit root in the nonlinear STAR framework, *Journal of Econometrics* 112, 359–379.

Kejriwal, M., and P. Perron. (2010) A sequential procedure to determine the number of breaks in trend with an integrated or stationary noise component, *Journal of Time Series Analysis* 31, 305–328.

Kendall, M. G. (1954) Note on the bias in the estimation of autocorrelation, *Biometrika* 41, 403–404.

Kew. H., and D. Harris. (2009) Heteroscedasticity-robust testing for a fractional unit root, *Econometric Theory* 25, 1734–1753.

Kılıç, K. (2011) Long memory and nonlinearity in conditional variances: a smooth transition FIGARCH model, *Journal of Empirical Finance* 18, 368–378.

Kim, T. H., Leybourne, S., and P. Newbold. (2002) Unit root tests with a break in innovation variance, *Journal of Econometrics* 109, 365–387.

Kim C. S., and P. C. B. Phillips (2000) Modified log periodogram regression, Yale University, mimeographed.

Kim C. S., and P. C. B. Phillips. (2006) Log periodogram regression: the nonstationary case, *Cowles Foundation Discussion Paper* No. 1587.

Kim, K., and P. Schmidt. (1993) Unit root tests with conditional heteroscedasticity, *Journal of Econometrics* 59, 287–300.

Klimko, L. A., and P. I. Nelson. (1978) On conditional least squares estimation for stochastic processes, *Annals of Statistics* 6, 629–642.

Kobayashi, M., and M. McAleer. (1999a) Tests of linear and logarithmic transformations for integrated processes, *Journal of the American Statistical Association* 94, 860–868.

Kobayashi, M., and M. McAleer. (1999b) Analytical power comparisons of nested and nonnested tests for linear and loglinear regression models, *Econometric Theory* 15, 99–113.

Krämer, W. (1998) Fractional integration and the augmented Dickey-Fuller test, *Economics Letters* 61, 269–272.

Krämer, W., and L. Davies. (2002) Testing for unit roots in the context of misspecified logarithmic random walks, *Economics Letters* 74, 313–319.

Künsch, H. R. (1987) Statistical aspects of self-similar processes, in Y. Prohorov and V. V. Sazanov (eds), *Proceedings of the First World Congress Vol. 1*, Utrecht: VNU Science Press, 67–74.

Kwiakowski, D., Phillips, P. C. B., Schmidt, P., and Y. Shin. (1992) Testing the null hypothesis of stationarity against the alternative of a unit root: how sure are we that economic time series have a unit root? *Journal of Econometrics* 54, 159–178.

Lee, J., and M. C. Strazicich. (2001). Break point estimation and spurious rejections with endogenous unit root tests, *Oxford Bulletin of Economics and Statistics* 63, 535–558.

Lee, O., and D. W. Shin. (2000) A note on stationarity of the MTAR process on the boundary of the stationarity region, *Economics Letters* 73, 263–268.

LeRoy, S. F. (1982) Expectations models of asset prices: a survey of the theory, *Journal of Finance* 37, 185–217.

Leybourne, S. J., McCabe, B. P. M., and T. C. Mills. (1996) Randomized unit root processes for modelling and forecasting financial time series: theory and applications, *Journal of Forecasting* 15, 253–270.

Leyboume, S. J., McCabe, B. P. M., and A. R. Tremayne. (1996) Can economic time series be differenced to stationarity? *Journal of Business and Economic Statistics* 14, 435–446.

Ling, S., and W. K. Li. (1998) Limiting distributions of maximum likelihood estimators for unstable autoregressive moving-average time series with general autoregressive heteroscedastic errors, *Annals of Statistics* 26, 84–125.

Ling, S., and W. K. Li. (2003) Asymptotic inference for unit root processes with GARCH(1,1) errors, *Econometric Theory*, 19, 541–564.

Ling, S., Li, W. K., and M. McAleer. (2002) Recent theoretical results for time series models with GARCH errors, *Journal of Economic Surveys* 16, 245–269.

Ling, S., Li, W. K., and M. McAleer. (2003) Estimation and testing for unit root processes with GARCH (1, 1) Errors: Theory and Monte Carlo Evidence, *Econometric Reviews* 22, 179–202.

Ling, S., Tong, H., and D. Li. (2007) Ergodicity and invertibility of threshold moving-average models, *Bernouilli* 13, 161–168.

Lo, A. W., and A. C. MacKinlay. (1989) The size and power of the variance ratio test in finite samples: a Monte Carlo investigation, *Journal of Econometrics* 40, 203–238.

Lobato, I. N., and C. Velasco. (2006) Optimal fractional Dickey–Fuller tests for unit roots, *Econometrics Journal* 9, 492–510.

Lobato, I. N., and C. Velasco. (2007) Efficient Wald tests for fractional unit roots, *Econometrica* 75, 575–589.

Lobato, I. N., and C. Velasco. (2008) Power comparison among tests for fractional unit roots, *Economics Letters* 99, 152–154.

Long, J. S., and L. H. Ervin. (2000) Using heteroscedasticity consistent standard errors in the linear regression model, *The American Statistician* 54, 217–224.

Lucas, A. (1995a) An outlier robust unit root test with an application to the extended Nelson-Plosser data, *Journal of Econometrics* 66, 153–173.

Lucas, A. (1995b) Unit root tests based on m estimators, *Econometric Theory* 11, 331–346.

Lumsdaine, R. L., and D. H. Papell. (1997) Multiple trend breaks and the unit root hypothesis, *The Review of Economics and Statistics* 79, 212–218.

Lundbergh, S., and T. Teräsvirta. (1999) Modelling economic high-frequency time series, Tinbergen Institute Discussion Papers 99-009/4.

Lundbergh, S., Teräsvirta, T., and D. van Dijk. (2003) Time-varying smooth transition autoregressive models, *Journal of Business & Economic Statistics* 21, 104–121.

Luukkonen, R. (1990) Estimating smooth transition autoregressive models by conditional least squares, in R. Luukkonen (ed.), *On Linearity Testing and Model Estimation in Nonlinear Time Series Analysis*, Helsinki: Finnish Statistical Society.

Luukkonen, R., Saikkonen, P., and T. Teräsvirta. (1988) Testing linearity against smooth transition autoregressive models, *Biometrika* 75, 491–499.

Lütkepohl, H. (1993) *Introduction to Multiple Time Series*, 2nd edition, Berlin: Springer Verlag.

Lütkepohl, H. (2006) *New Introduction to Multiple Time Series*, 2nd edition, Berlin: Springer Verlag.

MacKinnon, J. (1991) Critical values for cointegration tests, in R. F. Engle and C. W. J. Granger (eds), *Long Run Economic Relationships*, Oxford: Oxford University Press, 267–276.

MacKinnon, J. G., and A. A. Smith. (1998) Approximate bias correction in econometrics, *Journal of Econometrics* 85, 205–230.

MacKinnon, J. G., and H. White. (1985) Some heteroskedasticity consistent covariance matrix estimators with improved finite sample properties, *Journal of Econometrics* 29, 53–57.

Maki, D. (2009) Tests for a unit root using three-regime TAR models: power comparison and some applications, *Econometric Reviews* 28, 335–363.

Maki, D. (2010) Detection of stationarity in nonlinear processes: a comparison between structural breaks and three-regime TAR Models, *Studies in Nonlinear Dynamics and Econometrics* 14, 1–41.

Mandelbrot, B. B. (1982). *The Fractal Geometry of Nature*, New York: W.H. Freeman and Company.

Mandelbrot, B. B., and J. W. Van Ness. (1968) The fractional Brownian motions, fractional noises and applications, *SIAM Review* 10, 422–437.

Marinucci, D., and P. M. Robinson. (1999) Alternative forms of fractional Brownian motion, *Journal of Statistical Planning and Inference* 80, 111–122.

Martinez, W. L., and A. R. Martinez. (2002) *Computational Statistics Handbook with MATLAB*, London: Chapman and Hall.

McCabe, B. P. M., and R. J. Smith. (1998) The power of some tests for difference stationarity under local heteroscedastic integration, *Journal of the American Statistical Association* 93, 751–761.

McCabe, B. P. M., and A. R. Tremayne. (1995) Testing a time series for difference stationarity, *The Annals of Statistics* 23, 1015–1028.

McLeod, A. I., and A. W. Hippel. (1978) Preservation of the rescaled adjusted range: 1. A reassessment of the Hurst phenomenon. *Water Resources Reserch* 14, 491–508.

Miller, K. S., and B. Ross. (1993) *An Introduction to the Fractional Calculus and Fractional Differential Equations*, New York: John Wiley & Sons.

Mills, T. C. (2011) *The Foundations of Time Series Analysis*, Basingstoke: Palgrave-Macmillan.

Mills, T. C., and K. D. Patterson. (2006) *The Palgrave Handbook of Econometrics, Volume 1: Econometric Theory*, Basingstoke: Palgrave Macmillan.

Mills, T. C., and K. D. Patterson. (2009) *The Palgrave Handbook of Econometrics, Volume 2: Applied Econometrics*, Basingstoke: Palgrave Macmillan.

Mizon, G. E., and J-F. Richard. (1986) The encompassing principle and its application to non-nested hypotheses, *Econometrica* 54, 657–678.

Montañés, A., and M. Reyes. (1998) Effect of a shift in the trend function on Dickey Fuller unit root tests, *Econometric Theory* 14, 355–363.

Montañés, A., and M. Reyes. (1999) The asymptotic behaviour of the Dickey-Fuller tests under the crash hypothesis, *Statistics & Probability Letters* 42, 81–89.

Montañés, A., Olloqui, I., and E. Calvo. (2005) Selection of the break in the Perron-type tests, *Journal of Econometrics* 129, 41–64.

Moulines, E., and P. Soulier. (1999) Broadband log-periodogram regression of time series with long-range dependence, *The Annals of Statistics* 27, 1415–1439.

Murray, C., and E. Zivot. (1998) Inference on unit roots and trend breaks in macroeconomic time series, *Working Papers UWEC-2005-04*, University of Washington, Department of Economics.

Nelson, C., and C. I. Plosser. (1982) Trends and random walks in macroeconomic time series, *Journal of Monetary Economics* 10, 139–162.

Ng, S., and P. Perron. (1995) Unit root tests ARMA models with data dependent methods for the selection of the truncation lag, *Journal of the American Statistical Association* 90, 268–281.

Ng, S., and P. Perron. (2001) Lag length selection and the construction of unit root tests with good size and power, *Econometrica* 69, 1519–1554.

Nicolau, J. (2002) Stationary processes that look like random walks: the bounded random walk process in discrete and continuous time, *Econometric Theory* 8, 99–118.

Nielsen, M. O. (2009) A powerful test of the autoregressive unit root hypothesis based on a tuning parameter free statistic, *Econometric Theory* 25, 1515–1544.

Nucci, A. R. (1999) The demography of business closings, *Small Business Economics* 12, 25–39.

Nunes, L.C., Newbold, P., and C-M. Kuan. (1997) Testing for unit roots with breaks: evidence on the great crash and the unit root hypothesis reconsidered, *Oxford Bulletin of Economics and Statistics* 59, 435–448.

Oldham, J. B and J. Spanier. (1974) *The Fractional Calculus: Theory and Applications of Differentiation and Integration to Arbitrary Order*, New York: Academic Press.

Ozaki, T. (1985) Non-linear time series models and dynamical systems, in E. J. Hannan, P. R. Krishnaiah, and M. M. Rao (eds), *Handbook of Statistics*, Vol. 5, Amsterdam: Elsevier, 25–83.

Ozaki, T. (1992) A bridge between nonlinear time series models and nonlinear stochastic dynamical systems: a local linearization approach, *Statistica Sinica* 2, 113–135.

Park, J. Y., and M. Shintani. (2005) Testing for a unit root test against transitional autoregressive models, *Working paper No. 05-W10*, Vanderbilt University.

Parke, W. R. (1999) What is fractional integration? *The Review of Economics and Statistics* 81, 632–638.

Patterson, K. D. (2010) *A Primer for Unit Root Testing*, Basingstoke: Palgrave Macmillan.

Patterson, K. D., and S. M. Heravi. (2003) The impact of fat-tailed distributions on some leading unit root tests, *Journal of Applied Statistics* 30, 635–669.

Perron. P. (1989) The Great Crash, the oil price shock and the unit root hypothesis, *Econometrica* 57, 1361–1401.

Perron, P. (1990) Testing for a unit root in a time series with a changing mean, *Journal of Business & Economic Statistics* 8, 153–162.

Perron, P. (1997) Further evidence on breaking trend functions in macroeconomic variables, *Journal of Econometrics* 80, 355–385.

Perron, P. (2006) Dealing with structural breaks, chapter 8, in T. C. Mills and K. D. Patterson (eds), *The Palgrave Handbook of Econometrics: Volume 1: Econometric Theory*, Basingstoke: Palgrave Macmillan, 278–352.

Perron, P. and G. Rodriguez. (2003) GLS detrending, efficient unit root tests and structural change, *Journal of Econometrics* 115, 1–27.

Perron, P., and T. J. Vogelsang. (1992a) Testing for a unit root in a time series with a changing mean: corrections and extensions, *Journal of Business & Economic Statistics*, American Statistical Association 10, 467–470.

Perron, P., and T. J., Vogelsang. (1992b) Nonstationarity and level shifts with an application to purchasing power parity, *Journal of Business & Economic Statistics*, American Statistical Association 10, 301–320.

Perron, P., and T. J. Vogelsang. (1993) A note on the asymptotic distributions of unit root tests in the additive outlier model with breaks, *Revista de Econometria* 13, 181–201.

Perron, P., and T. Yabu. (2009) Testing for shifts in trend with an integrated or stationary noise component, *Journal of Business & Economic Statistics* 27, 369–396.

Petruccelli, J. D., and S. W. Woolford. (1984) A threshold AR(1) model, *Journal of Applied Probability* 21, 587–611.

Phillips, P. C. B. (1987) Time series regression with a unit root, *Econometrica* 55, 277–301.

Phillips, P. C. B. (1990) Time series regression with a unit root and infinite variance errors, *Econometric Theory* 6, 44–62.

Phillips, P. C. B. (1999a) Unit root log periodogram regression, *Cowles Foundation Discussion Paper* No. 1244.

Phillips, P. C. B. (1999b) Discrete fourier transforms of fractional processes, *Cowles Foundation Discussion Paper* No. 1243.

Phillips, P. C. B. (2007) Unit root log periodogram regression, *Journal of Econometrics* 138, 104–124.

Phillips, P. C. B., and P. Perron. (1988) Testing for a unit root in time series regression, *Biometrika* 75, 335–346.

Phillips, P. C. B., and K. Shimotsu. (2004) Local Whittle estimation in nonstationary and unit root cases, *Annals of Statistics* 32, 656–692.

Priestley, M. B. (1981) *Spectral Analysis and Time Series*, London: Academic Press.

Richardson, M., and J. H. Stock. (1989) Drawing inferences from statistics based on multiyear asset returns, *Journal of Financial Economics* 25, 323–348.

Robinson, P. M. (1991) Testing for strong serial correlation and dynamic conditional heteroskedasticity in multiple regression, *Journal of Econometrics* 47, 67–84.

Robinson, P. M. (1994a) Efficient tests of nonstationary hypotheses, *Journal of the American Statistical Association* 89, 1420–1437.

Robinson, P. M. (1994b) Time series with strong dependence, chapter 2, in C. A. Sims (ed.), *Advances in Econometrics, Sixth World Congress Volume I*, Cambridge: Cambridge University Press, 47–96.

Robinson, P. M. (1995a) Log-periodogram regression of time series with long range dependence, *Annals of Statistics* 23, 1048–1072.

Robinson, P. M. (1995b) Gaussian semiparametric estimation of long range dependence, *The Annals of Statistics* 23, 1630–1661.

Robinson, P. M. (2005) The distance between rival nonstationary fractional processes. *Journal of Econometrics* 128, 283–300.

Ross, S. (2003) *Probability Models*, 8th edition, London: Academic Press.

Rothenberg, T. J., and J. H. Stock. (1997) Inference in a nearly integrated autoregressive model with nonnormal innovations, *Journal of Econometrics* 80, 269–286.

Said, S. E., and D. A. Dickey. (1984) Testing for unit roots in autoregressive-moving average models of unknown order, *Biometrika* 71, 599–607.

Sayginsoy, O., and T. J. Vogelsang. (2011) Testing for a shift in trend at an unknown date: a fixed-B analysis of heteroscedasticity autocorrletion robust OLS-based tests, *Econometric Theory* 27, 992–1025.

Schmidt, P., and P. C. B. Phillips. (1992) LM tests for a unit root in the presence of deterministic trends, *Oxford Bulletin of Economics and Statistics* 54, 257–287.

Schoier, G. (1999) Strong consistency of conditional least squares estimators in multiple regime threshold models, *Statistical Methods and Applications* 8, 75–82.

Schwartz, J., and B. D. Rookey. (2008) The narrowing gender gap in arrests: assessing competing explanations using self-report, traffic fatality, and official data on drunk driving, 1980-2004, Criminology 46, 637–671.

Sen, A. (2001) Behaviour of Dickey-Fuller F-tests under the trend-break stationary alternative, *Statistics and Probability Letters* 55, 257–268.

Sen, A. (2003) On unit root tests when the alternative is a trend break stationary process, *Journal of Business & Economic Statistics* 21, 174–184.

Sen, A. (2007) On the distribution of the break-date estimator implied by the Perron-Type Statistics when the form of break is misspecified, *Economics Bulletin* 3, 1–19.

Seo, B. (1999) Distribution theory for unit root tests with conditional heteroskedasticity, *Journal of Econometrics* 91, 113–144.

Shimotsu, K. (2010) Exact local Whittle estimation of fractional integration with unknown mean and time trend, *Econometric Theory* 26, 501–540.

Shimotsu, K., and P. C. B. Phillips. (2000) Modified local Whittle estimation of the memory parameter in the nonstationary case, *Cowles Foundation Discussion Paper* No. 1265.

Shimotsu, K., and P. C. B. Phillips. (2005) Exact local Whittle estimation of fractional integration, *Annals of Statistics* 33, 1890–1933.

Shimotsu, K., and P. C. B. Phillips. (2006) Local Whittle estimation of fractional integration and some of its variants, *Journal of Econometrics* 103, 209–233.

Shin, D. W., and W. A. Fuller. (1998) Unit root tests based on unconditional maximum likelihood estimation for the for the autoregressive moving average, *Journal of Time Series Analysis* 19, 591–599.

Shin, D. W., and B. A. So. (2002) Recursive mean adjustment and tests for nonstationarities, *Economics Letters* 75, 203–208.

Shively, P. A. (2002) An exact invariant variance ratio test, *Economics Letters* 75, 347–353.

Skalin, J., and T. Teräsvirta. (1999) Another look at Swedish business cycles, 1861–1988, *Journal of Applied Econometrics* 14, 359–378.

Smith, J., and S. Yadav. (1994) Forecasting costs incurred from unit differencing fractionally integrated processes, *International Journal of Forecasting* 10, 507–514.

So, B. S., and D. W. Shin. (1999) Recursive mean adjustment in time-series inferences, *Statistics and Probability Letters* 43, 65–73.

Sowell, F. (1990) The fractional unit root distribution, *Econometrica* 58, 495–505.

Sowell, F. (1992a) Maximum likelihood estimation of stationary univariate fractionally integrated time series models, *Journal of Econometrics* 53, 165–188.

Sowell, F. (1992b) Modeling long-run behaviour with the fractional ARIMA model, *Journal of Monetary Economics* 29, 277–302.

Stern, D. I., and R. K. Kaufmann. (1999) Econometric analysis of global climate change, *Environmental Modelling and Software* 14, 597–605.

Stock, J., and M. Watson. (1999) A comparison of linear and nonlinear univariate models for forecasting macroeconomic time series, chapter 1, in R. F. Engle and H. White (eds), *Cointegration, Causality, and Forecasting – Festschrift in Honour of Clive W. J. Granger,* Oxford: Oxford University Press.

Su, J-J., and E. Roca. (2012) Examining the power of stochastic unit root tests without assuming independence in the error processes of the underlying time series, *Applied Economic Letters* 19, 373–377.

Sydsaeter, K., and P. Hammond. (2008) *Mathematics of Economics Analysis,* 3rd edition, London: Financial Times Press/Prentice Hall.

Tanaka, K. (1990) Testing for a moving average unit root, *Econometric Theory* 6, 433–444.

Tanaka, K. (1999) The nonstationary fractional unit root, *Econometric Theory* 15, 549–582.

Taniguchi, M. (1979) On estimation of parameters of Gaussian stationary processes, *Journal of Applied Probability* 16, 575–591.

Taniguchi, M. (1981) An estimation procedure of parameters of a certain spectral density model, *Journal of the Royal Statistical Society,* Series B 43, 34–40.

Taniguchi, M. (1987) Minimum contrast estimation for spectral densities of stationary processes, *Journal of the Royal Statistical Society,* Series B 49, 315–325.

Taniguchi, M., and K. J. van Garderen., and M. L. Puri. (2003) Higher order asymptotic theory for minimum contrast estimators of spectral parameters of stationary processes, *Econometric Theory* 19, 984–1007.

Taylor, M., Peel, D., and L. Sarno. (2001) Non-linear mean reversion in real exchange rates: toward a solution to the purchasing parity puzzles, *International Economic Review,* 1015–1042.

Teräsvirta. T. (1994) Specification, estimation, and evaluation of smooth transition autoregressive models, *Journal of the American Statistical Association* 89, 208–218.

Teräsvirta, T., and H. M. Anderson. (1992) Characterizing nonlinearities in business cycles using smooth transition autoregressive models, *Journal of Applied Econometrics* 7, S119–S136.

Thompson, S. B. (2004) Robust tests of the unit root hypothesis should not be 'modified', *Econometric Theory* 20, 360–381.

Tian, G., Zhang, Y., and W. Huang. (1999) A note on the exact distributions of variance ratio statistics, Peking University, mimeo.

Tse, Y. K., Ng, K. W., and X. Zhang. (2004) A small-sample overlapping variance-ratio test, *Journal of Time Series Analysis* 25, 127–134.

Tukey, J. W. (1967) An introduction to the calculations of numerical spectrum analysis, in B. Harris (ed.), *Advanced Seminar on Spectral Analysis of Time Series,* New York: John Wiley & Sons, 25–46.

Tweedie, R. L. (1975) Sufficient conditions for ergodicity and recurrence of Markov chains on a general state space, *Stochastic Processes and their Applications* 3, 385–403.

van Dijk, D., and P. Franses. (1999), Modeling Multiple Regimes in the Business Cycle, *Macroeconomic Dynamics* 3, 311–340.

van Dijk, D., Teräsvirta, T., and P. Franses. (2002) Smooth transition autoregressive models – a survey of recent developments, *Econometric Reviews* 21, 1–47.

Velasco, C. (1999a) Non-stationary log-periodogram regression, *Journal of Econometrics* 91, 325–371.

Velasco, C. (1999b) Gaussian semiparametric estimation of non-stationary time series, *Journal of Time Series Analysis,* 20, 87–127.

Velasco, C. (2000) Non-Gaussian log-periodogram regression, *Econometric Theory* 16, 44–79.

Vogelsang, T. J. (1997) Wald-type tests for detecting breaks in the trend function of a dynamic time series, *Econometric Theory* 13, 818–848.

Vogelsang, T. J., and P. Perron. (1998) Additional tests for a unit root allowing for a break in the trend function at an unknown time, *International Economic Review* 39, 1073–1100.

Wang, G. (2006) A note on unit root tests with heavy-tailed GARCH errors, *Statistics and Probability Letters* 76, 1075–1079.

Wang, G., and W-L. Mao. (2008) Unit root testing in the presence of heavy-tailed GARCH errors, *Australian Journal of Statistics* 50, 273–292.

Wang, W., van Gelder P. H. A. J. M., and J. K. Vrijling. (2005) Detection of changes in streamflow series in Western Europe over 1901–2000, *Water Supply* 5, 289–299.

White, H. (1980) A heteroskedastic-consistent covariance matrix estimator and a direct test of heteroskedasticity, *Econometrica* 48, 817–838.

White, H. (1982) Maximum likelihood estimation of misspecified models, *Econometrica* 50, 1–25.

Whittle, P. (1953) Estimation and information in stationary time series, *Arkiv för Matematik* 2, 423–432.

Wooldridge, J. M. (1990) An encompassing approach to conditional mean tests with applications to testing non-nested hypotheses, *Journal of Econometrics* 45, 331–350.

Wooldridge, J. M. (2011) *Introductory Econometrics: A Modern Approach*, 5th edition, Thomson Southwestern: Cengage Learning.

Wright, J. H. (2000) Alternative variance-ratio tests using ranks and signs, *Journal of Business & Economic Statistics* 18, 1–9.

Zivot. E., and D. W. K. Andrews. (1992) Further evidence on the Great Crash, the oil price shock and the unit root hypothesis, *Journal of Business & Economic Statistics* 10, 251–270.

Author Index

Subject Index